# Dogs Through Time: An Archaeological Perspective

Proceedings of the 1st ICAZ Symposium on the History of the Domestic Dog
Eighth Congress of the International Council for Archaeozoology (ICAZ98)
August 23-29, 1998
Victoria, B.C., Canada

Edited by

Susan Janet Crockford

BAR International Series 889
2000

Published in 2016 by
BAR Publishing, Oxford

BAR International Series 889

*Dogs Through Time: An Archaeological Perspective*

ISBN 978 1 84171 089 1



BAR Publishing is the trading name of British Archaeological Reports (Oxford) Ltd. British Archaeological Reports was first incorporated in 1974 to publish the BAR Series, International and British. In 1992 Hadrian Books Ltd became part of the BAR group. This volume was originally published by Archaeopress in conjunction with British Archaeological Reports (Oxford) Ltd / Hadrian Books Ltd, the Series principal publisher, in 2000. This present volume is published by BAR Publishing, 2016.

Printed in England

BAR
PUBLISHING

BAR titles are available from:

BAR Publishing
122 Banbury Rd, Oxford, OX2 7BP, UK
EMAIL info@barpublishing.com
PHONE +44 (0)1865 310431
FAX +44 (0)1865 316916
www.barpublishing.com

# CONTENTS

# CONTRIBUTORS

ANTONIAN, L.
Institute of Zoology, ANAS, P. Sevak st. 7., Yerevan 375014, Armenia

LÁSZLÓ BARTOSIEWICZ
Institute of Archaeological Sciences, Loránd Eötvös University, 1088 Muzeum krt 4/B, Budapest, Hungary
E-mail: h10459bar@iff.hu

LUMINITA BEJENARU
Faculty of Biology, University of Iasi, Bd. Copou 11, 6600 Iasi, Romania

I. LEHR BRISBIN, JR.
Savannah River Ecology Laboratory, PO Drawer E, Aiken, SC 29802, USA
E-mail: brisbin@srel.edu

MARYANN BURBIDGE
CBCB (Ornithology), Royal Ontario Museum, 100 Queen's Park, Toronto, Ontario, M5S 2C6, Canada

ASHLEY BYUN
407 – 611 University Dr., Saskatoon, Saskatchewan, S7N 3Z1, Canada

LOUIS CHAIX
Department of Archaeozoology, Museum of Natural History, 1 route de Malagnou, C.P. 6434 CH,
1211 Geneva, Switzerland.
E-mail: louis.chaix@mhn.ville-ge.ch

KATE M. CLARK
Centre for Human Ecology & Environment, University of Southampton, Southampton, Hampshire, SO17 1BJ, UK
E-mail: kate.clark@soton.ac.uk

JULIET CLUTTON-BROCK
Department of Zoology, The Natural History Museum, London, SW7 5BD, UK
E-mail: juliet.cb@btinternet.com

J.-M. CORDY
U.R. "E.V.E.H.", University of Liège, Institute L. Frédéricq (Bat. L1), Place Delcour, 17, 4020 Liège, Belgium

LESLIE CRAM
23 Boyers Orchard, Harby, Melton Mowbray, Leicestershire LE14 4 BA
E-mail: leslie.cram@talk21.com

SUSAN J. CROCKFORD
Pacific IDentifications Inc., 6011 Oldfield Road RR 3 Victoria, B. C. V9E 2J4, Canada.
E-mail: scrock@tnet.net

TAMAR DAYAN
Department of Zoology, Tel Aviv University, Tel Aviv 69978, Israel
E-mail: dayan@post.tau.ac.il

DODY FUGATE
Laboratory of Anthropology, Museum of New Mexico, PO Box 2087, Santa Fe, NM 87504-2087, USA
E-mail: dharma@compuserve.com or dfugate@lvr.state.nm.us

ALFRED GALIK
Institute for Palaeontology, Geozentrum University, Althanstr 14A, Vienna, a-1090, Austria
E-mail: alfred.galik@univie.ac.at

EHUD GALILI
Marine Archaeology Branch, Israel Antiquities Authority, PO Box 180, Atlit, Israel

BRENT M. HANDLEY
19 Fountain of Youth Blvd., Apt. A, St. Augustine, FL 32084, USA
E-mail: bmh1969@aol.com

HITOMI HONGO
Primate Research Institute, Kyoto University, 41 Kanrin, Inuyama, Aichi 484-8606, Japan
E-mail: hitomi@pri.kyoto-u.ac.jp

MARIE-PIERRE HORARD-HERBIN
UMR 6575 du C.N.R.S. University of Tours, 3 place Anatole France, 37000 Tours, France.
E-mail: galinie@univ-tours.fr

NAOTAKA ISHIGURO
Department of Veterinary Public Health, Obihiro University, Obihiro, Hokkaido, 080-8555, Japan
E-mail: ishiguro@obihiro.ac.jp

ANDREW C. KITCHENER
Department of Geology & Zoology, National Museums of Scotland, Chambers Street, Edinburgh,
EH1 1JF, UK
E-mail: a.kitchener@nms.ac.uk

BEN F. KOOP
Centre for Environmental Health, University of Victoria, PO Box 3020, Victoria, B.C., V8W 3N5, Canada
E-mail: bkoop@uvic.ca

JAMES K. MCINTYRE
South West Pacific Research Foundation, 631 Jarpon Ave., #6391, Fernandina Beach, FL 32034, USA
E-mail: jmcint6317@aol.com

JANICE KOLER-MATZNICK
NGSD Conservation Society, 5265 Old Stage Rd., Central Point, OR 97502, USA
E-mail: jkoler@iname.com

NINA L. MANASERIAN
Institute of Zoology, ANAS, P. Sevak st. 7., Yerevan 375014, Armenia
E-mail: c/o "Luba Balian" lbalian@aua.am

AKIRA MATSUI
Nara National Cultural Properties Research Institute, 2-9-1, Nijo-cho, Nara 630-8577, Japan.
E-mail: makira@nabunken.go.jp

JACOPO DE GROSSI MAZZORIN
Archaeological Institute of Rome, Laboratory of Archaeozoology, via S. Apollinare 8, 00186 Rome, Italy.
E-mail: hhhhjdgm@tin.it or degrossi@archeorm.arti.beniculturali.it

RICHARD H. MEADOW
Peabody Museum, Harvard University, 11 Divinity Ave., Cambridge, MA 02138, USA
E-mail: meadow@fas.harvard.edu

RUDOLF MUSIL
Department of Paleontology & Geology, Masaryk University, Kotlářská St. 2, 611 37 Brno, Czech Republic
E-mail: rudolf@sci.muni.cz

NAOHIKO OKUMURA
Society for Techno-Innovation of Agriculture, Forestry & Fisheries, 446-1 Ippaizuka, Kamiyokoba Tukuba,
Ibaraki, 305-0854, Japan
E-mail: okumura@gene.staff.or.jp

SANDRA L. OLSEN
Department of Anthropology, Carnegie Museum of Natural History, O'Neil Research Center, 5800 Baum Blvd.,
Pittsburgh, PA 15206, USA
E-mail: solsen+@pitt.edu

MAX G. PAVESIC
Department of Anthropology, Boise State University, Boise, Idaho 83725, USA

UTE RINK deceased
Centre for Environmental Health, University of Victoria, PO Box 3020, Victoria, B.C., V8W 3N5, Canada

MICHAEL L. RYDER
4 Osprey Close, Lord's Wood, Southampton, S016 8EX, UK

MARION SCHWARTZ
Department of Anthropology, Yale University, PO Box 208277, New Haven, CT 06520-8277, USA
E-mail: marion.schwartz@yale.edu

NOBUO SHIGEHARA
Primate Research Institute, Kyoto University, 41 Kanrin, Inuyama, Aichi 484-8606, Japan
E-mail: shigehar@pri.kyoto-u.ac.jp

ANTONIO TAGLIACOZZO
Museum of Prehistory and Ethnography - "L. Pigorini", Laboratory of Palaeontology and Archaeozoology, via le Lincoln 3, 00144 Rome, Italy.
E-mail: tagliant@tin.it

CARMEN TARCAN
Department of Archaeology, Simon Fraser University, Burnaby, B.C., Canada
E-mail: chriscu@sfu.ca

M. UDRESCU
U.R. "E.V.E.H.", University of Liège, Institute L. Fédéricq (Bat. L1), Place Delcour, 17, 4020 Liège, Belgium

RAUL VALADEZ AZUA
Institute of Anthropology, UNAM, Circuito Exterior, C.U. Delegación Coyoacan, C.P. 04510, Mexico D.F.
E-mail: stecnica@servidor.unam.mx

DANNY N. WALKER
Department of Anthropology, University of Wyoming, PO Box 3431, Laramie, WY 82071-3431, USA
E-mail: dnwalker@uwyo.edu

DIANE M. WARREN
Indiana University, Student Bldg., Rm 130, Bloomington, IN 47405, USA
E-mail: dmwarren@indiana.edu

BONNIE C. YATES
US Fish & Wildlife Service Forensic Laboratory, 1490 E. Main St., Ashland, OR 97520, USA
E-mail: bonnie_yates@fws.gov

ROBERT M. YOHE II
Department of Sociology and Anthropology, California State University, 9001 Stockdale Hwy., Bakersfield, CA 93311, USA
E-mail: Robert_Yohe@firstclass1.csubak.edu

# PREFACE

The remains of domestic dog, *Canis familiaris*, are found in archaeological sites around the world, providing an unexpected global link between archaeologists regardless of the cultures they study. Unfortunately, despite the personal appeal of their living members, dogs have been largely ignored archaeologically. The remains of dogs are rather esoteric components of archaeological faunal assemblages: since they only occasionally represent subsistence items (food), dogs do not fit into any faunal category that archaeozoologists routinely use to assess past human activities. As a consequence, dog remains from most sites are inconsistently reported and seldom given more than superfluous attention, even though numerically they often comprise a significant proportion of all vertebrate fauna. Increased awareness of the significance of dogs to both human history and evolutionary biology, however, has begun to arouse more interest in the study of prehistoric dogs.

Dogs were the first animal to establish a domestic relationship with humans and thus have the longest archaeological history of any domesticate. Dogs migrated from several points of origin, along with their human companions, to settle in every one of the world's habitable continents - they also occupied most islands and every imaginable environment, from arctic steppe to desert. Due to this wide-spread distribution over time and space, the dog is literally the *only* animal that prehistorians around the world have in common. Therefore, the questions which still need answering regarding the history of the dog are relevant to virtually all archaeologists regardless of where they work.

This collection of essays was compiled from papers presented at the 1st ICAZ Symposium on the History of the Domestic Dog, a special session held at the beginning of the Eighth International Congress of Archaeozoology (*ICAZ98)*, in Victoria, British Columbia, Canada (hosted by the Department of Anthropology, University of Victoria, August 23-28, 1998). *ICAZ* is the only international organisation of specialists who identify and analyse animal remains from archaeological sites and their meetings are held once every four years. For three days during the 1998 meeting, enthusiastic archaeozoologists from more than 20 countries gathered for the first time to exchange information about the history of the domestic dog. Many participants brought specimens of dogs to contribute to a "show and tell" comparative session held one afternoon. This fascinating and well-attended event allowed many of us currently working with dog remains to closely examine a variety of prehistoric types that we had only seen in pictures before that time. For our colleagues who had little practical experience with dog remains, it was a rare opportunity to expand their analytical horizons. Dody Fugate (Museum of New Mexico) volunteered her time to photograph these specimens, some of which appear in the concluding chapter. It is our hope that the presentation of these papers together in one book will not only enlighten our colleagues and interested non-professionals regarding what is presently known (and *not* known) about the history of dogs, but will also encourage more consistent and rigorous data collection and reporting of archaeological dog remains in the future.

This volume addresses essentially all of the issues which are raised during the analysis of dog remains recovered from archaeological contexts, primarily through the presentation of specific examples drawn from various regions. In the initial chapter, Juliet Clutton-Brock, curator emeritus of the Natural History Museum (London), provides a fascinating and quite personal overview of the history of research on dog evolution from her unique perspective as a principal participant. The remainder of the book is divided into seven parts by general theme.

Part One deals with dog evolution and evidence of early dogs. Two papers on the process of domestication are included: one is theoretical, presenting for the first time a testable, physiological hypothesis for how domestication occurs; the other is analytical, presenting compelling evidence for domestication associated with late Pleistocene (Magdalenian) horse hunters of northern Germany (*c.* 12,500-13,000BP). Two additional papers in this section discuss dogs of the Middle East, the first of which is a tribute to American archaeozoologist Barbara Lawrence. Lawrence's career dedication to canid studies included early dog remains from this region, for which she devised a comprehensive series of measurements that are defined here for the first time. The other paper deals with material from the Carmel Coast of Israel (*c.* 8,000BP), dogs that apparently had changed little over the *c.* 4,000 years since their first appearance in the area. Chapter six reports on a remarkable specimen of a small dog recovered from the French Alps (*c.* 10,500BP) that is older than any other truly small dog yet discovered in northern Europe from Neolithic times. The last paper of this section chronicles a remarkable reign of small Japanese dogs (*c.* 10,000 – 2,300BP) that remained essentially unchanged over 8,000 years.

Part Two contains contributions by several archaeozoologists who attempt to determine the roles that dogs played in ancient human societies: guardian, spiritual entity, companion, hunter, hauler, herder - or dinner? Two examples of interred dogs from *c.* 8,000BP are typical of archaeozoological dilemmas faced around the world (from Kasahkstan, northwestern Asia and western Idaho, USA): why were dogs often deliberately buried, alone or with humans, and what do these internment practices say about the spiritual roles dogs served in ancient cultures? Chapter ten examines skeletal pathologies on ancient dog remains, with a discussion about what such evidence tells us about the way dogs were treated or utilised. Two papers follow on the "dogs for dinner" theme. Cut marks left on bone as a result of butchering carcasses are commonly found on food animals (such as deer, cows, sheep, etc) but are noted on dogs only in selected areas, and often, only during restricted chronological periods. Consumption of dogs was common in the certain areas (eg the South Pacific and parts of southeast Asia), but the examples presented here come from specific time periods in the history of France, Belgium and Romania. One of these papers examines consumption evidence from cut marks alone, while the other looks at tooth wear in relation to ages of dogs at death in order to determine if specific patterns of age classes are represented in the butchered remains. Similar age-class analyses are often used in archaeozoology for inferring that certain bone assemblages represent the refuse resulting from culling managed populations of domestic livestock animals (such as horses and goats) rather than from the hunting of their wild ancestral species. The last paper in this section describes a rather large assemblage of dog remains from a unique site in Austria and the attempts to understand how and why the dogs came to be deposited there.

In Part Three, a suite of papers discuss the influence that the Romans had on European dogs of their time. The first paper of this section sets the stage by describing the changes in dog sizes and shapes that occurred in Italy, culminating in the regular occurrence of both large and very small, lap-type dogs by the end of the Roman era - interpreted as evidence of true dog breeding. Two papers follow, offering different perspectives on the effect that invasion by Romans had on native British dogs. The first of these papers is a very pertinent reminder that the archaeozoological record from Britain may be biased by factors that make it *appear* that variations in dogs were not present in Britain before the Romans arrived, when in fact that may not be true. The warning not to take the current "facts" at face value because of biases in types of sites excavated, reporting practices, and preservation factors, applies to dog remains from many regions outside Britain as well. The other British paper is an innovative analysis of dog footprints left in ceramic tiles, which presents another perspective on breed variation during Roman times. Lastly, a unique situation is described for the Roman period in ancient Hungary, where the invasion boundary is marked by the Danube River. Archaeological remnants of Roman settlements are thus found on one side of the Danube and those of native, rural communities on the other: dog remains are reported from sites in both regions.

Part Four covers the issue of skeletal variation in dogs from other locations. Were "breeds", as we define them today, present outside Roman Europe or do we merely see evidence of regional variation and subsequent migration? In North America, we finally have an answer for Mexico after many years of concerted archaeozoological research on dogs there, presented in Chapter eighteen. Another report presents results of a preliminary study on dogs from sites in the American Maritime Provinces, where there appears to be significant size variation. Chapter twenty is a comparative analysis of southeastern US and Mexican Mayan dogs, looking especially at their appearance in various art forms. The last contribution in this section is from Armenia, where some archaeozoologists are still wrestling with whether skeletal variation means "breeds" or different species origins, since this is a region where they feel jackals could have made a contribution to ancient dog stocks.

Part Five provides a refreshing break from skeletal analysis, with a comprehensive report on the distinctive qualities and current status of the extant New Guinea singing dog, a rare primitive dog type whose relationship to modern dogs is still unclear.

Part Six looks at various methods used in the analysis of archaeological remains. Although surviving only under rare conditions of prehistoric preservation, hair remains might contribute important clues to a dog's actual appearance or to breed origins (something we are otherwise left to guess about, when only bones remain). The first paper in this section provides some preliminary data on hair diameter and length as it relates to body size and type in modern dog breeds, a valiant start at creating a database of measurements against which archaeological finds might be compared. The next two papers deal with the important and little discussed issue of wolf/dog hybridisation. The first of these papers uses traditional multivariate analysis in an examination of a series of Arctic wolf skulls collected between the late 1800s and 1987, which reveal evidence of a short period of hybridisation. The second papers presents a brief, preliminary report on a simple method of distinguishing skulls of wolves, dogs and wolf/dog hybrids that doesn't require extensive measurements or complex computer analysis. Specimens of wolf, dog, and hybrid skulls were brought to the Dog Symposium for a "hands-on" demonstration of this method, which generated a lively discussion of the implications and evidence of ancient hybridisation events.

The last two papers of Part Six deal with molecular genetics. A Japanese team of geneticists and archaeozoologists present their analysis of mtDNA extracted from prehistoric dogs from Japan and a Canadian team do the same for prehistoric dogs from the Northwest Coast of North America. Some differences in methodology in these two molecular studies underlie the questions being asked and affect the scope of their conclusions. The Japanese team compare their extensive ancient dog data only to modern dogs (primarily those that are considered indigenous Japanese breeds), while the Canadian team combine a quite modest set of ancient dog data with an extensive set of all previously reported wolf and dog sequences plus some new ones. Therefore, the Japanese study draws conclusions *only* about the history of Japanese dogs, while the Canadian study is able to discuss the origins of North American dogs in relation to dog evolution in general.

Part Seven contains the concluding chapter, in which I attempt to synthesise various issues raised by other authors. One of these issues is the process of domestication, since some of the *a priori* assumptions associated with dog domestication (such as which subspecies or population of wolf was ancestral to the dog and under what conditions such a process might have taken place) clearly have a strong influence on the taxonomic identification of early dog material. Another important issue is recent and ancient hybridisation of the dog with its ancestral species (the grey wolf, *Canis lupus*), which presents taxonomic problems for both skeletal and genetic analyses of archaeological material. A third issue is one of clarifying regional variation in morphology (including size, shape, tooth anomalies, pathologies and tooth wear), since without consistent reporting practices it will not be possible to assess what (if anything) the morphological differences and similarities between populations tell us about the animals themselves and their associated human cultures. Lastly, the significance of particular similarities and differences related to the dog's role within human societies are discussed, such as burial practices, evidence of butchering, ritual sacrifice, and artistic representations. Comparative photographs of skeletal material brought by participants to the ICAZ Dog Symposium are used as illustrations here.

The second presentation in this final section is actually a comprehensive bibliography on dogs and wolves (stressing archaeological occurrences, hybridisation and the process of domestication), a unique and comprehensive listing of literature.

Together, these chapters provide a "sampler" of research on the history of the domestic dog world-wide and suggest where further attention is needed. Such a compilation of essays would not have been possible without the co-operation of all contributors in reviewing the papers of their colleagues, getting manuscripts submitted on time, and offering encouraging bits of enthusiasm for the finished product. Additional outside specialists assisted with reviewing particular manuscripts and provided further insight into the "problem" of interpreting archaeological dog remains. In particular, I would like to thank conference organiser Rebecca Wigen for accommodating our symposium into the 1998 ICAZ meeting; Juliet Clutton-Brock, Sandra Olsen, Robert Yohe and Danny Walker for taking on more than their share of reviewing responsibilities; Dody Fugate for photography; David Moyer for translating problematic computer files; and lastly, David Davison for patiently guiding me through the editorial process.

# Introduction

# INTRODUCTION

Juliet Clutton-Brock

The Natural History Museum
Department of Zoology
Cromwell Road
London SW7 5BD, UK

**Key words:** dog, evolution, archaeozoology, domestication, breeds, feral dogs

**Abstract**

As an introduction to this volume, a wide-ranging review will be given of the new and old evidence for the history of the domestication of the dog, which will begin with a discussion of the dog's place in nature. Topics to be covered, in an introductory form, will be theories on single vs. multiple origins, the process of domestication, the earliest identified finds of domestic dog world-wide, the development of breeds, and living, feral dogs as relics of prehistory.

## The development of ideas on dog origins

In this general introduction I don't want to discuss any detailed research that will be covered by other contributors. I am therefore going to describe the development of ideas about the origins of the dog, my views on the process of domestication, and my involvement with the archaeozoology of the dog over the last forty years.

I was given a puppy of my own when I was seven years old. This was in what was then Southern Rhodesia and the dog was a ridgeback cross who I called Zimba. I was passionately fond of Zimba and at the same time I discovered the excitement of archaeology with the finding of ancient pottery sherds on a hillock near the family house. I believe that my abiding interest in archaeozoology dates from that time.

For the past 40 years there has been a general consensus that the dog is descended solely from the wolf. This belief was not, however, arrived at easily; it had a long history of contention at many levels; I remember one visitor to the Natural History Museum who tried to convince me that the Irish wolfhound was descended from 'the hyaena'. On a more scientific level, it seems that the evidence from DNA analysis may also suggest anomalies which I shall discuss later.

In the 18th century, naturalists believed that every breed of dog had been separately created. Then came Darwin who at first was not at all clear of the taxonomic status of the family of canids and of the domestic dog in particular. It seems he believed that dogs could interbreed with the red fox (*Vulpes vulpes*), for in his pamphlet, *Questions about the breeding of animals* (1840), Darwin inquired: '...if a fox and a hound [hybrid] were crossed with pointer bitches what would the effect be in the first litter...?'

Twenty eight years later, in his *Animals and Plants Under Domestication* (1868), Darwin wrote as follows on the origins of the dog, and there was little discussion about this statement for the next hundred years:

> The chief point of interest is whether the numerous domesticated varieties of the dog have descended from a single wild species or from several. Some authors believe that all have descended from the wolf, or from the jackal, or from an unknown and extinct species. Others again believe, and this of late has been the favourite tenet, that they have descended from several species, extinct and recent, more or less commingled together.

Then came Lorenz who, in *Man Meets Dog* (1954) claimed 'two origins of fidelity' - the childish jackal dog and the proud lupus dog. Perhaps we all know the gossip which whispered that Lorenz only wrote this book because he wanted to buy a new Mercedes. Anyway, Lorenz later rescinded these quite unsubstantiated views on the dog's dual origin in his preface to *The Wild Canids* (Fox, 1975), with incisive comments about the vocalizations of jackals, which never resemble those of the dog.

For English speakers, Zeuner followed with his *A History of Domesticated Animals* (1963: 107) in which he tried to sort out the opposing views and he gave the following summary (in his own words):

> So far, then, there are four alternatives: The domesticated dog is descended (1) exclusively from the wolf, (2) exclusively from the jackal, (3) from both, with the dominance of a small race of wolf, or (4) from a wild dingo. Of these (4) is the best-founded

theory, (1) the second best, whilst (2) is no longer seriously considered probable.

Zeuner considered his 'wild dingo' to be a *Canis ferus*, that is a geographical race, very closely related to the Indian wolf (*Canis lupus pallipes*), with a distribution from Syria, eastwards to north India.

It was at this time that I began to look at the remains of dogs from archaeological sites while working for a Ph.D. under Professor Zeuner, and I published my first paper (1962) on the canid remains from Natufian sites in Palestine that had been identified as dog by Dorothea Bate (1937) and are held in The Natural History Museum, London.Believing that I was a modern scientist who was using up-to-date metrical analyses, I argued that these so-called dogs could not be distinguished from the small Arabian wolf (*Canis lupus arabs*) and that there was no osteological evidence for identifying them as dog.

I was, as we now know, quite wrong, Miss Bate was correct, and these canid remains from Kebarah, Mount Carmel, and Shukbah are part of an assemblage of dogs that has been added to recently, in some numbers, from Natufian sites in Israel, dated to around 12,000 years ago (Davis & Valla, 1978; Dayan, 1994; Tchernov & Valla, 1997).

It was in about 1960 that I first met Barbara Lawrence and Charles Reed who visited thc British Museum (Natural History) as it was then named, on their way to and from excavations in the Near East, and we had never-to-be forgotten discussions on bones, dogs, and the ancient peoples of western Asia.

Barbara Lawrence, amongst many other biologists, was now discounting the jackal as ancestor of the dog and in 1966 she took part in a symposium on the *Ecology and Behaviour of the Wolf* (Banks, Pimlott & Ginsburg, 1967) that became a seminal publication, not only in its innovative attitudes to the wolf but also because it included the now classic works of Schenkel and Scott on the behaviour of the domestic dog in relation to that of the wolf.

Ten years later in 1977, I had the privilege of visiting Barbara Lawrence at the Peabody Museum where she showed me the very large collection of dog remains that she was working on, in the footsteps of her predecessor, G. M. Allen. This was all part of my training in the infant subject of archaeozoology which has, over past decades, grown into an established, multidisciplinary field of research.

I will now move on to the process of domestication of the dog, as I have come to see it. There are two fundamental processes that combine to cause the evolution of a domestic species, and I use these words on purpose because I do believe that domestication is as much a part of evolution as the evolution of any wild species. The two processes are the biological and the cultural.

**Evolution of *Canis familiaris***

The biological process of domestication begins when a small number of animals are separated from the wild species and become habituated to humans. If these animals breed they form a founder group, which is changed over successive generations both in response to natural selection under the new regime of the human community and its environment, and by artificial selection for economic, cultural, or aesthetic reasons .

I see the biological process of domestication as an evolution that mimics that of island races of wild animals that become separated from their mainland ancestors. And I see the crucial factor in this differentiation of the new species as the chance genetic composition of the tiny group of founding animals, which is entirely responsible for all subsequent change. As discussed by Mayr as long ago as 1942, in his formulation of the founder effect, the genetic reorganization of this isolated population permits evolutionary changes that are many times more rapid than the changes within populations that are part of a continuous system. Adding to this, in the case of the dog, there may well have been innumerable separate tamings of wolves over time and over this carnivore's vast range which could have led to many small founder populations of domestic dogs, later to be subjected to both natural and artificial selection.

Once these founder populations of dog have come into being and their remains can be recognized in prehistoric sites we are faced with the problem of how to define and name them taxonomically. Are these canid remains still wolves, or are they perhaps descendants of some other unrecognized dingo-like ancestor, or are they a new species? The answer must surely depend on the definition of the species as a taxonomic unit. Today, the most commonly used definition is the biological species concept which holds that a species is a group of interbreeding natural populations that is genetically isolated from other such groups as a result of physiological or behavioural barriers (Mayr, 1966: 19).

There is a commonly held view that the separation of two species can be determined by whether or not they will produce fertile offspring when interbred, but on its own, the state of fertility of hybrid offspring is a quite inadequate means of defining a species. Many mammals that are normally considered to be good species will interbreed, although, because of a behavioural barrier, they may not usually do so in the wild, and their offspring will be fertile, as can occur with the dog, wolf, jackal, and coyote.

As far as I can see, the only reason for discounting domestication as an evolutionary process is the sociological view that any anthropogenic activity must be outside nature. However, I see no discontinuity between the human and animal worlds and I believe that domestication is an evolutionary process. I consider that if the domestic form of an animal is for all intents and purposes separated reproductively from the wild form then it should be classified as a separate species. It is for this reason that I argue that domestic animals should not be excluded from formal zoological nomenclature, and the use of Linnaean nomenclature, including *Canis lupus* for the wolf, *Canis familiaris* for the domestic dog, and *Canis dingo* for the dingo. I will discuss my views on the dingo later.

Once domestication has occurred and a new species is established, in this case the domestic dog, new breeds are produced from the domestic form by further reproductive isolation leading to genetic drift, as in the founder populations of new subspecies in the wild. The founders of the new breed contain only a small fraction of the total variation of the parent species, and it becomes a genetically unique population, which continues to evolve under natural and artificial selection. At any point the process can begin again and further new breeds can be developed by cross-breeding. The extreme phenotypic diversity present in modern breeds of dog would be the expected result of this combination of isolation, selection, and cross breeding.

The second fundamental process of domestication is the equally important cultural process, which affects both the human domesticator and the animal domesticate. Domestication of the dog had to end, if not to begin, with ownership. In order to be fully domesticated animals have to be incorporated into the social structure of a human community and become objects of ownership, inheritance, purchase, or exchange. The final relationship between human and wolf is the transformation from mutual antagonism with perhaps the occasional sharing of prey to total human control and domination.

Unlike the livestock animals, which probably evolved into domestic species with the biological and cultural processes impacting simultaneously on the wild species, the dog, like the brown rat, can be envisaged to have gone through an intermediate stage of symbiosis. By this, I mean that family groups of semi-tame wolves may have lived by scavenging the debris from around human hunting camps or settlements, just as pariah dogs do today in many parts of the world. Puppies from these packs would then have been gradually enfolded within the family life of the human social group and their progeny would become reproductively isolated from the wild wolves.

Support can be claimed for the theory that wolves were tamed and began to breed in this way in many different parts of the world by the large number of canid remains that can be identified as dog from sites of between 14,000 and 10,000 years ago. Although what is rather strange about these remains is their rather consistent morphology and their very marked differences in size and skull shape from the wolf, at least from the large European and North American wolf. This could be seen as evidence for diffusion of the earliest dogs from a few core areas such as western and eastern Asia. This second theory fits well with the osteological evidence that the early dogs are very similar to the present day small wolves from western Asia and with the work of Hemmer (1990) and others on the reduction of the cranial capacity as a direct result of the process of domestication. Hemmer has put forward the premise that it was the small subspecies of wolves and wild cats and those with smaller cranial capacities that were the progenitors of the domesticates.

On the other hand, there is no fossil, osteological evidence that the western Asiatic wolf at the end of the Pleistocene, 12,000 years ago, was indeed small, and the small size of the present wolves could be a quite recent development, as discussed by Dayan (1994).

Within the last few years the possible origins of the dog have received the attention of molecular biologists and the claim by Vilà *et al.* (1997) for the separation of dog from wolf to have occurred more than 100,000 years ago has received widespread publicity as well as viewed with considerable disbelief. Because this work of Vilà and his colleagues was so comprehensive it is worth summarising the results here, as they are bound to be used as a basis for much future discussion. The mitochondrial DNA was analysed of 140 modern dogs from 67 breeds and five crossbreeds plus 162 wolves from 27 populations in Europe, Asia, & North America and the authors claimed that:

1 The domestic dog is indeed closest in its origins to the wolf, *Canis lupus*, although, because mitochondrial DNA is maternally inherited, interbreeding between female dogs and male coyotes or jackals would not be detected.

2 The extreme phenotypic diversity of modern breeds of dog indicates that they had diverse origins.

3 All analyses supported the grouping of dog haplotypes into four distinct clades. A haplotype is the genotype of mitochondrial DNA; a clade is used here to mean a group that is defined by characters that are exclusive to all its members and distinguish this group from all others:

- CLADE 1 has 19 of the 26 dog haplotypes, comprising many common breeds such as the greyhound as well as the ancient breeds, dingo, New Guinea singing dog, and basenji.
- CLADE 2 has the haplotype from two Scandinavian breeds of dog, the elkhound and the jämthund, and is closely related to two haplotypes of wolves from Italy, France, Romania , and Greece.
- CLADE 3 has three haplotypes from a wide range of breeds for example the German shepherd, Mexican hairless, and Siberian husky.
- CLADE 4 has three dog haplotypes that were identical or very similar to the haplotype found in wolves from Romania and western Russia which suggests recent hybridization between dogs and wolves.

4 Dog sequences cluster with different groups of wolf haplotypes. Therefore, after the origins of dogs from a wolf ancestor, dogs and wolves probably continued to exchange genes. Most breeds show evidence of a genetically diverse

heritage because they contain different haplotypes.

5 The wolf haplotypes showed greater geographic restrictions than the dog haplotypes although three had a widespread distribution in wolves from eastern Europe and Saudi Arabia, while one was widespread throughout North America.

6 The dramatic conclusion of Vilà *et al.* was that, apparently, the wolf and the dog became separate breeding populations at least 135,000 years ago. This conclusion was arrived at from calculating the so-called molecular clock, which is based on the premise that mutation rates are stable. Thus, the molecular scientists argue that the coyote and wolf, which differ from each other by 20 substitutions are believed, on fossil evidence, to have diverged about one million years ago. Consequently the relative small sequence divergence of 12 substitutions between the most different genotypes found in dogs and wolves indicates that they could have separated as much as 135,000 years ago and certainly more than 14,000 years ago which is the time indicated by the archaeozoological record.

I can believe that wolves have been closely associated with hominids, from perhaps even earlier than the evolution of Homo sapiens, but it is hard to believe that these wolves were reproductively isolated from the wild species to the extent that they became genetically distinct so many thousand years ago. What is more, if a distinct race of dogs had evolved and was closely associated with humans, you would expect there to be depictions of them in the wonderful rock art of the French caves, say at Chauvet, dated to 30,000 years ago. But there are no artistic representations of dogs on rock art and no fossil remains of morphologically distinct dogs before around 15,000 years ago.

For the moment, this dramatic finding must remain a puzzle, although various explanations have been put forward. One is that, in fact, mutation rates are not stable and are in some way 'speeded up' by domestication, which would give a false timescale for molecular divergence, while another possibility is reversion to the idea that the dog is descended from a wild dog, now extinct, whose closest relative was the wolf but differed from it by the required 12 substitutions. It is of interest that, without any knowledge of the molecular divergence, Hemmer (1990: 42) speculated about just this possibility. If the world's populations of dogs were indeed descended from a separate lineage from modern wolves it would take us back to Zeuner and others' suggestion of a *Canis ferus* or dingo-like ancestor, and would have the merit that at least it would explain the widespread similar red coloration and conformation of the world-wide 'primitive' dogs, the dingo, Asian pariah, African village dogs, and the Native American feral dogs. There is no fossil evidence for such a species of wild dog, but surely these new uncertainties about the dog's true ancestor, provide an added important reason for attempting to preserve the pure lineages of the 'primitive' feral breeds. This applies particularly to those feral dogs that have had a long history of isolation, such as the New Guinea singing dog and the dingo.

I believe that these 'breeds' or indeed species, should be conserved and classified separately from all others. It is probable that a few domestic dogs were first taken to the island continent that is now called Australia, by boat, thousands of years ago by immigrant peoples.This must have been before the domestication of the pig, which was never taken to Australia as it was to New Guinea and to the Pacific islands in the early prehistoric period.However it must also have been later than 12,000 years ago, when Tasmania became separated from the Australian continent by the sea breaking through the Bass Straits, as there are no archaeozoological records of dogs on Tasmania. It can be speculated with some certainty that the first small founder group of domestic dogs soon began to hunt and feed on easily caught marsupial prey and to breed away from human control. The population of dogs would have expanded rapidly and soon spread widely over the continent. Today dingoes are in decline, but when they are pure bred they remain an ancient relic of early domestication and in my opinion they should have the species name *Canis dingo*, but I do not agree with Corbett (1985) that this taxon should include all the south east Asian dogs.

**Culture and consciousness**

Domestic dogs live in all the same diverse cultures as humans and their learned behaviour has to be responsive to a great range of different ways of life. Until very recently, many ethologists have believed it is 'unscientific' to describe the learned behaviour of animals as culture. But what after all is so special about the cultures of the human species? Every culture whether it is the ancient Egyptian or modern American is simply the product of learned behaviour passed down through the generations by the elders of the social group. I see the view that only humans can have a culture as a misguided relic of behaviourism and I have no problem in accepting that there is an unbroken continuity in the evolution of behaviour, and indeed consciousness, in all mammals including humans. Therefore I like to think that animals are just as capable of learning to live in a cultural milieu as humans and the perceptions of the dog are capable of almost unbounded adaptation, irrespective of whether it is the pampered companion of someone like me or the sledge dog of the Inuit. I cannot therefore, agree with Hemmer (1990: 91, 111) in his claim that the *Merkwelt*, or perceptual world, of the dog is reduced in correlation with the reduction in brain size that appears to occur in nearly all species of mammal that have been domesticated.

I should like to conclude by returning to the changes I have seen in the expansion of research in archaeozoology during the forty years of my academic career. What began with the listing of animal remains from archaeological sites has developed into a multidisciplinary subject with researchers world-wide producing data that are essential not only for archaeologists but also for agricultural historians, conservationists, and ethologists. Two other biological disciplines with practical applications have also been developed during this time, which I like to think I have been closely involved with. The first was the movement that has slowly spread world-wide for the conservation of endangered breeds of domestic animals, and this has included the dingo and other feral dogs. The second is the study of human and animal interactions, with the recently formed International Society for Anthrozoology. It is strange that it has taken so long for interest in the interface between the behaviour of humans and their animal companions to become a respectable part of scientific research and to be given the name anthrozoology.

I can think of many reasons for the late development of anthrozoology, but the main one must be the destructive power of the behaviourists. The theory of behaviourism postulated, in its most extreme form, that animal behaviour may be explained purely in terms of muscular and physiological or hormonal reactions to external and internal stimuli. Behaviourism started with human psychology but spread to studies of animal behaviour from the influence of B. F. Skinner (1904-90) who, in the 1940s, developed the Skinner box for experiments to condition rats, whereby they pressed a lever to gain a reward or avoid punishment. The greatest sin that a behaviourist could commit was to be anthropomorphic.

The influence of the behaviourists lasted for an amazingly long time but at last this era is almost over. I say 'almost' because, to me, behaviourism still has a legacy in the debate, which is increasing in momentum, on whether animals are not only sentient but have consciousness. A 'sentient being' is an organism that has senses and can 'feel' in response to a stimulus, but it is not necessarily 'conscious', that is self-aware. Many biologists and philosophers argue that only humans are self-aware; they not only know things but they know that they know them (Griffen, 1992: 11).

I believe, as do probably most archaeozoologists as well as dog owners, that there is an evolutionary continuum in the human and animal worlds and thereby in degrees of consciousness. In this I am helped by accepting that animal behaviour should be explained in anthropomorphic terms and by our knowledge of the unique partnership that began thousands of years ago between the two species of mammal, human and dog.

**References**

Banks, E. M., Pimlott, D. H. & Ginsburg, B. E. (1967). Ecology and behaviour of the wolf. *American Zoologist* **7**, 171-381.

Bate, D.M.A.(1937). Palaeontology: the fossil fauna of the Wady-el Mughara Caves. In (D. A. E. Garrod & D. M. A. Bate, Eds) *The Stone Age of Mount Carmel. Vol. 1, Part II.* Oxford: Clarendon Press, pp 139-237.

Clutton-Brock, J. (1962). Near Eastern canids and the affinities of the Natufian dogs. *Zeitschrift für Tierzüctungsbiologie* **76**, 326-33.

Corbett, L. K. (1985). Morphological comparisons of Australian and Thai dingoes: a reappraisal of dingo status, distribution and ancestry. *Proceedings Ecological Society Australia* **13**, 277-291.

Darwin C. (1840). Questions about the breeding of animals. *Society for the Bibliography of Natural History, Sherborn Fund Facsimile* **3**.

Darwin, C. (1868). *The Variation of Animals and Plants Under Domestication.* 2 vols. London: John Murray.

Davis, S. J. M. & Valla, F. R. (1978). Evidence for domestication of the dog 12,000 years ago in the Natufian of Israel. *Nature* **276**, 608-10.

Dayan, T. (1994). Early domesticated dogs of the Near East. *Journal of Archaeological Science* **21**, 633-640.

Griffen, D. R. (1992). *Animal Minds.* Chicago: University of Chicago Press.

Hemmer, H. (1990). *Domestication: the Decline of Environmental Appreciation.* Cambridge: Cambridge University Press.

Lorenz, K. Z. (1954). *Man Meets Dog.* London: Methuen.

Lorenz, K. (1975). Foreword. In (M. W. Fox, Ed) *The Wild Canids: Their Systematics, Behavioural Ecology and Evolution.* New York: Van Nostrand Reinhold.

Mayr, E. (1942). *Systematics and the Origin of Species.* New York: Columbia University Press.

Mayr, E. (1966). *Animal Species and Evolution.* Cambridge: The Belknap Press of Harvard University Press.

Tchernov, E. & Valla, F. F. (1997). Two new dogs and other Natufian dogs, from the southern Levant. *Journal of Archaeological Science* **24**, 65-95.

Vilà, C., Savolainen, P., Maldonado, J. E., Amorim, I. R., Rice, J. E., Honeycutt, R. L., Crandall, K. E., Lundeberg, J. & Wayne, R. K. (1997). Multiple and ancient origins of the dog. *Science* **276**, 1687-1689.

Zeuner, F. E. (1963). *A History of Domesticated Animals.* London: Hutchinson.

# Part 1

# Evolution & early dogs

# DOG EVOLUTION: A ROLE FOR THYROID HORMONE PHYSIOLOGY IN DOMESTICATION CHANGES

Susan J. Crockford
*Pacific IDentifications Inc.,*
6011 Oldfield Road RR #3
Victoria, B. C. V8X 3X1
Canada

**Key words:** dog, speciation, protodomestication, evolution, heterochrony, paedomorphosis, domestication, wolf, *Canis familiaris*

### Abstract

In this paper, I examine the evolution of the dog from an evolutionary perspective and present a model for domestication that does not require humans to have played an active role in the initial stage of the generation of primitive dogs. The model suggests that the first stage of domestication (*protodomestication*) was a natural colonisation process initiated by wolves themselves. Individual variation in the behavioural response of animals to stress is pivotal to this model, since there is an intimate connection between the adrenal stress-response system that mediates such behaviour and genetically-controlled thyroid hormone production. Thyroid hormone *also* mediates or controls foetal development, post-natal growth, pigment production in hair and skin, and reproduction. Since in essence all of the traits known to change as a result of domestication are controlled by thyroid hormone, such a physiological connection suggests that a simple biological mechanism may explain the physical and physiological changes that occurred in dog domestication.

I suggest that the anthropogenic environments associated with early human permanent settlements were attractive new habitats for colonising wolves but physiologically stressful due to the close and constant proximity of humans. Such habitats must have attracted stress-tolerant individuals as colonisers preferentially over less stress-tolerant variants. Thus, *only* the subset of animals which possessed a similar genetic phenotype for thyroid hormone physiology made up the founding population of incipient dogs. This subset of wolves which possessed a limited amount of variation for thyroid hormone production, interbreeding within a small isolated group, would have generated descendants with both morphological and behavioural differences relatively quickly because of the affects of thyroid hormone production on growth and development. Much later, human selection of physiologically similar subsets of dogs led to the creation of specific morphotypes and then breeds of morphologically and behaviourally distinct dogs.

This model provides a biological mechanism to explain how dog domestication could have occurred on more than one occasion from different wolf subspecies, as suggested by other evidence. While the hypothesis does not preclude the deliberate actions of people in the initial stage of domestication, it does not *require* such interference. The model also provides a testable scientific hypothesis that can be applied to speciation associated with colonisation in other vertebrates. The model thus contributes, for the first time, the theoretical framework necessary for understanding the initial stage of domestication as an evolutionary, rather than a cultural, process.

### Dedication

To Dr. Eitan Tchernov, with thanks, for unconscious inspiration. Eitan wanted very much to participate in the 1st ICAZ Dog Symposium but health problems intervened. His absence was keenly felt. I sincerely hope this paper represents the cusp of renewed interest in the study of domestication as an evolutionary process, a direction for scientific research I think he would be pleased to see taken.

## Introduction

Definitions abound for what constitutes a domestic animal but in general they encompass both cultural and biological criteria. Opinions vary regarding the extent and nature of cultural influences on domestication (eg, Crabtree, 1993; O'Connor, 1997) and a protracted debate on the subject is not attempted here. However, all definitions of domestication either state explicitly, or imply, that humans made a conscious choice to tame certain wild animals for their own use or benefit, requiring that *some animals be deliberately removed from the wild to form the source population* (Clutton-Brock, 1992a,b; Darwin, 1868; Price, 1984; Reed, 1984; Tchernov & Horwitz, 1991; Zohary, Tchernov & Horwitz, 1998). It is difficult, however, to reconcile this view with domestication in all taxa: quite simply, the similarities in biological changes that occur across taxonomic orders (see below) make one question how an explanation for domestication that depends on deliberate intent could have the same result for wolves as for sheep and pigs (see also Budiansky, 1992; O'Connor, 1997; Coppinger & Schneider,1995; Morey, 1990, 1992).

The morphological and physiological changes associated with domestication are surprisingly parallel even between distinct taxonomic lineages and ecologically diverse animals (Clutton-Brock, 1992a,b; Hemmer, 1990; Tchernov & Horwitz, 1991). In summary, these changes include:

- overall body size reduction
- shortening of the facial bones of the skull accompanied initially by tooth crowding and later by tooth size reduction
- reduction of horn size in animals with horns

- increased incidence of piebald coat colours (solid with white)
- increases in solid coat colours (red, yellow and black)
- increased docility/lessened fearfulness
- increased fecundity (larger litter sizes)
- lowered age of sexual maturity
- changes in reproductive timing (increases in frequency of oestrus (*polyoestrusness*) or shifts in its seasonal occurrence)

The biological differences between wild animals and their domestic descendants listed above can be clearly established for dogs, pigs, cattle, sheep, goats, and the lesser known Asian bovids - water buffalo and mithan (Mason, 1984). See Wayne (1986c) for a discussion of the changes associated with domestication in horses and cats.

Some of the distinctive differences between wild ancestors and their domestic descendants listed above have recently been shown to result from changes in developmental rates or in the timing of developmental events - the general evolutionary process known as *heterochrony* (in the dog, the particular process is called *paedomorphosis* or *neoteny*). For example, dogs have been shown to be paedomorphic in relation to wolves, based on several studies of behaviour and growth (Goodwin, Bradshaw & Wickens, 1997; Morey, 1992, 1994; Wayne, 1986a,b,c). This means that the most obvious differences between the wolf and dog result from the dog maturing at a stage of development equivalent to that of a juvenile wolf in both morphology and behaviour. These "juvenilized" traits seem to have been caused by a genetically-controlled change in growth rate that is implemented primarily during the early postnatal growth period and perhaps the late foetal stage (Wayne 1986a,b,c), an inherited pattern that may also be breed-specific (Coppinger & Schneider, 1995; Wayne, 1986c). Other traits (such as increases in fecundity, docile behaviour, piebaldness, and frequency of oestrus), appear to be associated *consequences* of paedomorphosis (Belyaev, 1979; Coppinger & Feinstein, 1991; Coppinger & Schneider, 1995; Hemmer, 1990; Morey, 1994).

Research on dogs and wolves therefore, suggests that changes in developmental rates and timing may account for *all* of the morphological, physiological, and behavioural differences between them - a pattern evident in many other domestic and wild animal lineages, including humans (Geist, 1986; Gould, 1977; McKinney & McNamara, 1991; Wayne, 1986c). As dogs represent an important and relatively well-understood example of developmental change (Gould, 1994; McKinney, 1998; McKinney & Gittleman, 1995), understanding how and why domestication occurred may shed critical insight into this general evolutionary phenomenon (eg Crockford, in press).

Archaeological skeletal remains are an essential source of evidence for unravelling the history of domestic species (Crabtree, 1993; Davis, 1987). However, the archaeological record does not show the clear intermediate stages between wild and domestic forms that we are led to expect from cultural explanations of domestication (Clutton-Brock, 1992b; Tchernov & Horwitz, 1991). The absence of intermediate forms in early dogs, for example, suggests that the physical changes associated with initial domestication were not gradual but abrupt (Hemmer, 1990; Morey, 1992, 1994), which in turn implies that the process must have been relatively rapid. This phenomenon is not unique to domesticates - similarly rapid speciation rates also appear to have occurred in non-domestic taxa. For example, the distinct "species flocks" of fish (groups of genetically similar but ecologically and morphologically distinct species) that inhabit some freshwater lakes in Africa have been postulated as having developed within as little as two hundred years (Meyer *et al.,* 1980; Owen, Crossley & Johnson, 1990).

Recognising early archaeological specimens of domestic taxa is difficult, in part because of *a priori* assumptions that have been made about domestication. For example, due to the small size of most early dog specimens in the Middle East, a small subspecies of wolf (similar in size to modern Israeli wolves) is presumed to be the ancestor of all domestic dogs (Clutton-Brock, 1995; Lawrence, 1967; Olsen, 1985). However, the assumption of a small wolf ancestor has been challenged recently by Dayan (1994), who pointed out that the wolf which lived in Israel just prior to the time when undisputed early dog remains are recovered in the Middle East (*c.* 12,000 BP) was not a small subspecies, but a very large one - typical of late Pleistocene fauna (see also Davis & Valla, 1978; Kurtén, 1968, 1988; Tchernov & Valla, 1997). Dayan suggested two possible explanations: 1) that if *c.* 12,000 year old Natufian specimens represent the earliest Israeli dogs, they must not have descended from a particularly small-sized subspecies of wolf. Consequently, early dogs found in regions which supported large-sized wolf subspecies, such as England, Germany and northern Iraq (Benecke, 1987; Clutton-Brock, 1995; Musil, this volume) could be indicators of *in situ* domestication from local wolves (ie suggesting that multiple domestication events are indeed possible); 2) alternatively, a canid specimen recovered from the period 14,000-24,000 BP (Dayan, 1994) may in fact be the earliest Israeli dog, suggesting that it likely descended from an earlier, small-sized wolf ancestor such as those found in 20,000-30,000 BP deposits. This latter hypothesis would uphold the assumption of a small-sized wolf ancestor for the Middle Eastern dog but push the date of the earliest archaeological evidence for dogs in Israel to at least *c.* 15,000 BP, and perhaps earlier. This debate is but one example of how the uncertainty surrounding our understanding of the process of dog domestication affects the interpretation of archaeological remains and illustrates why we need a critical reappraisal of the issues.

**An evolutionary perspective: "protodomestication"**

In considering domestication from an evolutionary perspective, I suggest that the process is actually composed of two distinct parts. The first stage (which I refer to as *protodomestication)* is in essence a natural speciation process whereby certain wild ancestors generate descendants with modified biological features, a process associated with the natural colonisation of human-dominated habitats. *Classic domestication* is the term I suggest be used to describe the processes of conscious and unconscious human selection (working in concert with natural selection) that modify any captive or commensal population, whether those animals are products of prior protodomestication or individuals deliberately removed from the wild. While these definitions contrast sharply with the traditional view of domestication (which collapses the two stages together for all taxa), I argue that many animals - especially dogs - have undergone both protodomestication and classic domestication while others are products of classic domestication only. Although the terms *protodomestication* and *classic domestication* might ultimately be replaced with other terminology, they suit the purposes of this discussion which intends only to distinguish between initial speciation events and those that came afterward.

Protodomestication is defined here as a colonisation process that occurred within human dominated or "anthropogenic" environments. An anthropogenic environment is a term used to describe a localised set of environmental conditions created by the impact of human populations, in this case the conditions resulting from the formation of permanent settlements first seen at the end of the Pleistocene (Horwitz & Tchernov, 1991; Tchernov, 1993a,b, 1998; Zohary *et al.,* 1998). Settlements need not have been absolutely permanent to have created anthropogenic effects. For example, Magdelanean hunters ( Musil, this volume) may have established large camps in the same location every year that were occupied over several seasons, since the wild horses they harvested were non-migratory animals. While not strictly permanent, such repeatedly occupied multi-season settlements may occasionally have attracted wolf colonisers.

An anthropogenic environment is also a habitat dominated by the continuous presence or proximity of humans. Colonisation of these new habitats was a choice animals could make, whether driven exclusively by the advantages offered (such as easily available food) or in concert with pressures exerted by changing conditions in their natural habitat. Individual animals with the highest tolerance to the associated stresses of these new environments probably provided the source population that ultimately became "domesticated". Figure 1 illustrates this colonisation model.

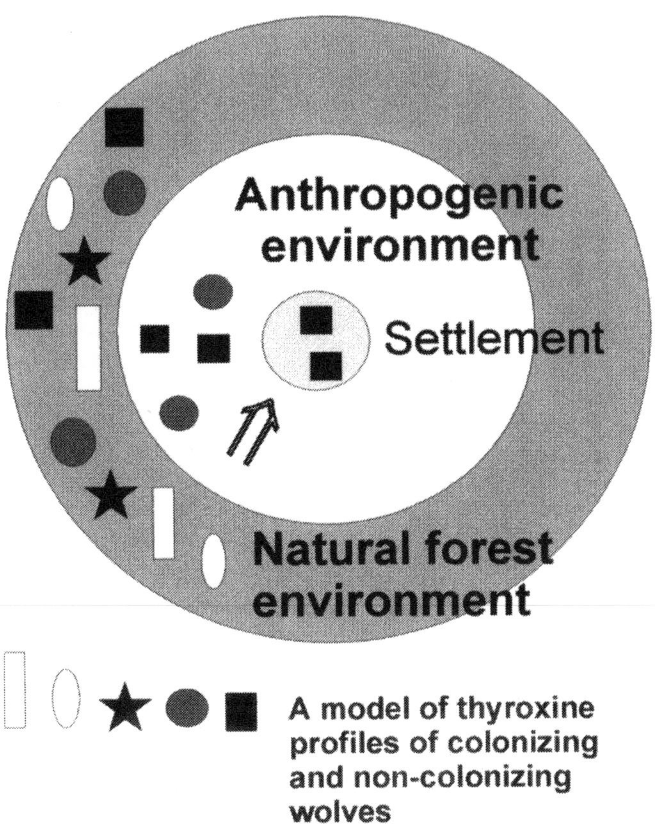


***Figure 1*** *A model for dog protodomestication through colonisation of anthropogenic environments. Wolves of varying physiological phenotypes in a wild population are represented by 5 different shapes (ie animals that possess slightly different thyroxine production profiles and thus show different behavioural responses to stress), only one of which is stress-tolerant enough to live and breed inside an anthropogenic environment.*

Individual variation in behavioural responses to stress is the key to explaining how wolves could have evolved rapidly into dogs without direct and deliberate human interference. There is an intimate connection between the adrenal stress-response system that mediates such behaviour and genetically-controlled thyroid hormone metabolism (Hadley, 1984; Boissy, 1995). This physiological connection suggests that a plausible biological mechanism may exist for the physical and physiological changes that occurred in all domesticates, because thyroid hormone also mediates the control of foetal development, post-natal growth, pigment production in hair and skin, and reproduction (Fig. 2). Since a physiological basis for domestication changes has been sought for at least 40 years, this is not a totally new concept. Thyroid hormone, however, has never been specifically implicated before now. A few experimental domestication programs have been undertaken, which attempted to demonstrate how the physical and behavioural changes associated with domestication could be generated. Before I embark on a discussion of the physiology of thyroid hormone, I summarise one of the experiments which successfully produced domestic animals (see also Hemmer, 1990).

**Experimental domestication**

A series of experiments on foxes which began in Siberia in the late 1950s (Belyaev, 1979, 1984) represent an elegant example of the intricate biological interactions pertinent to this discussion. The experiments began with a large population of farmed "silver" foxes (a naturally-occurring black morph of the red fox, *Vulpes vulpes*), that retained all the characteristics of the wild fox: it had one reproductive cycle per year, an annual fall moult and was generally timid of humans. The researchers assessed foxes from several fur farms and selected as an experimental population individuals which demonstrated noticeably less fearful behaviour towards humans (which I call *tolerant* ). After several generations of breeding and selecting only for tolerant behaviour, the timing of oestrus and the annual moult of many females had shifted to earlier in the season by several months. As the experiment continued, oestrus and moult receded further still until several females were experiencing two oestrous cycles annually. After 20 generations (20 years), some females were able to produce two litters per year and a number of other novel traits suddenly appeared, including a curled tail, drooping ears, a pattern of distinctive brown markings ("brown piebald"), and a more classic white piebald (black and white) pattern. The animals from this last generation had smaller adrenal glands associated with hormonal changes of several kinds and subsequent osteometric analysis of selected foxes found small changes had also occurred in cranial conformation (Trut *et al.,* 1991a,b).

Belyaev described the animals with these novel morphological and physiological traits as having remarkably "dog-like" behaviour: they barked and were quite unafraid of people. These behaviours were "dog-like" because they were juvenile, or paedomorphic, traits. Belyaev concluded that something in the selection for tolerant behaviour towards humans was not only causing paedomorphic changes in both morphology and behaviour but also disrupting the normally constrained timing of sexual reproduction. Subsequent investigations suggest that these changes were not caused by spontaneous mutations or other known mechanisms, which left investigators unable to explain the results (Belyaev, Ruvinsky & Trut, 1981; Trut, 1988, 1999; Trut & Osadchuk, 1997).

Coppinger & Schneider (1995: 41) assessed Belyaev's experiments in their discussion of neoteny (ie paedomorphosis) in relation to dog evolution and stated that:

> the transformation from wild fox into dog-like creature that Belyaev made in a few generations simply by selecting for tameness must mirror in some way the original transformation of wolf into dog. Belyaev thought that selection for tameness "destabilized" the genome in such a way as to create evolutionary novelties. As we learn more about gene action and biochemistry we find that he was probably not very far from the truth. The diestrous heat cycle and the piebald coat may be results of a neotenic process, *but they are not neotenic characters.*

My review of the biological literature suggests that selection for a single behavioural phenotype could have triggered the observed suite of well co-ordinated physiological and morphological traits in Belyaev's foxes because the common denominator among control mechanisms for these traits is thyroid hormone.

**Thyroid hormone metabolism**

Thyroid hormone is essential for early embryonic cell migration and differentiation; for both embryonic and postnatal growth (including brain development); hair production and manufacture of pigment for both skin and hair; adrenal gland function; and the development and function of the gonads (Hadley, 1984; McNabb & King, 1993). In essence, all of the traits which are known to change as a result of domestication, including piebaldness (Crockford, in press), are controlled by thyroid hormone.

Although thyroid hormone can be absorbed and utilised directly from food sources (Hadley, 1984) it is primarily a manufactured hormone. Present in all vertebrates, the thyroid gland (or glands, where it is a paired organ) produces two similar hormones, thyroxine ($T_4$) and its deiodinized, active form, triiodothyronine ($T_3$). Since the concentrations of $T_3$ and $T_4$ produced are interdependent, they are often referred to collectively by the term "thyroxine", which I use here for convenience.

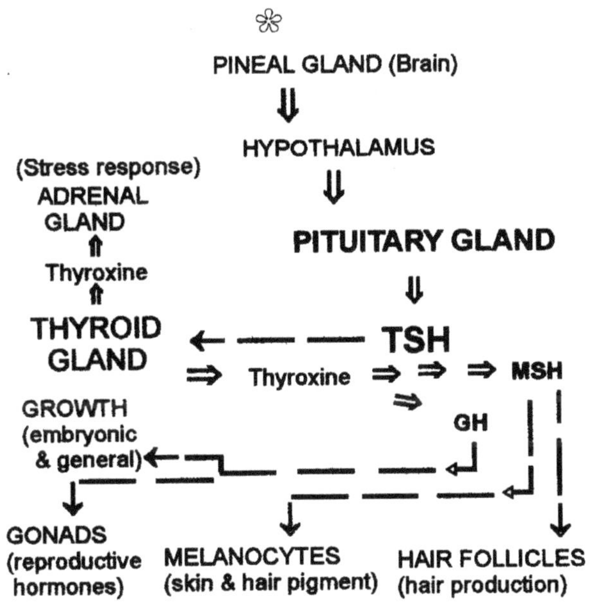


***Figure 2*** *The hierarchical influences of thyroxine on other body systems. TSH, thyroid stimulating hormone; MSH, melanocyte stimulating hormone; GH, growth hormone. Note that thyroxine affects embryonic and early foetal development directly (see text) and post-natal development via growth hormones.*

Thyroxine manufacture begins in the brain, due to stimulation of the pineal, hypothalamus and pituitary glands (Fig. 2). Thyroid stimulating hormone (TSH), released by the pituitary gland in response to intermittent stimulation from the pineal gland, initiates the production of thyroxine in distinct pulses of hormone. Thyroxine subsequently induces the production of several other essential hormones: growth hormone (GH), melanocyte stimulating hormone (MSH) and adrenal steroids (eg, adrenalin). MSH stimulates the production of pigment (*melanin*) in specialised cells that exist in skin and hair follicles, as well as stimulating growth of hair in hair follicles (*melanocytes*). GH stimulates general growth and development in various tissues, and triggers the release of reproductive hormones. Thyroxine itself initiates the production of adrenal steroids, which mediate the behavioural response to stress and stimuli (Boissy, 1995; Hadley, 1984; Windle *et al.*, 1998). As a consequence of the intermittent nature of thyroxine secretion, production of these critical hormones also fluctuates in a pulsatile manner (Thompson *et al.*, 1992). The pulsatile nature of this production is clearly apparent when blood concentrations of hormones are measured frequently and graphed over time. Such hormone profiles consist of repeated peaks (secretion phase) and troughs (non-secretion phase) that vary in height and breadth over the course of a day (Gancedo *et al.*, 1997; Windle *et al.*, 1998). Distinct patterns of thyroxine production shift somewhat with age and season in response to changing physiological demands (Kaptein, Hays, & Ferguson, 1994; Tomasi & Mitchell, 1994; Wilford, 1995).

Recent evidence suggests that thyroxine production patterns are both species-specific (Gancedo *et al.*, 1997; Harris, 1959; Schew, McNabb & Scanes, 1996; Silverin & Rudas, 1996) and breed-specific: differences in $T_4$ levels between breeds of dogs have been documented (Ferguson, 1994; Nunez *et al.*, 1970), as have differences in thyroxine-dependent hormones in dogs and rats (Arons & Shoemaker, 1992; Windle *et al.*, 1998). Of particular significance to this discussion is that because of the hierarchical relationship between thyroid and growth hormones, *species-specific patterns of thyroxine production results in species-specific growth and development.* In mammals, thyroid hormone regulates embryonic and early foetal growth directly (before the thyroid gland of the embryo is formed) through transfer of thyroxine across the placenta (Thommes & Wood, 1993). Late foetal growth is controlled by thyroid-induced growth hormone of the foetus, augmented by maternal contributions. At all stages of development, therefore, the pulsatile pattern of thyroxine secretion controls species-specific growth. Within the range of hormone patterns that define a species, these profiles also vary slightly in particular ways for each sex and for distinct morphotypes or breeds.

In addition to within and between species variation, it appears that thyroxine profiles also vary slightly between individuals. This individual variation is crucial to the role in which thyroid hormone metabolism plays in protodomestication. Precision in timing (frequency of pulses) and absolute amounts of thyroxine produced (amplitude of pulses) must be critical to certain target genes in various cells and organs that use thyroxine as a regulator. Very slight variations in the fluctuating pattern of production of thyroxine must produce small physiological differences among individuals within a population of any species, due to the profound influences which thyroxine has on other systems. These small physiological differences may be the ultimate cause of the noticeable morphological, reproductive and behavioural differences seen amongst individuals in virtually all mammals (Hayes & Jenkins, 1997).

I suggest, therefore, that slight differences in the pattern of production of thyroid hormone (which is itself, of course, genetically controlled) contributed the essential individual phenotypic variation within wolves required for protodomestication of dogs to occur. In other words, while staying within the boundaries of what is required for "wolf" development, slight differences in the height or frequency of thyroxine peaks during critical growth periods cause one wolf, for example, to be smaller or larger than its littermate at maturity, to react somewhat differently to stress of all kinds, to have different coloured fur and for onset of reproductive activity (*ie*, oestrus) to beginning slightly earlier or later. The resulting body size disparity between these individuals, permanently "recorded" in the skeleton, is the only manifestation of slight differences in thyroid physiology that archaeologists will ever see.

In summary, thyroxine metabolism is the body's physiological mechanism for adapting to short-term and long-term changes in environmental conditions, for co-ordinating life-long physiological responses and for implementing species-specific growth and development. If individual differences in thyroxine production patterns contribute strongly to individual variation within species, as I suggest must be the case, then population subdivision according to differences in thyroxine production patterns are a logical candidate for constituting the biological mechanism underlying all instances of protodomestication.

**Thyroxine production and protodomestication**

Individual variation in thyroid hormone production patterns is the key to proposing a scenario in which wolves could be exposed to the selection pressures necessary for the changes experienced by Belyaev's foxes to occur without deliberate human interference. I suggest that certain individual wolves within an ancient population would have been better able than others to colonise anthropogenic environments because they possessed the physiological ability to tolerate the increased stress and stimulus loads associated with human-dominated habitats. Colonising wolves capable of invading anthropogenic habitats would leave their stress-intolerant cohorts behind, effectively dividing the population non-randomly according to physiological phenotype. The invading group of stress-tolerant wolves would thus have a reduced amount of genetic variation in thyroid hormone production patterns. Interbreeding amongst these animals, within a restricted gene pool, would have resulted in the production of offspring with a limited complement of alleles for the genes controlling thyroxine production.

Within only a few generations, a new pattern of thyroxine production would have become established. Without the full range of variability for thyroxine production, descendants of this invading population would almost certainly have been behaviourally, reproductively and morphologically different from the ancestral wolf population after relatively few generations, due to the extensive influences of thyroxine. I suggest that this process could have produced a descendant population of primitive dogs with a well co-ordinated suite of distinctive traits within roughly the same short period of time as Belyaev's foxes (adjusted for the longer generation time for wolves, or *c.* 40 years). This extremely rapid change from one form to the other, even if slower by a factor of five (ie 200 years), would be archaeologically "invisible".

Protodomestication would have been complete once a morphologically-distinct animal, the primitive dog, emerged with its own unique thyroid hormone production profile. Protodomestication must be irreversible after a new thyroxine equilibrium is reached because some essential variation of the ancestor has been lost. Occasional hybridisation with wild cohorts has little effect, partly because paedomorphic domestic traits (genetically determined by the new thyroxine profile) tend to be inherited in dominant fashion (Belyaev, 1979; Scott, Fuller & King, 1959). More importantly, hybrid individuals with limited stress tolerance born in any generation would always tend to flee a human-dominated environment, removing them from the breeding pool of primitive dogs. Once wolves were biologically "domesticated", their distinctive traits (less fearful behaviour, increased tolerance to stress, smaller size, changes in reproductive seasonality, perhaps unique coloration) would have made them uniquely amenable to subsequent human subjugation and selection - whether that selection was deliberate or unconscious. In other words, once the speciation process of *protodomestication* was complete, *classic domestication* processes could begin. Even the first early primitive dogs, as described above, must have possessed behavioural and reproductive traits that encouraged them to mate with each other (Templeton, 1989), and thus qualify as a "true" species distinct from their wolf ancestors from that point onward (*cf* Mayr, 1982). This process constitutes evolution in every sense of the word: wolves and dogs are a natural species pair. There is no rationale for wolves and dogs to be considered "synonymous" entities, as Wilson & Reeder (1993: 281) have decreed: both ancestors and descendants deserve distinct scientific names.

Slight individual variations in thyroxine production profiles must have continued to exist within primitive dog populations and probably increased over time as natural mutations accumulated in the genes controlling thyroxine production. Individual variation in thyroxine profiles provided the raw material for both natural and human selection to implement further change. Some distinctive morphotypes eventually developed as a result of continued selection for traits linked to developmental processes. The intensely juvenile physical and behavioural features of some modern breeds reflects the continued impact that paedomorphosis has had on dog evolution through artificial human selection (Coppinger & Schneider, 1995; Goodwin *et al.,* 1997; Wayne, 1986c). As Ostrander and Giniger (1999) have recently remarked, the astounding diversity of phenotypically distinct and stable dog breeds that have been created within the last 250 years suggest that a surprisingly small number of key genes are responsible for "defining" breed characteristics of coat colour, morphology and behaviour. Recent reports that sighthounds have distinctly lower thyroxine levels than the average for dogs in general, taken in context of the multiple physiological and developmental roles for thyroxine, suggest that these key genes involve thyroxine production (Arons & Shoemaker, 1992; Ferguson, 1994; Nunez *et al.,* 1970). However, spontaneous mutations in other genes (such as those controlling albinism, dwarfism, and hairlessness) have obviously occurred as well, and have been perpetuated through controlled breeding (Hearing, 1993; Wilcox & Walkowicz, 1989).

This model presents a biological explanation for the suggestion made by Clutton-Brock (1995) and others, that protodomestication of the dog could have happened on more than one occasion from different wolf subspecies. Genetic analyses now support this view for dogs (Koop *et al.,* this volume; Vilà, Maldonado & Wayne, 1999; Wayne, 1993), as it has for cattle (Bailey *et al.,* 1996; Bradley *et al.,* 1998). It seems entirely possible, therefore, that a large-sized wolf subspecies could have generated domestic descendants, as discussed earlier in relation to Dayan's

(1994) paper. Clearly however, this model also provides an explanation for why protodomestication of dogs probably happened rarely overall, since the precise set of biological and ecological conditions required to precipitate colonisation of anthropogenic environments was unlikely to have occurred frequently.

## Potential for testing the hypothesis

The hypothesis that individual thyroid hormone phenotypes exist within wolf populations can be tested experimentally by direct measurement of thyroxine production patterns in individual wolves. Similarly, the correlated hypothesis - that a subset of wolves with particular thyroid hormone production patterns split off from the original population during, or as a consequence of, the protodomestication process - should be testable through measurement of thyroid hormone profiles in dogs. If thyroid hormone production patterns constitute the control mechanism by which paedomorphic changes are achieved in dogs, dog thyroxine profiles as a group should be significantly different from wolves. These profiles should also be constrained within certain limits for morphologically distinct breeds.

The dynamic nature of the endocrine system and its inherent sensitivity to stress of any kind will make such testing difficult (Boissy, 1995). However, as Windle *et al.* (1998) have recently demonstrated for adrenal steroids in rats, an automated blood sampling apparatus can be used to measure thyroid hormone levels within and between species frequently enough (every 10-15 minutes) for fine-scale patterns to be discerned (see also Gancedo *et al.*, 1997). Long-term breeding experiments, based on selection for animals of known thyroxine phenotype associated with specific behaviour patterns, should establish the effects of variable modes of thyroxine production or utilisation on developmental change. Ultimately, we should be able to identify the genetic sources of any significant differences.

## Discussion

The underlying physiological interactions between the traits which change predictably as a result of protodomestication suggest that an entirely natural, evolutionary explanation for initial domestication of the dog is both plausible and testable. If reduced fearfulness toward humans is a specific manifestation of the more general trait of *stress tolerance*, then protodomestication may be an entirely natural process involving nothing more than proximity to humans in the early stages. Humans provided the selection pressure that precipitated the division of existing populations of wolves into colonising and non-colonising subsets. The colonisation of anthropogenic environments by a population subset of individual wolves with a common behavioural phenotype could have resulted in the rapid accumulation of a suite of well co-ordinated physiological and morphological traits characteristic of dogs (without archaeologically-visible transition stages) because the common denominator among control mechanisms for these traits is thyroid hormone.

In fact, selection for any one of the traits linked through thyroid metabolism (as listed below in Fig.3) could change all other traits to some degree as a consequence. Therefore, deliberate human selection for wolves that were all white or black, for example, could perhaps have produced essentially the same result as natural colonisation of anthropogenic environments by stress tolerant individuals (if those rare colours were associated with stress-tolerant or "fear-less" behaviour).

| **Thyroid-mediated dog traits** | **Thyroid-mediated control hormones for these traits** |
|---|---|
| - overall body size<br>- snout & skull shape<br>- piebald coat pattern | THYROID HORMONE itself<br>(embryonic effects from maternal sources) |
| - barking<br>- reduced age at sexual maturity | GROWTH HORMONE<br>(foetal & post-natal effects from both maternal sources & the individual's own pituitary gland) |
| - fecundity (litter size)<br>- reproductive timing | SEX HORMONES<br>(reproductive glands) |
| - degree of docility/fearfulness (ie stress tolerance) | ADRENAL STEROIDS<br>(adrenal glands) |
| - solid coat colours<br>- moulting patterns<br>- hair length | MELANOCYTE STIMULATING HORMONE<br>(hair follicles & skin cells) |

***Figure 3*** *Summary of thyroid-mediated dog traits and the hormones that appear to control them.*

Most dog breeders now realise that selection for a single trait like head shape, for example, results in other physical, physiological and behavioural changes whether the breeder likes it or not (eg, Thomson, 1996). Selection for both single and multiple traits that are thyroid-linked, and subsequent crossing of established breeds with different thyroid-linked traits (Coppinger & Schneider, 1995), is probably responsible for a wide range of distinctive breed types. Traits controlled by chance mutations in genes not directly linked to thyroid metabolism further increase the variation available for artificial selection. The amalgamation of traits that are not thyroid-mediated to thyroid-mediated types could account for the existence of over 400 distinct modern breeds (Wilcox & Walkowicz, 1989).

Darwin (1868) appears to have suspected that understanding the process that produced domestic animals might reveal important insights into evolutionary change. However, more than one hundred years have passed since his two-volume treatise on domestic plants and animals first appeared, and real scientific research on domestication as an evolutionary process has so far been extremely limited. I suggest that protodomestication as defined here becomes an appropriate model for describing the truly dynamic relationship that clearly exists between individual variation, adaptation and speciation in all taxa (see Crockford, in press for a more comprehensive account). Specifically, the model provides a biological mechanism to explain how dogs could have evolved more than once from different wolf ancestors with essentially the same result and how morphologically and behaviourally distinct breeds could have been developed. Using dog evolution as an example, the concept of protodomestication provides the theoretical framework necessary for demonstrating experimentally how speciation through colonisation of new habitats actually works. This approach is a first step towards realising significant advances in understanding domestication as an evolutionary process.

**Acknowledgements**

The formulation of the ideas expressed in this essay emerged over several years of reading, writing and discussion with colleagues. I specifically thank Pacific ID partners Becky Wigen and Gay Frederick; colleagues at the University of Victoria (Margot Wilson, David Moyer, Nicholas Rolland); Michael McKinney (University of Tennessee, Knoxville); Lehr Brisbin, Jr. (Savannah River Ecology Laboratory, South Carolina); and Ken McNamara (Western Australian Museum, Perth). Sandra Olsen (Carnegie Museum, Pittsburgh), Juliet Clutton-Brock (Natural History Museum, London) and Quentin Mackie (University of Victoria) provided helpful comments on the manuscript.

**References**

Arons, C.D. & Shoemaker, W. J. (1992). The distribution of catecholamines and beta-endorphin in the brains of three behaviorally distinct breeds of dogs and their F1 hybrids. *Brain Research* **594**, 31-39.

Bailey, J. F., Richards, M. B., Macaulay, V. A., Colson, I. B., James, I. T., Bradley, D. G., Hedges, R. E. M. & Sykes, B.C. (1996). Ancient DNA suggests a recent expansion of European cattle from diverse wild progenitor species. *Proceedings of the Royal Society of London* B **263**, 1467-1473.

Belyaev, D. K. (1979). Destabilizing selection as a factor in domestication. *Journal of Heredity* **70**, 301-308.

Belyaev, D.K. (1984). Foxes. In (I. L. Mason, Ed) *Evolution of Domesticated Animals*. London: Longman Co., pp. 211-214..

Belyaev, D. K., Ruvinsky, A. O. & Trut, L. N. (1981). Inherited activation- inactivation of the star gene in foxes: its bearing on the problem of domestication. *Journal of Heredity* **72**, 267-274.

Benecke, N. (1987). Studies on early dog remains from northern Europe. *Journal of Archaeological Science* **14**, 31-45.

Boissy, A. (1995). Fear and fearfulness in animals. *Quarterly Review of Biology* **70**, 165-191.

Budiansky, S. (1992). *The Covenant of the Wild: Why Animals Chose Domestication*. London: Weidenfeld & Nicolson.

Bradley, D. G., Loftus, R. T., Cunningham, P. & MacHugh, D. E. (1998). Genetics and domestic cattle origins. *Evolutionary Anthropology* **6**, 79-86.

Clutton-Brock, J. (1992a). Domestication of animals. In (S. Jones, R. Martin & D. Pilbeam, Eds) *The Cambridge Encyclopedia of Human Evolution*, Cambridge: Cambridge University Press, pp. 380-385..

Clutton-Brock, J. (1992b). The process of domestication. *Mammalian Review* **22**, 79-85.

Clutton-Brock, J. (1995). Origins of the dog: domestication and early history. In (J. Serpell, Ed) *The Domestic Dog: Its Evolution, Behaviour and Interactions With People*. Cambridge: Cambridge University Press, pp.8-20.

Coppinger, R. & Feinstein, M. (1991). 'Hark! Hark! The dogs do bark...' and bark and bark. *Smithsonian* **21**, 119-129.

Coppinger, R. & Schneider, R. (1995). Evolution of working dogs. In (J. Serpell, Ed) *The domestic dog: its evolution, behaviour and interactions with people*. Cambridge: Cambridge University Press, pp. 21-47.

Crabtree, P. J. (1993). Early animal domestication in the middle East and Europe. In (M.B. Schiffer, Ed) *Archaeological Method and Theory*. Tuscon: University of Arizona Press, pp. 201-245.

Crockford, S. J. (in press) Animal domestication and heterochronic speciation: the role of thyroid hormone. In (K. McNamara & N. Minugh-Purvis, Eds) *Human Evolution Through Developmental Change.* Baltimore: Johns Hopkins University Press.

Darwin, C. (1868). *The Variation of Animals and Plants Under Domestication.* London: J. Murray.

Davis, S. J. M. & Valla, F. R. (1978). Evidence for domestication of the dog 12,000 years ago in the Natufian of Israel. *Nature* **276**, 608-610.

Davis, S. J. M. (1987). *The Archaeology of Animals.* New Haven: Yale University Press.

Dayan, T. (1994). Early domesticated dogs of the Near East. *Journal of Archaeological Science* **21**, 633-640.

Ferguson, D.C. (1994). Update on diagnosis of canine hypothyroidism. In (D.C. Ferguson, Ed) *Thyroid Disorders.* Veterinary Clinics of North America, Small Animal Practice **24.** Philadelphia: W. B. Saunders Co., pp. 515-539.

Gancedo, B., Alonso-Gomez, A. L., de Pedro, N., Delgado, M. J. & Alonso-Bedate, M. (1997). Changes in thyroid hormone concentrations and total contents through ontogeny in three anuran species: evidence for daily cycles. *General and Comparative Endocrinology* **107**, 240-250.

Geist, V. (1986). On speciation in ice age mammals, with special reference to cervids and caprids. *Canadian Journal of Zoology* **65**, 1067-1084.

Goodwin, D., Bradshaw, J. W. S. & Wickens, S. M. (1997). Paedomorphosis affects agonistic visual signals of domestic dogs. *Animal Behaviour* **53**, 297-304.

Gould, S. J. (1977). *Ontogeny and Phylogeny.* Cambridge: Harvard University Press.

Gould, S. J. (1994) *Eight Little Piggies.* New York, W.W. Norton and Co., pp. 382-395.

Hadley, M. E. (1984). *Endocrinology.* New Jersey: Prentice-Hall, Inc.

Harris, G. W. (1959). Neuroendocrine control of TSH regulation. In (A. Gorbman, Ed) *Comparative Endocrinology: Proceedings of the Columbia University Symposium on Comparative Endocrinology, 1958.* New York: J. Wiley and Sons, pp. 202-222.

Hayes, J.P. & Jenkins, S. H. (1997). Individual variation in mammals. *Journal of Mammalogy* **78,** 274-293.

Hemmer, H. (1990). *Domestication: The Decline of Environmental Appreciation.* Cambridge: University Press.

Kaptein, E. M., Hays, M. T. & Ferguson, D.C. (1994). Thyroid hormone metabolism: a comparative evaluation. In (D.C. Ferguson, Ed) *Thyroid Disorders.* Veterinary Clinics of North America, Small Animal Practice **24.** Philadelphia: W. B. Saunders Co., pp. 431-463.

Kurtén, B. (1968). *Pleistocene Mammals of Europe.* London: Weidenfeld & Nicolson.

Kurtén, B. (1988) *On Evolution and Fossil Mammals.* New York: Columbia University Press.

Lawrence, B. (1967). Early domestic dogs. *Zeitschrift fur Saugetierkunde* **32**, 44-59.

Mason, I. L. (1984). *Evolution of Domesticated Animals.* London: Longman Co.

McKinney, M. L. (1998). The juvenilzed ape myth–our "overdeveloped" brain. *Bioscience* **48**, 109-116.

McKinney, M. L. & Gittleman, J. L. (1995). Ontogeny and phylogeny: tinkering with covariation in life history, morphology, and behavior. In (K. J. McNamara, Ed), *Evolutionary Change and Heterochrony.* Chichester: J. Wiley and Sons, pp. 21-47.

McKinney, M. L. & McNamara, K. J. (1991). *Heterochrony – The Evolution of Ontogeny.* New York: Plenum Press.

McNabb, A. F. M. & King, D. B. (1993). Thyroid hormone effects on growth, development, and metabolism. In (M. P. Schreibman, C. G. Scanes & P. K. T. Pang, Eds) *The Endocrinology of Growth, Development, and Metabolism of Vertebrates.* New York: Academic Press, pp. 393-417.

Meyer, A., Kocher, T. D., Basasibwaki, P. & Wilson, A. C. (1990). Monophyletic origin of Lake Victoria ciclid fishes suggested by mitochondrial DNA sequences. *Nature* **347**, 550-553.

Morey, D. F. (1990). *Cranial allometry and the evolution of the domestic dog.* Ph.D. Thesis, University of Tennessee.

Morey, D. F. (1992). Size, shape and development in the evolution of the domestic dog. *Journal of Archaeological Science* **13**, 119-145.

Morey, D. F. (1994). The early evolution of the domestic dog. *American Scientist* **82**, 336-347.

Nunez, E. A., Becker, D. V., Furth, E. D., Belshaw, B. E.& Scott, J. P. (1970). Breed differences and similarities in thyroid function in purebred dogs. *American Journal of Physiology* **218**, 1337-1341.

O'Connor, T. P. (1997). Working at relationships: another look at animal domestication. *Antiquity* **71**, 149-156.

Olsen, S. J. (1985). *Origins of the Domestic Dog: The Fossil Record.* Tucson: University of Arizona Press.

Ostrander, E. A. & Giniger, E. (1997). *Semper fidelis*: what man's best friend can teach us about human biology and disease. *American Journal of Human Genetics* **61**, 475-480.

Owen, R. B., Crossley, R. & Johnson, T. C. (1990). Major low levels of Lake Malawi and their implications for speciation rates in cichlid fishes. *Proceedings of the Royal Society of London B* **240**, 519-53.

Price, E. O. (1984). Behavioral aspects of animal domestication. *Quarterly Review Biology* **59,** 1-32.

Reed, C. A. (1984). The beginnings of animal domestication. In (I. L. Mason, Ed) *Evolution of Domesticated Animals.* London: Longman Co., pp. 1-6.

Schew, W. A., McNabb, F. M. A. & Scanes, C. G. (1996). Comparison of the ontogenesis of thyroid hormones, growth hormone, and insulin-like growth factor-I in *ad Libitum* and food-restricted (altricial) European starlings and (precocial) Japanese quail. *General and Comparative Endocrinology* **101**, 304-316.

Scott, J. P., Fuller, J. L. & King, J. A. (1959). The inheritance of annual breeding cycles in hybrid basenji-cocker spaniel dogs. *Journal of Heredity* **50**, 254-261.

Tchernov, E. (1993a). The impact of sedentism on animal exploitation in the southern Levant. In (H. Buitenhuis & A. T. Clason, Eds) *Archaeozoology of the Near East: Proceedings of the First International Symposium on the Archaeozoology of South-Western Asia and Adjacent Areas.* Leiden: Universal Book Services, pp. 10-26.

Tchernov, E. (1993b). From sedentism to domestication-a preliminary review for the southern Levant. In (A.Clasen, S. Payne & H. P. Uerpmann, Eds) *Skeletons in her Cupboard: Festschrift for Juliet Clutton-Brock.* Oxford: Oxbow Books Monograph **34**, pp. 189-233.

Tchernov, E. (1998). Are late Pleistocene environmental factors, faunal changes and cultural transformations causally connected? The case of the southern Levant. *Paleorient* **23**, 209-228.

Tchernov, E. & Horwitz, L. K. (1991). Body size diminution under domestication: unconscious selection in primeval domesticates. *Journal of Anthropological Archaeology* **10**, 54-75.

Tchernov, E. & Valla, F.F. (1997). Two new dogs, and other Natufian dogs, from the Southern Levant. *Journal of Archaeological Science* **24**, 65-95.

Thommes, R. C. & Woods, J. E. (1993). Endocrine regulation of the growth/development of warm-blooded vertebrate embryo/fetuses. In (M. P. Schreibman, C. G. Scanes & P. K .T. Pang, Eds) *The Endocrinology of Growth, Development, and Metabolism of Vertebrates.* New York: Academic Press, pp. 495-518.

Thompson, Jr., D. L., Rahmanian, M. S., DePew, C. L., Burleigh, D. W., DeSouza, C. J. & Colborn, D. R. (1992). Growth hormone in mares and stallions: pulsatile secretion, response to growth hormone-releasing hormone, and effects of exercise, sexual stimulation, and pharmacological agents. *Journal of Animal Science* **70**, 1201-1207.

Thomson, K. S. (1996). The fall and rise of the English bulldog. *American Scientist* **84**, 220-223.

Tomasi, T. E. & Mitchell, D. A. (1994). Seasonal shifts in thyroid function in the cotton rat (*Sigmodon hispidus*). *Journal of Mammalogy* **75**, 520-528.

Trut, L. N. (1988). The variable rates of evolutionary transformations and their parallelism in terms of destabilizing selection. *Journal of Animal Breeding and Genetics* **105**, 81-90.

Trut, L. N. (1999). Early canid domestication: the fox-farm experiment. *American Scientist* **87**, 160-169.

Trut, L. N., Dzerzhinsky, F. Y. & Nikolsky, V. S. (1991a). A principal component analysis of changes in cranial characteristics appearing in silver foxes (*Vulpes vulpes* Desm.) under domestication (in Russian). *Genetika* **27**, 1440-1449.

Trut, L. N., Dzerzhinsky, F. Y. & Nikolsky, V. S. (1991b). The intercranial allometry and morphological changes in silver foxes (*Vulpes vulpes* Desm.) under domestication (in Russian). *Genetika* **27**, 1605-1611.

Trut, L. N. & Osadchuk, L. V. (1997). *Evolutionary-Genetic and Genetic-Physiological Aspects of Fur Animal Domestication: A Collection of Reports.* Oslo: IFASA/Scientifur.

Vilà, C., Maldonado, J. E. & Wayne, R. K. (1999). Phylogenetic relationships, evolution, and genetic diversity of the domestic dog. *Journal of Heredity* **90**, 71-77.

Wayne, R. K. (1986a). Limb morphology of domestic and wild canids: the influence of development on morphological change. *Journal of Morphology* **187**, 301-319.

Wayne, R. K. (1986b). Developmental constraints on limb growth in domestic and some wild canids. *Journal of Zoology, London* **210**, 381-399.

Wayne, R. K. (1986c). Cranial morphology of domestic and wild canids: the influence of development on morphological change. *Evolution* **40**, 243-261.

Wayne, R. K. (1993). Molecular evolution of the dog family. *Trends in Genetics* **9,** 218-224.

Wilcox, B. W. & Walkowicz, C. (1989). *The Atlas of Dog Breeds of the World.* Neptune City: T.F.H. Publications.

Wilford, C. L. (1995). A diagnostic dilemma: the enigmatic nature of hypothyroidism makes it difficult to distinguish from other diseases. *American Kennel Club Gazette* (Nov), pp. 66-71.

Wilson, D.E. & Reeder, D.M. 1993. *Mammal species of the world (2nd ed.): A taxonomic and geographic reference.* Washington: Smithsonian Institution Press.

Windle, R. J., Wood, S. A., Lightman, S. L. & Ingram, C. D. (1998). The pulsatile characteristics of hypothalamo-pituitary-adrenal activity in female Lewis and Fischer 344 rats and its relationship to differential stress responses. *Endocrinology* **139**, 4044-4052.

Zohary, D., Tchernov, E. & Horwitz, L. K. (1998). The role of unconscious selection in the domestication of sheep and goats. *Journal of Zoology, London* **245**, 129-135.

# EVIDENCE FOR THE DOMESTICATION OF WOLVES IN CENTRAL EUROPEAN MAGDALENIAN SITES

Rudolf Musil

Dept. of Geology and Palaeontology
Faculty of Science, Masaryk University,
Kotlárská St. 2, 611 37 Brno
Czech Republic

**Key words:** dog, domestication, Madgalenian, Pleistocene, wolf, archaeozoology, Germany

**Abstract**

At the end of the Pleistocene, there appear a great number of settlements of the Magdalenian culture in Central Europe (*c.* 12,000-9,000 BP). I have studied osteological material from many localities from this period in the Czech Republic and Germany. The three German localities (Kniegrotte, Oelknitz and Teufelsbruecke) are almost contemporaneous (*c.* 12,000-13,000 BP). In addition to a high quantity of horse remains, skeletal elements of *Canis* differing from "typical" wolf and displaying characteristics found in domestic dogs have been recovered as isolated finds in these sites.

What features divide wild wolves from domestic dogs? Generally reduced body size, shorter snout and facial parts of the skull, large teeth in relation to the maxilla and mandible, and a reduced brain volume are all seen in early dogs. It seems that these characters appear relatively quickly, within only a few generations. In the study of bones from the above mentioned localities I was surprised at their relatively small size. This refers not only to the fragments of long bones and tarsals but also to the mandibles. The teeth in the maxilla are compacted, with no diastema between C and $P_1$, while $P_3$ sits at an oblique angle to $P_2$ and $P_4$. I suggest this means that we have remains of domestic dogs in the Magdalenian of Germany.

The hunters of all these German localities, in contrast to other settlements in Moravia (Czech Republic), specialized in horse hunting. I believe that the methods of hunting taking place at that time are associated with dog domestication. The natural instinct of wolves, the ability of acknowledging man as leader of the pack and submitting to him may have been utilized by man in his hunting, for rounding up horses. It can therefore be suggested that the first domestication of wolves in these areas was associated with increasing the efficiency of hunting. Domestication of wolves can thus be expected everywhere where there was a relative specialization in the hunting of horses.

## Introduction

The Magdalenian is one of the last Palaeolithic cultures. Its seats are found all over central Europe, the eastern boundary being in Moravia (Czech Republic). Magdalenian settlements are not of long duration in central Europe, spanning a relatively short period between 12,000–9,000 BP (Feustel, 1980). The economic basis was still the hunting of animals, which was by that time highly developed. A number of very large prey animals typical of the preceding culture are, however, missing. Hunting thus concentrates on animals of "medium" and small size, predominantly reindeer, horses, hares and foxes.

In the past years I have had the opportunity of analysing the faunal remains from a number of Magdalenian strata (which occurred usually as only one cultural layer, occasionally two) both in Moravia (12 localities), and in Germany (3 localities). Not surprisingly, the species of game animals identified corresponds fully with the community of animals which lived in the surroundings of the localities studied and which could have been the object of hunting. Some animals occurred only sporadically, evidently hunted only occasionally, perhaps when hunting other species. Only where local conditions were optimal for some species do their numbers reach such a high level (as judged by relative counts of minimum number of individuals, MNI) that hunting appeared to be concentrated on that species. In eastern Germany at the time of the Magdalenian cultures, there were large herds of horses which clearly were the centre of interest of those people.

Let us compare the faunal remains from sites in two different environmental territories, where the qualitative and quantitative number of individual species can be expected to differ: Moravian Karst (Moravia) and east Germany. The fauna of the Magdalenian stations from the territory of the Moravian Karst is quite varied. The environment of that region did not allow the dominance of one species over others, so that hunters appear to have utilised all suitable animals living there. The percentage ratio of various species differs, of course, at individual localities. Species lists for a number of these locations are listed below, some with relative frequencies of identifiable taxa (from Musil, 1957, 1958).

**Adler's cave:** reindeer (most frequent), far behind it, hare; sporadically birds, pica, *wolf*, arctic fox, wild horse.

**Balcar's rock (four fireplaces):** reindeer, arctic fox and birds (most frequent); sporadically brown bear, pica, beaver, woolly mammoth, hyena, woolly rhinoceros.

**Kolíbky cave:** birds, arctic fox, hare, brown bear, woolly mammoth

**Kùlna cave:** reindeer, wild horse, hare, birds (most frequent); less frequent, arctic fox, auroch; isolated cases of *wolf*, wolverine and woolly mammoth.

**Ochoz cave:** wild horse (most frequent); sporadically hare, red deer, chamois or sheep, reindeer.

**Pekárna cave:** hare (36.8 %), reindeer (28.2 %), wild horse (19.0 %); birds (8.5 %); isolated cases of auroch, wolverine, red deer, arctic fox, red fox, woolly mammoth.

**Hadí cave:** reindeer and wild horse (most frequent); sporadically arctic fox, *wolf*, brown bear.

**Zitný cave:** reindeer (50 %), wild horse (27.7 %); isolated finds of hare, arctic fox, red deer.

In the above localities it is evident that the most frequently hunted species were reindeer and horses, in some localities also hares and foxes. As for birds, it is probable that only some of their finds comes from hunting. Those animals are thus represented that probably occurred in the surroundings in roughly equal proportions, so that it cannot be assumed that one species prevailed numerically over the others. As for the isolated finds of wolves, they were always typical, large sized wolves.

There is a somewhat different situation in the faunal remains from sites in the lowlands of Germany. The localities of Kniegrotte, Oelknitz and Teufelsbrücke are long-term settlements covering large areas (Musil, 1974, 1980, 1985). Magdalenian cultural layers within Teufelsbrücke (Alleröd IIa), Kniegrotte, and Oelknitz are all dated generally as belonging to the Dryas II period. In contrast, a site at Pekárna, although not discussed further here, possesses two Magdalenian layers which are classified within the earlier Dryas I (layer h) and Bölling (layer g) periods. The duration of these periods are as follow: Dryas III, 11,000–10,000; Alleröd, 12,000–11,000; Dryas II, a very short time, a little after 12,000; Bölling, 13,000–12,000; Dryas I, 14,000 (or 15,000) –13,000. More precisely, $^{14}$C dates (YBP) for these Magdalenian sites are as follows (Feustel 1980):

Kniegrotte: 12,230 ± 90, 13,585 ± 165

Oelknitz: 11,750 ± 250, 11,160 ± 85, 10,990 ± 85, 12,545 ± 80, 12,350 ± 85

(the last two from the bottom of the cultural layer).

Teufelsbrücke: 12.300 ± 85, 12.315 ± 100, 13.025 ± 85, 12.480 ± 90.

Although one finds the same species as in Moravia among the game, the steppe environment of the lowlands appears to have been particularly advantageous for some species. Horses in particular prevail in numbers above the others, suggesting that the Magdalenian people in these localities concentrated on them. Although it had previously been supposed that Magdalenian people were exclusively hunters of reindeer, the data presented here show that this view is not quite correct, that regional differences exist.

### Kniegrotte Cave

The first set of fauna from the lowlands which I analysed was from the site of Kniegrotte Cave (Musil, 1974), where the Magdalenian cultural layer was divided into four horizons. The division of the osteological material of one Magdalenian cultural layer into these horizons is not precise however, due to the methods used at the time of the excavation.

In the topmost horizon wild horse prevailed (35.9 %), followed by reindeer (17.9 %) and the two species of foxes (arctic fox 17.9 %, red fox 5.1 %). Other species were represented at lower frequencies: hare (7.7 %), brown bear (2.5 %), saiga (2.5 %) and birds (several species, 7.5 %). From the genus *Canis* only the proximal part of an ulna, a fragment of cranium, a calcaneus of smaller dimensions (max. length 60.7 mm, max. width 23.7 mm) and a scapula (cervicocaudal diameter of pars articularis 40.0 mm, laterocostal diameter of cavitas glenoidalis 24.2 mm) were found. All bones evidently belonged to one individual.

The maxilla fragment with teeth recovered from the upper layers of Kniegrotte Cave is illustrated in Figure 1. As the maxilla is fossilised in the same way as the other finds, there is no reason to suggest it is from a younger time. This canid was evidently a small but adult animal. The length $P^1$-$P^4$ of the alveolae = 62.3 mm; the length of $P^4$ = 25.8 mm; the length of $M^1$ = 16.9 mm; the height of $M^1$ = 20.7 mm; the length of $M^2$ = 9.0 mm; the width of $M^2$ = 13.3 mm. The total length of the row of teeth $P^1$-$M^2$ is 80.1 mm. The alignment and clustering of teeth in this maxilla fragment differs from typical wolves. The row of premolars and molars is relatively short and the alveolae are so close to each other that at $P^2$ and $P^3$ they partly overlap. This tooth crowding is considered a typical domestication feature.

The dimensions of this Kniegrotte Cave maxilla specimen are at the small end of the range of variation for "small" wolves from other localities. In Moravia there are three localities of approximately equivalent age that contained wolf remains: Dolní Vestonice, Pavlov and Predmostí. Dolní Vestonice has been dated to: 24,560 ± (660-610) to 29,300 ± (750 –690) BP, for the lower part of the cultural layer (19 dates). Pavlov has been dated to 25,160 ±130–26,650 ± 230 BP (6 dates), and Predmostí to 26,320 ±240–26,870 ± 250 BP (2 dates)(Jan van der Plicht, 1997).

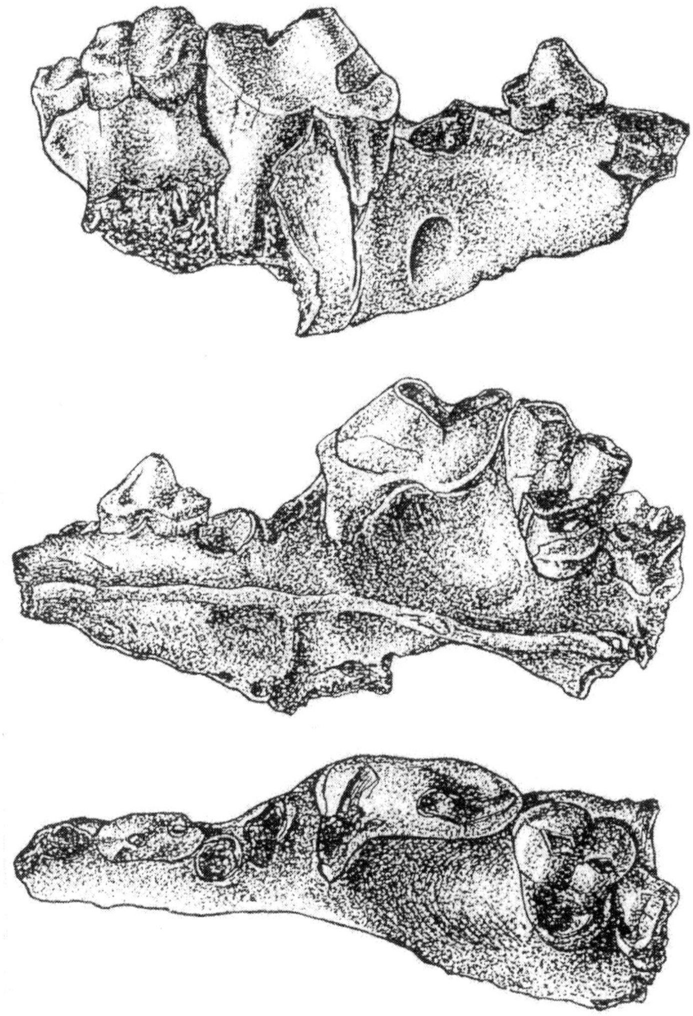

*Figure 1 The maxilla of Canis cf. familiaris from Kniegrotte Cave*

A few wolf maxilla remains are available for comparison to this specimen from Pavlov: $P^4$ ( length x width) = 27.0 x 15.0 mm; $M^1$ (length x width) = 17.1 x 21.0 mm; $M^2$ (length x width) = 10.0 x 15.0 mm.

In the middle horizon of Kniegrotte Cave the dominant species are hare (20.0 %), wild horse (18.2 %) and foxes (arctic fox 18.2 %, red fox 3.6 %). Other species are not as well represented: reindeer (5.4 %), saiga (3.6 %), red deer (1.8 %), brown bear (1.8 %), birds (several species, 12.6 %). In this horizon there were also *Canis* remains: a fragment of a metapodium and three fragments of mandibles with teeth from two individuals. The dimensions of these mandibles are listed in Table 1.

As for the maxilla fragment recovered from the upper layers, the dimensions of the mandibles listed in Table 1 indicate a relatively small animal standing at the lower limit of the range of variation for "typical" wolves found in the region. Although the remains of wolf bones from the cultural layers of Dolní Vestonice and Pavlov are very scarce and consist of small fragments only, they provide a few comparative measurements of mandibles (Tables 2 & 3). There are relatively more wolf mandible measurements from Predmostí material (Table 4 & 5), as well as some cranial measurements (Table 6). At Predmostí it was possible to divide all finds of wolves into groups of "large" and "small" animals, evidently representing sexual dimorphism. The dimensions of mandibles from the group of "small" animals from Predmostí, presumably female, are listed in Table 4.

| Mandible measurements | #1a | #1b | #2 |
|---|---|---|---|
| height of the mandible below C | 34.1 | - | - |
| height of the mandible, $P_4$ - $M_1$ | 14.7 | - | - |
| length of $M_2$ | 11.1 | - | - |
| length of $M_1$ | 27.0 | 27.5 | 27.5 |
| width of $M_1$ below the main cusp | 11.0 | 11.1 | 11.1 |
| length $P_1$ - $M_3$ in alveolae | 96.3 | - | - |
| length $P_1$ - $P_4$ in alveolae | 52.5 | - | - |

***Table 1*** *Mandible measurements (mm) of two putative Canis familiaris specimens from the middle layers of Kniegrotte Cave (#1a/b are same individual) (Musil ,1974)*

| Mandible measurements | #1 | #2 | #3 | #4 | range |
|---|---|---|---|---|---|
| height of mandible behind $M_2$ | 35.0 | 36.2 | 38.6 | 38.0 | 35.0-38.6 |
| height of mandible below main cusp of $M_1$ | 31.2 | 35.0 | 31.2 | 38.2 | 31.2-38.2 |
| width of mandible between $P_4$ - $M_1$ | - | 14.8 | 15.5 | 16.0 | 14.8-16.0 |
| length of $M_2$ | 11.5 | 11.2 | 10.5 | 11.2 | 10.5-11.5 |
| width of $M_1$ below main cusp | 12.8 | 13.2 | 13.1 | 12.1 | 12.1-13.2 |
| length of $M_1$ | 30.0 | 32.2 | 30.0 | 30.0 | 30.0-32.2 |
| length of $P_4$ | - | 17.2 | 17.7 | 16.7 | 16.7-17.7 |
| width of $P_4$ below main cusp | - | 9.0 | 8.6 | 8.3 | 8.3-9.0 |
| length of tooth row $P_1$ - $M_3$ | 98.6 | 99.0 | - | - | 98.6-99.0 |
| length of premolar row $P_1$ - $P_4$ | 50.3 | 53.0 | - | - | 50.3-53.0 |
| length of molar row $M_1$ - $M_3$ | 48.3 | 46.0 | - | 48.3 | 46.0-48.3 |

***Table 2*** *Mandible measurements of four wolf specimens from Dolní Vestonice (femur, total length = 246 mm also recovered) (Musil, 1959a)*

| Mandible measurements | 1 | 2 | 3 | 4 | 5 | "range" |
|---|---|---|---|---|---|---|
| height of mandible behind $M_2$ | - | 36.4 | - | 36.4 | 33.1 | 31.0-31.8 |
| height of mandible below main cusp of $M_1$ | - | 33.4 | - | 33.5 | 32.0 | 29.0-33.0 |
| width of mandible between $P_4$ and $M_1$ | - | 16.1 | - | 16.0 | 13.9 | 12.6-14.0 |
| length of $M_2$ | - | 12.0 | - | 11.4 | 10.6 | 11.0-12.0 |
| length of $M_1$ | - | 31.9 | 29.3 | - | 27.0 | 28.0-29.0 |
| width of $M_1$ below main cusp | - | 12.2 | - | 12.0 | 11.0 | 7.0-8.0 |
| length of $P_4$ | - | 17.8 | 17.0 | 17.0 | 15.2 | 16.0-17.0 |
| width of $P_4$ below main cusp | - | 8.4 | - | 8.4 | 7.2 | 7.0-8.0 |
| length of tooth row ($P_1$ -$M_3$), in alveolae | 98.5 | 100.8 | - | 100.0 | - | 96.0 |
| length of premolar row ($P_1$ _ $P_4$), in alveolae | 51.0 | 51.8 | 60.0 | 51.5 | - | 52.0 |
| length of molar row ($M_1$ – $M_3$) | 49.0 | 49.0 | - | 48.5 | 43.0 | 44.0 |

**Also found at this site were:**
Humerus (n=3), total lengths 209.0; 211.0; 229.0 mm;
Scapula (n=1), length & breadth of glenoid fossa 39.0 & 26.0 mm
Tibia (n=4), total lengths 218.0; 230.0; 240.0; 246.0 mm (n=2), breadth of distal epiphysis 28.2; 32.0 mm

***Table 3*** *Mandible measurements of wolf specimens from Pavlov, dimensions from Pavlov 1953 and "range" from Pavlov 1952 (Musil, 1959b)*

| Mandible measurements | range, in mm (n) | mean |
|---|---|---|
| height of the mandible below C | 28.0-36.0 (28) | 30.1 |
| height of the mandible between $P_4$ - $M_1$ | 13.0-15.0 (29) | 14.14 |
| length of $M_2$ | - | - |
| length of $M_1$ | 27.9-30.0 (30) | 29.31 |
| width of $M_1$ below the main cusp | 10.5-12.0 (27) | 11.45 |
| length $P_1$ - $M_3$ in alveolae | 91.0-102.0 (16) | 96.15 |
| length $P_1$ - $P_4$ in alveolae | - | - |

***Table 4*** *Mandible measurements of "small", presumably females wolves from the site of Predmostí (c. 26,500 BP) (van der Plicht, 1997)*

| Mandible measurements | large "type" range (mm) | n | mean | | small "type" range (mm) | n | mean |
|---|---|---|---|---|---|---|---|
| total length | 187-202 | 16 | 195.5 | | 162-183 | 9 | 175.8 |
| height of vertical ramus | 54-65 | 23 | 60.1 | | 45.0-55.9 | 16 | 53.3 |
| height of mandible behind $M_2$ | 34-40 | 42 | 35.8 | | 31-35 | 29 | 33.2 |
| height of mandible below main cusp of $M_1$ | 31-39 | 52 | 35.3 | | 28.0-37.3 | 36 | 31.3 |
| length of the symphysis | 40-52 | 22 | 47.6 | | 39-45 | 11 | 42.8 |
| width of mandible between $P_4$-$M_1$ | 13.5-18.5 | 53 | 15.3 | | 12.5-15.0 | 37 | 14.2 |
| length of tooth row, $P_1$-$M_3$ | 90.7-106.5 | 32 | 100.4 | | 91-102 | 21 | 95.4 |
| length of molar row, $M_2$-$M_3$ | 15.8-21.3 | 31 | 17.8 | | 15.8-19.0 | 19 | 17.1 |
| length of $M_1$ | 27.5-32.5 | 53 | 30.7 | | 27.9-30.0 | 38 | 29.3 |
| greatest breadth of $M_1$ | 11.0-13.3 | 51 | 12.1 | | 10.6-12.0 | 34 | 11.4 |
| length of $P_4$ | 16.2-18.2 | 46 | 17.3 | | 15.1-17.0 | 33 | 16.2 |
| greatest breadth of $P_4$ | 7.9-9.5 | 46 | 8.7 | | 7.0-8.5 | 32 | 7.8 |
| length of canine | 13.0-16.4 | 15 | 15.2 | | 13.0-15.5 | 11 | 13.9 |
| breadth of canine | 8.7-12.0 | 11 | 10.4 | | 8.5-10.5 | 8 | 9.7 |

***Table 5*** *Mandible measurements of Predmostí wolves. "Large" type, 55 specimens; "small" type, 38 specimens (Pokorný, 1951)*

| Cranial measurements | large "type" range (mm) | n | mean | | small "type" range (mm) | n | mean |
|---|---|---|---|---|---|---|---|
| total length **(1)** | 268.0-276.5 | 3 | 270.8 | | 239.0-251.0 | 3 | 246.3 |
| basal length **(3)** | 237.0-239.0 | 3 | 238.0 | | 210.0-223.5 | 4 | 217.0 |
| upper neurocranium length: akrokranion-frontal midpoint **(7)** | 142.0-149.5 | 4 | 145.0 | | 126.5-134.5 | 6 | 130.3 |
| neurocranium length: basion-nasion **(6)** | 63.5-65.5 | 3 | 64.7 | | 56.0-62.0 | 7 | 58.6 |
| greatest neurocranium breadth **(29)** | 80.5-83.0 | 3 | 81.8 | | 73.5-77.5 | 7 | 75.4 |
| least breadth of skull **(31)** | 40.5-48.0 | 3 | 44.5 | | 39.0-42.0 | 5 | 40.9 |
| greatest frontal breadth **(32)** | 71.0-75.0 | 3 | 73.3 | | 60.0-68.0 | 7 | 63.4 |
| least breadth between the orbits **(33)** | 49.8-55.0 | 3 | 52.8 | | 42.0-46.0 | 5 | 44.8 |
| greatest breadth, occipital condyles **(25)** | 50.0-55.0 | 3 | 52.7 | | 43.0-49.0 | 8 | 47.0 |
| height of the foramen magnum **(28)** | 19.0-21.0 | 4 | 20.0 | | 18.0-20.5 | 8 | 19.3 |
| greatest breadth, foramen magnum **(27)** | 24.0-26.0 | 4 | 25.5 | | 22.0-24.5 | 8 | 23.1 |
| "snout" length **(12)** | 133.0-136.0 | 3 | 134.8 | | 114.0-131.0 | 6 | 122.0 |
| palatal length **(13a)** | 121.0-130.5 | 7 | 127.6 | | 112.0-120.5 | 6 | 117.8 |
| breadth of snout between $P^4$ | 79.0-87.0 | 4 | 83.1 | | 76.5-81.0 | 6 | 78.8 |
| distance between infraorbital foramen | 54.0-58.5 | 4 | 56.7 | | 52.0-53.0 | 5 | 52.7 |
| breadth of snout between $P^2$ | 50.5-55.6 | 5 | 53.1 | | 47.0-53.0 | 6 | 49.7 |
| zygomatic breadth **(30)** | 140.0-154.0 | 3 | 146.7 | | 131.0-135.5 | 3 | 133.8 |

***Table 6*** *Cranial measurements of Predmostí wolves. "Large" type, 8 specimens; "small" type, 10 specimens (Pokorný 1951). Codes given by von den Driesch (1976) in brackets ()*

In the lowermost horizon of the Magdalenian cultural layer at Kniegrotte Cave, the following species were found: reindeer (66.6 %), wild horse (15.7 %), arctic fox (4.4 %), red deer (2.2 %), woolly mammoth (2.2 %), hare (2.2 %), saiga (2.2 %) and birds (2.2 %). In this horizon a fragment of an atlas belonging to a small wolf-like canid was also found.

The quantitative analysis of all Magdalenian layers indicates that reindeer (MNI 42.7/ 27.1%, the first number indicates the number of individual animals, the second one is the percentage) prevail as game, while horses are represented by somewhat fewer remains (MNI 33/21.3%). Arctic foxes have an overall MNI of 20 (12.9%) and birds an MNI of 18 (11.6%). Even if we admit the possible inconsistencies between the individual horizons due to recovery techniques, however, a gradual quantitative change in the game is evident from the base to the surface. The absolute numbers of horse remains change dramatically from bottom to top: layer 1 – MNI 2/NSP 4; layer 2 - MNI 21/NSP 62; layer 3 – MNI 23/NSP 276;

layer 4 – MNI 34/NSP 618. In the youngest horizon the remains of horse are dominant, while the oldest are dominated by reindeer and even contain isolated remnants of mammoth.

As far as small "wolves" are concerned, I met these for the first time at this location and was unsure how to classify the remains. Surprisingly, not only were the overall dimensions of the skeleton small, but the alignment of teeth in the maxilla were typical of dogs. The large premolars were crowded close together, with no gap between the canine and the first premolar, and both $P^2$ and $P^3$ twisted obliquely in the row. All of these features indicate a reduction of the facial part of the skull, a feature typical of dogs. To differentiate the finds from "typical" wolves, I identified them as *Canis sp. (cf. familiaris)*, the tentative nature of the classification reflecting the fact that they were the first canid finds with domestication features which I had seen in Magdalenian material. Due to the fact that in all Moravian Magdalenian localities there were always bones of "typical" wolves as well, I am assuming that the individuals showing signs of domestication must have been at the very early stages of the process.

### Teufelsbrücke Cave

The next German locality I encountered was Teufelsbrücke Cave (Musil, 1980), which has only one Magdalenian horizon. The fauna from this location was dominated by horses (MNI 30/NISP 546). Substantially less common were reindeer (MNI 2 or 3/NISP 70) and hare (MNI 6/NISP 57). Other species are represented by isolated remains: panther (NISP 1), arctic fox (NISP 7), fox sp. (arctic/red) (NISP 6), brown bear (NISP 4), auroch (NISP 1). The hunters of this Magdalenian station evidently also specialised in the hunting of horses, which must have been relatively common in that region.

I was surprised to find at this site "wolf" remains similar in size to those at Kniegrotte Cave, which convinced me that my earlier assessment of the canid material was legitimate. It am sure that it cannot be a coincidence. The small animal that I identified as *Canis sp. (cf. familiaris)* from Teufelsbrücke Cave is represented by a proximal fragment of a metapodial and a first phalanx whose length measurement (24.8 mm) is smaller than "typical" wolves. These remains probably would not have seemed particularly significant in themselves without the material from Kniegrotte Cave for comparison. In Teufelsbrücke Cave bones of "typical" wolves were also recovered: the dimensions of these elements (listed below) fall within the variation range of wolves at Dolní Vestonice and present-day wolves: Femur (distal part, max. width of distal epiphysis = 41.2 mm; max. diameter of distal epiphysis = 46.3 mm); Radius (distal part, width of epiphysis = 33.1 mm; diameter of epiphysis = 17.5 mm).

### Oelknitz

The last German Magdalenian site that I processed fauna from was Oelknitz (Musil, 1985). The people of that locality again seemed to specialise in hunting horses, as at the preceding localities. The game consisted of the following species: wild horse (MNI 127/61.35 %), hare (MNI 29/14.00 %), reindeer (MNI 27/13.04 %), arctic fox (MNI 12/5.79 %). Isolated remains of other species were also recovered: bovidae, bear, pig, woolly rhinoceros, lion, woolly mammoth, ibex and birds are all represented by one or a few elements only. Horses clearly dominate the list.

I expected now to find remains of small "wolves" in this assemblage. They were indeed present, and the fossilisation of their bones corresponded fully with the fossilisation of the other finds, suggesting that they could not come from temporally different sediments. The phalanges and metapodia were again smaller and, at the same time, slenderer than in "typical" wolves. The distal part of a humerus was also recovered (max. width of epiphysis = 42.6 mm; width of trochlea = 31.3 mm, diameter of the trochlea = 18.0 mm), again indicating a small animal. The same pattern is also seen in the distal part of a tibia (max. diameter of epiphysis = 20.3 mm; width of epiphysis = 29.0 mm). In this case I again classified the finds as *Canis cf. familiaris*.

Oelknitz thus became the third German locality characterised on the one hand by specialisation in hunting horses, and on the other by small "wolves" with features typically found in early dogs. The finds from these three localities, situated relatively near one another, suggest that the evidence is not coincidental and indicates perhaps the first attempts at the domestication of wolves. The first domestication of wolves may be thus connected with the hunting of horses. Even if one disregards the domestication features of the one specimen (shortened snout with crowded teeth), it might seem reasonable to suggest that wolves of smaller dimensions lived in the country at that time. But the finds of typical big wolves at Teufelsbrücke Cave suggest otherwise.

### Discussion

Magdalenian localities of east Germany represent the earliest finds to date of domesticated wolves in central Europe (Musil, 1970). This domestication appears to be connected with a specialisation in hunting horses. Magdalenian localities in which this specialisation does not exist always lack remains of small "wolves" with domesticated features and contain only typical big wolves.

It is necessary to realise that Magdalenian cultures span a relatively short period of time and therefore, one must assume that these first "domesticated" wolves must have still been very similar to their ancestors. Domestication certainly seems to depend on a particular subsistence pattern of humans, which evidently became established in the same culture but at different times in different regions.

In addition to the canid remains described here, dogs dating from the Mesolithic period have been found on the Danish Island of Seeland (Maglemose settlement) in large numbers. This settlement has been dated to the *Boreal* period, $C^{14}$ 6810 ± 70 years BC or stage V/VI (Degerbol, 1961). Still older are the finds from Star Carr in England, which belong to the end of the *Preboreal* (stage VI), dated to $C^{14}$ 7538 ± 350 years BC (9559 ± 210 BP and 9490 ± 350 BP) (Clutton-Brock & Noe-Nygaard, 1990). Of roughly the same age as some of the material described here (*c.* 9,000-10,000 BP) is the find described by Degerbol (1961) as "Senckenberg dog", from the moors near Frankfurt, Germany. In addition, Nobis (1979) reported on the mandible of a dog dated at *c.* 14,000 years BP from Oberkassel, near Bonn, Germany. These remains represent the oldest finds of dogs described in north-central Europe so far.

What evidence permit us to divide wolves into wild and domesticated ones, and what do these characteristics result from? The evidence is based on the following characteristics:

- generally reduced body size.
- shorter snout and facial parts of the skull.
- large teeth in relation to the size of the maxilla and mandible, so that they appear compacted and twisted obliquely within the tooth row, especially the premolars.
- reduced brain volume, associated with a reduced cranium size.

All of these characters can be considered as typical domestication features in mammals (Clutton-Brock, 1981) and they certainly appear to be in dogs. At the same time these characters appear relative quickly: it appears that not many generations are needed for their development. It is of interest to note that we also find these traits in wild wolves kept in captivity (Stockhaus, 1965). It appears that the influences of captivity, and subsequent changes in living conditions, can result in similar phenotypic changes in the first generation. Captivity and domestication thus may cause similar changes to individual parts of the skull and body proportions.

Some of the described domestication characteristics can be found in the finds from all of the Magdalenian finds reported here. In comparison, two dogs recovered from Maglemose, Denmark have been reported as having a gracile skeleton and crowded teeth (Degerbol, 1961). Dog skull remains from Star Carr, England are reported to possess only a small diastema between the canine and $P^1$, to have the entire tooth row shortened (the length from the posterior alveolus of C and the posterior alveolus $M^2$ measured 64 mm), and for $P^3$ and $P^4$ to overlap each other. On the basis of these distinctions Degerbol (1961) described the remains from Star Carr as belonging to *Canis familiaris*. In contrast to these dog remains, a wolf skull from the same deposits at Star Carr is reported as having a long facial portion and distinct gaps between the premolars.

I was surprised at the relatively small size of putative dog remains from Kniegrotte, Oelknitz and Teufelsbrücke caves. The overall body skeleton was less robust, which means that the animals in question were of smaller size. On the basis of Bergman's rules on the enlargement of dimensions in a cool climate, directly the opposite should be the case. The wolves in the Pleistocene, in fact, respond to this rule very well. The small size of the canid remains, therefore, must have had another cause than climatic size adaptation.

The maxilla recovered from Kniegrotte Cave put the entire question in an other light. The premolars are compacted, with no diastema between C and $P^1$, while $P^2$ and $P^3$ are positioned obliquely, suggesting that shortening of the facial bones of the skull must have occurred in addition to overall diminution. This specimen is so similar to dog remains identified from Star Carr discussed above that these Magdelanean finds were determined as belonging to *Canis familiaris*. Therefore, remains of domesticated wolves clearly existed during the Magdalenian period in central Europe, as predicted by the suggestion of Degerbol (1961) that Mesolithic dogs appear to have lived for a long period under conditions of domestication.

Even though I believe this assumption to be correct based on my experience in analysing the fauna of many Pleistocene localities, I consider it fair to mention possible objections to this view. It is generally known that the variability of dentition in wolves is very broad. Also, the domestication features discussed above are not typical of dogs only, but have also been reported in wolves, although rarely and developed to a lesser degree. Clutton-Brock (1970), for example, mentions oblique positioning of upper premolar teeth in two skulls of wolves. Unfortunately, it is not stated whether the reported skulls belonged to wolves living in zoos or possibly to animals which were actually hybrid crosses of wolf and dog.

In order to further investigate this "domestication" trait of tooth crowding, I examined the relatively numerous finds of wolves from Predmostí that should predate the occurrence of dogs (*c.* 26,000 BP) and I found that it was possible even in that set to find isolated cases in which the upper premolars of the skull were crowded and obliquely aligned. I also examined the skulls of recent wolves. In almost all specimens (n=30), there were gaps of varying width between the individual premolars, and the length of the diastema varied between 1 and 6 mm. In only three skulls were the premolars clustered close together, with $P^3$ and $P^4$ beginning to show an oblique overlap. It is clear that premolar crowding, described as a typical domestication feature, is occasionally found not only in wolves from the Last Glacial, but also in contemporary animals. However, this feature is not common and occurs in isolated cases only. In Pleistocene wolves from Predmostí, individuals with crowded teeth were always big animals and thus did not otherwise resemble dogs. In the recent specimens of wolves that I examined, which came from the time after World War II, the possibility of their being hybrid animals resulting from cross-breeding with dogs cannot, unfortunately, be excluded.

All that can be concluded from the above study of wolf skulls is that the short diastema associated with crowded premolars arranged in an oblique fashion can in isolated cases appear in wolves as well as dogs. So far this has only been observed in animals of large size; in other words, in typical wolves. Further study may verify if overall gracility and small body size must be taken into account when separating dogs from wolves based on premolar crowding.

It is clear that a simple explanation cannot account for the finds reported here, for example that smaller than average wolves occupied the part of east Germany where these sites are located. At Kniegrotte Cave, bones of typical large wolves were found along with the smaller animals classified as *Canis cf. familiaris* and the metric difference between the two size groups was considerable. If the putative dog remains had appeared at only one locality, it would be necessary to be cautious about classification. I contend that three separate finds of putative dog remains at three localities of the same age increases the probability that they represent domestication at its early stage.

### References


Clutton-Brock, J. (1970). The origins of the dog. In (D. Brothwell & E. Higgs, Eds), *Science in Archaeology*. New York: Praeger Publishers, pp. 303-309.

Clutton-Brock, J. (1981). *Domesticated Animals from Early Times*. London: British Museum (Natural History), Heinemann.

Clutton-Brock, J. & Noe-Nygaard, N. (1990). New osteological and C-isotope evidence on Mesolithic dogs: companions to hunters and fishers at Star Carr, Seamer Carr and Kongemose. *Journal of Archaeological Science* **17**, 643-653.

Degerbol, M. (1961). On a find of a preboreal domestic dog (*Canis familiaris* L.) from the Star Carr, Yorkshire, with remarks on other Mesolithic dogs. *Proceedings of the Prehistoric Society* **27**, 35-55.

Driesch, A. von den (1976). *A Guide to the Measurement of Animal Bones from Archaeological Sites*. Cambridge, MA: Harvard University Peabody Museum Bulletin **1**.

Feustel R. (1980). Magdalénienstation Teufelsbrücke. I. Archäologischer Teil. *Weimarer Monographien zur Ur-und Frühgeschichte Thüringens* **3**, 1-129.

Musil, R. (1957). Die Fauna des Magdalénien aus der Zitný-Höhle. *Acta Academy Societ Cechoslovenicae* **29**, 558-572.

Musil, R. (1958). Die Fauna der mährischen Magdalénienstationen. *Anthropozoikum* **7**, 7-26.

Musil R. (1959a). Poznámky k paleontologickému materiálu z Dolních Vestonic (Bemerkungen zum paläontologischen Material aus Dolní Vestonice Unterwisternitz). *Anthropozoikum* **8**, 73-82.

Musil R. (1959b). Osteologický materiál z paleolitického sídlište v Pavlove. Èást II. (Das osteologische Material aus der paläolithischen Siedlungsstätte in Pavlov. II. Teil. *Anthropozoikum* **8**, 83-106.

Musil, R. (1961). Die Fauna des Magdalénien der Hadí-Höhle. *Acta Musei Moraviae* **46**, 51-66.

Musil, R. (1970). Domestication of the dog already in the Magdalénien? *Antropologie* **8**, 87-88.

Musil, R. (1974). Tiergesellschaft der Kniegrotte. In (R. Feustel, Ed) *Die Kniegrotte*. Weimar: Weimarer Monographien zur Ur- und Frühgeschichte, pp. 30-95.

Musil, R. (1980). Die Großsäuger und Vögel der Teufelsbrücke. In (R. Feustel, Ed) *Magdalénienstation Teufelsbrücke*. Weimar: Weimarer Monographien zur Ur- und Frühgeschichte **3**, pp.5-69.

Musil, R. (1984). The first known domestication of wolves in Central Europe. *British Archaeological Reports International Series* **227**, 23-25.

Musil, R. (1985). *Die Fauna der Magdalénien-Siedlung Oelknitz*. Weimar: Weimarer Monographien zur Ur-und Frühgeschichte **17**, 3-87.

Nobis, G. (1979). Der älteste Haushund lebte vor 14.000 Jahren. *Umschau* **79**, 610.

Plicht, J. van der. (1997). The radiocarbon dating. In (I. Pavlov, Ed) *The Dolní Vestonice Studies* **4**, 427-436.

Pokorný, M. (1951). Príspìvek k paleontologii diluvia v Predmostí u Prerova (Contribution a l'etude paléontologique portant sur le Diluvium de Predmostí u Prerova). *Acta Musei Moraviae* **36**, 1-18.

Stockhaus, K. (1965). Metrische Untersuchungen an Schädeln von Wölfen und Hunden. *Zeitschrift für Zoologische Systematik und evolutionsforschung* **3**, 157-258.

# A PRELIMINARY LOOK AT SOME NEW DOMESTICATED DOGS FROM SUBMERGED NEOLITHIC SITES OFF THE CARMEL COAST

Tamar Dayan
Department of Zoology,
Tel Aviv University,
Tel Aviv 69978, Israel

Ehud Galili
Marine Archaeology Branch,
Israel Antiquities Authority,
P.O.Box 180, Atlit, Israel

**Key words**: Israel, Neolithic, dog, Atlit-Yam, Kfar-Galim, evolution, Natufian, PPNC, underwater archaeology

**Abstract**

We carried out a preliminary morphological analysis of some canid cranial remains from two submerged prehistoric sites off the coast of Israel: Atlit-Yam, a large Pre-Pottery Neolithic (PPNC) settlement (*c.* 8,100-7,500BP [all dates uncalibrated]), and Kfar-Galim, a Ceramic Neolithic settlement (*c.* 6,500-7,000BP). The Atlit-Yam remains include three parts of the same skull (braincase and two sides of the rostrum) with heavily worn dentition, a part of a cranium, a mandible bearing the molars and fourth premolar, and two premolars and two molars that may have all belonged to the same individual. The Kfar-Galim specimen is an almost complete cranium with heavily worn molars. The Atlit-Yam canids exhibit shortened mean upper toothrow lengths and reduced bullae as compared to the mean of recent Israeli wolves, but the difference is significant for only one specimen. Their upper carnassials are smaller than the mean for recent Israeli wolves, but not significantly so. The mandible and attached lower carnassial are both significantly shorter than the mean for recent wolves ($p<0.025$). The other lower carnassial is shorter than those of the mean for modern wolves, but not significantly so. The Kfar-Galim specimen is much smaller than the other specimens, is significantly smaller for all measured traits than the mean for recent wolves (GTL, CBL, upper toothrow, and bullar length; $p<0.01$), and its tympanic bullae are flattened. Although dogs may have been domesticated for over 4,000 years prior to the PPNC, the Atlit-Yam canids exhibit only slight metric changes in comparison to recent wolves of Israel, pointing to a fairly slow rate of morphological change at this stage.

## Introduction

The earliest recognizable domestic dogs were small animals, similar to small subspecies of wolf, the Indian (*Canis lupus pallipes*) or Arabian (*Canis lupus arabs*) races (eg Lawrence, 1967; Scott, 1968; Olsen, 1985) found in the Middle East, or the Chinese wolf (*Canis lupus chanco*) (Olsen & Olsen, 1977; Olsen, 1985). These races also possess the smallest cranial capacities that also make them suitable possible progenitors of domestic dogs (Hemmer, 1983). Therefore, there is widespread agreement that the dog was domesticated in southwest or southeast Asia.

Indeed many of the earliest recognizable domesticated dogs do originate from the Middle East. Canids from the Epi-Paleolithic of the Mount Carmel caves of Israel have been subject to several studies that yielded conflicting results (Bate, 1937; Degerbol, 1961; Clutton-Brock, 1962, 1969, 1979; Kurtén, 1965), but it now appears on morphological criteria that they are indeed domesticated dogs (Dayan, 1994). Other Natufian canids have been identified on both cultural and morphological criteria as early domestic dogs. These consist of remains of wolves or dogs buried with human skeletons (Davis & Valla, 1978; Valla, 1990; Tchernov & Valla, 1997). Clearly dogs were domesticated in the Near East by *c.* 12,500BP.

It has recently been suggested that the development of traits characterizing the process of domestication may have been a microevolutionary response to the development of new anthropogenic habitats generated by increasing sedentism during the Natufian period (Tchernov & Horwitz, 1991). This view conflicts with Vilà *et al.*'s (1997) time frame for dog domestication based on molecular evidence (*c.* 135,000 yrs BP), emphasizing the need for understanding the relationship between cultural, morphological, and molecular events.

Here we present some preliminary results of a study of some canid cranial remains from two submerged prehistoric sites off the coast of Israel: Atlit-Yam, a large Pre-Pottery Neolithic (PPNC) settlement (*c.* 8,100-7,500BP), and Kfar-Galim, a Ceramic Neolithic settlement (*c.* 6,500-7,000BP). These specimens help provide a point of reference to understanding the rate of morphological change in early domesticated dogs.

**The sites**

In the last 30 years numerous archaeological remains have been uncovered on the sea floor off the Carmel coast. These findings include submerged prehistoric settlements belonging to the Neolithic culture some 8,000 to 6,500 years BP (Fig. 1). Most of the finds are extremely well preserved due to the protective cover of sand that prevented damage by marine erosion. During the peak of the last glacial the world's ocean levels were some 100 m lower than at present (Emery & Aubrey, 1991) and the Carmel coastline was positioned *c.* 10 km to the west. The subsequently rising sea levels inundated prehistoric settlements situated near the ancient coastline (Galili *et al,* 1988).

Atlit-Yam: late Pre-Pottery Neolithic (PPNC)

The Atlit-Yam PPNC submerged village thrived some 8,100-7,450 years ago. It is located in the north bay of Atlit, submerged at a depth of 8-12 m, and is approximately 40,000 $m^2$ in area. Excavations have revealed foundations of rectangular stone structures, various round installations, a structure built of sandstone megaliths, anthropomorphic stone steles, and dozens of hearths with charcoal remains. The tool assemblages consist of stone, bone, and flint artifacts including axes, spearheads, sickle blades, and arrowheads. Dozens of human skeletons were uncovered buried in flexed positions, in and around the structures. Additional organic remains include animal bones, fish bones, and an abundance of charred and waterlogged plant remains (seeds, tree branches, and pollen). Floral and faunal remains suggest that the village economy was complex, based on several different food resources acquired through hunting, incipient herding, fishing, and farming (Galili *et al.*, 1993).

One of the most important finds is the water well that lies some 10.5 m below present-day sea level. The well is a cylindrical feature, 5.5 m deep and 1.5 m in diameter. The lowest 1.5 m of the well are cut into the kurkar (local term for carbonate-cemented quartz sandstone, mostly of aeolian origin) bedrock, and the upper sections are cut into clay sediments and built of undressed stones. The excavation of the well's fill was carried out with a dredging system and revealed two distinct zones: the lower, about 2 m thick, containing mostly kurkar stones of various sizes embedded in sandy clay, various flint, bone and stone artifacts, and a few animal bones; and the upper, some 3 m in thickness, containing various artifacts and large quantities of animal bones (including the dogs) embedded in brown clay (Galili & Nir, 1993).

Kfar-Galim: late Pottery Neolithic (Wadi Rabah culture)

A remarkable abundance and diversity of Pottery Neolithic remains were revealed in a narrow, almost continuous, submerged belt parallel to the Carmel coast (Galili & Weinstein-Evron, 1985; Galili & Sharvit, 1994-5; Galili *et al.*, 1997). This 15 km long stretch lies just seaward of the present shoreline, and is about 200 m wide. There are five submerged Neolithic sites in this zone, dated to *c.* seventh millenium BP. The submerged site Kfar-Galim is located 4 km south of Haifa, 0.5 to 4 m below present day sea level. Excavations and surveys at the site have revealed several rectangular water wells built of alternating layers of wooden branches and stones. A dog skull was found in one of these wells. A few round pits inlaid with undressed stones, and containing a small amount of potsherds and flint artifacts, and large amounts of plant remains (branches, seeds and pollen grains), were found. Stone basins and mortars of various shapes and sizes were found scattered around the pits.

**The specimens**

The Atlit-Yam specimens designated A-D were:

A Three parts of the same individual skull: braincase and the right and left parts of the rostrum. The sutures are closed and the dentition quite worn;

B Part of a cranium with no teeth;

C Mandible bearing $P_4$, $M_1$, and $M_2$;

D Several teeth that may have belonged to the same individual: upper carnassial and molars, and lower $P_3$, $P_4$, $M_1$, and $M_2$.

The single Kfar-Galim specimen was an almost complete cranium with closed sutures, bearing very worn molars, obviously an old individual. Its tympanic bullae were fairly flattened.

These canid specimens are too large to be golden jackals (*Canis aureus*). In general morphology they resemble either wolves (*Canis lupus*) or domesticated dogs. Early domesticated dogs are characterized by an overall size reduction, shortening of the facial region, and a lesser reduction of tooth size (eg, Olsen, 1985; Clutton-Brock 1993; Morey, 1992). In order to identify the Atlit-Yam and Kfar-Galim Neolithic canids, we took several measurements according to von den Driesch (1976):

- Greatest total length of the skull (GTL).

- Condylo-basal length of the skull (CBL).

- Length of the upper cheektooth row ($P^1$-$M^2$).

- Length of the upper carnassial, measured at the cingulum ($P^4$Le).

- Length of the lower premolar tooth row ($P_1$-$P_4$).

- Length of the lower carnassial measured at the cingulum ($M_1$L).

- Bullar length.

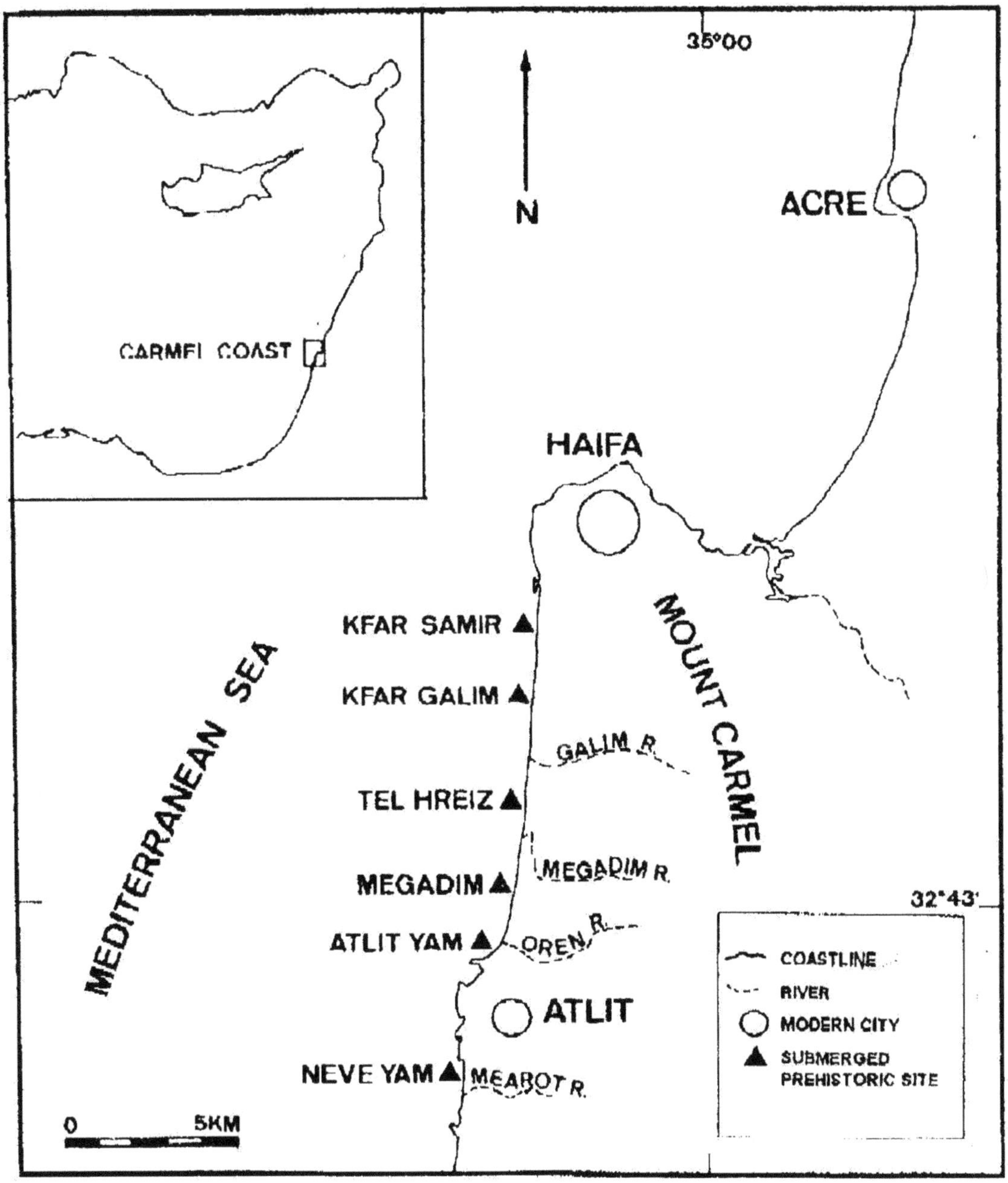


***Figure 1.*** *Map of the Carmel coast and locations of submerged Neolithic sites.*

| | N | mean | s.d. |
|---|---|---|---|
| GTL | 22 | 222.58 | 8.86 |
| CBL | 23 | 208.07 | 7.42 |
| Upper toothrow length | 24 | 76.24 | 2.79 |
| Upper carnassial length | 29 | 22.40 | 1.17 |
| Lower premolar row length | 24 | 47.03 | 2.35 |
| Lower carnassial length | 30 | 24.47 | 1.11 |
| Bullar length | 21 | 28.79 | 1.45 |

***Table 1*** *- Statistics of a sample of measurements of modern Israeli wolves (Canis lupus) (mm).*

Measurements were taken using digital calipers (Mahr) to 0.01 mm precision. We compared the results of our measurements with those of a sample of modern wolves from Israel (Table 1). Comparisons were made using a student's t-test (eg, Simpson *et al.*, 1960).

Specimen A is significantly shorter in toothrow length (70.20 mm; $p<0.025$) than the Israeli modern wolf sample. Specimen B (71.68 mm) is smaller too, but not significantly so. The mean for both is significantly ($p<0.01$) smaller than that of modern Israeli wolves ($p<0.05$). However, analyzing both specimens together assumes that they originate from the same population, an assumption that may be questionable.

In bullar length specimen A is significantly smaller than the mean for modern Israeli wolves (26.16 mm; $p<0.05$). Specimen B (26.46 mm) is smaller too, but not significantly so. The mean for both, however, is significantly smaller ($p<0.025$) than the mean for modern Israeli wolves ($p<0.025$), but this analysis should be considered with the same caveats as above.

The length of the upper carnassial of specimens B and C (20.88 & 21.19 mm, respectively) is smaller than the mean for modern Israeli wolves, but not significantly so, nor is the mean for both.

Specimen C is shorter in premolar row length (40.56 mm; $p<0.01$) than the mean for modern Israeli wolves. In carnassial length it is significantly smaller (22.16; $p<0.025$). The other lower carnassial (specimen D; 23.23 mm) is smaller than the mean for modern Israeli wolves, but not significantly so. However, the mean for both is significantly ($p<0.025$) smaller than the mean of modern Israeli wolves.

The smallest individual is the Kfar-Galim dog. Its skull length measurements (GTL, 183.60 mm; CBL, 172.20 mm) are significantly ($p<0.001$) smaller than those of modern Israeli wolves. The length of the cheektooth row is significantly shorter than the mean for modern Israeli wolves (59.48 mm; $p<0.001$). Its tympanic bullae are flattened, a characteristic of domestic dogs (see Harrison, 1973), and consequently they are shorter than those of modern wolves (24.69 mm; $p<0.01$). This morphological characteristic of domestic dogs was established based on a sample of modern dogs. Since canids in zooarchaeological contexts are seldom found with their bullae intact, this may be the first time that this characteristic is observed in ancient dogs.

## Conclusions

As a group, Atlit-Yam canids exhibit shortened jaws and reduced bullae as compared with recent Israeli wolves, and teeth that are reduced, but apparently to a lesser degree. These morphological traits are similar to those found in Pre-Pottery Neolithic dogs from Jericho (Zeuner, 1958; Dayan, 1994), and conform with the expected at earlier stages of domestication. Although dogs may have been domesticated for well over 4,000 years prior to the PPNC, the Atlit-Yam canids exhibit only slight metric changes in comparison with modern wolves of Israel, pointing to a fairly slow rate of morphological change at this stage.

## References


Bate, D.M.A. (1937). Palaeontology: the fossil fauna of the Wadi el-Mughara caves. In (D.A.E. Garrod & D.M.A. Bate, Eds) *The Stone Age of Mt. Carmel: excavations at the Wadi el-Mughara.*, Vol 1, Part 2. Oxford: Clarendon Press, pp.139-237.

Clutton-Brock, J. (1962). Near Eastern canids and the affinities of the Natufian dogs. *Zeitschrift für Tierzuchtung und Zuchtungsbiologie* **76**, 326-333.

Clutton-Brock, J. (1969). Carnivore remains from the excavations of the Jericho Tell. In (P.J. Ucko & G.W. Dimbleby, Eds) *The Domestication and Exploitation of Plants and Animals.* London: Duckworth, pp. 337-345.

Clutton-Brock, J. (1979). The mammalian remains from the Jericho Tell. *Proceedings of the Prehistoric Society* **45**, 135-157.

Clutton-Brock, J. (1993). Origins of the dog: domestication and early history. In (J. Serpell, Ed) *The Domestic Dog: its Evolution, Behaviour, and Interactions with People.* Cambridge: Cambridge University Press.

Davis, S.J.M. & Valla, F.R. (1978). Evidence for domestication of the dog 12,000 years ago in the Natufian of Israel. *Nature* **276**, 608-610.

Dayan, T. (1994). Early domesticated dogs of the Near East. *Journal of Archaeological Science* **21**, 644-640.

Degerbol, M. (1961). On a find of Preboreal domestic dog (*Canis familiaris* L.) from Star Carr, Yorkshire, with remarks on other Mesolithic dogs. *Proceedings of the Prehistoric Society* **27**, 35-55.

Driesch, A., von den. (1976). *A Guide to the Measurement of Animal Bones from Archaeological Sites.* Peabody Museum Bulletin I. Cambridge, Mass: Harvard University.

Emery, K.O. & Aubrey, D.G. (1991). *Sea Levels, Land Levels and Tide Gauges.* New York: Springer Verlag.

Galili, E. & Weinstein-Evron, M. (1985). Prehistory and paleoenvironments of submerged sites along the Carmel coast of Israel. *Paleorient* **11**, 37-52.

Galili, E., Weinstein-Evron, M. & Ronen, A. (1988). Holocene sea-level changes based on submerged archaeological sites off the northern carmel coast in Israel. *Quaternary Research* **29**, 36-42.

Galili, E., Weinstein-Evron, M., Hershkovitz, I., Gopher, A., Kislev, M., Lernau, O., Horwitz, L.K. & Lernau, H. (1993). Atlit-Yam: a prehistoric site on the sea floor off the Israeli coast. *Journal of Field Archaeology* **20**, 133.

Galili, E. & Nir, Y. (1993). The submerged Pre-Pottery Neolithic water well of Atlit-Yam, northern Israel, and its palaeoenvironmental implications. *The Holocene* **3**, 265-270.

Galili, E. & Sharvit, J. (1994-5). Evidence of olive oil production from the submerged site at Kfar Samir, Israel. *Mitekufat Haeven, Journal of the Israel Prehistoric Society* **26**, 122-133.

Galili, E., Stanley, D.J., Sharvit, J. & Weinstein-Evron, M. (1997). Evidence for earliest olive oil production in submerged settlements off the Carmel coast, Israel. *Journal of Archaeological Science* **24**, 1141-1150.

Harrison, D.L. (1973). Some comparative features of the skulls of wolves (*Canis lupus* Linn.) from the Arabian Peninsula and neighboring lands. *Bonn. Zool. Beitr.* **24**, 185-191.

Hemmer, H. (1983). *Domestikation Veramung der Merkwelt.* Braunschweig/Wiesbaden: Friedr.,Vieweg & Sohn.

Kurtén, B. (1965). The Carnivora of the Palestine caves. *Acta Zoologica Fennica* **107**, 3-74.

Lawrence, B. (1967). Early domestic dogs. *Zeitschrift fur Saugertierkunde* **32**, 44-59.

Morey, D.F. (1992). Size, shape, and development in the evolution of the domestic dog. *Journal of Archaeological Science* **19**, 181-204.

Olsen, S.J. (1985). *Origins of the Domestic Dog.* Tucson: University of Arizona Press.

Olsen, S.J. & Olsen, S.W. (1977). The Chinese wolf, ancestor of New World dogs. *Science* **197**, 533-535.

Scott, J.P. (1968). Evolution and domestication of the dog. *Evolutionary Biology* **2**, 243-275.

Simpson, G.G., Roe, A. & Lewontin, R. C. (1960) *Quantitative Zoology.* New York: Harcourt, Brace.

Tchernov, E. & Valla, F. R. (1997). Two new dogs and other Natufian dogs from the Southern Levant. *Journal of Archaeological Science* **24**, 65-95.

Tchernov, E. & Horwitz, L.K. (1991). Body size diminution under domestication: unconscious selection in primeval domesticates. *Journal of Anthropological Archaeology* **10**, 54-75.

Valla, F.R. (1990). Le Natufien: une autre facon de comprendre le monde? *Journal of the Israel Prehistoric Society* **23**, 171- 175.

Vilà, C., Savolainen, P., Maldonado, J. E., Amorim, I. R., Rice, J. E., Honeycutt, R. L., Crandall, K. A., Lundeberg, J. & Wayne, R. K. (1997). Multiple and ancient origins of the domestic dog. *Science* **276**, 1687-1689.

Zeuner, F.E. (1958). Dog and cat in the Neolithic of Jericho. *Palestine Exploration Quarterly* **1**, pp. 52-55.

# THE CONTRIBUTIONS OF BARBARA LAWRENCE TO THE STUDY OF DOGS, WITH A COMPREHENSIVE LISTING OF HER MEASUREMENT DEFINITIONS

Richard H. Meadow
Zooarchaeology Laboratory
Peabody Museum
Harvard University
11 Divinity Avenue
Cambridge, MA 02138
USA

**Key words**: dog, canids, osteometry, archaeozoology, cranial measurements, dental measurements, definitions, systematics, von den Driesch

**Abstract**

Barbara Lawrence (1909-1997) was a pioneer in the United States of the analysis of faunal remains from archaeological sites. One of her particular interests was the study of dogs and other canids. In line with this focus she assembled a large collection of modern canid skeletons and skins, and archaeological canid remains, in the Museum of Comparative Zoology, Harvard University, where she was Curator of Mammals from 1952 to 1976. Between 1967 and 1969 she published four seminal articles on the subjects of dog domestication and canid systematics in which she summarized the evidence as it stood at that time. To carry out her investigations, particularly on archaeological material, she defined a large number of potentially useful cranial and dental measurements which she then evaluated for descriptive and diagnostic value. Her measurement definitions have never been published in full, even though they are of particular value to the zooarchaeological community because they include detailed definitions for those parts of the skull that are most likely to be preserved in the archaeofaunal record. They are presented here compared to those of von den Driesch.

## Introduction

Barbara Lawrence (1909-1997) was an American pioneer in the field of Zooarchaeology (Meadow, 1997). One of her particular interests was the study of dogs and other canids. In line with this focus she assembled large collections of modern canid skeletons and skins and archaeological canid remains in the Museum of Comparative Zoology, Harvard University, where she was Curator of Mammals from 1952 to 1976. Between 1967 and 1969 she published four seminal articles on the subjects of dog domestication and canid systematics in which she summarized the evidence as it stood at that time (Lawrence, 1967, 1968; Lawrence & Bossert, 1967, 1969). To carry out her investigations particularly on archaeological material, she defined a large number of potentially useful cranial and dental measurements that she then evaluated for descriptive and diagnostic value (eg., Lawrence & Bossert 1967). Her measurement definitions have never been published in full, even though some are of particular value to the zooarchaeological community because they comprise definitions for those parts of the cranium and mandible that are most likely to be preserved in the archaeofaunal record. These definitions are presented here in Table 1 and Figures 1-6.

## Barbara Lawrence and the metrical study of *Canis* sp.

Barbara Lawrence's interest in the study of faunal remains from archaeological sites was sustained by her interest in dogs and dog domestication. From her first published zooarchaeological report (Lawrence, 1944) to her last (Lawrence & Reed, 1983), dog remains provided an important focus of research. Another focus was the systematics of the genus *Canis*. In these studies she collaborated with the statistician William Bossert to carry out early computer-aided multiple character analyses (discriminant function or linear discrimination analyses) to identify combinations of metrical characters that can be used to reliably separate the crania of wolves, coyotes, dogs, and hybrids (Lawrence & Bossert, 1967, 1969, 1975).

For the multiple character analyses, Lawrence took 42 measurements mostly on crania but with some on mandibles (Lawrence and Bossert, 1967: 231). These are presented here in Table 1 (L#1-47) and Figures 1-6.[1] Of

[1] The numbers used to designate the measurements in Tables 1-2 and Figures 1-6 are not those employed in Lawrence and Bossert (1967). They are designations that were established by Lawrence before the preparation of "The Dogs of Jarmo" (Lawrence and Reed, 1983) in which they are employed. That publication is said to have been "essentially completed by 1975" (ibid., footnote on page 485). In addition there are nine other measurements that must have been defined at that time (3, 15,

the measurements, 18 "were found not to be taxonomically reliable" (Table 1, symbol --) and 24 were selected as being possibly significant based on regression analysis (13 reflecting skull shape, 11 tooth form). Of these, 16 were found most diagnostic and thus useful for linear discrimination (Table 1, symbol **: 9 cranial, 7 tooth), and 8 were found not to be useful (Table 1, symbol *: 4 cranial, 4 tooth).

Also to be found in Table 1 and Figures 1-6 are the measurements defined by von den Driesch (1976). These are listed in that author's numerical order in Table 2. Only about one quarter of all the measurements in Tables 1 and 2 are commonly defined by Lawrence and von den Driesch. Thus, of 65 cranial measurements, there are 16 unique to Lawrence, 16 in common, and 33 unique to von den Driesch. Of 45 mandibular measurements, there are 20 unique to Lawrence, 13 in common, and 12 unique to von den Driesch. Overall, von den Driesch defines considerably more cranial measurements and Lawrence more mandibular measurements. In this regard, it is interesting to note that all of the 23 measurements added by Lawrence after the initial set of 42 (ie, L#48-72) are of the mandible. This probably reflects her efforts to deal with fragmented archaeological collections in which mandibles tend to be better represented than maxillary or other cranial remains.

Even the measurements that Lawrence defines for the mandible, however, are not exhaustive. A more complete sct of alveolar cheek tooth (including canine) measurements, for both the mandible and maxillary, are listed in Table 3. Crown measurements for individual cheek teeth (including canine) are also listed. In all, there are 49 maxillary and 60 mandibular measurement definitions noted. Of these, only 12 maxillary and 23 mandibular definitions are listed by Lawrence or von den Driesch or both. The reason for documenting a tooth row so exhaustively is to be able to compare highly fragmented material. Investigators, however, apparently have not found such an effort to be worthwhile or productive -- to judge from the fact that almost none have used this approach. The current author did use some of these measurements in describing the dog remains from the Neolithic site of Hajji Firuz Tepe in northwestern Iran (Meadow, 1983).

Lest the reader believes that the skull measurements listed in Tables 1-3 are exhaustive, it is important to note that other researchers have defined still other measurements. Examples include William Haag (1948)[2] and Darcy Morey (1986, 1990, 1992), Nobuo Shigehara and other Japanese researchers (eg, Shigehara, Onodera, & Eto, 1997), and Eitan Tchernov (eg, Tchernov & Valla, 1997). Many of these measurements, however, can be taken only on unbroken specimens, which limits their usefulness when dealing with archaeological remains.

37, 39, 40, 59, 61, 62, and 63) but which are not described in any available document. Finally, an additional measurement, 52a, was added sometime later.

[2] An interesting historical note is that neither Lawrence nor von den Driesch, when preparing their lists of measurements, seem to have been aware of the work of Haag (1948).

## Conclusion

Barbara Lawrence was an early proponent of the use of measurements in zooarchaeological research. She was able to use them effectively because she had a good understanding of the biology behind the morphology. She also understood the peculiar nature of archaeofaunal remains, modified as they have been by past human activities. Even so, she -- like most specialists -- was dependent upon the archaeologist for ensuring the integrity of the collections studied. This led her to accept at face value the antiquity of the Jaguar Cave (Birch Creek Valley, Idaho, USA) dogs on which she reported in the late 1960s (Lawrence, 1967, 1968). Subsequent AMS dating of two of the actual dog bones that she examined has shown that they are much less ancient than other animal bones from the site and than their previously presumed late Pleistocene/early Holocene age (Gowlett et al., 1987, pp. 145-146). When presented with these findings, she expressed no particular surprise or dismay. Instead she said that she was gratified that Science had developed techniques by which the antiquity of particularly significant specimens could be ascertained. She expressed confidence that, while specific results of her analyses might become obsolete, the approach to the study of canids that she employed would continue to be productive. And so it has proved to be.

## References Cited


Driesch, A. von den (1976). *A Guide to the Measurement of Animal Bones from Archaeological Sites.* Cambridge, MA: Harvard University Peabody Museum Bulletin **1**.

Gowlett, J.A.J., Hedges, R.E.M., Law, I.A., & Perry, C. (1987) Radiocarbon dates from the Oxford AMS system: archaeometry datelist 5. *Archaeometry* **29**(1), 125-155.

Haag, W.G. (1948). An Osteometric analysis of some aboriginal dogs. *The University of Kentucky, Reports in Anthropology* **7**(3), 107-264.

Lawrence, B. (1944). Bones from the Governador area. Appendix C in (E.T. Hall, Jr.) *Early Stockaded Settlements in the Governador, New Mexico* (Columbia Studies in Archeology and Ethnology, Vol. II, Part 1). New York: Columbia University Press, pp. 73-78.

Lawrence, B. (1967). Early domestic dogs. *Zeitschrift für Säugetierkunde* **32**(1), 44-59.

Lawrence, B. (1968). Antiquity of large dogs in North America. *Tebiwa* (The Journal of the Idaho State University Museum) **11**(2), 43-49.

Lawrence, B. & Bossert, W.H. (1967). Multiple character analysis of *Canis lupus, latrans,* and *familiaris* with a discussion of their relationships of *Canis niger*. *American Zoologist* **7,** 223-232.

Lawrence, B. & Bossert, W.H. (1969). The cranial evidence for hybridization in New England *Canis*. *Breviora* (Museum of Comparative Zoology) **330,** 1-13.

Lawrence, B. & Bossert, W.H. (1975) Relationships of North American *Canis* shown by a multiple character analysis of selected populations. In (M.W. Fox, Ed.) *The Wild Canids: Their Systematics, Behavioral Ecology and Evolution.* New York: Van Norstrand Reinhold, pp. 73-86.

Lawrence, B. & Reed, C.A. (1983). The dogs of Jarmo. In (L.S. Braidwood, R.J. Braidwood, B. Howe, C.A. Reed, & P.J. Watson, Eds.) *Prehistoric Archeology along the Zagros Flanks.* (Oriental Institute Publications 105). Chicago: The Oriental Institute of the University of Chicago, pp. 485-494.

Meadow, R.H. (1983). The vertebrate faunal remains from Hasanlu Period X at Hajji Firuz. Appendix G in (M.M. Voigt) *Hajji Firuz Tepe, Iran: The Neolithic Settlement.* (Hasanlu Excavation Reports, Volume I). Philadelphia: the University Museum, University of Pennsylvania, pp. 369-422.

Meadow, R.H. (1997). In Memoriam: Barbara Lawrence (30 June 1909 - 25 May 1997). *Archaeofauna* **6,** 145-148.

Morey, D.F. (1986). Studies on Amerindian dogs: taxonomic analysis of canid crania from the Northern Plains. *Journal of Archaeological Science* **13,** 119-145.

Morey, D.F. (1990). *Cranial allometry and the evolution of the domestic dog.* Ph.D. Dissertation. Knosville, TN: University of Tennessee. (Ann Arbor: University Microfilms International No. 9112869).

Morey, D.F. (1992). Size, shape and development in the evolution of the domestic dog. *Journal of Archaeological Science* **19,** 181-204.

Shigehara, N., Onodera, S., & Eto, M. (1997). Sex determination by discrimant analysis and evaluation of non-metric traits in the dog skeleton. In (S.J. Crockford) *Osetometry of Makah and Coast Salish Dogs.* Burnaby, B.C.: Archaeology Press: Simon Fraser University, pp. 113-133.

Tchernov, E. & Valla, F.F. (1997). Two new dogs, and other Natufian dogs, from the southern Levant. *Journal of Archaeological Science* **24,** 65-95.

***Table 1:*** *Measurements for canid skulls as defined by Barbara Lawrence (L) and Angela von den Driesch (D). Cr = cranium, Mn = mandible. All measurement definitions (L# and/or D#) are illustrated in Figures 1-6 except those noted as "n.i." Descriptions, modified from the unpublished notes of Barbara Lawrence and from von den Driesch (1976), are presented in the order defined by Barbara Lawrence. * and ** = tested for diagnostic value by Lawrence and Bossert (1967); ** = used in linear discrimination by same; -- = "found to be not taxonomically reliable" by same; + = used by Lawrence and Reed (1983).*

| L# | D# | Cr/Mn | Description |
|---|---|---|---|
| **+1 | - | Mn | Crown length P/4 (specify side(s)) |
| *2 | - | Mn | Greatest crown breadth P/4 at right angles to length (specify side(s)) |
| **4 | 1 | Cr | Total length of cranium from Akrokranion to Prosthion (tip of sagittal crest to labial alveolus of incisors) |
| --5 | 2 | Cr | Condylobasal length from aboral border of occipital condyles to Prosthion |
| --6 | 33 | Cr | Least interorbital breadth (Entorbitale to Entorbitale) |
| **7 | 30 | Cr | Zygomatic breadth (Zygion to Zygion) |
| **8 | - | Cr | Greatest breadth at parieto-temporal suture |
| --9 | 13 | Cr | Median palatal length from Staphylion to Prosthion |
| --10 | 19 | Cr | Alveolar length P4/ (specify side(s)) |
| **11 | - | Cr | Smallest length from aboral (distal) margin of alveolus of M2/ to depression between styloid process and inflation of bulla (specifiy side(s)) |
| **12 | - | Cr | Crown length from maxillary Canine to M2/ (specify side(s)) |
| --13 | - | Cr | Length of brain case from posterior frontal foramen in orbit to midpoint of foramen magnum (specify side(s)) |
| *14 | 32 | Cr | Breadth across postorbital processes (frontal breadth: Ectorbitale to Ectorbitale) |
| *16 | 25 | Cr | Breadth across occipital condyles |
| --+17 | 8 | Mn | Alveolar length from P/1 to M/3 (specify side(s)) |
| --18 | - | Mn | Length from condyle to distal alveolar margin of M/3 (specify side(s)) |
| *+19 | 13 | Mn | Crown length of M/1 from mesio-lingual corner to distal midpoint (specify side(s)) [also described as "parallel to main axis" and "parallel to inner margin"] |
| **20 | 18 | Cr | Crown length P4/ (specify side(s)) |
| --21 | 18a | Cr | Greatest crown breadth P4/ at right angles to length (specify side(s)) |
| **22 | - | Cr | Least crown breadth P4/ taken between roots (specify side(s)) |
| *23 | 12 | Cr | Length of rostrum from orbit to Prosthion (specify side(s) or mid-line) (mid-line = vdD#12) |
| *24 | 13a | Mn | Greatest crown breadth M/1 (specify side(s)) |
| **25 | - | n.i. Cr | Mesiodistal (Greatest) diameter of maxillary Canine at base of enamel (specify side(s)) |
| **26 | - | Cr | Greatest breadth across tooth rows at P or M (most lateral point of crowns) |
| --27 | 35 | Cr | Least palatal breadth, measured behind the canines / Least Breadth of Rostrum |
| --28 | - | Cr | Breadth across tips of nasals |
| --29 | - | Cr | Distance from tips of nasals in midline to medial projection of premaxillary between incisors |
| **30 | - | Cr | Least height from orbit to alveolar border of P4/ or M1/ (specify side(s)) |
| *31 | 37 | Cr | Greatest diameter of orbit from low point anteriorly to in front of postorbital process (specify side(s)) |
| **32 | - | Mn | Length of distal cusps of P/4 from notch to in line with distal edge of tooth (specify side(s)) |
| --33 | 15a | Cr | Aboral border of the alveolus of M3/ to oral border of the alveolus of C (specify side(s)) |
| --+34 | 1 | Mn | Greatest length of mandible from condyle to Infradentale (alveolus at I/1) (specify side(s)) |
| --35 | 7a | Mn | Alveolar length from mesial margin of /C to distal margin of M/3 (specify side(s)) |
| --36 | - | Mn | Crown length from mesial /C to distal M/3 (specify side(s)) |
| *38 | 39 | Cr | Height of braincase from basisphenoid to top of braincase at right angles to projection of tooth row and not including sagittal crest (specify side(s)) |
| **41 | 21a | Cr | Crown breadth M2/ (specify side(s)) |
| **42 | - | Cr | Breadth across tips of incisor crowns |
| --43 | - | Cr | Greatest crown depth (labial-lingual) I3/ (specify side(s)) |

| | | | |
|---|---|---|---|
| --44 | - | Cr | Height of bullae taken from horizontal line across bullae to basioccipital, measured on foramen magnum side of line |
| **45 | - | Cr | Least jugal height at right angles to long axis of skull |
| **46 | - | Cr | Breadth between inner margins of alveolus of P1/ |
| --47 | - | Cr | Aboral height of bony nares beside midline |
| +48 | 17 | n.i. Mn | Greatest breadth of horizontal ramus basal to M/1 (specify side(s)) |
| 49 | | Mn | Greatest height from alveolus between roots of M/1 to base of jaw (specify side(s)) |
| 50 | - | Mn | Greatest alveolar Diameter of /C (specify side(s)) |
| +51 | - | n.i. Mn | Greatest breadth at base of ascending ramus (specify side(s)) |
| 52 | - | n.i. Mn | Least breadth at symphysis with arm of calipers flat on medial side of symphysis (specify side(s)) |
| 52a | - | n.i. Mn | Breadth from symphysis to lateral alveolar margin of /C (specify side(s)) |
| 53 | 11 | Mn | Alveolar length P/1 to P/4 (specify side(s)) |
| 54 | - | Mn | Alveolar length mesial margin of P/4 to distal margin of M/3 (specify side(s)) |
| 55 | - | Mn | Alveolar length mesial margin of P/1 to distal margin of M/2 (specify side(s)) |
| 56 | 9 | Mn | Alveolar length P/2 to M/3 (specify side(s)) |
| 57 | 15 | Mn | Crown length M/2 (specify side(s)) |
| 58 | 15a | Mn | Crown breadth M/2 (specify side(s)) |
| 60 | 18 | Mn | Height of ascending ramus parallel to posterior margin (specify side(s)) |
| 64 | - | Mn | Alveolar length I/1 (Infradentale) to distal margin of P/2 (specify side(s)) |
| 65 | - | Mn | Alveolar length P/2 (specify side(s)) |
| +66 | - | n.i. Mn | Least breadth of jaw basal to Premolars (specify side(s)) |
| +67 | - | Mn | Alveolar length P/3 (specify side(s)) |
| +68 | 12 | Mn | Alveolar length P/2 to P/4 (specify side(s)) |
| +69 | | Mn | Least height of jaw in region of Premolars (specify side(s)) |
| [69] | 20 | Mn | Height of the mandible between P/2 and P/3, measured on the lingual side and at right angles to the basal border (specify side(s)) [Lawrence measures least height, which is sometimes equal to D#20] |
| +70 | - | Mn | Alveolar length mesial margin of P/2 to distal margin of M/2 (specify side(s)) |
| +71 | - | Mn | Alveolar length P/4 (specify side(s)) |
| 72 | - | Mn | Crown length mesial P/1 to distal M/3 (specify side(s)) |
| - | 3 | Cr | Basal length: Basion to Prosthion |
| - | 4 | Cr | Basicranial axis: Basion to Synsphenion (=Intersphenoid suture) |
| - | 5 | Cr | Basifacial axis: Synsphenion to Prosthion |
| - | 6 | n.i. Cr | Neurocranium length: Basion to Nasion (curved calipers only) |
| - | 7 | Cr | Upper neurocranium length: Akrokranion to Frontal midpoint |
| - | 8 | Cr | Viscerocranium length: Nasion to Prosthion |
| - | 9 | Cr | Facial length: Frontal midpoint to Prosthion |
| - | 10 | Cr | Greatest length of the nasals: Nasion to Rhinion |
| - | 11 | n.i. Cr | Length of braincase: Basion to cribriform plate (taken through foramen magnum) |
| - | 12 | Cr | "Snout" length: oral border of the orbits (median) to Prosthion |
| - | 13a | Cr | Palatal length: the median point of intersection of the line joining the deepest indentations of the Choanae to the Prosthion |
| - | 14 | Cr | Length of the horizontal part of the palatine: Staphylion to Palatinoorale |
| - | 14a | Cr | Length of the horizontal part of the palatine corresponding to v.d. D 13a |
| - | 15 | Cr | Alveolar length of cheektooth row (measured along alveoli on buccal side) (specify side(s)) |
| - | 16 | Cr | Alveolar length of molar row (measured along alveoli on buccal side) (specify side(s)) |
| - | 17 | Cr | Alveolar length of premolar row (measured along alveoli on buccal side) (specify side(s)) |
| - | 20 | Cr | Crown length M1/ (specify side(s)) |
| - | 20a | Cr | Crown breadth M1/ (specify side(s)) |
| - | 21 | Cr | Crown length M2/ (specify side(s)) |
| - | 22 | Cr | Greatest diameter of the auditory bulla (specify side(s)) |
| - | 23 | Cr | Greatest mastoid breadth = greatest breadth of the occipital triangle: Otion to Otion |
| - | 24 | Cr | Breadth dorsal to the external auditory meatus |
| - | 26 | Cr | Greatest breadth of the bases of the paraoccipital processes |
| - | 27 | Cr | Greatest breadth of the foramen magnum |
| - | 28 | Cr | Height of the foramen magnum: Basion to Opisthion |
| - | 29 | Cr | Greatest neurocranium breadth = greatest breadth of the braincase: Euryon to Euryon |
| - | 31 | Cr | Least breadth of skull = least breadth aboral of the supraorbital processes = breadth at postorbital constriction |
| - | 34 | Cr | Greatest palatal breadth: measured across the outer borders of the alveoli |
| - | 36 | Cr | Breadth at canine alveoli |
| - | 38 | Cr | Skull height. One arm of the slide calipers is placed basally on the basis of the skull (on the basioccipital) and the other dorsally on the highest elevation of the sagittal crest |
| - | 40 | Cr | Height of the occipital triangle: Akrokranion to Basion |
| - | 41 | n.i. Cr | Height of the canine (tip to apex) (specify side(s)) |
| - | 42 | n.i. Cr | Neurocranium capacity = capacity of the braincase (cc) [see v.d.D. 1976, pg. 43) |
| - | 2 | Mn | Length from the angular process to the Infradentale (specify side(s)) |
| - | 3 | Mn | Length from the indentation between the condyle process and the angular process to Infradentale (specify side(s)) |
| - | 4 | Mn | Length from the condyle process to the aboral\l (distal) border of the canine alveolus (specify side(s)) |
| - | 5 | Mn | Length from the indentation between the condyle process and the angular process to the aboral (distal) border of the canine alveolus (specify side(s)) |
| - | 6 | Mn | Length from the angular process to the aboral (distal) border of the canine alveolus (specify side(s)) |
| - | 7 | Mn | Length from the aboral (distal) border of the alveolus of M/3 to the aboral (distal) border of the canine alveolus (specify side(s)) |
| - | 10 | Mn | Alveolar length of molar row (specify side(s)) |
| - | 14 | Mn | Length of M/1 alveolus (specify side(s)) |
| - | 16 | Mn | Crown length M/3 (specify side(s)) |
| - | 16a | Mn | Crown breadth M/3 (specify side(s)) |
| - | 19 | Mn | Height of the mandible behind M/1, measured on the lingual side and at right angles to the basal border (specify side(s)) |
| - | 21 | Mn | Height of the canine (tip to apex) (specify side(s)) |

***Table 2:*** *Measurements for canid skulls as defined by Barbara Lawrence (L) and Angela von den Driesch (D). Cr = cranium, Mn = mandible. All measurement definitions (L# and/or D#) are illustrated in Figures 1-6 except those noted as "n.i." Descriptions, modified from the unpublished notes of Barbara Lawrence and from von den Driesch (1976), are presented in the order found in von den Driesch (1976: 42-46, 60-61).*

| L# | D# | Cr/Mn | Description |
|---|---|---|---|
| 4 | 1 | Cr | Total length of cranium from Akrokranion to Prosthion (tip of sagittal crest to labial alveolus of incisors) |
| 5 | 2 | Cr | Condylobasal length from aboral border of occipital condyles to Prosthion |
| - | 3 | Cr | Basal length: Basion to Prosthion |
| - | 4 | Cr | Basicranial axis: Basion to Synsphenion (=Intersphenoid suture) |
| - | 5 | Cr | Basifacial axis: Synsphenion to Prosthion |
| - | 6 | n.i. Cr | Neurocranium length: Basion to Nasion (curved calipers only) |
| - | 7 | Cr | Upper neurocranium length: Akrokranion to Frontal midpoint |
| - | 8 | Cr | Viscerocranium length: Nasion to Prosthion |
| - | 9 | Cr | Facial length: Frontal midpoint to Prosthion |
| - | 10 | Cr | Greatest length of the nasals: Nasion to Rhinion |
| - | 11 | n.i. Cr | Length of braincase: Basion to cribriform plate (taken through foramen magnum) |
| - | 12 | Cr | "Snout" length: oral border of the orbits (median) to Prosthion |
| 23 | 12 | Cr | Length of rostrum from orbit to Prosthion (specify side(s) or mid-line) (mid-line = D#12) |
| 9 | 13 | Cr | Median palatal length from Staphylion to Prosthion |
| - | 13a | Cr | Palatal length: the median point of intersection of the line joining the deepest indentations of the Choanae to the Prosthion |
| - | 14 | Cr | Length of the horizontal part of the palatine: Staphylion to Palatinoorale |
| - | 14a | Cr | Length of the horizontal part of the palatine corresponding to v.d. D 13a |
| - | 15 | Cr | Alveolar length of cheektooth row (measured along alveoli on buccal side) (specify side(s)) |
| 33 | 15a | Cr | Aboral border of the alveolus of M3/ to oral border of the alveolus of C (specify side(s)) |
| - | 16 | Cr | Alveolar length of molar row (measured along alveoli on buccal side) (specify side(s)) |
| - | 17 | Cr | Alveolar length of premolar row (measured along alveoli on buccal side) (specify side(s)) |
| 20 | 18 | Cr | Crown length P4/ (specify side(s)) |
| 21 | 18a | Cr | Greatest crown breadth P4/ at right angles to length (specify side(s)) |
| 10 | 19 | Cr | Alveolar length P4/ (specify side(s)) |
| - | 20 | Cr | Crown length M1/ (specify side(s)) |
| - | 20a | Cr | Crown breadth M1/ (specify side(s)) |
| - | 21 | Cr | Crown length M2/ (specify side(s)) |
| 41 | 21a | Cr | Crown breadth M2/ (specify side(s)) |
| - | 22 | Cr | Greatest diameter of the auditory bulla (specify side(s)) |
| - | 23 | Cr | Greatest mastoid breadth = greatest breadth of the occipital triangle: Otion to Otion |
| - | 24 | Cr | Breadth dorsal to the external auditory meatus |
| 16 | 25 | Cr | Breadth across occipital condyles |
| - | 26 | Cr | Greatest breadth of the bases of the paraoccipital processes |
| - | 27 | Cr | Greatest breadth of the foramen magnum |
| - | 28 | Cr | Height of the foramen magnum: Basion to Opisthion |
| - | 29 | Cr | Greatest neurocranium breadth = greatest breadth of the braincase: Euryon to Euryon |
| 7 | 30 | Cr | Zygomatic breadth (Zygion to Zygion) |
| - | 31 | Cr | Least breadth of skull = least breadth aboral of the supraorbital processes = breadth at postorbital constriction |
| 14 | 32 | Cr | Breadth across postorbital processes (frontal breadth: Ectorbitale to Ectorbitale) |
| 6 | 33 | Cr | Least interorbital breadth (Entorbitale to Entorbitale) |
| - | 34 | Cr | Greatest palatal breadth: measured across the outer borders of the alveoli |
| 27 | 35 | Cr | Least palatal breadth, measured behind the canines / Least Breadth of Rostrum |
| - | 36 | Cr | Breadth at canine alveoli |
| 31 | 37 | Cr | Greatest diameter of orbit from low point anteriorly to in front of postorbital process (specify side(s)) |
| - | 38 | Cr | Skull height. One arm of the slide calipers is placed basally on the basis of the skull (on the basioccipital) and the other dorsally on the highest elevation of the sagittal crest |
| 38 | 39 | Cr | Height of braincase from basisphenoid to top of braincase at right angles to projection of tooth row and not including sagittal crest (specify side(s)) |
| - | 40 | Cr | Height of the occipital triangle: Akrokranion to Basion |
| - | 41 | n.i. Cr | Height of the canine (tip to apex) (specify side(s)) |
| - | 42 | n.i. Cr | Neurocranium capacity = capacity of the braincase (cc) [see v.d.D. 1976, pg. 43) |
| 8 | - | Cr | Greatest breadth at parieto-temporal suture |
| 11 | - | Cr | Smallest length from aboral (distal) margin of alveolus of M2/ to depression between styloid process and inflation of bulla (specifiy side(s)) |
| 12 | - | Cr | Crown length from maxillary Canine to M2/ (specify side(s)) |
| 13 | - | Cr | Length of brain case from posterior frontal foramen in orbit to midpoint of foramen magnum (specify side(s)) |
| 22 | - | Cr | Least crown breadth P4/ taken between roots (specify side(s)) |
| 25 | - | n.i. Cr | Mesiodistal (Greatest) diameter of maxillary Canine at base of enamel (specify side(s)) |
| 26 | - | Cr | Greatest breadth across tooth rows at P or M (most lateral point of crowns |
| 28 | - | Cr | Breadth across tips of nasals |

| | | | |
|---|---|---|---|
| 29 | - | Cr | Distance from tips of nasals in midline to medial projection of premaxillary between incisors |
| 30 | - | Cr | Least height from orbit to alveolar border of P4/ or M1/ (specify side(s)) |
| 42 | - | Cr | Breadth across tips of incisor crowns |
| 43 | - | Cr | Greatest crown depth (labial-lingual) I3/ (specify side(s)) |
| 44 | - | Cr | Height of bullae taken from horizontal line across bullae to basioccipital, measured on foramen magnum side of line |
| 45 | - | Cr | Least jugal height at right angles to long axis of skull |
| 46 | - | Cr | Breadth between inner margins of alveolus of P1/ |
| 47 | - | Cr | Aboral height of bony nares beside midline |
| 34 | 1 | Mn | Greatest length of mandible from condyle to Infradentale (alveolus at I/1) (specify side(s)) |
| - | 2 | Mn | Length from the angular process to the Infradentale (specify side(s)) |
| - | 3 | Mn | Length from the indentation between the condyle process and the angular process to Infradentale (specify side(s)) |
| - | 4 | Mn | Length from the condyle process to the abora\l (distal) border of the canine alveolus (specify side(s)) |
| - | 5 | Mn | Length from the indentation between the condyle process and the angular process to the aboral (distal) border of the canine alveolus (specify side(s)) |
| - | 6 | Mn | Length from the angular process to the aboral (distal) border of the canine alveolus (specify side(s)) |
| - | 7 | Mn | Length from the aboral (distal) border of the alveolus of M/3 to the aboral (distal) border of the canine alveolus (specify side(s)) |
| 35 | 7a | Mn | Alveolar length from mesial margin of /C to distal margin of M/3 (specify side(s)) |
| 17 | 8 | Mn | Alveolar length from P/1 to M/3 (specify side(s)) |
| 56 | 9 | Mn | Alveolar length P/2 to M/3 (specify side(s)) |
| - | 10 | Mn | Alveolar length of molar row (specify side(s)) |
| 53 | 11 | Mn | Alveolar length P/1 to P/4 (specify side(s)) |
| 68 | 12 | Mn | Alveolar length P/2 to P/4 (specify side(s)) |
| 19 | 13 | Mn | Crown length of M/1 from mesio-lingual corner to distal midpoint (specify side(s)) [also described as "parallel to main axis" and "parallel to inner margin"] |
| 24 | 13a | Mn | Greatest crown breadth M/1 (specify side(s)) |
| - | 14 | Mn | Length of M/1 alveolus (specify side(s)) |
| 57 | 15 | Mn | Crown length M/2 (specify side(s)) |
| 58 | 15a | Mn | Crown breadth M/2 (specify side(s)) |
| - | 16 | Mn | Crown length M/3 (specify side(s)) |
| - | 16a | Mn | Crown breadth M/3 (specify side(s)) |
| 48 | 17 | n.i. Mn | Greatest breadth of horizontal ramus basal to M/1 (specify side(s)) |
| 60 | 18 | Mn | Height of ascending ramus parallel to posterior margin (specify side(s)) |
| - | 19 | Mn | Height of the mandible behind M/1, measured on the lingual side and at right angles to the basal border (specify side(s)) |
| [69] | 20 | Mn | Height of the mandible between P/2 and P/3, measured on the lingual side and at right angles to the basal border (specify side(s)) [Lawrence measures least height, which is sometimes equal to D#20] |
| - | 21 | n.i. Mn | Height of the canine (tip to apex) (specify side(s)) |
| 1 | - | Mn | Crown length P/4 (specify side(s)) |
| 2 | - | Mn | Greatest crown breadth P/4 at right angles to length (specify side(s)) |
| 18 | - | Mn | Length from condyle to distal alveolar margin of M/3 (specify side(s)) |
| 32 | - | Mn | Length of distal cusps of P/4 from notch to in line with distal edge of tooth (specify side(s)) |
| 36 | - | Mn | Crown length from mesial /C to distal M/3 (specify side(s)) |
| 49 | | Mn | Greatest height from alveolus between roots of M/1 to base of jaw (specify side(s)) |
| 50 | - | Mn | Greatest alveolar diameter of /C (specify side(s)) |
| 51 | - | n.i. Mn | Greatest breadth at base of ascending ramus (specify side(s)) |
| 52 | - | n.i. Mn | Least breadth at symphysis with arm of calipers flat on medial side of symphysis (specify side(s)) |
| 52a | - | n.i. Mn | Breadth from symphysis to lateral alveolar margin of /C (specify side(s)) |
| 54 | - | Mn | Alveolar length mesial margin of P/4 to distal margin of M/3 (specify side(s)) |
| 55 | - | Mn | Alveolar length mesial margin of P/1 to distal margin of M/2 (specify side(s)) |
| 64 | - | Mn | Alveolar length I/1 (Infradentale) to distal margin of P/2 (specify side(s)) |
| 65 | - | Mn | Alveolar length P/2 (specify side(s)) |
| 66 | - | n.i. Mn | Least breadth of jaw basal to Premolars (specify side(s)) |
| 67 | - | Mn | Alveolar length P/3 (specify side(s)) |
| 69 | | Mn | Least height of jaw in region of Premolars (specify side(s)) |
| 70 | - | Mn | Alveolar length mesial margin of P/2 to distal margin of M/2 (specify side(s)) |
| 71 | - | Mn | Alveolar length P/4 (specify side(s)) |
| 72 | - | Mn | Crown length mesial P/1 to distal M/3 (specify side(s)) |

Note that von den Driesch mandibular "measurements" 22 through 26 are not included here because they are dimensions calculated using various correction factors for basal length.

***Table 3:*** *Measurements for canid cheektooth rows that can be useful to document fragmented archaeological specimens. Those defined by Barbara Lawrence (L#) and Angela von den Driesch (D#) are noted in the first two columns and illustrated in Figures 3-6. Cr = cranium, Mn = mandible.*

| L# | D# | Cr/Mn | Description |
|---|---|---|---|
| 33 | 15a | Cr | Alveolar length mesial C/ to distal M2/ (specify side(s)) |
| | | Cr | Alveolar length distal C/ to distal M2/ (specify side(s)) |
| | 15 | Cr | Alveolar length mesial P1/ to distal M2/ (specify side(s)) |
| | | Cr | Alveolar length mesial P2/ to distal M2/ (specify side(s)) |
| | | Cr | Alveolar length mesial P3/ to distal M2/ (specify side(s)) |
| | | Cr | Alveolar length mesial P4/ to distal M2/ (specify side(s)) |
| | 16 | Cr | Alveolar length mesial M1/ to distal M2/ (specify side(s)) |
| | | Cr | Alveolar length mesial to distal M2/ (specify side(s)) |
| | | Cr | Alveolar length mesial C/ to distal M1/ (specify side(s)) |
| | | Cr | Alveolar length distal C/ to distal M1/ (specify side(s)) |
| | 17 | Cr | Alveolar length mesial P1/ to distal M1/ (specify side(s)) |
| | | Cr | Alveolar length mesial P2/ to distal M1/ (specify side(s)) |
| | | Cr | Alveolar length mesial P3/ to distal M1/ (specify side(s)) |
| | | Cr | Alveolar length mesial P4/ to distal M1/ (specify side(s)) |
| | | Cr | Alveolar length mesial to distal M2/ (specify side(s)) |
| | | Cr | Alveolar length mesial C/ to distal P4/ (specify side(s)) |
| | | Cr | Alveolar length distal C/ to distal P4/ (specify side(s)) |
| | | Cr | Alveolar length mesial P1/ to distal P4/ (specify side(s)) |
| | | Cr | Alveolar length mesial P2/ to distal P4/ (specify side(s)) |
| | | Cr | Alveolar length mesial P3/ to distal P4/ (specify side(s)) |
| 10 | 19 | Cr | Alveolar length mesial to distal P4/ (specify side(s)) |
| | | Cr | Alveolar length mesial C/ to distal P3/ (specify side(s)) |
| | | Cr | Alveolar length distal C/ to distal P3/ (specify side(s)) |
| | | Cr | Alveolar length mesial P1/ to distal P3/ (specify side(s)) |
| | | Cr | Alveolar length mesial P2/ to distal P3/ (specify side(s)) |
| | | Cr | Alveolar length mesial to distal P3/ (specify side(s)) |
| | | Cr | Alveolar length mesial C/ to distal P2/ (specify side(s)) |
| | | Cr | Alveolar length distal C/ to distal P2/ (specify side(s)) |
| | | Cr | Alveolar length mesial P1/ to distal P2/ (specify side(s)) |
| | | Cr | Alveolar length mesial to distal P2/ (specify side(s)) |
| | | Cr | Alveolar length mesial C/ to distal P1/ (specify side(s)) |
| | | Cr | Alveolar length distal C/ to distal P1/ (specify side(s)) |
| | | Cr | Alveolar length mesial to distal P1/ (specify side(s)) |
| | | Cr | Alveolar length mesial to distal C/ (specify side(s)) |
| | | Cr | Greatest diameter (GD) of C/ (specify side(s)) |
| | | Cr | Least diameter of C/ at same point as GD (specify side(s)) |
| | 41 | Cr | Height of C/ tip to apex (specify side(s)) |
| | | Cr | P1/ crown length (specify side(s)) |
| | | Cr | P1/ crown breadth (specify side(s)) |
| | | Cr | P2/ crown length (specify side(s)) |
| | | Cr | P2/ crown breadth (specify side(s)) |
| | | Cr | P3/ crown length (specify side(s)) |
| | | Cr | P3/ crown breadth (specify side(s)) |
| 20 | 18 | Cr | P4/ crown length (specify side(s)) |
| 21 | 18a | Cr | P4/ crown breadth (specify side(s)) |
| | 20 | Cr | M1/ crown length (specify side(s)) |
| | 20a | Cr | M1/ crown breadth (specify side(s)) |
| | 21 | Cr | M2/ crown length (specify side(s)) |
| 41 | 21a | Cr | M2/ crown breadth (specify side(s)) |
| 35 | 7a | Mn | Alveolar length mesial /C to distal M/3 (specify side(s)) |
| | | Mn | Alveolar length distal /C to distal M/3 (specify side(s)) |
| 17 | 8 | Mn | Alveolar length mesial P/1 to distal M/3 (specify side(s)) |
| 56 | 9 | Mn | Alveolar length mesial P/2 to distal M/3 (specify side(s)) |
| | | Mn | Alveolar length mesial P/3 to distal M/3 (specify side(s)) |
| | | Mn | Alveolar length mesial P/4 to distal M/3 (specify side(s)) |
| | 10 | Mn | Alveolar length mesial M/1 to distal M/3 (specify side(s)) |
| | | Mn | Alveolar length mesial M/2 to distal M/3 (specify side(s)) |
| | | Mn | Alveolar length mesial to distal M/3 (specify side(s)) |
| | | Mn | Alveolar length mesial /C to distal M/2 (specify side(s)) |
| | | Mn | Alveolar length distal /C to distal M/2 (specify side(s)) |

| | | | |
|---|---|---|---|
| 55 | | Mn | Alveolar length mesial P/1 to distal M/2 (specify side(s)) |
| 70 | | Mn | Alveolar length mesial P/2 to distal M/2 (specify side(s)) |
| | | Mn | Alveolar length mesial P/3 to distal M/2 (specify side(s)) |
| 54 | | Mn | Alveolar length mesial P/4 to distal M/2 (specify side(s)) |
| | | Mn | Alveolar length mesial M/1 to distal M/2 (specify side(s)) |
| | | Mn | Alveolar length mesial to distal M/2 (specify side(s)) |
| | | Mn | Alveolar length mesial /C to distal M/1 (specify side(s)) |
| | | Mn | Alveolar length distal /C to distal M/1 (specify side(s)) |
| | | Mn | Alveolar length mesial P/1 to distal M/1 (specify side(s)) |
| | | Mn | Alveolar length mesial P/2 to distal M/1 (specify side(s)) |
| | | Mn | Alveolar length mesial P/3 to distal M/1 (specify side(s)) |
| | | Mn | Alveolar length mesial P/4 to distal M/1 (specify side(s)) |
| | 14 | Mn | Alveolar length mesial to distal M/1 (specify side(s)) |
| | | Mn | Alveolar length mesial /C to distal P/4 (specify side(s)) |
| | | Mn | Alveolar length distal /C to distal P/4 (specify side(s)) |
| 53 | 11 | Mn | Alveolar length mesial P/1 to distal P/4 (specify side(s)) |
| 68 | 12 | Mn | Alveolar length mesial P/2 to distal P/4 (specify side(s)) |
| | | Mn | Alveolar length mesial P/3 to distal P/4 (specify side(s)) |
| 71 | | Mn | Alveolar length mesial to distal P/4 (specify side(s)) |
| | | Mn | Alveolar length mesial /C to distal P/3 (specify side(s)) |
| | | Mn | Alveolar length distal /C to distal P/3 (specify side(s)) |
| | | Mn | Alveolar length mesial P/1 to distal P/3 (specify side(s)) |
| | | Mn | Alveolar length mesial P/2 to distal P/3 (specify side(s)) |
| 67 | | Mn | Alveolar length mesial to distal P/3 (specify side(s)) |
| | | Mn | Alveolar length mesial /C to distal P/2 (specify side(s)) |
| | | Mn | Alveolar length distal /C to distal P/2 (specify side(s)) |
| | | Mn | Alveolar length mesial P/1 to distal P/2 (specify side(s)) |
| 65 | | Mn | Alveolar length mesial to distal P/2 (specify side(s)) |
| | | Mn | Alveolar length mesial /C to distal P/1 (specify side(s)) |
| | | Mn | Alveolar length distal /C to distal P/1 (specify side(s)) |
| | | Mn | Alveolar length mesial to distal P/1 (specify side(s)) |
| 50 | | Mn | Alveolar length mesial to distal /C (specify side(s)) |
| | | Mn | Greatest diameter (GD) of /C (specify side(s)) |
| | | Mn | Least diameter of /C at same point as GD (specify side(s)) |
| | 21 | Mn | Height of /C tip to apex (specify side(s)) |
| | | Mn | P/1 crown length (specify side(s)) |
| | | Mn | P/1 crown breadth (specify side(s)) |
| | | Mn | P/2 crown length (specify side(s)) |
| | | Mn | P/2 crown breadth (specify side(s)) |
| | | Mn | P/3 crown length (specify side(s)) |
| | | Mn | P/3 crown breadth (specify side(s)) |
| 1 | | Mn | P/4 crown length (specify side(s)) |
| 2 | | Mn | P/4 crown breadth (specify side(s)) |
| 19 | 13 | Mn | M/1 crown length (specify side(s)) |
| 24 | 13a | Mn | M/1 crown breadth (specify side(s)) |
| 57 | 15 | Mn | M/2 crown length (specify side(s)) |
| 58 | 15a | Mn | M/2 crown breadth (specify side(s)) |
| | 16 | Mn | M/3 crown length (specify side(s)) |
| | 16a | Mn | M/3 crown breadth (specify side(s)) |

Note that crown length measurements corresponding to the alveolar length measurements also can be taken if the teeth are present.

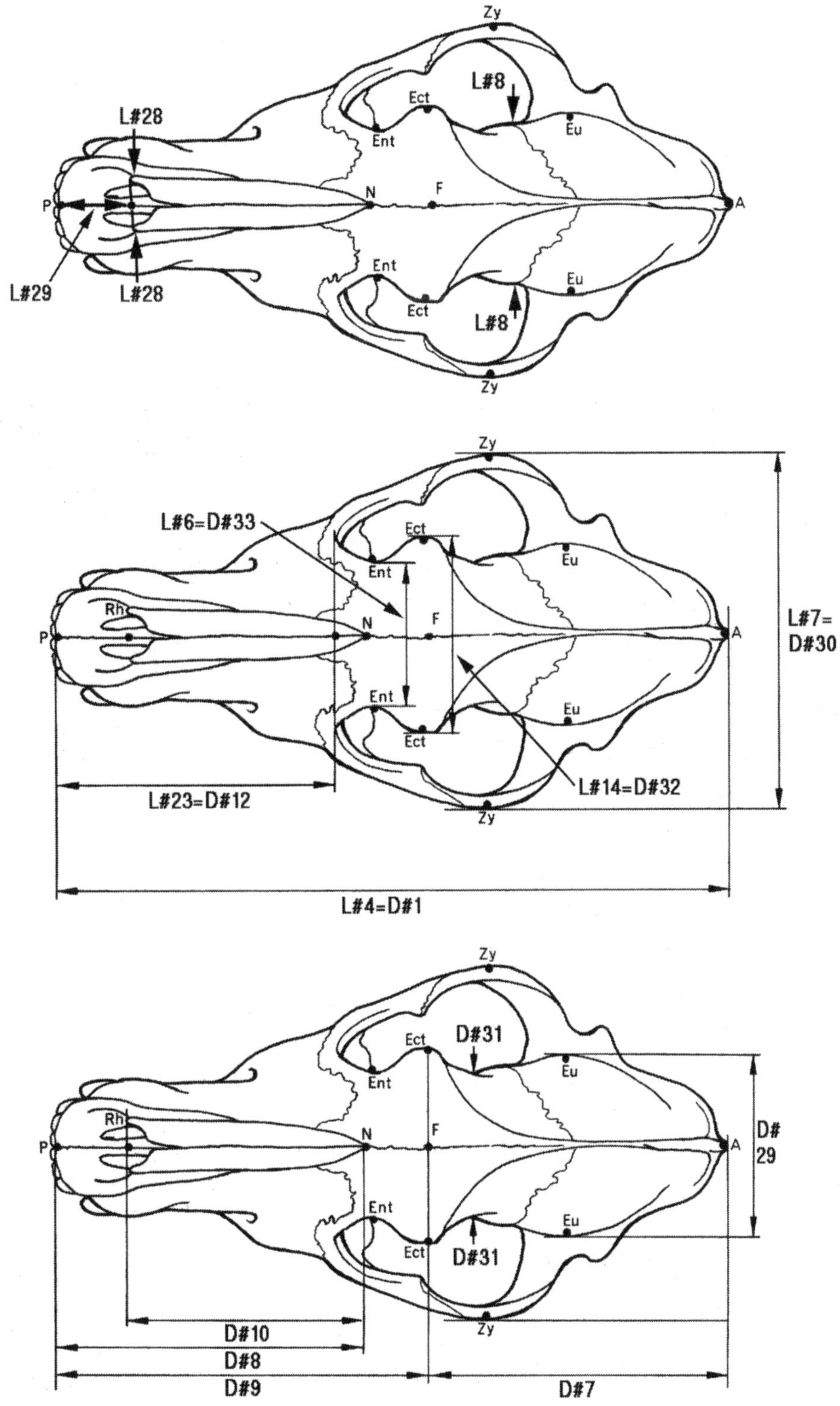


*Figure 1.* Canis *cranium, dorsal view, modified from von den Driesch (1976: Fig. 14a). L# = Barbara Lawrence measurement, D# = von den Driesch measurement. See Tables 1 and 2.*

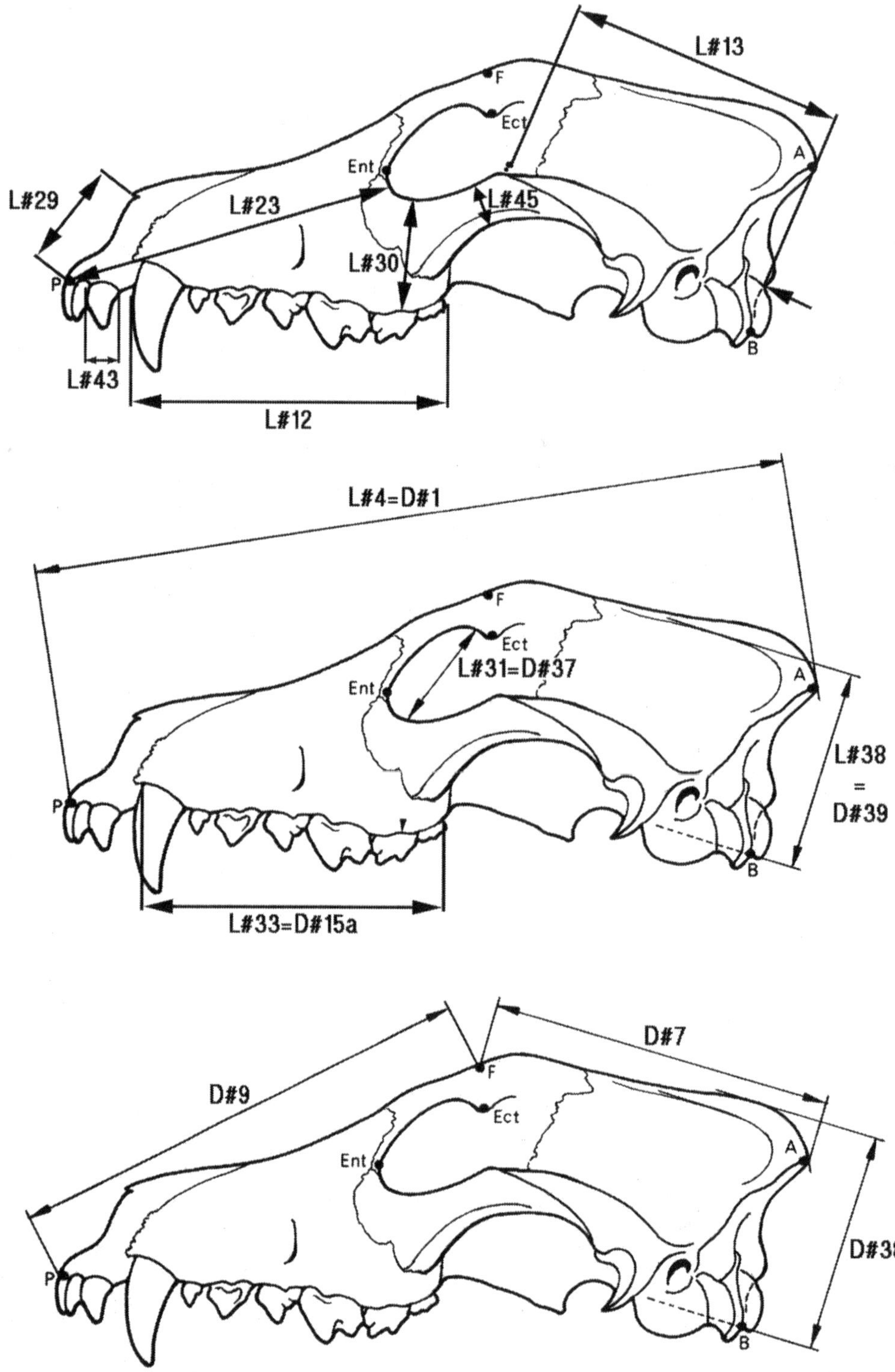


***Figure 2.*** Canis *cranium, left side view, modified from von den Driesch (1976: Fig. 14b). L# = Barbara Lawrence measurement, D# = von den Driesch measurement. See Tables 1 and 2.*

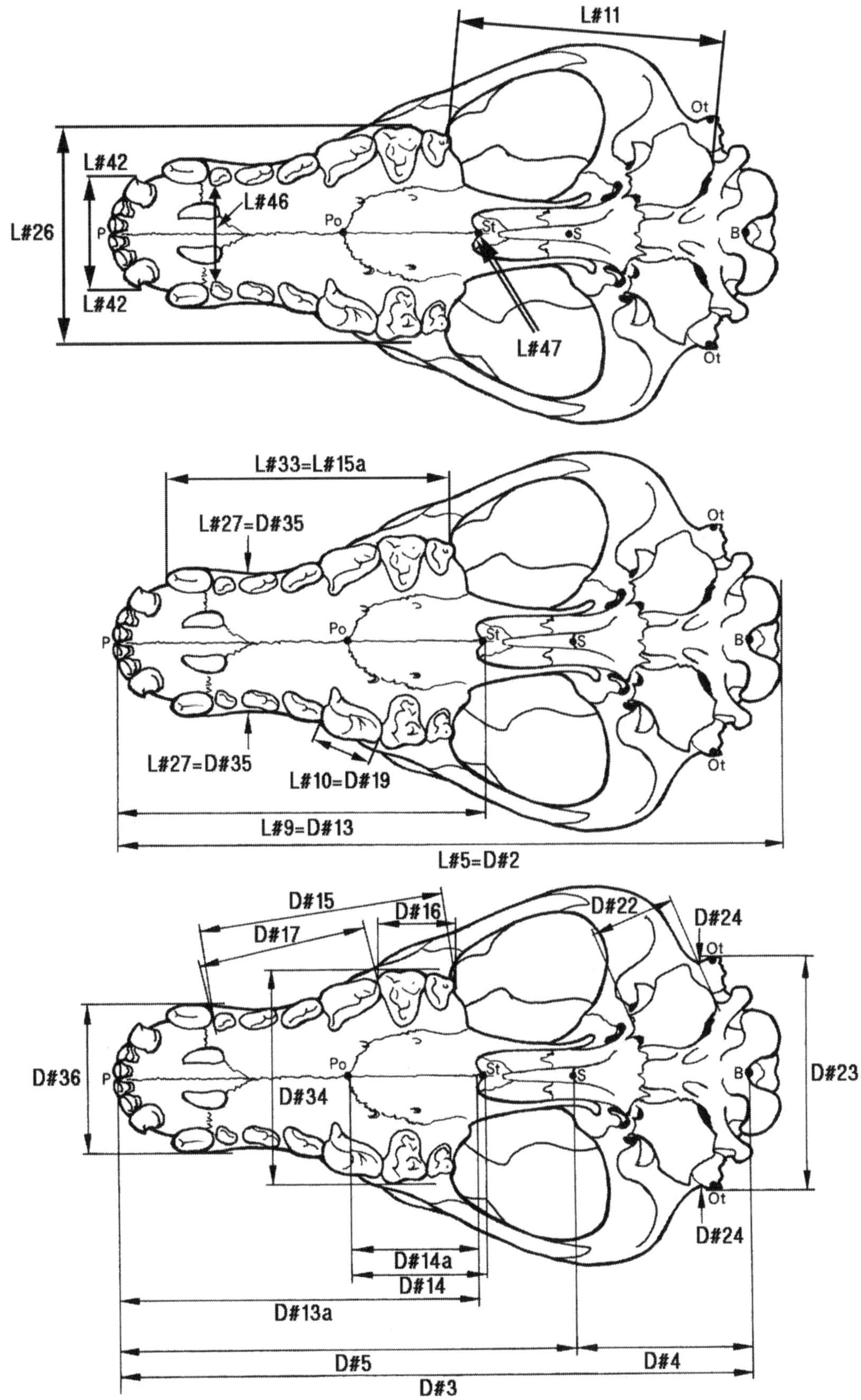


***Figure 3.*** Canis *cranium, basal view, modified from von den Driesch (1976: Fig. 14c). L# = Barbara Lawrence measurement, D# = von den Driesch measurement. See Tables 1 and 2.*

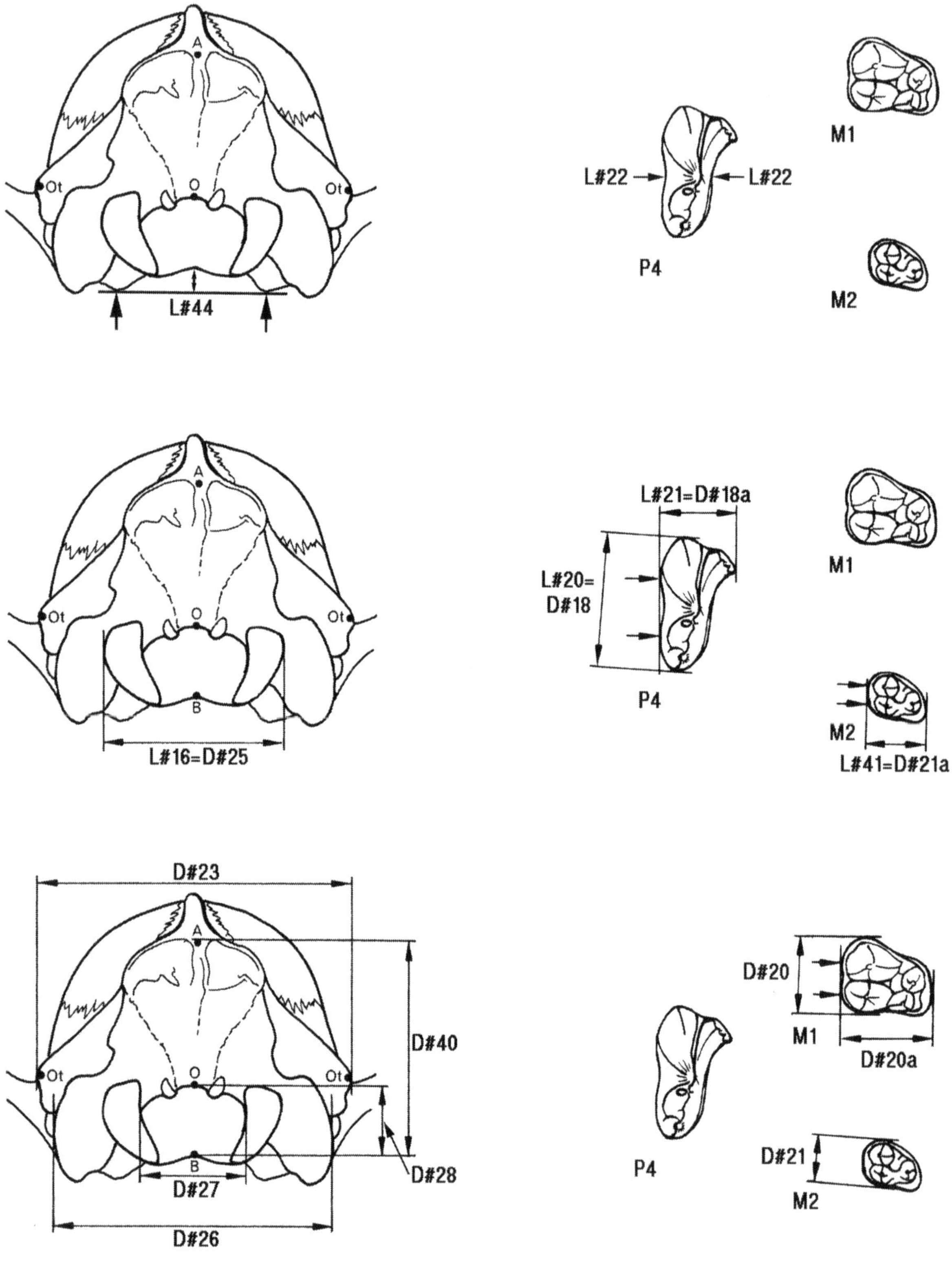


***Figure 4.*** Canis *cranium, nuchal view, and maxillary teeth, occlusal views, modified from von den Driesch (1976: Fig. 14d, 15a-c). L# = Barbara Lawrence measurement, D# = von den Driesch measurement. See Tables 1 and 2.*

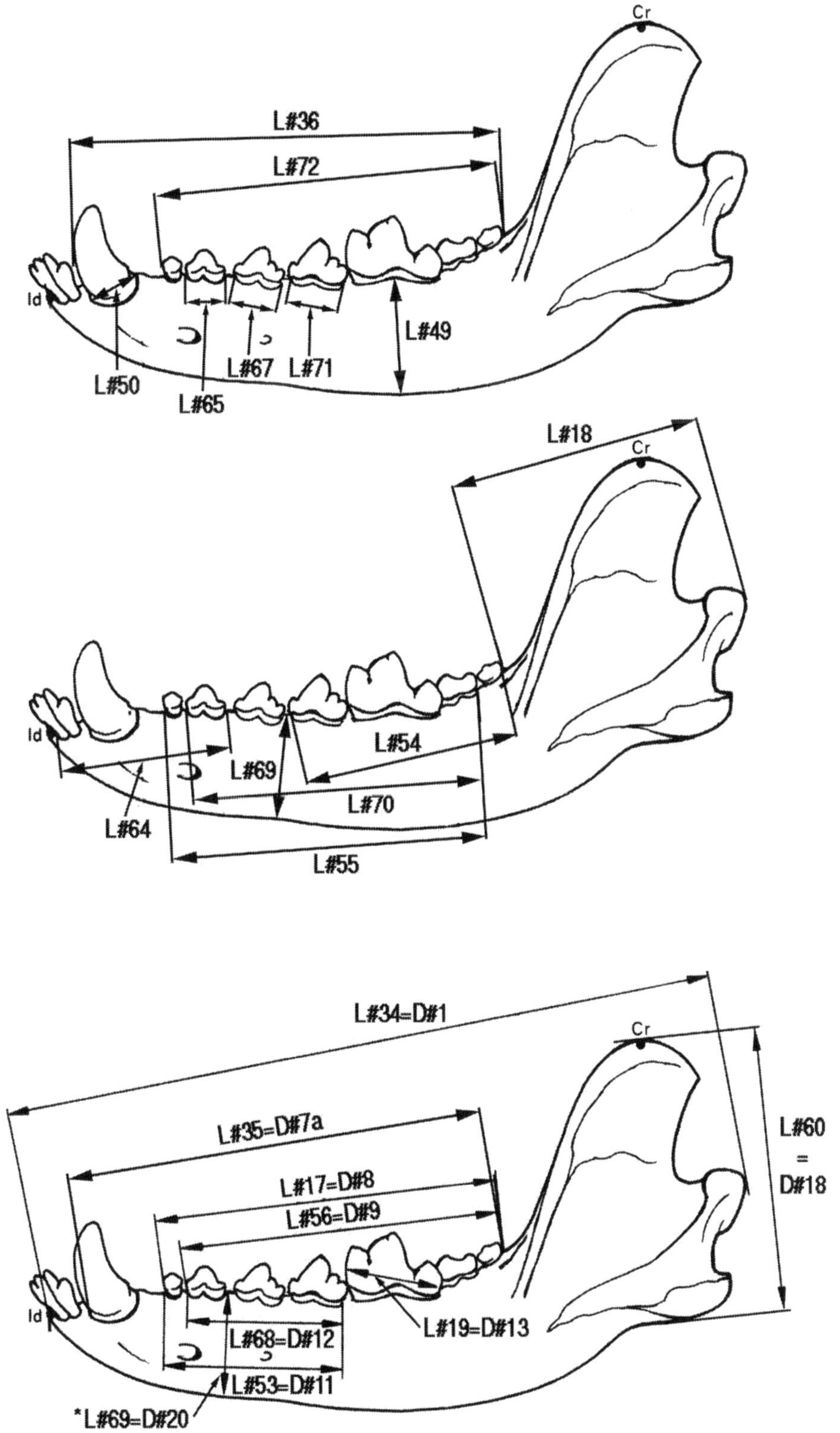


***Figure 5.*** Canis *mandible, lateral view of the left side, modified from von den Driesch (1976: Fig. 23a). L# = Barbara Lawrence measurement, D# = von den Driesch measurement. See Tables 1 and 2.*

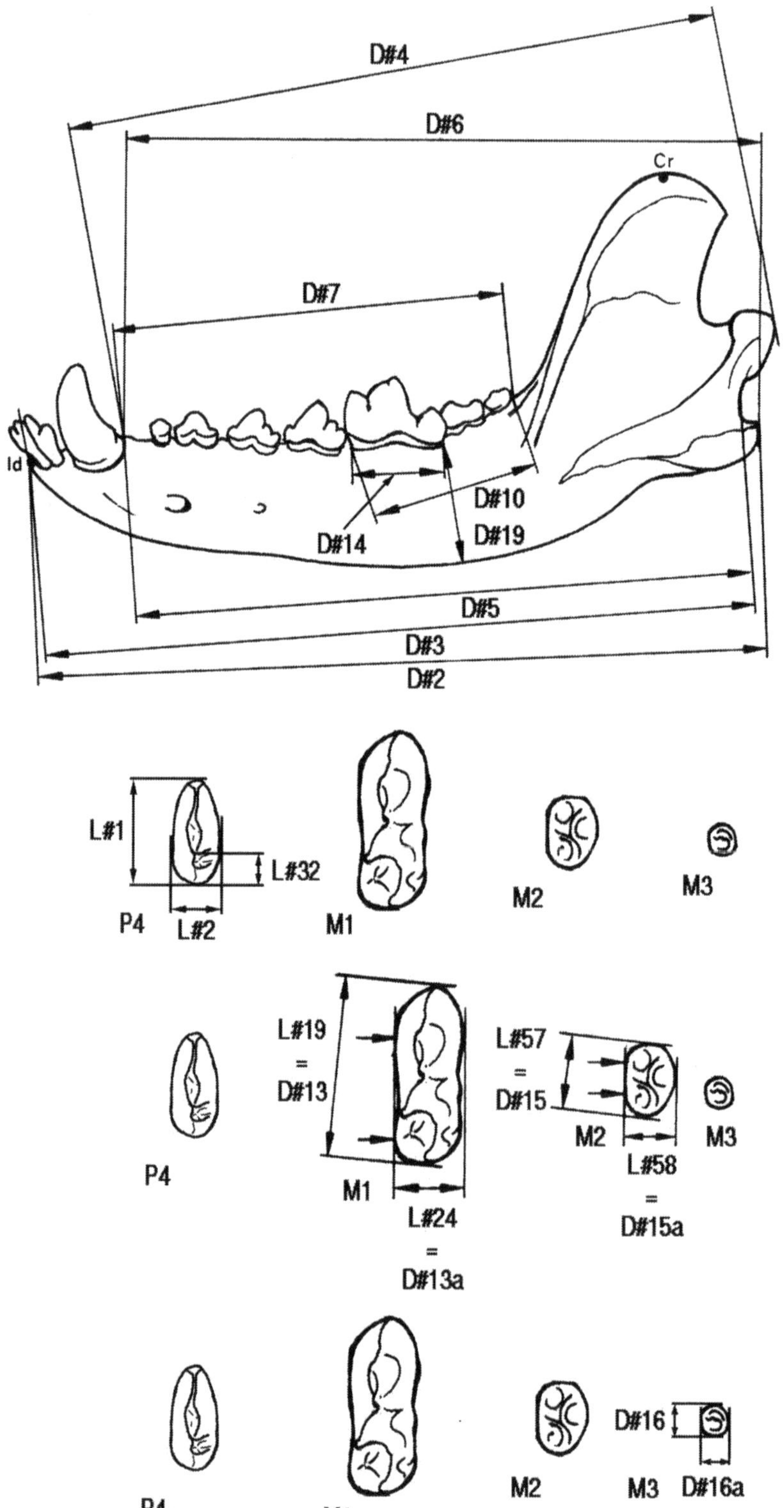


***Figure 6.*** Canis *mandible, lateral view of the left side, and mandibular teeth, occlusal views, modified from von den Driesch (1976: Fig. 23a-d). L# = Barbara Lawrence measurement, D# = von den Driesch measurement. See Tables 1 and 2.*

# A PREBOREAL DOG FROM THE NORTHERN ALPS (SAVOIE, FRANCE)

Louis Chaix
Département d'archéozoologie
Museum d'histoire naturelle
1 route de Malagnou
C.P. 6434 CH - 1211
Geneva, Switzerland

**Key words:** dog, Preboreal, France, Mesolithic dogs, Swiss Neolithic, archaeozoology

**Abstract**

The remains of a domestic dog were found in an Epipaleolithic layer from the rock-shelter of Saint-Thibaud-de-Couz in the French Alps.The stratigraphic position and the date of this animal fits with the transition between the Youngest Dryas and Early Preboreal, with a $^{14}C$ datation of 10,050 ± 100 BP. The dog is a small animal, compared with Mesolithic dogs from northern Europe. Its dimensions and morphology are very similar to those of the Swiss Neolithic. The lack of data about epipaleolithic wolves and our reduced knowledge of the domestication process in this geographical area do not allow more precise conclusions. In addition, the femur seems to indicate an osteodystrophical development of this dog.

## Introduction

The two rock-shelters of Saint-Thibaud-de-Couz are situated on the western slope from the mountainous massif of Chartreuse, in the French Alps (Fig. 1). The two sites, Jean-Pierre I and II presented an important stratigraphy (7 m) from the end of the Tardiglacial to the Late Holocene (Fig. 2). The chronological sequence is assured by a whole range of radiometric data, well correlated with the palynological biozones (Evin, 1994; Girard, 1994). The Bölling-Dryas 2 is characterised by an open environment and a contrasted and humid climate. The fauna shows the presence of arctic elements accompanied by steppical and alpine species. The Alleröd indicates a rapid increase of temperate and sylvatic taxa, when the Younger Dryas is weakly marked. During the Preboreal and Boreal, one observes an important development of the forest with broad-leaves trees and thermophilous plants. The fauna show a mixture of alpine and forest species.

In the rock- shelter Jean-Pierre I, the level 6A, beside the remains of a dog, delivered a fauna dominated by ibex (29.1%) and marmot (20.8%), followed by mountain hare, red deer, roe deer, brown bear and mustelids. From an archaeological point of view, this level is attributed to the Late Epipaleolithic. The lithic industry, very poor and monotonous, is characterized by the dominance of microliths (points and bladlets)(Bintz & Grünwald, 1996). Level 6A was dated, on animal bones, from 10,620 ± 210 years BP (Ly 1190).

Several dog remains were found in this level, in the northern area of Jean-Pierre I. The bones were found on a very small surface. The level is sealed by a compact and crusted sediment which prevents any contamination from the upper levels. For more security, an AMS date was made on the skull of the dog itself. The result was 10,050 ± 100 years BP (Ly-23 /Oxa 4405) which confirms the stratigraphic position of this canid and its attribution to the boundary between the end of the Younger Dryas and the beginning of the Preboreal.

The dog is known to be the first animal to have been domesticated in various regions of the globe. In Europe, evidence of Magdalenian dogs are documented from Spain (Altuna & Mariezkurrena, 1985), Switzerland (Morel & Müller, 1997) and Germany (Nobis, 1986; Musil, 1974, 1980, 1985). During the Preboreal, domestic dogs seem to be well represented, particularly in Mesolithic contexts from the northern part of Europe (Benecke, 1994). In France, the dog of Saint-Thibaud appears to be the oldest for this period. We can also observe that at the Mesolithic site of Noyen, in the Parisian Basin, dated between 8,000 and 7,400 BP, some wolves present characteristics (sagittal groove of the frontal , shortening of the snout), which probably indicate the beginning of domestication (Vigne & Vigne, 1988), similar to the observations made in the Late Magalenian site of Kniegrotte in Germany (Musil, 1974, & this volume).

## The specimen

The dog remains comprise the skull, right mandible, atlas, axis and the distal part of the left humerus (Fig.8, Plate 1). All of these bones were found together and belong to the same animal. A left femur, found in square 5, few meters from the other remains, in the northern sector also belongs to a dog. The femur, possibly from the same animal, is almost complete, but the great trochanter, yet unfused, is missing. A strong bite mark is visible on the anterior part

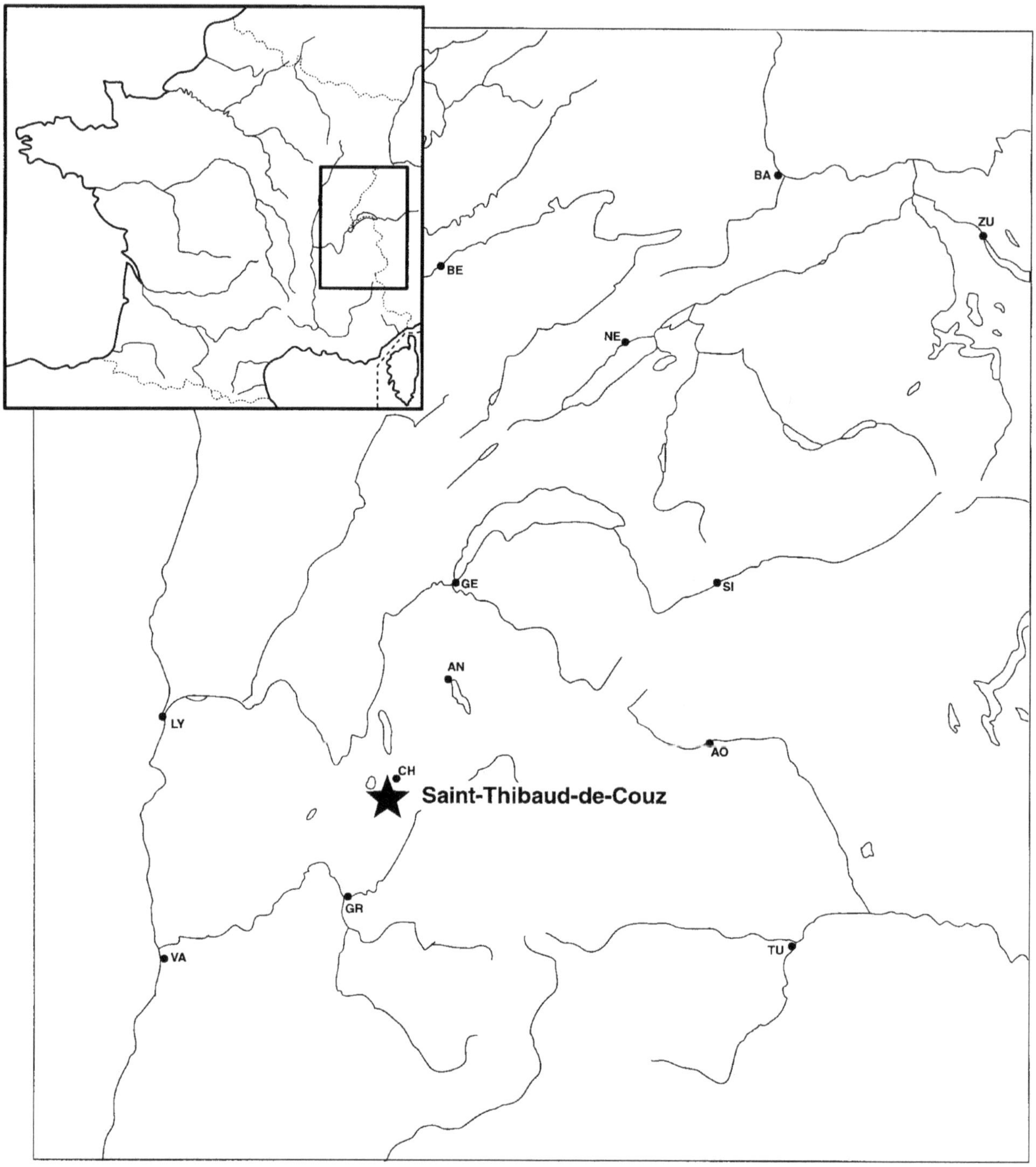


***Figure 1** Geographical situation of Saint-Thibaud-de-Couz (Savoie, France)*

of the distal shaft. The relation of this bone with the other elements is uncertain. Unfortunately, the skull has lost all the upper teeth except for a fragment of the left carnassial ($P^4$) and the right third incisor ($I^3$). The left mandible shows the presence of $P_3$, $M_1$ and $M_2$. All other teeth are missing. We can note the total absence of $M_3$ which seems uncommon amongst wild canids (Colyer, 1990). The atlas is almost complete while the axis is only represented by its proximal part. The humerus is attested only by its distal part.

Based on the stage of tooth wear and fusion stage of the femur we can ascribe an age of around 10 months to this dog (Habermehl, 1975). Some elements like the shape of the temporal crests, the poorly developed sagittal crest, and metrical data indicate the animal was probably a female (Fig.3). The withers height was estimated, from the length of the femur, to *c.* 40 cm (after Koudelka, 1885: 40.9 cm ; after Harcourt, 1974: 39.6 cm). This indicates a small sized dog compared to the individuals from the Early and Late Mesolithic of northern Europe, which have a range of estimated withers heights of 49-54.1 cm.

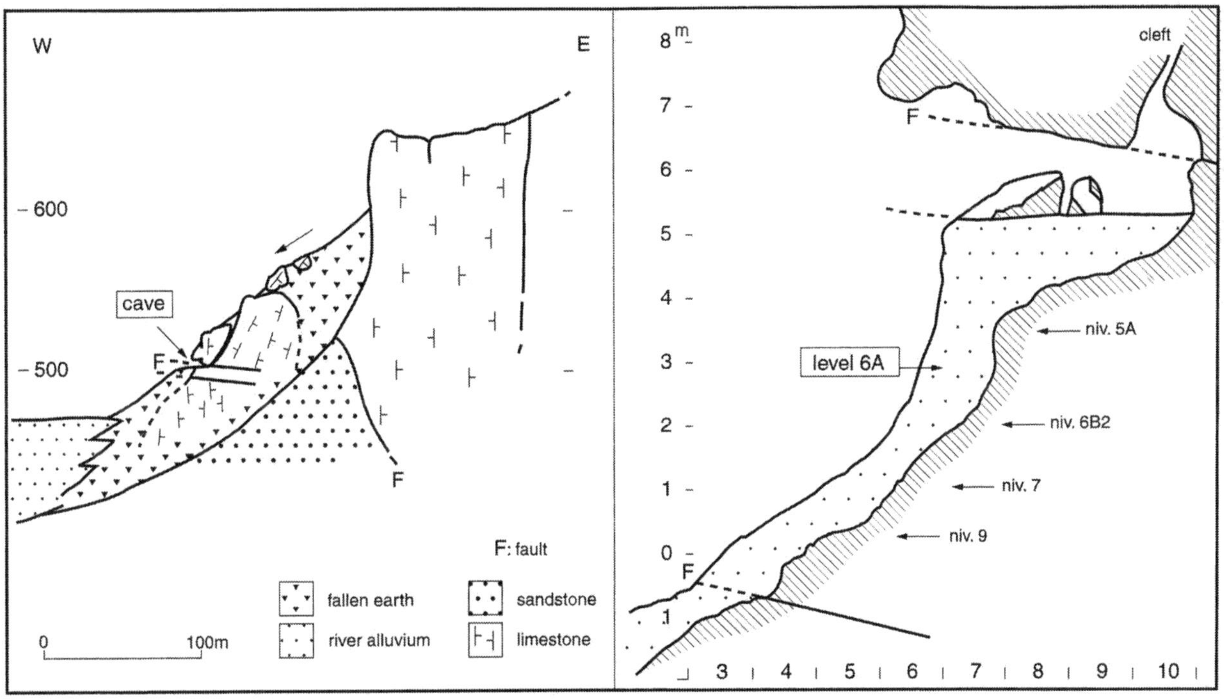


***Figure 2** Schematic stratigraphy of Jean-Pierre 1 rock-shelter (after Bintz, 1994). (a) geological cross-section (b) cross-section of the rock-shelter with the position of level 6A*

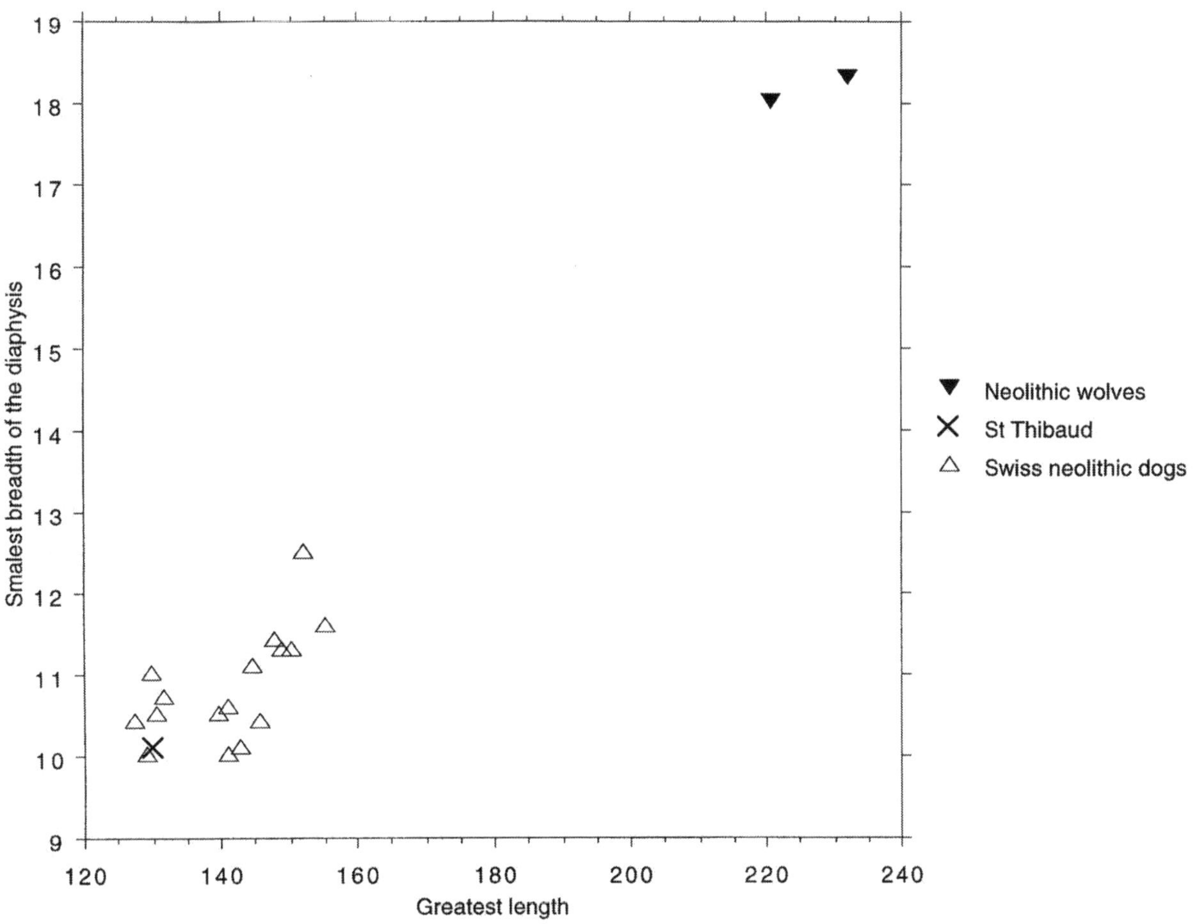


***Figure 3** Scattergram of the femur showing weak robusticity, suggesting that this individual is probably female*

| | | | |
|---|---|---|---|
| **SKULL** | | **MANDIBLE** | |
| | | | |
| Total length | 149 | Total length | 106.5 |
| Condylobasal length | 140.5 | Length between indentation-infradentale | 101.3 |
| Basal length | 132.5 | Length condyle-C post. | 91 |
| Basicranial axis | 43.4 | Length indentation - C post | 86.5 |
| Basifacial axis | 89.8 | Length C post - M3 post. | 58.9 |
| Neurocranium length | 79.5 | Length of the cheektooth row (P1-M3) | 55.5 |
| Upper neurocranium length | 71 | Length of the carnassial (cingulum) | 20.2 |
| Viscerocranium length | 76.5 | Breadth of the carnassial (cingulum) | 7.8 |
| Facial length | 89 | Height of the vertical ramus | 41.5 |
| Greatest length of the nasals | 52.5 | Basal length (Brinkmann 1924) | 124.7 |
| "Snout" length | 60 | Basal length (mean Brinkmann 1924) | 126.3 |
| Median palatal length | 73 | Basal length ( Dahr 1937) | 116.3 |
| Length of the horizonzal part of palatin | 23 | | |
| Greatest diameter of the auditory bulla | 21.9 | | |
| Greatest mastoid breadth | 53.6 | **ATLAS** | |
| Breadth dorsal to the ext. auditory mea | 46 | | |
| Greatest breadth of the occipital condyl | 31.3 | Greatest breadth over the wings | 56.5 |
| Greatest breadth of the paroccipital prc | 53.6 | Greatest length | 29.9 |
| Greatest breadth of the foramen magnu | 16.2 | Breadth of the cran. art. surf. | 32.5 |
| Height of the foramen magnum | 12.9 | Breadth of the caud. art. surf. | 25.5 |
| Greatest neurocranium breadth | 54 | Height | 21 |
| Zygomatic breadth | 83 | | |
| Least breadth of skull | 32.1 | | |
| Frontal breadth | 39.5 | **AXIS** | |
| Least breadth between the orbits | 27 | | |
| Greatest palatal breadth | 55 | Breadth of the cran. art. surf. | 22.8 |
| Least palatal breadth | 28.9 | Smallest breadth of the vertebra | 17.6 |
| Breadth at the canine alveoli | 22.4 | | |
| Greatest inner height of the orbit | 28 | | |
| Skull height | 46.5 | **HUMERUS** | |
| Height of the occipital triangle | 35.4 | | |
| Length of the cheektooth row | 49 | Greatest breadth of the distal end | 24.6 |
| Length of the carnassial alveolus | 16 | | |
| | | | |
| | | **FEMUR** | |
| | | | |
| | | Greatest length from caput femoris | 130.4 |
| | | Greatest depth of caput femoris | 15 |
| | | Smallest breadth of the diaphysis | 10.1 |
| | | Smallest circumference of the diaphysis | 32 |
| | | Greatest breadth of the distal end | 25.3 |

***Table 1*** *Metric data of the dog from Saint-Thibaud (after von den Driesch, 1976)*

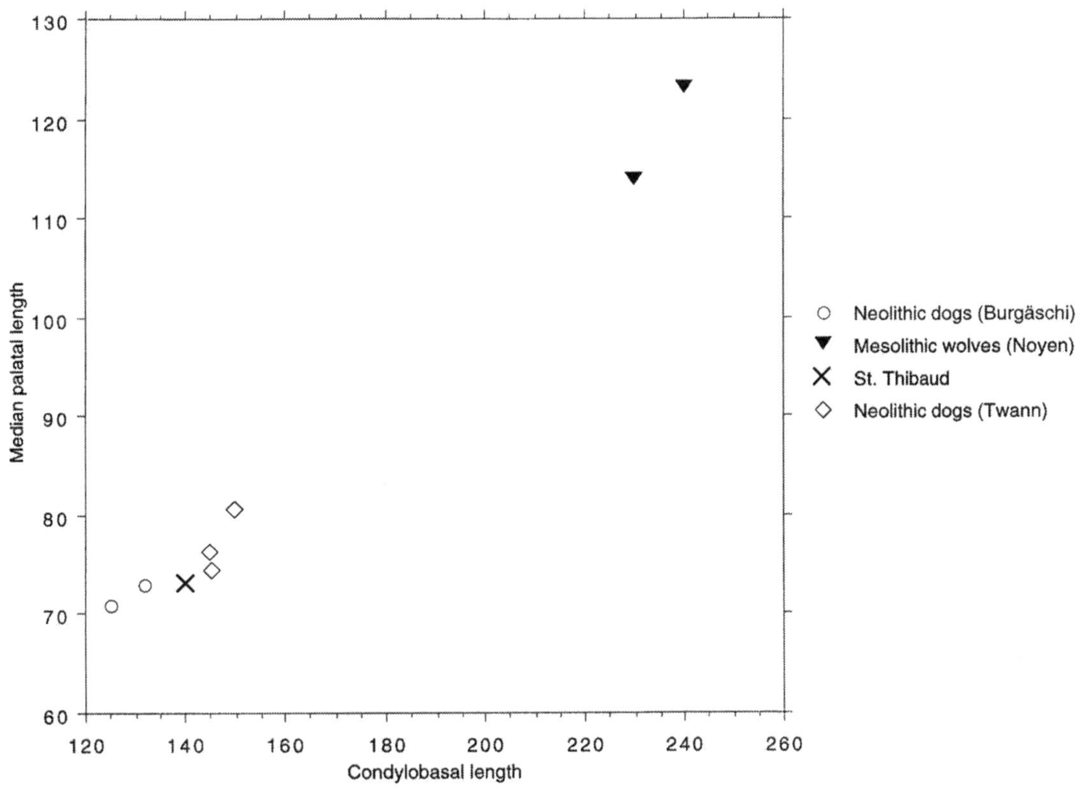


***Figure 4*** *Scattergram of the relation between skull's condylobasal length and median palatal length*

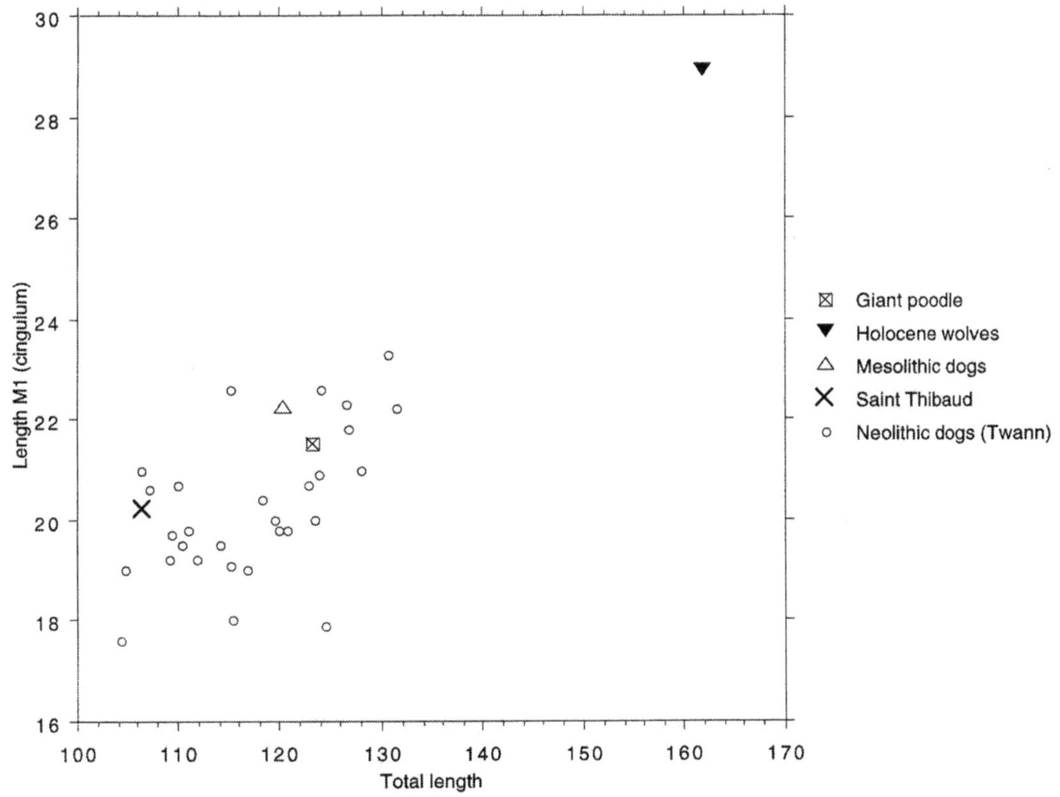


***Figure 5*** *Scattergram of the relation between the length of the mandible and the length of the carnassial (measured at the cingulum)*

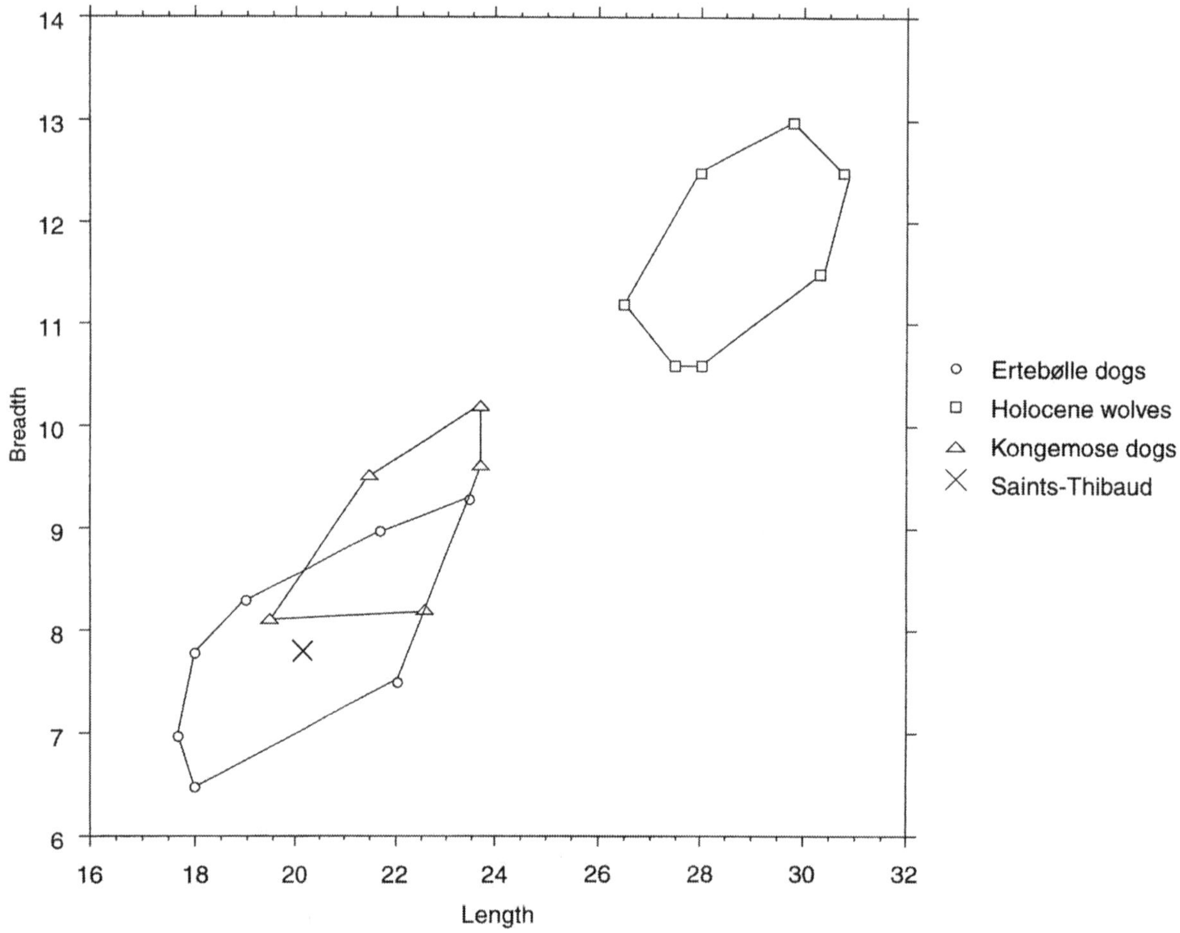


***Figure 6*** *Scattergram of the dimensions of the lower carnassial (at the cingulum) (after Benecke, 1994)*

The estimated height for the dog from Saint-Thibaud is more similar to that of the Middle Neolithic dogs, in example those of Switzerland, which have a mean of 43.6 cm and a range from 33.5-53.9 cm (Becker & Johansson, 1981).

All the measurements, after von den Driesch (1976) are given in the Table 1. From a morphological point of view, I note a number of points. The skull is well rounded, with smooth muscular crests and attachments. The frontal is flat, without the sagittal groove observed on the Mesolithic wolves from Noyen (Vigne & Marinval-Vigne, 1988). Consider the skull proportions, particularly the relation between the condylobasal length and the median palatal length: note that the specimen from Saint-Thibaud is very close to the Neolithic dogs of Switzerland (Fig. 4). There is no shortening of the facial part and the alveoli from the teeth show a total absence of crowding. Based on the mandible, the dog of Saint-Thibaud falls within the range of variation of Swiss Neolithic dogs (Fig. 5). Note that the alveolus of the second premolar ($P_2$) is clearly crowded, as the fourth premolar ($P_4$) is rather large. The dimensions of the carnassial ($M_1$) are also comparable to those of the Late Mesolithic dogs from the Ertebølle culture (Benecke, 1994) (Fig. 6 ).

The distal breadth of the humerus falls a new time near the mean of the Neolithic dogs fom Switzerland, below the range of variation of the Mesolithic dogs from northern Europe (Table 2). The femur shows a strong anterior curvature of the shaft, particularly in the distal part. This feature seems to be abnormal and possibly due to osteodystrophy (rickets or osteomalacy) (Baker & Brothwell, 1980; Moodie, 1923; Poplin, pers. com.). The relationship between the condylobasal length of the skull and the length of the femur, of domestic dogs from various contexts, shows a good correlation (0.811). (Fig.7). Therefore, it is probable that the femur fits with the dog remains found in area AW/5-6.

| HUMERUS | Distal breadth | | | |
|---|---|---|---|---|
| | N | mean | min | max |
| Saint-Thibaud | 1 | 24.6 | | |
| Northern mesolithic dogs | 15 | 32 | 26.5 | 35.7 |
| Swiss neolithic dogs | 95 | 25.6 | 21 | 32 |
| Giant poodles | 19 | 35.8 | 32 | 40.7 |
| Holocene wolves | 20 | 42.1 | 37.7 | 46.1 |

***Table 2*** *Humerus, distal portion. Comparative metric data*

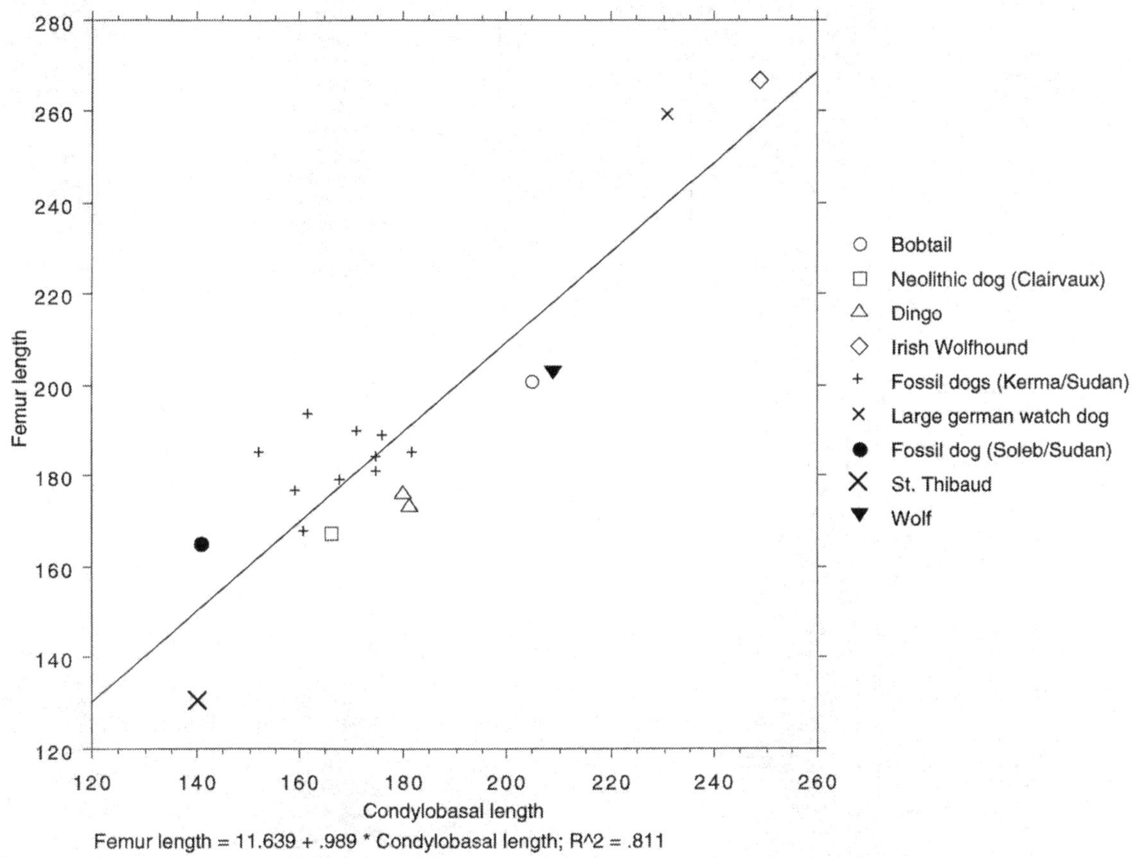


***Figure** 7 Scattergram: skull condylobasal length/femur length*

## Conclusion

The remains found in the cave of Saint-Thibaud certainly belong to a domestic dog (*Canis familiaris*). The dating of this animal can be attributed to the transition between the Younger Dryas and the Preboreal, in a level characterized by a typical Late Epipaleolithic industry. Compared with Mesolithic dogs from northern Europe, this animal is clearly smaller. Its morphology and osteometry place it in the lower part of the variation range of the Neolithic dogs of western Europe. The presence of a pathological femur seems to be linked to a dystrophic development of this animal. But our present knowledge of wild and domestic canids in this area is very limited and the strong polymorphism of the genus *Canis* adds complications to understanding the domestication process of our oldest companion.

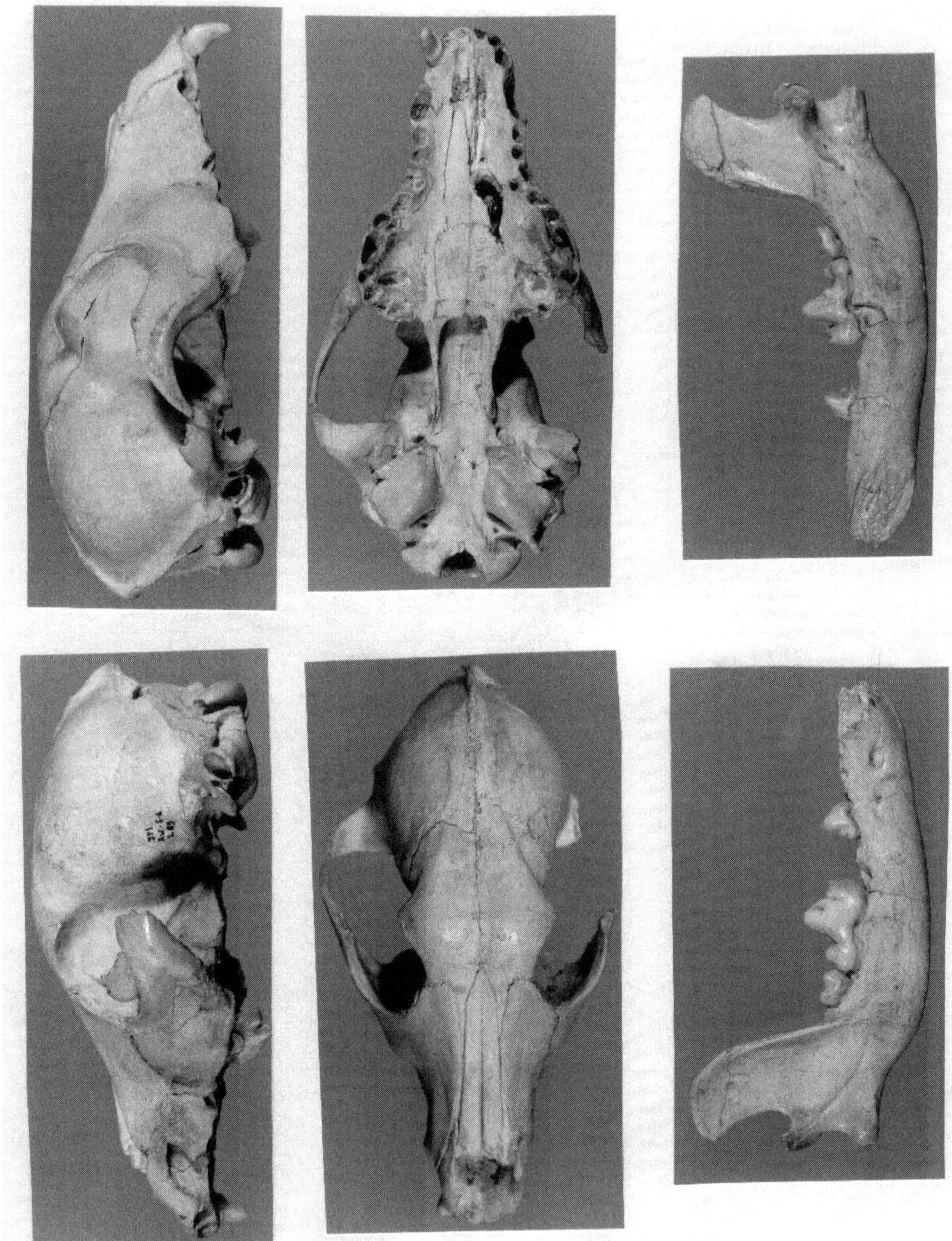

***Figure 8a*** *Plate 1. Skeletal remains from the dog of Saint-Thibaud*

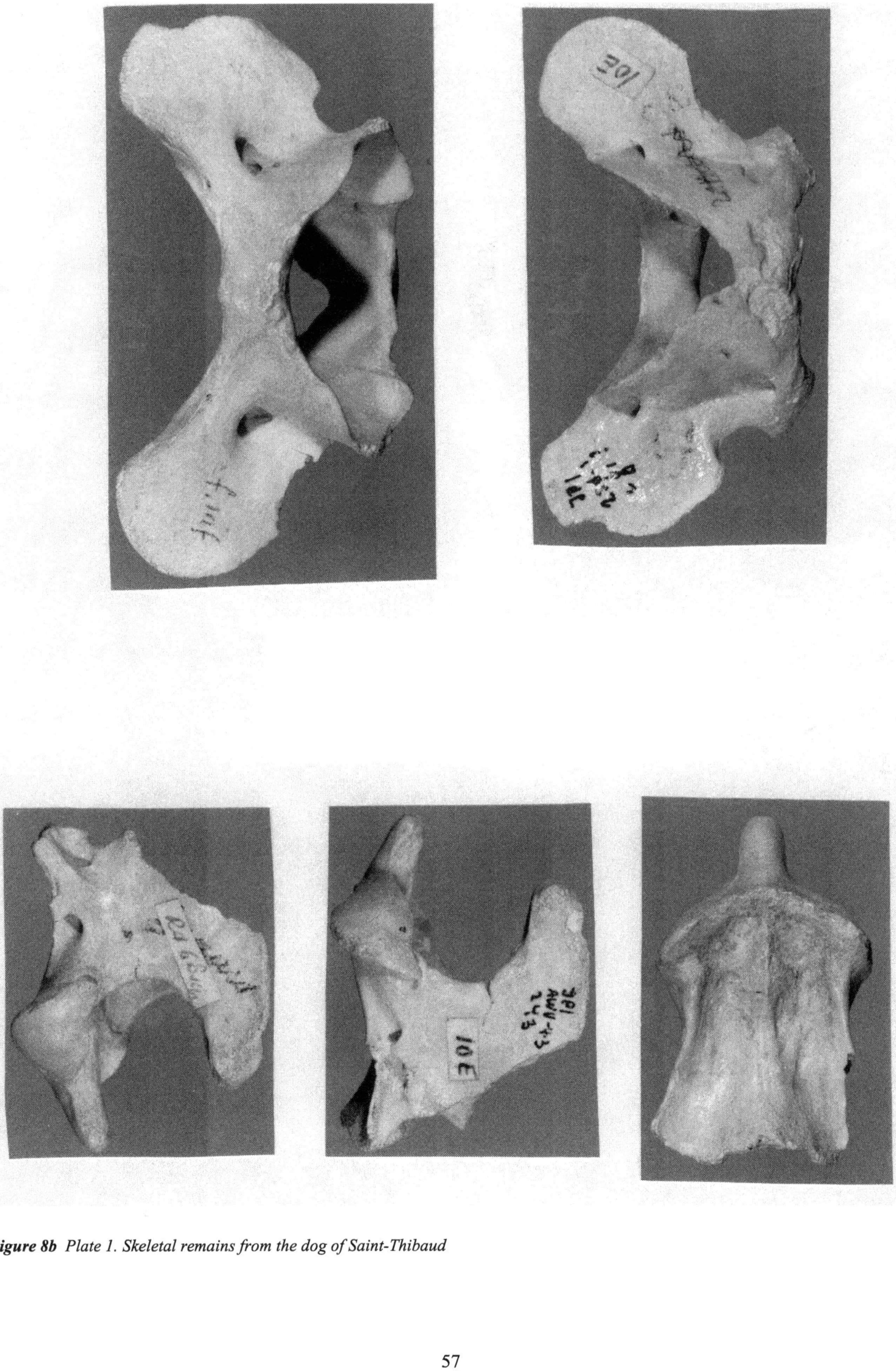

***Figure 8b*** *Plate 1. Skeletal remains from the dog of Saint-Thibaud*

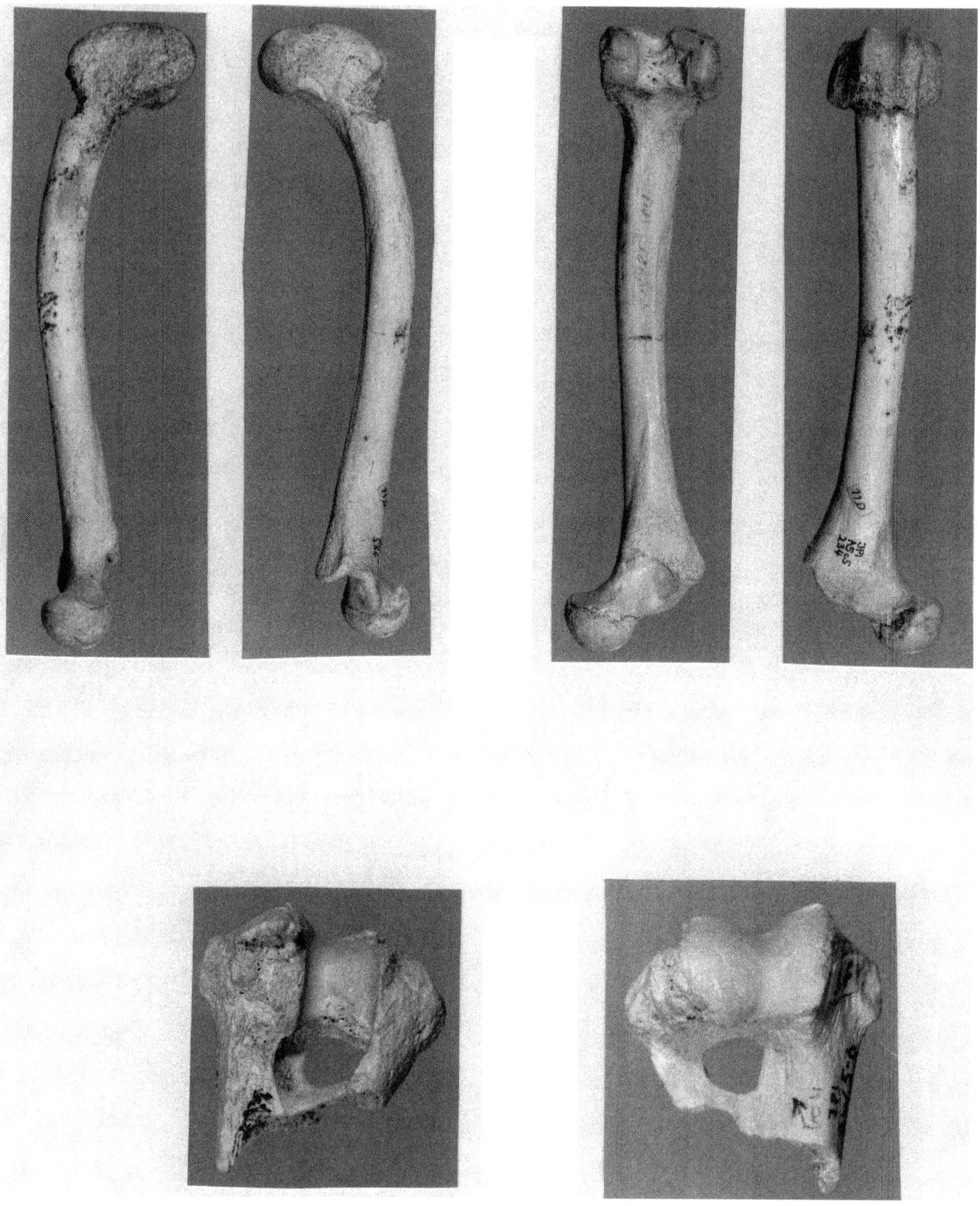

***Figure 8c*** *Plate 1. Skeletal remains from the dog of Saint-Thibaud*

## References


Altuna, J. & Mariezkurrena, K. (1985). Bases de subsitencia de los pobladores de Erralla. *Munibe* **37**, 87-117.

Baker, J. & Brothwell, D. (1980). *Animal Diseases in Archaeology*. London, Academic Press.

Becker, C. & Johansson, F. (1981). Tierknochenfunde. Zweiter Bericht. *Die neolithischen Ufersiedlungen von Twann* **11**. Bern: Staatlicher Lehrmittelverlag.

Benecke, N. (1994). Archäozoologische Studien zur Entwicklung der Haustierhaltung in Mitteleuropa und Südskandinavien von den Anfängen bis zum ausgehenden Mittelalter. *Schriften für Ur-und Frühgeschichte* **46**.

Bintz, P. & Grünwald, C. (1996). I. Analyse de l'industrie lithique. In (P. Bintz, dir) Les grottes Jean-Pierre 1 et 2 à Saint-Thibaud-de-Couz (Savoie). Paléoenvironnement et cultures du Tardiglaciaire à l'Holocène dans les Alpes du Nord. *Gallia-Préhistoire* **37**, 159-272.

Colyer, F. (1990). *Variations and Disease of the Teeth of Animals*. (Revised, A.E. W.Miles & C. Grigson, Eds). Cambridge: Cambridge University Press.

Driesch, A. von den (1976). *A Guide to the Measurement of Animal Bones from Archaeological Sites*. Cambridge, MA: Harvard University Peabody Museum Bulletin **1**.

Evin, J. (1994). Les datations radiométriques. In (P. Bintz, dir) Les grottes Jean-Pierre 1 et 2 à Saint-Thibaud-de-Couz (Savoie). Paléoenvironnement et cultures du Tardiglaciaire à l'Holocène dans les Alpes du Nord. *Gallia-Préhistoire* **36**, 183-188.

Girard, M. (1994). Les données de la palynologie. In (P. Bintz, dir) Les grottes Jean- Pierre 1 et 2 à Saint-Thibaud-de-Couz (Savoie). Paléoenvironnement et cultures du Tardiglaciaire à l'Holocène dans les Alpes du Nord. *Gallia-Préhistoire* **36**, 193-196.

Habermehl, K. H. (1975). *Die Altersbestimmung bei Haus- und Labortieren*. Berlin: Verlag Paul Parey.

Harcourt, R. A. (1974). The dog in prehistoric and early historic Britain. *Journal of Archaeological Science* **1**, 151-175.

Koudelka, F. (1885). Das Verhältnis der Ossa longa zur Skeletthöhe bei den Säugetieren. *Verhandlungen der Naturforschenden Verein Brünn* **24**, 127-153.

Morel, P. & Müller, W. (1997). Un campement magdalénien au bord du Lac de Neuchâtel. Etude archéozoologique (secteur 1). Hauterive-Champréveyres 11. *Archéologie neuchâteloise* **23**.

Moodie, R. L. (1923). *Palaeopathology : An Introduction to the Study of Ancient Evidences of Disease*. Urbana: University of Illinois Press.

Musil, R. (1974). Tiergesellschaft der Kniegrotte. In (R. Feustel, Ed) *Die Kniegrotte. Eine Magdalénienstation in Thüringen*. Weimar: Veröffentlichungen des Museums für Ur-und Frühgeschichte **5**.

Musil, R. (1980). Die Grosssäuger und Vögel der Teufelsbrücke. In (R. Feustel, ed) *Magdalénienstation Teufelsbrücke. II. Paläontologischer Teil*. Weimarer Monographien zur Ur-und Frühgeschichte **3**, 5-59.

Musil, R. (1985). Die Fauna der Magdalénien-Siedlung Oelknitz. *Weimarer Monographien zur Ur-und Frühgeschichte* **17**.

Nobis, G. (1986). Die Wildsäugetiere in der Umwelt des Menschen von Oberkassel bei Bonn und das Domestikationsproblem von Wölfen im Jungpaläolithikum. *Bonner Jahrbücher* **186**, 367-376.

Vigne, J. D. & Marinval-Vigne, M.C. (1988). Quelques réflexions préliminaires sur les Canidés mésolithiques de Noyen-sur-Seine (France) et sur la domestication du chien en Europe occidentale. *Archaeozoologia* **II**, 153-164.

# ANCIENT REMAINS OF JOMON DOGS FROM NEOLITHIC SITES IN JAPAN

Nobuo Shigehara
Primate Research Institute,
Kyoto Universit
41 Kanrin, Inuyama,
Aichi 484-8606
JAPAN

Hitomi Hongo
Primate Research Institute,
Kyoto University,
41 Kanrin, Inuyama,
Aichi 484-8606 JAPAN

**Key words:** dog, Japan, Jomon, Neolithic, archaeozoology, China

**Abstract**

The Jomon period began about 12,000 years ago and continued for about 10,000 years until 2,300 years ago. Jomon dogs migrated from the Asian mainland together with the ancestors of the Jomon people. The oldest dog remains in Japan were excavated from the Natsushima Shell-mound in Kanagawa Prefecture, dated to 9,300BP. Maximum cranial lengths of the Jomon dog is *c.*160 mm for males and *c.*150 mm for females. Cranial size gradually increased from the end of Jomon. Several dogs were intentionally buried. No attempt at breed improvement appears to have been made in the Jomon Period. Jomon dogs have the following primitive characteristics: both the skull and extremities are robust; the "stop" is relatively small; the width between the zygions is relatively narrow; the snout is relatively wide; periodontal diseases, such as discrepancy of teeth row, are rare; there is a high degree of tooth loss due to heavy attrition or breakage; the palate is of elongated triangular shape.

## Introduction

The Jomon period began about 12,000 years ago and continued for about 10,000 years (Table 1). Since the mode of subsistence during the Jomon period was based on hunting and gathering, dogs were probably very important as hunting companions. Although many dog skeletons have been excavated from archaeological sites of this period, few studies have examined the morphology of these Jomon dogs (Hasebe, 1924, 1952; Naora,1973; Shigehara, 1994; Nishimoto, 1981; Miyazaki, 1982). An overview of the history of research on Jomon dogs was presented by Ota (1980). Morphological changes of Japanese dogs from the Jomon period to the present have been studied and reported by Shigehara (1991, 1994). The morphology of present-day Shiba dog, considered to be a descendent of Jomon dogs, was studied by Shigehara, Onodera & Eto (1997), and a method of sex determination in excavated dog remains also described. Kaneko (1978) reports in his general overview of dog remains from the Jomon period that Jomon dogs have a single phylogenetic origin. Niwa (1982, 1987) studied the morphological characteristics of Jomon dogs and discussed the homogeneity of the population, as well as chronological and regional variations in the populations. The Niwa study (1982) on the cranial size of Jomon dogs reported that dogs were larger in northern Japan. Survey of excavated dog remains in Japan performed by Niwa (1987) revealed that dog skeletons were found from 282 Jomon sites. These sites include 22 Earliest, 45 Early, 53 Middle, 112 Late, and 50 Final Jomon period sites. In this study, regional variations are reported in the morphology of the caudal (posterior) margin of the hard palate. The present study uses primarily the measurement methods described by Saitoh (1963), Shigehara and Onodera (1984), and von den Driesch (1976), respectively.

| | | | |
|---|---|---|---|
| Jomon Period | 12,000 | - | 2,300 B.P. |
| Yayoi Period | 2,300 | - | 1,700 B.P. |
| Kofun Period | 1,700 | - | 1,300 B.P. |
| Nara • Heian Period | 1,300 | - | 800 B.P. |
| Medieval Period | 800 | - | 400 B.P. |
| Premodern Period | 400 | - | 130 B.P. |
| Modern Peridon | 130 | - | |

***Table 1*** *Japanese chronology*

## Origin of Jomon dogs

Native Japanese wolves are not considered to be the wild progenitor of Japanese dogs. Rather, dog are thought to have migrated from the Asian mainland together with the

| | Small | medium-small | medium | medium-larage | large | Driesch (1976) |
|---|---|---|---|---|---|---|
| Total L.(I-pr) | -155 | 156-170 | 171-185 | 186-200 | 201- | [1] |
| Cranial L.(na-I) | -83 | 84-93 | 94-103 | 104-113 | 114- | |
| Cranial Breadth (eu-eu) | -54 | 55-59 | 60-64 | 65-69 | 70- | [29] |
| Facial Length (pr.-na) | -76 | 77-84 | 85-92 | 93-100 | 101- | [8] |
| Snout Length | -64 | 65-72 | 73-80 | 81-88 | 89- | [12] |
| Maxillar Breadth | -52 | 53-57 | 58-62 | 63-67 | 68- | [34] |
| Mandibular Length | -113 | 114-124 | 125-135 | 136-146 | 147- | [M2] |
| Lower cheekteeth Length | -60 | 61-65 | 66-70 | 71-75 | 76- | [D15] |
| Humeral Length | -120 | 121-135 | 136-150 | 151-165 | 166- | [GL] |
| Radial Length | -115 | 116-130 | 131-145 | 146-160 | 161- | [GL] |
| Ulnala Length | -140 | 141-155 | 156-170 | 171-185 | 186- | [GL] |
| Femoral Length | -135 | 136-150 | 151-165 | 166-180 | 181- | [GL] |
| Tibial Length | -130 | 131-145 | 146-160 | 161-175 | 176- | [GL] |

***Table 2*** *Size categories of the dog (Hasebe, 1952). Last column shows the same measurements defined by von den Driesch (1976)*

ancestors of the Jomon people. Although several theories exist concerning the timing of human colonisation of the Japanese archipelago, the oldest human skeletal remains found in Japan thus far were excavated at the Minatogawa site in Okinawa. These remains were found to date back approximately 18,000 years. The oldest dog remains in Japan, excavated from the Natsushima Shell Mound in Kanagawa Prefecture, were found to date back to approximately 9,300BP. The second-oldest dog remains were excavated from the Kamikuroiwa Cave (Rock Shelter) site in Ehime Prefecture. These remains, which date back to 8,500-8,000BP, appear to have been intentionally buried.

Rice agriculture was introduced from the Asian mainland toward the end of the Final Jomon period and brought a drastic change in the diet and lifestyle of the Jomon people. As the Jomon people grew more dependent on agriculture, hunting became less important. Consequently, the dog as hunting companion became less valuable. The number of dogs found in archaeological sites declines sharply for later period sites, due primarily to less favorable burial condition for preservation of faunal remains. Skeletal remains from these periods, unlike those at Jomon sites, are not very well preserved.

## Morphological characteristics of Jomon dogs

Maximum cranial length of the Jomon dogs is *c.*160 mm for males and *c.*150 mm for females. The male Jomon dog is categorised as "medium-small sized", whereas the female is categorised as "small sized", according to the size classification suggested by Hasebe (1952) (Table 2). Jomon dogs have the primitive characters listed below:

- both the skull and extremities are robust (well built).
- the "stop" (concavity from the frontal region to the snout) is relatively small (Fig. 1).
- the width between the zygions (the most lateral points of the zygomatic arches) is relatively narrow.
- the snout is relatively wide.
- periodontal diseases, such as discrepancy of teeth row, are rarely observed among Jomon dogs.
- there is a high degree of tooth loss due to heavy attrition or breakage.
- the palate is relatively wide and is an elongated triangular shape (Fig. 2)

## Examples of Jomon dogs

Dogs from the Torihama Shell-mound

The Torihama Shell-mound is a waterlogged site of the Early Jomon Period (6,000- 4,500BP). Although a large number of faunal remains have been excavated from this site, only a few dog remains were found, including three complete crania (Table 3). Dogs were not intentionally buried at the Torihama Shell-mound, but rather were found to be scattered over the entire site. The skulls of the dogs from this site are relatively robust. The zygomatic arch is also heavily built, but the snout is relatively gracile. As was the case in other dogs of the Jomon Period, tooth loss often occurred in the premolars. The posterior margin of the medial (middle) part of the palate was observed to vary, taking on one of three shapes: posteriorly convex, straight, or concave. Thus, no morphological uniformity was found in this population. Although Niwa has suggested that the morphology of the palate is stable, and her study (1987) of the regional differences of dogs is based on this assumption, the high degree of variability at the Torihama Shell-mound presents evidence that contradicts this hypothesis. Except for a slightly longer face and wider zygomatic arch, the ratio diagram of the dogs found at the Torihama Shell-mound is similar to that of the dogs found at the Tagara Shell-mound (Miyagi prefecture) from the Late Jomon period.

Dogs from the Tagara Shell-mound

The Tagara Shell-mound site is a Late and Final Jomon period site located in the Tohoku district (Miyagi Prefecture) in northern Japan. At this site, the remains of 22 dogs, all of which were intentionally buried, were found

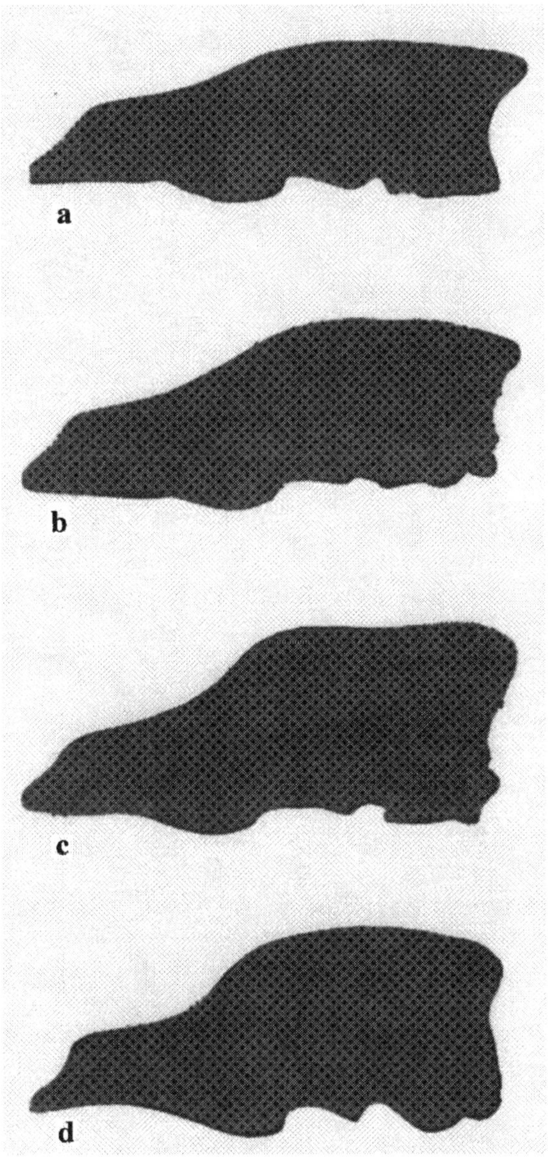


*Figure 1* *Silhouettes of lateral view of wolf and various dogs, same scale .(a) Canis lupus (b) Jomon dog (c) Medieval dog (d) typical recent Japanese Shiba dog*

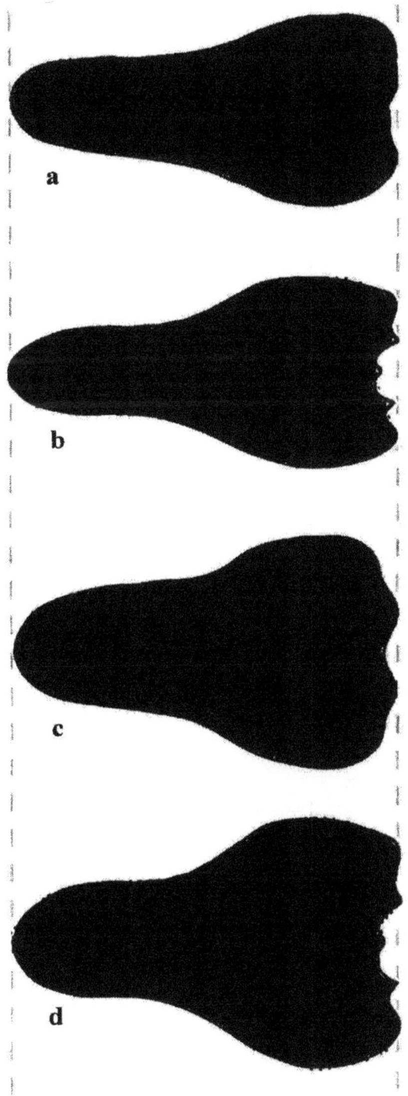


*Figure 2* *Silhouettes of hard palate of wolf and various dogs, same scale. (a) Canis lupus (b) Jomon dog (c) Medieval dog (d) typical recent Japanese Shiba dog.*

at this site (Shigehara & Onodera, 1984). Fifteen skulls, nine of which were males and six females, could be reconstructed. The maximum cranial length ranges from 157.6-171.3 mm (average 163.0 mm) in males and from 145.4-161mm (average 152.3 mm) in females (Table 3). These cranial size ranges are similar to those of present-day Shiba dogs. The bodies of these dogs are generally robust. Although the stop is rather large for Jomon dogs, the stop is still smaller than that of present-day Shiba dogs. Many of the teeth were broken or lost while the dogs were still alive. Thus, even though the body proportions of the dogs from this site are slightly different from other Jomon dogs, several of the features of these dogs are similar to those of the Jomon dogs. Despite being separated chronologically by 2,000 years, as well as the geographic separation between the Hokuriku District and the Tohoku District, these two dog populations have similar body size and skull proportions. The fact that dogs did not go through drastic morphological changes after being introduced to the Japanese Archipelago supports Kaneko's (1978) hypothesis that the Jomon dogs have a single origin.

### The history of Jomon dogs

The stable (uniform) morphological characteristics of Jomon dogs suggest that these dogs may share a single phylogenetic origin. If this is the case, dogs did not differentiate greatly after being introduced from the Asian mainland. Although the Yayoi people migrated from the Asian mainland, the morphology of the native Japanese dogs basically did not change, with the exception of the gradual enlargement of cranial size after the Yayoi period

| | jomon | | | | | | | |
|---|---|---|---|---|---|---|---|---|
| | Average ♂ | Average ♀ | Torihama early ♂ | Torihama early ♀(Ave) | Kaitori late ♂ | Tagara late-final ♂(Ave) | ♀(Ave) | No.2 ♂ |
| 1Maximum Cranial L.(pr-i) | 168.7 | 156.9 | 157 | 151.5 | 174 | 163.0 | 152.3 | 183.4 |
| 2:Condylobasal L.(pr- ) | 158.5 | 148.5 | 146 | 143.9 | 165.8 | 152.6 | 141.9 | 171.3 |
| 3:Zygomatic B.(zy-zy) | 95.4 | 87.5 | 96.0 | 87.7 | 97 | 88.3 | 85.2 | 95 |
| 4:nasion-inion L.(na-i) | 90.0 | 84.8 | 83.2 | 83.6 | | 87.2 | 82.8 | 94 |
| 5:nasion-basion(na-ba) | 86.5 | 80.8 | 80.5 | 80.1 | 91.2 | 84.2 | 79.0 | 91 |
| 6:G.neurocranium B.(eu-eu) | 52.5 | 51.6 | 53.6 | 50.7 | 51 | 52.4 | 51.6 | 51.5 |
| 7:Skull H.(br-ho) | 49.5 | 45.8 | 45 | 44.5 | 52 | 50 | 45 | 54 |
| 8:basion-bregma H.(ba-br) | 65.2 | 59.9 | 58.3 | 58.8 | 68.4 | 62.0 | 58.9 | 64.8 |
| 9:Least skull B.(ft-ft) | 31.7 | 31.4 | 32.3 | 29.8 | 31.1 | 31.7 | 30.0 | 32.8 |
| 10:Frontal B.(ect-ect) | 44.6 | 41.2 | 47 | 39.4 | 48 | 42.8 | 41.1 | 43.8 |
| 11:G.mastoid B.(ot-ot) | 61.6 | 58.8 | 57.9 | 57.6 | 62 | 59.5 | 57.7 | 64.4 |
| 12:B.dor.ext.aud.meatus(au-au) | 59.8 | 57.4 | 56.8 | 55.5 | 60.5 | 59.1 | 56.7 | 62.2 |
| 13:Least B.interorbit(ent-ent) | 31.1 | 28.2 | 29.4 | 27.4 | 33 | 28.7 | 27.6 | 32.4 |
| 14:Viscerocranium L.(pr-na) | 82.9 | 76.4 | 81 | 72.8 | 84.8 | 79.4 | 74.7 | 95.4 |
| 15:Snout L.(pr-o.a) | 72.2 | 67.5 | 66 | | 72.6 | 68.3 | 64.6 | 78 |
| 16:B.canine alveoli | 34.1 | 31.5 | 32.2 | | 34 | 34.5 | 30.8 | 37.3 |
| 17:Snout H.(na- ) | 39.7 | 36.8 | 40 | | 40 | 39 | 36 | 44 |
| 18:Nasal Curvature D. | 4.5 | 3.9 | 5.1 | | 5.2 | 5.0 | 3.4 | 4.8 |
| 19:Palatal L.(pr-sta) | 81.5 | 77.2 | 76 | | | 78.5 | 72.9 | 88.9 |
| 20:G.Palatal B. | 57.3 | 55.2 | 58.2 | | 58 | 57.4 | 53.7 | 62.3 |
| 21:Mandibular L.(1)(id-goc) | 123.5 | 113.3 | | | 128 | 118.5 | 112.2 | 134.4 |
| 22:Mandibular L.(2)(id-c.mid) | 123.1 | 113.7 | | | 129 | 117.4 | 110.6 | 133.7 |
| 23:H. vertical ramus | 48.8 | 43.4 | | | 52.5 | 44.3 | 44.3 | |
| 24:Min. B. ramus | 31.3 | 27.4 | | | 34 | 29.3 | 27.2 | |
| 25:Body H.(med. M1) | 23.6 | 21.1 | | | 24.1 | 22.4 | 20.5 | 24.7 |
| 26:Body Thickness(med. M1) | 11.0 | 9.9 | | | 10.6 | 10.9 | 9.8 | 12.1 |
| 27:Masseter fossa D. | 7.9 | 6.7 | | | 8.5 | 7.1 | 6.5 | 8 |
| Cranial index(1)(3/1) | 56.6 | 55.9 | 61.1 | 58.7 | 55.7 | 54.2 | 55.9 | 51.8 |
| Brain-case index(6/4) | 58.4 | 60.8 | 64.4 | 60.4 | | 60.1 | 62.3 | 54.8 |
| L-H index(8/1) | 38.6 | 38.1 | 37.1 | 38.3 | 39.3 | 38.0 | 38.7 | 35.3 |
| B-H index(8/6) | 124.3 | 116.2 | 108.8 | 115.9 | 134.1 | 118.3 | 114.1 | 125.8 |
| Trans. craniofacial index(3/6) | 181.8 | 170.2 | 179.1 | 173.1 | 190.2 | 168.5 | 165.1 | 184.5 |
| Postorbital index(9/6) | 60.5 | 60.9 | 60.3 | 58.7 | 61.0 | 60.5 | 58.1 | 63.7 |
| Facial index(14/3) | 86.9 | 87.2 | 84.4 | 81.9 | 87.4 | 89.9 | 87.7 | 100.4 |
| Snout index(15/1) | 42.8 | 42.6 | 42.0 | 42.2 | 41.7 | 41.9 | 42.4 | 42.5 |
| Nasal Curv.D. index(18/17) | 11.2 | 10.5 | 12.8 | 10.3 | 13.0 | 12.9 | 9.6 | 10.9 |
| Palatal index(20/19) | 71.1 | 70.3 | 76.6 | 64.4 | | 73.2 | 73.7 | 70.1 |
| Mandibular H-Th. index(26/25) | 47.1 | 46.7 | - | - | 44.0 | 48.5 | 47.8 | 49.0 |

***Table 3*** *Cranial measurements of some Jomon dogs*

(Shigehara, 1994). The direct ancestors of the present-day Japanese people did not bring many dogs when they migrated to Japan because they introduced rice agriculture, thus reducing the importance of hunting. From the Medieval period to the Edo Period, the size of these dogs continued to increase, and their skull proportions also changed gradually. The face and palate of these dogs became wider. The proportions of the upper cheek teeth also changed in accordance with these proportional changes. The bucco-lingual crown diameter also increased, and the relative width of the snout decreased greatly. In addition, the stoutness of these dogs decreased gradually.

In China, many dog burials have been excavated. However, few morphological studies have been performed concerning these dogs (Olsen, 1985; Wei Feng *et al.*, 1989; Shigehara *et al.*, 1998).

Shigehara *et al.* (1998) reported the morphological features of Neolithic dogs excavated in China. Wide morphological variations were found in Neolithic Chinese dogs, beyond any single dog size class used by Hasebe (1952). Dogs having a large stop have been determined to have been present in China around 3,000-4,000BP and 5,000-6,000BP (corresponding to the Middle-Late Jomon and Early-Middle Jomon period in Japan, respectively). Moreover, dogs much larger than Jomon dogs existed in central and northern China. However, in the Hemudu site, located in southern part of China, such large dogs were not found.

| Aoshima | Satohama | Maehama | Usuiso | Ootsubo | Shuridai | Tomisaki | Yoshigo | | recent Shiba | |
|---|---|---|---|---|---|---|---|---|---|---|
| final | final | final | final | early | late | late | final | final | | |
| ♂ | ♀ | ♂ | ♂ | ♀ | ♀ | ♂ | ♂ | ♀(Ave) | ♂(Ave) | ♀(Ave) |
| 164 | 163.1 | 171.8 | 177 | 162.2 | 156.6 | 166 | 155.9 | 155.8 | 155.5 | 145.3 |
| 154.6 | 157.5 | 161.4 | 167 | 152.6 | | 154.8 | 146.9 | 146.7 | 147.9 | 137.9 |
| 96 | 90 | 100 | 96 | 90.3 | 87.6 | 95 | 84 | 88.5 | 94.8 | 88.1 |
| 88.8 | 85.4 | 87.2 | 95.2 | 88.8 | 84.4 | | 83.1 | 86.1 | 86.1 | 80.8 |
| | | 87.2 | 94.6 | 84.6 | 80 | 86.5 | 81 | 80.5 | 84.2 | 79.1 |
| 52.8 | 53.6 | 53 | 54 | 51.2 | 52.4 | 54.1 | 49 | 50.5 | 50.1 | 48.3 |
| | 48 | | 52 | 47 | 44 | 50 | 44 | 45.8 | 48.5 | 45 |
| 63 | 63.6 | 64.2 | 70.8 | 61.8 | 58.1 | 64.1 | 57.9 | 59.1 | 65.0 | 61.4 |
| 32.7 | 35.2 | 33.7 | 29.8 | 31.9 | 29.5 | 32 | 29.3 | 30.7 | 29.5 | 28.2 |
| 43.6 | 45 | 46.1 | 46.1 | 42 | 40.1 | 43 | 38.8 | 40.5 | 43.5 | 39.3 |
| 64 | 61 | 59 | 63.2 | 59.6 | 59.2 | 61 | 55.1 | 59.0 | 55.2 | 51.6 |
| 61.7 | 60.3 | 59.1 | 62.5 | 58.7 | 57.4 | 60 | 53.9 | 56.0 | 54.2 | 50.7 |
| 30.5 | 30 | 34.7 | 32.5 | 29.3 | 27.8 | 29 | 26.6 | 27.4 | 28.7 | 25.4 |
| 79.6 | 82 | 89.2 | 86 | 75.6 | 76.6 | 80.6 | 76.7 | 71.3 | 74.6 | 69.4 |
| 70.4 | 68.6 | 77.7 | 78 | 70 | 66.3 | 70.6 | 67.7 | 63.1 | 65.0 | 60.1 |
| 36.3 | 35 | | 34.9 | 32.5 | 30 | 34 | 29 | 29.6 | 31.5 | 28.9 |
| 40 | 40 | 40 | 41 | 37 | 36 | | 35 | 36.0 | 34.2 | 32 |
| 3.5 | 3.8 | 4.3 | 4.2 | 4.6 | 3 | 4.6 | 4.6 | 4.4 | 6.4 | 5.5 |
| 77.8 | 81 | 85 | 85 | 79.7 | 76 | | 76.4 | 73.2 | 74.9 | 70.3 |
| 59.7 | 58.2 | 59 | 54 | 57.5 | 52 | | 53.8 | 53.3 | 59.2 | 55.2 |
| | | 125.1 | 128 | - | 114.9 | | 114.7 | 112.7 | 114.1 | 107.0 |
| 121.2 | 117 | 124.3 | 128 | - | 114.4 | | 114.5 | 112.8 | 114.2 | 107.0 |
| | | 47.4 | 49.4 | - | 43 | | 43.5 | 42.9 | 44.6 | 41.3 |
| 30.9 | 28.3 | 30.5 | 32.9 | - | 27.3 | | 27.3 | 26.7 | 27.7 | 25.5 |
| 23.4 | 21.9 | 24.1 | 25.4 | - | 20.8 | | 21.5 | 21.3 | 18.9 | 17.2 |
| 11.7 | 10.3 | 11.3 | 11.6 | - | 9.8 | | 9.5 | 9.5 | 9.1 | 8.2 |
| 7.8 | 6.6 | 7.8 | 8.5 | - | 7.2 | | 6.6 | 6.4 | 5.8 | 5.3 |
| 58.5 | 55.2 | 58.2 | 54.2 | 55.7 | 55.9 | 57.2 | 53.9 | 53.9 | 61.0 | 60.7 |
| 59.5 | 62.8 | 60.8 | 56.7 | 57.7 | 62.1 | | 59.0 | 59.4 | 58.2 | 59.8 |
| 38.4 | 39.0 | 37.4 | 40.0 | 38.1 | 37.1 | 38.6 | 37.1 | 37.5 | 41.9 | 42.3 |
| 119.3 | 118.7 | 121.1 | 131.1 | 120.7 | 110.9 | 118.5 | 118.2 | 117.2 | 129.9 | 127.3 |
| 181.8 | 167.9 | 188.7 | 177.8 | 176.4 | 167.2 | 175.6 | 171.4 | 172.6 | 189.4 | 182.7 |
| 61.9 | 65.7 | 63.6 | 55.2 | 62.3 | 56.3 | 59.1 | 59.8 | 61.2 | 59.0 | 58.3 |
| 82.9 | 91.1 | 89.2 | 89.6 | 83.7 | 87.4 | 84.8 | 91.3 | 91.3 | 78.8 | 78.8 |
| 42.9 | 42.1 | 45.2 | 44.1 | 43.2 | 42.3 | 42.5 | 43.4 | 43.6 | 41.8 | 41.4 |
| 8.8 | 9.5 | 10.8 | 10.2 | 12.4 | 8.3 | | 13.1 | 13.1 | 18.7 | 17.4 |
| 76.7 | 71.9 | 69.4 | 63.5 | 72.1 | 68.4 | | 70.4 | 72.9 | 79.2 | 78.6 |
| 50.0 | 47.0 | 46.9 | 45.7 | | 47.1 | | 44.2 | 44.7 | 48.3 | 47.7 |

Selective breeding of dogs, which was widely practised later in Europe, Egypt, and the Middle East, was not adopted in Japan. Breed improvement may not have been readily adopted due to the Japanese tendency to accept nature as it was. Selective breeding finally became common during the Edo Period, almost 2,000 years after the end of the Jomon period.

Efforts to preserve native Japanese dog breeds started in the 1930s. Through this movement (Ota, 1980), six breeds of native dogs were recognised and preserved: Hokkaido, Akita, Kai, Shiba, Kishu, and Shikoku breeds. The Akita breed is a large-sized dog and apparently has a history of hybridisation with some European breeds. The Hokkaido, Kishu and Shikoku breeds are of medium-size, and Kai breed is of medium-small size. The Shiba breed belongs to the small size category. Among these native breeds, the Shiba dogs is believed by several researches to be the direct descendant of the Jomon dogs (Kaneko, 1978; Ota, 1980).

**Conclusions**

- The ancestors of Jomon dogs may have migrated from the Asian mainland.
- The cranial size of Jomon dogs was found to be smaller than that of the later native dogs, except for the present-day Shiba dogs.
- Several dogs of the Jomon period were intentionally buried.

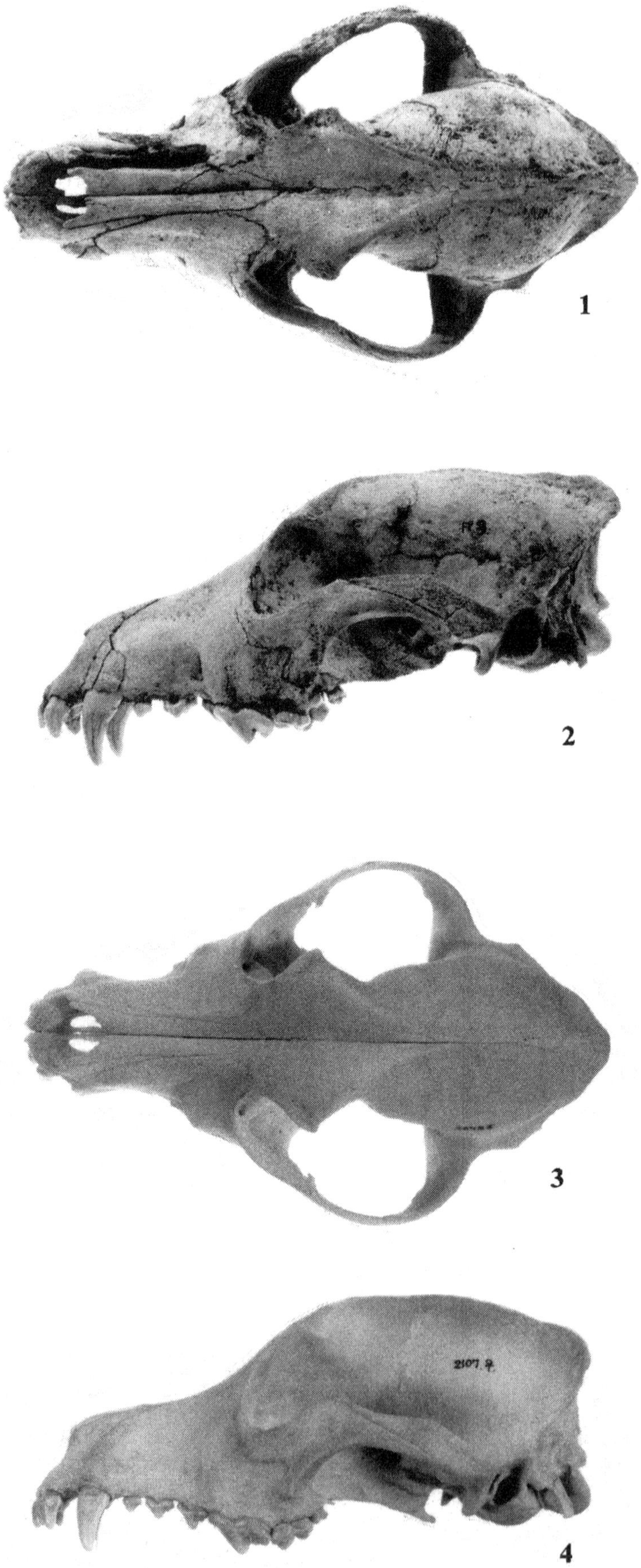


***Figure 3*** *Plate 1. The crania of a Jomon dog and a present-day Japanese Shiba dog. (1) upper view, Jomon dog (2) lateral view, Jomon dog (3) upper view, recent Shiba dog (4) lateral view, recent Shiba dog.*

- No attempt at breed improvement was made during the Jomon period.

- Morphological change in ancient native dogs started in the Yayoi period. In native Japanese dogs, the cranial size gradually increased from the Yayoi period. From the Medieval period, the relative facial breadth and palatal breadth increased gradually.

**Acknowledgements**

The authors would like to thank Professor Hiromasa Kaneko, who allowed us to view his collection of archaeological dog remains. The modern Shiba dog skeleton used for comparison is stored at the Dokkyo University School of Medicine. The authors also thank Susan Crockford, who organised the symposium in Victoria and provided us with useful comments. In addition, although individuals are not mentioned here, this paper could not have been completed without the help of several archaeologists at the Buried Cultural Resource Management offices.

**References**

Driesch, A. von den (1976). *A Guide to the Measurement of Animal Bones from Archaeological Sites.* Cambridge, MA: Harvard University Peabody Museum Bulletin **1**.

Hasebe, K. (1924). Über die Schadel und Unterkiefer von den steinzeitlich Japanischen Hunderassen. *Sendai* **10,** 1-33.

Hasebe, K. (1952): Dog bones. In (Bunkazai Hogo Iinkai). *Yoshigo Kaizuka [Yoshigo Shell Mound]*: 146-150. (In Japanese)

Kaneko, H. (1978). Animal remains from the archaeological sites of the Jomon Period (2). *Koukogaku Note* **7,** 1-18. (In Japanese)

Miyazaki, Y. (1982). *Kamei Site.* Report on file, Osaka Burial Assets Center, pp.205-230.

Naora, N. (1973). Ancient dog remains in and around Japan. In (Azekura Shobo, Ed) *Domestic Animal Remains Excavated From Ancient Archaeological Sites*. pp. 126-246, 248 (In Japanese).

Nishimoto, T. (1981): Animal remains: *Canis familiaris.* In (T.Ohba & H.Ooi, Eds) *Kabukai Site (2).* Tokyo: University of Tokyo Press, pp.415-424. (in Japanese)

Niwa, Y.(1982). Jomon people and their dogs. *Doubutsu to Shizen* **C12,** 18-21. (In Japanese).

Niwa, Y. (1987). To understand the real image of the ancient Japanese dogs. *Kiba* **2,** 5-29.

Olsen, S. J. (1985). *Origins of the Domestic Dog: The Fossil Record.* Tucson: University of Arizona Press.

Ota, K. (1980). The dog, its domestication and development, with special reference to the origin and the history of the Japanese native dog (a review). *Report of the Society for Researches on Native Livestock* **9,** 53-94. (Japanese, English summary)

Saito, H. (1963). *Osteometrie der Caniden.* Tokyo: Privately published.

Shigehara, N. & Onodera, S. (1984). Skeletal Remains of the domestic dogs from Tagara Shellmound, Miyagi Prefecture. *Journal of the Anthropology Society of Nippon* **92**, 187-210 (Japanese, English summary).

Shigehara, N. (1986). *Catalogue of the Ancient and Recent Canid Skeletons Collected by D. Kotondo Hasebe and preserved in the University Museum.* Tokyo: University of Tokyo Museum Material Report **13**.

Shigehara, N. (1991). Shape changes by era in the Japanese dog. *Bulletin of the National Museum of Japanese History* **29**, pp.89-101. (Japanese, English summary)

Shigehara, N. (1994). Morphological changes in Japanese ancient dogs. *Archeozoologica* **IV**, 78-94.

Shigehara, N., Onodera, S. & M. Eto (1997). Appendix A, Sex determination by discriminant analysis and evaluation of non-metrical traits in the dog skeleton. In (S. J. Crockford) *Osteometry of Makah and Coast Salish Dogs*. Burnaby, B. C: Simon Fraser University Archaeology Press **22**, pp. 113-126.

Shigehara N., Guoqin, Q., Komiya, H. & Jing, Y. (1998). Morphological study of the ancient dogs from three neolithic sites in China. *International Journal of Osteoarchaeology* **8**, 11-22 .

Wei feng, Wu Wcitang & Zhang Minhua (1989). The fauna from the neolithic site at Hemudu, Zhejiang. (Chinese with English Summary)

# Part 2

# Interpreting roles: early practical and ritual uses of dogs

# THE SECULAR AND SACRED ROLES OF DOGS AT BOTAI, NORTH KAZAKHSTAN

Sandra L. Olsen
Associate Curator of Anthropology
Carnegie Museum of Natural History
O'Neil Research Center
5800 Baum Blvd.
Pittsburgh, PA, USA 15206

**Key words**: dog, archaeozoology, interment, ritual, canid, Kazakhstan, Eurasian steppe

**Abstract**

The Eneolithic site of Botai, Kazakhstan, reveals much about both secular and sacred uses of dogs in the Eurasian steppe during the fourth millennium BC. By examining osteological and contextual data, hypotheses can be addressed regarding the various functions of dogs at Botai. Several salient characteristics have emerged.

Based on osteometrics, the Botai dogs were comparable in size and morphology to the modern Samoyed breed. The morphological information may shed light on genetic relationships among dog populations on the Siberian Plain and could someday elucidate the emergence of northern breeds. Reconstruction of the size range of the Botai dogs aids in determining the suitability of these dogs for certain tasks and in comparing them to prehistoric dogs in other regions.

Patterning in the placement of canine remains in threshold and foundation pits is interpreted here as evidence of ritual behavior. Botai dogs possessed a close connection with houses that may reflect their roles as guardians. The practice of burying dogs in pits to the west of houses at Botai also appears from England to China in the subsequent Bronze Age. According to later Indo-European and Indo-Iranian mythologies, dogs were the guardians of the gate to the Otherworld, which lay to the west.

In death, dogs were also closely associated with horses, the only other known domesticated animal of the Botai. This spiritual association may be analogous to a more secular relationship concerned with hunting or herding. There is no clear evidence that dogs were consumed at Botai, although they were decapitated and dismembered for ceremonial reasons.

Canine deposits at Botai suggest that many religious traditions of the Eurasian steppe have deep roots in northern Kazakhstan. Further exploration of pit features in the steppe should help clarify the origins of many rituals in this region. This study has demonstrated the need for more thorough analysis of animal deposits and illustrates the potential gain from more refined excavation and recording of such features.

## Introduction

The practice of burying dogs is almost as ancient as dog domestication itself (Beneke, 1987; Tchernov & Valla, 1997). Because dogs share a complex and unique relationship with humans, the variety of roles they fill poses some intriguing problems when attempting to interpret prehistoric dog remains (Morey & Wiant, 1992). There are two kinds of functions that may be reflected in dog interments: the sacred and the secular. Secular uses of dogs that are normally recognized include: dietary, fur-bearing, companionship, hunting, herding, security (guard dogs for humans or herds and watch dogs), haulage (by sled, travois, or packs) and military. The sacred is virtually limitless, but often there is an element in it that is a metaphorical representation of the dog's role in life. The sacred role, for example, may be for the dog to accompany its owner into the Afterlife as a companion or hunting aid. As White (1991: 14) has pointed out, "In a great number of cultures, then, the pastoral, cynegetic, and protective role of the dog is extended beyond the world of the living into the world of the dead." Whether certain dogs were regarded as sacred during their lifetime and therefore received special treatment prior to death is not easy to determine from the archaeological record. In many cases, it may be that the sacrificial ritual itself or the passage from living to dead conferred the sacred status on quite ordinary dogs as they entered the spirit world.

Ethnographic (Titcomb, 1969) and historical (White, 1991) studies have shown that there is rarely one simple type of human/dog relationship in any given society. Even among cultures that consume dogs, some individuals may be kept as pets, named, and even breast-fed by humans (Serpell, 1995; Titcomb 1969). Within a society, dogs may be divided into different functional groups, such as those that are considered edible and those that are companions. Likewise, the actual economic or other values assigned to dogs in a society may be in sharp contrast to their status and treatment in that society (Serpell, 1995). The final disposition of dog carcasses can reveal, at best, only part of the animal's functions in life or death. Hypotheses about the roles of dogs in prehistoric cultures are testable through the archaeological record only as far as the restricted parameters of burial and other evidence, like artistic representations, pathologies, etc. permit. Resulting interpretations are necessarily limited and often tentative. Despite all of these caveats, the unique roles dogs play in society make the pursuit of information about the prehistoric human/canine relationship extremely valuable (Clutton-Brock, 1981, 1984; White, 1991; Mair, 1998).

J.D. Hill (1996) has drawn attention to the problem of identifying ritual burials of animals as opposed to accidental taphonomic processes or actions taken by humans simply to discard carcasses. Hill sums it up with his opening question "Is it ritual or rubbish?" Other researchers, including Maltby (1985) and Wilson (1992), advocate caution in assigning ritual purposes to archaeological deposits. Hill's article sets out criteria for the Southern English Iron Age, but recommends that there can be no cross-cultural general criteria for establishing whether an animal deposit represents ritual or non-ritual behavior. A cautious stand would be to refer to an animal burial as "structured deposition" (Hill, 1996: 21), but while this minimizes criticism of the archaeologist describing the find, it does little to offer explanations for the original cause of such a deposit. Perhaps the best solution is to present the concrete evidence and follow it with possible interpretations of the data, without melding the two together. The reader is then free to make his or her own interpretations. In this work, Grant's (1984a: 533-534) definition of what constitutes an animal burial is applied, although with recognition that some might regard this as too liberal. According to Grant: "The definition of an animal burial for the purposes of this discussion is a group of bones, certainly or very probably found in articulation, representing a whole animal or at least a substantial part of a single animal, and usually showing no evidence of having been butchered." In addition, the placement of skulls, segregated from their postcranial skeletons, has been investigated in the Eurasian steppe, in light of compelling evidence that they appear to have been used for ceremonial purposes at Botai.

This study, an exploration of the roles that dogs may have filled in the Botai culture of northern Kazakhstan, is admittedly very preliminary. Although interpretation of the archaeological record always requires a certain degree of speculation, this research has shown critical ways in which the gathering of evidence can be substantially improved. For the benefit of future archaeologists, some standards should be set for recording any animal burial or possible ritual deposit. To aid in the reconstruction of both the sacred and the secular functions of animals in prehistoric societies, excavation of their interments should receive the same attention to detail afforded human burials. Observations should include the orientation of the animal; the position of its body; the body parts represented; the nature of skeletal articulation; and the association of the carcass with human interments, other animal remains, artifacts, and features. Attributes of the individual, such as age, sex, size, cranial morphology, pathologies, cutmarks, and taphonomic processes should also be recorded. Once detailed records from a large sample of animal deposits within a particular culture, tradition, or region have been established, patterning of the evidence should help resolve the sacred and secular roles of these animals. Until that time occurs in the Eurasian steppe, however, some initial groundwork can be laid by examining some of the evidence that currently exists.

**The Eneolithic Site of Botai, North Kazakhstan**

The site of Botai has had three dates made by the Oxford Accelerator Laboratory. All of these are from the Eneolithic occupation (Levine & Kislenko, 1997): OxA-4315, horse third tarsal, 3650BC; OxA-4316, horse first phalanx, 3650BC; OxA-4317, human cranium, 3650BC (all calibrated dates, with a 95.4% confidence level). Dates for the Mesolithic and Neolithic of this region have not been clearly defined. The Eneolithic (the period contemporaneous with the Copper Age, but lacking copper) of northern Kazakhstan covers a period beginning about 3600BC and ending with the arrival of bronze. The Bronze Age of northern Kazakhstan started at least as early as 2500BC, based on dates from the sites of Sergeevka and Balandino (Levine & Kislenko, 1997). The transition from the Bronze Age to the Iron Age took place between 1000-800BC (Koryakova, 1996).

A total of just three Eneolithic Botai culture sites have been identified to date and only the eponymous settlement has undergone significant excavation. The sites are distributed in the Ishim-Irtysh interfluve on or near small streams from about 52 to 55° north latitude and 66 to 70° east longitude, in the forest-steppe ecozone (Fig. 1). The Botai were located in the southern part of the West Siberian Plain, an area of thin saline soils of low fertility with an annual rainfall of 150 to 300 mm per year. Snow covers the ground 100 days a year on average between December and April and winter temperatures often plummet to -30° C. This combination of poor soil, short growing season, and harsh winters made this part of the steppes unsuitable for all but small bands of transient hunter-gatherers from Middle Paleolithic through Mesolithic times.

Covering nine hectares, the site of Botai is one of the largest known settlements in Kazakhstan prior to the Iron Age. Botai is situated on the Iman-Burluk river, a small tributary of the Ishim. The settlement consists of 158 documented pithouses of Neolithic and Eneolithic age. Mapping of the pithouses from the surface is possible because of their high visibility in the spring, when the depressions are filled with verdant vegetation (Fig. 2). To the east of the major concentration of houses was found a small Mesolithic camp. A few Bronze Age artifacts, primarily bone points, were sprinkled over the later ground surface overlying the settlement. The other two Botai

***Figure 1****. Map of Kazakhstan showing Botai site locations and neighboring cultures.*

***Figure 2****. The site of Botai, showing dark pithouse depressions on the surface.*

culture sites are Krasnyi Yar, just west of the modern city of Kokshetau, and Vasilkovka, between Krasnyi Yar and Botai. These sites are thought to be somewhat smaller than Botai, but their houses are not as readily visible on the surface and they have undergone very little testing.

Since the discovery of the site of Botai in 1980, 42 pithouses have been totally excavated and 18 more that were eroding out of the bank have been sampled by the University of North Kazakhstan under the direction of Victor Zaibert. The sizes of the Botai houses range on average from 8-12 m in diameter and 60-100 cm in depth. It is presently unclear how these structures were built because they reportedly lack postmolds for central support beams and no postmolds have been found around the walls. In settlements of the Tersek, a closely related culture to the west in the Turgay region, burnt pithouses yielded partially preserved fine woven sapling frameworks, suggesting a waddle and daub structure (Kalieva & Logvin, 1997). After Botai houses were abandoned and their roofs collapsed, they were usually filled in with refuse so that dense concentrations of faunal material can now be found overlying many houses.

Animal remains from Botai are estimated to total over 300,000 bones, of which 99% are from horse. Each of the species other than the horse and dog, is represented by fewer than 10 specimens. The list of taxa identified thus far includes: mollusk, fish, swan, goose, eagle, hare, souslik (*Spermophilus xerophilous*), marmot, beaver, red fox, wolf, bear, wolverine, saiga antelope, elk, red deer, roe deer, large bovid (*Bos/Bison*), wild boar, and kulan. Riverine resources were apparently of only minor importance, despite the settlement's proximity to the Iman-Burluk. The faunal collections from Botai were housed in a laboratory in the University of North Kazakhstan, in Petropavlovsk, Kazakhstan.

There is good evidence that wild horses were hunted by the Botai in the form of harpoon wounds in horse bones. Evidence for domestication is more circumstantial, but nonetheless compelling. The mortality pattern for the Botai horses, based on eruption and wear of dentition, indicates that 22% of the females were killed before reaching the age of sexual maturity, whereas 29% of the males were killed. The sample size is small, but this may show a slight preference for culling young domestic males. Unfortunately, the wild hunted population would dilute the probably much smaller domestic herds in terms of mortality patterns. One nearly complete horse fetus was found at the site, but this could have been carried in a wild mare that was hunted. Anthony and Brown (1998) claim to have found bit wear on lower premolars, but the microwear does not appear affected and beveling could be a natural phenomenon. Three lines of evidence are most convincing in arguing for horse domestication at Botai. The first is that horses were heavily involved in rituals, both in shrines or altars and in human burials. In later Eurasian steppe sites, domesticated animals are generally the subjects of such behavior, rather than wild species. Secondly, although there are not adequate quantitative data, it is obvious that the "Schlep Effect" (Perkins & Daly, 1968) was not significant at Botai. The finding of articulated vertebral columns, ribs, skulls, and foot bones means that either the Botai were killing horses very near the village or they had a means of hauling the heavy carcasses back home. Field dressing would be expected if the hunters were on foot; however, if they were riding horses to hunt wild horses, then they could carry much more back with them. Also, if some of the domestic horses were culled periodically in the village, then their skeletal remains should be relatively complete. Thirdly, the most common artifacts in the site were thong-smoothers made on horse mandibles. These notched tools show a very high polish and extremely fine striations sweeping over the notch. They were apparently used to straighten thongs made of leather or rawhide cut in a spiral fashion (Wilson, 1978). Thongs can serve many purposes, but thong-smoothers are not common artifacts in most sites. Thongs are very important for horsemen, since they can be made into bridles, whips, riding crops, pole-snares, lariats, hobbles, and other useful tackle. The combination of evidence, rather than any single piece, makes a fairly strong argument for incipient horse domestication at Botai. More work clearly needs to be done to secure this hypothesis, however.

The Botai culture has many characteristics that demonstrate a relationship with other cultures of the region from the Neolithic through the following Iron Age, at least. For the Eneolithic, it most closely resembles the Tersek culture (Kalieva & Logvin, 1997) in having cord-impressed, comb-impressed, and rocker-stamped pottery with geometric designs, similar pithouses, and a biface-dominated lithic assemblage. It has also been compared to the Surtanda culture in the Urals (Zaibert, 1993). Logvin (pers. comm., 1998) has even discussed collapsing all of these cultures and others in the Ural region into a single "Geometric Pottery Tradition." This tradition is indigenous, coming out of a Neolithic that ranged at least from the western Urals (Kerner, 1991) to Botai in the east. Clay funerary masks on human skulls, the sacrifice of horses in human burials, and specific pottery motifs, like hatched triangles and diamonds, link Botai to other Eurasian steppe cultures, including the subsequent, widespread Bronze Age Andronovo culture (Kadirbayev & Kurmankulov, 1992; Mallory, 1996). It therefore seems appropriate to compare dog deposits at Botai to ritual treatment of dogs in other sites in the Eurasian steppe and even to explore Indo-European or Indo-Aryan historical and religious documents, like the *Rig Veda* and *Avesta*.

### The Evidence for Reconstructing the Roles of Dogs at Botai

At Botai, dog remains are much rarer than horses, but they rank as the second-most common species. To date, at least 13 separate canine deposits representing at least 18 individuals (15 dogs and 3 wolves) have been recovered. In examining the dog remains from Botai, the following lines of evidence were addressed: the location and nature of the deposit, associated finds, body parts represented, age and sex, body size and proportion, pathologies and injuries, and cultural modifications of skeletal parts, as the data allow.

Location, Orientation, and Context of the Dog Remains and their Associated Finds

The distribution of dog remains and their occurrence in particular areas, structures, or features may provide crucial evidence for uncovering the roles of this domestic species in a prehistoric society. Table 1 summarizes this information for Botai in tabular form. The most recurrent patterns at the site involve the deposition of dogs (partial or whole) in distinct pits, either in the floors of houses, in niches in the walls (foundation pits), or just outside house walls. Because the early excavation records are not detailed and photographs and illustrations are generally not available, little can be deduced about the orientation of individual dog remains from Botai at this time.

One source of information that holds great potential for interpreting the secular and sacred roles of dogs is the assortment of artifacts and faunal remains that accompany a dog deposit. At this time, only a few provocative details have been documented for the Botai culture. Future excavations should produce more precise reports on the contents of burial pits, as well as regular storage pits and other features.

Remains of dogs were found in at least three separate pits to the west of houses. *Canis* 10, uncovered in a pit immediately west of House 35, consisted of the left and right forelimbs of one dog. The 60 cm deep pit was rectangular in plan, with its long axis running north /south. It was located in a large open area surrounded by houses and filled with numerous pits.

An irregularly-shaped pit to the west of House 139 contained the nearly complete skeleton (Fig. 3) of an adult male dog (*Canis* 15) and a small pile of horse bones, including a mandible, pelvis, and ribs. Based on Grant's (1984a) definition, this is classified as a dog burial, or intentional interment. The dog's skeleton was vaguely in anatomical position, but somewhat twisted, and articulated bones had drifted apart. The ribs, in particular, were all askew. The skull was partially inverted and fairly friable (Fig. 4). The dog's head was positioned toward the north wall of the pit and its nose pointed basically east. The taphonomic nature of the skeleton suggests that it may have lain in a pool of water that macerated the flesh before being covered with soil. These conditions could explain why the elements were not tightly articulated at the joints, had shifted, and spread apart without completely losing their orientation in the body. It could also have contributed to the poor condition of the skull. The horse bones were piled in a concentration against the east wall. *Green

| Botai | Species | Body Parts | Location | Feature | Associated Finds |
|---|---|---|---|---|---|
| 1 | *C. familiaris* | axial, fore, hind | House 32 | pit | MNI 5 dogs |
| 2 | *C. familiaris* | fore, hind | House 32 | pit | MNI 5 dogs |
| 3 | *C. familiaris* | skull, cervicals | House 32 | pit | MNI 5 dogs |
| 4 | *C. lupus* | skull | House 140 | pit | |
| 5 | *C. familiaris* | nearly complete | House 32 | pit | MNI 5 dogs |
| 6 | *C. familiaris* | axial, fore, hind | House 32 | pit | MNI 5 dogs |
| 7 | *C. familiaris* | axial, fore | House 32 | pit | MNI 5 dogs |
| 8 | *C. familiaris* | skull, cervicals | House 29 | pit | |
| 9 | *C. familiaris* | nearly complete | House 33 | pit | |
| 10 | *C. familiaris* | fore | west of H. 35 | pit | |
| 11 | *C. lupus* | mandibles | House 140 | pit | |
| 12 | *C. familiaris* | mandible | House 29 | pit | |
| 13 | *C. familiaris* | mandible | House 29 | pit | human skull cup, beaver, sm ruminant |
| 14 | *C. familiaris* | skull, cervicals | House 32 | pit | MNI 5 dogs |
| 15 | *C. familiaris* | nearly complete | west of H. 139 | pit | horse bones |
| 16 | *C. lupus* | fore paws, hind paws | House 32 | midden | |
| 17 | *C. familiaris* | skull | House 33 | pit | ochre, points, horse mandibles |
| 18 | *C. familiaris* | cranium | bank area | midden | horse cranium, stone flake |

***Table 1**. Location and context of canid remains.*

*Figure 3*. Canis 15 burial in pit to the west of House 139. Arrow points north.

*Figure 4*. Canis 15 skull. Scale is 5 cm.

(1992) has pointed out the recurrent placement of dogs in aquatic contexts in British Bronze Age, Celtic and Roman sites, including ponds, ditches, marshes, and wells. This she attributes to the chthonic (referring to underworld deities) roles that dogs played and equates such underwater deposits with underground placement in pits and shafts.

Another pit on the west margin of House 33 produced a single dog skull *(Canis 17)* on a possible altar. Like that outside House 35, the pit was rectangular with its long axis running north/south. The northeast corner of it protruded just inside the house, but most of the pit was located outside the west wall. The dimensions of the pit were 135 cm long by 65 cm wide by 40 cm deep, and it had vertical, slightly inclined walls, a flat bottom and a small step 15 cm high on the north side. The pit was one meter below ground surface. In the southern part of the pit on a stone slab with round stones placed around it were the skull of a dog and two anterior portions of horse mandibles. Also on the flat stone were two arrowheads, one of which was stuck into a piece of ochre. In addition, there were several other unidentified bones and a few pieces of flint.

On the basis of just three examples, it is not possible to say whether the orientation of extramural pits relative to houses was significant at Botai. However, because Indo-European and Indo-Aryan cultures saw the west as the gate to the Otherworld (White, 1991), there is good reason to investigate this trend further. According to Indo-European and Indo-Aryan mythology, two dogs served as the guardians to the gate (O'Flaherty, 1981; Jones-Bley, 1997). The positioning of dog deposits relative to houses should therefore be considered as important data to collect in the future for this region. Excavators should attempt to determine if there is any evidence for the location of house entrances in order to learn whether these could have been threshold offerings.

In addition to extramural dog deposits there were those placed in pits inside houses. Many of these were found in foundation pits that were placed half in and half outside the house and protruded through the subterranean wall at floor level. Pits in House 32 produced the remains of at least six and possibly seven dogs (*Canis* 1-3, 5-7, 14). Two of these individuals were juveniles and at least two were adult males. Unfortunately, there is no record of which pits contained the dogs or whether the dogs were combined in one pit or spread out amongst the three pits documented in the house. One of the features was an elongated pit that crossed from the interior floor of House 32, through the west wall and across an open area, linking that house to another house to the west of it. Two others were foundation pits dug into the north wall at floor level. According to one of the primary excavators, Alexander Kislenko (pers. comm., 1995), the dogs were found in a possible tunnel leading out of the house. This would suggest that they were deposited together in the western pit, but this could not be substantiated by the field notes. If so, this would make it the fourth case of dog remains placed on the west side of the house.

Merrifield (1987) distinguishes both foundation and threshold offerings in Indo-European settlements. Foundation offerings of humans, animals, and artifacts have been reported all across the Indo-European diaspora. These have been interpreted, in many cases, as reflections of "rites of commencement," performed by the builders prior to or during construction. At other times, they appear to signify renewal or termination. Threshold offerings are generally interpreted as "protection for the entrance," whether it be for a gate to a settlement or the door of a house. Here, it is evident that careful excavation and thorough recording could clarify the conditions under which dog deposits were made. If the pit was dug as a niche from inside the house at or below floor level, then it would be classified as a foundation pit. If it was dug down from the ground surface outside the house, it might be seen as a threshold pit. The distinction should be obvious during excavation in most cases. The soil in pits at Botai, as elsewhere, is generally much darker than surrounding sterile soil. In the case of the dog burial to the west of House 139, the pit was clearly external and was dug down from ground surface, so it was clearly not a foundation offering. Whether it was a threshold offering depends on where the doorway into the pithouse was. Doorways have not been identified definitively at Botai at this time. It is possible that the houses were entered through a central smoke hole by ladder, as was done by some Siberian tribes (Zolotarev 1938: 19) and the Lapps (Bosi 1960) in the last century.

House 32 also produced a highly unusual array of other species. Three marmot (*Marmota bobak*) mandibles (probably used as chisels, based on wear on the incisors), two beaver (*Castor fiber*) right humeri, a brown bear (*Ursus arctos*) carpal, a wolf (*Canis lupus*) innominate and femur, a red fox (*Vulpes vulpes*) femur, and a saiga antelope (*Saiga tatarica*) horn core, along with the remains of at least 23 horses, were represented in the house contents. Given that over 99% of the faunal material at Botai (Olsen, 1996) is from horse and that the number of elements from other species is very low there, this house is quite remarkable. Kislenko (1993) reported that most of the osteological material was concentrated in a deposit of clay from the collapsed wall in the northwest quadrant of the foundation. He also stated that part of the bones were discovered in the fill of the three foundation pits there and that those bones were better preserved than the ones found in midden fill. The occurrence of a diverse fauna associated with dogs is similar to a ritual found at the famous Iron Age site of Danebury, in England. There, a pit from the earliest phase of the settlement produced two dogs in close association with 20 very carefully chosen bones of seven other (domestic and wild) species (Green 1992, 1997).

An oval foundation pit in the southern wall of House 33 contained a nearly complete skeleton of a dog (*Canis* 9), along with six horse skulls and several artifacts in a foundation pit in the south wall. The skull of a dog, its articulated vertebral column, scapulae, one humerus, and

***Figure 5***. *Canis 18, dog cranium resting on a horse cranium in a midden near the river bank. Arrow points north.*

***Figure 6***. *Fox skull with a piece of quartz resting in the nasal region. Arrow points north.*

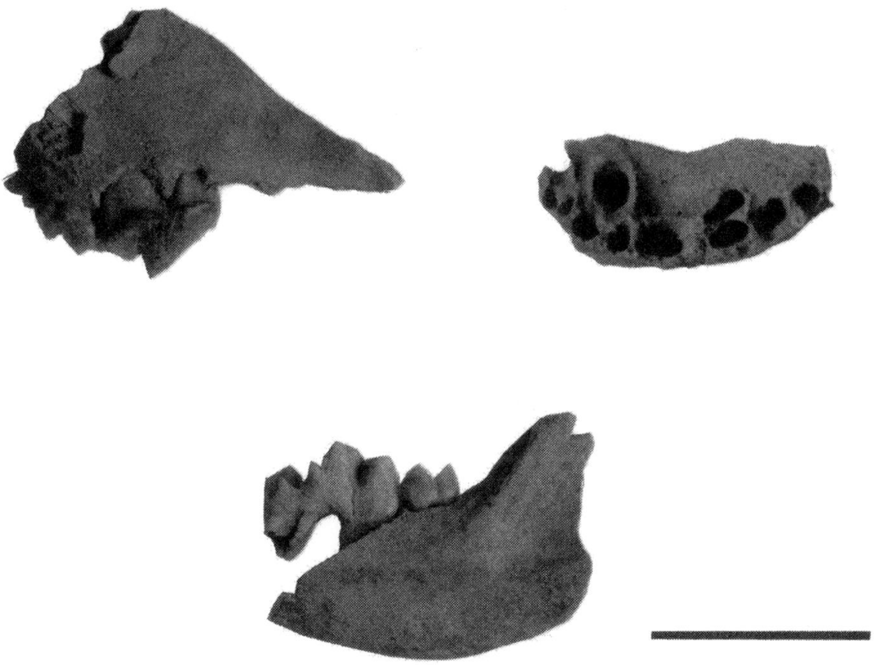

***Figure** 7. Fragments of wolf maxilla and mandible (Canis 4) from Neolithic House 140 pit. Scale is 5 cm.*

most of the hind limbs were found on the floor of the pit. Above the floor were the articulated bones of both hind feet, some artifacts made on unidentified long bones, and some potsherds. The placement of the six horse skulls relative to the dog burial is not described in the field notes, unfortunately.

| Element | 1 | 2 | 3 | 4 | 5 | 6 | 7 | 8 | 9 | 10 | 11 | 12 | 13 | 14 | 15 | 16 | 17 | 18 |
|---|---|---|---|---|---|---|---|---|---|---|---|---|---|---|---|---|---|---|
| Cranium | | | 1 | 1 | 1 | | | 1 | | | | | | 1 | 1 | | 1 | 1 |
| Mandible | | | | L | B | | | | | | B | L | R | | B | | B | |
| Hyoid | | | | | 1 | | | | | | | | | | 1 | | | |
| Atlas | 1 | | 1 | | 1 | | | 1 | 1 | | | | | 1 | 1 | | 1 | |
| Axis | 1 | | 1 | | 1 | | | 1 | 1 | | | | | 1 | 1 | | 1 | |
| Other cervical | 5 | | 1 | | 3 | 4 | 5 | | 3 | | | | | 1 | 5 | | X | |
| Thoracic | 13 | | | | 5 | 3 | 3 | | | | | | | | 12 | | X | |
| Rib | 9 | | | | 3 | 1 | | | 4 | | | | | | 17 | | | |
| Sternebra | 3 | | | | | | | | | | | | | | 1 | | | |
| Lumbar | 4 | | | | 3 | | | | 6 | | | | | | 6 | | | |
| Sacrum | | | | | | | | | 1 | | | | | | 1 | | | |
| Baculum | | | | | | 1 | | | | | | | | | 1 | | | |
| Caudal | | | | | 4 | 3 | | | | | | | | | 5 | | | |
| Scapula | B | | | | B | B | | | B | L | | | | | B | | | |
| Humerus | B | B | | | B | B | B | | L | L | | | | | B | | | |
| Ulna | B | | | | B | B | | | | | | | | | B | | | |
| Radius | B | B | | | B | B | | | | R | | | | | B | | | |
| Carpal | | | | | | 2 | | | | | | | | | ? | | | |
| Metacarpal | B | B | | | B | B | | | | | | | | | L | R | | |
| Innominate | | | | | B | | | | B | | | | | | B | | | |
| Femur | R | | | | B | B | | | B | | | | | | B | | | |
| Patella | | | | | | | | | | | | | | | B | | | |
| Tibia | R | | | | B | B | | | B | | | | | | B | | | |
| Fibula | | | | | R | R | | | | | | | | | B | | | |
| Calcaneus | | | | | | B | | | B | | | | | | B | | | |
| Astragalus | | | | | L | B | | | R | | | | | | B | | | |
| Tarsal | | | | | | 4 | | | | | | | | | ? | | | |
| Metatarsal | | | | | B | B | | | B | | | | | | R | L | | |
| 1st Phalanx | 6 | 13 | | | 9 | 9 | | | | | | | | | 7 | | | |
| 2nd Phalanx | 2 | 9 | | | 1 | 4 | | | | | | | | | 4 | | | |
| 3rd Phalanx | | 1 | | | | 2 | | | | | | | | | 4 | | | |

*Table 2. Canid element frequencies.*Canis *1-18. L=left; R=right; B=both; X=present, number unknown; ?=presence uncertain; number indicates the quantity of those elements found.*

Botai resembles other contemporary and later sites in the steppe (Telegin, 1986) as well as Bronze and Iron Age sites across Europe (Grant 1984b; Merrifield, 1987; Green, 1992) in exhibiting a close association between the horse and dog. In addition to examples described above, a dog cranium (*Canis* 18) was found resting on that of a horse in a midden near the river bank (Fig. 5). Given the density and dominance of horse bones over all other cultural material at Botai, it is predictable that dog remains should occur in close proximity to some horse elements, but the fact that these associations are most often with horse crania and mandibles appears to be significant. It might be expected that there were connections between the dog and horse at Botai. If horses were domesticated, as this author believes, then dogs and horses were, for practical purposes, the only domesticated animals there. A scattering of a few bovine bones have yet to be identified as bison, aurochs, or domestic cattle. If domesticated horses were used for riding during hunts, then dogs could have been used alongside the horses. The teaming of horses and dogs in hunts could provide a stimulus for sacred associations, as well. In Scythian art, dogs are sometimes shown in close association with horses. A silver-gilt ritual vessel of Scythian form, but probably made by a Greek artisan, was found in the Solokha barrow, near Zaporizhzhya, Ukraine (Piotrovsky *et al.*, 1987). The vessel shows dogs with men on horseback having an encounter with a lion. The Greek author Flavius Arrianus (born in AD96) wrote a treatise on dogs called *Cynegeticus*, the same title as a book written by Xenophon 550 years earlier. In the second book, Arrianus describes Celts hunting hares with hounds while the men followed on horseback (Hull, 1964).

The horse/dog associations at Botai are reminiscent of the cult stallion skull and pair of dogs found at Dereivka, Ukraine (Telegin, 1986), now thought to be an Iron Age intrusion. Recently, a tooth removed from the Dereivka stallion mandible was AMS dated to 790-200BC, calibrated (OxA 7185: 410-200BC, Ki 6962: 790-520BC) (Anthony and Brown 2000). This would make it considerably younger than Botai and means that it is not a component of the Copper Age Sredny Stog occupation at Dereivka as once thought (Telegin, 1986).

The dog remains at Dereivka were found below and to the sides of the stallion skull. One dog's remains consisted of a west-facing skull and a coiled, articulated vertebral column. The second dog was represented by a southeast-facing skull and articulated bones of the left and right forelimb paws. Mythology may come into play at Dereivka. Two dogs served as the companions of the lord of the dead, Yama, in the Indic religious text, the *Rig Veda* (O'Flaherty, 1981: 43), or Yima in the Iranian text, the *Avesta.* The *Avesta* makes reference to a yellow dog with four eyes and a white dog with yellow ears that were companions to Yima (Jones-Bley, 1997: 206). In both the *Rig Veda* and *Avesta,* the dogs were considered the guardians to the gate of the kingdom of the dead. This recurrent theme also appears among mythological references in Latin, Greek, Germanic, Armenian, and Celtic texts (Lincoln, 1979). As White (1991:39) says, "Indo-European dogs of the dead generally moved around in pairs, however: when they were not twin brothers, they could be represented as two dogs in one, as in the well-known case of Cerberus." Merrifield (1987) also provides numerous examples of pairs of dogs in ritual deposits. Whether examples of paired dogs in prehistoric settings can be linked directly to that mythology is difficult to say, but such patterns should be recorded with the hope that trends will eventually shed light on prehistoric rituals. Pairing of dogs in ritual deposits was not a recurrent pattern at Botai, but the expressions of rituals pertaining to dogs seems to have been very flexible or diverse there, as it was throughout Eurasia.

Another interesting, although perplexing, finding is that canid skulls are sometimes accompanied by a small stone at Botai. One fox skull found in the bank midden had a piece of flint lying just in front of the maxilla (Fig. 6). A dog cranium from the same midden (*Canis* 18) had a piece of quartz lying in its left orbit. Whether these are just coincidental or represent some ritual practice is uncertain at present, but perhaps future discoveries will help determine if a pattern is evident.

Two wolves, represented only by a cranium and mandibular fragments of one (*Canis* 4, Fig. 7) and mandibles of another (*Canis* 11), were found in a pit in House 140, a Neolithic dwelling. The fore and hind feet of a wolf (*Canis* 16) were found in the northwest quadrant of the foundation of House 32, possibly in a pit, but no further data are available. The foot bones were probably the residuals of a wolf skin.

Body Part Representation

Caution must be exercised when attempting to extrapolate from the representation of body parts at Botai because the conditions of excavation and curation, as well as the numerous times the material was moved from one laboratory to another, subjected the dog remains to considerable post-excavation loss and potential mixing. Of the material examined, skulls, cervical vertebrae, and limb bones predominated (Table 2). The finding of partial skeletons cannot be totally disregarded as the result of post-excavation practices, because this pattern has been observed elsewhere in the Eurasian steppe (Telegin, 1986).

Although no cutmarks were observed on the Botai canines, it is clear that dog heads were often separated from their bodies and handled independently. The finding of isolated dog skulls in pits and middens is difficult to interpret, but it should be noted that horses and several other species, including three foxes, two wolves, and a wolverine, were treated similarly at Botai. The prehistoric peoples of the Eurasian steppe demonstrated a strong pattern of treating skulls of both humans and other animals in special ways. Two human skulls at Botai were from decapitated individuals. One of those was skinned, covered with a clay mask and drilled at the top for suspension before reaching its final resting place in a trench running northeast from the exterior of House 44 (Rikushina & Zaibert, 1984). The second was made into a bowl. Perhaps animal skulls were carried as fetishes, kept in skins and worn as parts of costumes, or buried for ritual purposes.

To illustrate the kind of ritual that might be represented at Botai, an example is given about a group of people that lived in Manchuria during the tenth century. The Turko-Mongol Khi-tan people held a celebration on the eighth day of the eighth month to honor their dog ancestor, known as Nai-ho, or "dog skull" (White, 1991: 133). On the day of the Dog's Head Festival, the local ruler was to kill a white dog outside his tent. He then buried the dog with its muzzle protruding from the earth. At the end of seven days, he moved his tent so that it rested on top of the buried dog's head. The Khi-tan were identified with other Asian groups that were reported to have dog-headed men descended from the union of warrior women and their dogs.

The recurrence of isolated dog skulls in the site, sometimes clearly in ritual context, is interesting in light of the mythology of cynocephali reported by European, Indian, and Chinese observers. The Franciscan missionary John of Plano Carpini reported a tale told by his Mongol hosts in 1247 of the dog-men that lived north of them (White, 1991: 131). Similar accounts of the Kingdom of Dogs were attributed to the Armenian King Het'un I, also in the thirteenth century, and to a Chinese traveler Hu Chiao in the tenth. Based on repeated tales of dog-headed men associated with Amazon women (female warriors), White (1991: 124) has concluded through a kind of triangulation that the Kingdom of the Dog originated in Central Asia. To quote White, "As we have intimated, there is a certain intersection in these three traditions of the geographic localizations of the Dog-Man races: the vortex of all three would be the vast region of Central Asia or Turkestan." He includes among these Altaic people: the Mongols, Tungus, Uighurs, Khi-tans, Turks, Tibetans, Koreans, Ainus, Eskimos, Hungarians, Finns, and Estonians, as well as the Indo-Iranian Scythians. Davis-Kimball (1997) has interpreted the presence of weapons in female graves at the cemetery of Pokrovka (east of the Ural river on the northern Kazakh border) as evidence for the existence of Sauromatian and Early Sarmatian Amazons in the Iron Age.

The occurrence of dog, fox, and wolf skulls, as well as articulated foot bones, at Botai, is reminiscent of head and hoof burials of horses, sheep, and cattle, that may represent skins, in subsequent Bronze and Iron Age contexts in the Eurasian steppe and beyond (Piggott, 1962; Vasiliev *et al.*, 1994). Keeping the bones in the hide helps retain some of the form of the living animal and gives it support, although the same can be achieved by stuffing the skin as a partial taxidermy mount. Leaving the bones in place is simpler, however, and after the skin is dried, they will not fall out the way straw or other stuffing might. More importantly, by leaving the skull in place in dog or wolf skins, the frightening teeth may be displayed.

The early accounts of dog-headed warriors could be based on men wearing dog or wolf skin capes over their head and shoulders. Ancient art can provide supplemental evidence for the use of canid skin capes. One example is a bas-relief depicting two warriors carrying lances on a bronze helmet plate from Torlunda, Sweden, (sixth or seventh century AD) (Mallory, 1996). The second warrior is wearing a canid skin over his head and shoulders that appears to still harbor teeth in the head, indicating that the skull was still inside the skin. The forelimbs are wrapped around the man's neck and the tail hangs down the man's back. Such costumes may have originated as hunting disguises or clan emblems. Perhaps when men wearing such costumes first appeared in foreign lands, they were mistaken for mythological beasts--half dog and half man.

A second example depicting possible dog-men is a silver goblet from the Bronze Age kurgans of Trialeti, in the Ktsiya-Khrami river valley 110 km from Tbilisi, Georgia (Mongait, 1959). The upper of two horizontal bands of chased designs consists of a procession of animal-headed men with tails drinking from beakers. They are walking toward a figure on a throne next to a tree, two altars, and sacrificial animals. The ancestral family of the Indo-Iranian Achaemenidian Dynasty (559-480BC) was known as *saka haumavarka*, or "those who change themselves into wolves in the ecstasy brought on by haoma" (White, 1991:215). Haoma, also known as soma, was a sacred beverage made by pressing an uncertain species of fungus or plant that may have possessed hallucinogenic properties. Perhaps the animal-headed men on the goblet from Trialeti were supposed to have been drinking haoma.

The myth of cynanthropy could have enhanced fear and alienation on the part of foreign societies the warriors contacted. Based on historical accounts, this seems to have been the case. During the tenth century, the Chinese traveler Hu Chiao, described the Khi-tan and their homeland: "Further to the north is the Kingdom of Dogs [Guo-guo], where the inhabitants have the bodies of men and the heads of dogs. They have long hair, they have no clothes, they overcome wild beasts with their bare hands, their language is the barking of dogs. Their women have a human form and can speak Chinese; when they give birth to males, they have the form of dogs; when they give birth to females, these have the form of humans. They intermarry, live in caves and eat their food raw; their women and girls are man-eaters" (White 1991: 133). The various historical accounts of cynocephali show how broad the swath of the dog-headed men was, as they made themselves known to people from Europe to India to China.

The idea that canid skins were worn as headdresses or parts of costumes by men should be explored further in the Eurasian steppe. The rarity of canid remains in burials makes this difficult, but if canid skulls are found in other ritual contexts with articulated feet and possibly caudal vertebrae (minus other elements), then it is likely that those deposits represent hides.

| Botai | Age | U. Dent. | L. Dent. | Scap. | Hum. | Rad. | Ulna | MC | Inn. | Fem. | Tib. | Calc. | Phal. |
|---|---|---|---|---|---|---|---|---|---|---|---|---|---|
| 1 | 5 to 8 | | | | <12 | <8 | <15 | | | <18 | <18 | | >5 |
| 2 | >15 | | | | >12 | >6 | | >5 | | | >14 | >15 | |
| 3 | adult | | | | | | | | | | | | |
| **4** | >6 | >6 | >5 | | | | | | | | | | |
| 5 | >18 | >5 | >5 | | >12 | >18 | >15 | | >18 | >18 | >18 | | |
| 6 | >18 | | | | >12 | >18 | >15 | | | >18 | >18 | >15 | |
| 7 | <12 | | | | <12 | | | | | | | | |
| 8 | adult | | | | | | | | | | | | |
| 9 | >18 | | | | >6 | | | | >6 | >18 | >18 | >15 | |
| 10 | <8 | | | <8 | | <8 | <15 | | | | | | |
| **11** | >5 | | >5 | | | | | | | | | | |
| 12 | >6 | | >6 | | | | | | | | | | |
| 13 | >6 | | >6 | | | | | | | | | | |
| **14** | adult | | | | | | | | | | | | |
| 15 | >18 | | | >8 | >12 | >18 | >15 | >5 | >18 | >18 | >14 | | >5 |
| **16** | 5 | | | | | | | 5 | | | | | |

***Table 3**. Age distribution, in months for* Canis *1-16. Bold specimen numbers indicate wolves.*

Age and Sex

The age distribution of the domestic dogs (*Canis familiaris*) can be seen in Table 3. With four individuals under a year and ten dogs over fifteen months, the percentage of immature dogs is about 29%. The sample size at present is too small to determine if culling was conducted at Botai. In any case, ritual, rather than population control, may have been the primary cause of death in most of these specimens. Two dogs could only be classified as over six months, so they could fit into either the juvenile or adult category. Two cases of exostosis on adult elements may have been caused by arthritis and could indicate senility, but the fact that the exostoses are restricted to just one element in each skeleton suggests that the pathologies are due to work-related strain.

Only three individuals could be sexed as male based on the presence of a baculum. Two bacula came from the pit in House 32, where the remains of at least six dogs were mixed together. The other was found with *Canis* 15, the small adult male in the pit outside House 139. Because of the partial nature of carcasses in ritual deposits, coupled with curation problems, it is not wise to speculate that any of the dogs missing a baculum was female.

Size and Proportions

When the Botai dog skeletal remains were measured, they were found to compare fairly well with the modern Samoyed breed in size and form. Five Samoyed skeletons (three adult females and two adult males) in the Carnegie Museum of Natural History Mammal Collection (Specimen numbers CM 59583, 59585, 59936, 59935, and 61408) were measured and compared to the Botai specimens. This is not meant to imply that the Eneolithic dogs of Botai belonged to a specific breed or that the modern Samoyed breed as it appears today extends back in time to the occupation of Botai. However, given the proximity of the antecedents to Samoyed people in this region during the fourth millennium, it is not surprising to find that the Botai dogs are more similar to the Samoyed breed than they are to other breeds of great antiquity like the Egyptian Greyhound (Zeuner, 1963), Tibetan hunting dog (White, 1991), Mesopotamian Molossian (Hull, 1964), or Greek Laconian hound (Hull, 1964).

The Samoyed (or Nentsy) people, who are so closely associated with the Samoyed dog breed, speak a Uralic language. According to Hajdú (1963), the ancient home for the Uralic language was the west side of the Ural Mountains, in the vicinity of the bend in the Volga, along the Kama and Vyatka rivers, and eastward. By about 3,000BC the Samoyed language had diverged from Finno-Ugrian (a language group also derived from Uralic and including Finnish and Hungarian) when the Samoyed people spread out across the forest zone north of 57° latitude. At the time they split off, the Samoyed people had domestic dogs; and linguistic analysis demonstrates that they were already acquainted with the sled. Finding dogs similar in proportion and body size to the Samoyed breed just south of the location where Samoyed people crossed around 3,000BC could be interpreted as evidence that an ancestral stock of the Samoyed dog breed and its relatives already existed in this region by the fourth millennium. This is also supported by the fact that the spitz and similar breeds found in Finland today are considered to have come from the same stock as the Samoyed breed.

Apart from any possible genetic connections between the Botai and Samoyed dogs, it is productive to compare the Eneolithic canines to a living breed in order to assess the potential strength, speed, endurance and other qualities of the prehistoric dog's physique. Although it is dubious to recreate the temperament or coat of prehistoric dogs from skeletal material, the Samoyed may provide a range of comparable physical capabilities that can be applied to the Botai canines. Differences between the Samoyed and Botai dogs will be described here and Tables 4-6 provide the metric evidence for such divergences.

The skull of the Samoyed is described as broad, flat, wedge-shaped, and strong (Canadian Kennel Club, 1982). The stop (where the nasal bones meet the frontal) is visible, but not very pronounced (Table 3). The muzzle is medium in length, the jaws are robust, and the face tapering. The overall shape of the Botai skulls is very similar to that of Samoyeds (Table 4, Figs. 14-17). The total length of the skulls are comparable and the size ratio of the neurocranium to the muzzle is about the same. The height of the forehead is practically identical. The mandibles are approximately the same length. Spacing and size of the dentition is quite comparable between the archaeological and modern specimens.

| Meas. Code | Botai 3 | Botai 5 | Botai 8 | Samoyed M 61408 | Samoyed M 59935 | Samoyed F 59936 | Samoyed F 59585 | Samoyed F 59583 |
|---|---|---|---|---|---|---|---|---|
| 1 | | 192 | | 181.6 | 183.7 | 178 | 168 | 175 |
| 7 | | 101 | | 86.6 | 86.5 | 85.5 | 81 | 81.2 |
| 8 | | 96 | | 94.3 | 94 | 94.5 | 85.8 | 88.6 |
| 9 | | 108 | | 108.5 | 111.4 | 105.8 | 98 | 106 |
| 17 | | 53.8 | | 50.3 | 48.2 | 49 | 45.8 | 46.6 |
| 22 | | 24.2 | 29.2 | 21.8 | 21.8 | 22.4 | 22 | 21.9 |
| 23 | | 75.4 | 81.2 | 60.8 | 61 | 59.3 | 57.9 | 59.6 |
| 25 | 50 | 44.3 | 46.6 | 37 | 38.7 | 37.3 | 34.7 | 36.8 |
| 26 | | 60 | 66 | 50.5 | 53.6 | 52.2 | 48.6 | 48.3 |
| 27 | 24.1 | 21.7 | 23 | 20.3 | 21 | 21.3 | 19.6 | 20.1 |
| 28 | | 18.5 | 16.3 | 17.6 | 16.2 | 15 | 15 | 16.7 |
| 29 | | 54.6 | | 52.4 | 54.8 | 55.9 | 55.5 | 53.6 |
| 32 | | 59.4 | | 52.6 | 52 | 46.8 | 48.7 | 48.5 |
| 33 | | 39 | | 37.4 | 36.3 | 35.5 | 36.2 | 34 |
| 38 | | 67.3 | 66.4 | 68.8 | 69.7 | 70.2 | 66 | 65.8 |
| 40 | | 53.7 | 54 | 50.6 | 50.8 | 47.3 | 44.6 | 47.5 |
| P4/ GB | | 10.7 | | 9.9 | 9.3 | 9.1 | 10.1 | 8.9 |
| P4/ L | | 18.2 | | 18.3 | 17.7 | 17.1 | 17.3 | 18.1 |

***Table 4**. Cranial measurements (mm) of Botai dogs and modern Samoyeds. M=male; F=female.Measurement codes from von den Driesch (1976): 1. Total length; 7. Upper neurocranium length; 8.Viscerocranium length; 9. Facial length; 17. Length of premolar row (buccal); 22. Greatest diameter of auditory bulla; 23. Greatest mastoid breadth; 25. Greatest breadth of occipital condyles; 26. Greatest breadth of bases of paroccipital processes; 27. Greatest breadth of foramen magnum; 28. Height of foramen magnum; 29. Greatest neurocranium breadth; 32. Frontal breadth; 33. Least breadth between orbits; 38. Skull height; 40. Height of occipital triangle; P4 GB. Greatest breadth of upper fourth premolar; P4 L. Greatest length of upper fourth premolar.*

| Element | Meas. Code | Botai 3 | Botai 5 | Botai 9 | Botai 14 | Botai 15 | Botai Average | Samoyed Average |
|---|---|---|---|---|---|---|---|---|
| Atlas | GL | 47.8 | 41.7 | 39.3 | 39.2 | 41.4 | 41.9 | 38 |
| | GB | 76.3 | 86 | 82 | 72.9 | 84.3 | 80.3 | 74.6 |
| | BFcr | 37.6 | 33.3 | 30.9 | 41 | 31.3 | 34.8 | 38.2 |
| Axis | LaPa | | 55 | | | 55 | 55 | 54 |
| | H | | 41.1 | 39 | | 44 | 41.4 | 37 |
| Sacrum | GL | | | | | 43.1 | 43.1 | 34.9 |
| | GB | | | 48 | | 45.6 | 46.8 | 45.1 |
| | PL | | | | | 37.8 | 37.8 | 34.5 |

***Table 5**. Axial measurements (mm) of Botai dogs and the average of the five modern Samoyeds. Measurement codes from von den Driesch (1976): GL: Greatest length; GB: Greatest breadth; BFcr: Greatest breadth of cranial articular surface; LaPa: Greatest length of arch; H: Greatest height;PL: Physiological length.*

| Element | Meas. Code | Botai 2 | Botai 5 | Botai 6 | Botai 9 | Botai 15 | Botai Average | Samoyed Average |
|---|---|---|---|---|---|---|---|---|
| Scapula | HS | | | 13.7 | | | 13.7 | 13.5 |
| | SLC | | 31.4 | | 27.5 | 27 | 28.6 | 23.5 |
| | GLP | | 34.4 | 34.2 | 29.3 | 29.9 | 31.9 | 28.3 |
| | LG | | 35 | 34.5 | 29.6 | 25.7 | 31.2 | 24.7 |
| | BG | | 21 | 22.1 | 18.5 | 18.4 | 20 | 17.1 |
| Humerus | GL1 | | | | | 148 | 148 | 154.9 |
| | GLC | | | | | 143 | 143 | 152.4 |
| | DP | 39.1 | 43 | 44.5 | | 39.5 | 41.5 | 38.2 |
| | SD | 15.2 | 14.3 | 15 | 13.1 | 12.9 | 14.1 | 11.5 |
| | BD | | 27.5 | 27 | 21.6 | 32.8 | 27.2 | 32.9 |
| | BT | | | | | 22 | 22 | 24.6 |
| Radius | GL | | | | | 150 | 150 | 154.6 |
| | BP | 21.6 | 20.3 | 20 | | 17.6 | 19.9 | 16.9 |
| | SD | 16.1 | 13.9 | 14.7 | | 12.7 | 14.3 | 12.4 |
| | BD | | 21.7 | 20.5 | | 24.1 | 22.1 | 23.5 |
| Ulna | GL | | | | | 177 | 177 | 184.5 |
| | DPA | | 27.3 | 29.1 | | 25.1 | 27.2 | 25.3 |
| | SDO | | 24.2 | 24.9 | | 21.7 | 23.6 | 18.8 |
| | BPC | 18 | 20 | 20 | | 17.7 | 18.9 | 16.8 |
| Innominate | GL | | | | 118.7 | 147 | 133 | 143.1 |
| Femur | GL | | | | | 160 | 160 | 168.9 |
| | BP | | | | | 36.7 | 36.7 | 36.6 |
| | DC | | 21.1 | 21.1 | | 18.8 | 20.3 | 17.6 |
| | SD | | 14.8 | | 13.5 | 13 | 13.8 | 11.8 |
| | BD | | 34.4 | | 30 | 30.8 | 31.7 | 31.5 |
| Tibia | BP | | 36.5 | 35 | 32.1 | | 34.5 | 34.1 |
| | SD | 15.2 | 14.3 | 15.8 | 13.2 | 12.9 | 14.3 | 11.8 |
| | BD | 26.8 | 24.5 | 24.4 | 22.1 | 22.4 | 24 | 23.3 |
| Astragalus | GL | 34.3 | 28.6 | 31.6 | 31.4 | | 31.5 | 27 |
| Calcaneus | GL | 55.5 | | 47.5 | 43 | 42.5 | 47.1 | 44.1 |
| | GB | 20.9 | | 20 | 17.8 | 18.3 | 19.2 | 17.2 |

***Table 6**. Appendicular measurements (mm) of Botai dogs and the average of the five modern Samoyeds. Measurement codes from von den Driesch (1976): HS: Height along spine (scapula); SLC: Smallest length of neck (scapula); GLP: Greatest length of glenoid process (scapula); LG: Length of glenoid cavity (scapula); BG: Breadth of glenoid cavity (scapula); GL1: Greatest length of lateral part (humerus); GLC: Greatest length from caput or head (humerus); GL: Greatest length; DP: Depth of proximal end; SD: Smallest breadth of diaphysis; BD: Greatest breadth of distal end; BT: Greatest breadth of trochlea; DPA: Depth across Processus anconaeus (ulna); SDO: Smallest depth of olecranon (ulna); BPC: Greatest breadth of proximal articular surface (ulna).*

There are a few minor differences between the Botai dog skulls and those of Samoyeds that are worth noting. The sagittal and nuchal crests are more pronounced in the Botai specimens than in both the male (Fig. 8) and female Samoyeds. There is a more caudally-extended overhang at akrokranion, "the most aboral point of the vertex of the cranium in the median plane" (von den Driesch, 1976). The stop is less abrupt and the muzzle more gradually sloping in the Botai skulls. The frontal breadth (from ectorbitale to ectorbitale) is slightly greater in the Botai dogs (Fig. 9, 10). The greatest neurocranial breadth (ie, the greatest breadth of the braincase), is somewhat larger in the Botai dogs. The zygomatic arches are broader and the occipital condyles are larger in the Botai specimens. The mandibles of the Botai dogs are more robust than those of the Samoyed. This is most easily seen at the symphysis, which is higher and thicker in the Botai specimens (Fig. 11). In general, the Botai dogs are cranially somewhat more robust than Samoyeds, as they are post-cranially. The Botai dogs would have had stronger neck and chewing muscles than Samoyeds. This may reflect their functions in the prehistoric culture, but is more likely a retention of the primitive condition in these probable descendants of the wolf.

The atlas, axis, and sacrum on average are slightly larger in the Botai specimens than in the modern Samoyeds (Table 5). The clearest differences are in the greatest breadth of the atlas across the wings and the greatest length of the sacrum.

The lengths of the limbs of *Canis* 15 are similar to those of Samoyeds (Table 6). Based on a number of measurements, however, it is clear that the Botai dogs were generally more heavily built than modern Samoyeds. This implies that they were also stronger and heavier, although their stride may have been about the same.

To calculate the difference in limb proportions, a robusticity index (R) (Table 7) was devised. This index is presented as a ratio of the Botai limb bone robusticity over that of the modern Samoyed. To calculate the robusticity of limb bones, the smallest shaft diameter (SD) of the bone was divided by the bone's greatest length (GL) (von den Driesch, 1976). This figure was then multiplied by 100. The robusticity index formula is as follows: R = [Bot. SD/GL x 100] / [Sam.SD/GL x 100]. This index was calculated for the humerus, radius, ulna, and femur by comparing *Canis* 15 to the average of the five modern Samoyeds. *Canis* 15, an adult male, ranged from 6-20% more robust than the average of three female and two male Samoyeds. Given that *Canis* 15 was somewhat more gracile in all of the limb bones than other specimens from Botai (Table 6), the average Botai dog must have been at least 20% more robust than the average Samoyed. The weight ranges for modern Samoyeds are 20-30 kg for males and 17-25 kg for females (Fiorone, 1970). The average Botai dogs probably would have weighed in at the upper end of the scale for Samoyed males or occasionally even slightly heavier.

The shoulder height of *Canis* 15 (male) was in the upper range of modern Samoyed females and less than modern male Samoyeds. Based on Harcourt's (1974) formula for reconstructing shoulder height from limb bone length, *Canis* 15 stood at *c.* 49.4 cm. This figure was obtained by calculating Harcourt's individual formulae for the radius (49.6), ulna (49.8), and femur (48.9) and averaging those

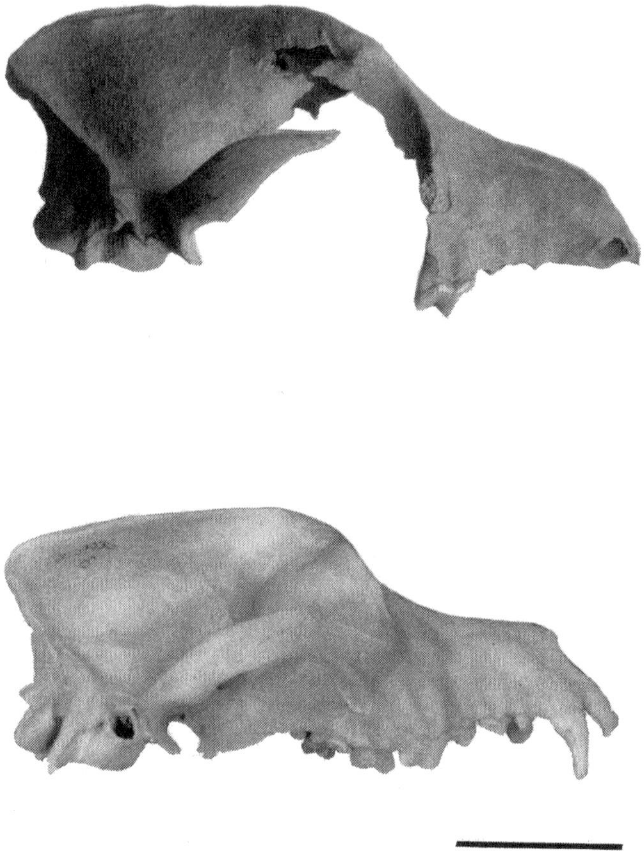

*Figure 8*. Profile of crania of Botai Canis 5 and Samoyed male (CMNH 59935). Scale is 5 cm.

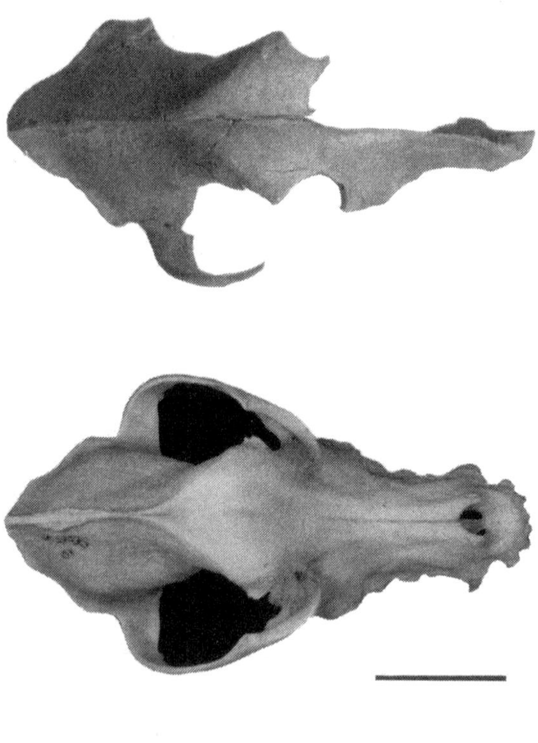

*Figure 9*. Top view of crania of Botai Canis 5 and Samoyed male (CMNH 59935). Scale is 5 cm.

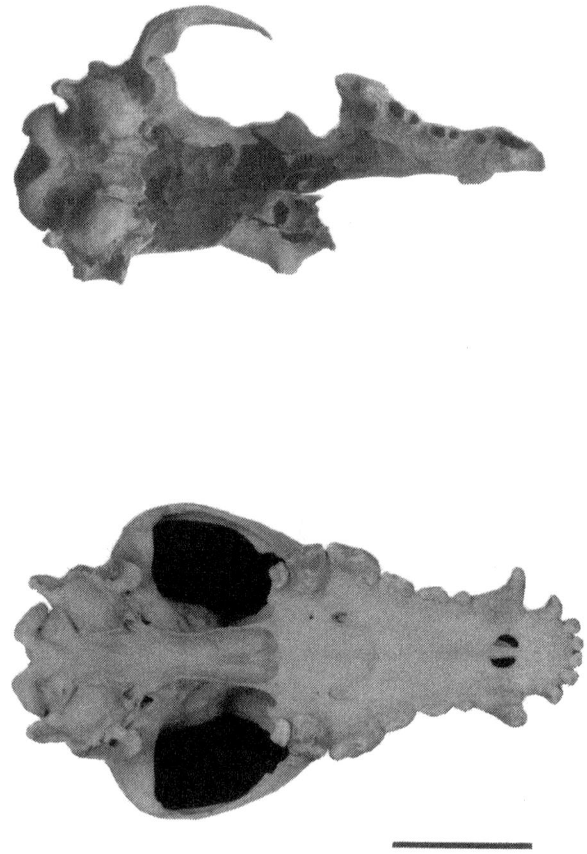

*Figure 10*. Basicranial view of crania of Botai Canis 5 and Samoyed male (CMNH 59935). Scale is 5 cm.

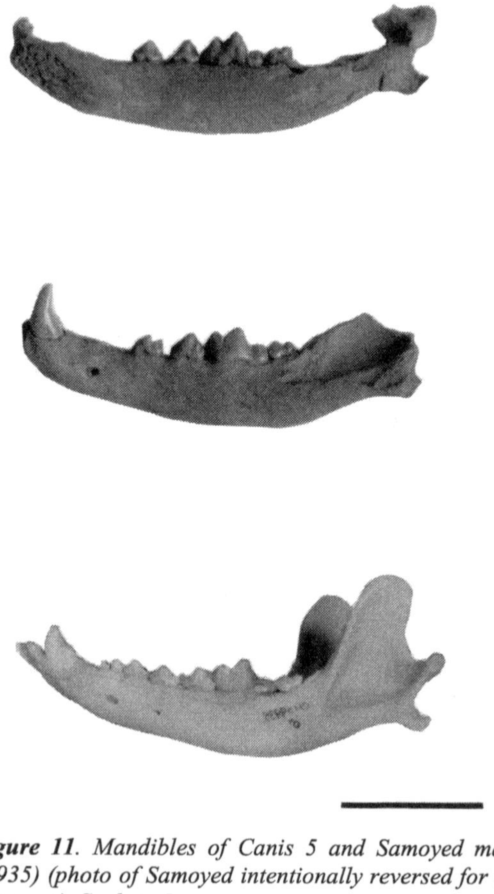

*Figure 11*. Mandibles of Canis 5 and Samoyed male (CMNH 59935) (photo of Samoyed intentionally reversed for comparison purposes). Scale is 5 cm.

three figures (Table 8). Male Samoyeds are 51-56 cm at the shoulder and females are 45-51 cm (Fiorone, 1970). Because of its modest size compared to the other dog remains, *Canis* 15 might have been presumed to be female, except for the presence of the baculum. Measurements suggest that *Canis* 15 was atypically small for a Botai male.

| Element | Botai 15 SD | Botai 15 GL | Samoyed SD | Samoyed GL | Botai/Samoyed |
|---|---|---|---|---|---|
| Humerus | 13 | 148.5 | 11.5 | 154.9 | 1.2 |
| Radius | 12.7 | 150 | 12.4 | 154.6 | 1.1 |
| Ulna | 21.7 | 177 | 18.8 | 184 | 1.2 |
| Femur | 13 | 160 | 11.8 | 168.9 | 1.2 |

***Table** 7. Robusticity Index, comparing Botai Canis 15 to the average for the five Samoyeds (mm).*

| Element | Total Length (mm) | Harcourt's formula* | Shoulder Height (cm) |
|---|---|---|---|
| Radius | 150 | (3.18 x 150) + 19.51 | 49.6 |
| Ulna | 177 | (2.78 x 177) + 6.21 | 49.8 |
| Femur | 160 | (3.14 x 160) - 12.96 | 48.9 |
| Average | | | 49.4 |

*(Harcourt 1974: 154)

***Table 8**. Shoulder height estimate for Canis 15.*

Sexual dimorphism is slight in modern Samoyeds, because they have been bred primarily for sledding. Pairing of males and females in harness reduces aggressive behavior on dog teams, but it means that sexual dimorphism must be negligible so that gaits can be matched on teams (Coppinger & Schneider, 1995). Sexual dimorphism was demonstrated to be minimal in the five Samoyed skeletons examined in this study. The limb measurements for female Samoyeds in the Carnegie collection ranged between 88-100% of those of males. The average female measurement was 92% of the male size (taking into consideration both breadth and length data). Because of the fragmentary nature of the Botai skeletons, sexual dimorphism could not be ascertained on the basis of the metric analysis.

Comparisons between the body size and proportions of the Samoyed and Botai dogs are very informative regarding their relative physical capabilities. The Samoyed is classified as a working dog that is both strong and agile. The ideal Samoyed has a body of medium-length that is muscular, with a deep chest, straight front, and strong neck and loins (Canadian Kennel Club, 1982). The legs are moderately long, partly because of the deep chest. The hindquarters should be well-developed. The Samoyed is intermediate between the Alaskan or Siberian Husky, the swiftest of the sled dogs, and the heavier freight hauler, the Alaskan Malamute. The Alaskan Husky has greater endurance than heavier breeds. Larger dogs use more energy transporting their own body weight and have more difficulty dissipating heat when exerting themselves (Coppinger & Schneider, 1995). The Botai dogs would have been stronger and more capable of hauling heavy freight than Samoyeds, but would have had less speed and endurance in long hauls. The Samoyed people used their dogs to transport their children and goods (mostly dried fish) via sleds (Donner, 1954). Given their moderately robust physique, the Botai dogs also could have been used to pull a sled or travois or to carry packs. If horses at Botai were domesticated, however, the dogs may not have been needed for hauling except with sleds in the winter. At present, there is no evidence for or against the Botai dogs being used to haul freight or pull sleds.

The Samoyed is a breed that has survived several centuries in northern Siberia, demonstrating that its body form is very well-adapted to cold temperatures. Samoyeds were even used on Sir Robert Scott's second expedition to the South Pole (Canadian Kennel Club, 1982). Body size and proportions are also important for the Botai dogs, given the harsh winters experienced on the Siberian Plain, where temperatures often drop to -30° C. The wind chill factor also plays an important role in the open terrain of northern Kazakhstan, where winds can exceed 125 kmph at times throughout the year. The weight of the Samoyed is at the low end of the range that is optimal for dogs that must survive extremely low winter temperatures. Dogs native to the Arctic normally weigh between 32-36 kg (Coppinger & Schneider, 1995). The Botai dogs were somewhat heavier than the Samoyed, making them even better adapted to cold winters. Thinner dogs or dogs with longer limbs, according to both Bergman's and Allen's Rules (Feldhamer *et al.,* 1999), have less volume to keep their bodies warm. More robust dogs, like those found at Botai, are therefore better adapted to this climate.

If dogs were used for hunting at Botai, they probably could have been trained to track and run wild horses in relay fashion until the horses were exhausted. Based on their measurements, the Botai dogs would not have been extremely swift, as were the more slender Egyptian Greyhounds (Zeuner, 1963). They were also less rangy than the Scythian dogs depicted in artistic representations. One such example is a gold torc from the Tolstaya Mogila barrow near Dnepropetrovsk, Ukraine, made by Bosporan craftsmen who worked in the Graeco-Barbarian style (Piotrovski *et al.*, 1987). Perhaps not coincidentally, these canines closely resembled the long-legged Greek Laconian breed used primarily for hunting hares (Hull, 1964). It is also doubtful that the Botai dogs would have been as powerful as the Mastiffs or Molossians used by the Assyrian king Assurbanipal to hunt wild onagers (Clutton-Brock, 1996). Modern male Mastiffs stand at least 76 cm at the shoulder and females are about 70 cm tall (Canadian Kennel Club, 1982). Mastiffs are also considerably heavier (70-85 kg) than Samoyeds and, therefore, the Botai dogs. Based on their size and morphology, the Botai dogs may have been a well-adapted compromise - an animal of moderate strength, endurance, and speed, but one adequately-suited to the harsh Siberian winters. Used in relay fashion, these dogs could probably exhaust wild horses in time. It is well-established that wolves can bring down horses. That skill could still have been present in the Botai dogs, although, as domesticates, they either may have been trained from actually killing their prey or such aggressive tendencies may have been bred out of the dogs.

It is useful to compare the Botai dogs to other archaeological specimens from a similar time period. A dog from Grimes Graves, England (Burleigh *et al.*, 1977), dating to 2280-2450BC (based on antler found in close proximity), is somewhat more recent, but is a good comparison. The skull of the Grimes Graves dog has an even greater overhang at akrokranion than the Botai dogs and a less pronounced stop. In these regards it is more wolf-like in appearance and very different from a modern Samoyed. The Grimes Graves skull, however, is smaller than the Botai skulls. The zygomatic arches and the frontal breadth are narrower in the Grimes Graves specimen. In these regards, the Grimes Graves skull is more similar to the Samoyed than to the Botai dogs, but probably only because it is smaller overall. The Grimes Graves dog was estimated to stand *c.* 52.1 cm at the shoulder. This means that it was taller than Botai *Canis* 15, but may have been about the same height as the larger Botai dogs, for which height has not been estimated. The Grimes Graves dog is very similar to *Canis* 15 in breadth of long bone articular condyles and shaft diameter. It is, therefore, generally at the gracile end of the range of Botai dogs, but still well within that range. Given its height, this means that the Grimes Graves dog was somewhat lankier, with a longer stride than *Canis* 15 and a lighter-weight body than the average Botai dog. It probably would have been swifter, but less strong.

Pathologies and Injuries

The Botai dog remains exhibit few indications of illness or trauma, healed or otherwise. For the most part, the dogs appear to have been healthy and none was recognizably senile. Two pathologies were observed: an exostosis on the proximal end of an ulna of *Canis* 5 and one on a calcaneus of *Canis* 15. Diagnosis by a veterinarian might help to determine whether the pathologies were caused by injury, work-related stress, or simply arthritis from extreme old age. Arthritis might be likely to show up at other joints, as well, unless those particular joints had undergone more than the usual amount of stress in life. All that can be said about the ages of the two individuals with exostosis is that they were both over 18 months. Healed fractures can reflect either mistreatment or veterinary care, but no such examples were observed. There is also no indication of the cause of death in any of these animals, but the remains are rarely complete and the conditions of their burials are usually not recorded in detail.

In this case, the sample size is too small to expect to find many pathologies, but it can be helpful to consider how the roles of the dog in a society might impact on its overall health. Dogs that habitually pull heavy loads might show hypertrophied muscle attachments, arthritis of the shoulder or other joints, and other indications of strain. It is possible that the two examples of exostosis in limb joints were due to excessive strain in the line of duty. Dogs used in hunting large game, particularly wild horses, might exhibit various kinds of traumas caused by the defensive kicks of the prey. Those that act like pariahs, skulking around the midden scrounging scraps of garbage, might be expected to have injuries associated with rock-throwing or beating, such as indentations on their skulls.

Cultural Modifications

There is no sound indication that the Botai dogs were consumed by humans. There are no cutmarks, spiral fractures, or signs of burning, and very few postcranial remains have been found in midden fill. It must be acknowledged, however, that the discovery of isolated skulls and many partial skeletons demonstrates that dogs were indeed dismembered. The dismemberment may reflect ritual activity rather than dietary preparation, however. Carnivore heads belonging to several species have been found in middens. These are perplexing in nature and also may have been ritual deposits (see above discussion of body part representations).

The absence of evidence for marrow extraction is indicated by the fact that virtually all the limb bones were over 90% complete. Only the scapulae, which have little or no marrow but are very fragile, were more extensively fragmented. The extent of articulation of vertebral columns and limbs also argues against a pattern of exploitation of canines for dietary purposes.

**Interpretations of the Botai Data**

The process of interpreting dog deposits at Botai involved looking at the locations and nature of the remains and associated finds, the elements present, the ages and sexes of the individuals, their sizes and proportions, pathologies, and cultural modifications to the skeletons. In addition to the Botai data, parallel evidence from sites belonging to a wide range of cultures and periods across the Eurasian steppe and elsewhere in Europe were considered. Because the Botai dogs resembled the Samoyed breed in size and skull morphology, the use of that breed by the Samoyed people was explored. This at least provided an indication of the physical capabilities and limitations of the prehistoric Botai dogs.

At the most fundamental level, the features containing dog remains at Botai can be said to be "structured deposits" (Hill, 1996) that show dogs to have been repeatedly interred in small pits, either in houses or just outside them. These pits contained whole dog carcasses, parts of carcasses, or multiple carcasses, as well as associated horse bones and other materials, at times. Dog skulls were sometimes placed in middens, but postcranial remains were rarely discarded in trash fill. It is, of course, possible that some of the deposits represent nothing more than a customary way of discarding the remains of a deceased animal by placing it in an abandoned storage pit.

Interpretation is difficult because the deposits do not reflect a single routine or possible ritual that was rigidly defined by a universal standard dictating the precise placement, association of objects, or qualities of the dog itself. In other words, if the deposits represent rituals, then they may not be strictly prescribed according to specific codes. Instead, the deposits may reflect beliefs or traditions expressed variously, according to a family's or individual's own preferences. Or, more likely, they may depict several different kinds of ritual acts involving dogs. There are strong indications that dogs were sometimes offered on altars, as in the deposit containing dog and horse remains, ochre and projectile points on a stone slab.

There appears to have been a connection between horses and dogs in the belief system at Botai. The frequency with which dogs and horses are linked in Eurasian rituals may reflect their close association in life. Evidence abounds for the use of dogs with horses in pursuing game in later cultures (Hull, 1964). Pastoralists also often ride horses while using dogs to round up or navigate their herds across the landscape. This secular pairing could be transferred easily into a sacred association of these two species.

At Botai, whole dogs, partial carcasses, and skulls were placed in foundation pits. Foundation and building offerings were widespread in the Indo-European realm and beyond. Thresholds were also frequently provided with ritual deposits in order to protect the entrance of the structure. According to Merrifield (1987: 52), "A dog was buried under the sill beam of a timber-framed house in Newgate Street, London, when this was built after a fire of around AD130-135 (pl 16), and another was buried in Winchester before a masonry building was built over it at about the same date." In China, far across the continent, at the Neolithic village of Pan-p'o (which began about 5000BC) dogs were buried under posts and around the foundations of dwellings (Mair, 1998). This example probably predates any contact with the northern "barbarians" from Central Asia, but later examples in the Shang may be the result of such exchange of ideas as are thought to have occurred with the nomads. At the famous capital of the Shang Dynasty, An-yang, dogs and horses were buried together in tombs (Mair, 1998).

The occurrence of dog deposits or burials on the west side of houses at Botai is provocative. According to Indo-European mythology, the gate to the Otherworld was to the west and was guarded by two dogs (or in the case of Cerberus, a two- or three-headed dog) (O'Flaherty, 1981; White, 1991; Jones-Bley, 1997). Ritual burials of dogs at western gates is found from England to China. The late Roman British fortified town of Chester-le-Street, dating to the late third to fourth century, produced a large pot containing a small dog under a masonry floor of the north tower of the town's west gate (Merrifield, 1987). Twelve dogs, plus bones of horses, cattle, and other animals were found with fourth century beakers and a bowl in the northwest corner tower of the city wall at Caerwent (Merrifield, 1987).

It is apparent that the Botai dogs had a close association with houses. Perhaps the best interpretation for the strong connection between dogs and houses at Botai is that dogs were buried just outside or in foundation pits because one of their primary secular functions in life was to guard the household. If this charge was transferred to the spiritual world, then the dog might be buried in or near the home to safeguard the inhabitants spiritually. This role of spiritual guardian could then be a metaphorical shift from the more pragmatic secular role of guard dogs.

Although it is very speculative to attempt any reconstruction of ancient dog breed behavior, it might be noted that members of the Samoyed breed are considered exceptional guard dogs (Fiorone, 1970). The Samoyed people take them wherever they travel. The dogs are tied to poles in the camp, where they bark readily, especially when strangers approach (Donner, 1954; Fiorone, 1970).

Throughout Indo-European and Indo-Iranian texts, there is a recurrent theme of dogs as guardians (Green, 1997). A vivid reference to guard dogs in the home can be found in a Vedic hymn entitled "Varuna Provoked to Anger" (*Rig Veda* 7.86). The hymn describes the sage Vasistha being attacked by a guard dog when he entered the house of Varuna, the sky god who watched over people's deeds. The dog barked, ran toward the man, and tried to bite him (O'Flaherty, 1981: 212). Vasistha put the dog to sleep with the words of a sleeping spell (*Rig Veda* 7.55). The second and third verses especially refer to the guard dog in the house:

> 2 White and tawny son of Saramá, when you bare your teeth they gleam like spears in your snapping jaws. Fall fast asleep!
>
> 3 Bark at the thief or at the marauder, as you run up and back again, O son of Saramá. But you are barking at those who sing Indra's praises; why do you threaten us? Fall fast asleep!

According to White (1991: 39), "In Rome, the Lares, the ancestral divinities who protected the home and hearth, were conceived of as dogs or as men clad in dogs' skins." The Persians considered fire, the dog, and the cock as the three protectors (Rudenko, 1970).

The author does not have access to a vast amount of comparative information for canids in domestic contexts in the Eurasian steppe, partly because many of the later nomadic cultures are better known from cemeteries than settlements. At Dereivka, a Sredny Stog village of Early Kurgan tradition in Ukraine, a nearly complete dog skeleton was found lying on its side under the north wall of House 2 in a large foundation pit (Telegin, 1986). Telegin interprets it as an example of "a cult associated with a belief in guardian animals." Several stones were found about the animal's body and its head pointed north, both patterns found at Botai. According to Telegin (1986: 23), "ritual interments of dogs are known fairly well in the cultures of the Copper-Bronze Age, as, for example, in the late Tripolye culture (Tudorovo), the Yamnaya culture (Nerushai), the Corded Ware culture (Stzhizhkov), etc."

Because of the recurrent appearance of dogs in or around houses at Botai, the roles of real and spiritual guardians may be the most visible canine functions reflected in the archaeological record. This does not eliminate the possibility of additional canine roles, however. The stature and build of the Botai dogs would have made them capable sled-dogs or pack animals. Although they would not have been as swift as other breeds, like the Laconian or Greyhound, or as powerful as the Mastiff or Molossian, they may still have been able to hunt large game in relay fashion. The association of dogs and horses at Botai could support hypotheses that dogs were used for either hunting or herding. The Botai dogs would have been in the size range for herding dogs, but were probably somewhat small to have been guard dogs for domestic horses. The Samoyed breed is reportedly used to guard and drive reindeer herds (Canadian Kennel Club, 1982), but the Samoyed people preferred to guard the livestock themselves (Donner, 1954).

The question of whether the dogs were sacrificed or died of natural causes has yet to be addressed. At this stage, it is difficult to know which was the case, but the occurrence of at least six dogs in one pit in House 32 suggests that sacrifice may have been conducted. The co-occurrence of dog remains and horse skulls may provide another clue. Grant (1984b: 222) has argued that "the likelihood of several animals of different species dying of natural causes at the same time is relatively small and suggests that at least some of the animals mentioned were deliberately killed." Dog sacrifices were sanctioned in Vedic text, as when dogs were offered to Indra in sacrifice and when Gautama sacrificed and may have eaten a dog (White, 1991).

There is little to indicate that the dogs of Botai were being eaten. Their postcranial remains are extremely rare in midden fill, no cutmarks have been found, their bones are not smashed open to extract marrow, and there are no signs of their bones being burnt from being thrown on the fire after the flesh was consumed.

According to the *Rig Veda,* eating a dog was one of the most polluting things one could do to one's body (O'Flaherty, 1981: 145). In "The Birth and Childhood Deeds of Indra," (*Rig Veda* 4.18), verse 13, Indra says, "Because I was in desperate straits, I cooked the entrails of a dog."

There are several references in the Vedic texts indicating that dogs were not suitable for eating. Food touched by dogs was impure and utensils that were touched by dogs had to be cleansed or discarded (Simoons, 1994).

Some evidence of the consumption of dogs has been recorded in the Neolithic, Bronze Age and later sites in Europe (Cunliffe, 1991; Simoons, 1994; Bökönyi, 1974). The Bronze Age site of West Row (Olsen, 1994) and the Iron Age settlement of Danebury (Grant, 1984a) contained dog remains that probably represent food refuse. These exhibited cutmarks and, at West Row, breakage typical of marrow extraction. The frequencies of apparent consumption of dogs varies through time in different regions. Bökönyi (1974: 320) notes that "The essential change that had taken place by the end of the Bronze Age in the use of the dog is that by that period man in Europe had ceased eating dog meat."

A survey of the literature reveals that dogs in the Eurasian steppe held a very different status in human funerary ritual than livestock. Horses, cattle, and sheep are often found in burial chambers or beside them, but dogs are generally not. In a review of ritual treatment of horses in the Early Kurgan tradition, Mallory (1981) reported only two examples of dog remains with human burials in the Pontic-Caspian region. One consisted of a dog skull and long bone found in a human burial pit with the radius, tibia, and pelvis of a horse, and a horn-core and tibia of a cow at Khutor Veselyj 3 (Artamonov, 1949). Another was a dog skull buried with a horse "hoof" in a Catacomb interment at Troitskoe 4/24 (Klejn, 1960). These associations between horse remains and dog skulls may reflect rituals similar to ones performed at Botai that were incorporated in human burials.

At the famous Bronze Age cemetery of Sintashta, in the Trans-Ural steppe (near Chelyabinsk, Russia), horses, cattle, sheep, and more rarely goats were buried with humans inside burial chambers or adjacent to them around the end of the second millennium BC (Gening, 1979). About 40 graves containing 65 humans were excavated. Dogs were found only in the backfill overlying the burial chamber. In three such cases, dogs were laid on their side with their legs extended. In another, a calf and a dog were positioned thus. At Botai, the only human remains were associated with horse sacrifices, not dogs. In a large pit in one house, four humans were buried with fourteen horses (Rikushina & Zaibert, 1984).

Exclusion of dogs from human burial chambers was also the rule during the Iron Age of the Eurasian steppe (Davis-Kimball, pers. comm., 1999), although other animals like the ram, horse, and more rarely cattle or pigs (Barbarunova, 1995; Davis-Kimball & Yablonsky, 1995; Moshkova, 1995) might be included as food offerings. Dogs were also absent from the burial complex at Pazyryk (Rudenko, 1970). Dog remains have been reported in a "sacrificial complex" within a kurgan belonging to the Iron Age Sargat culture in the Gayevo cemetery, east of Ekaterinburg, Russia (Koryakova & Daire, 1997). Details were not provided, so it is not clear whether the dog bones were actually in the burial chamber or not. The rarity of dogs in human interments has interesting implications for the Eurasian steppe. Many of the animals ritually sacrificed for the funeral may have been killed to accompany the deceased because of their importance as food and in the case of the horse and cattle as possible transportation or traction. The herbivores, particularly horses and cattle, were likely to have been considered more valuable than dogs because of their expensive upkeep in terms of labor and land. In Vedic India, the Brahmanian texts rank the value of sacrificial victims in descending order as follows: humans, horses, cattle, sheep and goats. Dogs are omitted entirely.

In reviewing the historic texts of Indo-European cosmologies, the dog has a dual personality of both auspicious and dangerous character. In this aspect, the sacred world has direct parallels to the secular. Whereas this useful creature may serve as companion, hunting aid, herder, guardian, draft animal, and so forth, it may at times also be regarded as an impure carrion-eater and vicious attacker. The Homeric and Indian sources identify a small pariah dog that fed on the bodies of soldiers after battles (White, 1991). According to O'Flaherty (1981: 43) Yama's guardian dogs "are regarded (like many Vedic gods) as dangerous because they kill you...but also as potentially benevolent, because they lead you to heaven." The idea of frightening hounds being connected with the Otherworld was widespread and has been recognized in the mythologies of many of the Indo-European and Indo-Iranian cultures. Lincoln (1979: 285), in his linguistic investigation of mythological dogs in Indo-European cultures, states: "The Indo-European hellhound, then, had a name which was drawn from the sound of his snarl or growl. To a certain extent, this may be due to the role of dogs as devourers of corpses, noted so often in the literature of various Indo-European peoples, something which naturally gave them strong funerary associations." Davidson (1998: 49) summarizes: "These sinister dogs appear to have been associated with the malicious hosts of the unquiet dead, and at times the soul of the newly dead on its way to the Otherworld was their quarry, while they might also attack the living who were unfortunate enough to encounter them." This ambiguous profile makes the dog a somewhat risky inclusion in the burial chamber of a loved one and may explain why dogs were rarely incorporated in the actual chamber.

As briefly described above, according to the *Rig Veda* , the god Yama, lord of the kingdom of the dead, had two dogs as his constant companions (O'Flaherty, 1981:43). The dogs guarded the gates of the Otherworld and functioned as "psychopomps" (conductors of souls to the Otherworld) (Mair, 1998). In the funerary hymn about Yama and the Fathers *(Rig Veda* 10.14), the dual roles of the dog as a guardian that can lead you to heaven and as a dangerous threat to human life are demonstrated. Verses 10 through 12 of the hymn have been translated as follows (O'Flaherty, 1981: 43):

> 10 (To the dead man:) Run on the right path, past the two brindled, four-eyed dogs, the sons of Saramá, and then approach the fathers, who are easy to reach and who rejoice at the same feast as Yama.
>
> 11 Yama, give him over to your two guardian dogs, the four-eyed keepers of the path, who watch over men. O king, grant him happiness and health.
>
> 12 The two dark messengers of Yama with flaring nostrils wander among men, thirsting for the breath of life. Let them give back to us a life of happiness here and today, so that we may see the sun.

The finding of dog burials in the mound fill above the burial chamber at Sintashta may reflect the idea of Yama's dogs serving as the guardians to the Otherworld. Because dogs could be evil and eat human flesh, their bodies may have been deliberately kept at a distance from the body, outside the burial chamber. Placement in the overlying soil might symbolically represent the idea that the human spirit had to go to or hurry past Yama's guardian dogs so that they could guide it into the Afterlife.

Variation can be found in the beliefs of Indo-Iranians, however. According to the *Avesta* (*Videvdat VIII*), the yellow, four-eyed dog that accompanied Yima was supposed to be brought to glance at the corpse in order to repel the evil forces of the demon Nisu, who would try to enter a man's body as his soul left (Gening, 1979). The Persians (White, 1991) and later followers of the Zoroastrian religion (Gening, 1979; Malandra, 1983) believed the corpse should be eaten by raptors or dogs as the preferred funerary practice in order to avoid polluting the fire, water, or earth.

Later Indo-European cultures do have occasional references to dogs sacrificed in human burials, but they are certainly not as common as cattle, sheep, and horse sacrifices. Homer (Iliad, 23:175) stated that nine dogs were sacrificed for the funeral of Patroclus, along with sheep, horses, and men. Dogs were included in Hittite graves of ordinary citizens, along with cattle, sheep, pigs, horses, and donkeys (MacQueen, 1975; Jones-Bley, 1997). Ireland has produced dogs in human burial tombs from the Neolithic onward (McCormick, 1985/6; Jones-Bley, 1997). The Late Bronze or Early Iron Age of Greece, Cyprus, and Crete witnessed the burial of dogs in tombs with humans or in pits adjacent to human burials in cemeteries (Simoons, 1994). Dog burials are found with human graves in the Elbe-Saale region of Germany at Slavic sites at approximately AD400-600 (Simoons, 1994). Ibn Fadlan described a Viking chieftain's burial on the Volga river, in AD922, in which a dog, two horses, two cows, a cock, and a hen were sacrificed, along with servants and female slaves (Brønsted, 1965; Jones-Bley, 1997). Finally, some Prussians were buried with both male and female human sacrifices, horses and hunting dogs (Jones-Bley, 1997).

Other attempts have been made to reconstruct the role of dogs in a prehistoric society based on burial data. Cantwell (1980) has described a situation for the Middle Woodland of Illinois parallel to that of Botai. There, dogs were also buried in pits in house floors or in other domestic situations, but not with humans. Preceding and subsequent human burials sometimes contained dogs, however. According to Cantwell, this marked shift indicated a change in the roles of dogs in society, which may have been reversed later. The shift from burying dogs in separate pits in houses or above the human burial chamber in the Eurasian steppe to directly placing dogs with humans in later Indo-European sites may be another example of a change in the perception of the dog's role in society through time.

## Conclusions

The study of the canid remains from the Eneolithic site of Botai demonstrates how meaningful even small samples with little contextual information can be for interpreting prehistoric behavior. At the same time, it provides incentive for archaeologists to be considerably more diligent in recording information about animal deposits in much greater detail in the future. If such scant data as those available for Botai can indicate ritual behavior and the use of dogs as both secular and sacred guardians, then comprehensive records of animal deposits should produce even more information about the belief system of a society and the functions of animals in that culture. Dogs are unique in their relationship to humans and their roles are highly varied and complex. Although it is too optimistic to think that all the interactions between dogs and humans will be expressed in the archaeological record, it is not too hopeful to search for enlightening patterns among their remains.

References to the ritual treatment of dogs in later Indo-European context are varied and numerous. Green (1997:82) states that "In both Classical and Celtic mythology and religion, dogs possessed composite symbolism in which the animals could represent hunting, healing and death." They occur in Celtic European art quite frequently as protectors of gods and goddesses. The dog held a chthonic role, or link to the Underworld, over much of Eurasia, as it appeared in ritual deposits from as far west as Great Britain (Green, 1992, 1997; White, 1991) to as far east as China (Mair, 1998). Interpretation of the sacred, as well as the secular roles of dogs at Botai is by no means straightforward. It is apparent, however, that dogs were involved in the local mythology. There are strong indicators that one of the roles that the dog performed was guardian or protector. The close association of dogs and horses may also reflect the use of dogs in hunting, or perhaps horse herding. The emphasis on dog skulls in rituals may be connected to the myths of cynocephali in Central Asia. Variation in ritual expression at Botai, in terms of the body parts represented, the placement in pits both inside and to the west of houses, and the associated artifacts, indicates that there must have been numerous ceremonies prescribing the use of dogs. Based on the results from Botai, more close scrutiny of the context of animal remains and associated finds in the future is strongly recommended.

## Acknowledgements

I wish to offer my gratitude to my colleagues in Kazakhstan, particularly Victor Zaibert and Alexander Kislenko, without whose cooperation and assistance this research would not have been possible. I am also indebted to the National Geographic Society, the National Science Foundation (grant #BCS-9816476), and the Carnegie Museum of Natural History M. Graham Netting Research Fund and Edward O'Neil Research Fund, all of which have provided funding for this research. As highly-valued assistants in Kazakhstan, Jacqueline Payette and Barbara Pitman carefully organized and measured all the canid material from Botai. The Carnegie Museum of Natural History Section of Mammals generously provided access to their dog collections, as well as the friendly assistance of their collection manager, Suzanne McLaren. I am especially grateful to the reviewers of this manuscript, Katheryn Linduff, Victor Mair, Mary Littauer, John Wible, Mary Dawson, and Chris Beard, for their insight and critical evaluation. For organizing the stimulating symposium that brought together this remarkable group of canid researchers and for her dedication and perseverance in the editing of this volume, I would like to thank Susan Crockford.

## References

Anthony, D. & Brown, D. (1998). Bit wear, horseback riding and the Botai site in Kazakstan. *Journal of Archaeological Science* **25**: 331-347.

Anthony, D. & Brown, D (2000). Eneolithic horse exploitation in the Eurasian steppes: diet, ritual, and riding. *Late Prehistoric Exploitation of the Eurasian Steppe,* Papers presented for the symposium in Cambridge, England, held 12 January-16 January, 2000. Vol. 1: 1-12.

Artamonov, M. (1949). Kurgan excavations on the Manyjche river in 1937. *Soviet Archaeology* **11**: 305-336 (Russian).

Barbarunova, Z. (1995). Early Sarmatian culture. In (J. Davis-Kimball, V. Bashilov, & L. Yablonsky, Eds) *Nomads of the Eurasian Steppes in the Early Iron Age*. Berkeley, CA: Zinat Press, pp. 121-135.

Benecke, N. (1987). Studies of early dog remains from northern Europe. *Journal of Archaeological Science* **14**: 31-49.

Bökönyi, S. (1974). *History of Domestic Mammals in Central and Eastern Europe.* Budapest, Hungary: Akadémiai Kiadó.

Bosi, R. (1960). *The Lapps.* New York, NY: Frederick A. Praeger.

Brønsted, J. (1965). *The Vikings.* London: Penguin Books (Translated by Kalle Skov).

Burleigh, R., Clutton-Brock, J., Felder, P. & Sieveking, G. 1977. A further consideration of Neolithic dogs with special reference to a skeleton from Grime's Graves (Norfolk), England. *Journal of Archaeological Science* **4**: 353-366.

Canadian Kennel Club (1982). *The Canadian Kennel Club Book of Dogs*. Toronto: Canadian Kennel Club.

Cantwell, A.-M. (1980). Middle Woodland dog ceremonialism in Illinois. *Wisconsin Archeologist* **61**: 480-496.

Clutton-Brock, J. (1981). *Domesticated Animals from Early Times.* Austin, TX: University of Texas Press.

Clutton-Brock, J. (1984). Dog. In (I. L. Mason, Ed) *Evolution of Domesticated Animals.* London: Longman, pp. 198-211.

Clutton-Brock, J. (1996). Horses in History. In (S. L. Olsen, Ed) *Horses Through Time.* Boulder, CO: Roberts Rinehart, pp. 83-102.

Coppinger R. & Schneider, R. (1995). Evolution of working dogs. In ( J. Serpell, Ed) *The Domestic Dog: Its Evolution, Behaviour and Interactions with People.* Cambridge, UK: Cambridge University Press, pp. 21-47.

Cunliffe, B. (1991). *Iron Age Communities in Britain.* London: Routledge.

Davidson, H. (1998). *Roles of the Northern Goddess.* London: Routledge.

Davis-Kimball, J. (1997). Warriors Women of the Eurasian Steppes. *Archaeology* **50**: 44-48.

Davis-Kimball, J. & Yablonsky, L. (1995). *Kurgans on the Left Bank of the Ilek: Excavations at Pokrovka, 1990-1992.* Berkeley, CA: Zinat Press.

Donner, K. (1954). *Among the Samoyed in Siberia.* New Haven, CT: Human Relations Area Files.

Driesch, A. von den (1976). *A Guide to the Measurement of Animal Bones from Archaeological Sites.* Cambridge, MA: Harvard University Peabody Museum Bulletin **1**.

Feldhamer, G., Drickamer, L., Vessey, S. & Merritt, J. (1999). *Mammalogy: Adaptation, Diversity, and Ecology.* Boston, MA: McGraw-Hill.

Fiorone, F. (1970). *The Encyclopedia of Dogs.* New York, NY: Thomas Y. Crowell Co.

Gening, V. F. (1979). The cemetery at Sintashta and the early Indo-Iranian peoples. *Journal of Indo-European Studies* **7**: 1-30.

Grant, A. (1984a). Ritual behaviour: the special bone deposits. In (B. Cunliffe, Ed) *Danebury: An Iron Age Hillfort in Hampshire, Vol. 2: The Excavations 1969-1978.* Council for British Archaeology Research Report **52,** pp. 533-543.

Grant, A. (1984b). Survival or sacrifice? A critical appraisal of animal burial in Britain in the Iron Age. In (C. Grigson & J. Clutton-Brock, Eds) *Animals and Archaeology: 4. Husbandry in Europe.* Oxford, UK: BAR International Series **227**, pp. 221-227.

Green, M. (1992). *Animals in Celtic Life and Myth.* London: Routledge.

Green, M. (1997). *Dictionary of Celtic Myth and Legend.* London: Thames & Hudson.

Hajdú, P. (1963). *The Samoyed Peoples and Languages.* Bloomington, IN: Indiana University Publications Uralic and Altaic Series **14**.

Harcourt, R. (1974). The dog in prehistoric and early historic Britain. *Journal of Archaeological Science* **1**: 151-175.

Hill, J. D. (1996). The identification of ritual deposits of animals: a general perspective from a specific study of 'special animal deposits' from the Southern English Iron Age. In (S. Anderson & K. Boyle, Eds) *Ritual Treatment of Human and Animal Remains.* Oxford, UK: Oxbow Books, pp. 17-32.

Hull, D. B. (1964). *Hounds and Hunting in Ancient Greece.* Chicago, IL: University of Chicago Press.

Jones-Bley, K. (1997). Defining Indo-European burial. In *Varia on the Indo-European Past: Papers in Memory of Marija Gimbutas.* Washington, DC: Indo-European Studies Monograph **19**, pp. 194-221.

Kadirbayev, M. & Zh. Kurmankulov (1992). *The Culture of Ancient Stock-breeders and Metallurgists of Sary-Arka.* Alma Ata, Kazakhstan: A. X. Margulana Institute of Archaeology, Kazakhstan Academy of Science. (Russian).

Kalieva, S. & V. Logvin (1997). *Turgay Stock-breeders of the Third Millennium BC.* Kustanay, Kazakhstan: A. X. Margulana Institute of Archaeology, Kazakhstan Academy of Science. (Russian).

Kerner, A. (1991). The Isetskoe Pravoberehnoye settlement. In *The Neolithic Sites of the Urals.* Sverdlovsk, USSR: Academy of Science of the USSR. pp. 46-67. (Russian).

Kislenko, A. (1993). Experimental reconstruction of an eneolithic house. In *Problems of Economic and Technological Reconstruction from Archaeological Data.* Petropavlovsk, Kazakhstan: A. X. Margulana Institute of Archaeology, Kazakhstan Academy of Science. (Russian).

Klejn, L. S. (1960). The kurgans at Troits'koe. *Archaeological Sites* **8**: 141-163. (Russian).

Koryakova, L. (1996). Complex societies in central-northern Eurasia (the second-first millennia BC). In (G. Afanas'ev, S. Cleuziou, J. Lukacs & M. Tosi, Eds) *Colloquia of the XIII International Congress of Prehistoric and Protohistoric Sciences: The Prehistory of Asia and Oceania.* Forlì, Italy: A.B.A.C.O. Edizioni, pp. 17-23.

Koryakova, L. & M.-Y. Daire. (1997). Summary (in English). In (L. Koryakova & M.-Y. Daire, Eds) *The Culture of Trans-Uralian Cattle and Horse Breeders on the Turn of an Era; The Gayevo Burial Ground of the Sargat Community: Anthropological Research.* Ekaterinburg, Russia: Ekaterinburg Publishers, pp. 156-172. (Russian).

Levine, M. & A. Kislenko (1997). New eneolithic and early Bronze Age radiocarbon dates for north Kazakhstan and south Siberia. *Cambridge Archaeological Journal* **7**: 297-300.

Lincoln, B. (1979). The Hellhound. *Journal of Indo-European Studies* **7**: 273-285.

MacQueen, J.G. (1986). *The Hittites and Their Contemporaries in Asia Minor.* London: Thames & Hudson.

Mair, V. (1998). *Canine Conundrums: Eurasian Dog Ancestor Myths in Historical and Ethnic Perspective.* Philadelphia, PA: University of Pennsylvania Sino-Platonic Papers **87**.

Malandra, W. (1983). *An Introduction to Ancient Iranian Religion.* Minneapolis: University of Minnesota Press.

Mallory, J. (1981). The ritual treatment of the horse in the Early Kurgan Tradition. *Journal of Indo-European Studies* **9**: 205-226.

Mallory, J. (1996). *In Search of the Indo-Europeans.* London: Thames & Hudson.

Maltby, M. (1985). Patterns in faunal assemblage variability. In (G. Barker & C. Gamble, Eds)

*Beyond Domestication in Prehistoric Europe*. London, UK: Academic Press, pp. 33-74.
McCormick, F. (1985/6). Animal bones from prehistoric burials. *Journal of Irish Archaeology* **III**: 37-48.
Merrifield, R. (1987). *The Archaeology of Ritual and Magic*. New York, NY: New Amsterdam Books.
Morey, D.F. & M. D. Wiant. (1992). Early Holocene domestic dog burials from the North American Midwest. *Current Anthropology* **33**: 224-229.
Mongait, A. (1959). *Archaeology in the USSR*. Moscow, Russia: Foreign Languages Publishing House.
Moshkova, M. (1995). Late Sarmatian Culture. In (J. Davis-Kimball, V. Bashilov, & L. Yablonsky, Eds) *Nomads of the Eurasian Steppes in the Early Iron Age*. Berkeley, CA: Zinat Press, pp. 149-163.
O'Flaherty, W. 1981. *The Rig Veda*. London, UK: Penguin Books.
Olsen, S. L. (1994). Exploitation of mammals at the early Bronze Age site of West Row Fen (Mildenhall 165), Suffolk, England. *Annals of Carnegie Museum* **63**: 115-153.
Olsen, S. L. (1996). Prehistoric adaptation to the Kazak steppes. In (G. Afanas'ev, S. Cleuziou, J. Lukacs & M. Tosi, Eds) Colloquia of the XIII International Congress of Prehistoric and Protohistoric Sciences: The Prehistory of Asia and Oceania. Forlì, Italy: A.B.A.C.O. Edizioni, pp. 49-60.
Perkins, D. & Daly, P. (1968). The potential of faunal analysis: An investigation of the faunal remains from Suberde, Turkey. *Scientific American* **219**: 96-106.
Piggott, S. (1962). Head and hoofs. *Antiquity* **36**: 110-118.
Piotrovski, B., Galanina, L. & Grach, N. (1987). *Scythian Art*. Oxford, UK: Phaidon.
Rikushina, G. & Zaibert, V. (1984). Preliminary report about human skeletal remains of the Eneolithic settlement of Botai. In *The Bronze Age of the Uralo-Irtish Region*. Chelyabinsk, Russia: Bashkir State University, pp. 121-135. (Russian).
Rudenko, S. (1970). *Frozen Tombs of Siberia: The Pazyryk Burials of Iron-Age Horsemen*. Berkeley, CA: University of California Press (Translated by M. W. Thompson).
Serpell, J. (1995). From paragon to pariah: Some reflections on human attitudes to dogs. In (J. Serpell, Ed) *The Domestic Dog: Its Evolution, Behaviour, and Interactions with People*. Cambridge, UK: Cambridge University Press, pp. 245-256.
Simoons, F. (1994). *Eat Not This Flesh: Food Avoidances from Prehistory to the Present*. Madison, WI: University of Wisconsin Press.
Tchernov, E. & Valla, F. (1997). Two new dogs, and other Natufian dogs, from the southern Levant. *Journal of Archaeological Science* **24**: 65-95.
Telegin, D. (1986). *Dereivka: A Settlement and Cemetery of Copper Age Horse Keepers on the Middle Dnieper*. Oxford, UK: BAR International Series **287**.
Titcomb, M. (1969). *Dog and Man in the Ancient Pacific*. Honolulu, HI: B. P. Bishop Museum Special Publication **59**.
Vasiliev, I., Kuznetsov, P. & Semenova, A. (1994). *Potapovski Kurgan Burials of the Indo-Iranian Tribe on the Volga*. Samara, Russia: Samara Institute. (Russian).
White, D. (1991). *Myths of the Dog-Man*. Chicago, IL: University of Chicago Press.
Willis, M. B. (1995). Genetic aspects of dog behaviour with particular reference to working ability. In (J. Serpell, Ed) *The Domestic Dog: Its Evolution, Behaviour and Interactions with People*. Cambridge, UK: Cambridge University Press, pp. 51-64.
Wilson, B. (1992). Considerations for the identification of ritual deposits of animal bones in Iron Age pits. *International Journal of Osteoarchaeology* **2**: 341-349.
Wilson, G. (1978). *The Horse and Dog in Hidatsa Culture*. [Reprinted from: Anthropological Papers of The American Museum of Natural History, New York, 1924.] Lincoln, NE: J & L Reprint Co., Reprints in Anthropology **10.**
Zaibert, V. (1993) *The Eneolithic of the Ural-Irtysh Interfluve*. Petropavlovsk, Kazakhstan: A. X. Margulana Institute of Archaeology, Kazakhstan Academy of Sciences. (Russian).
Zeuner, F. (1963). *A History of Domesticated Animals*. New York, NY: Harper & Row.
Zolotarev, A. (1938). The ancient culture of north Asia. *American Anthropologist*, n.s. 40: 13-23.

# EARLY ARCHAIC DOMESTIC DOGS FROM WESTERN IDAHO, USA

Robert M. Yohe II
Department of Sociology & Anthropology (as of Aug.1/99)
California State University
9001 Stockdale Hwy.,
Bakersfield, CA
93311
USA

Max G. Pavesic
Department of Anthropology
Boise State University
Boise, Idaho 83725

**Key words:** dog, Idaho, archaeozoology, burials, Archaic, interments

**Abstract**

During excavations at the Braden site (10-WN-117) near Weiser, Idaho, in the late 1960s, field crews from various institutions unearthed part of an early Archaic cemetery, recently re-dated to 6,600BP Associated with a dozen human burials at this site were the remains of two dogs (*Canis familiaris*). This study presents the descriptive and osteological analysis of the canid remains and a comparison of these data with other dogs that have been encountered in the American West.

## Introduction

Dog burials are relatively uncommon in the early archaeological record of North America (Kerber, 1997; Schwartz, 1997). This is especially true in the American West, where domestic dog remains, in general, are infrequently reported and intentional dog interments rarer still. A recent discussion of prehistoric dog remains from the Great Basin presented by Lupo & Janetski (1994) is the most comprehensive comparative analysis of domestic canids for the interior West to date. Prehistoric dog interments are known from the Northwest Coast, some associated with human burials (Cybluski, 1992), though many dog remains recovered from numerous sites on Vancouver Island and coastal British Columbia are contextually problematic (see Crockford, 1997). Only a few dog burials from the southern Plateau are known (Haag, 1983; Switzer, 1992). Evidence for at least one intentional dog burial has been described from an Early Horizon site (CA-SJo-68) in coastal California (Haag & Heizer, 1953), and dog sacrifice and burial with owners is known from the ethnographic record from the northern coastal area (Elsasser, 1978; Lapena, 1978), the Gabrieleno of southern California (Bean & Smith, 1978), and the Yokuts of the southern San Joaquin Valley (Wallace, 1978). Ethnographic dog sacrifice at death of owner is also known from the Plateau (Switzer, 1992). Such practices have a long history world-wide (Schwartz, 1997).

More than 30 years ago, two dog burials were exhumed from what was reported as a "mass grave" consisting of seven human burials surrounded by at least five additional human interments on the Braden farm in Weiser, western Idaho (Butler, 1980; Harten, 1980). A recent direct radiocarbon assessment of *Canis familiaris* bone from this feature suggests that the interments took place nearly 6,600 years ago (Yohe & Pavesic, 1996), making these canid remains among the oldest in western North America, and perhaps the earliest known intentional dog burials directly associated with human burials in the Americas. The report that follows is a descriptive analysis of these remains and a comparison of the Braden dogs with others in western North America, particularly the Great Basin and Northwest.

## The Braden site

The Braden site (10WN117) is located on a long, undulating dune above the modern course of the northward flowing Snake River in western Idaho. The site was accidentally discovered during a farm irrigation project in 1908 but remained unknown to archaeologists until the late 1960s and 1970s when a series of excavation samples were secured (Pavesic, 1985). Since its initial discovery, the deposit has yielded a spectacular array of artifactual materials, later proving to be associated with human skeletal remains. Controlled excavations have yielded a

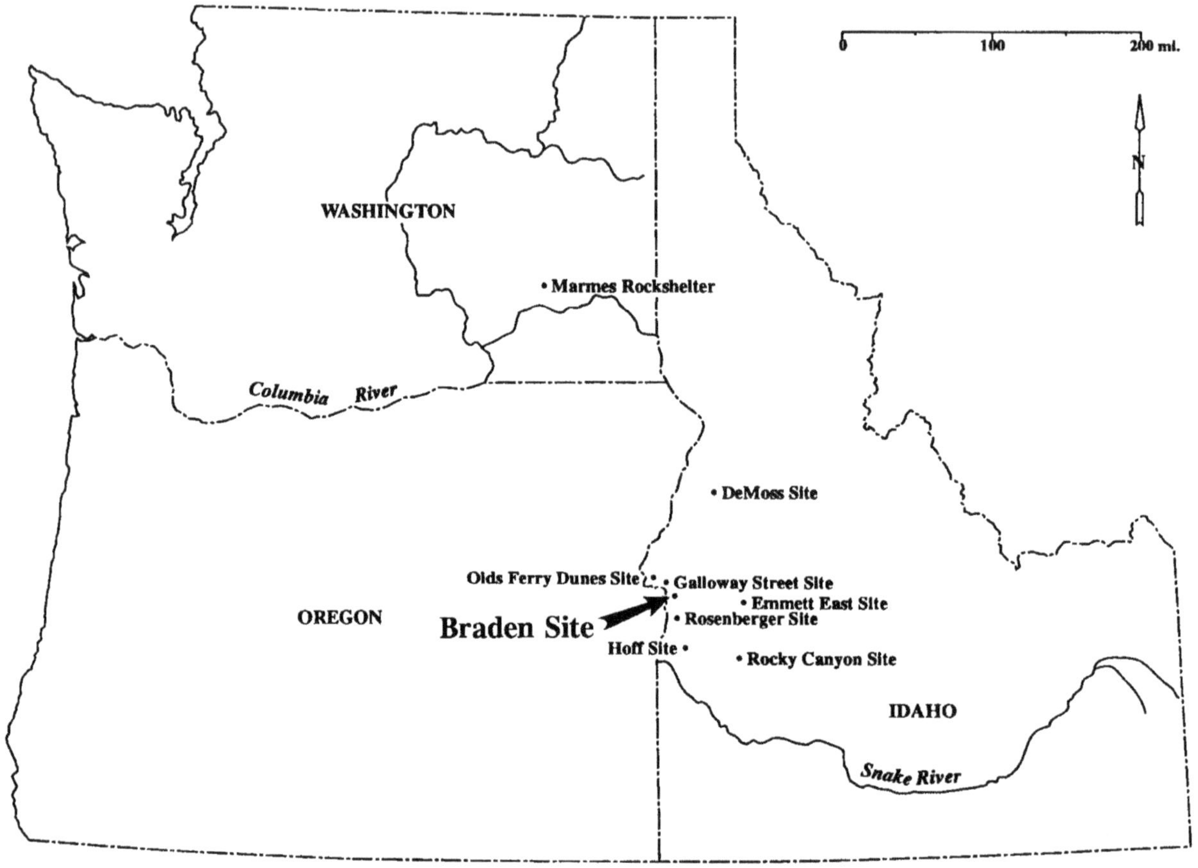


***Figure 1*** *The Braden site (10-WN-117) and related western Idaho Archaic burial localities.*

minimum sample of a dozen human burials of mixed age and sex, along with two domesticated dogs (*Canis familiaris*). Excavations have also revealed caches of funerary objects, the largest of which yielded 165 formed artifacts, including 153 obsidian preforms or blanks (Butler, 1980: 123, Fig. 9, upper).

At Braden, the primary feature identified to date has been the exposure of seven human burials and two domestic dogs with artifacts in what was originally interpreted as a "mass burial" episode (Butler, 1980; Harten, 1975, 1980). Subsequent research has revealed that this feature is the result of multiple prehistoric interments and not a single event (Pavesic, 1979, 1985). The initial age of 5,790±120 radio carbon years BP (WSU-1487) obtained from this feature was derived from a composite sample of bones collected from several individual burials along with miscellaneous bone fragments (Harten, 1975: 72-73). During the current study, a sample of dog bone (metapodials) was submitted for analysis and provided a date of 6,590±90BP (Beta-90555) (Yohe & Pavesic, 1996). Of the two recovered dog burials, one clearly represents a primary burial, whereas the second had been disturbed by later human interments.

The Braden site is the keystone in the reconstruction of the *Western Idaho Archaic Burial Complex*, which includes a total of eight culturally and ritually related prehistoric sites in western Idaho along with the Cascade phase burials from the Marmes Rockshelter in southwestern Washington (Green *et al.*, 1986; Pavesic, 1985; Pavesic *et al.*, 1993) (Fig. 1). The "complex" assemblage includes large turkey-tail points and cache blades, side-notched projectile points, Cascade points, triangular obsidian preforms, *Olivella* shell beads, pipes, and micaceous schist shaft straighteners. The artifacts, freshly produced specifically for inclusion with the burials (Madsen, 1997), were placed among flexed or semi-flexed inhumations, liberally sprinkled with red ochre and usually located in high sandy knolls overlooking streams. Overall, the western Idaho pattern shares many traits common to North American burial complexes, including interments segregated from habitations, purposely selected locations, ritual treatment of the dead, and distinctive, special use artifacts (Pavesic, 1992; Penney, 1993).

## The "mass grave" and the Braden dogs

Despite the fact that the dog remains from the Braden site were exhumed more than three decades ago, they had not been subjected to formal analysis prior to this study. Although the postcranial specimens were recognized as canid during the analysis of the human skeletal material (Harten, 1975: 48), the skulls had been misplaced and did not resurface for several years. Another problem that complicated the analysis of the skeletons until recently was the question of direct context. With the exception of the dog skulls mentioned in one excavator's notes (Harten, 1975: 13), the remains were not always clearly identified as canid in the field (frequently only as "animal" or "cat") and only some of the "animal" specimens were mapped in with the human remains. Only through careful laboratory analysis and the study of several field photographs of the burial feature was it possible to reconstruct the specific provenience of the dog specimens.

The "mass grave" feature containing the human/dog interments was excavated by student volunteers from Boise State College (later to become Boise State University) in 1967. The Boise State effort was the first systematic investigation at the Braden site, which had been dug into previously by non-archaeologists. At the beginning of the 1967 excavation, a grid was established consisting of an 8 x 10 m rectangle divided into twenty 2 x 2 m units or "pits." Of this number, twelve pits were excavated completely during that season. A total of twelve human burials and the two dogs were recovered, six of the interments and the two dogs clustered in the "mass grave" (designated as Feature 3) in units 13, 14, 18, and 19 at a depth of approximately one m. Idaho State University, collaborating with Boise State College, returned to the site for an additional season in 1968, concentrating on digging immediately north of the previous year's excavation (Butler, 1980). Several biface caches, a partial child burial, assorted human bone fragments, and numerous flaked stone artifacts were recovered during this season, but nothing approached the complex burial features encountered the year before.

Pavesic (1979) returned to the Braden site in 1978 to verify previously reported stratigraphic associations in Horizon II, the artifactual and burial bearing lens (Butler, 1980: 121). This work clearly demonstrated the reported "mass burial" was the result of multiple interment episodes rather than a single event. Additionally, a close examination of previously published photographs (Harten, 1975: Figs. 2 and 3), revealed "jumbled postcranial skeletal remains, a few bone fragments in a perpendicular position, and several chipped stone blades on edge" (Pavesic, 1985: 64), again suggesting prehistoric disturbance.

Assisted by photographs and excavation maps of Feature 3, it was possible to locate several dog remains, some of which could be matched to specific canid elements in the collection (Fig. 2). The ability to recognize individual elements provided significant aid in the differentiation between the two dogs (designated Dog 1 and Dog 2). Since the more complete dog skeleton (Dog 1) proved to be slightly larger than the other, more fragmentary individual (Dog 2), it was possible to sort many of the postcranial remains based on size.

Figure 3 represents the reconstruction of Feature 3 based on the photographs and feature maps. The configuration of the various interments suggests only one primary inhumation (Burial 2) and five secondary bundle-type burials. As shown in Figure 3, Dog 1 (on the left) is

***Figure 2*** *Photo, view northeast, of Feature 3 at the Braden site (10-WN-117) with arrow showing location of dog remains.*

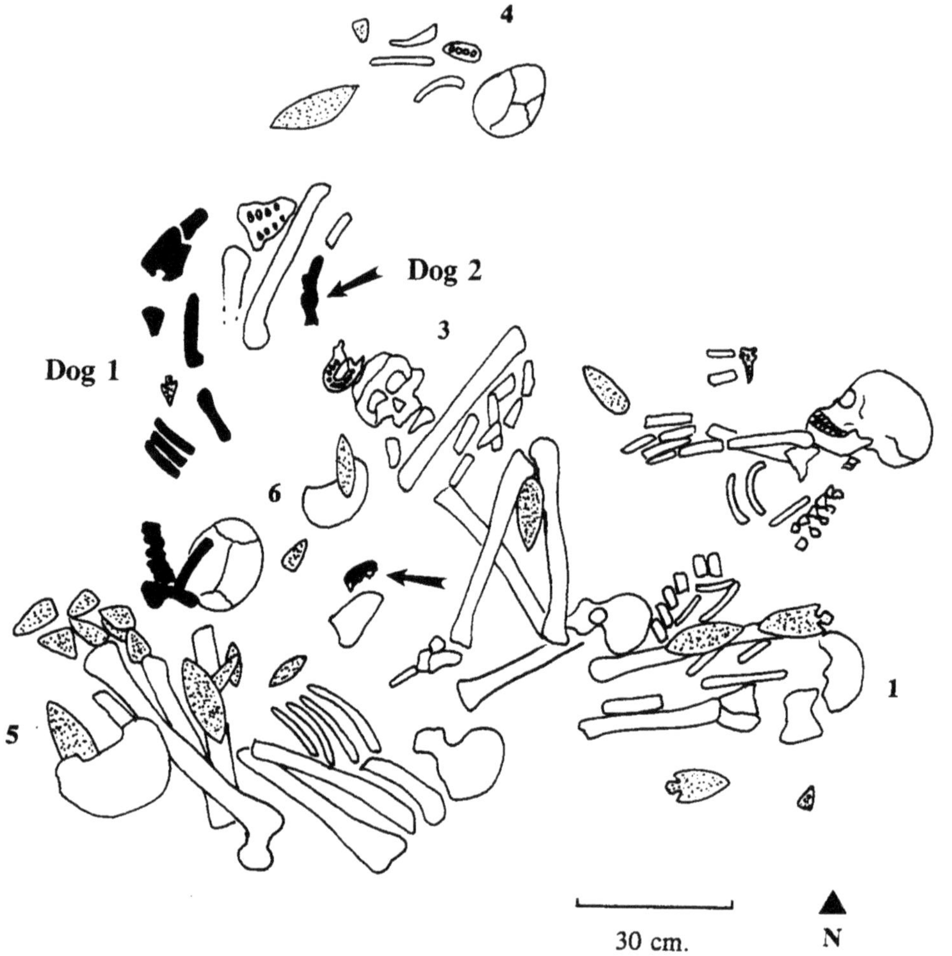


***Figure 3*** *Plan view of multiple interment feature showing postioning of canid remains with respect to Burial 2 and five subsequent secondary inhumations. Numbers represent field-numbered human burials; stippled objects are flaked stone artifacts.*

represented fairly completely *in situ*, while only two elements (an innominate and a mandible) were noted in place for Dog 2 in either the photograph or on the feature record. In attempting to reconstruct the burial sequence of Feature 3, it appears that both dogs may have been buried with Burial 2 (since at least Dog 1 appears to be a primary interment), with Dog 1 oriented with its head to the north and Dog 2 with its head to the south (if the position of the innominate and maxilla for Dog 2 is correct and not the result of jumbling). The prehistoric excavation of Burials 3 and 6 appears to have intruded upon and displaced much of Dog 2, as can be seen in the figure. This likely explains the more incomplete inventory of Dog 2 elements as well as the greater degree of fragmentation.

## Canid skeletal analysis

As noted above, sorting of the postcranial remains of the two dogs was based on both the field photographs and the slight size difference between Dogs 1 and 2. Certain elements could not be specifically attributed to one animal or the other (some rib fragments, tarsals, carpals, metapodials, phalanges), but most postcranial elements could. Most of the ribs are believed to belong to Dog 1, since "dog ribs" were included on one of the level records for Feature 3 for this dog. Figure 4 provides a visual inventory of the fragmentary elements present for each individual. Ironically, the most complete of the two crania belongs to Dog 2 ; the skull of the more complete skeleton of Dog 1 is very fragmented (Figs. 5 & 6). It is possible that the ancient excavation of the grave for Burial 6 might have damaged and displaced the right front portion of the skull of Dog 1 in addition to severely disturbing most of the postcranial skeleton of Dog 2.

The condition of the bones varies from good to fair. Many elements were covered with a thin coating of calcium carbonate which had been manually removed during laboratory cleaning when the specimens were originally catalogued. Unfortunately, this left many nicks and grooves in the bone that were observed when the specimens were examined for intentional cutting or butchering marks. However, the fresh appearance of this damage made it obvious in most cases that it was not the result of a pre-interment event. No observable butchering or non-recent cut marks were seen on any of the bones. The bones of the Dog 2 skull were slightly distorted due to post-depositional processes and had been coated with white glue (*Elmer's*) in the field. In fact, many of the elements appear to have been

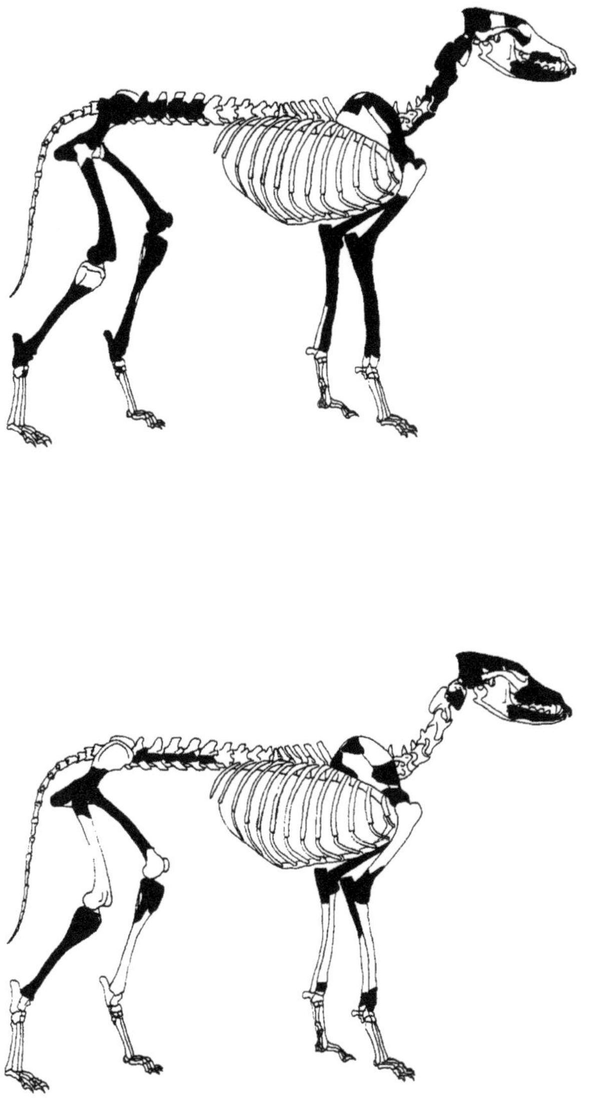

***Figure 4*** *Visual representation of elements preserved from Dogs 1 (top) and 2 (bottom), Braden site.*

***Figure 5*** *Left and right lateral views of representation of cranial and mandibular fragments recovered from Dog 1.*

coated with this substance, presumably as a preservative.

Both dogs are adult animals with all epiphyses completely closed. The tooth wear, both maxillary and mandibular, for Dog 1 is pronounced, whereas that for Dog 2 is negligible (Fig. 7). This suggests that Dog 1 may have been considerably older than Dog 2. No gross skeletal pathologies were apparent during the examination of the bone.

Measurements of all the canid skeletal elements were made in accordance with the standards of von den Driesch (1976), using digital sliding calipers and an osteometric board. Table 1 lists all of the cranial measurements for both individuals; Table 2 mandibular measurements; and Table 3 the postcranial metrical values. Estimated measurements were not made for incomplete long bones, but in cases where bilateral symmetry allowed (ie the frontal bone), such estimates were made.

| Attribute | Dog 1 | Dog 2 |
|---|---|---|
| Length of carnassial ($P^4$) | 19.2 | 19.9 |
| Greatest breadth of carnassial ($P^4$) | 10.2 | 10.3 |
| Length of carnassial alveolus | -- | 18.1 |
| Length of $M^1$ | -- | 13.3 |
| Breadth of $M^1$ | -- | 15.0 |
| Length of $M^2$ | -- | 5.9 |
| Breadth of $M^2$ | -- | 9.6 |
| Greatest diameter of auditory bulla | -- | 27.3 |
| Greatest mastoid breadth | 64.0 (est.) | 62.9 |
| Breadth dorsal to the external auditory meatus | -- | 58.4 |
| Greatest breadth of the foramen magnum | 18.0 | -- |
| Height of foramen magnum | 14.0 | 15.6 |
| Frontal breadth | 50.0 | 48.0 (est.) |
| Skull height | -- | 55.0 |
| Height of the occipital triangle | 43.0 | 40.0 (est) |

***Table 1*** *Cranial measurements* of Canis familiaris *from the Braden site (mm)*

| Attribute | Dog 1 | Dog 2 |
|---|---|---|
| Length of premolar row, $P_2$ - $P_4$ | 33.1 | -- |
| Length of carnassial ($M_1$) | 21.4 | 22.8 |
| Breadth of carnassial | 9.1 | 9.2 |
| Length of $M_2$ | 9.1 | 8.4 |
| Breadth of $M_2$ | 6.6 | 6.5 |
| Greatest thickness of body of jaw | 13.2 | -- |
| Height of mandible behind $M_1$ | 22.9 | -- |
| Height of mandible between $P_2$ - $P_4$ | 20.6 | 20.6 |

***Table 2*** *Mandibular measurements of* Canis familiaris *from the Braden site (mm)*

| Attribute | Dog 1 | Dog 2 |
|---|---|---|
| **Scapula** | | |
| Height | 127.8 (r) | -- |
| Smallest length of collum scapulae | 23.5 (r) | -- |
| Greatest length of the glenoid process | 29.6 (r) | -- |
| Length of glenoid cavity | 21.6 (r) | 20.1 |
| Breadth of glenoid cavity | 18.3 | -- |
| **Humerus** | | |
| Greatest length | 162.0 (l) | -- |
| Greatest length from caput | 159.0 (l) | -- |
| Depth of the proximal end | 40.9 (l) | -- |
| Smallest breadth of diaphysis | 11.9 (l) | 11.5 (r) |
| Greatest breadth of distal end | 31.9 (l) | 29.3 (l) |
| **Radius** | | |
| Greatest length | 156.0 (l) | -- |
| Greatest breadth of proximal end | 17.9 | 17.1 |
| Smallest breadth of diaphysis | 11.5 | -- |
| Greatest breadth of distal end | 24.8 | -- |
| **Ulna** | | |
| Depth across the processus anconaeus | 25.2 | 24.0 |
| Smallest depth of olecranon | 20.9 | 20.5 |
| Greatest breadth across coronoid process | 17.2 (r) | -- |
| **Pelvis** | | |
| Greatest length of one half | 150.0 (r) | -- |
| Length of acetabulum on rim | 20.5 | 19.4 |
| Smallest height of shaft of ilium | 8.3 | 8.1 |
| Smallest breadth of shaft of ilium | 19.6 | 19.0 |
| Inner length of obturator foramen | 30.0 (r) | -- |
| **Femur** | | |
| Greatest breadth of the proximal end | 37.4 | -- |
| Greatest depth of caput femorus | 17.8 (r) | 17.6 (r) |
| Smallest breadth of diaphysis | 11.6 | 11.8 |
| Greatest breadth of the distal end | -- | 32.4 |
| **Tibia** | | |
| Greatest length | -- | 160.0 |
| Greatest breadth of proximal end | 34.0 (l) | 33.3 (r) |
| Smallest breath of diaphysis | 12.1 (l) | 11.4 (r) |
| Greatest breadth of distal end | 22.5 | 20.8 |
| Greatest depth of distal end | 16.5 | 15.3 |

***Table 3*** *Postcranial measurements of* Canis familiaris *from the Braden site (mm). Right or left of paired elements measured indicated as (r) or (l)*

## Results and Discussion

Based on the above analysis, the dogs from Braden seem to be two very different individuals. Dog 1 is slightly larger, much older, and maybe longer-faced than Dog 2. Shoulder height (SH) was calculated employing the formulae proposed by Harcourt (1974) for both dogs using the humerus for Dog 1 and the tibia for Dog 2. When the value for Dog 1 humeral length (162.0 mm) is used in the equation, the resulting estimated shoulder height is 52.9 cm. When the bones of two limbs are present, such as the humerus and radius, a more accurate determination of shoulder height can be assessed using the combined-limb formula found in Harcourt (1974). Only Dog 1 has two complete single-limb elements, the left humerus and radius, producing an estimated shoulder height of 52.0 cm. For Dog 2, the only complete limb bone is the tibia. Using the tibial formula presented by Harcourt (1974), Dog 2 has an estimated shoulder height is 47.7 cm. This suggests a 10% size difference between the two animals, which is not inconsistent with the apparent size differential seen in the elements of the these animals (Fig. 8).

Clark (1995) has proposed a method of determining the body length of a dog by using total pelvic length (PL) with the a regression formula ( BL(cm)=0.47 X GL[PL(mm)]-15.7). The total pelvic length of Dog 1 is 150.0 mm, providing a total body length value of 54.8 cm. According to Clark (1995), dogs with shoulder heights less than estimated body lengths are classified as "long bodied" dogs, which seems to be the case for Dog 1.

The sex of both dogs is very tentatively determined as male. No baculae were noted in the bones from Feature 3, which might suggest that the dogs are both female, but the fact that the faunal assemblage from the site has not been analyzed (and hence, it may still be mixed with other bone) and the disturbed nature of the Dog 2 interment could account for the missing penis bones. Observations of certain non-metric, sexually dimorphic traits, such as development of the sagittal and nuchal crests (especially the flaring of the nuchal crest in the Braden specimens), suggest both dogs may be male, though the margin of error in such cases is high (Shigehara *et al.*, 1997).

The relative age determination of the dogs is based on tooth wear: the occlusal attrition of Dog 1 is significantly greater than Dog 2. Recent studies of $M_1$ attrition in a large sample of known-age late Iron Age dogs in France suggest a series of age-specific wear stages (Horard, 1997; Horard-Herbin, this volume). Significant wear on the entoconid, metaconid, and hypoconid may suggest an age greater than 48 months (Horard-Herbin, this volume), which is consistent with the wear observed in Dog 1. Using these same criteria, Dog 2 could be as young as 24 months, since there is no notable wear on the maxillary or mandibular teeth. This is consistent with the epiphyseal data. However, such age assessments would be contingent on the assumption that attrition rates in Archaic Idaho domestic canids are comparable to Iron Age French dogs, which is non-demonstrable at this point. Therefore, the more parsimonious interpretation is that Dog 1 is considerably older than Dog 2, but to what degree is indeterminate at this time.

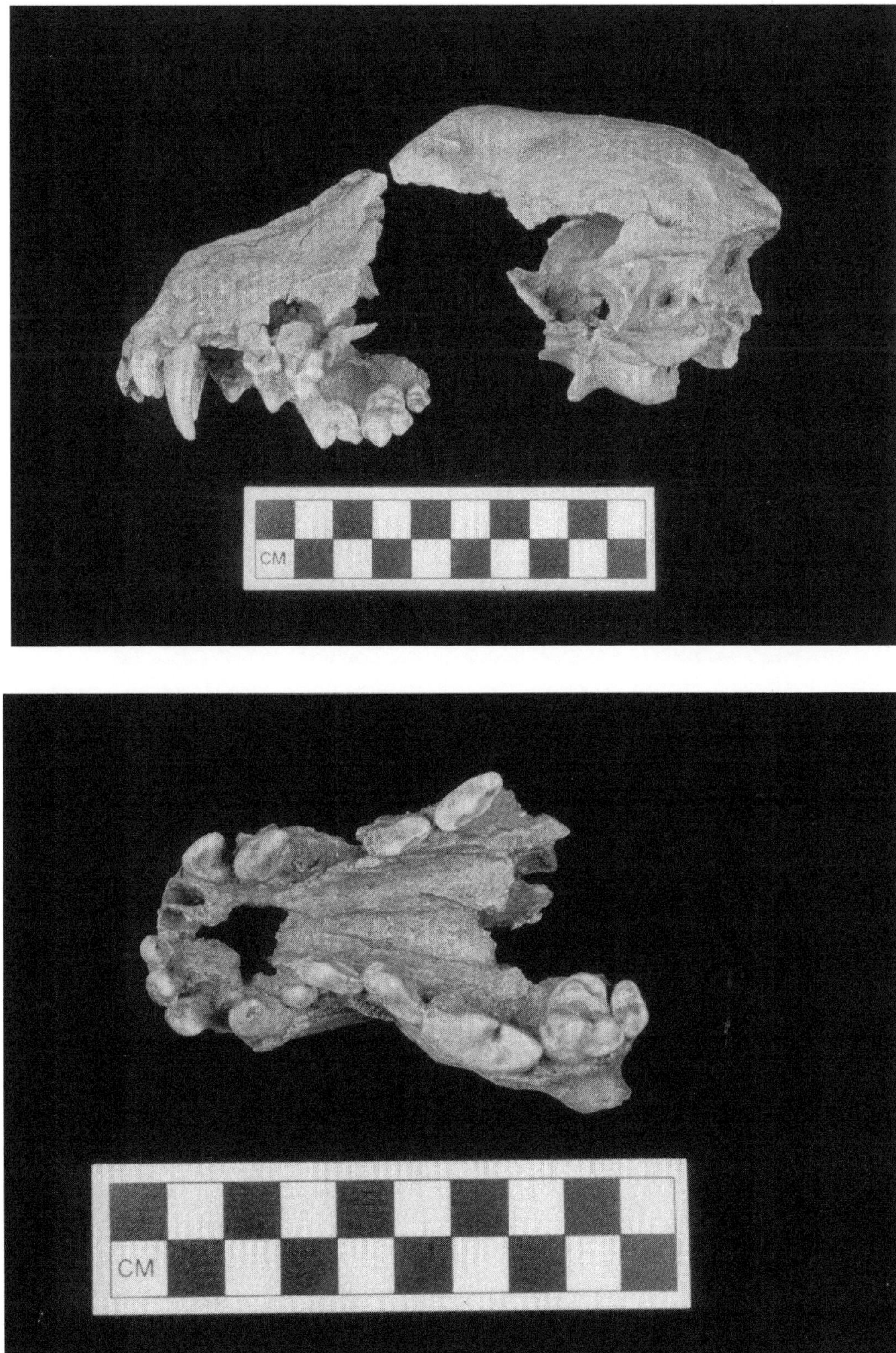


***Figure 6*** *Skull of Dog 2, left lateral view (top) and palatal view (bottom).*

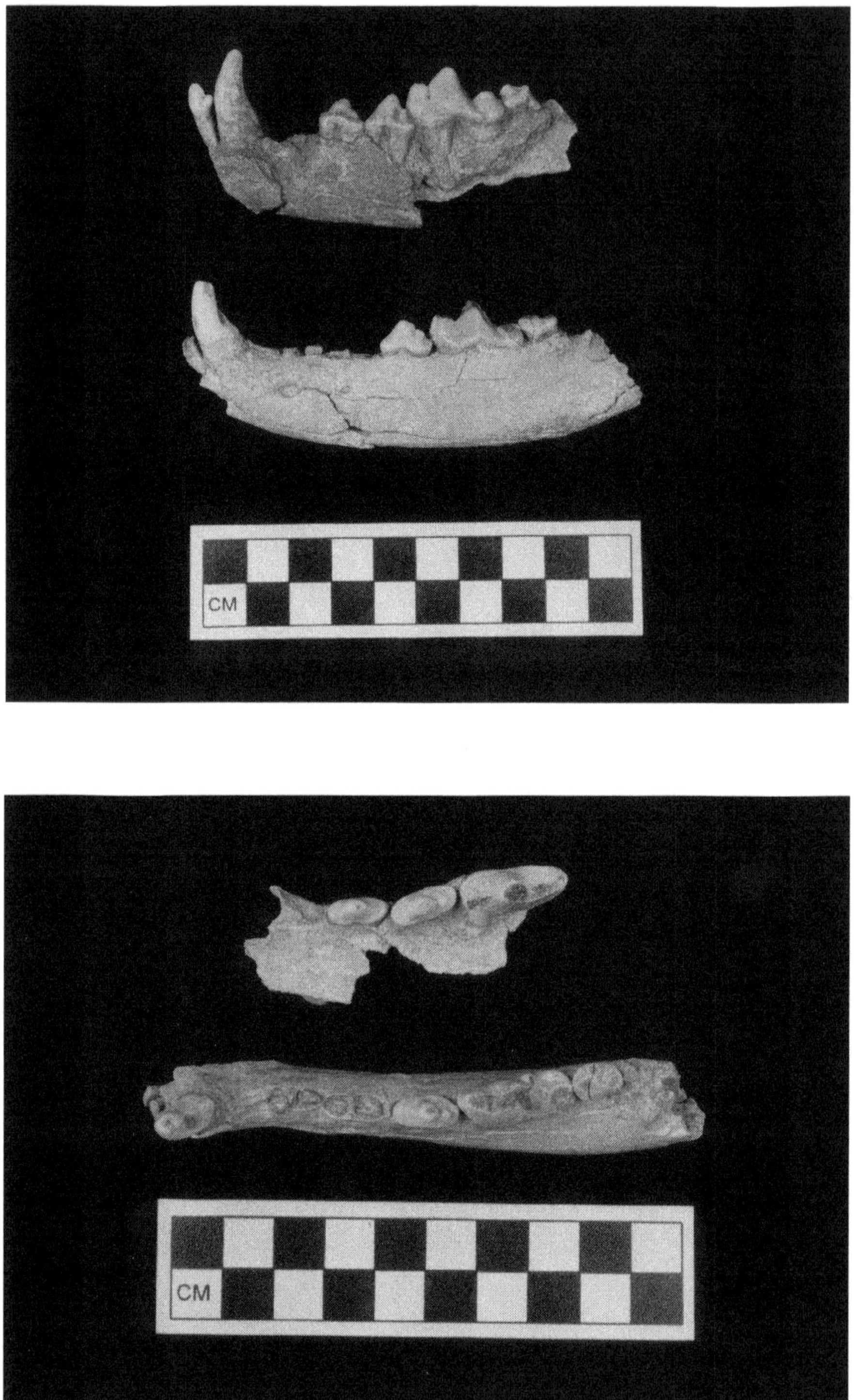


***Figure*** *7 Tooth wear for Dogs 1 and 2, Braden. Top photograph, occlusal view of left mandible and right maxilla fragment of Dog 1. Bottom photograph, lateral views of left mandible fragments of Dog 2 (upper) and Dog 1 (lower).*

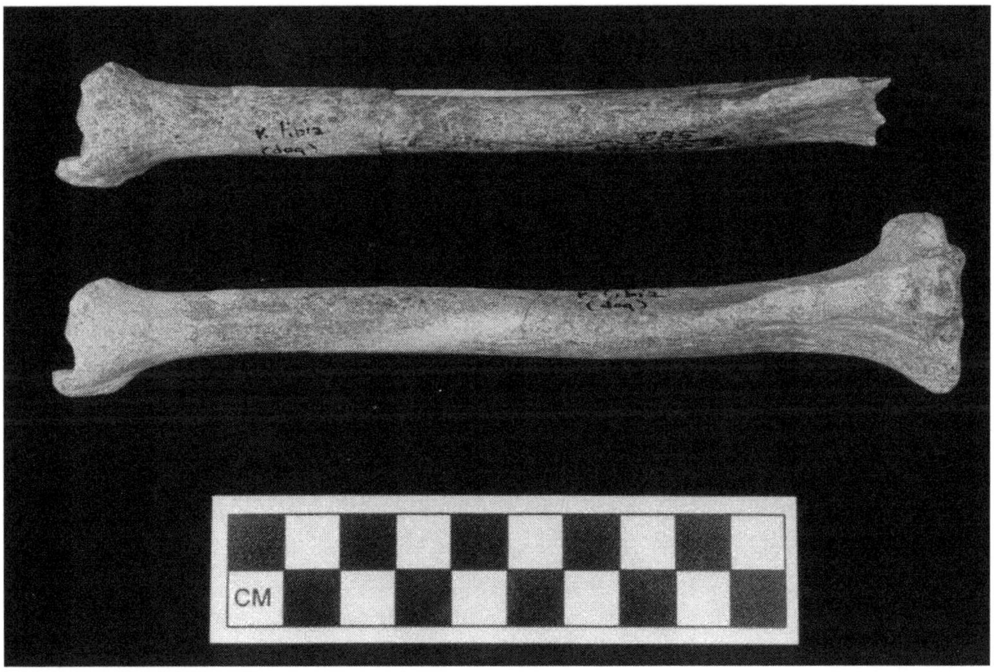


***Figure 8** Size differential between postcranial elements of Dogs 1 and 2, Braden.*

Both dogs have a congenital absence of $P_1$, a condition that was observed in 82% of a large sample of prehistoric dog mandibles (n = 68) from the Northwest Coast (Crockford, 1997) and has been commonly observed elsewhere (Allen, 1920, Haag, 1948, but see Switzer, 1992). Interestingly, the larger dog has slightly smaller teeth than Dog 2, even though it has a more massive mandible.

There was no apparent cause of death based on the analysis skeletal material, although the dogs were likely sacrificed at the death of their owner, which we have posited is the primary inhumation designated Burial 2. Since dogs have obviously served as a food source to Native Americans in the past (*cf* Snyder, 1991), the bones of both dogs were scrutinized for apparent butchering marks, even though it seemed very likely, given the positioning of Dog 1, that it was buried intact. As noted above, no butchering marks or cuts indicative of sacrifice were observed.

In summary, the Braden dogs appear to represent the remains of two adult animals, one much older (Dog 1) than the other and about 10% larger. Each stood about 50 cm. at the shoulder, about the size of a modern Dalmatian. Both dogs may be male, and each exhibit a congenital absence of $P_1$, but possess no apparent skeletal pathologies.

## Comparison with other prehistoric dogs: Northwest Coast; Interior Plateau; Great Basin

When the Braden dogs are compared with other prehistoric dog material from the Northwest Coast, Plateau, and Great Basin, they seem to fall well within the upper size range for all dogs from these areas. Figure 9 is a comparison of maxillary and mandibular carnassial lengths for the Braden dogs, mean values of these measurements for nine dogs from the Wildcat Canyon site (35-Gm-9) of north-central Oregon (Haag, 1983), the mean values of a large sample of Northwest Coast dogs (Crockford, 1997), three specimens from the Birch Creek Valley (Jaguar Cave) in Idaho (Lawrence, 1967, 1968), and a large dog from a burial at the Mosida site (42-Ut-808) in Utah (Janetski *et al.*, 1992). The same holds true when postcranial skeletal elements are compared (Fig. 10) among Braden, Wildcat Canyon, and Crockford's (1997) 20-site Northwest Coast sample. The Utah specimen, one of the larger known dogs from western North America (see Lupo & Janetski, 1994, Table 3; Crockford, 1997, Table 6-3), is only slightly larger than Dog 1, further supporting the contention that the Braden animals are above average in size. The Mosida site specimen is also one of the older dogs (4,720 ± 20BP, see Janetski *et al.*, 1994) known from the west, but still almost 2,000 younger than the Braden dogs.

## The antiquity of the Braden dogs

At nearly 6,600BP, the Braden dogs appear to be the oldest deliberate dog burials associated with human remains from the Northwest and possibly western North America. This is the second time that Idaho has made a claim for ancient canids, the first time being with the dog remains from Jaguar Cave in the Birch Creek Valley of eastern Idaho (Lawrence, 1967, 1968). These dog remains, consisting of two paired mandibles found in the same stratum as a "hearth" dated to 10,370BP and a maxillary fragment slightly stratigraphically superior (Lawrence, 1967), were later directly dated with AMS at 3,220 ± 80BP (OxA-922)

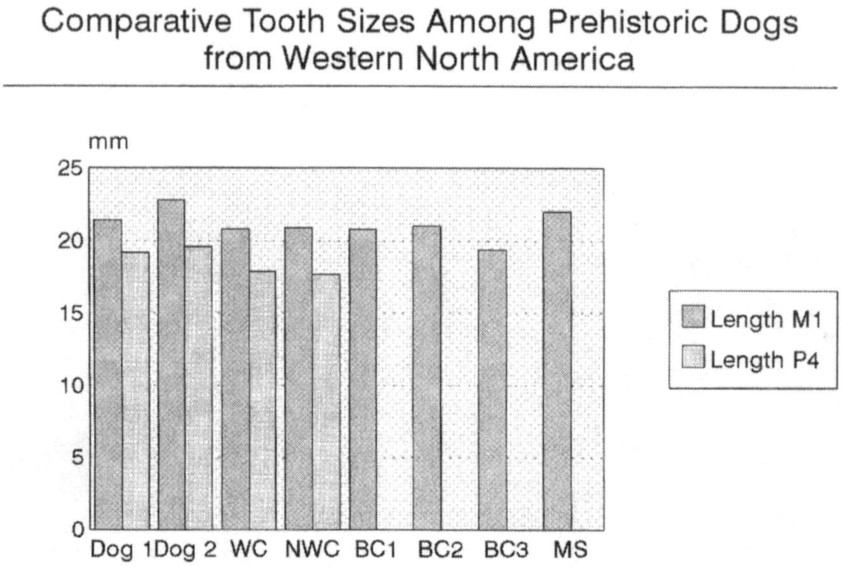


***Figure 9*** *Tooth size comparisons of prehistoric domestic dogs from western North America: Braden (Dog 1 and Dog 2); Wildcat Canyon (WC) (Haag, 1983:202; mean values for $P^4$ [n = 7] and $M_1$ [n = 7]); Northwest Coast of North America (NWC)(Crockford, 1997:Tables 4-2b and 52-b; mean values for $P^4$ [n = 17] and $M_1$ [n = 36]); Jaguar Cave, Birch Creek Valley, Idaho (BC1, BC2, BC3)(Lawrence, 1967, 1968); the Mosida site, north-central Utah (Janetski et al., 1992; Figure 7).*

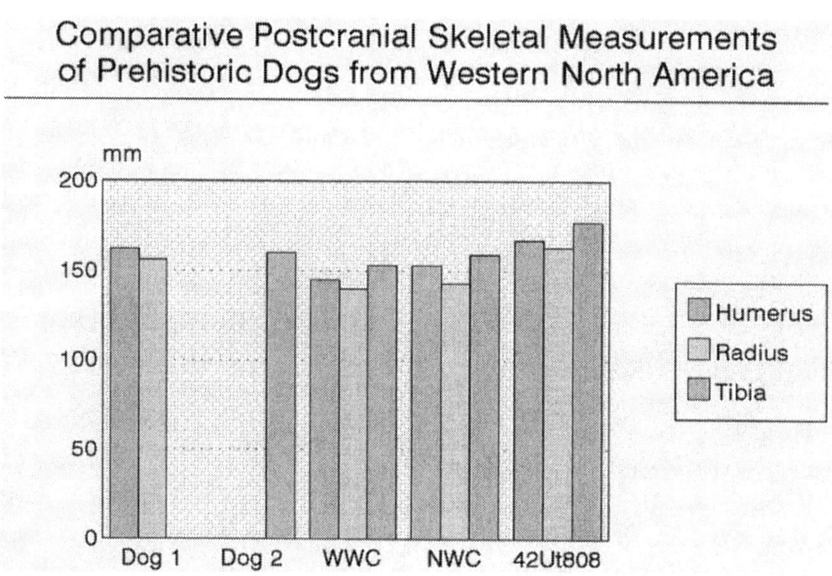


***Figure 10*** *Comparative postcranial greatest length (GL) measurements from prehistoric domestic dogs from western North America: Braden (Dog 1 and 2); Wildcat Canyon (WCC)(Haag, 1983:202; mean values for humerus [n = 8], radius [n = 8], and tibia [n = 7]); Northwest Coast of North America (Crockford, 1997:Tables 6.3, 6.5, & 7.4; mean for humerus (n = 29), radius [n = 21], and tibia [n = 24); Mosida site (42Ut808) (Lupo & Janetski 1994:211[Table 3]).*

(mandible), and 940 ± 80BP (OxA-923) (maxilla) (Gowlett *et al.*, 1987). The oldest known dogs in the Great Basin were identified by Grayson (1988) from the recent reanalysis of vertebrate fauna from Danger Cave (42-T0-13) in northwestern Utah. Represented by mandibles and skull fragments, these dogs date to between 9,000 and 10,000BP. Also from northwestern Utah are dog remains (mandibles) dating from 7,450-8,000BP from Hogup Cave (Haag, 1970) recovered from general cave fill. Dog remains dating to 6,500BP also have been claimed for Ventana Cave in southern Arizona (Pferd, 1987). However, none of these early domesticated canids were recovered from burial contexts. Other than the Braden dogs, only the dog from the Mosida burial can be demonstrated to be a deliberate dog interment from the early-mid Archaic (4,700BP) (Janetski *et al.*, 1992; Lupo & Janetski, 1994).

Earlier deliberate dog interments than Braden have been described from other North American sites; ancient dog burials dating to 8,500-7,000BP are known from the Koster site in Illinois (Morey & Wiant, 1992), and a possible 7,500BP small dog burial has been reported from Rodgers Shelter in southern Missouri (McMillan, 1970). However, at this time, the Braden specimens appear to represent the earliest known example of *intentional dog burial in direct association with human interments* in North America.

**Acknowledgments**

We thank Danny Walker for his original sorting and assistance with the Braden dog material; Amy Dansie for discussions relating to ancient dogs in the American West; and Leslie Quintero for her comments on a draft of the original manuscript of this paper.

**References**

Allen, G. M. (1920). Dogs of the American aborigines. *Harvard University Museum of Comparative Zoology Bulletin* ***63***, pp.431-517.

Bean, L. & Smith, C. R. (1978). Gabrieleno. In (R. F. Heizer, Ed.) *Handbook of North American Indians, Vol. 8 (California).* Washington: Smithsonian Institution, pp. 538-549.

Butler, B. (1980). The 1968 excavations at the Braden burial site (10-WN-117), an early archaic cemetery in western Idaho. In (L. Harten, C. Warren & D. Tuohy, Eds.) *Anthropological Papers in Honor of Earl Swanson, Jr.* Pocatello: Special Publication of the Idaho Museum of Natural History, pp. 117-129.

Clark, G. R. (1995). The Kuri in prehistory: a skeletal analysis of an extinct Maori dog. Master's Thesis, University of Otago, New Zealand.

Crockford, S. J. (1997). *Osteometry of Makah and Coast Salish Dogs*. Burnaby: Simon Fraser University Archaeology Press **22**.

Cybluski, J. S. (1992). A Greenville burial ground: human remains and mortuary elements in British Columbia coast prehistory. *Archaeological Survey of Canada Mercury Series Paper* ***146***. Ottawa: Canadian Museum of Civilization.

Driesch, A. von den (1976). *A Guide to the Measurement of Animal Bones from Archaeological Sites, Peabody Museum Bulletin* ***1.*** Cambridge: Peabody Museum of Archaeology and Ethnology.

Elsasser, A. B. (1978). Mattole, Nongatl, Sinkyone, Lassik, & Wailaki. In (R. F. Heizer, Ed.) *Handbook of North American Indians, Vol. 8 (California).* Washington: Smithsonian Institution, pp. 190-204.

Gowlett, J.A., Hedges, R. E. M, Law, I. A. & Perry, C. (1987). Radiocarbon dates from the Oxford AMS system: archaeometry datelist 5. *Archaeometry* **29**, 125-155.

Green, T. J., Pavesic, M. G., Woods, J. C. & Titmus, G. L. (1986). The DeMoss burial locality: preliminary observations. *Idaho Archaeologist* **9**, 31-40.

Grayson, D. K. (1988). Danger Cave, Last Supper Cave, and Hanging Rock Shelter: The Faunas. *Anthropological Papers of the American Museum of Natural History* **66**.

Haag, W.G. (1948). An osteometric analysis of some aboriginal dogs. *University of Kentucky Reports in Anthropology* *7*, 107-264.

Haag, W. G. (1970). Dog remains from Hogup Cave. In (C. M. Aikens, Ed.) Hogup Cave, pp. 274-274. *University of Utah Anthropological Papers* **84**.

Haag, W.G. (1983). Dog remains from the Wildcat Canyon site. Appendix G in (D. E. Dummond & R. Minor, Eds) *Archaeology in the John Day Reservoir: the Wildcat Canyon site, 35-GM-9.* University of Oregon Anthropological Papers **30**.

Haag, W.G. & Heizer, R. F. (1953). A dog burial from the Sacramento Valley. *American Antiquity* **18,** 263-264.

Harcourt, R. A. (1974). The dog in prehistoric and early historic Britain. *Journal of Archaeological Science* **1,** 151-174.

Harten, L. (1975). The osteology of the human skeletal material from the Braden site, 10- WN-117, in western Idaho. Master's Thesis, Idaho State University.

Harten, L. (1980). The osteology of the human skeletal material from the Braden site in western Idaho. In (L. Harten, C. Warren & D. Tuohy, Eds.) *Anthropological Papers in Honor of Earl Swanson, Jr.* Pocatello: Special Publication of the Idaho Museum of Natural History, pp. 180-200.

Horard, M.-P. (1997). *Le village celtique des Arènes à la Levroux (4).* 12$^{e}$ suppl. à la Revue Archéologique du Centre de la France.

Janetski, J., Lupo, K., McCullough, J. & Novak, S. (1992). The Mosida site: A middle Archaic burial from the eastern Great Basin. *Journal of California and Great Basin Anthropology* **14**, 180-200.

Kerber, J. E. (1997). *Lambert Farm: Public Archaeology and Canine Burials Along Narragassett Bay.* Fortworth: Harcout, Brace & Company.

Lapena, F. R. (1978). Wintu. In (R. F. Heizer, Ed.) *Handbook of North American Indians, Vol. 8 (California).* Washington: Smithsonian Institution, pp. 324-340.

Lawrence, B. (1967). Early domestic dogs. *Zeitscheift fur Saugetier Kunde* **32**, 44-59.

Lawrence, B. (1968). Antiquity of large dogs in North America. *Tebiwa* **11**, 43-49.

Lupo, K. D. & Janetski, J. C. (1994). Evidence of domesticated dogs and some related canids in the eastern Great Basin. *Journal of California and Great Basin Anthropology* **16**, 199-220.

Madsen, C. (1997). Microwear analysis of the lithic assemblage at the Rosenberger site. *Journal of California and Great Basin Anthropology* **19**, 116-123.

McMillan, R. B. (1970). Early canid burial from western Ozark Highlands. *Science* **167,** 1246-1247.

Morey, D. F. & Wiant, M.D. (1992). Early Holocene domestic dog burials from the North American Midwest. *Current Anthropology* **33**, 224-229.

Pavesic, M. (1979). Recent findings at the Braden burial site, western Idaho. Presented paper, 44th annual meeting, Society of American Archaeology, Vancouver.

Pavesic, M. (1985). Cache blades and turkey tails: piecing together the western Idaho archaic burial complex. In (M. Plew, J. Woods & M. Pavesic, Eds.) *Stone Tool Analysis: Essays in Honor of Don E. Crabtree*. Albuquerque: University of New Mexico Press, pp. 55-89.

Pavesic, M. (1992). Death and dying in the western Idaho archaic. In (A. Goldsmith, S. Garvie, D. Selin & J. Smith, Eds.) *Ancient Images, Ancient Thought: The Archaeology of Ideology*. Calgary: Archaeological Association, University of Calgary, pp. 289-293.

Pavesic, M., Miller, S., Gamel, P. & Green, T. (1993). The DeMoss site: a material culture and faunal update. *Idaho Archaeologist* **16,** 3-15.

Penney, D. (1985). The late Archaic period. In (D. Brose, J. Brown & D. Penney, Eds.) *Ancient Art of the American Woodland Indians*. New York: Harry N. Abrams, Inc., pp. 15-37.

Pferd, W. (1987). *Dogs of the American Indians*. Fairfax: Denlinger's Publishing.

Snyder, L. M. (1991). Barking mutton: ethnohistoric, ethnographic, archaeological, and nutritional evidence pertaining to the dog as a Native American food resource on the Plains. In (J.R. Purdue, W. E. Klippel & B. W. Styles, Eds) *Beamers, Bobwhites, and Blue-Points: Tributes to the Career of Paul W. Parmalee*. Illinois State Museum Scientific Papers **23**, pp. 361-378.

Schwartz, M. (1997). *A History of Dogs in the Early Americas*. New Haven: Yale University Press.

Shigehara, N., Onodera, S. & Eto, M. (1997). Sex determination by discriminate analysis and evaluation of non-metric traits in the dog skeleton. Appendix A in (S. Crockford) *Osteometry of Makah and Coast Salish Dogs*. Burnaby: Simon Fraser University Archaeology Press **22**, pp. 113-126.

Switzer, L. L. (1992). *An osteometric comparison of Northwest aboriginal dogs*. Master's Thesis, Washington State University.

Wallace, W. (1978) Southern Valley Yokuts. In (R. F. Heizer, Ed.) *Handbook of North American Indians, Vol. 8 (California)*. Washington: Smithsonian Institution, pp. 448-461.

Yohe, R. & Pavesic, M.. (1996). Of snails and puppy dog tales: new data relating to the western Idaho Archaic burial complex. Presented paper, 25th Great Basin Anthropological Conference, Kings Beach, California.

# PALAEOPATHOLOGY OF ARCHAIC PERIOD DOGS FROM THE NORTH AMERICAN SOUTHEAST

Diane M. Warren
Department of Anthropology
Student Building Room 130
Indiana University
Bloomington, IN 47405,
USA

**Key words:** dog, pathology, osteology, Archaic period, osteoarthritis, osteophytosis, bone fractures, US Southeast, archaeozoology

**Abstract**

One hundred and ninety-four domestic dogs from twelve Archaic (hunting-gathering) period sites in the American Southeast were examined. All are thought to date between 7,000 and 3,000BP, and the majority are deliberate burials. The dogs are small to medium in size and range in age from subadult to old adult. High frequencies of axial skeleton fractures were observed at all sites. Vertebral osteoarthritis and marginal osteophytosis are common at some sites, although significant differences in frequency occur. Explanations for these differences include disparate uses of dogs at the sites and differential exposure to environmental pathogens.

## Introduction

Although only a small fraction of an individual's experiences are recorded in its skeleton, osteology and palaeopathology are often the only source of information about particular aspects of life in the past. Our understanding of previous human activities, health, nutrition, and social and biological relationships has been greatly increased by the investigation of the types and patterns of pathology observable on skeletal remains. Dog skeletal studies (eg Kerber, 1997; Snyder, 1991; Wapnish & Hesse, 1993; Wing, 1991) are similarly valuable for illuminating the varieties of interaction between these animals and humans over time and across cultures.

Deliberate burial of domestic dogs is common at Middle and Late Archaic period (8,000-3,000BP) sites in the North American Southeast. Haag (1948), Smith (1975) and others have suggested that this practice may reflect the dogs' functional importance. Because Archaic societies are seasonally mobile while hunting-gathering, a hunting and transport function of the dogs has been proposed. Dog burials decrease in frequency after the Archaic and are rare at Mississippian (1,100-400BP) and historic period sites in the region. This decline is accompanied by a rise in the occurrence of butchered and disarticulated dog remains at sites, a phenomenon attributed to a changing role of dogs in increasingly sedentary and agriculture-dependent societies (Haag, 1948; Smith, 1975). By early historic times Native groups in the Southeast used dogs for hunting only occasionally, and the use of dogs as pack animals or for traction is not reported (Swanton, 1946).

Despite their numbers, the dogs buried at Archaic sites in the Southeast have been largely ignored by archaeozoologists, and their use and treatment in life remains a subject of speculation. Previous detailed studies of dogs from this region have been descriptive (Skaggs, 1946; Morey, 1994) or have focussed on osteometry, particularly as it relates to taxonomic or domestication issues (Haag, 1948; Colton, 1970; Morey & Wiant, 1992). No comprehensive palaeopathological analyses have been undertaken.

This study investigates aspects of the skeletal biology and palaeopathology of dogs from Archaic sites in the North American Southeast. Age at death, size, skeletal fractures and joint disease are compared among the sites. This will allow for a better understanding of the range of variation among these dogs. Possible relationships between pathology and dog use and treatment are explored. Results are preliminary and further interpretation is limited until controlled comparisons of the Archaic dogs and other samples can be made.

## Materials

Dogs were examined from twelve Southeastern Archaic period sites: Perry, Alabama; Carlston Annis, Read, Barrett, Butterfield, Kirkland, Ward, Bowles, Chiggersville, and Indian Knoll, Kentucky; and Eva and Cherry, Tennessee. To maximize sample sizes geographically proximal sites were grouped, producing five site clusters (Table 1): Alabama (AL), Butler County, Kentucky (KYBT), McLean County, Kentucky (KYMCL), Ohio County, Kentucky (KYOH), and Benton County, Tennessee (TN). Locations of the clusters are shown in Figure 1.

A minimum of 194 dogs is included in the sample. One hundred and eighty-nine of these consist of multiple skeletal elements and were likely deliberate interments of complete or nearly complete animals. The remaining dogs are represented by isolated crania, mandibles or postcranial bones. None of the isolated bones in the sample have cutmarks, fire damage or other evidence of butchery, and their relationship to the burials is unclear. Burials are typically of a single animal, but burials including multiple dogs and burials with humans also occur.

Ceramics recovered from Perry, Alabama indicate a sizeable post-Archaic component at the site, and many of the more than one hundred dogs present are of ambiguous affiliation. To minimize the number of post-Archaic dogs in the Alabama sample, only dogs from the lowest strata (5' & 6' levels) are included, ceramics being concentrated in the upper two feet of the site (Webb & DeJarnette, 1942;

| Cluster and site | Dogs | References |
|---|---|---|
| Alabama (AL) | | |
| Perry (1Lu25) | 23 | (Webb & DeJarnette, 1942; Webb & DeJarnette, 1948) |
| Butler Co., Kentucky (KYBT) | | |
| Carlston Annis (15Bt5) | 21 | (Webb, 1950; Marquardt & Watson, 1983) |
| Read (15Bt10) | 25 | (Webb, 1950) |
| McLean Co., Kentucky (KYMCL) | | |
| Barrett (15McL4) | 12 | (Webb & Haag, 1947) |
| Butterfield (15McL7) | 7 | (Webb & Haag, 1947) |
| Kirkland (15McL12) | 12 | (Webb & Haag, 1940; Webb & Haag, 1947) |
| Ward (15McL11) | 24 | (Webb & Haag, 1940) |
| Ohio Co., Kentucky (OHKY) | | |
| Bowles (15Oh13) | 7 | (Marquardt & Watson, 1983) |
| Chiggersville (15Oh1) | 12 | (Webb & Haag, 1939) |
| Indian Knoll (15Oh2) | 23 | (Webb, 1946) |
| Tennessee (TN) | | |
| Eva (40Bn6) | 21 | (Lewis & Lewis, 1961) |
| Cherry (40Bn84) | 7 | (Magennis, 1977) |

***Table 1** Sites and site clusters included in this study*

***Figure 1** Map of the southeastern US, indicating location of the site clusters: (1) McLean County, Kentucky; (2) Ohio County, Kentucky; (3) Butler County, Kentucky; (4) Tennessee; (5) Alabama*

Webb & DeJarnette, 1948). Dogs at the other sites are generally accepted as Middle to Late Archaic, and probably date to between 7,000 and 3,000BP (Morey, 1992; see also references Table 1).

### Sample Composition

Sex

Thirty-nine of the 194 dogs are identified as male based on the presence of a baculum (Table 2). Several females are inferred from the inclusion of fetal pups with an adult dog. Attempts to use the morphological sexing methods of The & Trouth (1976) and Onodera, Shigehara & Eto (1987) were unsatisfactory due to fragmentation of the remains and my initial unfamiliarity with the techniques. Other morphological sexing methods were not applied. Given the high proportion of dogs of indeterminate sex and the overall small number of dogs in each site cluster males and females are combined.

Age

Dogs were assigned to one of six age categories based on skeletal maturity, dental eruption and tooth wear (Arnall, 1961; Kremenak, 1967; Silver, 1963; Shabestari *et al.*, 1967). *Fetal* dogs, identified on the basis of long bone length (Andersen, 1970), are not considered here. *Subadults* are less than one year of age and have many unfused epiphyses; older subadults possess a permanent dentition. Dogs estimated to be more than one year of age are considered adult, sexual maturity in modern dogs typically occurring between six and eighteen months (Mandeville & Sidewater, 1985: 678-679). Adult dogs are subdivided by skeletal maturity and tooth wear. *Young adults* have light tooth wear and may have some unfused epiphyses. *Middle adults* are skeletally mature and have moderate tooth wear. Heavy to extreme tooth wear indicates an *old adult*. *Indeterminate adult* dogs cannot be assigned to a more precise category. The age distributions of the dogs in each site cluster are presented in Table 3.

Chi-square tests show the age distributions are not significantly different among the site clusters. This is true whether the adult dogs of indeterminate age are considered ($x^2 = 19.88$, $p = 0.228$) or ignored $x^2 = 9.98$, $p = 0.618$). Adult dogs predominate, comprising between 82 and 93 percent of the individuals in each cluster. In all but the Alabama cluster, young, middle and old adults are approximately equally represented, with each category contributing between 25 and 41 percent of determinate age adults. In the Alabama cluster old adults comprise only 8 percent of the determinate adults.

Subadults are underrepresented in all site clusters (although not necessarily at all sites). A high proportion of dogs in this age category is expected in both attritional and catastrophic samples. The L-shaped mortality curve of a catastrophic sample reflects the age distribution of the living population, which typically includes a large number of subadults (Klein & Cruz-Uribe, 1984). Almost 30 percent of the free-ranging Chilean dogs observed by Rodolfo Martin *et al.* (1977) were one year of age or younger. In an attritional sample the greater susceptibility to natural death of sub- and old adults produces a U-shaped mortality profile (Klein & Cruz-Uribe, 1984). Beck (1973) estimated that approximately half of the feral urban dogs he observed died during their first year. The small percentage of subadults in the clusters could be due to a preference for adult dogs for burial; it could also be the result of preservation or excavation bias.

The high proportion of young and middle adults is inconsistent with attritional mortality and suggests that the buried dogs are not merely an unbiased sample of natural deaths. A clearly catastrophic age distribution is seen among the adult Alabama dogs. Such a distribution could result from random selection for interment of dogs from the living population, perhaps through sacrifice. It could also result from preferential selection for burial of particular age classes of dogs dying of natural causes. In the other clusters the high proportion of old versus young adults is not what would be expected in a living population, suggesting old dogs were more desirable for burial. A combination of sacrifice and selective burial of natural deaths could also have occurred.

Although dog sacrifice could have produced the age distributions seen at the sites, the remains themselves provide little direct evidence for this practice. Many of the dogs have fractured crania but postmortem breaks make perimortem injuries difficult to assess. One dog from the Ohio County cluster (Indian Knoll) has a possible prehistoric cutmark on its sternum. Dogs may have been sacrificed but in ways which did not affect the skeleton, for instance by strangulation.

### Osteometry

Twenty-seven measurements of the skull, teeth, and postcranial skeleton were attempted on each dog. Of these, eight are used to estimate size; the remainder could be taken on too few individuals for inclusion in this study. Only elements in which epiphyseal union was initiated and mandibles in which all permanent teeth had erupted were measured. When possible measurements were taken from the left side; if not, right side values were substituted. Measurements are not available from Bowles site in the Ohio County, KY cluster or from Cherry site in the Tennessee cluster.

Small to medium sized dogs are seen in all site clusters (Table 4). Discrete size classes are not apparent, nor are grossly discernible skull types. T-tests reveal no significant differences (0.05 level) in mean measurement values among the dogs from Butler County, KY, McLean County, KY, and Tennessee. The Alabama dogs have significantly larger postcranial elements and a longer $M_3$-$P_2$ cheek tooth row than dogs from some but not all other clusters (Table 4). Mean cheek tooth row lengths in the Ohio County, KY cluster are significantly shorter than those observed in all clusters except McLean County.

Shoulder height was estimated from humerus length using Harcourt's (1974) formula. Average values range from 37.1 cm at Ohio County, KY to 40.2 cm in the Alabama cluster (Table 5). The estimated shoulder heights of the smallest and largest dogs in the sample are 30.6 cm (Ohio County cluster) and 45.7 cm (McLean County and Tennessee clusters).

**Pathology**

All dogs were examined macroscopically for the presence of 1) skeletal fractures, 2) osteoarthritis , and 3) vertebral marginal osteophytosis. Frequency and patterning of these pathologies may reflect habitual activity or treatment by humans.

| | Site cluster | | | | | |
|---|---|---|---|---|---|---|
| **Sex** | **AL** | **KYBT** | **KYMCL** | **KYOH** | **TN** | **Total** |
| Male | 8 | 10 | 4 | 14 | 3 | 39 |
| Female | 0 | 1 | 1 | 0 | 0 | 2 |
| Indeterminate | 15 | 35 | 50 | 28 | 25 | 153 |

***Table 2.*** *Sex distribution*

| | Age category[1] | | | | | |
|---|---|---|---|---|---|---|
| **Sites by cluster** | **SA** | **YA** | **MA** | **OA** | **IA** | **Total** |
| Alabama (AL) | | | | | | |
| Perry | 2 | 5 | 7 | 1 | 8 | 23 |
| Butler Co, KY (KYBT) | | | | | | |
| Carlston Annis | 3 | 8 | 4 | 5 | 1 | 21 |
| Read | 2 | 8 | 6 | 8 | 1 | 25 |
| McLean Co., KY (KYMCL) | | | | | | |
| Barrett | 4 | 0 | 3 | 2 | 3 | 12 |
| Butterfield | 2 | 0 | 2 | 1 | 2 | 7 |
| Kirkland | 0 | 7 | 2 | 2 | 1 | 12 |
| Ward | 4 | 3 | 5 | 8 | 4 | 24 |
| Ohio Co., KY (KYOH) | | | | | | |
| Bowles | 2 | 2 | 1 | 0 | 2 | 7 |
| Chiggersville | 2 | 1 | 0 | 5 | 4 | 12 |
| Indian Knoll | 1 | 9 | 6 | 4 | 3 | 23 |
| Benton Co., TN (TN) | | | | | | |
| Eva | 2 | 3 | 5 | 6 | 5 | 21 |
| Cherry | 0 | 4 | 2 | 1 | 0 | 7 |

[1]Age category: SA=subadult, YA=young adult, MA=middle adult, OA=old adult, IA=indeterminate adult

***Table 3.*** *Age Distributions*

| | AL | KYBT | KYMCL | KYOH[a] | TN[b] |
|---|---|---|---|---|---|
| **Measurement** | **Range (mean); N** | **Range (mean); N** | **Range (mean); N** | **Range (mean); N** | **Range (mean); N** |
| Humerus | | | | | |
| Length | 11.7-13.9 (12.5); 143[e] | 9.9-13.5 (11.8); 25 | 10.0-14.1 (11.9); 17 | 9.7-13.7 (11.6); 24 | 10.8-14.1 (12.1); 15 |
| Head diameter | 3.1-4.0 (3.4); 135 | 2.9-3.8 (3.3); 29 | 2.8-3.8 (3.3); 18 | 2.7-3.7 (3.2); 24 | 2.8-3.8 (3.3); 16 |
| Radius length | 11.3-12.6 (12.0); 104[e] | 10.3-13.4 (11.8); 27 | 9.8-13.0 (11.4); 16 | 9.9-13.8 (11.5); 24 | 10.7-14.2 (12.0); 13 |
| Femur length | 13.0-14.8 (14.0); 83[d, e] | 11.2-14.5 (13.2); 22 | 10.8-15.0 (12.9); 19 | 10.9-15.2 (12.9); 25 | 12.2-16.0 (13.4); 15 |
| Tibia length | 13.4-15.8 (14.4); 113[d, f] | 11.2-15.7 (13.5); 28 | 11.1-15.1 (13.3); 21 | 11.4-15.6 (13.3); 27 | 12.5-15.2 (13.6); 17 |
| $P_4$ length | 1.48-1.83 (1.68); 11 | 1.48-1.83 (1.67); 33 | 1.45-1.87 (1.67); 32 | 1.52-1.82 (1.66); 27 | 1.45-1.83 (1.66); 17 |
| $M_1$ length | 1.76-2.07 (1.93); 11 | 1.71-2.10 (1.88); 39 | 1.66-2.09 (1.88); 35 | 1.69-2.02 (1.87); 25 | 1.68-2.06 (1.88); 16 |
| Tooth row, $P_2$-$M_3$ | 5.8-6.3 (6.1); 95 | 5.4-6.5 (6.0); 245 | 5.0-6.5 (5.9); 15 | 5.3-6.3 (5.8); 19 | 5.6-6.6 (6.1); 125 |

[a]Bowles site not included.
[b]Cherry site not included.
[c]Significantly greater (P > 0.05) than KYBT.
[d]Significantly greater (P > 0.05) than KYMCL.
[e]Significantly greater (P > 0.05) than KYOH.
[f]Significantly greater (P > 0.05) than TN.

***Table 4.*** *Osteometrics. Measurements (cm) after Haag (1948)*

| Site cluster (n) | Mean humerus length, cm (range) | Mean shoulder height, cm |
|---|---|---|
| AL (14) | 12.5 (11.7-13.9) | 40.2 |
| KYBT (25) | 11.8 (9.9-13.5) | 37.8 |
| KYMCL (17) | 11.9 (10.0-14.1) | 38.2 |
| KYOH[a] (24) | 11.6 (9.7-13.7) | 37.1 |
| TN[b] (15) | 12.1 (10.9-14.1) | 38.8 |

[a]Bowles site measurements unavailable
[b]Cherry site measurements unavailable

***Table 5.** Estimated shoulder*

Incidence of each type of pathology will have an age component due to cumulative risk exposures over time. Ideally, only dogs of the same age are compared; however, the small number of dogs per age category required that young, middle, old and indeterminate adults be combined. Similar age distributions among the clusters should minimize the effect of grouping the adults. Subadults are too poorly represented for a meaningful comparison across site clusters, and are omitted from this study.

Fractures

Skeletal fractures were recorded by location on the bone and degree of healing. Only fractures in which some healing had taken place are included. Fragmentary bones and functional units (eg forepaw, thoracic vertebrae, ribs) which are less than fifty percent complete are excluded. Fractures of multiple bones in a functional unit are treated as a single occurrence. Left and right sides are scored separately; the frequencies are then combined.

Fractures of the axial skeleton are most common in all clusters (Table 6). Vertebrae (Fig. 2a) and ribs are particularly prone. The frequency of fracture of the spinous, transverse and other processes of the thoracic and lumbar vertebrae approaches 75 percent in the McLean County and Tennessee clusters. Moderate frequencies (9-15 percent) of compression fractures of the endplates of the vertebral bodies were observed in the lumbar region of dogs in all but the Alabama cluster (Fig. 2b). These fractures are similar to Schmorl's nodes and are likely the result of degenerative or traumatic displacement of the intervertebral disk into the vertebral body. Affected vertebrae typically exhibit marginal osteophytes (see below). Intrusion of the intervertebral disk into the vertebral body (rather than into the vertebral foramen) is very rare in modern clinical samples (Palmer, 1993: 155); the frequencies seen among the Archaic dogs are curious.

Appendicular fractures are less common among the Archaic dogs. In all but the Alabama cluster the os coxae are the most frequently fractured bone in the appendicular skeleton; healed fractures are seen in over 10 percent of os coxae at Tennessee. Fractures of the shafts of the major limb bones (humerus, radius, femur, tibia) are relatively rare, affecting less than three percent of the elements observed in any cluster. Chi-square analysis reveals no significant differences (0.05 level) among the five clusters in frequency of fracture of any skeletal element or unit.

Osteoarthritis

Osteoarthritis is the bony endproduct of non-inflammatory degeneration of the articular cartilage found in synovial joints. Cartilage degeneration has a variety of causes including joint, muscle, and skeletal trauma, aging, developmental defects, and nutritional disorders (Ornter & Putschar, 1985; Palmer, 1993: 247; Roberts & Manchester, 1995). Because of its relationship to repetitive joint use, osteoarthritis patterning within the skeleton can reflect habitual activities to a limited extent. Affected joint surfaces may show pitting, eburnation, marginal lipping, or some combination of these responses. Inflammatory joint disease, or septic arthritis, is omitted from this study.

A joint was considered for analysis if at least one contributing articular surface was available. Location, severity and contributing factors such as fracture were recorded for all joints exhibiting osteoarthritis. Because use of the left and right sides of the body is balanced in most dog activities, contra-lateral elements are treated as a single structure. Functional units (eg hindpaw) and series of repeated joints (eg costovertebral) are also considered a single structure.

The highest incidence of osteoarthritis in all clusters is in the costo- and intervertebral (zygopopheseal) joints (Table 7). Intervertebral osteoarthritis is most common in the upper thoracic region (TV 1-9). Significantly less osteoarthritis in this region is seen in the dogs from McLean County than in those from Ohio County ($x^2$ = 8.94, P = 0.003) and Alabama ($x^2$ = 5.29, P = 0.022). The higher incidence of lumbar intervertebral osteoarthritis in the Ohio County cluster versus the Alabama, McLean County and Tennessee clusters is also statistically significant ($x^2$ = 19.03, P = 0.000).

Arthritic changes in the paw and elbow are moderately frequent in all clusters. Beyond this no clear patterns of osteoarthritis in the appendicular skeleton emerge. Frequencies of osteoarthritis in the non-vertebral joints are not significantly different (0.05 level) across the five clusters.

Vertebral marginal osteophytosis

Marginal osteophytosis (MO) involves the production of bony spurs on the margins of the vertebral bodies. Like osteoarthritis, this condition has multiple etiologies, including degeneration, trauma and inflammation of the intervertebral disks (Ornter & Putschar, 1985; Palmer, 1993). These may be indistinguishable macroscopically. Spondylosis, the degenerative form, is often age-related and is common in modern clinical samples ((Morgan *et al.*, 1967; Palmer, 1993: 157). Using radiographs, Morgan *et al.* (1967) observed what they interpret as age-related marginal osteophytes in over 60 percent of a sample of English dogs of assorted breeds. Inflammatory marginal osteophytosis, or diskospondylitis, typically occurs in response to bacteremia, fungal infection or migrating foreign bodies (Palmer, 1993: 177).

Location and severity of osteophytes were recorded. Also noted were factors that may have contributed to osteophyte formation, such as osteitis or compression fractures of the vertebral endplates. Vertebral regional units which were less than fifty percent complete were excluded. Multiple osteophytes in a region are scored as a single occurrence.

In all clusters marginal osteophytes are most commonly located in the lower thoracic and lumbar vertebrae (Table 8). Modern clinical samples exhibit the same pattern, which is attributed to the increased pressure on the disks in areas of greatest spinal curvature (Palmer, 1993: 157). Chi-square tests reveal significantly higher frequencies (0.05 level) of lower thoracic and lumbar MO among the Alabama, Tennessee and McLean County dogs. Startling numbers of dogs are affected; 93% of the Alabama dogs have marginal osteophytes of the lumbar vertebrae. The Alabama dogs also have significantly more upper thoracic MO than dogs from all other clusters, although frequencies at Benton and McLean Counties surpass those at Tennessee. No significant differences were observed among the clusters in frequencies of cervical and sacral marginal osteophytosis.

**Discussion and Conclusion**

The data presented here suggest the Archaic dogs from Kentucky, Tennessee and northern Alabama are of broadly similar skeletal morphology. All dogs are small to medium in size and show no unequivocal evidence of discrete size classes or head shapes. Measurement ranges overlap among all clusters. The Alabama dogs are significantly larger in some respects than those from the other site clusters, yet only 3.1 cm separate the mean shoulder height of these dogs (40.2 cm) from that of the Ohio County dogs (37.1 cm). Such small differences in mean measurement values are difficult to interpret, but may be due to a deliberate selection for larger dogs at the Alabama site or to genetic drift. Based on a comparison of size among eleven Alabama and eleven Kentucky Archaic sites, Haag (1948: 211-212) concluded that the Alabama and Kentucky dogs were different enough to be considered separate varieties. This is not borne out by the more limited sample of Alabama dogs in the present study. Interestingly, there is a trend towards size reduction over time at the Alabama site considered here, with dogs from the upper, ceramic-bearing component of the site having significantly smaller (0.05 level) mean humeral length, humeral head diameter, $M_1$ length and cheek tooth row length than their Archaic predecessors. This contradicts Haag's (1948) observations of increasing dog size over time in the Southeast and requires further exploration.

Frequency and patterning of skeletal fractures are also similar across the site clusters. Rib, vertebrae and skull fractures predominate, and could be due to blows from humans. Alternately, Wobeser (1992) points out that skeletal fractures among Canadian wolves and coyotes may result from interspecies conflict, particularly from encounters with large prey animals. Dogs could similarly clash with larger animals when hunting either on their own or with humans. Ethnohistoric accounts from the Southeast describe the use of dogs in hunting bear and deer (Swanton, 1946). Vertebral process fractures could also result from heavy packs or travois-like poles placed on the back. Mutliple etiologies for the fractures are likely.

Despite their similarities, the Archaic dogs exhibit some significant differences in frequency of vertebral osteoarthritis and marginal osteophytosis. Apparently, disparate levels of risk for vertebral joint disease occurred across the clusters. This may have resulted from the use of dogs as pack animals at some sites, although the pattern of differences is difficult to interpret from a functional standpoint. For example, the Alabama, Butler County and McLean County dogs have high levels of lumbar marginal osteophytosis relative to those seen among the Ohio County and Tennessee dogs. Lumbar osteoarthritis, on the other hand, was not observed among the Alabama, Butler and McLean dogs but is seen at moderate levels in the Ohio County and Tennessee clusters. This apparent negative relationship between marginal osteophytosis and osteoarthritis is puzzling, since both disorders can result from disk degeneration or trauma. It must be emphasized that in the thoracic vertebrae the relationship between osteoarthritis and marginal osteophytosis is less clear. Possibly the the high levels of marginal osteophytosis at some sites are due to inflammation rather than to disk degeneration or trauma, an interpretation supported by the absence of both osteoarthritis and compression fractures of the vertebral body endplates in the lumbar region of the

| Element or unit | AL f/n[1] | AL % | KYBT f/n | KYBT % | KYMCL f/n | KYMCL % | KYOH f/n | KYOH % | TN f/n | TN % |
|---|---|---|---|---|---|---|---|---|---|---|
| Cranial vault[2] | 0/10 | 0.0 | 1/19 | 5.3 | 2/17 | 11.8 | 1/25 | 4.0 | 1/11 | 9.1 |
| Face[3] | 1/11 | 9.1 | 5/29 | 17.2 | 1/21 | 4.8 | 2/26 | 7.7 | 2/14 | 14.3 |
| Mandible | 0/22 | 0.0 | 3/68 | 4.4 | 0/59 | 0.0 | 0/51 | 0.0 | 0/32 | 0.0 |
| Scapula | 0/29 | 0.0 | 3/42 | 7.1 | 0/47 | 0.0 | 1/57 | 1.8 | 1/24 | 4.2 |
| Humerus shaft | 0/34 | 0.0 | 0/69 | 0.0 | 2/67 | 2.3 | 1/63 | 1.6 | 0/38 | 0.0 |
| Humerus other | 0/34 | 0.0 | 0/69 | 0.0 | 0/67 | 0.0 | 0/63 | 0.0 | 1/38 | 2.6 |
| Radius shaft | 0/28 | 0.0 | 0/67 | 0.0 | 1/61 | 1.6 | 0/58 | 0.0 | 1/35 | 2.9 |
| Ulna shaft | 1/31 | 3.2 | 2/66 | 3.6 | 0/62 | 0.0 | 1/58 | 1.7 | 1/35 | 2.9 |
| Ulna other | 0/31 | 0.0 | 1/66 | 1.5 | 0/62 | 0.0 | 1/58 | 1.7 | 1/35 | 2.9 |
| Forepaw[4] | 1/20 | 5.0 | 3/49 | 6.1 | 1/42 | 2.3 | 1/54 | 1.9 | 1/31 | 3.2 |
| Os coxae | 1/22 | 4.5 | 5/47 | 10.6 | 1/29 | 3.4 | 3/57 | 5.3 | 2/27 | 7.4 |
| Femur shaft | 0/32 | 0.0 | 1/72 | 1.4 | 0/57 | 0.0 | 0/63 | 0.0 | 0/45 | 0.0 |
| Femur other | 0/32 | 0.0 | 1/72 | 1.4 | 0/57 | 0.0 | 0/63 | 0.0 | 0/45 | 0.0 |
| Tibia shaft | 0/34 | 0.0 | 1/73 | 1.4 | 0/58 | 0.0 | 0/65 | 0.0 | 0/40 | 0.0 |
| Fibula shaft | 0/14 | 0.0 | 3/35 | 8.6 | 1/34 | 2.9 | 1/43 | 2.3 | 0/23 | 0.0 |
| Hindpaw[5] | 1/21 | 4.8 | 1/48 | 2.1 | 1/45 | 2.2 | 2/51 | 3.9 | 0/32 | 0.0 |
| Cervical vertebrae | | | | | | | | | | |
| Atlas (CV1) | 1/9 | 11.1 | 3/29 | 10.3 | 0/22 | 0.0 | 3/29 | 10.3 | 1/13 | 7.7 |
| Axis (CV2) | 0/8 | 0.0 | 1/24 | 4.2 | 2/22 | 9.1 | 2/28 | 7.1 | 0/15 | 0.0 |
| CV 3-7[6] | | | | | | | | | | |
| body | 0/12 | 0.0 | 1/25 | 4.0 | 0/22 | 0.0 | 0/30 | 0.0 | 0/14 | 0.0 |
| processes | 0/12 | 0.0 | 0/25 | 0.0 | 1/22 | 4.5 | 1/30 | 3.3 | 2/14 | 14.3 |
| Thoracic vertebrae | | | | | | | | | | |
| TV 1-9[6] | | | | | | | | | | |
| body | 0/14 | 0.0 | 0/24 | 0.0 | 0/21 | 0.0 | 0/28 | 0.0 | 0/14 | 0.0 |
| processes | 5/14 | 35.7 | 17/24 | 70.8 | 11/21 | 52.4 | 13/28 | 46.4 | 9/14 | 64.3 |
| TV 10-13[6] | | | | | | | | | | |
| body | 0/12 | 0.0 | 1/20 | 5.0 | 0/18 | 0.0 | 0/27 | 0.0 | 0/16 | 0.0 |
| processes | 4/12 | 33.3 | 14/20 | 70.0 | 13/18 | 72.2 | 11/27 | 40.7 | 7/16 | 43.8 |
| Lumbar vertebrae[6] | | | | | | | | | | |
| body | 0/14 | 0.0 | 4/27 | 14.8 | 2/22 | 9.1 | 7/29 | 24.1 | 2/18 | 11.1 |
| processes | 5/14 | 35.7 | 20/27 | 74.1 | 15/22 | 68.2 | 16/29 | 55.2 | 7/18 | 38.9 |
| Sacrum | 0/9 | 0.0 | 2/22 | 9.1 | 0/22 | 0.0 | 0/27 | 0.0 | 0/16 | 0.0 |
| Ribs6 | 5/13 | 38.5 | 10/22 | 45.5 | 5/13 | 38.5 | 9/29 | 31.0 | 6/13 | 46.2 |

[1]Fractured/observed
[2]Cranial vault = parietals, temporals, occipitals as a unit
[3]Face = maxillae, zygomatics, frontals as a unit
[4]Forepaw = carpals, metacarpals as a unit
[5]Hindpaw = tarsals, metatarsals as a unit
[6]Considered as a unit.

***Table 6.*** *Fracture incidence, adult dogs. Left and right sides are combined*

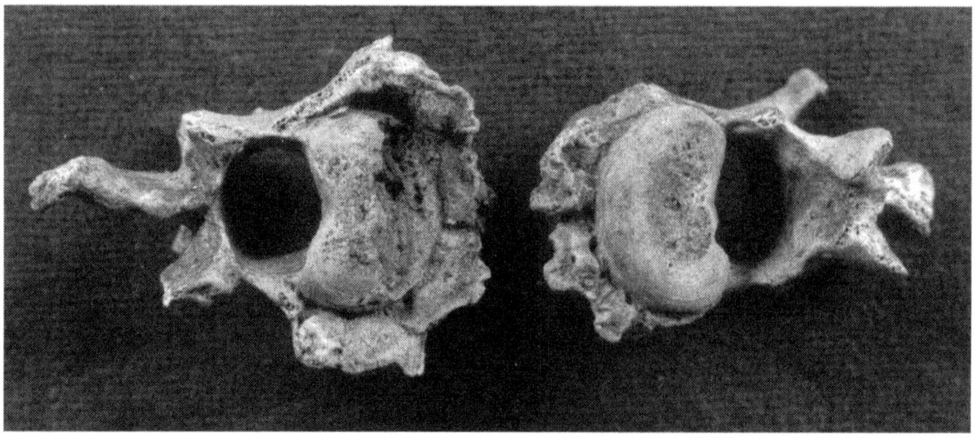

*Figure 2* *Vertebral pathologies. Cranial (left) and caudal (right) views of contiguous lumbar vertebrae showing healed fracture of the spinous processes and moderate-to-severe vertebral marginal osteophytosis (15Oh1, Dog 1-50/1-57)*

*Figure 3* *Cranial view of lumbar vertebra showing compression fracture of the body endplate (40Bn6, dog 16)*

| Joint | AL p/n[a] | AL % | KYBT p/n | KYBT % | KYMCL p/n | KYMCL % | KYOH p/n | KYOH % | TN p/n | TN % |
|---|---|---|---|---|---|---|---|---|---|---|
| Jaw | 1/10 | 10.0 | 1/34 | 2.9 | 0/25 | 0.0 | 1/25 | 4.0 | 0/15 | 0.0 |
| Shoulder | 0/16 | 0.0 | 1/35 | 2.8 | 0/33 | 0.0 | 2/35 | 5.7 | 0/17 | 0.0 |
| Elbow | 2/18 | 11.1 | 1/38 | 2.6 | 2/39 | 5.1 | 4/35 | 11.4 | 0/23 | 0.0 |
| Forepaw | 2/13 | 15.4 | 2/30 | 6.7 | 0/25 | 0.0 | 3/30 | 10.0 | 0/18 | 0.0 |
| Hip | 1/16 | 6.2 | 1/34 | 2.9 | 0/29 | 0.0 | 1/34 | 2.9 | 0/21 | 0.0 |
| Knee | 0/19 | 0.0 | 1/40 | 2.5 | 0/34 | 0.0 | 0/34 | 0.0 | 0/23 | 0.0 |
| Hindpaw | 2/13 | 15.4 | 0/28 | 0.0 | 1/25 | 4.0 | 0/30 | 0.0 | 1/18 | 5.6 |
| Intervertebral[b] | | | | | | | | | | |
| CV | 1/12 | 8.3 | 2/25 | 8.0 | 1/22 | 4.5 | 4/30 | 13.3 | 0/14 | 0.0 |
| TV 1-9 | 6/14 | 42.8[e] | 8/24 | 33.3 | 2/21 | 9.5 | 14/28 | 50.05 | 5/14 | 35.7 |
| TV 10-13 | 3/12 | 25.0 | 2/20 | 10.0 | 0/18 | 0.0 | 4/27 | 14.8 | 2/16 | 12.5 |
| LV | 0/14 | 0.0 | 0/27 | 0.0 | 0/22 | 0.0 | 8/29 | 27.6[c,d,e] | 3/18 | 16.7 |
| Sacral | 0/9 | 0.0 | 0/22 | 0.0 | 0/22 | 0.0 | 0/27 | 0.0 | 0/16 | 0.0 |
| Costovertebral | 5/13 | 38.5 | 11/22 | 50.0 | 6/13 | 46.2 | 15/29 | 51.7 | 10/13 | 76.9 |

[a]pathologic joints or units/observed joints or units
[b]CV =Cervical, TV = Thoracic, LV=Lumbar
[c]Significantly greater ($P > 0.05$) than AL.
[d]Significantly greater ($P > 0.05$) than KYBT.
[e]Significantly greater ($P > 0.05$) than KYMCL.

*Table 7* *Osteoarthritis incidence, adult dogs. All degrees of severity are combined*

| | AL | KYBT | KYMCL | KYOH | TN |
|---|---|---|---|---|---|
| Region[1] | p/n[2] (%) | p/n (%) | p/n (%) | p/n (%) | p/n (%) |
| CV | 2/12 (16.7) | 1/25 (4.0) | 0/22 (0.0) | 0/30 (0.0) | 0/14 (0.0) |
| TV 1-9 | 9/14 (64.3)[3,4,5,6] | 7/24 (29.2)[6] | 5/21 (23.8)[6] | 3/28 (10.7) | 0/14 (0.0) |
| TV 10-13 | 8/12 (66.7)[5,6] | 15/20 (75.0)[5,6] | 13/18 (72.2)[5,6] | 6/27 (22.2) | 1/16 (6.3) |
| LV | 13/14 (92.9)[3,5,6] | 17/27 (63.0)[5,6] | 19/22 (86.4)[5,6] | 10/29 (34.5) | 4/18 (22.2) |
| SV | 0/9 (0.0) | 1/22 (4.5) | 1/22 (4.5) | 2/27 (7.4) | 0/16 (0.0) |

[1]Region: CV = Cervical, TV = Thoracic, LV = Lumbar, SV = Sacral
[2] pathological units/observed units
[3]Significantly greater ($P \leq 0.05$) than KYBT.
[4]Significantly greater ($P \leq 0.05$) than KYMCL.
[5]Significantly greater ($P \leq 0.05$) than KYOH.
[6]Significantly greater ($P \leq 0.05$) than TN.

***Table 8*** *Vertebral marginal osteophytoses. All severeties combined*

Alabama dogs. Varying levels of exposure to bacteria, fungi or other environmental pathogens rather than different activity patterns may underlie the differences in vertebral joint disease among the Kentucky, Tennessee and Alabama dogs.

Future studies of these dogs will consider dental pathologies and skeletal pathologies not addressed here. Controlled comparison with samples of dogs of known activity (eg traction) will allow for a more meaningful interpretation of the palaeopathology of the Southeastern dogs as it relates to their use and treatment.

**Acknowledgements**

I am grateful to the following individuals for allowing me access to the dogs in their collections: Nancy O'Malley and Mary Lucas Powell (University of Kentucky Museum of Anthropology, Lexington), Eugene Futato (University of Alabama, Alabama Museum of Natural History, Division of Archaeology, Moundville, Alabama), and Jefferson Chapman (Frank H. McClung Museum, Knoxville, Tennessee). Della Cook, Lorena Havill, Elizabeth Pennefather-O'Brien, Staffan Peterson and two anonymous reviewers provided insightful comments, criticisms and support. Finally, heartfelt thanks to Susan Crockford, who organized this symposium and expertly assembled this volume.

**References**

Andersen, A. C. (1970). Skeletal system. A. Osteogenesis. In (A. C. Andersen & L. S. Good, Eds.) *The Beagle as an Experimental Dog*. Ames, Iowa: Iowa State University Press, pp. 149-158.

Arnall, L. (1961). Some aspects of dental development in the dog--II. Eruption and extrusion. *Journal of Small Animal Practice* **1,** 259-267.

Beck, A. M. (1973). *The Ecology of Stray Dogs*. Baltimore: York Press.

Colton, H. S. (1970). The aboriginal southwestern Indian dog. *American Antiquity* **35,** 153-159.

Haag, W. G. (1948). An osteometric analysis of some aboriginal dogs. *University of Kentucky Reports in Anthropology* **7,** 1-264.

Harcourt, R. A. (1974). The dog in prehistoric and early historic Britain. *Journal of Archaeological Science* **1,** 151-175.

Kerber, J. E. (1997). *Lambert Farm: Public Archaeology and Canine Burials Along the Narragansett Bay*. Fort Worth: Harcourt Brace & Co.

Klein, R. G. & Cruz-Uribe, K. (1984). *The Analysis of Animal Bones from Archaeological Sites*. Chicago: University of Chicago Press.

Kremenak, C. R. (1967). Dental exfoliation and eruption chronology in Beagles. *Journal of Dental Research* **46,** 687-93.

Lewis, T. M. N. & Lewis, M. K. (1961). *Eva: An Archaic Site*. Knoxville: University of Tennessee Press.

Magennis, A. L. (1977). Middle and Late Archaic mortuary patterning: An example from the Western Tennessee Valley. Master's Thesis, University of Tennessee.

Mandeville, J. J. & Sidewater, A. (1985). *The Complete Dog Book*. New York: Howell Book House, Inc.

Marquardt, W. H. & Watson, P. J. (1983). The Shell Mound Archaic of western Kentucky. In (J. L. Phillips & J. A. Brown, Eds.) *Archaic Hunters and Gatherers in the American Midwest*. New York: Academic Press, pp. 323-339.

Morey, D. F. (1992). Size, shape and development in the evolution of the domestic dog. *Journal of Archaeological Science* **19,** 181-204.

Morey, D. F. (1994). *Canis* remains from Dust Cave. *Journal of Alabama Archaeology* **40,** 160-69.

Morey, D. F. & Wiant, M. D. (1992). Early Holocene domestic dog burials from the North American Midwest. *Current Anthropology* **33,** 224-229.

Morgan, J. P., *et al.* (1967). Spondylosis deformans (vertebral osteophytosis) in the dog. *Journal of Small Animal Practice* **8,** 57-66.

Onodera, S., Shigehara, N. & Eto, M. (1987). Discriminant analysis of the sexual differences of the skeletons of Shiba dogs (*Canis familiaris*). *Acta Anatomica*

*Nipponica* **62,** 19-32. (in Japanese, English summary).
Ornter, D. J. & Putschar, W. G. J. (1985). *Identification of Pathological Conditions in Human Skeletal Remains*. Washington, DC: Smithsonian Institution Press.
Roberts, C. & Manchester, K. (1995). *The Archaeology of Disease*. Ithaca, NY: Cornell University Press.
Rodolfo Martin, M., Francisco Marin, L. B. & Miguel Rivera, M. (1977). Estudio demografico de la poblacion canina en localidades urbanas menores de 8.500 habitantes de la provincia de Valdivia. *Arch. Med. Vet.* **9,** 29-35.
Shabestari, L., Taylor, G. N. & W. Angus (1967). Dental eruption pattern of the beagle. *Journal of Dental Research* **46,** 276-278.
Silver, I. A. (1963). The ageing of domestic animals. In (D. Brothwell, E. Higgs & G. Clark, Eds.) *Science in Archaeology*. NY: Basic Books, Inc., pp. 250-272.
Skaggs, O. (1946). A study of the dog skeletons from Indian Knoll with special reference to the coyote as progenitor. *University of Kentucky Reports in Anthropology and Archaeology* **4,** 341-355.
Smith, B. (1975). Middle Mississippian exploitation of animal populations. *University of Michigan Museum of Anthropology Anthropological Papers* **57**.
Snyder, L. M. (1991). Barking mutton: Ethnohistoric, ethnographic, archaeological, and nutritional evidence pertaining to the dog as a Native American food resource on the Plains. In (J. R. Purdue, W. E. Klippel & B. W. Styles, Eds.) *Beamers, Bobwhites, and Blue-points: Tributes to the Career of Paul Parmalee*. Springfield: Illinois State Museum, pp. 359-378.
Swanton, J. R. (1946). *The Indians of the Southeastern United States*. Washington, DC: Smithsonian Institution Press.
The, T. L. & Trouth, C. O. (1976). Sexual dimorphism in the basilar part of the occipital bone of the dog (*Canis familiaris*). *Acta Anatomica* **95,** 565-71.
Wapnish, P. & Hesse, B. (1993). Pampered pooches or plain pariahs? The Ashkelon dog burials. *Biblical Archaeologist* **56,** 55-80.
Webb, W. S. (1946). Indian Knoll, Site Oh2, Ohio county, Kentucky. *University of Kentucky Reports in Anthropology and Archaeology* **4**.
Webb, W. S. (1950a). The Carlston Annis Mound, Site 5, Butler County, Kentucky. *University of Kentucky Reports in Anthropology* **7**.
Webb, W. S. (1950b). The Read shell midden. *University of Kentucky Reports in Anthropology* **7**.
Webb, W. S. & DeJarnette, D. L. (1942). An archaeological survey of the Pickwick Basin in the adjacent portions of the states of Alabama, Mississippi and Tennessee. *Smithsonian Institution Bureau of American Ethnology Bulletin* **129**.
Webb, W. S. & DeJarnette, D. L. (1948). The Perry site, Lu°25. *Geological Survey of Alabama, Alabama Museum of Natural History Museum Paper* **25**.
Webb, W. S. & Haag, W. G. (1939). The Chiggersville site. *University of Kentucky Reports in Anthropology* **4,** 1-62.
Webb, W. S. & Haag, W. G. (1940). Cypress Creek Villages Sites 11 and 12, McLean county, Kentucky. *University of Kentucky Reports in Anthropology* **4,** 67-110.
Webb, W. S. & Haag, W. G. (1947). Archaic sites in McLean county, Kentucky. *University of Kentucky Reports in Anthropology* **7**.
Wing, E. S . (1991). Dog remains from the Sorce site on Vieques Island, Puerto Rico. In (J. R. Purdue, W. E. Klippel & B. W. Styles, Eds.) *Beamers, Bobwhites, and Blue-points: Tributes to the Career of Paul W. Parmalee*. Springfield: Illinois State Museum, pp. 379-386.
Wobeser, G. (1992). Traumatic, degenerative, and developmental lesions in wolves and coyotes from Saskatchewan. *Journal of Wildlife Diseases* **28,** 268-75.

# DOG MANAGEMENT AND USE IN THE LATE IRON AGE : THE EVIDENCE FROM THE GALLIC SITE OF LEVROUX (FRANCE).

Marie-Pierre Horard-Herbin

UMR 6575 du C.N.R.S.
Universite de Tours,
3 place Anatole France,
37000 Tours, France

**Key words:** dog, Iron Age, dog consumption, archaeozoology, age determination, tooth wear

**Abstract**

Age estimation of dogs, based either on epiphyseal fusion or dental data, pose a problem since an individual's age can not generally be determined beyond the first two years of life. These restrictions are especially frustrating when considering the role of dogs in late Iron Age settlement contexts: were these animals solely a source of meat or were they used for other purposes as well ? The ability to construct age profiles and analyse kill-off patterns for this period is thus very important.

Here I propose an age estimation method based on the occlusal surface wear patterns of the first mandibular molar. This particular tooth appears most suitable for this type of research, owing to its conformation and ease of identification, in addition to being relatively abundant and well preserved in faunal assemblages containing dog remains. Application of this method is demonstrated through a case study analysis of dog remains excavated from the Gallic village of Levroux, France.

## Introduction

Age estimation of dogs, based either on epiphyseal fusion or dental data poses a problem since an individual's age can not generally be determined beyond the first two years of life. Indeed, the time of post-cranial epiphyseal fusion gives no information beyond the age of 24 months (fusion of the iliac crest) and dental eruptions end at an even earlier age of seven months with the appearance of the last molar. Looking at incisor wear allows for estimates beyond five years but these tiny teeth easily fall out of the jaw and rarely make their way to the zooarchaeologist's bench. Therefore, the analysis of ages is very limited in as much as we have very little data concerning dogs over two years of age.

These restrictions are especially frustrating when considering the role of dogs in late Iron Age settlement contexts : were these animals solely a source of meat or were they used for other purposes as well? The ability to construct age profiles and analyse kill-off patterns for this period is thus very important.

Here I propose an age estimation method based on the occlusal surface wear patterns of the first mandibular molar. This particular tooth appears most suitable for this type of research, owing to its conformation and ease of identification, in addition to being relatively abundant and well preserved in faunal assemblages containing dog remains. Application of this method is demonstrated through the case study analysis of dog remains excavated from the Gallic village of Levroux, France (Fig. 1).

Dental wear of the first lower molar in dogs was examined from two archaeological assemblages. The first comprise 111 mandibles from a Gallic period village in Indre, France (Levroux), dated to 200BC-AD50 (Buchsenschutz *et al.*, 1988, 1993, 1994). Isolated and fragmented dog mandibules were recovered from domestic refuse pits that appeared to have been randomly filled. The dog remains (2,662 specimens out of a total of 48,488 identified faunal remains) were mixed with those of other domestic animal foods as well as various residues of ceramics, metal, and glass.

The second assemblage contains 70 mandibles from a Roman period sanctuary, in Côte-d'Or, France (Vertault), dated to the first century AD. In contrast to those from Levroux, these mandibules come from 150 complete or nearly complete dog skeletons (where complete skeletons are males only). In 70 of the pits, the dogs appear to have been sacrificed and deposited on their left side, head facing west. Some of the dogs may have been killed by a blow on the cranium and subsequently slaughtered, but the remainder of the skeletons showed no butchery marks or other evidence of having been eaten (Méniel, 1992).

Wear patterns on the teeth were codified and seven different wear stages were defined (listed below), which include several degrees of observed variability (Fig. 2).

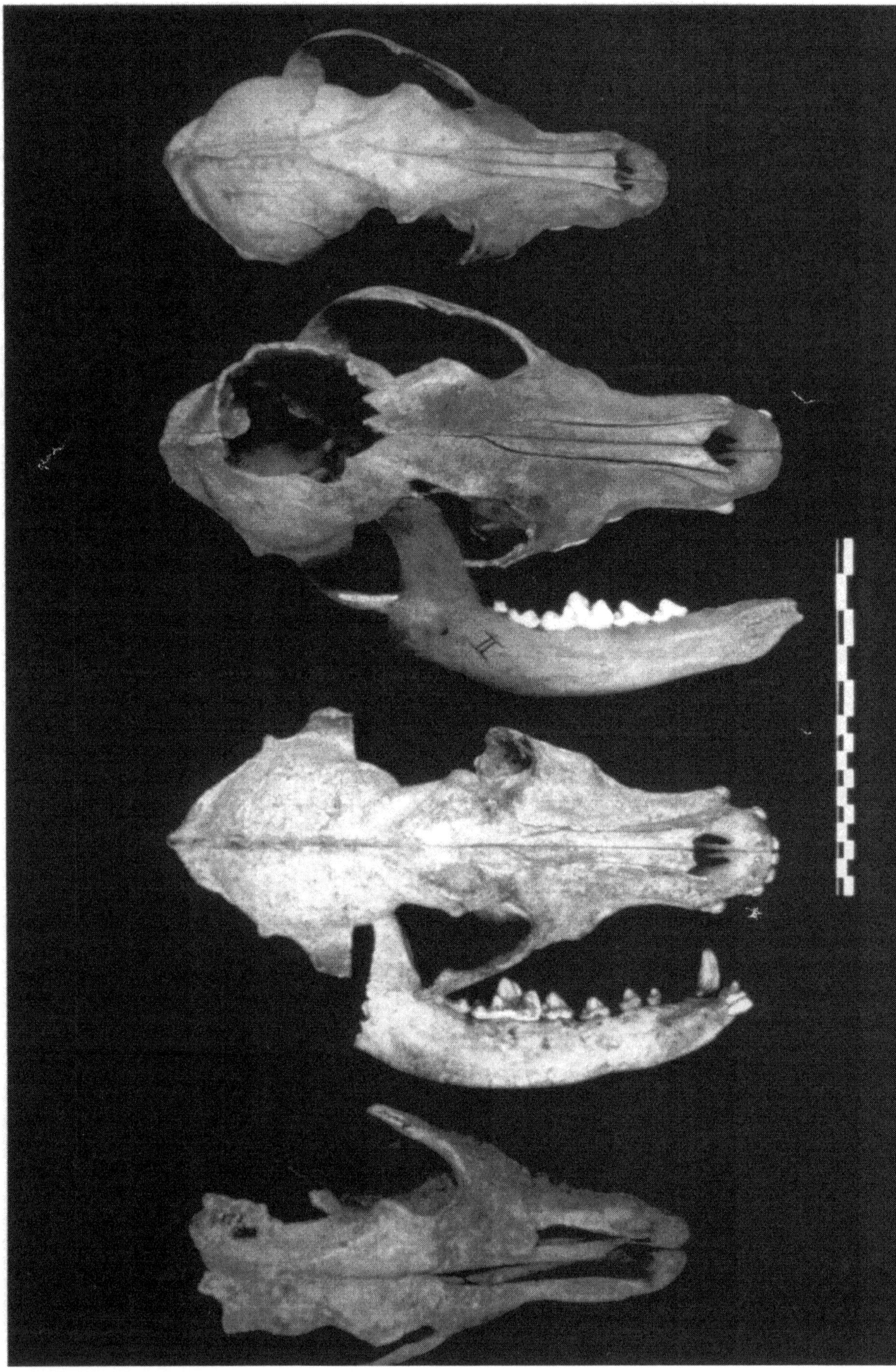

*Figure 1 Dogs from the Gallic site of Levroux (Indre, France)*

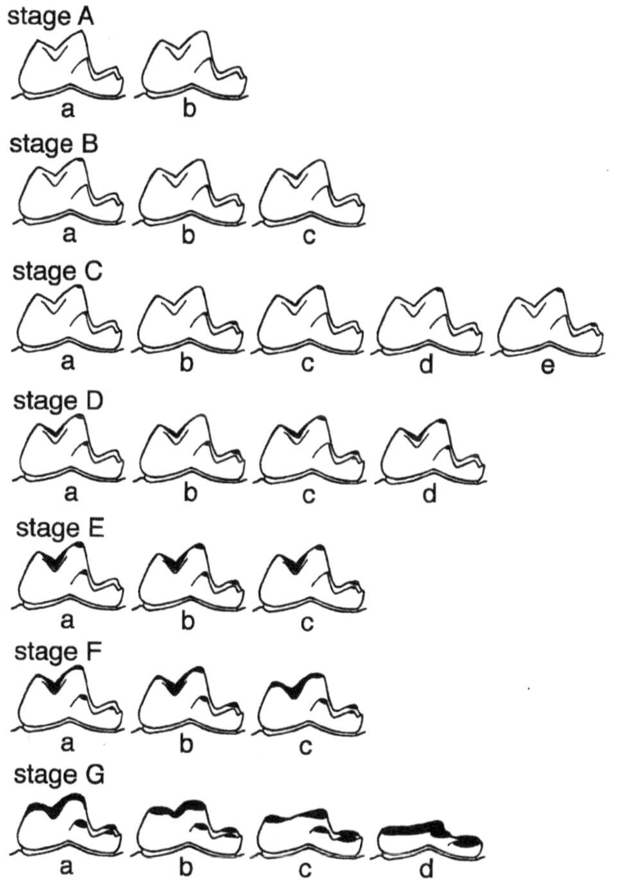


***Figure 2** Sequence of dental attrition for the dog. Lingual view of right lower permanent first molar. Wear facets are shaded in black*

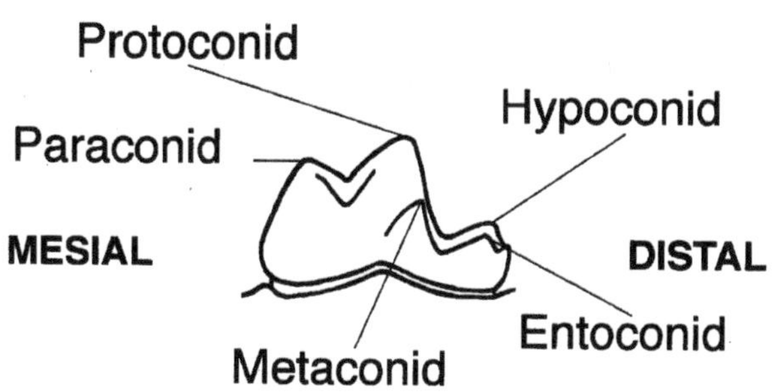


***Figure 3** Nomenclature of the trigonid and the talonid of the lower permanent first molar*

**Wear stages**

A) At stage A the lower permanent first molar is new and fully erupted (a) or starts to wear but without exposed dentine (Fig.2b).

B) At stage B one island of slight wear is observed, with exposed dentine on the protoconid (a), the metaconid (b), or at the junction of the paraconid and the protoconid (c) (for nomenclature, see Figure 3).

C) At stage C two islands of wear are observed: on the protoconid and the metaconid (a), on the metaconid and the hypoconid (b), on the protoconid and the hypoconid (c), on the protoconid and the entoconid (d), or at the junction of the paraconid and the protoconid and on the protoconid (e).

D) At stage D exposed dentine appears systematically at the junction area between the paraconid and the protoconid with, as the case may be, the metaconid and the hypoconid (a), the protoconid and the metaconid (b), the protoconid, the metaconid and the hypoconid (c), or the protoconid, the entoconid and the hypoconid (d).

E) Stage E corresponds to a new amplification of wear at the junction area between the paraconid and the protoconid, as well as on the protoconid, which stays separated. Wear is also apparent on the metaconid (a), the hypoconid (b), the metaconid and the hypoconid (c), or on the entoconid and the hypoconid (d).

F) At stage F dentine of the paraconid and the protoconid join up, while wear continues to increase on the metaconid and the entoconid (a), on the metaconid and the hypoconid (b) or on the metaconid, the entoconid and the hypoconid (d).

G) At stage G all cusps are heavily worn, and the occlusal surface is flat. The wear-surface between the paraconid and the protoconid is very large (a), crown height is reduced by a third (b). At stage G (c), the wear-surface joins up the paraconid and the protoconid, and at stage G (d), the wear-surface joins the protoconid and the metaconid. The tooth is then reduced by two thirds.

After codifying and determining wear stages, I attempted to assign true ages by fitting comparative data from an archaeological collection. The archaeological collection of dogs from the Roman period sanctuary of Vertault (Méniel, 1992) was chosen for the following reasons: it is an exceptional collection of 70 complete dog skeletons, comprising various age groups, thus assuring statistic validity; the morphology and size of the dogs are homogeneous within the collection - no breeds are apparent; the same zootechnical context can be assumed; the animals are assumed to have been fed in a similar way, on a rustic diet. Owing to these characteristics, this collection is considered to be appropriate for this study, since a relatively constant rate of tooth wear can be assumed. Despite the problems of using an archaeological assemblage (where true ages cannot be known), it would have been very difficult to find a series of dog skeletons from a modern population that could be used to accurately relate the dental rate of attrition with the age of dogs by conventional zooarchaeological methods.

**Age determination**

Each individual from this Roman collection was assigned an age based on skull and post-cranial epiphyseal data (Barone, 1986; Silver, 1969), as well as on the dental eruption sequence (Hillson, 1986), and incisor wear (Pierard, 1972). After examining the attrition of the first permanent lower molar of each dog, it was possible to relate true age with the wear-stage observed (Table 1 & Fig. 4).

| AGE IN MONTHS | 6-10 | 10-15 | 15-24 | 24-36 | 36-48 | 48-72 | > 72 |
|---|---|---|---|---|---|---|---|
| n | 26 | | 25 | 12 | 4 | 1 | 2 |
| Stage A | 46% | | 12% | | | | |
| Stage B | 23% | | 16% | | | | |
| Stage C | 31% | ? | 20% | 17% | | | |
| Stage D | | | **48%** | 17% | | | |
| Stage E | | | 4% | **67%** | 25% | | |
| Stage F | | | | | 75% | | |
| Stage G | | | | | | 100% | 100% |

***Table 1*** *Results of relationship of toothwear stage to age. Significant differences in* ***bold***

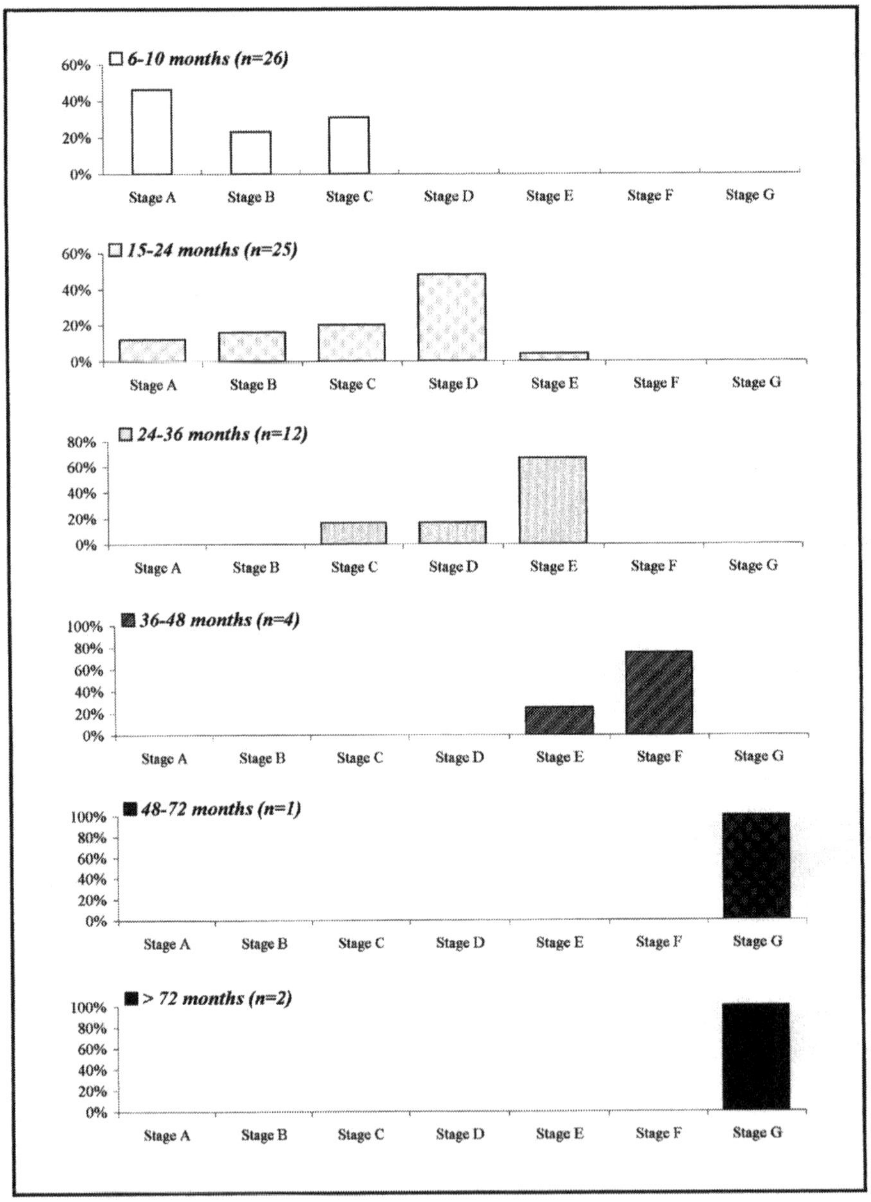


***Figure 4*** *Wear stages and variability observed in each of the age-classes defined*

The results of this analysis suggests that this collection of 70 complete dog skeletons is insufficient for defining age parametres based on tooth wear. Young adults aged between 10 and 15 months were absent from the sample and it was therefore impossible to refine the hierarchical classification of stage B and C. In addition, there is also an insufficient number of old dogs (only seven individuals more than 36 months). These gaps will have to be filled with data from additional analyses of complete dog skeletons.

Although the small sample size does not allow an assignment of absolute age to individuals, it is nevertheless possible to observe a relationship between the rate of dental attrition and age. The dogs can be classified into three major age groups: young (stages A, B and C), intermediate (D, E and F) and old (G). These age-classes were then used to construct histograms for the study of kill-off patterns in dogs from other assemblages.

The age estimation method developed here were used to analyse kill-off patterns of dogs from the Gallic village of Levroux. In this particular site, 2,662 dog remains have been identified but only 1,026 were subjected to a detailed analysis (Horard-Herbin, 1997). Anatomical elements consisted primarily of cranial fragments and bones from the upper limbs whereas ribs, vertebrae and lower limb bones were poorly represented. This skeletal element distribution is similar to that of pig and caprines remains from the same site (27,311 and 7,155 elements, respectively).

Butchery marks were observed on some of the dog bones (17.5%): either heavy cuts made by axes or cleavers (associated with the first stages in the cutting of the carcass), or fine cuts made by knives (associated with the final preparation and/or consumption). Skins appear to have been systematically removed, and some bones were burnt (2.9%). Butchery patterns observed for the dog were analogous to those of other domestic species (pig, goat and cattle), whose carcasses appear to have been processed in the same general way.

Thus, on the Levroux site, consumption of dogs during the period 200-80BC can be attested without ambiguity based on the presence of disarticulated dog remains among food residues, the frequency and type of damage to the bones, and to the the similarites in skeletal part representation and butchery patterns observed for both dog and other domestic animal remains. This phenomenon is also observed in other habitation sites of containing Late Iron Age fauna (Méniel, 1987).

Age data for each isolated mandible and tooth were combined according to the chronological sequence of the site, from which two histograms were constructed (Fig. 5). As this analysis is based on the study of the first lower molar which erupts between three and five months (Hillson, 1986), information concerning very young dogs is missing. It is estimated that about 20% of the population would have been composed of puppies not represented in the first two histograms [Dog mandibles from Phase 1 (200-150BC) were recovered from 10 features whereas those from Phase 2 (150-100 BC) were recovered from 15 different features. Only one feature was closely examined within each phase: feature L154 comprising 15 of the 29 adult mandibles (52%) for Phase 1; and feature R110 comprising 22 mandibles out of 52 for Phase 2 (42%). In addition to the mandibles from adult dogs, feature L154 also contained mandibles from four puppies (two aged three months, one aged four months and one aged five months) out of a total of 19 individuals ; and feature R110 contained mandibles from five puppies (two aged three months, two aged four months, and one aged five months) out of a total of 27 individuals. These very young dogs, who also exhibit butchery marks, thus represent between 18.5 and 21% of the total number of individuals, which enables us to estimate the proportion of puppies whose ages are not considered in the histograms at about 20%].

The kill-off patterns for the first two chronological phases (200-150BC and 150-100BC) are quite similar. In both cases, between 72 and 69% of the dogs were slaughtered at a relatively young age (stages A, B, and C) and between 21 and 27% were slaughtered at an intermediate age (stages D, E and F), while the proportion of old dogs is very small (between 7 and 4%, stage G).

These age profiles, in which a majority of dogs are slaughtered at the same age, are quite similar to those observed for pigs at the same period. Slaughter appears to have been primarily targeted at young individuals intended for meat production, while a few adults were kept for breeding purposes and to assure renewal of the flock.

The pattern of dog exploitation remains relatively constant during the whole of the Gallic period at Levroux. A majority of dogs were killed young, and some very young dogs, aged between three to five months, were also slaughtered. Dogs do not appear to have been kept mainly as a secondary food source. Rather, the kill-off patterns suggest that they were the object of a standardized meat production management, carried out in a very systematic fashion.

It thus seems that dogs in this village were raised more for their meat than for the other purposes they might serve, such as hunting, guarding or companionship. Alternatively, individuals that filled those services roles may not have been eaten and could have been disposed of and buried somewhere other than the middens, such as remains recovered from sites at sanctuaries or necropolis (Méniel, 1992).

This study on interpretation of age profiles for dogs was hindered by a methodological problem: inappropriate assumptions may have been incorporated due to the necessity of basing the comparison on mortality profiles built for understanding husbandry practices of the more common meat producing animals. Ideally, the interpretation should take into consideration the natural mortality profile of dogs as well as a model designed more specifically to understand meat production in dogs. I am still looking for this type of data in the ethnographic literature.

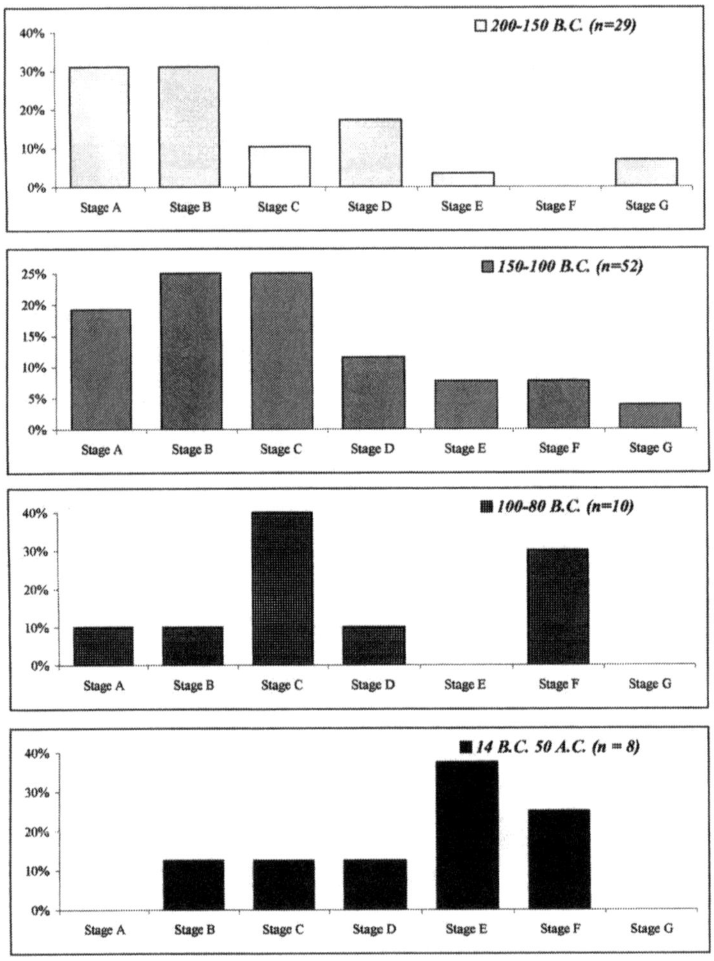


***Figure 5*** *Kill-off patterns in dogs from the Gallic site of Levroux (Indre, France)*

## Conclusion

This study establishes a method for estimating age of dogs by reference to dental attrition stages for both associated and isolated lower permanent first molars. The method enables the construction of age profiles for a species in which the dental eruption sequence and epiphyseal fusion occur at an early age. Dogs can thus be classified into major age-classes (young, intermediate and old) which can be used to build histograms for the study of mortality or kill-off patterns. This method of age estimation for the dog opens up new possibilities in understanding the role of this species at the end of the Iron Age, a period when dogs are meat-producing animals but also play an important role in social and religious life.

## Acknowledgements

I am most grateful to Patrice Méniel for placing this material at my disposal.
Evelyne Cossette and Virginia Elliot were kind enough to translate the text into English.

## References

Barone, R. (1976). *Anatomie comparée des mammifères domestiques*, 3 tomes. Paris : Vigot.

Buchsenschutz, O., Colin, A., Krausz, S., Levéry, M. & Soyer, C. (1988). *Levroux 1 : L'évolution du canton de Levroux d'après les prospections et les sondages archéologiques*. Supplément à la Revue Archéologique du Centre de la France. Levroux : ADEL.

Buchsenschutz, O., Colin, A., Krausz, S., Levéry, M. & Soyer, C. (1993). *Levroux 2 : Le village celtique des Arènes à Levroux. Description des structures*. Revue Archéologique du Centre de la France 8ème supplément. Levroux : ADEL.

Buchsenschutz, O., Colin, A., Krausz, S., Levéry, M. & Soyer, C. (1994). *Levroux 3 : Le village celtique des Arènes à Levroux. Description du mobilier*. Revue Archéologique du Centre de la France, 10e supplément. Levroux : ADEL.

Hillson, S. (1986). *Teeth*. Cambridge: Cambridge University Press.

Horard-Herbin M. P. (1997). *Levroux 4 : L'élevage et les productions animales dans l'économie de la fin du second Age du Fer à Levroux (Indre).* 12e supplément de la Revue Archéologique du Centre de la France. Levroux: ADEL.

Méniel, P. (1987). *Chasse et élevage chez les Gaulois (450-52 av J.-C.).* Paris: Errance.

Méniel, P. (1992). *Les sacrifices d'animaux chez les Gaulois.* Paris: Errance.

Silver, I. A. (1969). The ageing of domestic animals. In (D. Brothwell & E. Higgs, Eds) *Science in Archaeology.* Thames & Hudson, London, pp. 283-302.

# CONSOMMATION DE LA VIANDE DE CHIEN: LE *VICUS* DE BRAIVES (BELGIQUE) ET LES SITES GETO-DACES ET ROMAINS DE ROUMANIE

# (BUTCHERY EVIDENCE ON DOG FAUNAL REMAINS FROM ROMAN PERIOD SITES IN BELGIUM (BRAIVES) AND ROMANIA)

Carmen Tarcan
Department of Archaeology
Simon Fraser University
Burnaby, B.C. V5A 1S6
Canada

J.-M. Cordy
U.R. "E.V.E.H.", Université de Liège,
Institute L. Frédéricq (Bat. L1),
Place Delcour, 17, 4020 Liège - Belgium

Luminita Bejenaru
Faculté de Biologie,
Université de Iasi,
Bd. Copou 11, 6600 Iasi, Roumanie

M. Udrescu
U.R. "E.V.E.H.", Université de Liège,
Institute L. Frédéricq (Bat. L1),
Place Delcour, 17, 4020 Liège - Belgium

**Key words :** dog, butchering, Roman, Romania, Belgium, archaeozoology, dog comsumption, Pre-Roman

**Abstract**

This study examines butchery traces shown by the dog assemblage from a Gallic-Roman site from Braives (first-fourth centuries AD), located in Belgium. Some comparisons are made with collections from Geto-Dacic (fourth century BC - first century AD) and Daco-Roman (first-third centuries AD) sites, located in Romania. The butchery marks found on the dog remains from Braives show that the dog was eaten. Its share in the daily diet of the inhabitants was not significant in comparison to the other consumed species. We suggest that the consumption of dog at Braives was related more to some local habits than with food necessities. The dog assemblages from the pre-Roman period sites in Romania show only indirect evidence of consumption: the constant presence of burned and fragmented bones of dog among the culinary refuse. In contrast, the analyses of dog remains from later, Roman period sites in Rumania have not provided any evidence of dog consumption.

## Introduction

Cinq campagnes de fouilles (I à V) ont été menées dans le *vicus* gallo-romain de Braives (Province de Liège, Belgique). Plusieurs zones d'habitat s'étalant généralement entre le Ier et le IVe siècles après J.C. ont été mises à jour: la zone centrale du *vicus* ou "Braives I" (BV I), le quartier des potiers ou "Braives II" (BV II), la zone périphérique occidentale ou "Braives III" (BV III), la zone centre-ouest ou "Braives IV" (BV IV) et la fortification du Bas-Empire, ainsi qu'un puits ou "Braives V" (BV V). Dans les cinq zones d'habitat, près de 5300 pièces ostéologiques ont été identifiées (Table 1).

L'étude des matériaux ostéologiques des ces cinq campagnes de fouilles n'a permis que peu de constatations quant à une éventuelle utilisation alimentaire des chiens. Ainsi, c'est dans Braives IV que, pour la première fois, une possible utilisation alimentaire du chien est envisagée: "ces animaux (le cheval et le chien) n'étaient donc pas - ou peu - destinés à la boucherie" (I. Trabert, 1990, p.213). Dans le Braives V, sur la base de deux fragments d'os coxal, les auteurs ont conclu que "le chien ...a sans doute été consommé occasionnellement..." (Yernaux, Udrescu & Cordy, 1993 : 237). Enfin, dans le puits C 13 (Braives V), qui est très riche (presque 6 000 pièces ostéologiques), les auteurs signalent qu' "aucun os de chien de Braives...ne porte de telles marques (traces de découpage)" (Lentacker, Van Neer & Desender, 1993 :327).

| **Especes** | **Braives I** | | **Braives II** | | **Braives III** | | **Braives IV** | | **Braives V** | |
|---|---|---|---|---|---|---|---|---|---|---|
| | **N** | **%** | **N** | **%** | **N** | **%** | **N** | **%** | **N** | **%** |
| Bos taurus | 511 | 97.15 | 434 | 66.98 | 55 | 45.08 | 740 | 50.24 | 1030 | 48 |
| Equus caballus | 2 | 0.38 | 35 | 5.4 | 11 | 9.02 | 241 | 16.36 | 158 | 7.36 |
| Ovis aries+Capra hircus | 3 | 0.57 | 126 | 19.44 | 23 | 18.85 | 175 | 11.88 | 451 | 21.02 |
| Sus domesticus | 5 | 0.95 | 51 | 7.87 | 31 | 25.41 | 125 | 8.49 | 445 | 20.74 |
| Canis familiaris | 1 | 0.19 | 1 | 0.15 | | | 191 | 12.97 | 51 | 2.38 |
| Felis catus | | | | | | | | | 1 | 0.05 |
| | | | | | | | | | | |
| Bos primigenius | | | 1 | 0.15 | 2 | 1.64 | | | | |
| Cervus elaphus | 2 | 0.38 | | | | | | | 2 | 0.09 |
| Capreolus capreolus | | | | | | | 1 | 0.07 | | |
| Sus scrofa | 2 | 0.38 | | | | | | | | |
| Vulpes vulpes | | | | | | | | | 1 | 0.05 |
| Lepus europaeus | | | | | | | | | 6 | 0.28 |
| Talpa europaea | | | | | | | | | 1 | 0.05 |
| **Total des mammiferes** | 526 | 100 | 648 | 100 | 122 | 100 | 1473 | 100 | 2146 | 100 |
| | | | | | | | | | | |
| Homo sapiens sapiens | 19 | 3.5 | | | 1 | 0.8 | 2 | 0.13 | 24 | 0.97 |
| Oiseaux | | | 2 | 0.3 | 1 | 0.8 | 25 | 1.7 | 271 | 10.98 |
| Batraciens | | | | | | | | | 1 | 0.04 |
| Mollusques | 1 | 0.2 | | | | | | | 26 | 1.05 |
| **Total divers** | 20 | 3.7 | 2 | 0.3 | 2 | 1.61 | 27 | 1.83 | 322 | 13.04 |
| **Total general** | 546 | | 650 | | 124 | | 1500 | | 2468 | |

***Tableau 1***: *Decompte des restes et proportions des especes identifiees dans les differentes secterurs fouilles dans le vicus de Braives (cf Yernaux, Udrescu & Cordy, 1993).*

Pour la période gauloise (Ve - Ier siècles av. J.-C.), il n'y a pas de doute quant à la consommation de viande de chien. Ainsi par exemple, Meniel constatait que, chez les gaulois, "la viande de chien a donc été consommée et si la découverte de squelettes dépourvus de telles traces montre que cette pratique n'est pas systématique, leur petit nombre ne permet pas de mettre en cause le statut d'animal de boucherie" (Meniel, 1987: 29). Ce fut d'ailleurs le statut du chien partout en Europe avant la conquête et l'avis des différents auteurs est quasi unanime à ce sujet.

Chez les gallo-romains (Ier siècle av. J.-C. - IVe siècle après J.-C.), Meniel (1992: 16) est d'avis que "l'influence du monde romain est d'ailleurs tout à fait perceptible après la conquête, cheval et chien ne faisant alors plus partie des animaux consommés en Gaule". Par contre, un passage en revue de quelques données d'archéozoologie sur la période romaine, principalement en Allemagne, a permis aux auteurs de l'étude des ossements du puits $C^{13}$ de Braives, de constater que "certains sites romains ont livré des os de chiens avec des traces de découpage..." et que "les auteurs croient que ces chiens étaient consommés" (Lentacker, Van Neer & Desender, 1993: 327). Comme on peut facilement s'en rendre compte, pour la période qui suit la conquête romaine, l'avis des auteurs sus-mentionés est partagé.

## Nos observations

Jusqu'à présent, nous avons étudié les traces de "boucherie" sur les os de caprinés, de cheval et de chien de Braives I à V et dans un premier temps dans la présente publication, nous n'allons présenter que les cinq pièces ostéologiques de chien qui portent de telles traces:

1. Fragment d'os coxal droit, la région ischiatique, Braives V, no. 43-12 (Fig. 1).

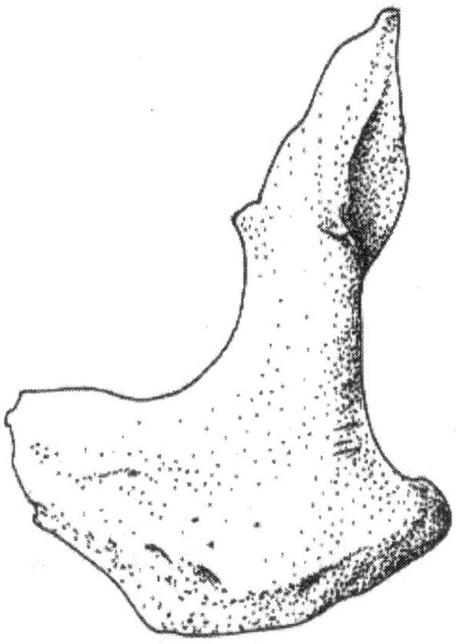

***Figure** 1 Fragment d'os coxal droit de chien (*Canis familiaris*) avec des entailles (Braives V, 43-12).*

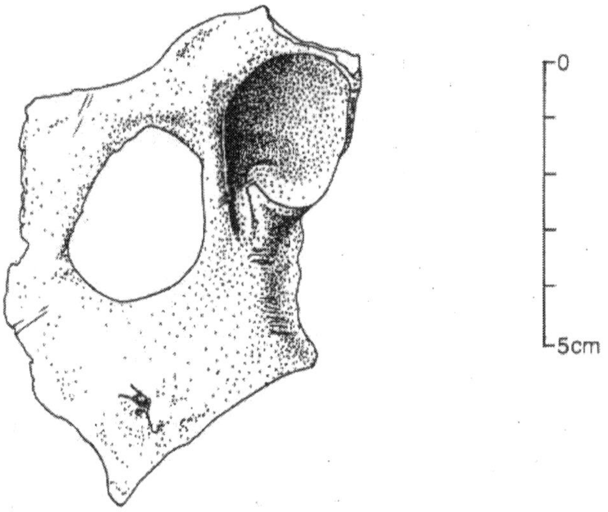


***Figure** 2 Fragment d'os coxal gauche de chien (*Canis familiaris*) avec de nombreuses entailles (Braives V, 43-13).*

On constate trois entailles parallèles sur la petite échancrure sciatique, placées plutôt vers la tubérosité ischiatique. Les trois traces ont 3 mm de longueur et sont peu profondes; chaque entaille est le résultat de plusieurs passages du tranchant de couteau et elles sont un peu en oblique par rapport au diamètre longitudinal de l'os coxal.

2. Fragment d'os coxal gauche, région acétabulaire, Braives V, no. 43-13, (Fig. 2).

On observe plusieurs courtes entailles au niveau du bord de l'acetabulum, qui atteignent même la surface semi-lunaire. Sur la partie présente de l'ischium, il y a plusieurs entailles parallèles, assez courtes (2-3 mm longueur) et peu profondes. Quelques entailles sont visibles entre l'éminence ilio-pubienne et la symphyse pelvienne. Enfin, on constate plusieurs entailles (7-8 mm longueur) au niveau de la symphyse pelvienne, obliques vers le foramen obturé.

Les particularités morphométriques des deux fragments d'os coxal nous font croire qu'ils proviennent du même bassin (Fig. 3)

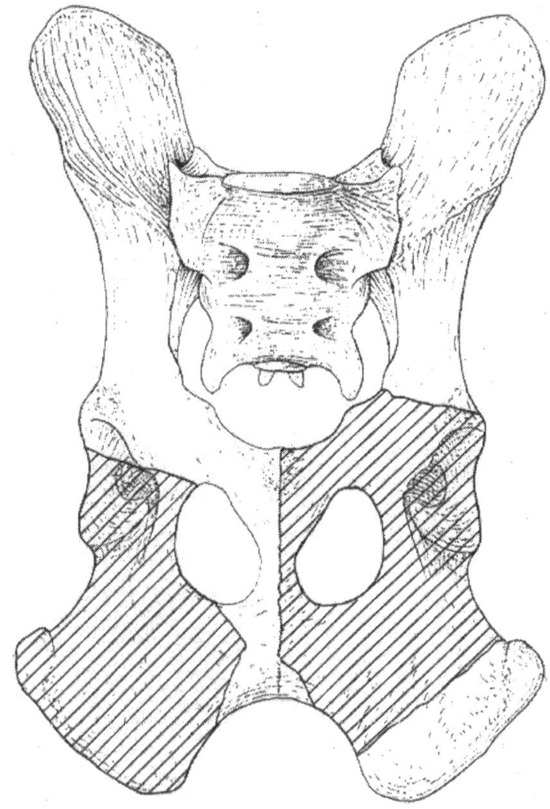

***Figure** 3 Schéma de l'emlacement des deux fragments d'os coxal (Braives V, 43-12 et 43-13) sur un bassin de chien (*Canis familiaris*)*

3. Moitié distale d'un fémur droit, Braives IV, no. 87-30 (Fig. 4)

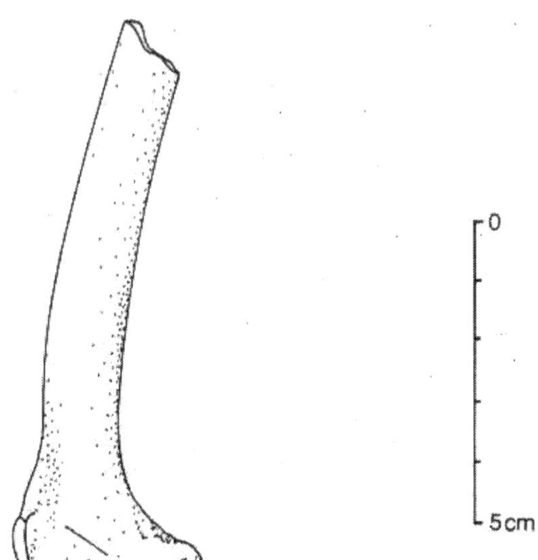


***Figure** 4 Moitié distale de fémur droit de chien (*Canis familiaris*) avec des entailles (Braives IV, 87-30); vue médiale.*

Sur la face médiale de l'épiphyse distale, on remarque une trace assez grande de couteau que relie presque la lèvre médiale de la trochlée au condyle médial; à cela s'ajoutent encore trois entailles au niveau du même condyle.

4. Tibia droit, Braives IV, no. 31-4 (Fig. 5)

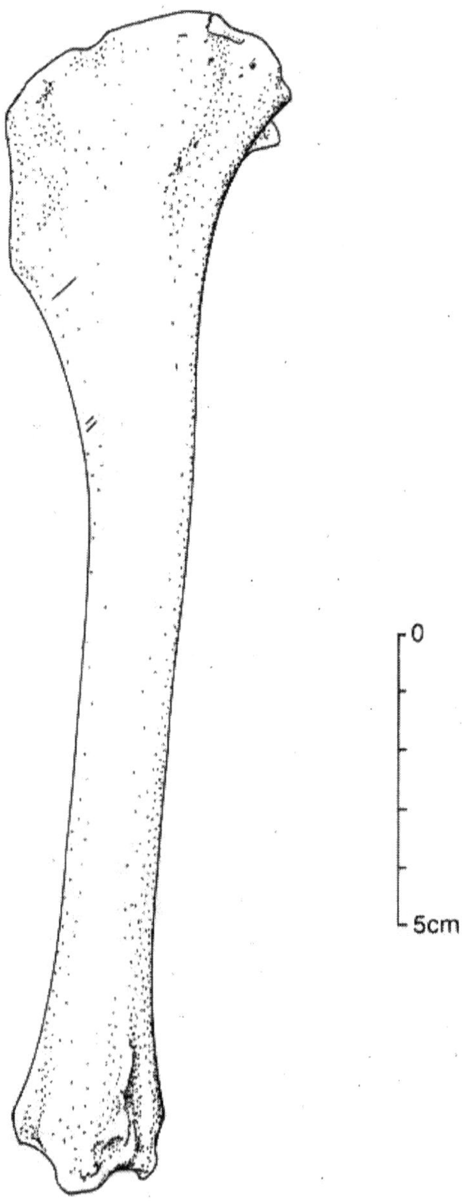


***Figure 5*** *Tibia droit de chien (*Canis familiaris*) avec des entailles (Braives IV, 31-4) ; vue médiale.*

Partant de la base de la crête tibiale, oblique vers la partie proximale de l'os, se trouve une entaille unique de 6 mm longueur; 20 mm plus bas, deux entailles plus courtes sont orientées dans la même direction.

5. Tibia gauche, Braives V, no.63-9 (Fig. 6)

Sur la face caudale du corps de l'os, un peu plus haut que le foramen nourricier, on observe quatre fines entailles, parallèles et presque perpendiculaires à la longueur de l'os. Une des entailles se prolonge sur le bord médial du tibia; elle est doublée à ce niveau par plusieurs encoches parallèles, résultat de plusieurs passages du tranchant du couteau.

Les entailles observées au niveau de toutes les pièces ostéologiques décrites ont été faites avec le même type d'outil coupant, avec le tranchant très fin, de type couteau.

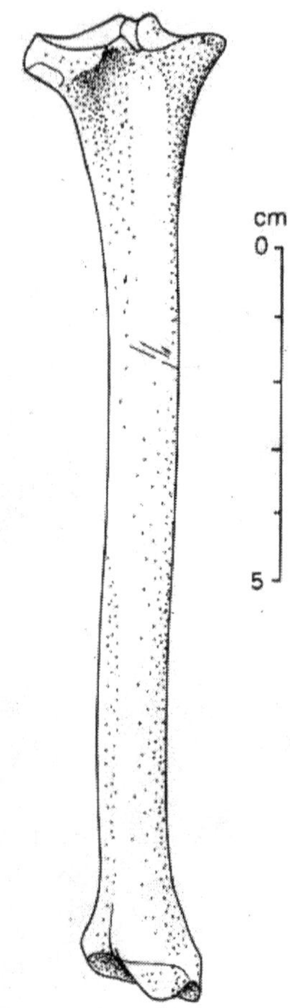


***Figure 6*** *Tibia gauche de chien (*Canis familiaris*) avec des entailles (Braives V, 63-9) ; vue caudale.*

## Discussion

Une première estimation de la proportion de pièces ostéologiques présentant des traces de décarnisation et/ou de découpe au niveau des caprinés, du cheval et du chien montre, pour les deux premiers, des pourcentages semblables qui sont supérieurs à celui du chien (Table 2). Il n'y a pas de différence entre les entailles répertoriées sur les os de chien et celles observées au niveau des os de caprinés et du cheval, ce qui nous fait penser qu'elles ont été faites avec le même type d'objet tranchant. Par contre, pour les caprinés et surtout le cheval, nous avons répertorié une plus grande variété de traces qui prouvent l'utilisation de plusieurs types d'outils tranchants. Une chose semble pourtant être claire: le chien a été consommé dans le site gallo-romain de Braives; son importance alimentaire était probablement beaucoup plus réduite par comparaison aux espèces domestiques utilisées habituellement comme nourriture (boeuf, mouton, cochon). De toute façon, même un pourcentage si réduit de pièces qui portent des traces de écarnisation chez le chien ne nous permet plus de mettre en doute son statut d'animal consommé à Braives.

| ESPECE | BRAIVES I – V | | |
|---|---|---|---|
| | Nombre de pièces examinées | Nombre de pièces avec des traces | % |
| *Canis familiaris* | 96 * | 4 | 4.2 |
| *Ovis - Capra* | 678 | 62 | 9.1 |
| *Equus caballus* | 390 | 39 | 10.0 |

* Nous n'avons compté que les pièces ostéologiques isolées

***Tableau 2:*** *Nombre des pièces présentant des traces de boucherie chez les caprinés, le cheval et le chien*

L'économie animalière pendant la période gallo-romaine satisfaisait largement le besoin en protéines animales, par l'utilisation d'espèces "classiques" comme nourriture. Dans ce contexte, continuer à consommer de la viande de chien tenait probablement plutôt des habitudes alimentaires locales, que des nécessités.

Si à Rome et pour la haute aristocratie la viande de chien était "immonde", dans les campagnes et, certainement, dans les provinces assez éloignées de Rome, les habitudes alimentaires ont pu se perpétuer. Même dans les *vicus*, des centres fortement romanisés, comme à Braives, la partie "basse" de la population a pu garder certaines de ses habitudes alimentaires antérieures à la conquête. Nous sommes d'avis qu'à l'époque gallo-romaine, il y avait des communautés très "romanisées" dans lesquelles la consommation de viande de chien n'était plus envisagée, mais qu'il y en avait d'autres, plus petites et probablement plus « éloignées » de centres politico-administratifs, dans lesquelles la consommation de viande de chien continuait comme avant la conquête.

D'ailleurs, de telles habitudes alimentaires se sont gardées en Europe, dans des régions proches de la Belgique, jusqu'à très récemment. Ainsi, en Allemagne, la fermeture des boucheries dans lesquelles on vendait de la viande de chien s'est faite à peine en 1986: "After 1930 it was frequently called for abolishing the inspection order for dogs of 1900 or for the prohibition of dog slaughterings. Although four draft laws of the years 1954, 1959, 1963, 1985 have existed, the slaughter prohibition enacted to law in the year 1986." (Geppert, 1990: 201).

## Comparaison avec la Roumanie

Les avis concernant l'utilisation du chien aussi comme nourriture pendant la période géto-dace (IVe siècle av. J.-C. - Ier siècle après J.-C.) et daco-romaine (IIe - IIIe siècles après J.-C.), si peu nombreux soient-ils, sont assez contradictoires. Ainsi, Haimovici écrit: "Un autre problème est de savoir si les géto-daces ont utilisé les chiens dans l'alimentation. Nous avons tendance à répondre par la négative..." (Haimovici, 1987: 152). Par contre, C. Hriscu et collab. écrivent que "La présence des os de chien parmi les déchets ménagers dans les habitations mêmes ... la présence à peu près constante de pièces ostéologiques brûlées ... nous fait croire que l'espèce était utilisée ... aussi dans l'alimentation chez les géto-daces" (Hriscu, Bejenaru & Udrescu, 1996: 135).

On a parfois l'impression que les archéozoologues ont "contourné" délibérement ou involontairement le "sujet"; des ressentiments "actualistes" les empêchent peut-être de reconnaître une telle éventualité: nos ancêtres géto-daces ont mangé de la viande de chien, quelle hantise! Il y a pourtant suffisamment d'arguments en faveur des deux points de vue. De toute façon, même si la viande de chien a été consommée, le faible pourcentage de pièces ostéologiques attribuées au chien dans tous les établissements géto-daces et daco-romains nous fait croire qu'une telle utilisation a été tout à fait subsidiaire. En échange, il était employé assez fréquemment dans les manifestations rituelles et funéraires; on l'utilisait certainement pour la garde des troupeaux domestiques et des habitations et, peut-être, pour la chasse.

A notre connaissance, les études archéozoologiques pour la période daco-romaine ne mentionnent pas une éventuelle utilisation du chien dans alimentation. Une des priorités de la recherche archéozoologique en Roumanie est de résoudre ce problème, soit par une nouvelle analyse des échantillons déjà étudiés, soit par une analyse attentive de matériaux futurs.

## Conclusions

Le chien a été consommé dans le site gallo-romain de Braives comme en témoignent les cinq pièces ostéologiques portant des entailles. Toutefois, l'importance alimentaire de cet animal était certainemenft beaucoup plus réduite par rapport aus espèces domestiques utilisées habituellement comme nourriture dans les sites gauloises et gallo-romains.

Dans les sites géto-daces et daco-romains, l'identification des os des chien parmi les déchets ménagers dans les habitations même ainsi que la présence constante de pièces ostéologiques brûlées, confirment également l'hypothse de la consommation de la viande de chien de l'antiquité. Par contre, dans tous ces sites, le faible pourcentage de pièces ostéologiques attribuées au chien nous fait croire à nouveau qu'une telle utilisation a été tout à fait accessoire.

**Acknowledgements**

Les dessins de cette publication ont été réalisés par Feli Giraldo-Martin du projet PRIME no. 30170, asbl Paléontologie et Archéologie Karstique, rue Lamarck 121 à 4000 Liège (Belgique).

**Bibliographie**

Cordy, J.-M. (1981). Archéozoologie. In (R. Brulet *et alii, Eds*) *Braives gallo-romain I. La zone centrale, Publ. Hist. Art et Archeol. Univ. Cath. Louvain* **24**, 191-200.

Cordy, J.-M. et Stassart, M. (1983). Archéozoologie. In (R.Brulet *et alii, Eds*) *Braives gallo-romain II. Le quartier des potiers, Publ. Hist. Art et Archeol. Univ. Cath. Louvain,* 190-200.

Cordy, J.-M. et Rapaille, A. (1985). Archéozoologie. In (R. Brulet *et alii, Eds*) *Braives gallo-romain III. Le quartier des potiers, Publ. Hist. Art et Archeol. Univ. Cath. Louvain* **46**, 137-147.

Geppert, P. (1990). *Hundeschlachtungen in Deutschland im 19. Und 20. Jahrhundert unter besonderer Berücksichtigung der Verältnisse.* München: Thèse.

Haimovici, S. (1987). Cresterea animalelor la geto-daci (sec IV i.e.n - sec I e.n.) din Moldova si Muntenia. *Thraco-Dacica* **VIII**, 144-153.

Hriscu, C., Bejenaru, L. & Udrescu, M. (1996). Materialul osteologic din asezarile geto-dace (sec IV-I i. Hr.) din zona Cascioarele-Greaca-Prundu. Date zooarheologice. In (V. Sirbu *et alii, Eds*) *Asezari din zona Cascioarele-Greaca-Prundu, mileniile I i.Hr. - I d.Hr,* 134-138. Braila.

Lentacker, A., Van Neer, W. & Desender, K. (1993). Archéozoologie. In (R. Brulet *et alii, Eds*) *Braives gallo-romain V; La fortification du bas-empire, Publ. Hist. Art et Archeol. Univ. Cath. Louvain* **83,** 284-339.

Meniel, P. (1987). *Chasse et élevage chez les Gaulois (450-52 av J.C.).* Paris: Ed. Errance.

Meniel, P. (1992). *Les sacrifices d'animaux chez les Gaulois*. Paris: Ed. Errance.

Trabert, I. (1990). Archéozoologie. In (R. Brulet *et alii, Eds*) *Braives gallo-romain IV. La zone Centre-Ouest, Publ. Hist. Art et Archeol. Univ. Cath. Louvain* **77,** 203-214.

Yernaux, G., Udrescu, M. & Cordy, J.-M. (1993), Archéozoologie. In (R. Brulet *et alii, Eds*) *Braives gallo-romain V; La fortification du bas-empire, Publ. Hist. Art et Archeol. Univ. Cath. Louvain* **83,** 225-241.

# DOG REMAINS FROM THE LATE HALLSTATT PERIOD OF THE CHIMNEY CAVE DUREZZA, NEAR VILLACH (CARINTHIA, AUSTRIA).

Alfred Galik
Institute for Palaeontology
Geozentrum University
Althanstr 14A,
Vienna, a-1090,
Austria

**Key words:** dog, archaeozoology, Austria, Carinthia, Hallstatt period, burials, Iron Age

**Abstract**

The site reported on here is located between the Tscheltschnigkogel and the Dobratsch massif near Villach in southern Carinthia, Austria. The Durezza cave is the vertical continuation of a doline, which was originally filled up by a conical pile of sediment. As cave explorers started to dig there, they discovered many human and animal bones, and few bronze ornaments. The remaining sediment was scientifically excavated in cooperation with the cave explorers, to document the finds and to ascertain their horizontal and vertical distribution.

The bones were scattered and not in anatomical position. Dog skulls lay next to human skulls, for example. Characteristic bronze ornaments date the complex of finds to a relative short period of time, *c.* 200 years during the latter part of the Iron Age "Hallstatt" period (*c.* 700-450BC). Among the animal remains, domestic taxa are predominant. Most of these remains are represented by dog bones. The dog bones show a biased distribution, caused by the excavation methods used by the cave explorers. Large, visible dog bones are dominant: the distribution shows a paucity of smaller elements. However, some small bones were recovered and both very small caudal vertebrae as well as complete skulls are represented.

More than 2,800 dog specimens could be identified and they represent a minimum number of 45 individuals. About 1,400 elements out of all dog specimens were complete. More than 500 dog bones were rearticulated and more than 300 pairs of elements were matched. Te rearticulations generally do not include the vertebrae, although it was possible to rearticulate two almost complete spinal columns. Pairs and rearticulated elements represent every body part: trunk, limbs and head. Although cut and tooth marks were found on a few bones, it seems that complete dog bodies were thrown into the cave. More than 2,500 of the bones of dog remains came from adult individuals, apparently in the prime of life.

The cave was presumably used by a local human population to throw people and animals into, for whatever reason. However, careful investigations of the dog remains allow us a very close look at the dog population that lived during the Hallstatt period in southern Carinthia. It is possible to make statistically well defined statements about the size variations, body form, diseases and nutrition of these dogs. The results will be valuable for comparisons with other dog populations or for inferring breeding strategies.

## Introduction

In 1989 cave explorers discovered the Durezza cave in a small valley near Villach in Austria (Galik & Pacher, 1997). Unfortunately they excavated about 2/3 of the cave sediment and deposited it outside the cave. However, the remaining sediment was scientifically excavated in 1996 (Galik, 1997a) and washed through sieves with a mesh widthf two millimetres. The sediment heaps outside the cave became re-excavated and a lot of small bones like metapodials and phalanxes were recovered. The cave sediment contained a few artifacts (bronze ornaments, glass beads, weaving weights, and pottery fragments), as well as numerous human and animal bones. Characteristic bronze fibulae and bronze hair rings were found on the top as well as at the base of the sedimentary complex. Therefore, the whole complex can be dated to the latter part of the Iron Age "Hallstatt" (*c.* 700-450BC) period (D2 to La Tène A), with a maximum time range of 200 years assumed for the whole complex (Gleirscher, 1997a). Two $C^{14}$ dates of human bones confirmed the archaeological dating (Fuchs, 1996).

The majority of the animal bones belong to domestic animals, which are predominated by dogs (Fig.1). The aim of this work was to enlarge the knowledge about dogs from this period, using these closely dated dog bones from this unique site in Austria. The material contains a high number of complete bones, which are useful for quantitative analysis.

## Methods

This work consists of statistical analysis of dog elements, in two parts. The first part consists of quantitative analysis, aimed at revealing what conditions may have caused the dogs to be placed in the cave. The second part entails morphometric analysis, which is necessary to show any morphological homogeneity or differences within the sample.

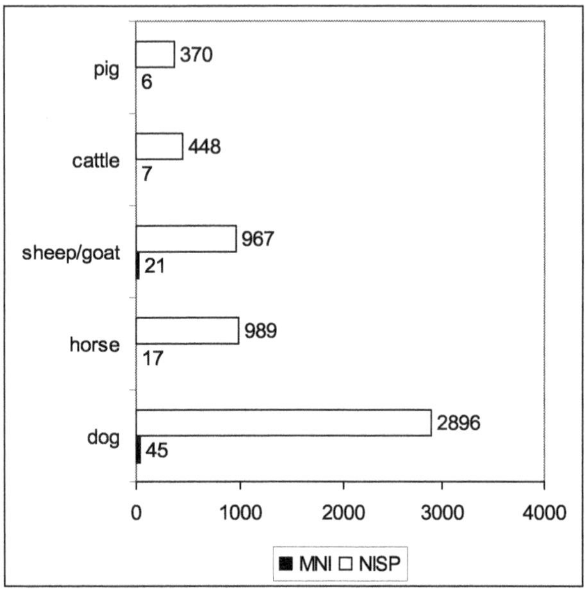


*Figure 1 Minimum Individual Number (MNI) and Number of Identified Specimens (NISP) of domestic animals from the Durezza cave*

Univariate statistics were done to analyse the body dimensions of the Durezza dogs. Univariate or bivariate statistics often do not discriminate efficiently enough to reveal differences or homogenities in skeletal material. Cluster and discriminant analysis were used as tools to test if different morphometric groups could be discerned among the canid sample (*cf* Benecke, 1987; Morey, 1990). These statistics were useful for condensing many metric traits into single steps of analysis for morphometrical description of elements. The resulting morphologies of skulls and mandibles are discussed in detail. The metric traits are listed in Tables 7 and 8.

As many bone fragments as possible were refitted to original complete bones. The frequencies of specimens are listed as "NISP" (Number of Identified Specimens) in Table 3 (Ringrose, 1993; Grayson, 1984; Lyman, 1994). Specimens are complete bones (**Ec**, or "complete elements"), as well as bone fragments from a skeleton. Useful fracture types were defined for fragmented limb bones to prevent biased frequencies for several statistics, specifically MNE (Minimum Number of Elements) and MNI (Minimum

| Skull | Code | Mandible |
|---|---|---|
| (1) Total length | v1 | (1) Total length |
| (2) Condylobasal length | v2 | (8) Length of the teeth row ($P_1 - M_3$) |
| (3) Basal length | v3 | Reconstructed basal length |
| (4) Basicranial axis | v4 | (2) Length: processus angularis-infradentale |
| (5) Basifacial axis | v5 | (3) Length: saddle between p. condylaris and p. angularis to infradentale |
| (7) Height of the Neurocranium triangle | v6 | (4) Length: proc. condylaris to the aboral alveolus edge of the canine |
| (8) Facial length | v7 | (5) Length: saddle between p. condylaris and p. angularis to the alveolus edge of the canine |
| (9) Viscerocranium length | v8 | (6) Length: p. angularis to the aboral alveolus of the canine |
| (13) Median palatal length | v9 | (7) Length: the aboral border of the alveolus of $M_3$ to the aboral border alveolus of the canine |
| (15) Length of the cheektooth row | v10 | (10) Length of the molar row |
| (19) Length of the carnassial alveolus | v11 | (17) Greatest thickness of the body jaw |
| (21) Breadth of $M^2$ | v12 | (19) Height of the mandible behind $M_1$ |
| (22) Greatest diameter of the bulla ossa | v13 | (20) Height of the mandible between $P_2$ an $P_3$ |
| (23) Greatest mastoid breadth | v14 | (14) Greatest length of the $P_4$-alveolus |
| (29) Greatest neurocranium breadth | v15 | (13a) Greatest length of $M_1$ |
| 30) Zygomatic breadth | v16 | (13b) Greatest breadth of $M_1$ |
| (31) Least breadth of the skull | v17 | (15a) Greatest length of $M_2$ |
| (32) Frontal breadth | v18 | (15b) Greatest breadth of $M_2$ |
| (33) Least breadth between the orbits | v19 | |
| (34) Greatest palatal breath | v20 | |
| (35) Least palatal breath | v21 | |
| (36) Breadth at the canine alveoli | v22 | |
| (37) Greatest height in the orbita | v23 | |
| (38) Skull height | v24 | |
| 40) Least breadth of the neurocranium | v25 | |

*Table 1 Measurements of skull and mandible. The numbers in parenthesis refer to codes of von den Driesch (1976), v1 to v25 are codes for measurements used in this work*

| | | |
|---|---|---|
| Shoulder height | Humerus<br>Radius<br>Ulna<br>Femur<br>Tibia | 3.43 * TL - 26.54<br>3.18 * TL + 19.51<br>2.78 * TL + 6.21<br>3.14 * TL - 12.69<br>2.92 * TL + 9.41 |
| Skull Indices | Cephalic Index (CI):<br>Snout Index (SI):<br>Snout Width Index (SWI): | Max. zygomatic width * 100/TL<br>Nasion to the alveolar * 100/TL<br>Snout width between the canines * 100/Nasion to Prostion |
| Mandible | Basal Length (BL): | Length of the teeth row ($p_1$-$m_3$) * 2.8-44 |

***Table 2** Formulas used to calculate: Cephalic Index, Snout Index, Snout Width Index (Harcourt, 1974); skull basal length (BL) from mandible dimensions (Dahr, 1937); shoulder height (Harcourt, 1974) from the total lengths (TL) of limb bones*

Number of Individuals) (Lyman, 1994; Galik, 1997b). Five fracture types were categorised for incomplete limb bones: proximal fragments; proximal + shaft; shaft; shaft + distal; distal fragments. For example, a proximal fragment, a shaft fragment and a distal fragmented specimen represent one element, whereas a proximal + shaft fragment and a shaft + distal fragment represent two elements. The MNE consists of the frequencies of left and right skeletal elements. For calculating the MNI, the specimens were differentiated into paired and unpaired bones. In paired elements, the specimens with the higher frequency (left or right) were counted for the MNI. The different quantitative techniques complement each other and provide a good overview of the bone material.

Another point of interest was the preservation of the dog bones. The bones were carefully examined for natural and artificial modifications and pathologies (*cf* Sadek–Kooros, 1975; Lyman, 1994; Galik, 1997b), listed in Table 4. Two kinds of fracture types were noted: green bone fractures which occur in bones containing collagen, producing spiral fracture types with a smooth fracture surface; fractures which occur in bones that have lost collagen during deposition in sediment, producing fractures that are perpendicular to the long axis of the bone with a crumbly fracture surface. The dog bones were repaired as much as possible to get information about the completeness of the dog's skeletons. The individual ages of dogs were estimated after Habermehl (1975).

The measurements for the morphometrical analysis were taken according to von den Driesch 1976 (Table 1), on complete or at least almost complete skeletal elements.

The Cephalic Index, Snout Index and Snout Width Index were calculated from formulae given by Harcourt (1974) to characterise dog skulls (Table 2). The Cephalic Index gives the width of the skull relative to its length, while the Snout Index indicates the length of the snout relative to the whole head. The Snout Width Index shows the width of the muzzle relative to the length of the nose.

In addition to the indices mentioned above, univariate statistics were carried out to assess body dimensions of the dogs. For this reason the basal lengths were measured from the canid skulls. According to Dahr (1937), the basal length of the skull can be estimated from mandibles as well and the formula used for this calculation is listed in Table 2. The resulting values were compiled into univariate statistics (Table 5). The shoulder heights were calculated from the total lengths (TL) of the long bones (Harcourt, 1974) , using the formulae listed in Table 2.

All data were logarithmically transformed before being used in statistical analyses. These metric traits were standardised and subsequently subjected to a classification by hierarchical cluster analysis (Ward's Method). The proposed group membership by cluster analysis was identified by linear discriminance analysis. Missing values were substituted using the means for the predictor variables during the classification, so that these cases could also be classified. The resulting discriminant scores are expressed in Tables 7 and 8, together with the group memberships and metric data from the bones.

## Results

The NISP frequencies in Table 3 differ considerably among the skeletal elements. The highest frequencies were encountered in elements which are predominant in a dog skeleton and can be combined into more general categories like vertebrae, phalanges and metapodials. Numerous specimens came from intensively fragmented bones like skulls and ribs. Long bones, such as the humerus or femur, are rather similarly distributed in their frequencies of NISP. Small bones such as sesamoids, sternal elements, and astragali, or the more fragile fibulae, are poorly represented (Table 3). The 156 skull specimens represent 45 individuals. The MNI frequencies of skull, femur, humerus and tibia are conspicuously similar. The poor recovery of small bones is confirmed in the MNI statistics.

The comparison of NISP and Ec can be considered as an expression of bone fragmentation (Table 3). The closest correspondence between Ec and NISP occurs in the small and solid bones like calcaneus or phalanges. However, many complete elements were found among the more fragile elements, too. Perpendicular fractures characteristic of post-depositional breakage are predominant among the well preserved bones, whereas bones with green bone fractures are remarkably underrepresented (Table 4). Cut

| | astragalus | calcalneus | costa | femur | fibula | humerus | mandible | metapodial | os penis | os pisiforme | pelvis | phalanx | radius | sacrum | scapula | sesamoid | skull | sternalia | tibia | ulna | vertebra | total |
|---|---|---|---|---|---|---|---|---|---|---|---|---|---|---|---|---|---|---|---|---|---|---|
| NISP | 28 | 48 | 320 | 96 | 43 | 94 | 79 | 251 | 5 | 1 | 75 | 664 | 45 | 23 | 52 | 7 | 156 | 18 | 89 | 83 | 719 | 2896 |
| Ec | 27 | 45 | 16 | 34 | 1 | 42 | 38 | 226 | 3 | 1 | 8 | 660 | 25 | 12 | 16 | 7 | 19 | | 27 | 18 | | 1389 |
| MNI | 17 | 34 | | 43 | 21 | 44 | 30 | 21 | 5 | 1 | 35 | 14 | 22 | 19 | 35 | 1 | 45 | | 43 | 36 | | 45 |
| MNE | 27 | 45 | | 82 | 33 | 79 | 49 | 475 | 5 | 1 | 61 | 661 | 40 | 19 | 57 | 1 | 45 | | 85 | 66 | | 1786 |

***Table 3*** *Numbers of Identified Specimens (NISP), Minimum Numbers of Individuals (MNI), complete Elements (Ec) and Minimum Numbers of Elements (MNE) in the recovered sample.*

| | skull | mandible | stylopodium | zygopodium | autopodium | ribs | vertebrae | pelvis | total |
|---|---|---|---|---|---|---|---|---|---|
| Cut marks | | 2 | | 4 | 3 | 3 | 1 | 1 | 14 |
| Carnivore gnawings | 1 | 2 | 8 | 8 | 21 | 23 | 6 | 2 | 71 |
| Perpendicular fractures | 118 | 50 | 93 | 159 | 36 | 386 | 191 | 67 | 1100 |
| Green bone fractures | 7 | 4 | 10 | 29 | 7 | 44 | 8 | | 109 |
| Exostosis | | | 2 | 2 | 24 | 2 | 29 | | 59 |
| Healed fractures | | | | 5 | | 4 | 2 | | 6 |
| Deformations | | | | | 2 | | 19 | | 21 |
| Intravital lost teeth | 15 | 30 | | | | | | | 45 |
| Irregular arranged teeth | 2 | 4 | | | | | | | 6 |
| Traces of inflammations | 5 | 3 | | | | | | | 8 |

***Table 4*** *Frequencies of natural and artificial modifications, and pathologies on skeletal elements in the recovered sample*

marks appeared only on 14 bones and 71 specimens showed traces of carnivore gnawing. Most of the carnivore punctures were on ribs and elements of the autopodium (Table 4). Cranial elements, limb bones and vertebrae were affected by pathologies. Jaws showed intravital loss of teeth. Some of the alveoli are completely closed, or small root stumps jut out of the bone structure. Irregularly arranged teeth were found as well as traces of inflammations of the gum. Elements of the autopodium and vertebrae were mainly affected by exostosis (Table 4). Bone deformations appeared in the form of bent processes of vertebrae and two crooked metapodials. Few bones showed healed fractures.

A considerable number of bones (421) were paired and rearticulated. Only a few of the vertebrae have been re-associated into individual spinal columns. Many mandible pairs could be matched and craniums fitted to paired and to single mandibles. Elements of the fore limbs (humerus, ulna and radius) were often found together, even as pairs, as were elements of the hind limbs (femur and tibia, sometimes with their fitting calcaneus and astragalus). A number of autopodium elements, mainly metapodials were also rearticulated and paired.

The majority of the sample, 2500 specimens, came from adult individuals and almost all of the teeth showed no wear. Only three individuals were not adult: one was an infantile individual, one individual younger than 12 months; and another younger than 24 months. An exact determination of sex in dog individuals was only possible by association with *os penii*. Five male individuals were thus identified among the Durezza dog remains (Table 3).

The Cephalic Index values have a mean of 56 and indicate skulls with fairly broad zygomatic arches. The mean of the Snout Index is about 48 and the Snout Width Index lies at about 42 (Table 5). The skulls, therefore, can be said to have long snouts with fairly broad muzzles. The basal lengths range in size from 16-20 cm (Table 5), producing a bimodal distribution in the frequency histogram (Fig.2).

Figure 3 shows a frequency histogram of the reconstructed shoulder heights from humerus, radius, ulna, femur and tibia. Heights calculated from intact, paired elements were very similar and appear mostly in the same bars but shoulder heights calculated from repaired elements sometimes differed in their results. The greatest differences are in the size range of a few centimetres and they are probably not of significant importance to the analysis. The

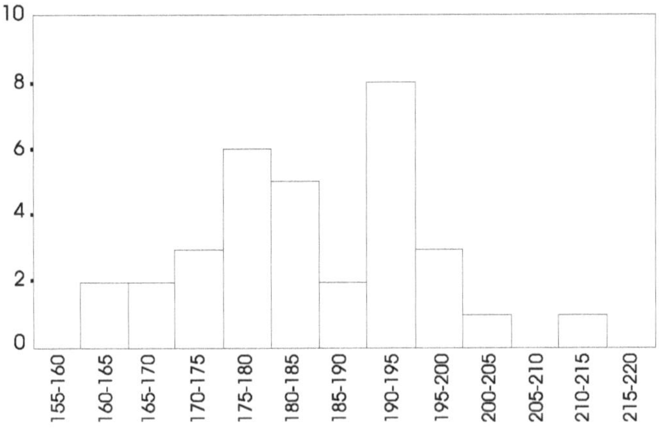


*__Figure 2__ Absolute frequencies of basal length in mm*

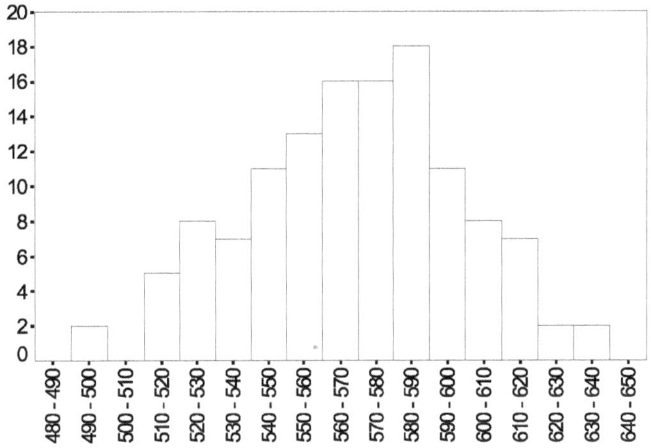


*__Figure 3__ Absolute frequencies of shoulder heights in mm*

| | N | Min. | Max. | Mean | S.d. |
|---|---|---|---|---|---|
| Shoulder height (mm) | 126 | 491 | 635 | 569 | 29.69 |
| Basal length (mm) | 27 | 163 | 202 | 184 | 9.84 |
| Cephalic Index | 6 | 54 | 59 | 56 | 2.27 |
| Snout Index | 8 | 42 | 54 | 48 | 4.74 |
| Snout Width Index | 9 | 34 | 48 | 42 | 4.2 |

*__Table 5__ Size variation of the shoulder heights, basal lengths, Cephalic Index, Snout Index and Snout Width Index for the recovered sample*

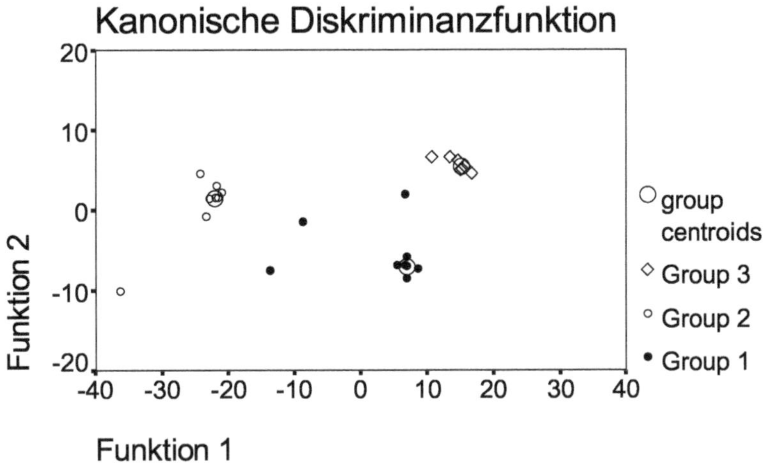


*__Figure 4__ Scatter plot produced by with discriminance scores of mandibles using two discriminance functions*

dogs represented in the sample are seen to have had a shoulder height that varied from 49 cm–63.5 cm, producing a unimodal distribution in the frequency histogram.

Sixteen variables were used to classify the mandibles by cluster analysis. The analysis identified three possible groups. Two significant separating canonical discriminance functions confirmed the mandible groups and all the individuals were correctly classified (Table 6). The smallest mandibles are in group one, whereas group three consists of the largest bones. The best morphological homogeneity is represented in group three, which contains three pairs of mandibles (Fig. 3). Many elements of group one as well as group two cluster very close to the group centroid, although they contain only one pair of mandibles (Fig. 4).

The skulls clustered into two groups. Twenty five metric traits were used in the discriminant analysis to provide the separation. Both groups were significantly discriminated (Table 6) and correctly classified. Group one contains three larger skulls and in group 2 are the smaller specimens. A closer examination of the discriminant scores shows that the three skulls of group one, and the three skulls of group two, are almost identical (*cf* Table 8). In group two are three "pairs" of skulls, which are very similar in their metrical dimensions (*cf* Table 8).

| Element | Discriminant Function | Eigenvalue | canonical correlation | Wilk`s Lambda | $Chi^2$ | P |
|---|---|---|---|---|---|---|
| Skull | 1 | 124.976 | .996 | .08 | 21.76 | .010 |
| Mandible | 1 | 305.768 | .998 | .000 | 74.020 | .000 |
| | 2 | 33.001 | .985 | .029 | 28.211 | .013 |

***Table 6** Results of the discriminant function analysis for three groups of mandibles and two groups of skulls*

| Mand | G | F1 | F2 | R | v1 | v2 | v3 | v4 | v5 | v6 | v7 | v8 | v9 | v10 | v11 | v12 | v13 | v14 | v15 | v16 | v17 | v18 |
|---|---|---|---|---|---|---|---|---|---|---|---|---|---|---|---|---|---|---|---|---|---|---|
| **1185** | . | . | . | | . | 80 | | | | | | | | | | | | | | | | |
| **2760** | 1 | 6.9311 | -8.524 | | 145 | 76 | 176 | 140 | 133 | 126 | 118 | 126 | 76 | 37 | 12 | 25 | 21 | 23 | 22 | 9 | 10 | 7 |
| **2761** | 1 | 6.7005 | -6.887 | | 146 | 77 | 179 | 144 | 137 | 127 | 120 | 127 | 83 | 37 | 12 | 24 | 20 | 22 | 21 | 10 | 10 | 7 |
| **2899** | . | . | . | | | 71 | 163 | | | | | | | | | | | | | | | |
| **3062** | . | . | . | | | 81 | 192 | | | | | | | | | | | | | | | |
| **3949** | . | . | . | | | 88 | 210 | | | | | | | | | | | | | | | |
| **3950** | . | . | . | | | 82 | 194 | | | | | | | | | | | | | | | |
| **4385** | 2 | -36.23 | -10.08 | | 147 | 75 | 173 | 144 | 139 | 129 | 122 | 128 | 84 | 33 | . | . | . | 20 | 22 | 10 | 9 | 7 |
| **5156** | 2 | -21.77 | 3.0680 | | 170 | 83 | 196 | 161 | 154 | 142 | 134 | 138 | 90 | 37 | 13 | 30 | 24 | 23 | 23 | 9 | 9 | 7 |
| **5159** | 2 | -21.87 | 1.6003 | | 163 | 81 | 191 | 160 | 154 | 142 | 133 | 139 | 89 | 37 | 14 | 29 | 24 | 23 | 24 | 10 | 9 | 7 |
| **5185** | 2 | -22.89 | 1.5040 | | 157 | 81 | 192 | 156 | 151 | 138 | 131 | 137 | 90 | 39 | 13 | 27 | 24 | 23 | 23 | 11 | 10 | 7 |
| **5187** | 1 | 6.6153 | 2.0153 | | 169 | 80 | 188 | 160 | 157 | 144 | 133 | 137 | 89 | 38 | 14 | 29 | 22 | 21 | . | . | . | . |
| **5188** | 3 | 15.312 | 5.1091 | | 170 | . | . | 163 | 157 | 147 | 135 | 141 | 92 | 41 | 14 | 30 | 25 | 25 | 26 | 11 | 10 | 8 |
| **5190** | 1 | -13.65 | -7.518 | | 150 | 75 | 174 | 145 | 143 | 124 | 117 | 120 | 81 | 36 | 13 | 24 | 22 | 21 | . | . | 9 | 7 |
| **5197** | . | . | . | | | 78 | 181 | | | | | | | | | | | | | | | |
| **5382** | 1 | 6.9433 | -5.813 | | 150 | 79 | 184 | 145 | 139 | 136 | 128 | 130 | 87 | 38 | 12 | 25 | 20 | 23 | 24 | 9 | 10 | 7 |
| **5383** | . | . | . | | | 73 | 168 | | | | | | | | | | | | | | | |
| **5157a** | 1 | 6.9070 | -7.021 | 1 | 147 | . | . | 147 | 140 | 126 | 118 | 126 | 81 | 38 | 13 | 25 | 21 | 23 | 24 | 10 | 9 | 8 |
| **1654** | 1 | -8.762 | -1.446 | 2 | . | 81 | 191 | 150 | 149 | . | 128 | . | 89 | 37 | 15 | 30 | 23 | . | 23 | 9 | 9 | 7 |
| **1331** | . | . | . | 3 | . | 83 | 194 | | | | | | | | | | | | | | | |
| **1332** | 2 | -21.49 | 1.6548 | 3 | 154 | 83 | . | 157 | 145 | 138 | 129 | 138 | 90 | 37 | 12 | 27 | 20 | 22 | 22 | 9 | 9 | 7 |
| **4072** | 2 | -23.42 | -.7636 | 4 | 159 | 82 | . | 154 | 148 | 139 | 128 | 134 | 89 | 38 | 13 | 26 | 21 | 23 | 23 | 10 | 9 | 7 |
| **5200** | . | . | . | 4 | | 82 | 194 | | | | | | | | | | | | | | | |
| **5158** | 1 | 5.5392 | -6.841 | 5 | 147 | . | . | 145 | 139 | 127 | 119 | 125 | 85 | 39 | 14 | 23 | 20 | 23 | 24 | 10 | 10 | 8 |
| **5186** | 1 | 8.6680 | -7.251 | 5 | 147 | 77 | 181 | 144 | 139 | 128 | 120 | 126 | 84 | 39 | 14 | 25 | 20 | 23 | 24 | 10 | 10 | 7 |
| **5160** | 2 | -21.01 | 2.2444 | 6 | 156 | 84 | 201 | 154 | 146 | 136 | 128 | 136 | 90 | 41 | 13 | 25 | 23 | 23 | 23 | 10 | 10 | 7 |
| **5189** | 2 | -24.30 | 4.6340 | 6 | 154 | . | . | 154 | 146 | 136 | 127 | 135 | 89 | 41 | 13 | 24 | 23 | 23 | 23 | 9 | . | . |
| **5191** | 3 | 14.986 | 5.0369 | 7 | 161 | 80 | . | 160 | 148 | 137 | 139 | 137 | 86 | 40 | 14 | 28 | 22 | 23 | 24 | 10 | 11 | 8 |
| **5199** | 3 | 14.639 | 6.1793 | 7 | 162 | 81 | . | 159 | 154 | 137 | 128 | 135 | 86 | 40 | 14 | 29 | 23 | 23 | 24 | 9 | 11 | 8 |
| **5192** | 3 | 16.675 | 4.5785 | 8 | 160 | 83 | 197 | 155 | 150 | 139 | 129 | 134 | 89 | 39 | 14 | 26 | 24 | 22 | 24 | 10 | 11 | 11 |
| **5194** | 3 | 13.433 | 6.6633 | 8 | 160 | 85 | . | 154 | 148 | 142 | 131 | 137 | 91 | 39 | 14 | 28 | 25 | 23 | 24 | 9 | 11 | 11 |
| **1101** | 3 | 10.682 | 6.6547 | 9 | 158 | . | . | 149 | . | 139 | 129 | 130 | . | 39 | 13 | 30 | 24 | 24 | 23 | 10 | 11 | 8 |
| **1102** | 3 | 15.717 | 5.4626 | 9 | 158 | 81 | 192 | 149 | 147 | 138 | 130 | 130 | 89 | 39 | 13 | 30 | 24 | 24 | 23 | 10 | 11 | 8 |

***Table** 7 Group memberships (**G**), discriminant scores from functions **F1** and **F2**, and measurements taken from mandibles. Row **R** lists additional specimen #'s assigned to mandibles that could be matched to skulls (see Table 8). Measurements **v1** to **v18** defined in Table 1*

| Skull | G | F1 | R | v1 | v2 | v3 | v4 | v5 | v6 | v7 | v8 | v9 | v10 | v11 | v12 | v13 |
|---|---|---|---|---|---|---|---|---|---|---|---|---|---|---|---|---|
| 1107 | 1 | -4.35994 | 9 | 203 | 189 | 180 | 33 | 130 | 87 | 77 | 134 | 101 | 68 | 19 | 10 | 24 |
| 1174 | 1 | -7.05404 | | 195 | 174 | 163 | 42 | 123 | 92 | 80 | 110 | 95 | 69 | 18 | 11 | 23 |
| 1175 | 1 | 10.33574 | | 210 | 191 | 181 | 46 | 134 | 97 | 86 | 124 | 98 | 71 | 19 | 11 | 24 |
| 1194 | 1 | -7.86747 | 2 | 212 | 200 | 190 | 51 | 137 | 95 | 107 | 130 | 105 | 72 | 19 | 10 | 23 |
| 1755 | 1 | -6.07185 | | 198 | 183 | 172 | 39 | 134 | 98 | 84 | 117 | 99 | 102 | 20 | 8 | 23 |
| 3063 | 1 | 0.22139 | | 205 | 195 | 183 | 49 | 144 | 96 | 84 | 117 | , | 74 | 21 | 9 | 26 |
| 3938 | 1 | -5.68307 | | 215 | 208 | 196 | 45 | 145 | 98 | 86 | 120 | 106 | 72 | 22 | 11 | 24 |
| 3942 | 1 | -6.09194 | | 199 | 183 | 171 | 46 | 126 | 102 | 99 | 109 | 99 | 63 | 19 | 10 | 24 |
| 4477 | 1 | -13.5261 | | 202 | 187 | 177 | 45 | 132 | 96 | 105 | 121 | 104 | 65 | 21 | 12 | 22 |
| 4677 | 1 | -5.31709 | | 200 | 187 | 176 | 44 | 133 | 94 | 101 | 115 | 101 | 85 | 19 | 11 | 23 |
| 3936 | 1 | -4.25698 | | 206 | 192 | 179 | 46 | 134 | 104 | 99 | 112 | 100 | 71 | 19 | 11 | 25 |
| 5203 | 1 | -7.09324 | | 193 | 178 | 169 | 41 | 129 | 104 | 94 | 104 | 98 | 66 | 19 | 11 | 23 |
| 5214 | 2 | 16.66986 | | 255 | 240 | 209 | 60 | 149 | 123 | 103 | 136 | 114 | 75 | 21 | , | 26 |
| 5212 | 2 | 16.12205 | 7 | 200 | 186 | 177 | 44 | 132 | 101 | 96 | 113 | 101 | 69 | 19 | 11 | , |
| 5213 | 2 | 16.74673 | | 235 | 222 | 196 | 56 | 140 | 108 | 94 | 126 | 104 | 73 | 20 | , | 6 |

| Skull | G | F1 | R | v14 | v15 | v16 | v17 | v18 | v19 | v20 | v21 | v22 | v23 | v24 | v25 |
|---|---|---|---|---|---|---|---|---|---|---|---|---|---|---|---|
| 1107 | 1 | -4.35994 | 9 | 69 | 51 | 115 | 42 | 60 | 44 | 68 | 37 | 41 | 32 | 61 | 45 |
| 1174 | 1 | -7.05404 | | 69 | 56 | 105 | 40 | 49 | 35 | 64 | 32 | 35 | 30 | 56 | 39 |
| 1175 | 1 | 10.33574 | | 68 | 54 | , | 36 | 59 | 43 | 68 | 33 | , | , | 65 | 37 |
| 1194 | 1 | -7.86747 | 2 | 73 | 56 | 124 | 41 | 58 | 44 | 74 | 42 | 45 | 32 | 67 | 45 |
| 1755 | 1 | -6.07185 | | 68 | 56 | 113 | 38 | 53 | 36 | 69 | 40 | 42 | 33 | 60 | 48 |
| 3063 | 1 | 0.22139 | | 72 | 57 | , | 41 | 60 | , | , | , | , | , | , | , |
| 3938 | 1 | -5.68307 | | 78 | 55 | 115 | 44 | 73 | 50 | 77 | 45 | 46 | 32 | 69 | 53 |
| 3942 | 1 | -6.09194 | | 70 | 56 | , | 35 | 43 | 36 | , | 36 | 38 | , | 55 | 47 |
| 4477 | 1 | -13.5261 | | 71 | 56 | 108 | 39 | 58 | 43 | 70 | 39 | 42 | 31 | 61 | 45 |
| 4677 | 1 | -5.31709 | | 69 | 57 | 116 | 36 | 48 | 38 | 68 | 38 | 40 | 31 | 59 | 48 |
| 3936 | 1 | -4.25698 | | 75 | 60 | 122 | 38 | 56 | 40 | 76 | 43 | 44 | 32 | 60 | 41 |
| 5203 | 1 | -7.09324 | | 65 | 52 | 108 | 36 | 64 | 34 | 66 | 34 | 36 | 31 | 55 | 37 |
| 5214 | 2 | 16.66986 | | 82 | 65 | 129 | 43 | 69 | 48 | 78 | 43 | 48 | 40 | 66 | 59 |
| 5212 | 2 | 16.12205 | 7 | 69 | 57 | , | 36 | 49 | 38 | 68 | 39 | 40 | 30 | 59 | 43 |
| 5213 | 2 | 16.74673 | | 77 | 62 | 122 | 42 | 63 | 44 | 75 | 40 | 43 | 38 | 62 | 55 |

***Table 8** Group membership (G), discriminant scores from function **F1**, and measurements taken from skulls. Row **R** lists mandibles which could be matched to skulls (see mandible #'s, Table 7). Measurements **v1** to **v25** defined in Table 1*

## Discussion

The Durezza cave, with its closely dated and extensive bone material, is an important and unique site in Austria. Dog bones from the Hallstatt and La Tène period have previously been found in rural contexts. The canids from these periods have been described as moderately large animals (Bökönyi, 1974; Benecke, 1994; Petters, 1997). A high frequency of dog bones is unusual from such sites in Austria. The Durezza dog bones fit quite well into the size range of previously reported Hallstatt and La Tène period dogs, even though the larger Durezza individuals exceed slightly the size of dogs from other Hallstatt sites (Pucher, 1986,1987, 1996; Riedel, 1986, 1998). Large canid remains are known from a similar site, the Grotta delle Ossa (Škocjan), a chimney cave in Northern Italy, but those remains were identified as wolves (Riedel, 1977). The bones classified as dog from the Italian site have smaller dimensions than the mean of values of Durezza dogs (Riedel, 1977).

The bimodal distribution of skull basal lengths for the Durezza sample shows a big gap between the two peaks. Therefore, the distribution pattern does not allow a separation into female and male individuals based on size. However, the body dimensions of the Durezza dogs do not show an extraordinary size variation. The variation of basal lengths is about 4 cm from largest to smallest, and the estimated shoulder heights differ by about 14 cm. Although groups were significantly discriminated by multivariate statistical methods, the importance of the results was the confirmation of morphological homogeneity of the bone groups. The analysis results suggest specimens of almost identically formed skulls and mandibles were present. This unusual morphological homogeneity of skulls and mandibles could be considered as an indication of a very closely related population (*cf* Benecke, 1994: 219- comparing Neolithic dogs from Twann, Kt. Bern in Switzerland). The archaeological dating that indicates a short period of time for deposition may be supported by a close relationships among the dogs. The "breeding strategies" of the Durezza people, if they existed, were probably to obtain moderately large to large dogs with fairly broad and long skulls. Such animals were quite common in these periods of time and are usually considered as "unspecialised" working dogs (used for herding, hunting or guarding) (Bökönyi, 1974; Benecke, 1994).

The dogs represented in the sample appeared to have been rather healthy adult animals. Pathologies show no unusual diseases. A look at the fracture types indicates that postdepositional destruction was the main cause of breakage. The obvious lack of small bones can be explained by the rough excavation methods of the cave explorers. The MNIs of skull and some long bones are conspicuously similar. However, the number of elements that could be re-associated into individuals and the presence of all skeletal elements support the assumption that at least corresponding body parts were thrown into the cave.

Consumption of dog meat was quite common during the Hallstatt and La Tène periods (Bökönyi, 1974; Benecke, 1994). Evidence of dog meat exploitation in rural contexts from Austrian Hallstatt periods exist for Gars Thunau (Kanelutti, 1993), Stillfried (Pucher, 1992), Inzersdorf ob der Traisen (Pucher, 1989), Schleinbacher Ziegelei in Mistelbach (Pucher, 1996) and Böheimkirchen (Riedel, 1997) in Lower Austria. The few cut marks which occurred on the Durezza dog bones are probably indications of butchering.

Dog bones from sites within rural contexts were sometimes affected by carnivore gnawing. Few of the Durezza dog bones showed gnawing marks and it was usually possible to match skeletal elements from single individuals with tooth marks. If the carcasses were not gnawed by carnivores before getting thrown into the cave, they must have been scavenged inside the cave by dogs or foxes which had been thrown in alive and survived the fall.

From an archaeological point of view the majority of the Durezza remains must be interpreted as cultic and/or sacrificial offerings (Gleirscher, 1997b; Behm-Blancke, 1976). The very low frequencies of ornaments and the large number of animal bones are atypical for regular burials. Dead people were usually burnt and buried afterwards in cemeteries during this time period. However, burials from a cemetery near the Tscheltschnigkogel appear to have stopped just about the time when the input of people and animals into the Durezza cave started (Gleirscher, 1997b). The anthropological implications of the Durezza cave analysis suggest the possibility of burials because the death profiles of human remains from the cave are similar to profiles from grave yards (Reuer-Fabrizzi & Reuer, 1997).

Dogs appear to have played an important role in everyday life during the Metal ages.

Occasionally dogs are known from Hallstatt and La Tène human burials (Behrens, 1964; Benecke, 1994; Leskovar, 1996). Dogs were buried or added to human burials as complete animals or as body parts. Mythical, religious and sociological reasons are the most common explanations for dogs in burial contexts (Behrens, 1964; Blänkle, 1986; Green, 1992; Leskovar, 1996), which can also be stated for the Durezza dogs. The results derived from archaeozoological investigations of dog bones unfortunately do not provide conclusive evidence that Durezza cave was used for atypical and irregular burials, or alternatively, if the dogs and the other animals represented by skeletal remains were sacrificed.

**References**

Behm-Blancke, G. (1976). Zur Funktion bronze- und eisenzeitlicher Kulthöhlen im Mittelgebirgsraum. *Ausgrabungen und Funde* **21**, 1-4.

Behrens, H. (1964). Die neolithischen bis frühmetallzeitlichen Tierskelettfunde der Alten Welt. *Veröffentlichungen des Landesmuseum für Vorgeschichte in Halle* **19**, 1-135.

Benecke, N. (1987). Studies on early dog remains from Northern Europe. *Journal of Archaeological Science* **14**, 31-49.

Benecke, N. (1994). *Der Mensch und seine Haustiere.* Stuttgart: Theiss Verlag. Blänkle, P. H. (1986). Eine mittel- bis spätlaténezeitliche Hundebestattung aus Offenbach am Main. *Rumpenheim. Archäologisches Korrespondenz-blatt* **16**, 165-185.

Bökönyi, S. (1974). *History of Domestic Animals in Central and Eastern Europe.* Budapest: Akadémiai Kiado.

Dahr, E. (1937). *Studien über Hunde aus primitiven Steinzeitkulturen in Nordeuropa.* Lund: Lunds Univ. Arskrift NF Avd. 2,32 **Nr. 4**.

Driesch, A., von den (1976). *A Guide to the Measurement of Animal Bones from Archaeological Sites.* Cambridge: Peabody Museum Bulletin **1**.

Fuchs, M. 1996. Radiometrische Datierungen der Menschenknochen aus der Durezzahöhle. *Fundberichte aus Österreich* **35**, 446.

Galik, A. & Pacher, M, (1997). The Durezza cave near Villach (Carinthia, Austria): cave with sacrifical offerings from the Hallstatt period? Proceedings of the 12th International Congress of Speleology, August 10-17, 1997, La Chaux de Fonds, Switzerland.

Galik, A. (1997a). Zur Grabungsmethodik und Fundsituation in der Durezza Schacht-Höhle. Neues aus Alt-Villach. *Jahrbuch des Stadtmuseums* **34**, 9-17.

Galik, A. (1997b). Die hallstattzeitlichen Tierknochen aus der Schachthöhle Durezza am Tscheltschnigkogel bei Villach in Kärnten. Neues aus Alt-Villach. *Jahrbuch des Stadtmuseums* **34**, 31-55.

Gleirscher, P. (1997a). Hallstattzeitliche Kleinfunde aus der Durezza–Schachthöhle. Neues aus Alt-Villach. *Jahrbuch des Stadtmuseums* **34**, 17-31.

Gleirscher, P. (1997b). Überlegungen zur Deutung der Durezza-Schachthöhle. Neues aus Alt-Villach. *Jahrbuch des Stadtmuseums* **34**, 125-213.

Grayson, D. K. (1984). *Quantitative Zooarchaeology*. Orlando: Academic Press.

Green, M. (1992). *Animals in Celtic Life and Myth*. London: Routledge.

Habermehl, K. H. (1975). Altersbestimmung bei Haus- und Labortieren. Unpublished manuscript.

Harcourt, R. A. (1974). The dog in prehistoric and early historic Britain. *Journal of Archaeological Science* **1**, 151-175.

Kanelutti, E. (1993). *Archäozoologische Untersuchungen am Schanzberg von Gars/Thunau.* In (H. Friesinger, F. Daim, E. Kanelutti & O. Cichocki, Eds) Bioarchäologie + Frühgeschichtsforschung. Archaeologia Austriaca-Monographien **2**, 169-184.

Leskovar, J. (1996). *Frühkeltische Siedlung und ein Frauengrab mit Hund und Schwein.* Linzer Archäologische Forschungen, Sonderheft **XVII**, 1-32.

Lyman, L. R. (1994). *Vertebrate Taphonomy*. Cambridge: Cambridge University Press.

Morey, D. F. (1990). *Cranial allometry and the evolution of the domestic dog*. Ph.D. Thesis, University of Tennessee.

Petters, J. (1997). Der Hund in der Antike aus archäozoologischer Sicht. *Anthropozoologica* **25-26**, 511-523.

Pucher, E. (1982). Tierknochenfunde aus Stillfried an der March (Niederösterreich). Dissertation Wien, unpubl.

Pucher, E. (1986). Bronzezeitliche Knochen vom Buchberg, OG. Wiesing, Tirol. *Fundberichte aus Österreich* **23**, 209-220.

Pucher, E. (1987). Tierknochen aus der Bronzezeit des Buhubergs (Niederösterreich). *Wissenschaftliche Mitteilungen aus dem Niederösterreichischen Landesmuseum* **4**, 11-35.

Pucher, E. (1989). Der Knochenabfall einer hallstatt-latènezeitlichen Siedlung bei Inzersdorf ob der Traisen (Niederösterreich). In (P. C. Ramsl, Ed) *Inzersdorf–Walpersdorf, Studien zur späthallstatt-bis latènezeitlichen Besiedlung im Traisental, Niederösterreich.* Fundberichte aus Österreich **A6**, 56-67.

Pucher, E. (1996). Die tierknochenfunde aus der Schleinbacher Ziegelei, Bezirk Mistelbach, Niederösterreich (Grabung 1981-1986). *Annales Naturhistorisches Museum Wien* **97A**, 21-54.

Reuer-Fabrizzi, S. & Reuer, E. (1997). Anthropologische Untersuchungen an 138 skeletten aus der Durezza–Schachthöhle bei Villach. Neues aus Alt-Villach. *Jahrbuch des Stadtmuseums* **34**, 111-125.

Riedel, A. (1977). I resti animali della grotta delle ossa (Skocjan). *Estratto dagli Atti del Museo Civico di Storia Naturale-Trieste*, **Vol. XXX**, Fasc. 2, Nr. 7, 126-195.

Riedel, A. (1986). Ergebnisse von archäozoologischen Untersuchungen, im Raum zwischen Adriaküste und Alpenhauptkamm (Spätneolithikum-Mittelalter). *Padusa* **22**, 1-220.

Riedel, A. (1998). Archäozoologische Untersuchungen an den Tierknochenfunden aus der Veterov Kultur von Böheimkirchen (Niederösterreich). *Annales des Naturhistorischen Museum Wien* **99A**, 341-374.

Ringrose, J. T. (1993). Bone counts and statistics: a critique. *Journal of Archaeological Science* **20**, 121-157.

Sadek-Kooros, H. (1975). Intentional fracturing bone: description of criteria. In (A. Clason, Ed) *Archaeozoological Studies*. Amsterdam: American Elsevier, pp. 139-150.

# Part 3

# Interpreting skeletal variation: the Roman influence on breed development

# MORPHOLOGICAL AND OSTEOLOGICAL CHANGES IN THE DOG FROM THE NEOLITHIC TO THE ROMAN PERIOD IN ITALY

Jacopo De Grossi Mazzorin
Soprintendenza Archeologica di Roma,
Laboratorio di Archeozoologia,
via S. Apollinare 8, 00186 Roma, Italy.

Antonio Tagliacozzo
Soprintendenza Speciale al Museo Preistorico Etnografico
"L. Pigorini",
Laboratorio di Paleontologia e Archeozoologia,
viale Lincoln 3, 00144 Roma, Italy.

**Key words**: dog, Italy, osteometry, Neolithic, Bronze Age, Iron Age, Roman period, breeds, archaeozoology

**Abstract**

The scope of this work is to examine the evolution of the domesticated dog in Italy from the Neolithic Age (sixth millennium BC) to the end of the Roman period (sixth century AD). Its aim is to evaluate whether, in the period under examination, the development and changes in the morphology and size of canine remains reflect the emergence of different breeds and varieties of *Canis familiaris* and if so, where and when these new strains occurred.

The results presented involve the preliminary analysis of data from about 2000 specimens of skeletal dog remains taken from approximately 200 archaeological excavations. The majority of the data are from protohistoric sites in north and central Italy, in particular the Bronze Age. From the data collected, it is possible to make some remarks regarding the modifications in morphology and size of dogs in Italy in ancient times.

From the analysis of the morphology of the skeletal remains of Italian dogs it is possible to say that a fairly good size variability is present from the Neolithic period onward. It increases in the Bronze Age, but morphological differences detected appear to be the result of accidental cross-breeding rather than real selection. Large dogs are present only starting with the Iron Age, when there seems to be a selection for bigger animals. Starting with the Iron Age it is possible that there is the first real selection leading to real breeds. Greater size variability can be found in the Roman period. At this time, similarly to other European Roman provinces, lap-dogs appear, characterized by a particularly reduced stature.

## Introduction

This work is the continuation of a study whose preliminary results were presented at the previous International Congress of Archaeozoology (ICAZ) in Konstanz (De Grossi Mazzorin & Tagliacozzo, 1997). The aim of this research is to examine the evolution of the domestic dog in Italy from the Neolithic to the end of the Roman period. The goal is to evaluate whether the development and changes in morphology and size of canid remains reflects the emergence of different kinds and varieties of *Canis familiaris* and if so, where and when these new strains have occurred. When possible comparisons with modern breeds were made in order to evaluate which characters were being modified and which ones were selected. Clearly, it is not our intent to refer the findings to modern breeds which are of recent selection, but we shall use these breeds only to give an idea of the different size and morphology of the animals.

The results presented here involve the analysis of data from over 2000 specimens of dog coming from approximately 250 archaeological sites (Table 1). This osteological sample, while numerically significant, shows several limits which influence the final conclusions. In particular, only a small part of the sample was analysed directly by the authors, while the majority of the data were obtained from the literature. A problem that became clear during the research was the disparity of information available for the different chronological periods and the distribution of sites throughout Italy: the majority of the data are from Bronze Age sites in North and Central Italy.

## Methodology

Only a small portion of the remains were studied directly by the authors and the majority of the data was obtained from the literature. All available data on the dimensions of the specimens under investigation were collected and recorded on computer. Measurements were taken following von den Driesch (1976) and refer to adult animals. Data on the sex of specimens was not available and so not are considered in this paper. Quantitative and statistical analyses were then carried out on the main bone

| Site | **Area** | **Date** | **References** |
|---|---|---|---|
| Abisso Cesca | Friuli V.G. | I A | Riedel, 1973 |
| Acquarossa | Lazio | I A | Tagliacozzo, 1994a |
| Altino | Veneto | E R | Riedel, 1985a |
| Ancona - C.le dei Cappuccini | Marche | B A | Wilkens, 1990 |
| Appiano | Trentino - Alto Adige | B A | Riedel, 1985b |
| Attiggio di Fabriano | Marche | N - E | Wilkens, 1990 |
| Barche di Solferino | Lombardia | B A | Riedel, 1976a |
| Bari – S.Maria | Puglia | B A | Wilkens, 1991 |
| Belverde di Cetona | Umbria | B A | Tagliacozzo, in prep. |
| Bologna - S. Caterina | Emilia Romagna | I A | Farello, 1995a |
| Bolsena | Lazio | E R | Tagliacozzo, 1995 |
| Braida Roggia | Friuli V.G. | I A | Riedel, 1981 |
| Broglio di Trebisacce | Calabria | B A | Tagliacozzo, 1994b |
| Casale del Dolce | Lazio | N - E | Fiore & Tagliacozzo, 1997a, b |
| Casale Rivalta | Emilia Romagna | I A | Farello. 1990a |
| Case Vandelli | Emilia Romagna | I A | Farello. 1995a |
| Castelliere degli Elleri | Friuli V.G. | B A - I A | Riedel. 1976b |
| Castelliere di Monte d'Oro | Friuli V.G. | I A | Riedel. 1973 |
| Castelliere di Nivize | Friuli V.G. | I A | Riedel. 1968a |
| Castelrotto | Veneto | I A | Riedel. 1985c |
| Castenaso | Emilia Romagna | I A | Farello. 1994 |
| Castiglione | Lazio | B A | De Grossi Mazzorin. in prep. |
| Cattinara | Friuli V.G. | I A | Riedel. 1974 |
| Cavalzara | Veneto | B A | Riedel. 1979 |
| Caverna Cotariova | Friuli V.G. | ? | Riedel .1976c |
| Cerveteri | Lazio | I A | Clark. 1989 |
| Cisano | Veneto | B A | Riedel. 1990 |
| Classe | Emilia Romagna | L R | Farello. 1995b |
| Coccioli | Abruzzo | B A | De Grossi Mazzorin. in prep. |
| Colle S. Agata | Lazio | I A | Tagliacozzo. in prep. |
| Colle S. Magno | Lazio | B A | Tagliacozzo. in prep. |
| Colognola ai Colli | Veneto | I A | Riedel. 1984a |
| Colombare di Negrar | Veneto | E | Riedel. 1976d |
| Comacchio | Emilia Romagna | E R | Farello. 1990b |
| Coppa Nevigata | Puglia | B A | Siracusano. 1990; 1991 |
| Coppetella | Marche | N | Wilkens. 1988a |
| Cornuda | Veneto | N | Riedel. 1988 |
| Cretoncini | Lazio | I A | De Grossi Mazzorin. 1995a |
| Cures | Lazio | I A | Ruffo. in prep. |
| Este (necropoli) | Veneto | I A | Tagliacozzo. in prep. |
| Feniletto | Veneto | B A | Riedel. 1982a |
| Ficana | Lazio | B A - I A | De Grossi Mazzorin. 1997; in prep. |
| Fidene (necropoli) | Lazio | E R | De Filippis & De Grossi Mazzorin, in press |
| Fidene U.P.F. | Lazio | I A | De Grossi Mazzorin, in prep. |
| Fidenza | Emilia Romagna | I A | Farello, 1995a |
| Fiorano Modenese | Emilia Romagna | I A | Farello, 1989a |
| Fondo Paviani | Veneto | B A | Riedel, 1979 |
| Frattesina | Veneto | B A | De Grossi Mazzorin, in prep. |
| Ghisalba | Lombardia | E R | Fusco, 1981 |
| Giovinazzo | Marche | B A | Cataldo, Radina & Wilkens, 1989-90 |
| Gorzano | Emilia Romagna | B A | De Grossi Mazzorin, 1994 |
| Grammichele | Sicilia | B A | Villari, 1995 |
| Grotta Bella | Umbria | B A | Guerreschi *et al.*, 1992 |
| Grotta Calafarina | Sicilia | E | Villari, 1995 |
| Grotta Continenza | Abruzzo | N | Wilkens, 1987 |
| Grotta del Pino | Campania | B A | Tagliacozzo, in prep. |
| Grotta della Chiusazza | Sicilia | E | Villari, 1995 |
| Grotta delle Gallerie | Friuli V.G. | N - E | Riedel , 1968b |
| Grotta delle Ossa | (Slovenia) | B A - E R | Riedel, 1977 |
| Grotta dell'Uzzo | Sicilia | N | Tagliacozzo, 1993a; in prep. |
| Grotta n. 1745/4558 | Friuli V.G. | I A | Riedel, 1973 |

| | | | |
|---|---|---|---|
| I Camponi | Veneto | B A | Riedel, 1992a |
| Isolone sul Mincio | Lombardia | B A | Riedel, 1975 |
| La Marmottta | Lazio | N | Cassoli & Tagliacozzo, 1995 |
| La Starza | Campania | B A | Albarella, in prep. |
| Le Paludi di Celano | Abruzzo | B A | De Grossi Mazzorin, 1991 |
| Ledro | Trentino - Alto Adige | B A | Riedel, 1976e |
| Luni sul Mignone | Lazio | B A | Lepiksaar, 1975 |
| Maddalena di Muccia | Marche | N | Wilkens, in prep. |
| Madonna del Petto | Puglia | B A | Curci, 1995 |
| Marzabotto | Emilia Romagna | I A | Farello, 1995a |
| Matrensa | Sicilia | N | Villari, 1995 |
| Megara Hyblaea | Sicilia | B A | Villari, 1995 |
| Mezzocorona | Trentino - Alto Adige | E R | Riedel & Rizzi, 1994 |
| Mirandola loc. Arginone | Emilia Romagna | I A | Farello, 1992 |
| Modena | Emilia Romagna | E R | De Grossi Mazzorin, 1988 |
| Moie di Pollenza | Marche | I A | Wilkens, 1988b |
| Moletta | Trentino - Alto Adige | N | Riedel, 1984b |
| Montale | Emilia Romagna | B A | De Grossi Mazzorin, 1994 |
| Monte Casale | Sicilia | B A | Villari, 1995 |
| Monte Castellaccio | Emilia Romagna | B A | De Grossi Mazzorin, 1996a |
| Monte Rovello | Lazio | B A | Caloi & Palombo, 1986 |
| Monte Zoppega | Veneto | B A - I A | Riedel, 1979 |
| Monteriggioni "Campassini" | Toscana | I A | Bartoloni, Cianferoni & De Grossi Mazzorin 1997 |
| Mulino S. Antonio | Campania | N | Albore Livadie *et al.*, 1987-88 |
| Musarna | Lazio | E R | Tagliacozzo, in press |
| Oderzo - via dei Mosaici | Veneto | I A - E R | Tagliacozzo & Fiore, 1992 |
| Olbia | Sardegna | E R | Manconi, 1996 |
| Osimo | Marche | I A | Wilkens, 1997 |
| Osteria dell'Osa | Lazio | I A | De Grossi Mazzorin, 1992 |
| Padova - largo Europa | Veneto | I A - E R | Tagliacozzo, 1993b |
| Padova - via Giustiniani | Veneto | E R | Tagliacozzo, in prep. |
| Peschiera | Veneto | B A | Riedel, 1982b |
| Piano Donatelli | Marche | N | Wilkens, 1988c |
| Piercastello | Calabria | E R | Albarella, 1989 |
| Poggio Rusco | Lombardia | B A | Catalani, 1984 |
| Poviglio - S. Rosa | Emilia Romagna | B A | Riedel, in prep. |
| Pozzuolo del Friuli | Friuli V.G. | I A | Riedel, 1983-84 |
| Pyrgi | Lazio | I A | Caloi & Palombo, 1980 |
| Quarto del Tormine | Veneto | B A | Riedel, 1987 |
| Razza di Campegine | Emilia Romagna | N | Cazzella *et al.*, 1976 |
| Redù | Emilia Romagna | B A | De Grossi Mazzorin, 1994 |
| Ripa Tetta | Puglia | N | Wilkens, 1988d |
| Ripabianca di Monterado | Marche | N | Wilkens, in prep. |
| Riparo La Roberta | Toscana | B A | Bigini, 1985 |
| Ripoli | Abruzzo | N | De Grossi Mazzorin, in prep. |
| Roma - Aqua Marcia | Lazio | E R | De Grossi Mazzorin, 1996b |
| Roma - Domus Parthorum | Lazio | L R | Tagliacozzo, in prep. |
| Roma - Meta Sudans | Lazio | L R | De Grossi Mazzorin, 1995b |
| Roma - Palatino | Lazio | I A | De Grossi Mazzorin, in prep. |
| Rubiera Ca' del Cristo | Emilia Romagna | B A | Farello, 1989b |
| S. Ambrogio | Emilia Romagna | B A | De Grossi Mazzorin, 1994 |
| S. Candido | Trentino - Alto Adige | E R | Riedel, 1983 |
| S. Claudio | Emilia Romagna | I A | Farello, 1990c |
| S. Giacomo degli Schiavoni | Molise | L R | Albarella, Ceglia & Roberts, 1993 |
| S. Giorgio Valpolicella | Veneto | I A | Riedel, 1992b |
| S. Giuliano | Emilia Romagna | B A | De Grossi Mazzorin, 1996c |
| S. Maria in Selva | Marche | N | Wilkens, 1987 |
| S. Paolina di Filottrano | Marche | B A | Wilkens, 1988e |
| Santorso | Veneto | I A - E R | Cassoli & Tagliacozzo, 1991 |
| Sestino | Toscana | E R | De Grossi Mazzorin, in prep. |
| Sonnenburg | Trentino - Alto Adige | B A | Riedel, 1984c |

| | | | |
|---|---|---|---|
| Stentinello | Sicilia | N | Villari, 1995 |
| Stufels - hotel Dominik | Trentino - Alto Adige | I A - E R | Riedel, 1986 |
| Sulmona | Abruzzo | I A | De Grossi Mazzorin, 1995c |
| Tabina di Magreta | Emilia Romagna | B A | De Grossi Mazzorin, in prep. |
| Tarquinia | Lazio | I A | Bedini, in prep. |
| Terramare varie | Emilia Romagna | B A | De Grossi Mazzorin, 1994 |
| Terranegra | Veneto | I A | Riedel, 1979 |
| Thapsos | Sicilia | B A | Villari, 1995 |
| Torre Mordillo | Calabria | B A | Tagliacozzo, in press |
| Udine | Friuli V.G. | L R | Riedel, 1993 |
| Vejano - Borgo | Lazio | B A | Cosentino & De Grossi Mazzorin, in press |
| Veroli | Lazio | I A | Tagliacozzo, in prep. |
| Verucchio | Emilia Romagna | I A | Farello, 1997 |
| Viverone | Piemonte | B A | Fedele, 1983 |
| Volano | Trentino - Alto Adige | E R | Riedel & Scarpa, 1988 |

***Table 1*** *List of the archaeological sites with at least one useful measurement (N, Neolithic; E, Eneolithic; BA, Bronze Age; IA, Iron Age; ER, Early Roman; LR, Late Roman)*

measurements. Our work used as models similar research conducted on dogs by Harcourt (1974) and Clark (1995) for Great Britain, Benecke (1987) for Northern Europe, Peters (1994) for Central Europe, Morey (1992) for North America, and Shigehara (1994) for Japan. The metric data of modern breeds were taken from Wagner (1930). Data have been divided into the following chronological periods: Neo-eneolithic (second half of sixth millennium-third millennium BC); Bronze Age (twentieth-tenth century BC); Iron Age (ninth-third century BC); early Roman (third century BC-second century AD); and late Roman period (third-sixth century AD).

Cranial dimensions were initially determined using von den Driesch's system of 40 measurements, but on the basis of our experience with previous work and the above mentioned research, only 13 have been selected for analysis: these appear sufficient to express the real size of the skulls. This substantially reduced the osteological sample, but at the same time allowed more valid comparisons. Following the coded numbering system established by von den Driesch (1976), the cranial measurements selected are defined below.

**Cranial measurements**

1 - Total length
2 - Condylobasal length
7 - Upper neurocranium length
8 - Viscerocranium length
13 - Median palatal length
15 - Length of cheek-tooth row ($P^1$-$M^2$)
18 - Length of the carnassial
29 - Greatest neurocranium breadth
31 - Least breadth of skull
32 - Frontal breadth
33 - Least breadth between the orbits
34 - Greatest palatal breadth
36 - Breadth at the canine alveoli

In addition to absolute size, other features, such as the shape of the skull, the length of the muzzle and the width of the snout, the neurocranium and palatal breadth, may shed light on the morphological character of a specimen. Therefore, some indices were calculated. Such indices express, as a percentage, the ratios between some of the measurements described above and reflect the different relationships among the parts of the cranium. Seventeen different indices were calculated, some of them similar to those of Harcourt (1974), Morey (1992), and Clark (1995). Other indices were tentatively modified by us in order to adapt them to our specimens: these were most often incomplete and it was therefore not always possible to take all measurements. Of these indices only six were actually selected because they describe better the morphological differentiation of the skulls. The cranial indices used are defined below.

**Cranial indices**

Ind. 1 - Cephalic index (29) * 100 / (1)
Ind. 2 - Cephalic index (29) * 100 / (2)
Ind. 3 - Snout index (15) * 100 / (1)
Ind. 4 - Snout index (13) * 100 / (2)
Ind. 5 - Snout width index (34) * 100 / (15)
Ind. 6 - Snout width index (36) * 100 / (34)

The same method was used for the mandible, where nine measurements were selected and nine indices calculated. Later the indices were reduced to six. Mandible code numbers and measurement definitions, are defined below along with index definitions for mandibles.

**Mandible measurements**

1 - Total length
6 - Length: angular process-aboral border of the canine alveolus
8 - Length of cheek-tooth row ($P_1$-$M_3$)
10 - Length of the molar row
12 - Length of the premolar row
13 - Length of the carnassial
17 - Greatest thickness of the body of jaw
18 - Height of the vertical ramus
19 - Height of the mandible behind $M_1$

**Mandible indices**

Ind. 1 - (18) * 100 / (1)
Ind. 2 - (19) * 100 / (1)
Ind. 3 - (18) * 100 / (8)
Ind. 4 - (19) * 100 / (10)
Ind. 5 - (17) * 100 / (1)
Ind. 6 - (17) * 100 / (19)

For the postcranium, measurements of humerus, radius, femur, tibia, and metapodials were selected. In particular the total length, the proximal breadth, shaft breadth, and distal breadth were measured. Sometimes also the anterior-posterior diameters of the proximal and distal end were considered.

**Analysis of the data**

Skulls (crania)

There were 167 more or less complete skulls for which at least one of the 13 selected measurements could be taken: 23 were Neolithic, 73 of the Bronze Age, 33 of the Iron Age, 27 of the early Roman period and 11 of the late Roman period. The sample was drawn from 68 sites. Skulls were classified according to Hasebe (1952): skulls with a total length of less than 155 mm, as small dogs; skulls between 156 and 185 mm, as medium-small dogs; between 171 and 185 mm, as medium dogs; between 186 and 200 mm, as medium-large dogs; and above 200 mm, as large dogs.

Classification of skulls by total length (Fig. 1a) (n=33) show that the smallest specimen (115 mm.) came from the late Roman period at Classe near Ravenna, where three other skulls of less than 155 mm were also found. However, small crania are also present earlier: in the Bronze Age (at Cetona, 132 mm; Bagno di Rubiera, 142 mm; Barche di Solferino, 4 skulls, between 127 and 147.5 mm); in the Iron Age (at Veroli, 153 mm); and in the early Roman period (at Padova, 153 mm; Fidene, 147 mm). In general none of the skulls examined reaches the size of modern dwarf breeds, like the Pekinese or the dwarf pinscher whose skulls do not reach 100 mm. The only Neolithic skull for which total length could be taken is from Ripoli and belongs to a medium-small dog (164 mm). The largest skull is from the Iron Age context of Pyrgi and, following the Hasebe's classification, is the only complete cranium belonging to a truly large sized dog (213.5 mm). Medium-large dogs, with dimensions above 186 mm, are present in all periods (Terramare, Cetona, M. Castellaccio, and S. Giuliano in the Bronze Age; Case Vandelli in the Iron Age; Padova in the Roman period).

Comparison of the length of the upper cheektooth row (n=57), or the length of the upper carnassial (n=132) were also useful, as often skull length was not available (Fig. 1b-c). Cheektooth rows less than 55 mm long and carnassials of less than 16 mm are considered as belonging to small dogs while measurements above 67 mm and 20 mm respectively are referred to large dogs. These data only partially confirmed previous findings. A prevalence of medium-small sized dogs can be noted in all periods including the Neolithic. Dogs classifiable as large are very rare: one in the Neolithic, two in the Bronze Age and four in the Iron Age. However, we noted an increase in skull size in the Iron Age compared to earlier periods.

The different morphologies of the most complete skulls were investigated by applying several indices and logarithmic comparisons of some measurements. Forty-two skulls, reported in Table 2, allowed the calculation of at least one such index.

It was possible to apply the first Cephalic Index (ind.1) to 23 skulls from each period with the exception of the Neolithic (Fig. 2). This index shows the presence, as early as the Bronze Age (Cetona and Rubiera), of short skulls with fairly broad neurocrania (index about 37%). More rounded skulls were found at Classe in the Late Roman period (indices from 39 to 40%). Particularly narrow skulls are present in the Bronze Age at Cetona and the Terramare settlements, and also in the Late Roman period at Classe. Comparison of this index for specimens from the Bronze Age site of Cetona, it is possible to note that two different morphologies exist in the breadth-length ratio, varying from 26 to 37%. In the Late Roman period this tendency is somewhat accentuated, as the specimens from Classe demonstrate (indices 28.3-39.8%).

Comparison of the second Cephalic Index (ind.2) confirms in general these first evaluations. Other indices (ind.3 and 4) allow the evaluation of different morphologies by comparing facial length: total cranial length (Fig. 3). An example of a dog with a particularly short muzzle was found at the Bronze Age site of S. Ambrogio (ind.3=30.3%; ind.4=52.5%). The skull, for its absolute dimensions and only for this ratio, seems very similar to a modern Bulldog (ind.3=31.6%; ind.4=52%). In the Bronze Age, small sized dogs with a long muzzle are also present. Dogs with a short muzzle become more common in the Roman period. In the Iron Age, a medium sized skull had a particularly long muzzle (Veroli). A group of Bronze Age dogs is similar to a modern dingo, while a large Iron Age skull from Pyrgi shows features close to a Doberman.

Indices 5 and 6 (which reflect relative width of the muzzle) show that as early as the Bronze Age (Cetona) there are dogs which show both broadening and shortening of the muzzle. Unfortunately, it was only possible to use the cheektooth row length instead of the palatal length for most specimens. Skulls with broad muzzles were recovered from the early Roman period in Padova, and the late Roman period in Classe. In all periods, dogs with a narrow muzzle coexist (Cetona, Padova and Classe). Again, for this ratio the dog from Pyrgi resembles a Doberman. Little dogs with long pointed muzzles, like a modern dwarf pinscher, are present only from the Late Roman Period.

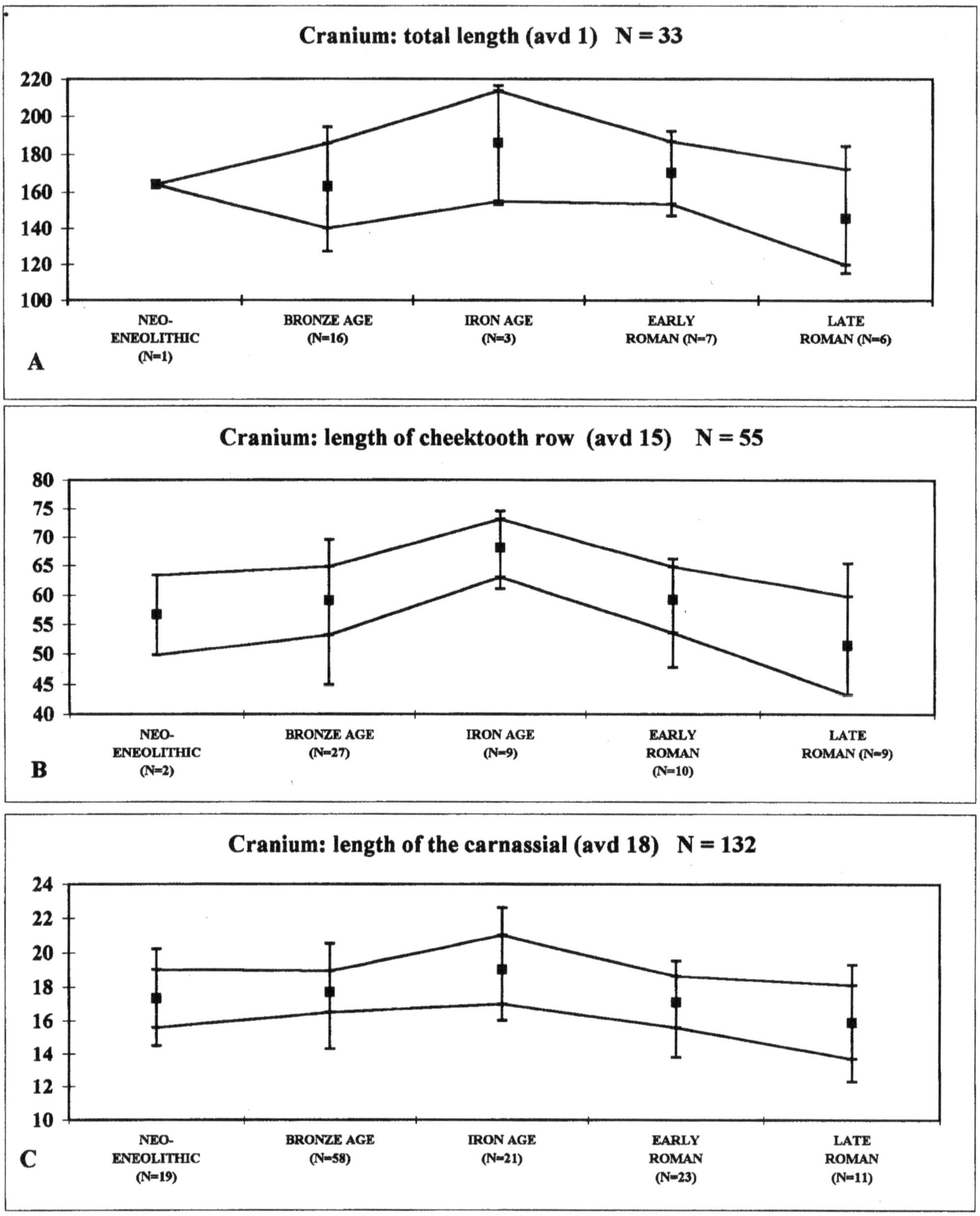


***Figure 1*** *Skull measurements in the different periods analyzed: A) total length, B) length of the cheek-tooth row, C) length of P4 (length in mm; after von den Driesch 1976)*

| Site | dat. | Date | 1 | 2 | 13 | 15 | 29 | 32 | 34 | 36 | ind.1 | ind.2 | ind.3 | ind.4 | ind.5 | ind.6 |
|---|---|---|---|---|---|---|---|---|---|---|---|---|---|---|---|---|
| **S.Maria in Selva** | n | N | | | | 63.3 | 58.0 | 41.2 | 61.7 | | | | | | 97.5 | |
| **C. del Dolce** | n-e | N-E | | | 85.6 | | 53.9 | | 56.7 | 32.4 | | | | | | 57.1 |
| **Terramare** | b m r | BA | | | 91.5 | 58.5 | | | 66.0 | 40.0 | | | | | 112.8 | 60.6 |
| **Terramare** | b m r | BA | 189.0 | 181.0 | 93.0 | 64.5 | 51.0 | 50.0 | | 37.7 | 27.0 | 28.2 | 34.1 | 51.4 | | |
| **S. Ambrogio** | b m r | BA | 170.0 | 157.0 | 82.4 | 51.5 | 53.0 | 50.3 | 57.5 | 30.0 | 31.2 | 33.8 | 30.3 | 52.5 | 111.7 | 52.2 |
| **Ancona** | b f | BA | 179.2 | 171.2 | | 63.1 | 57.2 | 45.2 | 62.6 | 37.4 | 31.9 | 33.4 | 35.2 | | 99.2 | 59.7 |
| **Belverde Cetona** | b a m r | BA | 176.0 | 166.0 | 89.1 | 62.5 | 48.3 | 41.6 | 56.7 | 33.5 | 27.5 | 29.1 | 35.5 | 53.7 | 90.7 | 59.0 |
| **Belverde Cetona** | b a m r | BA | 161.8 | 154.0 | 77.8 | 57.0 | 50.3 | 47.5 | 57.2 | 32.7 | 31.1 | 32.7 | 35.2 | 50.5 | 100.3 | 57.2 |
| **Belverde Cetona** | b a m r | BA | 132.6 | 124.0 | 64.0 | 50.0 | 48.8 | 32.2 | 50.8 | 30.5 | 36.8 | 39.3 | 37.7 | 51.6 | 101.6 | 60.1 |
| **Belverde Cetona** | b a m r | BA | 194.2 | 185.0 | 102.5 | 69.5 | 51.7 | 51.3 | 60.4 | 34.9 | 26.6 | 28.0 | 35.8 | 55.4 | 86.9 | 57.8 |
| **M. Castellaccio** | b m | BA | 186.0 | 175.5 | 90.4 | 65.0 | 57.2 | | 58.9 | 33.9 | 30.8 | 32.6 | 34.9 | 51.5 | 90.6 | 57.6 |
| **Bagno Rubiera** | b m | BA | 142.0 | 138.0 | 75.4 | 51.8 | 52.8 | 38.5 | 48.6 | | 37.2 | 38.3 | 36.5 | 54.6 | 93.8 | |
| **S. Giuliano** | b r | BA | 190.5 | 178.0 | 94.0 | 65.5 | 61.5 | 52.6 | 62.0 | 37.2 | 32.3 | 34.6 | 34.4 | 52.8 | 94.7 | 60.0 |
| **Ledro** | b a m | BA | 173.4 | | 83.2 | 61.0 | | 53.5 | 61.2 | 33.8 | | | 35.2 | | 100.3 | 55.2 |
| **Ledro** | b a m | BA | | | 82.3 | 60.0 | | | 62.6 | 34.5 | | | | | 104.3 | 55.1 |
| **Ledro** | b a m | BA | | | 92.1 | 63.2 | | | 60.7 | 35.2 | | | | | 96.0 | 58.0 |
| **Barche** | b a m | BA | 147.5 | | | | | 42.7 | 54.7 | 30.4 | | | | | | 55.6 |
| **Barche** | b a m | BA | 160.5 | | | 59.2 | | 41.8 | 61.0 | | | | 36.9 | | 103.0 | |
| **Barche** | b a m | BA | 127.0 | | 65.2 | 44.9 | | 43.2 | 47.9 | 24.7 | | | 35.4 | | 106.7 | 51.6 |
| **Barche** | b a m | BA | 130.7 | | 69.2 | 49.6 | | 43.5 | 47.0 | 26.4 | | | 37.9 | | 94.8 | 56.2 |
| **Barche** | b a m | BA | | | | 58.0 | | 35.0 | 56.0 | | | | | | 96.6 | |
| **Osteria Osa** | I fe | IA | | | | 69.0 | | | 60.0 | 36.5 | | | | | 87.0 | 60.8 |
| **Veroli** | VII-VI a.C. | IA | 153.0 | | | 63.0 | | | | 32.0 | | | 41.2 | | | |
| **Case Vandelli** | V a.C. | IA | 191.0 | 182.3 | | 68.3 | 58.0 | | | | 30.4 | 31.8 | 35.8 | | | |
| **Pyrgi** | IV-III a.C. | IA | 213.5 | 201.0 | 108.0 | 74.6 | | | 67.3 | 39.8 | | | 34.9 | 53.7 | 90.2 | 59.1 |
| **Padova Lg. Europa** | I-II d.C. ? | ER | 152.8 | 146.5 | 78.0 | 55.1 | 55.4 | 39.9 | | 30.7 | 36.3 | 37.8 | 36.1 | 53.2 | | |
| **Padova Lg. Europa** | I-II d.C. ? | ER | 172.3 | | | 57.7 | 58.5 | 33.7 | 65.7 | 36.5 | 34.0 | | 33.5 | | 113.9 | 55.6 |
| **Padova Lg. Europa** | I-II d.C. ? | ER | 186.7 | 179.0 | 97.0 | 60.8 | 57.0 | | 60.0 | 37.0 | 30.5 | 31.8 | 32.6 | 54.2 | 98.7 | 61.7 |
| **Padova Lg. Europa** | I-II d.C. ? | ER | 192.0 | 182.0 | 96.0 | 64.0 | 58.0 | | | | 30.2 | 31.9 | 33.3 | 52.7 | | |
| **Fidene t. 13** | II d.C. | ER | 166.0 | | 87.3 | 59.8 | | 42.0 | 57.5 | 33.0 | | | 36.0 | | 96.2 | 57.4 |
| **Fidene t. 14/III** | II d.C. | ER | | 169.0 | 88.0 | 63.0 | 58.0 | 47.1 | 61.5 | 35.3 | | 34.3 | | 52.1 | 97.6 | 57.4 |
| **Fidene t. 29/I** | II d.C. | ER | 147.0 | 136.5 | 72.6 | 47.8 | 51.5 | 39.1 | 48.7 | 28.9 | 35.0 | 37.7 | 32.5 | 53.2 | 101.9 | 59.3 |
| **Fidene t. 29/II** | II d.C. | ER | 174.0 | | 88.6 | 54.2 | 54.0 | | 56.5 | 33.0 | 31.0 | | 31.1 | | 104.2 | 58.4 |

| | | | | | | | | | | | | | | | | |
|---|---|---|---|---|---|---|---|---|---|---|---|---|---|---|---|---|
| **Classe** | IV-V d.C. | LR | 115.0 | 109.0 | 59.0 | 43.3 | 45.8 | 33.5 | 42.6 | 22.8 | 39.8 | 42.0 | 37.7 | 54.1 | 98.4 | 53.5 |
| **Classe** | IV-V d.C. | LR | 128.0 | 120.0 | 62.2 | 44.5 | 50.3 | 42.5 | 48.5 | 25.5 | 39.3 | 41.9 | 34.8 | 51.8 | 109.0 | 52.6 |
| **Classe** | IV-V d.C. | LR | 135.0 | 126.0 | 66.0 | 46.0 | 44.5 | 39.0 | 47.8 | 24.2 | 33.0 | 35.3 | 34.1 | 52.4 | 103.9 | 50.6 |
| **Classe** | IV-V d.C. | LR | | | | 44.6 | 52.0 | 36.1 | 52.8 | 28.1 | | | | | 118.4 | 53.2 |
| **Classe** | IV-V d.C. | LR | 143.0 | 134.0 | 67.0 | 49.0 | 47.4 | 49.6 | 50.7 | 28.4 | 33.1 | 35.4 | 34.3 | 50.0 | 103.5 | 56.0 |
| **Classe** | IV-V d.C. | LR | | | | 50.5 | 51.5 | 38.8 | 54.0 | 31.4 | | | | | 106.9 | 58.1 |
| **Classe** | IV-V d.C. | LR | 170.0 | 161.0 | 87.2 | 59.4 | 54.3 | 53.5 | 57.8 | 31.2 | 31.9 | 33.7 | 34.9 | 54.2 | 97.3 | 54.0 |
| **Classe** | IV-V d.C. | LR | 184.0 | 173.0 | 90.7 | 61.1 | 52.1 | 50.1 | 60.3 | 33.3 | 28.3 | 30.1 | 33.2 | 52.4 | 98.7 | 55.2 |
| **Udine** | IV-VII d.C. | LR | | | 92.1 | 65.4 | | | 61.0 | | | | | | 93.3 | |
| | | **N** | 31 | 23 | 30 | 39 | 26 | 28 | 35 | 34 | 23 | 22 | 30 | 21 | 34 | 31 |
| | | **Min** | 115.0 | 109.0 | 59.0 | 43.3 | 44.5 | 32.2 | 42.6 | 22.8 | 26.6 | 28.0 | 30.3 | 50.0 | 86.9 | 50.6 |
| | | **Max** | 213.5 | 201.0 | 108.0 | 74.6 | 61.5 | 53.5 | 67.3 | 40.0 | 39.8 | 42.0 | 41.2 | 55.4 | 118.4 | 61.7 |
| | | **Mean** | 164.0 | 158.7 | 83.1 | 57.8 | 53.1 | 43.3 | 56.7 | 32.4 | 32.4 | 34.3 | 35.0 | 52.8 | 100.1 | 56.7 |
| | | **s. d.** | 24.5 | 24.9 | 12.8 | 8.0 | 4.3 | 6.4 | 6.1 | 4.4 | 3.7 | 4.0 | 2.1 | 1.4 | 7.6 | 2.9 |
| **Modern breed** | | | **1** | **2** | **13** | **15** | **29** | **32** | **34** | **36** | **ind.1** | **ind.2** | **ind.3** | **ind.4** | **ind.5** | **ind.6** |
| **Shephard** | | | 238.3 | 207.2 | 113.1 | 74.7 | 58.6 | 68.4 | 78.0 | 47.6 | 24.6 | 28.3 | 31.3 | 54.6 | 104.4 | 61.0 |
| **Bulldog** | | | 161.5 | 138.8 | 72.2 | 51.0 | 57.2 | 63.4 | 73.8 | 42.4 | 35.4 | 41.2 | 31.6 | 52.0 | 144.7 | 57.5 |
| **Bassett** | | | 142.5 | 122.6 | 69.4 | 51.6 | 52.2 | 42.9 | 51.9 | 28.5 | 36.6 | 42.6 | 36.2 | 56.6 | 100.6 | 54.9 |
| **Boxer** | | | 177.0 | 152.0 | 80.0 | 54.3 | 57.8 | 60.5 | 79.5 | 46.2 | 32.7 | 38.0 | 30.7 | 52.6 | 146.4 | 58.1 |
| **St. Bernard** | | | 253.5 | 223.0 | 123.8 | 76.8 | 60.5 | 75.5 | 86.3 | 56.8 | 23.9 | 27.1 | 30.3 | 55.5 | 112.4 | 65.8 |
| **Borzoi** | | | 221.2 | 198.2 | 110.8 | 78.6 | 55.3 | 55.5 | 62.5 | 34.5 | 25.0 | 27.9 | 35.5 | 55.9 | 79.5 | 55.2 |
| **Doberman** | | | 214.4 | 190.4 | 103.9 | 72.8 | 56.5 | 58.3 | 67.4 | 38.8 | 26.4 | 29.7 | 34.0 | 54.6 | 92.6 | 57.6 |
| **Pointer** | | | 205.0 | 186.1 | 102.2 | 71.7 | 57.6 | 57.1 | 73.2 | 42.2 | 28.1 | 31.0 | 35.0 | 54.9 | 102.1 | 57.7 |
| **Dingo** | | | 192.0 | 169.3 | 93.2 | 67.0 | 54.2 | 51.3 | 59.8 | 35.0 | 28.2 | 32.0 | 34.9 | 55.1 | 89.3 | 58.5 |
| **Pekinese** | | | 90.2 | 79.2 | 43.0 | 29.5 | 50.2 | 45.5 | 48.5 | 26.5 | 55.7 | 63.4 | 32.7 | 54.3 | 164.4 | 54.6 |
| **Pinscher** | | | 97.0 | 84.1 | 46.4 | 33.0 | 46.7 | 35.4 | 39.4 | 20.0 | 48.1 | 55.5 | 34.0 | 55.2 | 119.4 | 50.8 |
| **Wolf** | | | 243.0 | 225.0 | 113.0 | 81.5 | 64.0 | 69.4 | 71.2 | 42.0 | 26.3 | 28.4 | 33.5 | 50.2 | 87.4 | 59.0 |

***Table 2*** *Measurements of the skulls that allowed the calculation of at least one index (N, Neolithic; E, Eneolithic; BA, Bronze Age; IA, Iron Age; ER, Early Roman; LR, Late Roman). Measurement codes after von den Driesch (1976)*

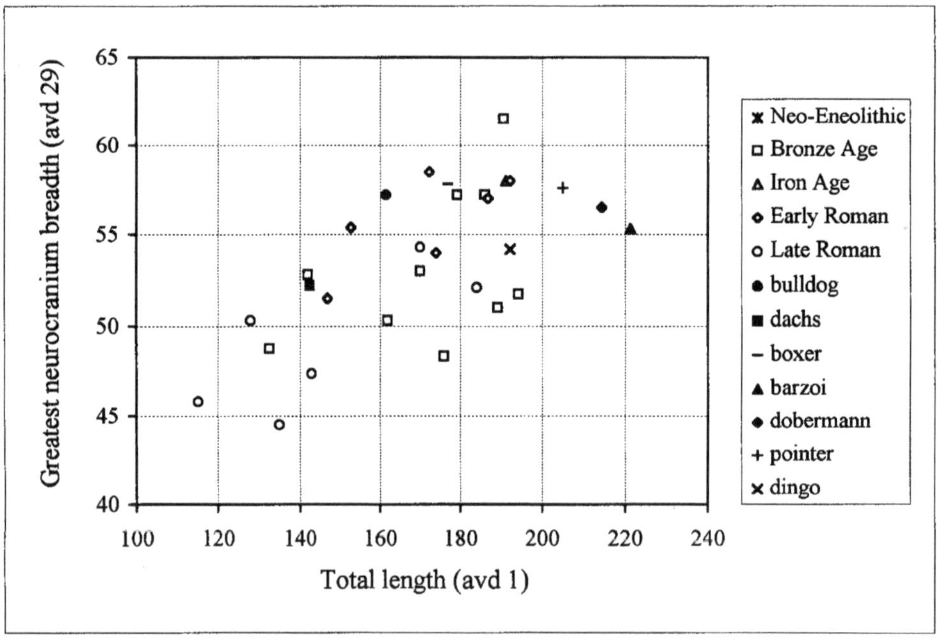


***Figure 2*** *Skull: scatter diagram (meas. 1 to meas. 29) (length in mm; avd, von den Driesch 1976)*

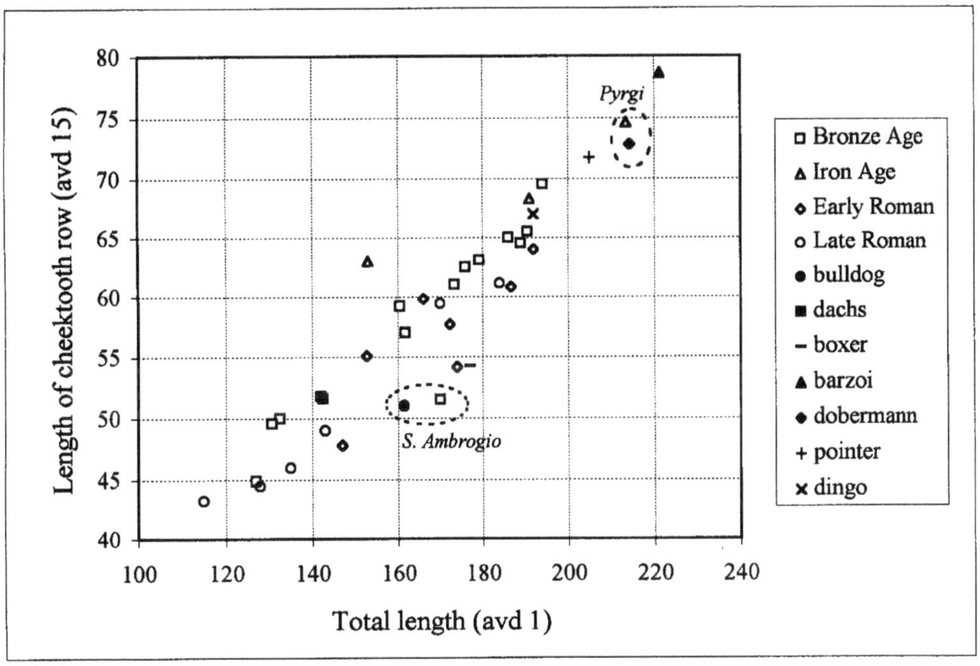


***Figure 3*** *Skull: scatter diagram (meas. 1 to meas. 15) (length in mm; avd, von den Driesch 1976)*

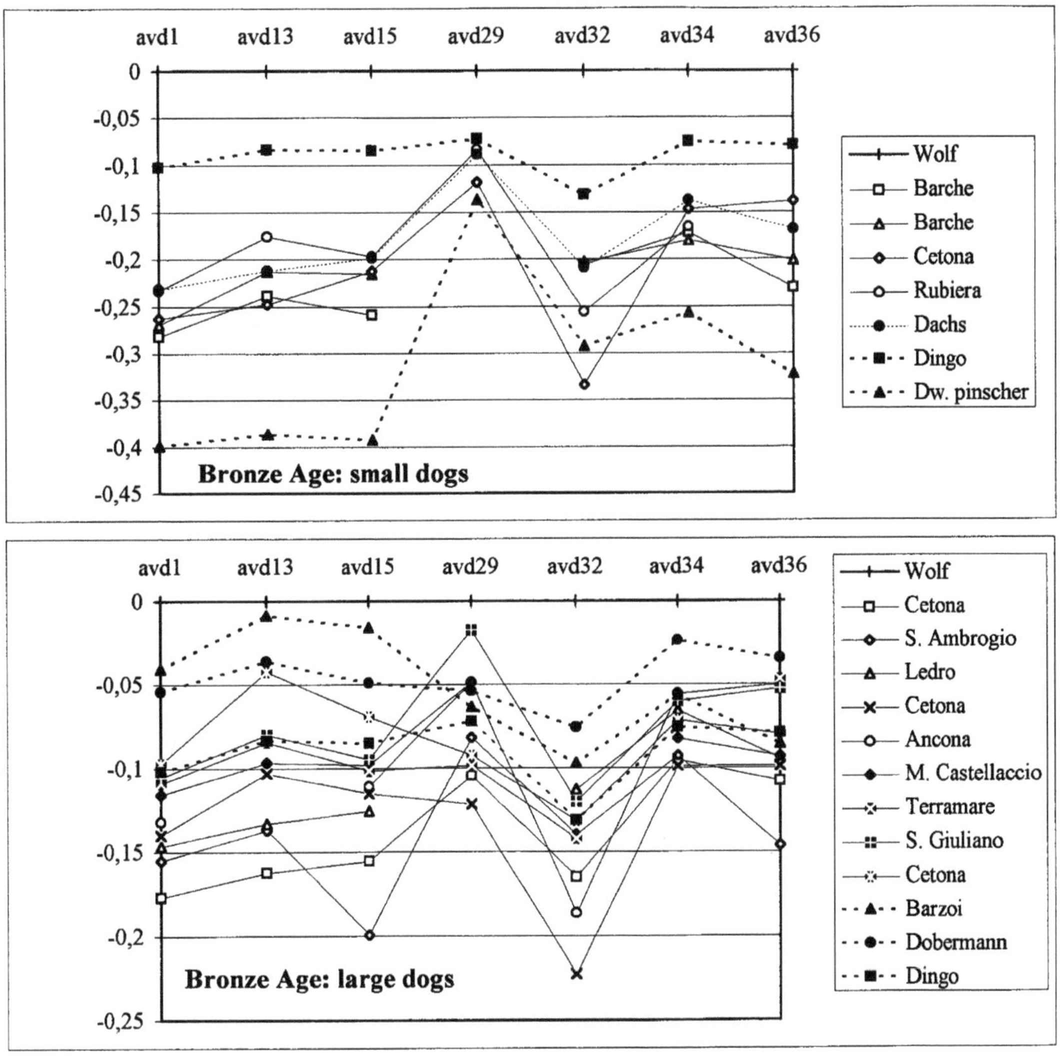


***Figure 4*** *Logarithmic difference of Bronze Age skulls compared to a modern wolf and to some modern breeds (avd, von den Driesch 1976)*

Six measurements of the skull were compared to a modern wolf and those of modern breeds. The few Neolithic skulls are dimensionally close to the dingo although from a morphological point of view they resemble a dachshund. The extreme variability and the high number of Bronze Age skulls suggested two graphs can be drawn on the basis of these dimensions (Fig. 4). In the Bronze Age, smaller skulls show a strong dimensional and morphological similarity to a modern dachshund. Only one skull (Belverde di Cetona) with a particularly wide muzzle and a narrow frontal is quite different. Larger variability can be seen in the medium-large sized skulls, some with a long and narrow muzzle like a greyhound or a Doberman, and others with a short, wide muzzle. However, on the whole, the morphology of the Bronze Age skulls is similar to that of a dingo. In the Iron Age the skull from Pyrgi confirms the similarity to a Doberman. In the Roman Age the presence in the same site of skulls with different morphologies is confirmed. In the Late Roman period strong dimensional differences are evident.

In conclusion, as regards the skull, it is possible to say that a notable dimensional variety, from medium-large to small, has been present since the Neolithic and became greater in the Bronze Age. However, large dogs are present only since the Iron Age, when there seems to be an increase in the average dimensions. An even greater dimensional variability occurs in the late Roman period, when very small dogs also appear. Substantial morphological differences, both in cranial length and muzzle shape, do occur in the Bronze Age but they appear to be the result of accidental cross-breeds rather than real selection, since at the same sites there are skulls with mixed features. Starting in the Iron Age, it is possible that the first real selection (probably for large dogs) leading to real breeds occurs. Only in the Roman period does selection become of greater importance and real breeds of different size and conformation appear.

Mandibles

There were 394 more or less complete mandibles for which at least one of the nine selected measurements could be taken. Among theses, 261 allowed the calculation of at least one of the indices. The indices employed in the study permitted an evaluation of the morphological modifications related to shortening of the snout. The ratios between the absolute length of the mandible (or the cheektooth rows) and the height of the body of the vertical ramus (or *thickness* of the body) were used.

The mandible specimens were also classified according to total length, following Hasebe (1952): less than 113 mm, as small; between 114 and 124 mm, as medium-small; between 125 and 135 mm, as medium; between 136 and 146 mm, as medium-large; and above 147 mm, as large. For specimens from the Neolithic period, the absolute length shows a relatively low variability, with all dogs classified as medium sized (Fig. 5a). For the Bronze Age there is a larger variability in the absolute dimensions of the mandible (101 and 148 mm), although medium (125 and 135 mm) and medium-small (114 and 124 mm) sized ones are most prevalent. In the Iron age, although less data were available, animals with larger mandible dimensions are represented (121 and 153mm) and there is at least one dog that can be classified as large. In the early Roman period (107 and 151mm) and in the late Roman period (83 and 135mm) the greatest mandibular length variability is evident.

These data from mandible total length are confirmed by the measurement of the lower cheektooth row (Fig. 5b) and of the lower carnassial (Fig. 5c). This latter measurement shows that small varieties of dog were already present in the Neolithic (La Marmotta). On the whole, the data agree and confirm a substantial reduction in size between the Iron Age and the Roman period.

The indices used in the analysis of the mandible sample demonstrate the relationship between the absolute length of the mandible to the length of the cheektooth row, and to the height of the horizontal ramus of the mandible (Table 3 & Fig. 6). Comparison of these figures indicates that as early as the Neolithic there existed a large amount of variability, demonstrated by the presence of dogs with massive mandibles along with those having slender ones of the same absolute length. In particular, the sample from S. Maria in Selva demonstrates this phenomenon, since although most of the mandibles fall within the range of medium length dimensions, mandibles with a horizontal ramus that is very thick or very thin in relation to the length of the cheektooth row are also present.

For the Bronze Age sample, comparison of the mandibular indices demonstrates the existence of mandibles of similar length but with different shapes, especially those from the five settlements with the highest number of remains: Ledro, Frattesina, Monte Castellaccio, Isolone, and Cetona. In particular, the samples from Ledro and Cetona show the greatest variability in the absolute length (Ledro=111-141 mm; Cetona=112-148 mm) and in the ratios between the different measurements, suggesting the presence of several kinds of dogs. Monte Castellaccio demonstrates less dimensional variability, but strong morphological differences. Finally, Isolone and Frattesina show less dimensional variety and only in a few specimens can morphological differentiation be detected.

In the Iron Age, the evidence suggests a substantial morphological uniformity in mandibles, possibly due to the scarcity of diagnostic specimens. Only at Sulmona do two different morphologies seem to coexist. In contrast, the Roman Period sample from Padova seems to indicate the coexistence of at least three different types of dog.

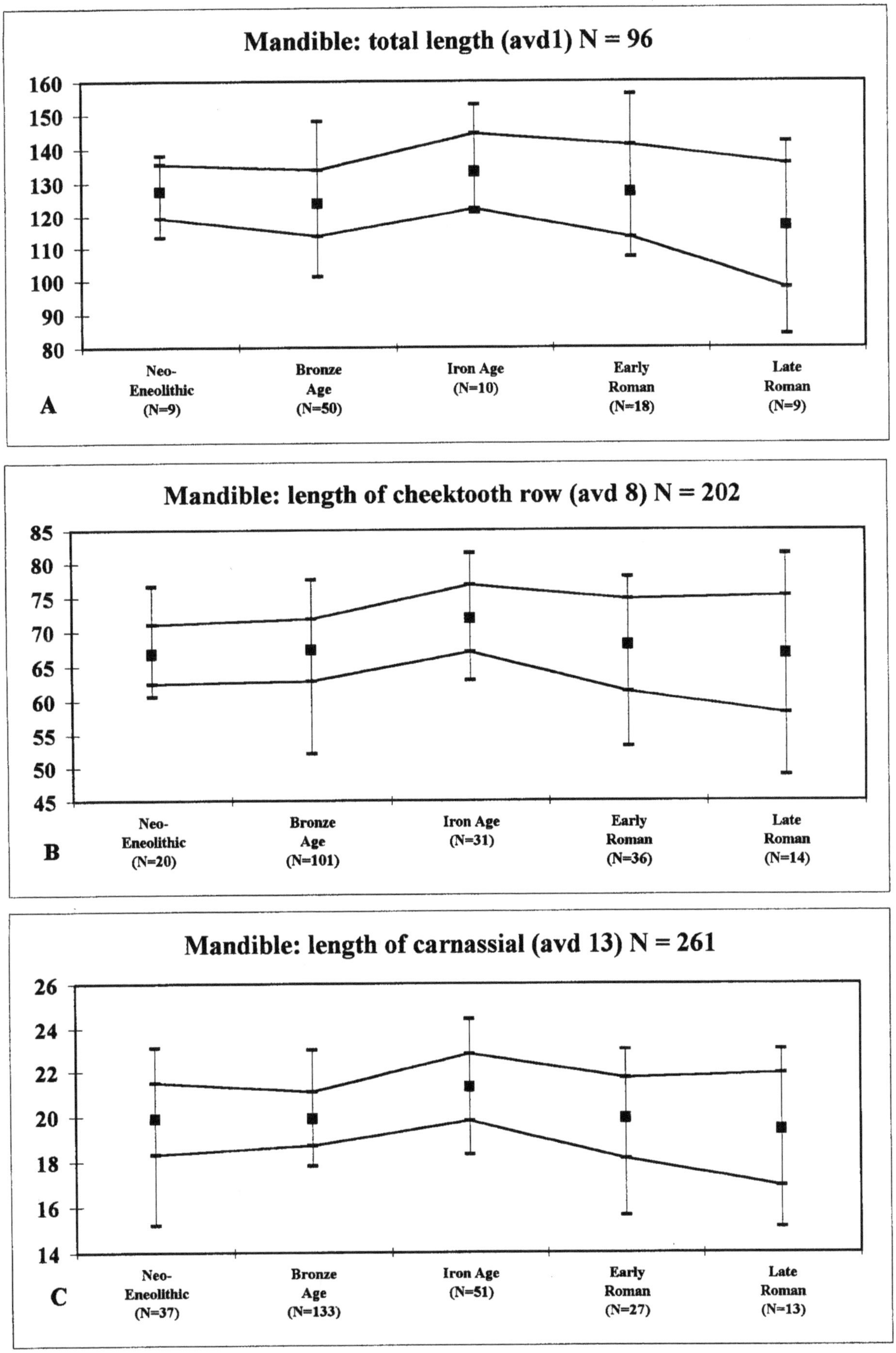


***Figure 5*** *Mandible measurements in the different periods analyzed: A) total length,B) length of the cheek-tooth row, C) length of M1 (length in mm; after von den Driesch 1976)*

| Site | Date | 1 | 8 | 10 | 17 | 18 | 19 | ind.1 | ind.2 | ind.3 | ind.4 | ind.5 | ind.6 |
|---|---|---|---|---|---|---|---|---|---|---|---|---|---|
| **Ripoli** | N | 126 | 69 | 32 | 10.6 | | 20.7 | | 16.4 | 30.0 | 64.7 | 8.4 | 51.2 |
| **Stentinello** | N | 118 | 65.8 | 32.3 | | 45.5 | 21.3 | 38.6 | 18.1 | 32.4 | 65.9 | | |
| **Gallerie** | N-E | 135 | | 37 | 12 | 51 | 24 | 37.8 | 17.8 | | 64.9 | 8.9 | 50.0 |
| **Gallerie** | N-E | 128 | 66 | | 9 | 49 | 20 | 38.3 | 15.6 | 30.3 | | 7.0 | 45.0 |
| **Gallerie** | N-E | 138 | 76 | | 12 | | 25 | | 18.1 | 32.9 | | 8.7 | 48.0 |
| **C. del Dolce** | N-E | 113.3 | 66.3 | 32.2 | 12.1 | 46.2 | 21.9 | 40.8 | 19.3 | 33.0 | 68.0 | 10.7 | 55.3 |
| **C. del Dolce** | N-E | 126.6 | 68.9 | 31.8 | | 49.1 | 20.4 | 38.8 | 16.1 | 29.6 | 64.2 | | |
| **C. del Dolce** | N-E | 126.7 | 68.8 | 32.2 | 10.6 | | 19.6 | | 15.5 | 28.5 | 60.9 | 8.4 | 54.1 |
| **Terramare** | BA | 128.8 | 68 | 33.5 | 11 | | 20.5 | | 15.9 | 30.1 | 61.2 | 8.5 | 53.7 |
| **Terramare** | BA | 119 | 64.5 | 33.5 | 10.5 | 48 | 20 | 40.3 | 16.8 | 31.0 | 59.7 | 8.8 | 52.5 |
| **Terramare** | BA | 127.5 | 70 | 34.3 | 11.5 | 50 | 21 | 39.2 | 16.5 | 30.0 | 61.2 | 9.0 | 54.8 |
| **Redù** | BA | 134.4 | 69.2 | 34 | 12 | 50.3 | 22 | 37.4 | 16.4 | 31.8 | 64.7 | 8.9 | 54.5 |
| **Feniletto** | BA | 119.2 | 65.5 | 31.7 | 10 | 47.7 | 20 | 40.0 | 16.8 | 30.5 | 63.1 | 8.4 | 50.0 |
| **Isolone** | BA | 114.2 | 62.2 | 31.2 | 10.5 | 42.9 | 20 | 37.6 | 17.5 | 32.2 | 64.1 | 9.2 | 52.5 |
| **Isolone** | BA | 113.6 | 63.6 | 32.8 | 10.3 | 41 | 20.7 | 36.1 | 18.2 | 32.5 | 63.1 | 9.1 | 49.8 |
| **M. Rovello** | BA | 143 | 73.8 | 36.8 | 12.2 | | 26.5 | | 18.5 | 35.9 | 72.0 | 8.5 | 46.0 |
| **Ledro** | BA | 111 | 64.7 | 33 | 10.8 | 46.6 | 21.4 | 42.0 | 19.3 | 33.1 | 64.8 | 9.7 | 50.5 |
| **Ledro** | BA | 114.5 | 65.8 | 30.3 | 9 | 45.3 | 18.2 | 39.6 | 15.9 | 27.7 | 60.1 | 7.9 | 49.5 |
| **Ledro** | BA | 114.8 | 65.3 | 32 | 9.3 | 45.6 | 18.3 | 39.7 | 15.9 | 28.0 | 57.2 | 8.1 | 50.8 |
| **Ledro** | BA | 115.5 | 52.1 | 30.7 | 9.5 | 44.2 | 20 | 38.3 | 17.3 | 38.4 | 65.1 | 8.2 | 47.5 |
| **Ledro** | BA | 117.4 | 68 | 34.7 | 9.3 | | 19.7 | | 16.8 | 29.0 | 56.8 | 7.9 | 47.2 |
| **Ledro** | BA | 125.2 | 66.3 | 32.3 | 10.5 | | 23.7 | | 18.9 | 35.7 | 73.4 | 8.4 | 44.3 |
| **Ledro** | BA | 125.7 | 69.2 | 33.6 | 10.1 | 49.6 | 21.7 | 39.5 | 17.3 | 31.4 | 64.6 | 8.0 | 46.5 |
| **Ledro** | BA | 127 | 70 | 33 | 10.4 | 50.8 | 21.8 | 40.0 | 17.2 | 31.1 | 66.1 | 8.2 | 47.7 |
| **Ledro** | BA | 130.7 | 70.8 | 32.5 | 10.8 | 52.3 | 23.7 | 40.0 | 18.1 | 33.5 | 72.9 | 8.3 | 45.6 |
| **Ledro** | BA | 133.3 | 69.5 | 32.9 | 10.7 | 53.2 | 23.2 | 39.9 | 17.4 | 33.4 | 70.5 | 8.0 | 46.1 |
| **Ledro** | BA | 134.7 | 71 | 31.8 | 10.7 | 51.6 | 23 | 38.3 | 17.1 | 32.4 | 72.3 | 7.9 | 46.5 |
| **Ledro** | BA | 137.4 | 73.6 | 34.7 | 10.5 | 54.7 | 21.9 | 39.8 | 15.9 | 29.8 | 63.1 | 7.6 | 47.9 |
| **Ledro** | BA | 140.7 | 75.4 | 35.3 | 11.8 | 58.6 | 24.1 | 41.6 | 17.1 | 32.0 | 68.3 | 8.4 | 49.0 |
| **Ficana 2** | BA | 131 | 71 | 33.5 | 10.5 | | 21.6 | | 16.5 | 30.4 | 64.5 | 8.0 | 48.6 |
| **Frattesina** | BA | 129.5 | 68.5 | 33.7 | 11.5 | 50.5 | 23.3 | 39.0 | 18.0 | 34.0 | 69.1 | 8.9 | 49.4 |
| **Frattesina** | BA | 136.7 | 73 | 35 | 10.8 | | 23.8 | | 17.4 | 32.6 | 68.0 | 7.9 | 45.4 |
| **Frattesina** | BA | 134.5 | 68.5 | 33 | 11.4 | 56 | 23.3 | 41.6 | 17.3 | 34.0 | 70.6 | 8.5 | 48.9 |
| **M. Castellaccio** | BA | 126 | 66.4 | 32.7 | 10.4 | | 22.1 | | 17.5 | 33.3 | 67.6 | 8.3 | 47.1 |
| **M. Castellaccio** | BA | 109 | 60.2 | 31.5 | 9.9 | 41.5 | 17 | 38.1 | 15.6 | 28.2 | 54.0 | 9.1 | 58.2 |
| **M. Castellaccio** | BA | 123.8 | 67.5 | 33.5 | 9.9 | 46.2 | 18.9 | 37.3 | 15.3 | 28.0 | 56.4 | 8.0 | 52.4 |
| **Montevenera** | BA | 114 | 64 | 29.9 | 9.5 | 45.5 | 20.1 | 39.9 | 17.6 | 31.4 | 67.2 | 8.3 | 47.3 |
| **Belverde Cetona** | BA | 128.79 | 63.05 | 27.02 | 12.3 | 50.87 | 20.4 | 39.5 | 15.8 | 32.4 | 75.5 | 9.6 | 60.3 |
| **Belverde Cetona** | BA | 148.16 | 77.62 | 37.32 | 12.3 | 59.86 | 23.3 | 40.4 | 15.7 | 30.0 | 62.4 | 8.3 | 52.8 |
| **Belverde Cetona** | BA | 112.7 | 61.24 | 30.26 | 9.2 | 48.89 | 16.85 | 43.4 | 15.0 | 27.5 | 55.7 | 8.2 | 54.6 |
| **Belverde Cetona** | BA | 120.5 | 67.33 | 32.97 | 11.2 | 44.47 | 18.76 | 36.9 | 15.6 | 27.9 | 56.9 | 9.3 | 59.7 |
| **Belverde Cetona** | BA | 122.61 | 64.26 | 31.11 | 10.6 | 45.44 | 19.23 | 37.1 | 15.7 | 29.9 | 61.8 | 8.6 | 55.1 |
| **Belverde Cetona** | BA | 111.7 | 59.63 | 30.11 | 9.1 | 45.02 | 17.65 | 40.3 | 15.8 | 29.6 | 58.6 | 8.1 | 51.6 |
| **Belverde Cetona** | BA | 101.3 | 59.02 | 29.24 | 9.6 | | 16.27 | | 16.1 | 27.6 | 55.6 | 9.5 | 59.0 |
| **Belverde Cetona** | BA | 121.33 | 63.52 | 30.92 | 10.5 | 46.14 | 19.19 | 38.0 | 15.8 | 30.2 | 62.1 | 8.7 | 54.7 |
| **Belverde Cetona** | BA | 130.04 | 66.01 | 30.68 | 12.1 | | 22.3 | | 17.1 | 33.8 | 72.7 | 9.3 | 54.3 |
| **Belverde Cetona** | BA | 128 | 67.42 | 32.59 | 11.2 | | 20.33 | | 15.9 | 30.2 | 62.4 | 8.8 | 55.1 |
| **S. Giuliano** | BA | 132.5 | 69.5 | 31.5 | 10.2 | 53 | 22.2 | 40.0 | 16.8 | 31.9 | 70.5 | 7.7 | 45.9 |
| **Gr. del Pino** | BA | 140.8 | 75.9 | 35.2 | 11.5 | 58.4 | 25.7 | 41.5 | 18.3 | 33.9 | 73.0 | 8.2 | 44.7 |
| **Sulmona** | IA | 120.9 | 66.5 | 32.6 | 9 | | 18.6 | | 15.4 | 28.0 | 57.1 | 7.4 | 48.4 |
| **Gr. 1745/4558** | IA | 123.2 | 70.8 | | 12 | | 22 | | 17.9 | 31.1 | | 9.7 | 54.5 |
| **Gr. 1745/4558** | IA | 122.7 | 66.8 | | 12.1 | | 21.5 | | 17.5 | 32.2 | | 9.9 | 56.3 |
| **Ficana 3b-c** | IA | 129 | 72.5 | 34 | 11 | 53 | 22 | 41.1 | 17.1 | 30.3 | 64.7 | 8.5 | 50.0 |
| **Castenaso** | IA | 123.8 | 67.6 | 35.1 | 10.8 | 49.7 | 23.2 | 40.1 | 18.7 | 34.3 | 66.1 | 8.7 | 46.6 |
| **Pyrgi** | IA | 153 | 81 | 39 | 13.1 | 61 | 28 | 39.9 | 18.3 | 34.6 | 71.8 | 8.6 | 46.8 |
| **Colognola** | ER | 138 | 76.2 | 38.8 | 12.8 | | 24.6 | | 17.8 | 32.3 | 63.4 | 9.3 | 52.0 |
| **Padova lg. Europa** | ER | 107.5 | 55 | 25.73 | 9.1 | 39.8 | 15.3 | 37.0 | 14.2 | 27.8 | 59.5 | 8.5 | 59.5 |
| **Padova lg. Europa** | ER | 144.6 | 75.7 | 30.6 | 12.2 | 50.9 | 20.3 | 35.2 | 14.0 | 26.8 | 66.3 | 8.4 | 60.1 |
| **Padova lg. Europa** | ER | 150.6 | 73.8 | 34.18 | 11.1 | 55.7 | 23.25 | 37.0 | 15.4 | 31.5 | 68.0 | 7.4 | 47.7 |
| **Padova lg. Europa** | ER | 130.4 | 70.1 | 30.3 | 12.5 | 50.9 | 20.4 | 39.0 | 15.6 | 29.1 | 67.3 | 9.6 | 61.3 |
| **Olbia** | ER | 110.7 | 61.6 | | 10.7 | | 21.3 | | 19.2 | 34.6 | | 9.7 | 50.2 |

| | | | | | | | | | | | | | |
|---|---|---|---|---|---|---|---|---|---|---|---|---|---|
| **Olbia** | ER | 125.2 | 65.8 | | 10.7 | | 21.5 | | 17.2 | 32.7 | | 8.5 | 49.8 |
| **Fidene t. 13** | ER | 124.7 | 66.6 | 31.9 | 10.7 | 53.1 | 21.6 | 42.6 | 17.3 | 32.4 | 67.7 | 8.6 | 49.5 |
| **Fidene t. 13** | ER | 126.2 | 66.8 | 31.4 | 10.9 | 51.6 | 23.6 | 40.9 | 18.7 | 35.3 | 75.2 | 8.6 | 46.2 |
| **Fidene t. 21/b** | ER | 111.5 | 60.7 | 30.5 | 10 | 43.5 | 17.9 | 39.0 | 16.1 | 29.5 | 58.7 | 9.0 | 55.9 |
| **Fidene t. 14/II** | ER | 131.5 | 66.8 | 32 | 11.4 | | 22.6 | | 17.2 | 33.8 | 70.6 | 8.7 | 50.4 |
| **Fidene t. 14/III** | ER | 132.8 | 71 | 32.5 | 11 | 50 | 21.2 | 37.7 | 16.0 | 29.9 | 65.2 | 8.3 | 51.9 |
| **Fidene t. 14/III** | ER | 132.5 | 71.5 | 32.8 | 11 | 50.4 | 21.4 | 38.0 | 16.2 | 29.9 | 65.2 | 8.3 | 51.4 |
| **Fidene t. 29/I** | ER | 108.8 | 57.6 | 30.1 | 10.7 | 43.2 | 17.1 | 39.7 | 15.7 | 29.7 | 56.8 | 9.8 | 62.6 |
| **Fidene t. 29/II** | ER | 121.2 | 62.2 | 31.5 | 10.3 | 52.2 | 21 | 43.1 | 17.3 | 33.8 | 66.7 | 8.5 | 49.0 |
| **Classe** | LR | 83.8 | 48.8 | | 7.5 | 30.3 | 13.6 | 36.2 | 16.2 | 27.9 | | 8.9 | 55.1 |
| **Classe** | LR | 123.4 | 63.6 | 29 | 10 | 49.6 | 19.6 | 40.2 | 15.9 | 30.8 | 67.6 | 8.1 | 51.0 |
| **Classe** | LR | 94.8 | 57.9 | 25.1 | 8.8 | 36.3 | 14.2 | 38.3 | 15.0 | 24.5 | 56.6 | 9.3 | 62.0 |
| **Classe** | LR | 115 | 62.3 | 29.3 | 10.4 | 47.7 | 21.2 | 41.5 | 18.4 | 34.0 | 72.4 | 9.0 | 49.1 |
| **Classe** | LR | 124 | 65.7 | 30.6 | 9.8 | 50 | 19.8 | 40.3 | 16.0 | 30.1 | 64.7 | 7.9 | 49.5 |
| **Classe** | LR | 105 | 63.8 | 30.9 | 10.1 | 37.5 | 15.1 | 35.7 | 14.4 | 23.7 | 48.9 | 9.6 | 66.9 |
| **Classe** | LR | 127 | 66.3 | 30.1 | 9.8 | 49.6 | 19.3 | 39.1 | 15.2 | 29.1 | 64.1 | 7.7 | 50.8 |
| **Classe** | LR | 135 | 70.2 | 31.1 | 10.2 | 50.2 | 20.4 | 37.2 | 15.1 | 29.1 | 65.6 | 7.6 | 50.0 |
| | | | | | | | | | | | | | |
| | **N** | 70 | 70 | 65 | 70 | 53 | 70 | 53.0 | 70.0 | 70.0 | 65.0 | 70.0 | 70.0 |
| | **Min.** | 83.8 | 48.8 | 25.1 | 7.5 | 30.3 | 13.6 | 35.2 | 14.0 | 23.7 | 48.9 | 7.4 | 44.3 |
| | **Max.** | 153.0 | 81.0 | 39.0 | 13.1 | 61.0 | 28.0 | 43.4 | 19.3 | 38.4 | 75.5 | 9.9 | 66.9 |
| | **Mean** | 124.4 | 66.8 | 32.2 | 10.6 | 48.7 | 20.8 | 39.3 | 16.7 | 31.1 | 64.6 | 8.6 | 51.6 |
| | **s. d.** | 12.5 | 5.8 | 2.5 | 1.1 | 5.8 | 2.7 | 1.9 | 1.2 | 2.7 | 5.8 | 0.6 | 4.9 |

***Table 3*** *Measurements of the mandibles that allowed the calculation of at least four indexes (N, Neolithic; E, Eneolithic; BA, Bronze Age; IA, Iron Age; ER, Early Roman; LR, Late Roman). Measurement codes after von den Driesch (1976)*

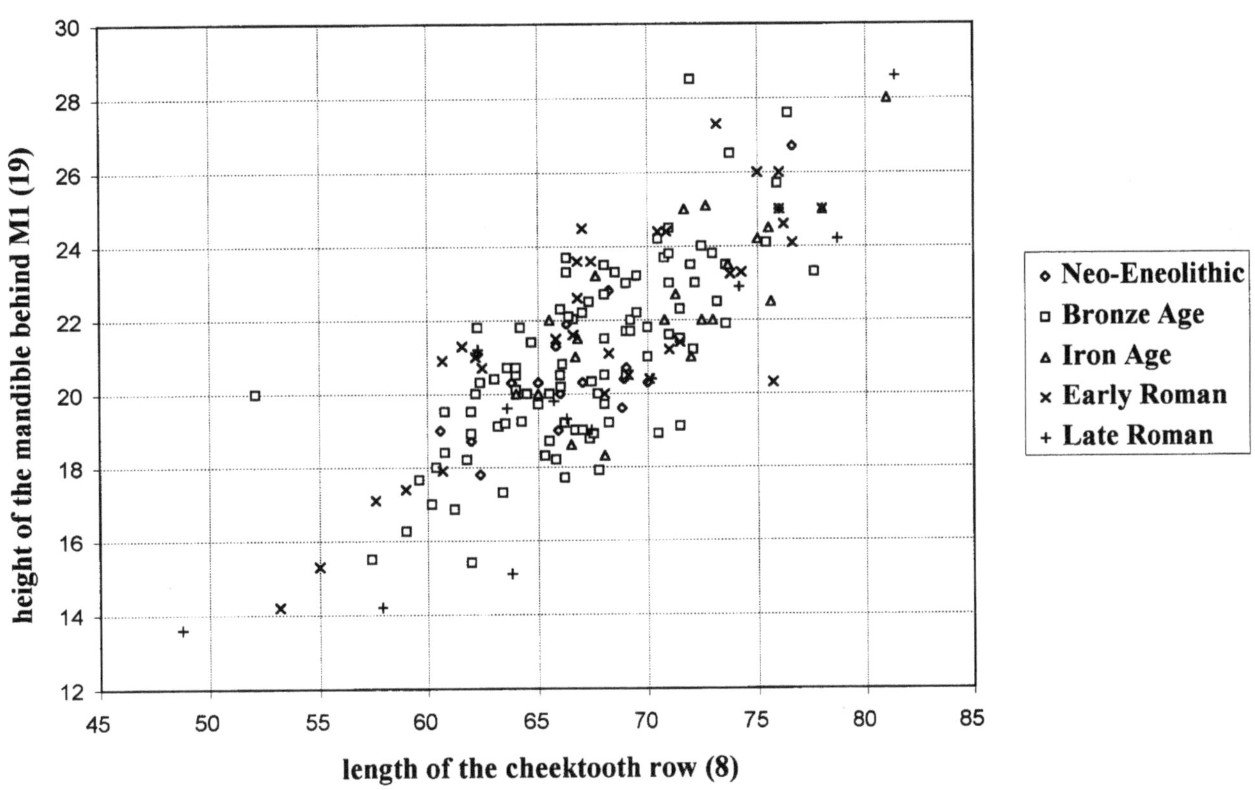


***Figure 6*** *Mandible: scatter diagrams (meas. avd 19 to meas. avd 10), (meas. avd 19 to meas. avd 8) (length in mm; avd, von den Driesch 1976)*

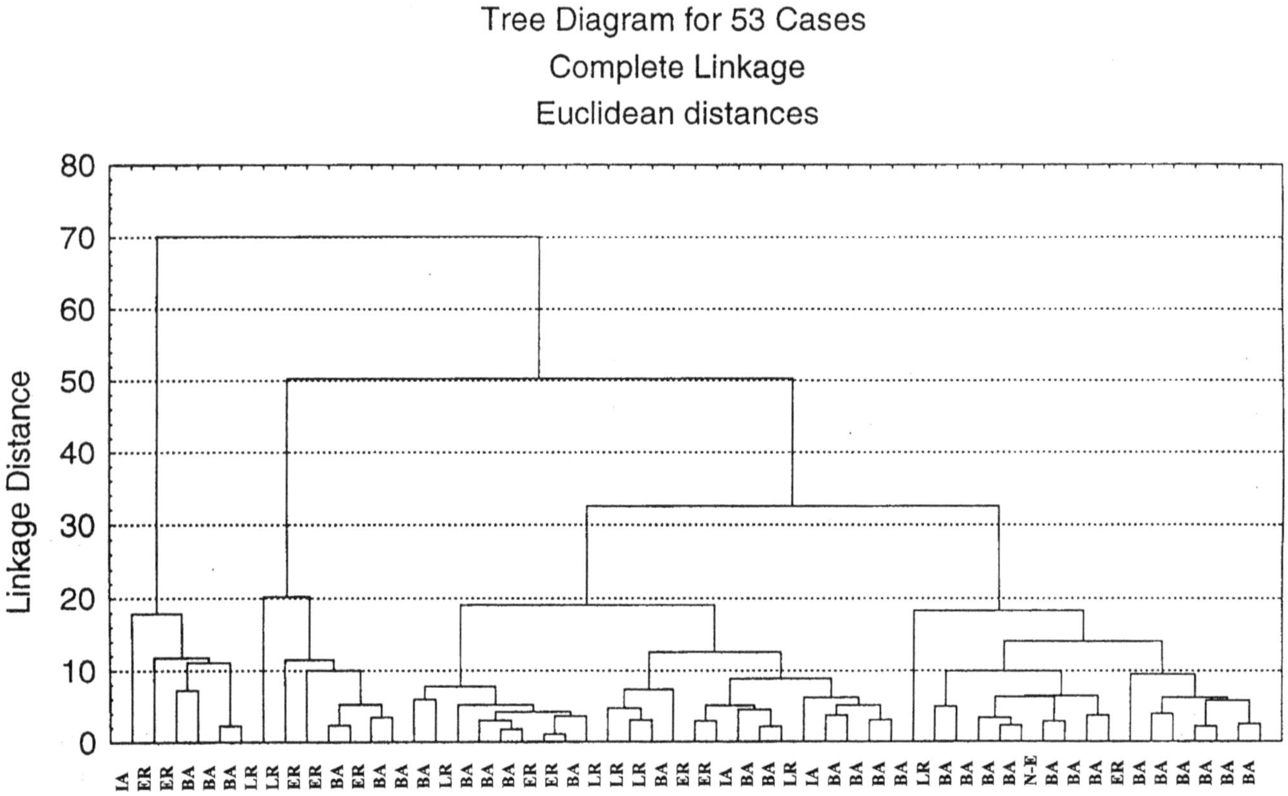


***Figure** 7 Cluster analysis of the mandibles with the measurements 1, 8, 10, 17, 18, and 19 ( avd, von den Driesch 1976)*

Finally, a greater variability in the mandibles can be noted, especially in the small forms, in the dogs of the Late Roman period.

Analysis of variance applied to nine metric traits of the mandible demonstrated a significant dimensional heterogeneity among the samples for all variables, with the exception of the length between the angular process and the aboral border of the canine alveolus. Moreover, analysis of the heterogeneity of the variances (Levene's Test) shows that seven out of nine variables have a significant difference in the variability within the samples. Therefore, the dimensional variability within the chronological series investigated is different.

Cluster analysis applied to this mandible sample (Fig. 7), based on the standardized Euclidean distance and using the algorithm of aggregation of the complete linkage, isolates four groupings that do not show any chronological or geographical aggregation. In particular, the mandibles from the same site are present in three (Ledro, Padova & Classe) or four (Cetona) groups. Only the fourth group seems to be made mainly of Bronze Age dogs.

The mandible analysis confirms the conclusions based on analysis of skulls: that dogs of variable size, from small to medium-large, existed as early as the Neolithic and the Bronze Age, but large dogs are only present from the Iron Age onward, when a increase in the average dimensions is confirmed. An even greater dimensional variability occurs in the late Roman period when even very small dogs appear. Substantial morphological differences, evident especially in the height of the horizontal ramus of the mandible and the shortening of the snout, occur as early as the Bronze Age, but real breed selection appears to start in the Iron Age and continues through the Roman period.

Limbs

A total of 1124 limb bone remains were analysed in this study. From these, it was possible to calculate the withers height for 473 cases. Dividing the sample into the different chronological periods, the estimate of withers height is known for 34 dogs of the Neo-Eneolithic, 137 of the Bronze Age, 87 of the Iron Age, 137 of the early Roman period, and 78 of the late Roman period. Compared to our previous work (De Grossi Mazzorin & Tagliacozzo, 1997), the sample of data increased greatly (especially for the Neo-Eneolithic and the early Roman period) both because of new excavations and because, after the research conducted by K. M. Clark (1995), the measurements of the metapodials, not previously considered, could also be included in the analysis. Comparison of the withers height estimates indicates a consistent variability in Neolithic, with an average increase in stature from the Bronze Age to the Iron Age (Table 4; Fig. 8). In the Roman period, large dogs with withers height above 62 cm are represented.

| Period | N | min. | max. | mean | s. d. |
|---|---|---|---|---|---|
| N-E | 34 | 35.3 | 55.6 | 45.3 | 4.3 |
| BA | 137 | 36.0 | 62.1 | 47.4 | 5.1 |
| IA | 87 | 37.4 | 60.8 | 51.5 | 5.6 |
| ER | 137 | 29.4 | 64.0 | 44.2 | 7.8 |
| LR | 78 | 26.1 | 69.7 | 47.8 | 12.5 |

***Table 4** Dog withers height estimates for chronological periods in Italy, following Harcourt (1974) & Clark (1995), in mm (N, Neolithic; E, Eneolithic; BA, Bronze Age; IA, Iron Age; ER, Early Roman; LR, Late Roman); s.d. is standard deviation*

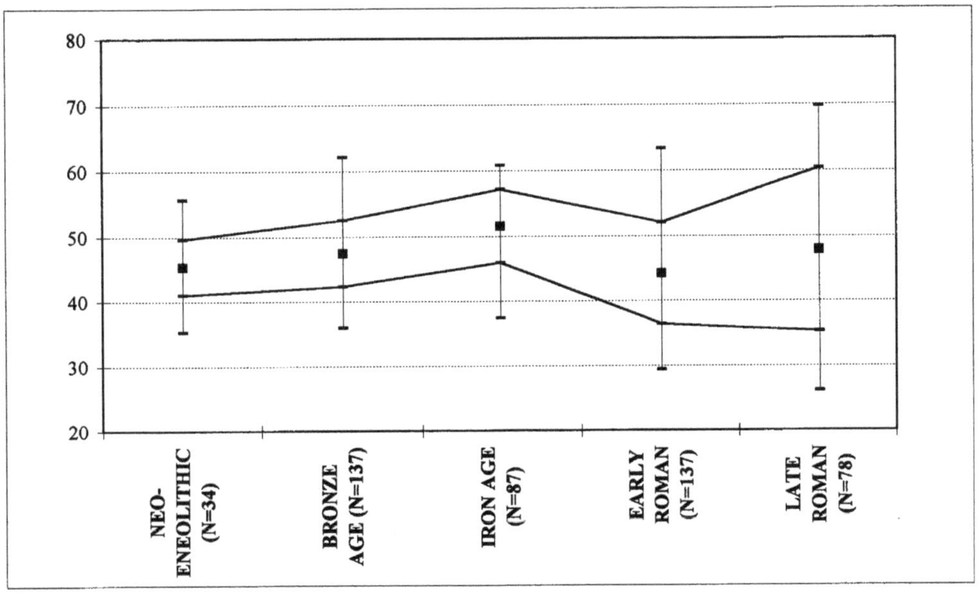


***Figure 8** Withers heights (in cm)*

From a comparison of the relationship between total bone length (GL) and the width of the distal end (Bd) of major limb elements recovered from each chronological period, some scatter plots were prepared (Fig. 9). The samples are not equally balanced. Nevertheless, the diagrams demonstrate a number of correlations that lead to a series of suggestions useful in analysing the issue of changes over time. In the graphs, the different ranges referable to Hasebe's size classification are also apparent (Shigehara, 1994). As regards the Neolithic and Eneolithic periods, the remains indicate in general medium-small sized dogs, with the exception of a very small dog at La Marmotta and a large one at Scamuso. In the Bronze Age, within a wide range of variability, the majority of the dogs are of medium size. In this period, however, large dogs were also present. In the Iron Age, there was a smaller range of variability, but a general increase in size is particularly evident, especially at Pyrgi, Nivize and Padova. During the Roman period, however, the range of variability increases substantially, especially in the final phase. All of the long bones confirmed the existence in the Roman period of small dogs with slightly brachymelic limbs. The "brachymelia" is a mutation that results in short-legged animals, probably due to a prenatal evolutionary disturbance, an endochondral ossification disorder (Hilzheimer, in Teichert, 1987). Brachymelic dogs are characterized by short, thick and twisted legs.

Withers heights have also been compared to the slenderness index of the femur, for all dogs from the analyzed sites and to the means of this index for some modern breeds (Fig. 10). An increase in dimensions can be seen from the Bronze to the Iron Age. Protohistoric dogs are in general of medium size (eumetric) with a tendency towards slender limbs. In the two phases of the Roman period a wider distribution of canid forms can be seen: from little dogs (hypometric) with size between 25 and 35 cm to hypermetric dogs higher than 60 cm (like a Doberman or a pointer).

In order to better demonstrate the distribution of small sized dogs, those from archaeological sites are compared to some small sized modern breeds (Fig. 11). Some dogs with brachymelic limbs are documented in Rome (*Meta Sudans*, Olbia, and Classe): these are probably lap-dogs, the size of a modern Pekinese. Other small sized dogs, withers height between 35 and 40 cm, but with more harmonious morphology, are present at Padova and Fidene. Those of Padova are close in dimensions to a modern spitz or a small fox-terrier, while those of Fidene, like some small Bronze Age dogs, are closer to a fox-terrier or a whippet in conformation.

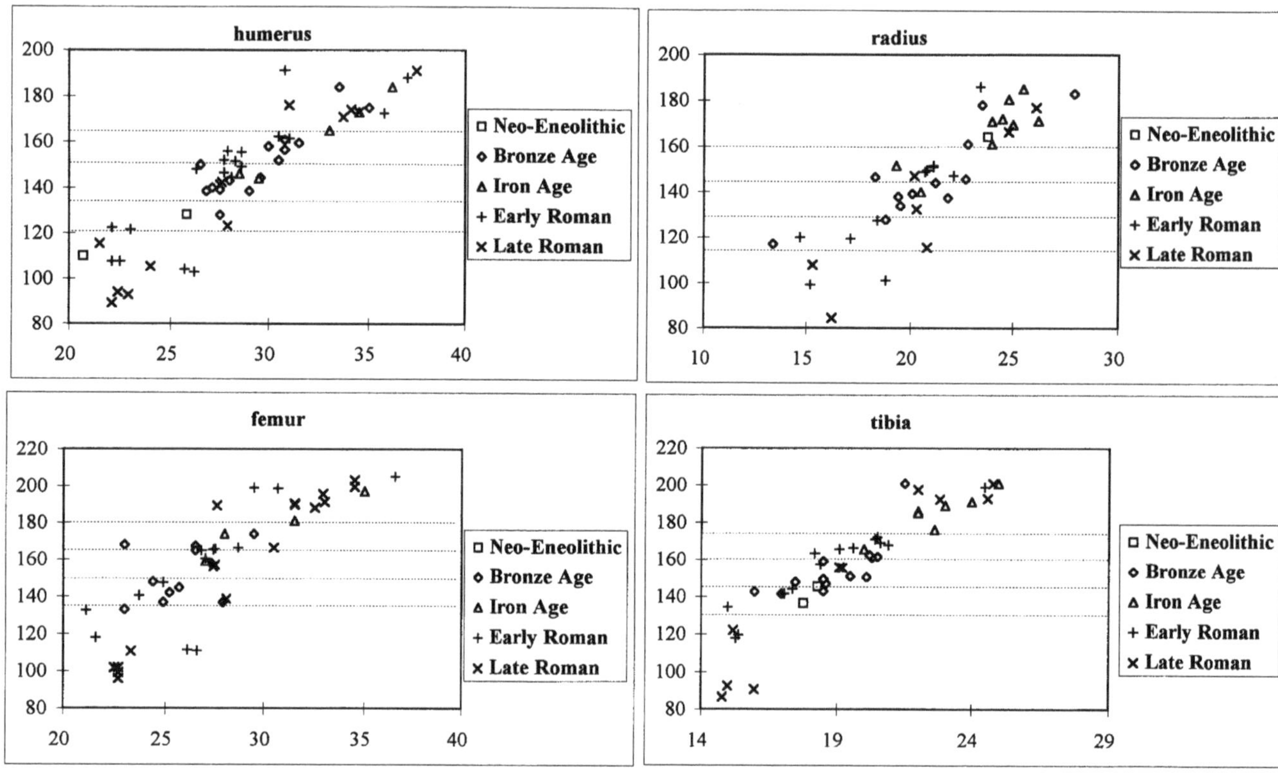


***Figure* 9** *Long bones: scatter diagrams. ordinate: greatest length (GL); abscissa greatest breadth of distal end (distal breadth, Bd) (in mm)*

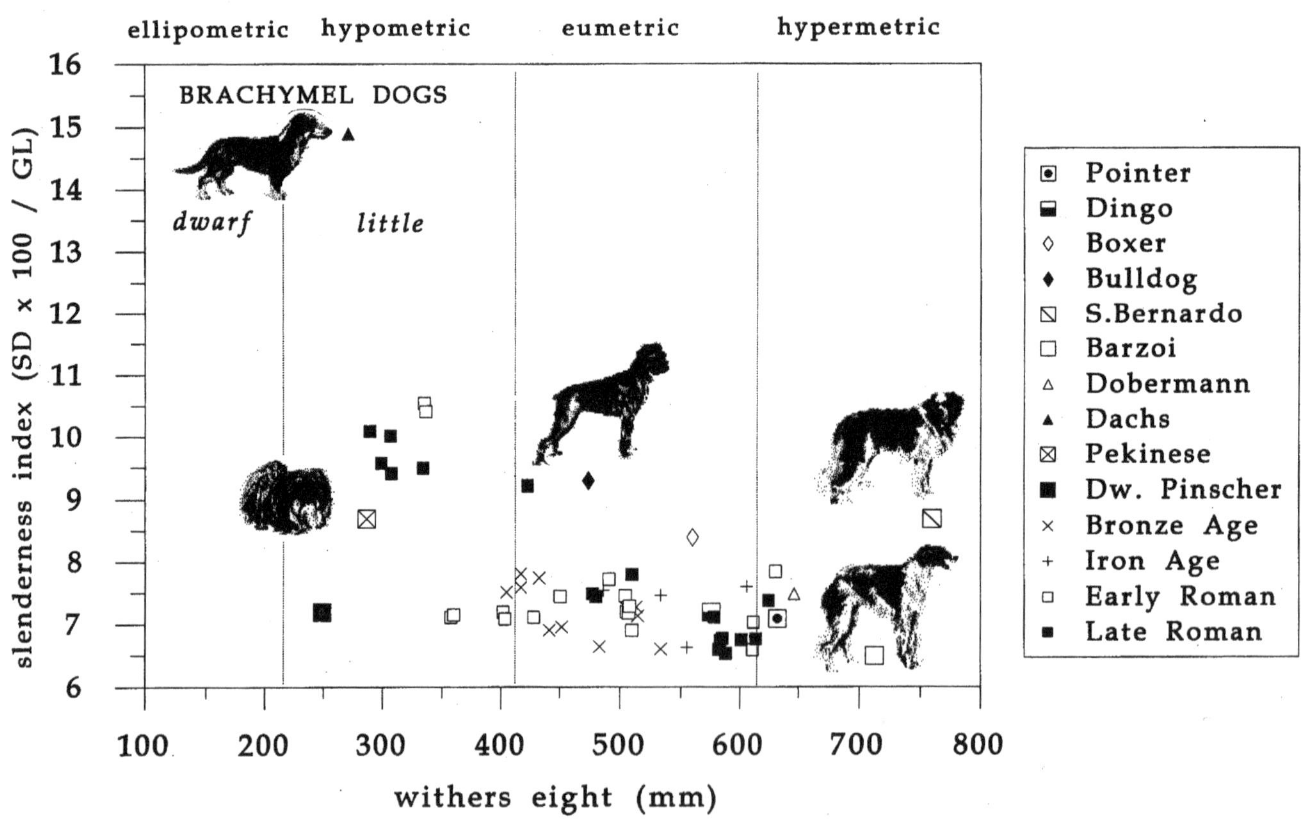


***Figure 10*** *Femurs from the Italian archaeological sites compared to those of some modern breeds. (slenderness index to withers height)*

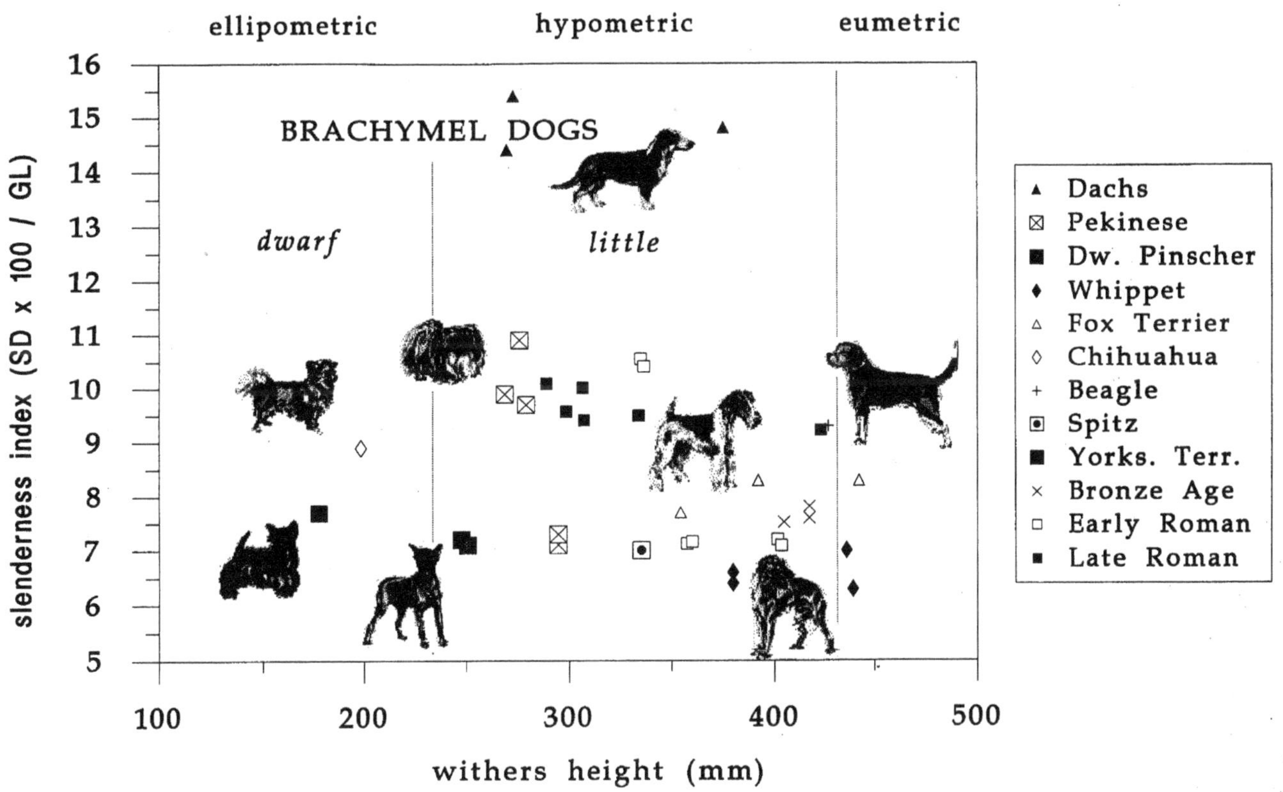


***Figure 11*** *Femurs of the small dogs from the Italian archaeological sites compared to those of some modern small and dwarf breeds. (slenderness index to withers height)*

## Conclusions

From the analysis of the morphology of the skeletal remains of Italian dogs it is possible to conclude that a fairly wide dimensional variability is present from the Neolithic period onward. Variability increases in the Bronze Age, but the morphological differences detected appear to be the result of accidental cross-breeding of randomly occurring variants rather than real selection, since in the same site skulls with mixed features were found. Large dogs are present only from the Iron Age onwards, when there seems to be selection for larger animals. In the Iron Age it is possible that the first real selection begins, producing dogs of greater size. Even greater size variability can be found in the Roman period in Italy. At this time, similar to other European Roman provinces (Altuna & Mariezkurrena, 1992; Teichert, 1987), lap-dogs appear, characterized by a particularly reduced stature.

## References:

Albarella, U. (1989). Hipponion-Vibo Valentia: necropoli in località Piercastello: la fauna. *Annali della Scuola Normale Superiore di Pisa* **19**, 825-832.

Albarella, U., Ceglia, V. & Roberts, P. (1993). S. Giacomo degli Schiavoni (Molise): an early fifth century AD deposit of pottery and animal bones from central Adriatic Italy. *Papers of the British School at Rome* **61**, 157-230.

Albore Livadie, C., Fedele, F., Albarella, U., de Matteis, F., Esposito, E. & Federico, R. (1987-88). Ricerche sull'insediamento tardo-neolitico di Mulino Sant'Antonio (Avella). *Rivista di scienze preistoriche* **41**, 65-103.

Altuna, J. & Mariezkurrena, K. (1992). Perros enanos en yacimientos romanos de la peninsula iberica. *Archaeofauna* **1**, 83-86.

Bartoloni, G., Cianferoni, G. C., De Grossi Mazzorin, J. (1997). Il complesso rurale di Campassini (Monteriggioni): considerazioni sull' alimentazione nell'Etruria settentrionale nell'VIII e VII secolo a.C. In *Aspetti della Cultura di Volterra etrusca fra l'età del Ferro e l'età ellenistica e contributi della ricerca antropologica alla conoscenza del Popolo Etrusco, Atti del XIX Convegno di Studi Etruschi ed Italici* (Volterra, ottobre 1995), 93-127.

Benecke, N. (1987). Studies on early dog remains from northern Europe. *Journal of Archaeological Science* **14**, 31-49.

Bigini, I. (1985). Reperti faunistici. *Rassegna d'Archeologia* **4**, 143-145.

Caloi, L.& Palombo, M.R. (1980). Il cane domestico di Pyrgi. *Studi Etruschi* **58**, 293-328.

Caloi, L. & Palombo, M.R. (1986). La fauna di Monte Rovello (XII-IX sec. a.C.): origine delle forme domestiche-Elenco del materiale. *Notiziario del Museo di Allumiere* **7**, 49-82.

Cassoli, P.F. & Tagliacozzo, A. (1991). La fauna degli scavi 1983-1986 a Santorso, Vicenza (Età del Ferro). *Preistoria Alpina* **25**, 165-216.

Cassoli, P.F. & Tagliacozzo, A. (1995). I reperti ossei faunistici dell'area della piroga. In (M.A. Fugazzola Delpino & M. Mineo, Eds) *La Piroga Neolitica del Lago di Bracciano ("La Marmottal")*. Bullettino di Paletnologia Italiana **86**, pp. 267-288.

Catalani, P. (1984). Poggio Rusco (MN): la fauna. *Preistoria Alpina* **20**, 203-210.

Cataldo, L., Radina, F. & Wilkens, B. (1989-90). L'insediamento protostorico di Giovinazzo (Bari): considerazioni sulla facies protoappenninica e appenninica nell'Italia sud-orientale. *Rivista di Scienze Preistoriche* **42**, 171-240.

Cazzella, A., Cremaschi, M., Moscoloni, M.& Sala, B. (1976). Siti neolitici in località Razza di Campegine (Reggio Emilia). *Preistoria Alpina* **12**, 79-126.

Clark, G. (1989). A group of animal bones from Cerveteri. *Studi Etruschi* **55**, 253-269.

Clark, K.M. (1995). The later prehistoric and protohistoric dog: the emergence of caninc diversity. *Archaeozoologia* **7**, 9-32.

Cosentino, S. & De Grossi Mazzorin, J. (in press). Archeozoologia e paleobotanica dell'insediamento protostorico del Borgo di Vejano (VT). In (N. Negroni Catacchio, Ed) *Preistoria e Protostoria in Etruria.* Milano: Atti del IV Incontro di Studi, pp. 1-12.

Curci, A. (1995). I resti faunistici dell'insediamento dell'età del Bronzo di Madonna del Petto, scavi 1977. *Taras* **15**, 205-216.

De Filippis, M. & De Grossi Mazzorin, J. (in press). Riti particolari in un sepolcreto a Fidene. *Atti del Convegno "Culto dei morti e costumi funerari romani"*, Roma (aprile 1998).

De Grossi Mazzorin, J. (1988). Il sondaggio stratigrafico nella sede centrale della Cassa di Risparmio di Modena (1985-1986): I resti faunistici. *Modena dalle origini all'anno mille, Studi di Archeologia e Storia* **1,** 449-455.

De Grossi Mazzorin, J. (1991). I resti faunistici dell'insediamento dell'Età del Bronzo finale delle Paludi di Celano: campagne di scavo 1986-1989. *Atti del Convegno di Archeologia "Il Fucino e le aree limitrofe nell'antichità"*. Roma , pp.165-173.

De Grossi Mazzorin J. (1992). I resti ossei animali rinvenuti nella necropoli di Osteria dell'Osa. In (A.M. Bietti Sestieri, Ed) *La necropoli laziale di Osteria dell'Osa.* Roma, pp. 487-489.

De Grossi Mazzorin, J. (1994). I resti faunistici provenienti dalle Terramare esposti nel Museo Civico di Modena: alcune considerazioni. *Quaderni del Museo Archeologico-Etnologico di Modena, 1, Studi di Preistoria e Protostoria* **1**, 145-152.

De Grossi Mazzorin, J. (1995a). Ricerche zooarcheologiche in alcuni insediamenti protostorici dell'Etruria meridionale. In (N. Negroni Catacchio, Ed), *Preistoria e Protostoria in Etruria.* Milano: Atti del II Incontro di Studi, pp.17-26.

De Grossi Mazzorin, J. (1995b). La fauna rinvenuta nell'area della Meta Sudans nel quadro evolutivo degli animali domestici in Italia. Atti del I Convegno Nazionale di Archeozoologia, *Padusa Quaderni* **1**, 309-318.

De Grossi Mazzorin, J. (1995c). Sepolture con cani nella necropoli pre-romana di Sulmona (AQ). Atti del I Convegno Nazionale di Archeozoologia, *Padusa Quaderni* **1,** 375-376.

De Grossi Mazzorin, J. (1996a). Archeozoologia delle "ossa di bruti" provenienti dagli scavi della stazione preistorica sul Monte Castellaccio presso Imola. In (M. Pacciarelli, Ed) *La collezione Scarabelli 2.* Imola: Musei civici di Imola preistoria, pp.181-218.

De Grossi Mazzorin, J. (1996b). Resti faunistici. In (R. Volpe, Ed) *Aqua Marcia.* Firenze: Lo scavo di un tratto urbano, pp. 203-214.

De Grossi Mazzorin, J. (1996c). Analisi dei resti faunistici dell'insediamento protostorico di San Giuliano di Toscanella. In (M. Pacciarelli, Ed) *La collezione Scarabelli 2.* Imola: Musei civici di Imola preistoria, pp. 308-312.

De Grossi Mazzorin, J. (1997). Analisi dei resti faunistici dall'abitato di Ficana (zone 3b-c). In (J.R. Brandt) *Ficana* vol. II., part 1. Roma: Ist. Pol. Stato, pp.405-423.

De Grossi Mazzorin, J. & Tagliacozzo, A. (1998). Dog remains in Italy from the Neolithic to the Roman period. *Anthropozoologica* **25-26**, 429-440.

Driesch, A. von den (1976). A guide to the measurement of animal bones from archaeological sites. *Peabody Museum Bulletin* **1**.

Fedele, F. (1983). Le macrofaune antropiche dell'età del Bronzo piemontese. *Rivista Piemontese di Storia Naturale* **4**, 85-109.

Fusco, V. (1981). Osservazioni sui resti faunistici di Ghisalba. *Annali Benacensi* **7**, 205-210.

Farello, P. (1989a). Fiorano modenese: reperti faunistici. In (G. C., Ambrosetti, R. Macellari & L. Malnati, Eds) *Rubiera.* Reggio Emilia: "Principi" etrusci in Val di Secchia, pp.179-184.

Farello, P. (1989b). Rubiera, Ca' del Cristo: reperti faunistici. In (G. C., Ambrosetti, R. Macellari & L. Malnati, Eds) *Rubiera.* Reggio Emilia: "Principi" etrusci in Val di Secchia, pp.139-142.

Farello, P. (1990a). Casale di Rivalta: reperti faunistici. In (G. C., Ambrosetti, R. Macellari & L. Malnati, Eds) *Vestigia Crustunei.* Reggio Emilia: Insediamenti etruschi lungo il corso del Crostolo, pp. 241-255.

Farello, P. (1990b). Reperti faunistici. In (R. Berti, Ed) *Fortuna Maris.* La nave romana di Comacchio, pp.118-130.

Farello, P. (1990c). San Claudio: reperti faunistici. In (G. C., Ambrosetti, R. Macellari & L. Malnati, Eds) *Vestigia Crustunei.* Reggio Emilia: Insediamenti etruschi lungo il corso del Crostolo, pp. 133-140.

Farello, P. (1992). Mirandola loc. Arginone: reperti faunistici. In (M. Calzolari & L. Malnati, Eds) *Gli Etruschi nella Bassa Modenese.* Modena: Nuove scoperte e prospettive di ricerca in un settore dell'Etruria padana, pp. 273-283.

Farello, P. (1994). L'insediamento di Castenaso: allevamento, caccia e pesca in un sito villanoviano. Insediamenti della prima età del Ferro, *Studi e documenti d'Archeologia Quaderni* **5**, 218-223.

Farello, P. (1995a). L'Emilia dal VI al V secolo a.C.: caccia e allevamento. Atti del I Convegno Nazionale di Archeozoologia, Rovigo (marzo 1993). *Padusa Quaderni* **1**, 209- 217.

Farello, P. (1995b). I cani tardo- antichi rinvenuti in un condotto fognario di Classe (RA). Atti del I Convegno Nazionale di Archeozoologia, Rovigo (marzo 1993). *Padusa Quaderni* **1**, 295- 308.

Farello, P. (1997). Reperti faunistici da Verucchio (RN)-Scavo 1963. *Archeologia dell'Emilia Romagna* **I/1**, 41-49.

Fiore, I. & Tagliacozzo, A. (1997a). Analisi dei resti ossei faunistici di una struttura neolitica a fossa (116) dell'area E. In (A. Zarattini & L. Petrassi, Eds) *Casale del Dolce. Ambiente, economia e cultura di una comunità preistorica della Valle del Sacco.* Roma, pp. 228-248.

Fiore, I. & Tagliacozzo, A. (1997b). I reperti osteologici animali delle tombe 1 e 8 (area A) e 4 (area C) della necropoli eneolitica: il contributo della tafonomia. In (A. Zarattini & L. Petrassi, Eds) *Casale del Dolce. Ambiente, economia e cultura di una comunità preistorica della Valle del Sacco.* Roma, pp.200-228.

Guerreschi, G., Catalani, P., Longo, G. & Iannone, A. (1992). Grotta Bella (Terni). Una sequenza stratigrafica dal Neolitico inferiore all'età imperiale: i livelli preistorici. *Bullettino di Paletnologia Italiana* **83**, 143-228.

Harcourt, R.A. (1974). The dog in prehistoric and early historic Britain. *Journal of Archaeological Science* **1**, 151-175.

Hasebe, K. (1924) *Uber die Schadel und Unterkiefer von den steinzeitlich Japanischen Hunderassen.* Sendai: Arbeiten der Anatomische Institution der Kaiserlich-Japanische Universität.

Lepiksaar, J. (1975). Animal remains. In (L. Mignone, Ed) *The Zone of the Large Iron Age Building.* Acta Instituti Romani Regni Sueciae 4/XXVII, pp. 77-86.

Manconi, F. (1996). Olbia, su cuguttu 1992: i reperti faunistici. In (A. Mastino & P. Ruggeri, Eds) *Olbia in età antica.* Sassari, pp. 447-460.

Morey, D.F. (1992). Size, shape and development in the evolution of the domestic dog. *Journal of Archaeological Science* **14**, 181-204.

Peters, J. (1994). Le chien dans l'antiquité. *Société Francophone de Cynotechnie*, Séminaire des 25 et 26 Mars 1994, 1-30

Riedel, A. (1968a). I mammiferi domestici del Castelliere di Nivize nel Carso triestino. *Atti e Memorie della Commissione Grotte Eugenio Boegan* **8**, 125-144.

Riedel, A. (1968b). Nuovo contributo alla conoscenza della fauna domestica neo-eneolitica della Grotta delle Gallerie. *Atti e Memorie della Commissione Grotte Eugenio Boegan* **8**, 111-123.

Riedel, A. (1973). I mammiferi domestici della Grotta n. 1745/4558 V.G. e di faune oloceniche minori. *Atti e Memorie della Commissione Grotte Eugenio Boegan* **13**, 53-90.

Riedel, A. (1974). La fauna dell'insediamento preistorico di Cattinara. *Atti e Memorie della Commissione Grotte Eugenio Boegan* **14**, 79-92.

Riedel, A. (1975). La fauna del villaggio preistorico di Isolone della Prevaldesca. *Bollettino del Museo Civico di Storia Naturale di Verona* **2**, 355-414.

Riedel, A. (1976a). La fauna del villaggio preistorico di Barche di Solferino. *Atti del Museo Civico di Storia Naturale di Trieste* **29**, 215-318.

Riedel, A. (1976b). La fauna del Castelliere degli Elleri (Trieste). *Atti del Museo Civico di Storia Naturale di Trieste* **29**, 105-122.

Riedel, A. (1976c). Resti faunistici preistorici della Caverna Cotariova (Trieste) (Scavi B. Lonza). *Atti del Museo Civico di Storia Naturale di Trieste* **29**, 69-104.

Riedel, A. (1976d). La fauna del villaggio eneolitico delle Colombare di Negrar (Verona). *Bollettino del Museo Civico di Storia Naturale di Verona* **3**, 205-238.

Riedel, A. (1976e). La fauna del villaggio preistorico di Ledro. *Studi Trentini di Scienze Naturali*, **53**, 3-120

Riedel, A. (1977). I resti animali della Grotta delle Ossa (Skocjan). *Atti del Museo Civico di Storia Naturale di Trieste* **30**, 125-208.

Riedel, A. (1979). La fauna di alcuni insediamenti preistorici del territorio Veronese. *Atti del Museo Civico di Storia Naturale di Trieste* **31**, 41-73.

Riedel, A. (1981). La fauna di Braida Roggia a Pozzuolo del Friuli. *Atti dei Civici Musei di Storia ed Arte di Trieste* **12**, 121-131.

Riedel, A. (1982a). Die fauna von Feniletto (Verona). *Rivista d'Archeologia* **6**, 28-30.

Riedel, A. (1982b). Die fauna einer bronzezeitlichen Siedlung bei Peschiera am Gardsee. *Rivista d'Archeologia* **6**, 23-27.

Riedel, A. (1983). Tierfunde einer römischen Fundstätte von Innichen. *Padusa*, **19**, 3-18.
Riedel, A. (1983-84). The fauna of the excavations of Pozzuolo del Friuli (1980-1983) (Castelliere dei Ciastei-Trench 1). *Atti dei Civici Musei di Storia ed Arte di Trieste* **14**, 215-276.
Riedel, A. (1984a). The fauna of the excavations of Colognola ai Colli (Verona, Northern Italy: Iron Age). *Bollettino del Museo Civico di Storia Naturale di Verona* **11**, 277-318.
Riedel, A. (1984b). Die Fauna der vorgeschichtlichen Fundstätte von Moletta bei Arco im Trentino. *Preistoria Alpina* **20**, 131-138.
Riedel, A. (1984c). Die Fauna der Sonnenburger Ausgrabungen. *Preistoria Alpina*, **20**, 261-280.
Riedel, A. (1985a). Die fauna von Altino (Venetien) im verhältnis zu den faunen nordostitaliens und der Alpenländer, Razprave IV. *Razreda SAZU* **26**, 131-146.
Riedel, A. (1985b). Die fauna einer bronzezeitlichen Siedlung bei Eppan. *Rivista d'Archeologia* **9**, 9-25.
Riedel, A. (1985c). The fauna of the Iron Age site of Castelrotto (Verona). *Padusa* **21**, 55-97.
Riedel, A. (1986). Die fauna von zwei römischen Fundstätten im Brixner Gemeindegebiet. *Der Schlern* **58**, 455-498.
Riedel, A. (1987). Mozzecane, Quarto del Tormine: la fauna. *Quaderni di Archeologia del Veneto* **3**, 117-118.
Riedel, A. (1988). The neolithic animal bones deposit of Cornuda (Treviso). *Annali dell'Università di Ferrara* **1**, 71-90.
Riedel, A. (1990). La fauna dell'età del Bronzo di Cisano (Verona). In (L. Salzani, Ed) *Nuovi scavi nella palafitta di Cisano.* Comune di Bardolino, pp. 53-59.
Riedel, A. (1992a). The Bronze Age animal bone deposit of Nogarole Rocca I Camponi (Verona). *Padusa* **28**, 87-104.
Riedel, A. (1992b). Considerazioni sulla fauna dell'età del Ferro. In (P. Brugnoli & L. Salzani, Eds) *San Giorgio di Valpolicella.* Verona: Scavi archeologici e sistemazioni museali, pp. 81-87.
Riedel, A. (1993). Tierknochenfunde aus den Ausgrabungen im Bereich des Schlosses von Udine (Friaul). *Aquileia Nostra* **64**, 70-106.
Riedel, A. & Rizzi, J. (1988). Resti faunistici domestici e selvatici: produzione e consumo. In (E. Cavada, Ed) *Archeologia a Mezzocorona: Documenti per la storia del popolamento rustico di età romana nell'area atesina.* Patrimonio storico e artistico del Trentino **15**, pp. 233-246.
Riedel, A. & Scarpa, G. (1988). Resti animali di un complesso produttivo di età tardoromana a Volano. *Annali dei Musei Civici di Rovereto* **4**, 37-54.
Shigehara, N. (1994). Morphological changes in Japanese ancient dogs. *Archaeozoologia* **6**, 79-94.
Siracusano, G. (1990). Un commento sugli ultimi ritrovamenti faunistici del Subappenninico di Coppa Nevigata. *Atti 11° Convegno Nazionale sulla Preistoria-Protostoria Storia della Daunia, San Severo*, pp. 117-127.
Siracusano, G. (1991). Prima valutazione delle misurazioni osteometriche dei resti faunistici di Coppa Nevigata. *Scienze dell'Antichità Storia Archeologia Antropologia* **5**, 265-289.
Tagliacozzo, A. (1993a). Archeozoologia della Grotta dell'Uzzo, Sicilia. Da un'economia di caccia ad un'economia di pesca ed allevamento. *Supplemento al Bullettino di Paletnologia Italiana,* serie II, Roma **84**.
Tagliacozzo, A. (1993b). Le ossa animali. In (C. Balista & A. Ruta Serafini, Eds) Saggio stratigrafico presso il muro romano di Largo Europa a Padova. *Quaderni di Archeologia del Veneto* **9**, pp. 95-111.
Tagliacozzo, A. (1994a). I resti faunistici di Acquarossa. *Acta Instituti Romani Regni Sueciae* 4°, XXXVIII, **VII**, 156-165.
Tagliacozzo, A. (1994b). I dati archeozoologici: economia di allevamento e caccia a Broglio di Trebisacce. *Enotri e Micenei nella Sibaritide.* Taranto, pp. 587-652.
Tagliacozzo, A. (1995). I resti ossei faunistici. In (M-H. Et J. Santrot, Eds) *Bolsena VII. La citerne 5 et son mobilier: Production importasions et consommation. IIIe siecle/début Ier siècle Av. J-C. et deuxième tiers du Ier siècle ap. J.- C., Fuilles de l'École française de Rome à Bolsena, Tome VII.* Mèlanges d'archeologie et d'histoire Supplément **6**, pp. 323- 347.
Tagliacozzo, A. & Fiore, I. (1992). Oderzo, via dei Mosaici. I resti faunistici. *Quaderni di Archeologia del Veneto* **8**, 166-173.
Teichert, M. (1987). Brachymel dogs. *Archaeozoologia* **1**, 69-75.
Villari, P. (1995). *Le faune della tarda preistoria nella Sicilia orientale.* Siracusa.
Wagner, K. (1930). Rezente Hunderassen. Eine osteologische Untersuchung Skriften utgitt av Det Norske Videnskaps. *Akademi i Oslo, I, Mat.-Naturv. Kl.* **9**, 1-157.
Wilkens, B. (1985). La fauna dei livelli neolitici ed eneolitici di Attiggio. *Picus. Studi e ricerche sulle Marche nell'antichità* **5**, 203-214.
Wilkens, B. (1987). La fauna del villaggio neolitico di S. Maria in Selva. *Picus. Studi e ricerche sulle Marche nell'antichità* **7**, 169-194.
Wilkens, B. (1988a). La fauna di Coppetella (Iesi, Marche). *Atti della Società Toscana di Scienze Naturali* Memorie Serie A **95**, 363-375.
Wilkens, B. (1988b). La fauna di Moie di Pollenza. In *Le Marche. Archeologia, Storia, Territorio, Arcevia-Sassoferrato*, pp. 137-144.
Wilkens, B. (1988c). I resti faunistici di Pano Donatelli (neolitico). In *Le Marche. Archeologia, Storia, Territorio, Arcevia-Sassoferrato*, pp. 327-364.

Wilkens, B. (1988d). La fauna del villaggio di Ripa Tetta (Lucera): nota preliminare. *Taras* **8**, 55-61.

Wilkens, B. (1988e). S. Paolina di Filottrano (Marche): i resti faunistici. *Atti della Società Toscana di Scienze Naturali* Memorie Serie A **95**, 345-361.

Wilkens, B. (1990). La fauna del villaggio del Colle dei Cappuccini (Ancona). *Rassegna d'Archeologia* **9**, 237-252.

Wilkens, B. (1991). I resti faunistici dell'abitato dell'età del Bronzo di Bari (S. Mara del Buon Consiglio). *Taras* **11**, 85-93.

Wilkens, B. (1997). I resti faunistici dei livelli piceni di Osimo. *Rassegna d'Archeologia* **14**, 237-252.

# DOGGED PERSISTENCE: THE PHENOMENON OF CANINE SKELETAL UNIFORMITY IN BRITISH PREHISTORY.

Kate M. Clark
Centre for Human Ecology & Environment
University of Southampton
Southampton, Hampshire,
SO17 1BJ
UK

**Key words:** dog, breeds, Britain, archaeozoology, Roman period, osteometry, metric archives

**Abstract**

In 1974 Ralph Harcourt published a review of the metrical data available at that time for archaeological dog remains excavated from British sites. His work illuminated the proliferation of skeletal variety, in size and morphology, that appeared in the 400 years of the Romano-British period and suggested a marked homogeneity of form prior to the first century AD. In the quarter century following his digest the techniques of faunal analysis in British archaeological research have developed a wider focus. The preoccupation with relative representations of consumable animals to identify dietary preference or contingency is now recognised as scientifically flawed because of its reliance on subjective assessment in the absence, in most cases, of supporting environmental data, evidence of contemporary cultural attitude, or understanding of basic principles of animal function in domesticated circumstances.

The 25 years that separate us from Harcourt's review coincide with the existence of the English Heritage Faunal Remains Unit at Southampton, and in the last four years Southampton has also supported the Centre for Human Ecology and Environment which predominantly undertakes research into man/animal relationships. In addition, these organisations maintain a literature bank of all published and much unpublished British material available since 1974. It is estimated that over these 25 years, colleagues at these two organisations themselves examined and reported on over 1.5 million animal bone specimens, and can access directly the results of other workers who have been active during that time.

This paper examines the impact of this increased faunal data set on our understanding of the prehistoric dog in Britain.

## Introduction

This paper examines the question of skeletal uniformity in prehistoric British dogs. In particular it reassesses the trends visible in the later prehistoric, particularly the later Iron Age populations. A number of questions are raised about the apparent uniformity of dogs prior to the Romano-British period, and the paper also discusses whether archaeologists are optimising the data available to us in British archaeological assemblages.

There are two problematic aspects of the study of dogs in Britain. One is the tendency to turn suggestions into rules, and thereafter seek only to reinforce them. And the other is the difficulty of using certain osteometric techniques which perhaps are not generally applicable to the study of all species found on archaeological sites. The case discussed here is whether archaeologists may be losing valuable information on prehistoric dogs because of inappropriate measurement strategies.

In 1974 Ralph Harcourt published a review of the metrical data available at that time for archaeological dog remains excavated from British sites. His review highlighted the proliferation of skeletal variety, in size and morphology, that appeared in the 400 years of the Romano-British period. His samples also suggested a marked degree of uniformity prior to the first century AD. He comments on the small range of variability in Iron Age skulls, and on the very low frequency of smaller animals. Although there appeared to be more variability in the later prehistoric dog than in the earlier periods, the early data are very sparse. Certainly there appears nothing in the Iron Age data to predict the explosion of variability that had occurred by the second century AD (Clark, 1996a).

In the 25 years following Harcourt's digest, the techniques of faunal analysis in British archaeological research have developed a wider focus. The preoccupation with relative proportions of remains of consumable animals to identify dietary preference or contingency is now recognised as scientifically flawed. It relies essentially on subjective assessment in the absence, in most cases, of supporting environmental data, evidence of contemporary cultural attitude, or understanding of basic principles of animal function in domesticated circumstances.

In Britain, legislation pertaining to the impact on archaeological remains by construction and development works has been a major driver in assemblage recovery. An increased amount of animal bone has been recovered from a wide range of site type and size since 1974, and the material has been subjected to a more rigorous and standardised procedure; however, the need to recover has superseded the need to develop long term research designs. Nevertheless, the statutory requirements have resulted in an accumulation of data which should give access to a far wider range of hypotheses than just those concerning consumption. The environmental and domestic implications of the presence of wild fauna and commensals have become active areas of research, for example, together with the considerations of pathological manifestations and, most significantly, the development and variability of skeletal characteristics within individual species. Of course archaeologists have to consider to what extent the increased number and size of assemblage outweighs the biases inherent in recovery from mainly urban areas, but have to accept that the potential in the osteological material archive is now very high.

The 25 years that separate us from Harcourt's review coincide with the existence of the English Heritage Faunal Remains Unit at Southampton, and in the last four years Southampton has also supported the Centre for Human Ecology and Environment which predominantly undertakes research into man/animal relationships. In addition, these organisations maintain a literature bank of all published and much unpublished British material available since 1974. It is estimated that over these 25 years workers at Southampton have examined and reported on over 1.5 million animal bone specimens, and can now access directly the results of other workers who have been active during that time. Archaeologists should therefore expect an enhanced data set for prehistoric dogs, even though some analysts may have been more interested in gnawing marks on consumable domesticates than with the species which inflicted them.

Reality of course falls below expectation, and some of the disappointment is due to a bias towards the proto-historic and later periods excavated during modern urban development subjected to recent archaeological planning conditions. Even so, given the enormous data resource British archaeozoologists should be able to claim a degree of objective re-analysis of Harcourt's observations for prehistoric dogs. In fact, archaeozoologists could claim that such an exercise should, by its results, confirm that faunal analysis has moved on from the simplistic, "supermarket survey" view of "diet = relative percentage of species = status" to the more fundamental investigation of species modification consequent upon permutations of environmental change, shift in cultural priorities, and development of husbandry skills.

## Current View

The "consumer" approach is bad news for dog studies. Assemblage statistics will usually indicate the presence of dogs, but with an incidence often less than 1%, and the animal is frequently relegated to the category of "other species" or treated as an undiscussed incidental. Measurements and notes frequently do not make it out of the archive. Dogs in special circumstances - interred in human graves, buried as part of a ritual suite - do fare better than their disarticulated contemporaries scattered amongst domestic debris, but until fairly recently even these prize specimens were given only cursory metrical treatment even though in many cases they were accorded the significance of an *in situ* site photograph.

In Britain this unsystematic approach has been compounded by the difference in assemblage size and frequency between the prehistoric periods and the subsequent, clearly defined Romano-British era. In southern Britain in particular the predominance of Romano-British material is illustrated by the metrical database held in Southampton University, in which measurements from assemblages funded by public and developer money in this area since 1975 have been collated (Centre for Human Ecology & Environment 1995) (Table 1). Although not part of the argument of this paper, I include here the Medieval data to show the remarkable drop in measured dogs for that period.

| | ***Measurements of all species*** | ***Measurements of dog*** |
|---|---|---|
| Neolithic to end Iron Age | 4960 | 329 |
| Romano-British | 9648 | 899 |
| Medieval | 36142 | 395 |

***Table 1*** *Animal bone metrical archive (1975-1995), mainly southern Britain*

Equally telling is the number of dog measurements taken in Southampton since 1996, when the metrical archive was uploaded, to the present, and I compare this to the measurement total of the preceding 20 years (Table 2).

| | ***Animal bone metrical archive (1975-1995)*** | ***Centre for Human Ecology (1996-1998)*** |
|---|---|---|
| Neolithic | 12 | 37 |
| Bronze Age | 39 | 75 |
| Iron Age | 278 | 884 |
| Romano-British | 899 | 1313 |

***Table 2*** *Measurements taken on dog bones from British sites*

## Data: availability and restrictions

The concern of this paper is that British prehistory ended in the mid first century AD, and within a very few generations the character of a domestic species population had undergone an extraordinary transformation. There is a tacit acceptance that pre-Romano British dogs were "plain" dog, lacking in size and morphological variability except for the appearance of a few smaller animals in the Late Iron Age, which Teichert (1997) has claimed cannot be attributed to human interference in the breeding of these animals.

This simplistic scenario is difficult to accept at face value. It invokes necessary explanations for which we have no archaeological evidence - such as immediate introduction of a new genetic resource, rapid and massive shift of cultural attitude towards the species, and new husbandry practices to which dogs make a significant contribution. Any or all of these aspects may apply, but none have been demonstrated sufficiently to explain the appearance of such an interface in the species, and certainly no theoretical discussion has taken place to explore their possibilities. For the present, therefore, archaeologists should pursue the null hypothesis that the development of dog variability in Britain is, at best, exponential. We could claim a null hypothesis that the intraspecific development of variability is constant, but results from other prehistoric British species, notably cattle, indicate that inter-period shifts in size and morphology are driven by ecological and/or economic factors supported by independent archaeological evidence (eg Thomas, 1989; Smith, 1984).

The term "pursue" is paramount here. To test the argument British archaeologists require robust samples across time, and these we do not have. As faunal analysts our embarrassment is that the deficiency in sample is unlikely to be a lack of specimens, but is a disinterest in a domestic species which does not contribute consistently over time to diet.

On an international, evolutionary, scale this might be seen as a little local difficulty in British archaeology. However, there may be more to this question because faunal analysts in Britain are having to deal with an explosion of specific variability within a very tight chronology, and if it happens in Britain it may have occurred elsewhere. To address the problem our academic investment in faunal analysis dictates that we reassess our analytical strategy and recognise the deficiencies in the data pool. Archaeologists also need to recognise that some of those deficiencies are a consequence of our misconceptions about what is at the root of human/animal relationships. Eating animals is doubtless a primary driving force. Living with those animals produces the complex range of forces which we see in effect in the wider archaeological evidence, and it is the interaction of those forces which give us the greater insight into antique societies. The interaction of dog and human is one small window into those societies, but will never be disclosed by strategies designed to identify "diet" and "status". Evidence of life with dogs is ubiquitous, whether it is the dog remains themselves or the effects of dog teeth on the bones of other species, but evidence of consumption of dogs is very occasional in Britain.

**Some results and indications**

So far I have only considered the metrical analysis of dogs, but archaeozoologists need to take account also of non-metric characteristics. The presence of the following traits currently precludes a specimen from being "prehistoric":

- Dental anomaly. No absence of mandibular $M_3$ has yet been recorded in British material of prehistoric date. This condition is almost certainly brought about by diminution of jaw length where osteological reduction proceeds faster than tooth reduction and so there can be too little space for the eruption of $M_3$. Possible indication of similar jaw reduction in prehistoric cattle however is shown in loss of aboral pillar of mandibular $M_3$ resulting in overgrowth and pointed appearance of maxillary $M^3$. This has been noted in British cattle from the Bronze Age onwards (Legge, 1991), and on mainland Europe from the Neolithic onwards (eg Förster, 1974). In cattle, of course, faunal analysts are able to observe the diminution of the tooth itself, whereas in dogs we observe either presence or absence. However, it is the consistency of the underlying mechanism that we are seeking.
- Tooth crowding. This has not been satisfactorily recorded in prehistoric dogs. Again, it is likely to be a result of accelerated diminution of jawbone.
- Anomalous limb bone conformation. No bowed limb bones are reported prior to the Romano-British period. Bowed limb bones probably are a congenital deformity where there is an imbalance in the growth centre of the bone in utero. In animals of otherwise normal conformation it results in a shortened but relatively more robust appearance.
- The saggital crest - no lack of crest has been reported in prehistoric dogs.
- Cephalic indices. No bracycephalic or dolicocephalic skull types have been described for the prehistoric period. Round-headed bracycephalic dogs do not have saggital crests, and this conformation is most visible in very small animals. Essentially they retain paedomorphic characteristics and in the live animal this is exhibited as large eyes, well spaced within a domed head. Dolicocephalic skulls present, in the live animal, as the typical greyhound, a type recognised early in mainland Europe and Asiatic material but not identifiable in British prehistoric specimens. In Romano-British remains, however, archaeologists have incontrovertible evidence for this type and the model from Lydney in Gloucestershire is a classic example (Toynbee, 1973).
- Pathological evidence. Very advanced degenerative osteoarthritis, at the stage which would render the animal very slow and possibly incapacitated, has not been recognised in prehistoric dogs. Such indications of

- advanced age have, however, been well described in Romano-British material (Harcourt, 1967).

Non-metric characteristics such as those listed above give archaeologists important insights into the populations, but they do not define them. The primary evidence has to be osteometric. This is the only level at which faunal analysts can hope to achieve representative samples, and at which we can monitor the quality of observation. However, in Britain the osteometric strategies of faunal analysts can be criticised because they have not been designed to optimise the data available in fragmented remains. There has been a tendency to take measurements from which a conclusion can be drawn immediately regarding the skeleton in which the element originated. The classic examples are the overall lengths of long bones, from which to calculate shoulder height, measurements of the dentition in complete mandibles and maxillae, and the components of skull indices such as total length of the skull and zygomatic breadth (Harcourt, 1974). These are all highly valuable in themselves, but they are limited in their contribution to a strategy which should encompass the widest range of incomplete remains. It is understandable that workers do not spend valuable time measuring proximal or distal ends of longbones, or recover perhaps two measurements from a skull fragment, because for these limited data they have often have no *comparanda* against which to set their results.

Nevertheless, archacologists have to be aware that complete elements are most often recovered from contexts which reflect particular types of deposition, or particular taphonomic circumstances. The most common circumstance for dogs which incurs both these situations is discrete burial. It is clear from the majority of assemblage reports that the incidence of dog remains is due mainly to the incorporation of dog bones in refuse deposits, and these bones are commonly disarticulated, fragmented, and subject to the level of attrition normally associated with consumption debris. A greater interest needs to be developed in the measurement of those fragments to which we can assign a minimum relative maturity, for instance the depth of the fused femoral caput and the depth of the head of the humerus.

However, to review this question of persistent homogeneity in the British prehistoric dog faunal analysts must use the osteometric data accumulated to date, recognising that there may be an inherent bias introduced by circumstances of deposition rather than specific incidence.

One of the most popular measurements is of the total length of the limb bones, because there are the factors calculated by Harcourt on which to estimate shoulder height. I could argue here that short legged dogs will have an equal chance of recognition in debris in any period because the shorter the bone the less likely it is to fragment. I have looked at the results published in 1974 by Harcourt (Fig. 1). The paucity of data for periods earlier than the British Bronze Age, and the disparity of sample sizes between the prehistoric and the Romano-British period, mean that faunal analysts should only look at the range of long bone measurements and not attempt to apply even the simplest descriptive statistics. However, ranges should be treated with caution and ideally archaeologists should aim to develop sample sizes on which we can demonstrate clearly the degree of symmetry in distribution.

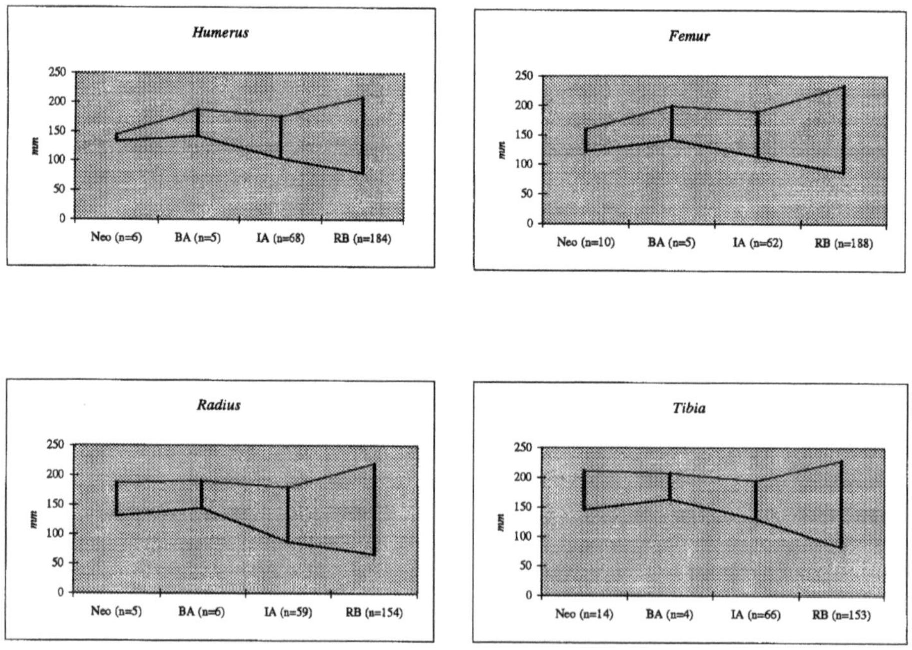


***Figure 1*** *Longbone measurement ranges, from Harcourt, 1974 (Neo, Neolithic; BA, Bronze Age; IA, Iron Age; RB, Romano-British)*

The widening of the length range in the Romano-British period is evident in all four longbone elements. All four elements show the presence of shorter dogs during the Iron Age, compared with the earlier periods. The humerus alone shows an increased presence of tall dogs in the Iron Age. But of course the sample sizes preclude claiming significant variability between the Iron Age and the preceding era. But there is perhaps a perceived trend which suggests the presence of shorter legged dogs in the Iron Age.

I then added the data available to archaeologists since 1975 (Fig. 2). This is salutary, and shows the constraints imposed by considering only complete elements. The Iron Age sample has increased by around 38%, and the Romano-British sample by some 60%. The earlier prehistoric samples are scarcely affected. All I can say is that the indication of the presence of shorter dogs in the Iron Age is reinforced.

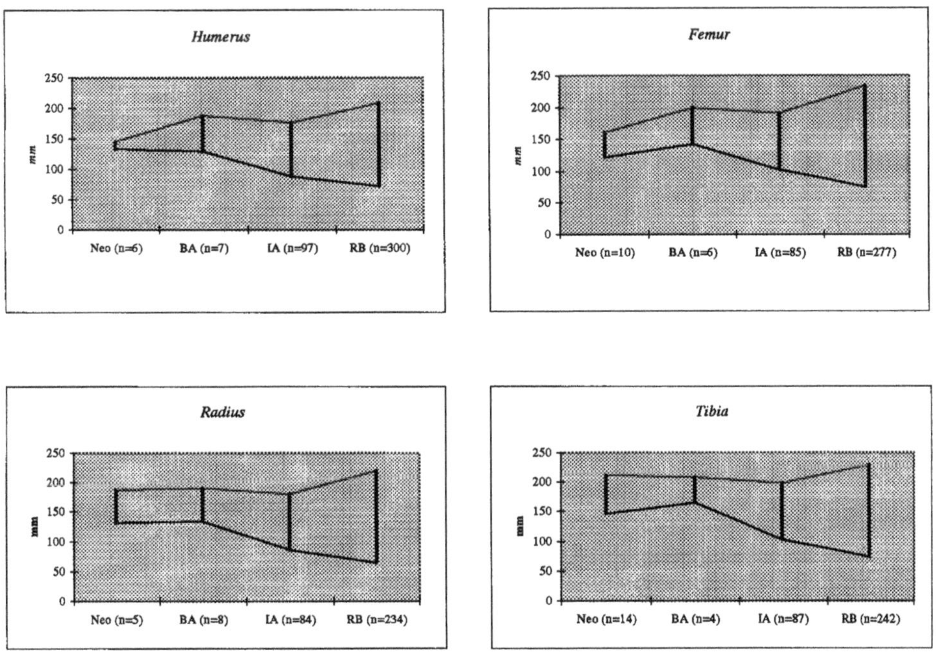


***Figure** 2 Longbone measurement ranges, all data available to 1998 (Neo, Neolithic; BA, Bronze Age; IA, Iron Age; RB, Romano-British)*

The sample problem returns me to an earlier argument in this paper. Although the length measurements are popular, complete limb bones on which to take them are far more rare than epiphyseal fragments and they tend to occur most frequently where whole or partial skeletons are recovered. So there is both a bias towards dogs from particular circumstances, and a diminished sample from other deposits because there is selection for completeness in otherwise fragmentary assemblages.

Faunal analysts are accustomed to viewing our osteometric data in bar graphs, and the humerus length shown in Figure 3 has the classic shapes that we have come to recognise. The bimodal distribution of the Romano-British, the more restricted Iron Age distribution with the marked skew to the right, and the early prehistoric specimens clumped in the middle with the occasional large outlier. To explain the samples illustrated here, the Romano-British group are those measurements collected since 1975; the Iron Age and earlier samples also include Harcourt's results.

If I abandon any claim to sample size validity I suggest that the shapes for the prehistoric be examined more closely. Here are the questions that faunal analysts should be asking. Is there a marked shift to larger dogs in the Bronze Age; is there really a major interface in dog morphology between the Middle and the Late Iron Age?

I examined another element that should produce information about morphological change. The mandible produces a suite of potential measurements involving the dentition, the robustness of the jaw, and its overall length. Popular measurements amongst workers are the lengths of the cheektooth row and the length of the carnassial. This last measurement has been shown in studies on modern dogs to be of very limited use in defining size (Clark, 1996b), and the first requires the recovery of complete mandibles. Of more use are measurements of the molar and premolar rows, which can be taken on partial mandibles. As single measurements on adult jaws they will give some indication of relative skull size.

I have no previous study of these dimensions on which to build, and so the data presented here (Fig. 4) arc those held on the Animal Bone Metrical Archive together with further specimens analysed since 1996. Again, the paucity of data in the earlier prehistoric is clear - no measurements are

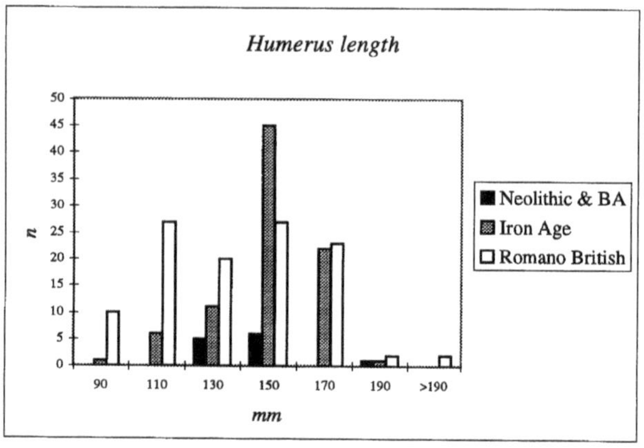


*Figure 3* *Length of humerus for chronological periods*

available for the Neolithic and only four for the Bronze Age. But even in the restricted sample for the Iron Age the presence of small dogs is apparent, and indeed there appear to be some dogs which may be smaller than their Romano-British counterparts in this sample. On all other evidence this latter observation can be considered rather anomalous and may very well be a result of the low sample size.

I have checked the "pedigree" of the small Iron Age specimens, identified both in longbones and mandibles, and it is impeccable. All specimens were identified by one or other of two eminent and experienced faunal analysts under the aegis of English Heritage, and the accidental incorporation of juvenile mandibles is highly unlikely.

## Implications

I have quite consciously not presented any data beyond indicating ranges. I have deliberately avoided even the simplest descriptive statistics on these samples. My aim in this paper is to raise the issue of curiosity about an extraordinary species and question the analytical strategies applied to the understanding of its development in British prehistoric society.

Twenty-five years ago Harcourt produced a seminal work which, on the whole, has proved incontrovertible in its results. That work produced a series of graphically illustrated datum lines against which subsequent British observations have been measured, and it has proved inordinately successful time and time again. This endowment has, however, left archaeologist with some problems. Harcourt identified the most diagnostic planes of measurement and tested them against his known *comparanda*. For his samples this has produced a standard which is unlikely to be assailed. However, his diagnostic planes are a base line, drawn up from analysis of complete skeletons. The archaeozoological task is to develop archaeologically valid osteometric bases on which to sample dog populations. Zygomatic breadth, length of cranium, and length of longbones are not valid osteometric data for the sampling of a species whose remains will be incorporated within domestic debris from prehistoric sites.

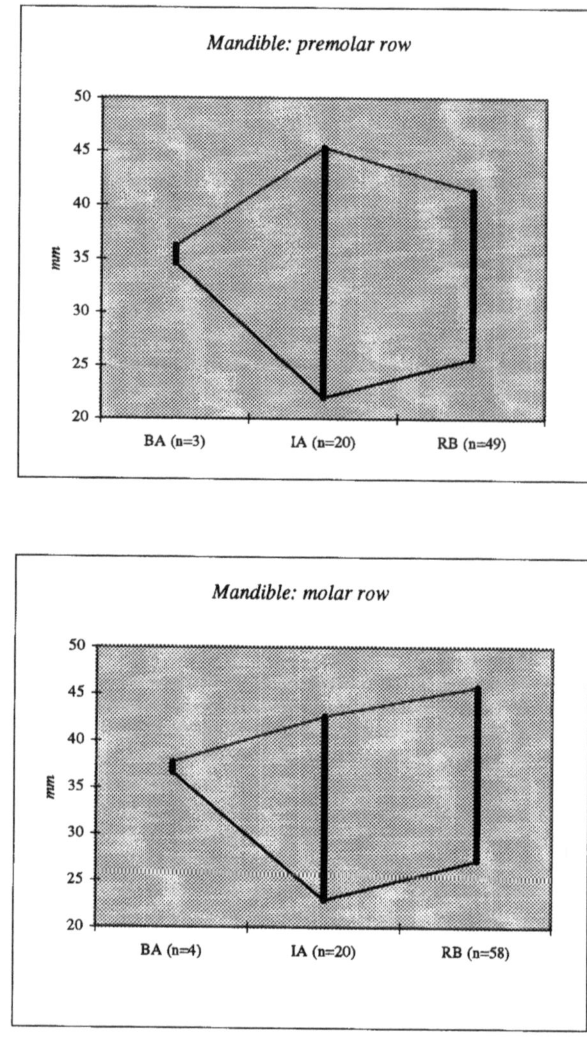


*Figure 4* *Mandibular measurement ranges (BA, Bronze Age; IA, Iron Age; RB, Romano-British)*

The curiosity of faunal analysts should be more aroused than it is, and we should be working towards accumulating appropriate data to feed it. This has not been done. Since 1974, when Harcourt pointed up so clearly the problem, British faunal analysts have done little to advance his work. We have merely used his results as a convenient datum level and abdicated any responsibility for advancing understanding of the species in the context of British archaeology. There are so many further questions that archaeozoologists should be asking, but we need to accumulate the appropriate type of data for interrogation.

## References

Centre for Human Ecology & Environment (1995). Animal bone metrical archive project: draft report on the project phase for English Heritage. University of Southampton: on file.

Clark, K. M. (1996a). The later prehistoric and protohistoric dog: the emergence of canine diversity. *Archaeozoologia* **VII/2**, 9-33.

Clark K. M. (1996b). Neolithic dogs: a reappraisal based on evidence from the remains of a large canid deposited in a ritual feature. *International Journal of Osteoarchaeology* **6**, 211-219.

Förster W. (1974). Die Tierknochenfunde aus der neolithischen Station Feldmeilen-Vorderfeld am Zürichsee. II. Die Wiederkäuer. Diss. Munchen.

Harcourt R. A. (1967). Osteoarthritis in a Romano-British dog. *Journal of Small Animal Practice* **8**, 521-522.

Harcourt, R. A. (1974). The dog in Prehistoric and Early Historic Britain. *Journal of Archaeological Science* **1**, 151-175.

Legge A. J. (1991). Animal remains from Neolithic and Bronze Age sites at Down Farm. In (J. Barrett *et al.*, Eds) *Papers on the Prehistoric Archaeology of Cranborne Chase* **11**. Oxford: Oxbow, pp. 54-100.

Smith R. W. (1984). The ecology of Neolithic farming systems as exemplified by the Avebury region of Wiltshire. *Proceedings of the Prehistoric Society* **50**, 99-120.

Teichert M. (1987). Brachymel dogs. *Archaeozoologia* **I**, 69-75.

Thomas R. N. W. (1989). Cattle and the Romano-British economy: a metrical analysis of size variation. Ph.D. thesis. University of Southampton.

Toynbee J. M. C. (1973). *Animals in Roman Life and Art.* London: Thames & Hudson.

# VARIETIES OF DOG IN ROMAN BRITAIN

Leslie Cram
Reading Museum Service, The Town Hall
Blagrave Street, Reading
RG1 1QH
UK

**Key words:** Britain, dog, footprints, Roman period, osteometry, archaeozoology, tile making

**Abstract**

Early excavations of Roman sites in Britain usually only described dog bones as of a type similar to a modern breed. The first metrical study of Roman dog bones was in 1974 and used long bone lengths translated into shoulder heights to show a range from toy dog size to dogs as large as wolves. A 1995 study on bones then showed that toy dogs were more frequent in the later Roman period. Recent work on Roman dog footprints (tracks) on tile has shown a range in animal size similar to sizes from bones. These tracks were made in wet clay while the tiles were drying laid out on the ground. The tracks of modern breeds of dog show toy dogs to be as small as Roman, but large modern breeds such as a greyhound of wolf size have tracks smaller than wolf. Bones and footprints of dog from Castleford, an army settlement in Yorkshire, show bones ranging from toy dog size but not up to the largest Roman dogs. The tracks show that the commonest foot impression was of human beings followed by dog and that the smallest toy dogs were not found in the tile kilns. Bones of dog from Silchester, a civilian settlement in Hampshire, show a similar range of animal sizes. The tracks show that dogs were the creature which most frequently trod on the drying tiles, followed by human beings and a range of farmyard animals, and that toy dogs did live around the tile kilns. Brief reference is given to accounts of dogs in Britain by Roman writers and illustrations and statues of dogs produced in Britain.

## Introduction

"Footprints?"

"Mr Holmes, they were the footprints of a gigantic hound!"

*The Hound of the Baskervilles* (Conan Doyle, 1901-02)

Extensive excavations of Roman sites in Britain began in the later years of last century. The excavation reports recorded the discoveries of dog bones, occasionally referring to the context in which they were found or whether the bones were from a complete skeleton. Usually any description was a reference to the size of the animals in relation to a modern breed of dog.

In this paper care is taken to use the word breed only for modern dogs where reproduction is controlled by humans with the knowledge of genetics that became available early in the twentieth century. The word variety is used for populations of dogs in earlier times (Clutton-Brock, 1987).

A quarter of a century ago the results of these years of excavation were brought together for the first time. The work was done by Ralph Harcourt (1974) on dog bones held in museum collections. He took measurements of dog bones from the whole of Britain according to the main divisions of the archaeological record from Mesolithic to Anglo-Saxon.

Two measurements were taken on the long bones, the total length and the mid-shaft diameter. Harcourt's measurements are the same as those later codified by von den Driesh (1976) apart from the femur which he measured between the proximal and distal articular extremities and the shaft diameter which he measured at mid shaft whereas von den Driesch measures as the minimum. Skull measurements were also taken by Harcourt which accord with von den Driesch. Harcourt pooled his measurements from the whole of Britain and from each archaeological period. He then translated bone measurements into shoulder heights.

The results from the Iron Age (500BC-AD43) are relevant to the study of the dogs of the following Roman period. Early Iron Age dogs differed little from the dogs of earlier times (Clark, this volume) but in the later Iron Age Harcourt found small animals which were of a size of what are described by the British Kennel Club as "toy dogs" (sometimes called "lap dogs") with a shoulder height of some 290 mm. These are explained as imports from the Roman Empire flourishing then across the English Channel, along with other luxury goods shown in the archaeological record by high status pottery. The largest Iron Age dogs were of a shoulder height of some 580 mm.

The Roman invasion of Britain extended the Empire based in Italy from the eastern Mediterranean and North Africa to the borders of Scotland. The Emperor Claudius brought elephants to Britain to celebrate his conquest of AD 43 and other animals were introduced to be a permanent part of the economy. Harcourt included 1156 Roman dog bones in his study. His measurements show toy dogs of even smaller size than the Iron Age with shoulder height 230 mm, comparable to a modern Pomeranian. The form of these toy dog long bones was with shafts as slender in proportion to their length as the larger dogs. Thus faunal analysts interpret the evidence to show that the toy dogs continued to be imported from mainland Europe. He also found dogs larger than the Iron Age at the other extreme, which archaeologists often account for by introduced large varieties. These large dogs were of shoulder height up to 720 mm. His statistics indicate "*two distinct but overlapping populations*". Harcourt's most frequent long bone measurement, where his measurement is the same as that of von den Driesch, is the humerus where he has 184 specimens ranging from 79 mm to 208 mm in length and a standard deviation of 30.6 mm (Fig.1). He noted rounded skulls with little sagittal crest.

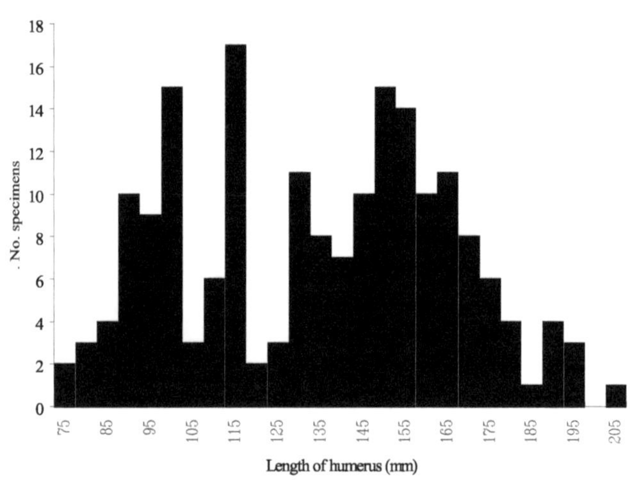


***Figure 1*** *Roman dogs, humerus length from Harcourt (1974), redrawn in Clark (1995).*

The work of Harcourt was taken further by Kate Clark in 1995. Clark took bones available since 1974, pooling measurements from 206 specimens from the whole of Britain, but allocating them to four stages of a century each within the Roman period. She expressed her results in shoulder heights, comparing these to the humerus length of Harcourt. Her results confirmed those of Harcourt and went on to show that the toy dog was more common in the later Roman period (Fig. 2).

Looking back over more than a century of discoveries of skeletal remains of dog, Clark discussed the value of complete skeletons compared to isolated bones or semi-articulated skeletons. To get large enough samples there is need to use the most commonly remaining bones and she proposes that the measurements of the frontal bone above the orbits will give valuable and more abundant data than relying on the less frequent complete skulls. Complete skeletons are often found in burials showing that the dogs had some status in Roman society, which may have separated them from interbreeding with the more common dogs of the streets in the towns and army camps whose remains are of isolated and broken bones.

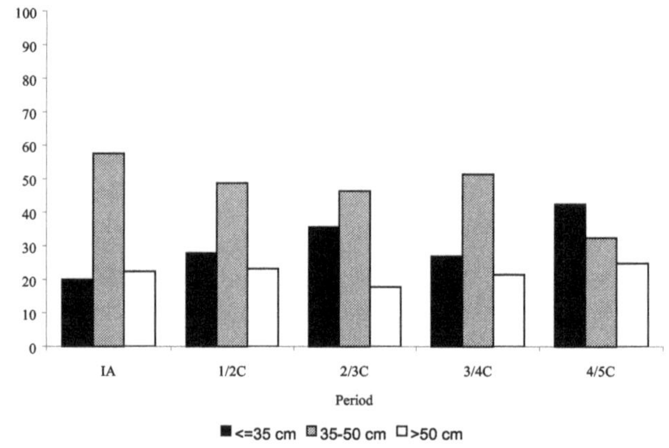


***Figure 2*** *Iron Age and Romano-British dogs, shoulder heights, post 1974 data from Clark (1995: fig. 5).*

Increasing attention is now being given to dog bones and especially to complete skeletons (Baxter, in press) and the results of Harcourt and Clark are being extended. A problem in defining the range of sizes of Roman dogs is where the larger dogs reach the size of the smaller wolves. Harcourt reported that the largest of his Roman dogs overlapped the bottom of the range of wolves and pointed out that it is not possible to tell dog and wolf apart by the post-cranial skeleton, only the mandible and skull from the size and crowding of teeth and the skull shape will show the distinction. The largest dog he could prove was one with a tibia of 229 mm, shoulder height 680 mm, shown to be dog by the associated skull. He refers to wolf skeletons smaller than this. More recently excavated collections of post-cranial bones, for instance shown in Clark's bibliography, usually show a range up to some 600 mm shoulder height with no immediately clear pattern of whether the larger specimens come from particular types of site such as large settlements or isolated villas. But occasional sites such as the villa at Shakenoak (Cram, 1978) have given limb bones showing shoulder heights approaching 800 mm. Are these from unusually large guard dogs or from wolves killed to protect livestock? Remains where limbs and head are found together are vital for telling us the size of the largest dogs.

## Dog bones from Britain vs the Roman Empire

Study of archaeological dog remains from Italy confirm the evidence from Britain. Mazzorin and Tagliacozzo (this volume) report early Roman (third century BC to second century AD) dogs with shoulder heights between 294 mm and 640 mm, and late Roman (third century AD to fourth century AD) shoulder heights between 261 mm and 697 mm. Toy dogs are thus shown to have been bred in Italy before their first importation into Roman Britain. It is interesting that the dogs of Roman Britain were both smaller and larger than the contemporary dogs in Italy.

## Dog footprints from Roman Britain

The evidence represented by the bones of dogs are of the animals after their death and incorporation into an archaeological deposit. Another source of information in the Roman period is from the animals when they were alive. This evidence comes from footprints made by dogs walking over wet brick and tile while laid out to dry. The evidence brought together by McWhirr (1979) shows that tile making was introduced into Britain by the Roman armies which initially produced brick and tile in kilns operated by themselves. Soon the techniques were learnt by the civilian population and kilns were constructed at the nearest source of brickclay for the settlement where the tiles were to be used. Footprints therefore come from the dogs living by the tile kilns although the tiles are found in the archaeological context after being used in buildings. However, we may presume that the civilian kilns were close to the settlements and the dog population extended from settlement to kiln.

Study of the foot impressions shows the frequency of dogs near the tile kilns compared to the other animals and human beings that left impressions in the wet clay, and also the sizes of the footprints (Cram, 1984, 1985). Measurements can be taken of the *track* (the individual footprint) and the *trail* (the succession of tracks). The track of a dog has four toe pads (digital pads) in a semicircle in front of the main pad (interdigital pad), each toe pad having a corresponding claw mark. Cat tracks have four toe pads but these lack claw marks. The fox has four toe pads and claw marks but the two central toe pads are positioned further forward than in the dog. The front paw of the dog is larger than the hind both in length and width, and the main pad has a concave rear compared to the hind main pad which is convex (Bang & Dahlstrom, 1974). For the dog the measurement is taken excluding any claw marks as these are so variable, and are of the length of the track from the front of the toe pads to the back of main pad and of the maximum width across the toe pads. Measurement can also be taken of the length of the toe pads excluding the main pad. This is the only measurement of length available when the main pad is indistinctly shown or lacking as is often the case, but is less useful as it does not indicate fore or hind. The most commonly available measurement is the track width.

Caution must be taken when using measurements from tracks left in wet clay. Firstly the clay and therefore the track contracts by some 10% in drying. Secondly the track gives no indication of the age of the animal. Tracks of a puppy are larger in proportion to the shoulder height than they are when the animal reaches adult size. It must also be remembered that the size of the track left by the paw of an individual dog varies according to the terrain and the gait. The male dog is slightly larger than the female and may therefore be expected to have a slightly larger track.

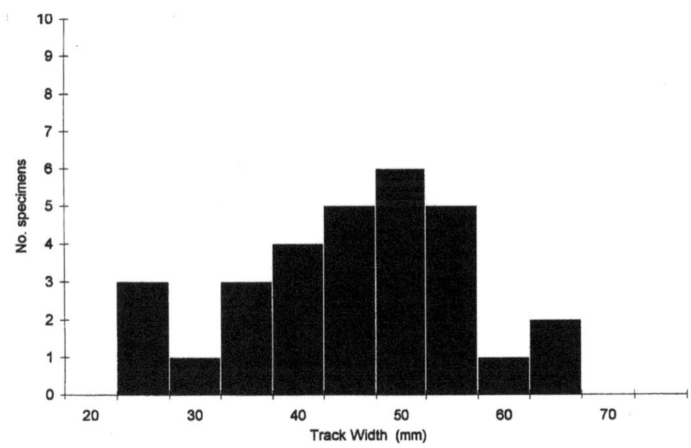


***Figure 3*** *Romano-British dogs, forefoot track width.*

Measurements of dog track width have been pooled from the whole Roman period and the whole of Roman Britain to compare with Harcourt's long bone results. The collections are mainly from Castleford (Cram, 1998), Colchester (unpublished), and Silchester (Cram,1979, 1999). The foretrack has a range of 25 mm to 69 mm, a mean of 47.8 mm and a standard deviation of 11.2 mm (Fig. 3) . The hindtrack has a range from 21 mm to 65 mm, a mean of 40.9 mm and a standard deviation of 9.3 mm. To translate track sizes into shoulder heights, comparative tracks were obtained from known breeds of modern dogs (Table 1). A Chihuahua of shoulder height 130 mm has a foretrack width of 23 mm, on the other extreme a German Shepherd with shoulder height of 625 mm has a foretrack measurement of 65 mm. The track sizes show the range of Roman dogs being equally as small as shown by the bones. The tracks do not reach the size of wolf, of 100-110 mm for the fore foot and 70-80 mm for the hind (Bang & Dahlstrom, 1974), despite the largest bones being of wolf size. However modern breeds of dog have smaller tracks in proportion to their shoulder height than does the wolf. Table 1 shows a greyhound of wolf shoulder height but a track half the size of a wolf. This reduction in paw size in proportion to shoulder height may have already taken place in the Roman period.

At present there are not enough measurements of dog tracks dated within the Roman period to give reliable evidence to show any variation in sizes over time.

## Literary and artistic evidence for dogs

Quick reference can be made to two other sources of information on dogs in Roman Britain. Contemporary

| Breed | Shoulder height in mm | | Track width in mm | |
|---|---|---|---|---|
| | Male | Female | Fore | Hind |
| Chihuahua[a] | 130 | 130 | 23 | 22 |
| Jack Russell terrier[b] | | 290 | 30 | 27 |
| Jack Russell terrier[b] | 315 | | 45 | 40 |
| Bearded collie[a] | 530-560 | 510-530 | 57 | |
| Border collie[a] | 530 | slightly less | 57 | 40 |
| Flat coated retriever[a] | 580-610 | 560-590 | 76 | 65 |
| Lercher puppy[b] | | 380 | 35 | 32 |
| Lercher adult[b] | | 560 | 54 | |
| German shepherd[a] | 625 | 575 | 65 | 56 |
| Greyhound[a] | 710-760 | 680-710 | 49 | 47 |
| Wolf[c] | 700-800 | - | 100-110 | 70-80 |

(a) Shoulder heights from the Kennel Club (1992). (b) Measurement of shoulder heights were taken at the Canine Defence League kennels at Newbury. (c) Shoulder heights and track sizes from Bang & Dahlstom (1974: 56). All track measurements apart from the wolf were taken at Newbury Canine Defence League kennels.

*Table 1 Shoulder height and track width of modern breeds of dogs.*

Roman writers described dogs in Britain and referred to the export of hunting dogs (Alcock, 1996 & pers. comm.).
Toy dogs are known from classical writings and illustrations before the Roman conquest of Britain, being shown on the vases of Greece before the Roman Empire (Ash, 1927).

No exhaustive search has yet been made for all depictions and statues of dog made in Roman Britain, nor any classification of these into varieties to compare with the evidence from dog skeletal remains. But the following references, drawn up by Joan Alcock, give an introduction to this evidence. There are statues showing toy dogs held by women (Jenkins, 1957: plate 7; Toynbee, 1962: plate 76), dogs associated with chasing hares sculpted in stone (Clifford, 1938: 301; Toynbee, 1962: plate 79; Toynbee, 1964: plate 38b) and often shown on pottery (Gillam, 1957: 229; Toynbee, 1962: plates 178-179), dogs hunting deer on a mosaic (Alcock, 1996: 57), on pottery and other media (Toynbee, 1962: plate 262; Wright & Richmond, 1955: plate 36), bronze statues of long limbed dogs (Toynbee, 1964: plates 34b, 34c, 79b) and stocky dogs (Toynbee, 1962: plates 62 & 64), dogs associated with the hunter goddess Diana (Toynbee, 1962: plate 68; Royal Commission, 1928: plate 12; Wright & Richmond, 1955: plate 35) and miscellaneous other dogs (Cunliffe, 1969: plate 64; Smith, 1852: plate 25; Toynbee, 1962: plates 73 & 177; Toynbee, 1964: plate 45a; Wheeler & Wheeler, 1932: plate 26). There are many other representations of dogs found on objects in Britain but made elsewhere in the Roman Empire, in particular on Samian pottery (Oswald, 1936-7: plates 77-79 & 82), but these were obviously not modelled on native British dogs.

### Varieties of dogs from an army settlement

The final part of this paper compares the dog population as shown by both bones and footprints from two distinct Roman settlements. Castleford in west Yorkshire was a Roman fort with an adjoining civilian settlement or vicus supplying services (Cool & Philo, 1998). It was excavated between 1974 and 1988 with modern techniques, including recovery and recording of all tile and animal bone. The tiles of both the fort and vicus are thought to have been made in kilns nearby, with the exception of bathhouse tiles made in an army tile kiln which are not considered here. For this study it has been assumed that the dogs formed one interbreeding population between fort, vicus and kilns. However the majority of the dog bones and foot impressed come from the vicus at a date when the fort was not operating. Despite the extensive nature of the excavations archaeologists might wish that the samples recovered were larger.

The number of dog bones recovered was small, 394 specimens of which 26 are complete mature long bones giving a measurement (Berg, in press; Cram, 1998). No complete skeleton was found. The measurements come mainly from the vicus and show a range of sizes from a tibia of length 80 mm (below the smallest recorded by Harcourt) but not reaching the size of Harcourt's longest measurements.

Approximately 1,000 excavated complete and fragmentary tiles were examined from which 25 were seen to have foot impressions (Cram, 1998). Fifteen of these were human, eight dog and two, small ruminant. The dog fore-foot track widths varied from 27 mm to 57 mm. These support the evidence from the bones of dogs ranging from small but not to the largest sizes. A hind-foot track length of 62 mm is noteworthy.

### Varieties of dogs from a civilian settlement

The Roman town of Silchester in north Hampshire, known in Roman times as Calleva Atrebatum, contrasts with Castleford. The settlement was civilian, the county town of the Atrebatres Tribe, with little or no unrest or military presence during the whole Roman period. Excavations at Silchester were extensive between 1890 and 1909 when the whole area within the walls was explored on a site which had been abandoned after Roman times (Boon, 1974). These excavations have given the most complete plan of a Roman town in Europe.

The early excavations were standard for their times. The material selected to be placed in museums tended to be the best or most complete specimens and there is little or no recording with these apart from occasional writing on the specimen such as a pit or house number. Excavation techniques did not recognise the nuances of stratigraphy that are recorded today. So in the case of animal remains there is usually no knowledge of any date within the four hundred years of the Roman occupation of the town. The site is at present being investigated by area excavation over a five year programme. The excavation director considers that the unrecorded bones from the early collection may date from any time between the first century BC Iron Age settlement and the post Roman settlement up to the fifth to sixth century AD (Fulford, pers. comm.) Indeed it may be argued that dog bones may have been interred in later times into the Roman levels. For this paper it is assumed that all dog bones are Roman. Perhaps it should be said here that this possibility of contamination is relevant to many other early museum collections of dog bones.

However, animal bone was kept and in quantities rivalling the pottery. The excavation reports recognised the potential of the collection but gave no measurements or detailed description. The report of the 1905 discoveries record that "*The large number of canine remains from Silchester would be of great interest to anyone working out the different varieties of dogs known to the Romans*" (St John Hope, 1906). Eight hundred and ninety-one bones of dog are held in the collections of Reading Museum Service and a further 60 in the Natural History Museum in London. Later excavations have used increasingly sophisticated techniques but so far have found little dog material (Hamilton-Dyer, 1997; Maltby, 1984).

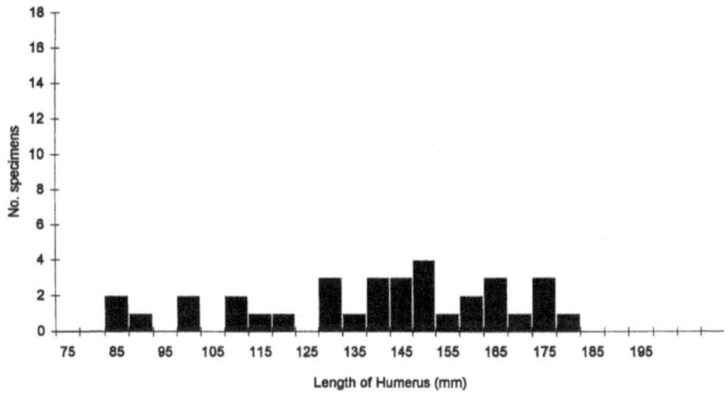


***Figure 4*** *Silchester dogs, left humerus length.*

There are 34 left humeri which give measurements which can be compared with the work of Harcourt and Clark. The lengths range from 84 mm to 176 mm with mean of 165 mm and a standard deviation of 26.8 mm (Fig. 4). Measurement of the femur gives a shortest length according to von den Driesch of 86 mm, of the size of the

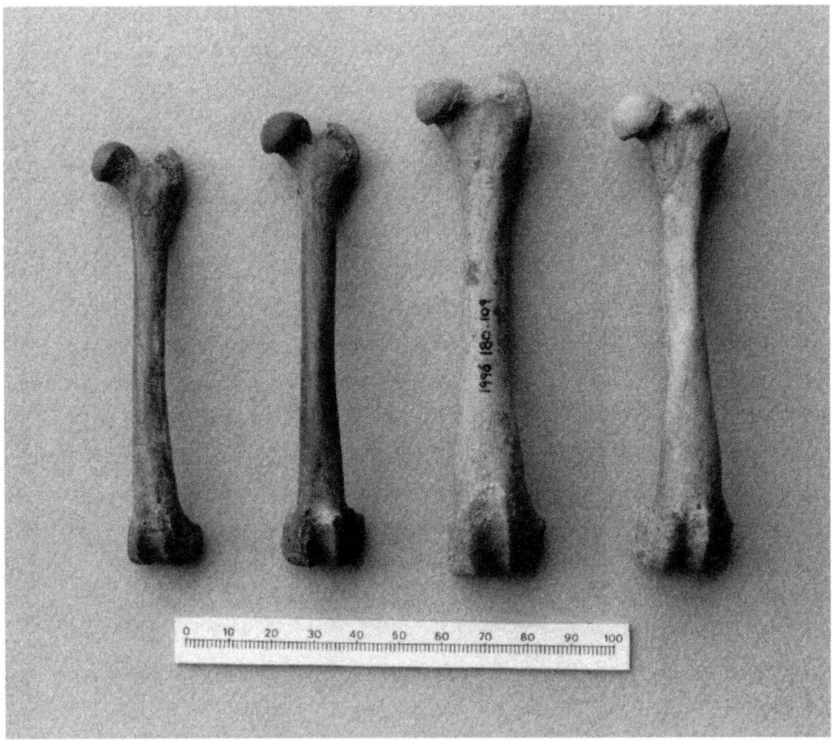


***Figure 5*** *Silchester dogs, femur anterior view, left to right REDMG 1996.180.1864 and 1862, greater trochanter lower than the head, the shaft is bowed forward and slender, REDMG 1996.180.109 and 684, greater trochanter higher than the head, with the shaft stout, bowed backward and expanded at the base. Scale in mm.*

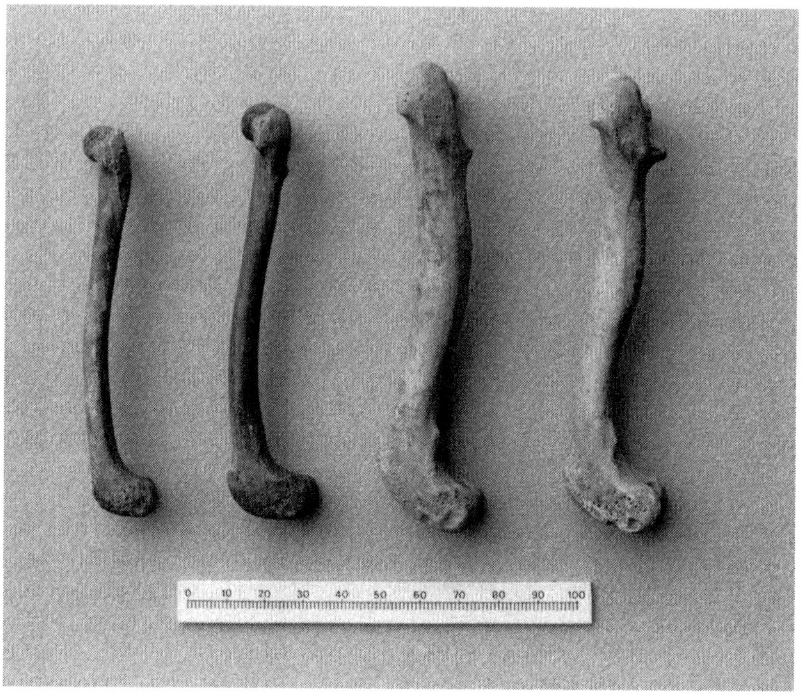


***Figure 6** Silchester dogs, femur medial view, left to right REDMG 1996.180.1864 and 1862, greater trochanter lower than the head, the shaft is bowed forward and slender REDMG 1996.180.109 and 684 greater trochanter higher than the head, with the shaft stout, bowed backward and expanded at the base. Scale in mm.*

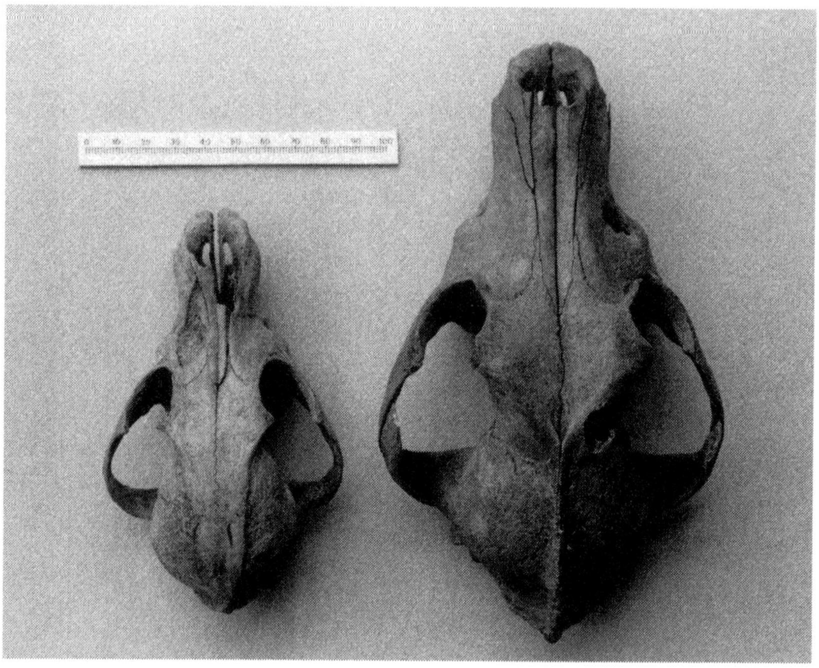

***Figure 7** Silchester dogs, varieties of skull, left to right REDMG 1996.180.1653 one of smallest complete skulls with healed fracture on right maxilla and REDMG 1996.180.1646 one of the largest complete skulls with healed fracture on left maxilla. Scale in mm.*

smallest reported by Harcourt although his measurements were on different points of the bone. The femora range up to the longest at 203 mm . Two distinct forms of femur can be seen. One form has the greater trochanter higher than the head, with the shaft bowed backward and a stout shaft expanded at the base. This form makes up 17 % of the femora. The other form has the greater trochanter lower than the head, the shaft is bowed forward and slender (Fig. 5 & 6). Other long bones, especially the humerus and tibia, show these characteristics of stout extremities and twisted shafts and they can be likened to the brachymel dogs described by Teichert (1987) resulting from breeding from a gene for dwarfism.

The 54 skull lengths vary from 110 mm to 198 mm with a standard deviation of 18.3 mm. Most lack development of the sagittal crest (Fig.7). An interesting feature of the collection of dog skulls from Silchester is the frequency of damage to the head. Twenty-two percent of the skulls have healed fractures on the frontal, maxilla or nasal bones. These fractures might be explained as coming from human blows to the head of the dog while it carried out its function as a guard dog. However these healed fractures are as common in smaller dogs as in the larger and therefore suggest general ill-treatment of the animals.

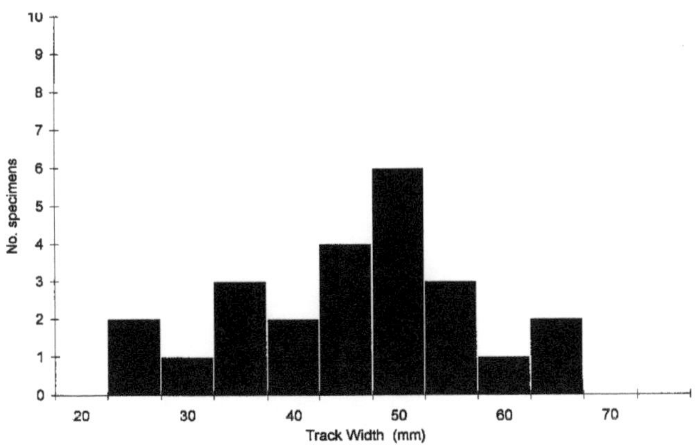


***Figure 8** Silchester dogs, forefoot track widths.*

With one exception none of the Silchester bones reaches the largest sizes found by Harcourt, the exception being a tibia with a length of 241 mm. The other Silchester tibiae range up to 208 mm, while Harcourt gives measurements up to 229 mm. This Silchester tibia is therefore taken to be of wolf.

A fuller report on the Silchester animal remains with discussion of the possibilities from contamination by later material is forthcoming (Clark & Cram, in preparation).

Excavations at Silchester have produced an exceptionally large number of foot impressed tiles. One hundred and fifteen remain from the early excavations (Cram & Fulford, 1979). The excavations under Professor Michael Fulford over the last 20 years have produced another 87 (Cram, 1999). It must be acknowledge here that the smallest of the dog tracks were not recognised as being of that species in the work of 1979 because it was not realised that Roman dogs could be so small.

One tile has the stamp of the emperor Nero and a dog pawprint over this. It was made in the earliest days of the Roman settlement by the army and has not been included in this study. The other tiles are considered to have all been made in kilns near to the town and which shared the same dog population as represented by the bones found within the walls. The species represented on the tiles differ from Castleford. Most common are dog footprints. After this come human beings, sheep and goat, chickens and other birds, cats and cattle. These species suggest that the

***Figure 9** Roman tiles from Silchester showing the largest dog track on the left REDMG 1995.98.26 and smallest on the right REDMG 1995.98.30. Scale in mm.*

tile kilns and drying sheds were positioned next to farmyards. The frequencies of the species contrast with the army kilns of Castleford with their predominantly human impressions followed by dog.

The dog tracks as measured by the track width range from 25 mm to 69 mm, mean 48.4 mm and standard deviation of 11.3 mm for the 24 measured from the forefoot and 21 mm to 65 mm, mean 40.7 mm, standard deviation 10.4 mm for the 17 measured from the hindfoot (Fig.8 & 9).

The combined evidence from bones and tracks from Silchester shows toy dogs in the town and farmyard tilemaking sites as small as found anywhere else in Roman Britain. The dog bones however do not reach the sizes of the largest Roman dogs. Footprints from dogs of the farms and tile kilns around Silchester are as large as have been found anywhere else in Roman Britain.

To summarise the comparisons of the two settlements; Silchester civilian settlement bone sizes are comparable with those from the army camp at Castleford. The Castleford dog tracks show animals as large as at Silchester but toy dogs are not recorded from Castleford tile making yards as small as have been found at Silchester.

***Figure 10*** *Roman floor tile from Silchester with multiple tracks of dog, REDMG 1995.98.17. Scale in mm.*

***Figure 11*** *Roman floor tile from Silchester showing a dog track with leaf impression, REDMG 1995.98.19. Scale in mm.*

A final comment on the Silchester tiles is the intimate detail they give of the life of the dogs. One tile is entirely covered by dog paw prints and suggests that the animal, with a hindfoot track width of 65 mm, was tethered nearby so walking repeatedly over the tile while it dried (Fig.10). Another has the impression of a leaf that was lying on the wet tile. In what year the tile was made we do not know, but the dog walked over the tile in the autumn. (Fig.11).

**Acknowledgements**

I thank Richard Sabin and the Natural History Museum, London for access to Silchester material in its collections, David Berg of West Yorkshire Archaeology Service for access to data on dog bones in advance of publication, the Canine Defence League kennels at Newbury for help in obtaining tracks of modern breeds of dog, and Kate Clark and Juliet Clutton-Brock for discussions over the content of the paper. I am grateful to Joan Alcock for references to literature and illustrations on Roman dogs in Britain by classical writers (The export of hunting dogs from Britain - Nemesianus, Cynegetica 225-226; Strabo, 4. 5. 2; Symmachus, Epistolae II 77. Description of Agassian dogs - Oppian, Cynegetica I 468-480. Description of dogs that could break the neck of a bull - Claudian, De Consulatu III 301). I thank David Pearson for oversight of the statistical data on the Silchester collection and Charity Gordon for preparation of the statistical figures.

**References**

Alcock, J. (1996). *Life in Roman Britain*. London: B. T. Batsford/English Heritage.

Ash, E. C. (1927). *Dogs, Their History and Development.* London: Ernest Bess Limited.

Bang, P. & Dahlstrom, P. (1974). *Animal tracks and signs.* London: Collins.

Baxter, I. L. (in press). A tale of two dogs: dog burials on Roman cemetery sites. In (C. Duhig, Ed) *Taphonomy in Human and Animal Bones*. Proceedings of the 7th conference of the Osteological Research Group. Oxford: Oxbow Books.

Berg, D. (in press). The mammal bones. In ( P. Abramson, D. Berg & M. R. Fossick) *Roman Castleford – Excavations 1974-85, Volume 2, The Structural and Environmental Evidence.* Wakefield: Yorkshire Archaeology **5.**

Boon, G. C. (1974). *Silchester - The Roman Town of Calleva Atrebatum*. Newton Abbot: David & Charles.

Clark, K. M. (1995). The later prehistoric and protohistoric dog: the emergence of canine diversity. *Archaeozoologia* **7**, 9-32.

Clifford, E. (1938). Roman altars in Gloucestershire. *Transactions of the Bristol and Gloucestershire Archaeological Society* **60,** 297-307.

Clutton-Brock, J. (1987). *A Natural History of Domesticated Mammals*. Cambridge: Cambridge University Press.

Conan Doyle, A. (1901-02), *The Hound of the Baskervilles.* London: The Strand Magazine.

Cool, H. E. M. & Philo, C. (1998). *Roman Castleford - excavations 1974-85, volume 1, The Small Finds.* Wakefield: Yorkshire Archaeology **4**.

Cram, C. L. (1978). Animal bones. In (A. C. C. Brodribb, A. R. Hands & D. R. Walker) *Excavations at Shakenoak Farm, near Wilcote, Oxfordshire, Part V Sites K and E*. Oxford: British Archaeological Reports, pp. 117-60; 171-78.

Cram, C. L. & Fulford, M. J. (1979). Silchester tile making - the faunal environment. In (A. McWhirr, Ed.) *Roman Brick and Tile.* British Archaeological Reports International Series **68**, pp. 201-209.

Cram, C. L. (1984). Footprints in the sands of time. *British Archaeological Reports International Series* **227**, 229-235.

Cram, C. L. (1985). Empreintes sur des tuiles romanes. *Les Dossiers Histoire et Archeologie* **90**, 88 - 96.

Cram, C. L. (1998). Human and animal foot impressed tiles. In (H. E. M. Cool & C. Philo, Eds) *Roman Castleford - Excavations 1974-85, Volume 1, The Small Finds.* Wakefield: Yorkshire Archaeology **4**, pp. 232-238.

Cram, C. L. (1998). *Roman dog bones from Castleford excavations 1987 and 1988.* Manuscript held at West Yorkshire Archaeological Service, Wakefield.

Cram, C. L. (1999). Foot impressed tiles. In (M. G. Fulford and J. R. Timby) *Late Iron Age and Roman Silchester: Excavations on the Forum-Basilica, 1977 and 1980-86.* Britannia Monograph **15**.

Cunliffe, B. (1969). *Roman Bath.* London: Reports of the Research Committee of the Society of Antiquaries of London **24**.

Driesch, A., von den (1976). *A Guide to the Measurements of Animal Bones from Archaeological Sites.* Harvard: Peabody Museum Bulletin **1.**

Hamilton-Dyer, S. (1997). The animal bones. In (M. Fulford, S. Rippon, S. Ford, J. Timby & B. Williams) Silchester: excavations at the north gate, on the north walls, and in the northern suburbs 1988 and 1991-3. *Britannia* **28,** 131 - 135.

Harcourt, R. (1974). The dog in prehistoric and early historic Britain. *Journal of Archaeological Science* **1**, 151-175

Jenkins, F. (1957). The role of the dog in Roman-British religion. *Latomus* **16**, 60-76.

Kennel Club (1992). *Illustrated Breeds Standards*. London: Cresset Press.

Maltby, M. (1984). The animal bones. In (M. J. Fulford) *Silchester Defences 1974-80*. Britannia Monograph **5**, 199 - 212.

*McWhirr, A. (1979). Roman* Brick and Tile. British Archaeological Reports International Series **68**.

Oswald, F. (1936-7). *Index of Figure-Types on Terra Sigillata (Samian Ware)*. Annals of Archaeology & Anthropology, supplement. Liverpool: Liverpool University Press.

Royal Commission on Historical Monuments, England. (1928). *An Inventory of the Historical Monuments in London*: *Volume 3, Roman*. London: HMSO.

St. John Hope, W. H. (1906). Excavations on the site of the Roman City at Silchester, Hants in 1905. *Archaeologia* **60**.

Smith, H. E. (1852). *Reliquiae Isuriane: The Remains of Isurium (Aldborough)*. London: William Hilton.

Teichert, M. (1987). Brachymel dogs. *Archaeozoologia* **1**, 69-75.

Toynbee, J. M. C. (1962). *Art in Roman Britain*. London: Phaidon Press.

Toynbee, J. M. C. (1964). *Art in Britain under the Romans*. Oxford: Clarendon Press

Wright, R. P. & Richmond, I. A. (1955). *Catalogue of the Roman inscribed and sculptured stones in the Grosvenor Museum, Chester*. Chester: Chester and North Wales Archaeological Society.

# METRIC VARIABILITY IN ROMAN PERIOD DOGS IN PANNONIA PROVINCE AND THE BARBARICUM (HUNGARY)

László Bartosiewicz
Institute of Archaeological Sciences,
Loránd Eötvös University,
Budapest, Hungary

**Key words:** dog, breeds, Roman, Barbaricum, Hungary, archaeozoology, craniometry, Carpathian Basin

**Abstract**

During the Period of the Roman Empire (AD first-fifth centuries), the political boundary between the province of Pannonia Inferior and the Barbaricum, the *limes*, was marked by the Danube river. The Roman province occupied its right bank, while the Great Hungarian Plain, extending from the left bank, was inhabited by various pastoral peoples of eastern origins. This dichotomy is also mirrored in animal husbandry. Hypothetically, the idea of conscious animal breeding spread beyond the borders of Italy within the Roman Empire. Early population movements of the Migration Period, on the other hand, brought peoples to the Carpathian Basin whose animal keeping was mostly rooted in empirical traditions. Dog skulls from the Roman urban settlement of Tác-Gorsium in Pannonia displayed considerable variability, possibly indicative of breed formation. Meanwhile, bones from large "working dogs" of the pastoral Sarmatians found at the eastern site of Gyoma 133 appear more homogeneous, at least in terms of their cranial morphology. Significant differences between the dog skulls from these two disparate settlements were, however, found only in the mean values of some neurocranium measurements.

## Introduction

During the period of the Roman Empire (AD first-fifth centuries), the Carpathian Basin was politically divided along the Danube river. The Roman province of Pannonia Inferior occupied the right bank (Transdanubia, ie modern day western Hungary), while the left bank, also known as the "Barbaricum", was inhabited by various, usually pastoral peoples of eastern origins.

It has been suggested that this geopolitical distinction was mirrored in animal husbandry traditions. Hypothetically, the idea of conscious animal breeding, well known from the works of Cato, Columella and Varro, was widely spread beyond the borders of Italy within the Roman Empire. Early waves of people in the Migration Period, on the other hand, brought groups to the Carpathian Basin whose animal keeping was mostly rooted in an empirical tradition that had developed in the Eurasian steppe.

The domestic dog, an ubiquitous and multiparous species with a short generation interval, seems to be a good indicator of the aforementioned differences in animal breeding. It has been hypothesized that dog remains found in complex urban settlements in Pannonia, which display considerable variability, potentially indicate Roman Period "breed" formation. Meanwhile, the "working dogs" of pastoral groups such as the Sarmatians, are expected to look more homogeneous, at least in terms of size as reflected by their osteomorphology.

## Material and Methods

By the time of the Roman Period, evidence for the consumption of dog flesh in the Carpathian Basin disappears (Bökönyi, 1974: 320). General domestic animal husbandry practices had become well established by the end of prehistoric times, so that the opportunistic exploitation of dogs for meat became unnecessary (Bartosiewicz, 1990: 291). Consequently, the bones of these animals are less likely to occur in ordinary settlement refuse. In Pannonia Province, 40 more-or-less complete dog skulls were selected for study from the large urban settlement of Tác-Gorsium. Bökönyi (1984: 15) identified 47,941 animal bones in the hand-collected material from this site. This high NISP value in and of itself shows the relatively low concentration of dog remains in this Roman provincial town which developed from a military fort around the middle of the AD second century and which covered the greatest area by the middle of the fourth century. Such a settlement would most likely have been inhabited by a truly Romanized population as opposed to local Celtic villages.

Sarmatian dogs were represented by 24 measurable skulls from the AD second-third century Sarmatian site of Gyoma 133. These were selected from a set of 8,555 hand-collected identifiable bones (Bartosiewicz, 1996: 380). The many times greater contribution of dog skulls to this assemblage of identifiable bones (0.281 % in comparison with the 0.083 % at Tác-Gorsium) results from the presence of (sometimes multiple) dog burials at this rural settlement.

Hypothetically, the two settlements were treated as distinct geopolitical units. In addition to the different ways they functioned, they were both located relatively far from the Danube which formed the *limes* of the Roman Empire in this area. Tác-Gorsium lay in the interior of the province, some 35 km west of this line, while Gyoma was located in what corresponds to the heartland of Sarmatian territories on the Great Hungarian Plain, at least 140 km east of the

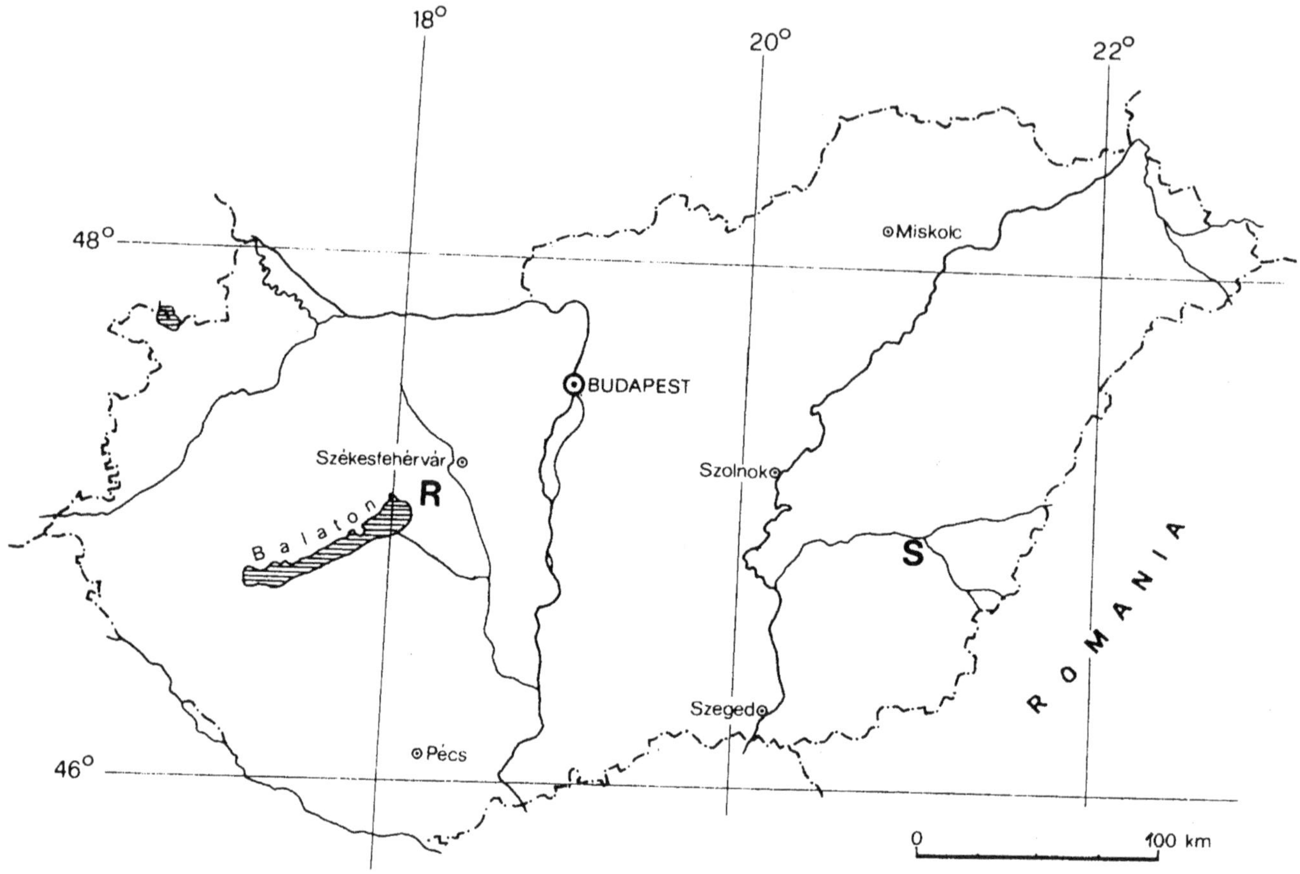


***Figure 1*** *Map of Hungary showing the location of Tác-Gorsium (R) and Gyoma 133 (S)*

| | Name of measurement | Abbreviation | Code in Figure 2 |
|---|---|---|---|
| 1 | total length | A-P | 1 |
| 2 | basal length | B-P | 3 |
| 3 | height of the occipital triangle | B-A | 40 |
| 4 | basion-staphylion length | B-St | not listed for dog, "99" |
| 5 | median palatal length | St-P | 13 |
| 6 | breadth at the canine alveoli | C-C | 36 |
| 7 | zygomatic (greatest) breadth | Zyg-Zyg | 30 |
| 8 | frontal breadth | Ect-Ect | 32 |
| 9 | greatest neurocranium breadth | Eu-Eu | 29 |
| 10 | greatest mastoid breadth | Ot-Ot | 23 |
| 11 | greatest breadth of the occipital condyles | Cond. occ. | 25 |

***Table 1*** *Cranial measurements used in the comparison of Roman provincial and Sarmatian dogs (measurement codes from von den Dreisch, 1976).*

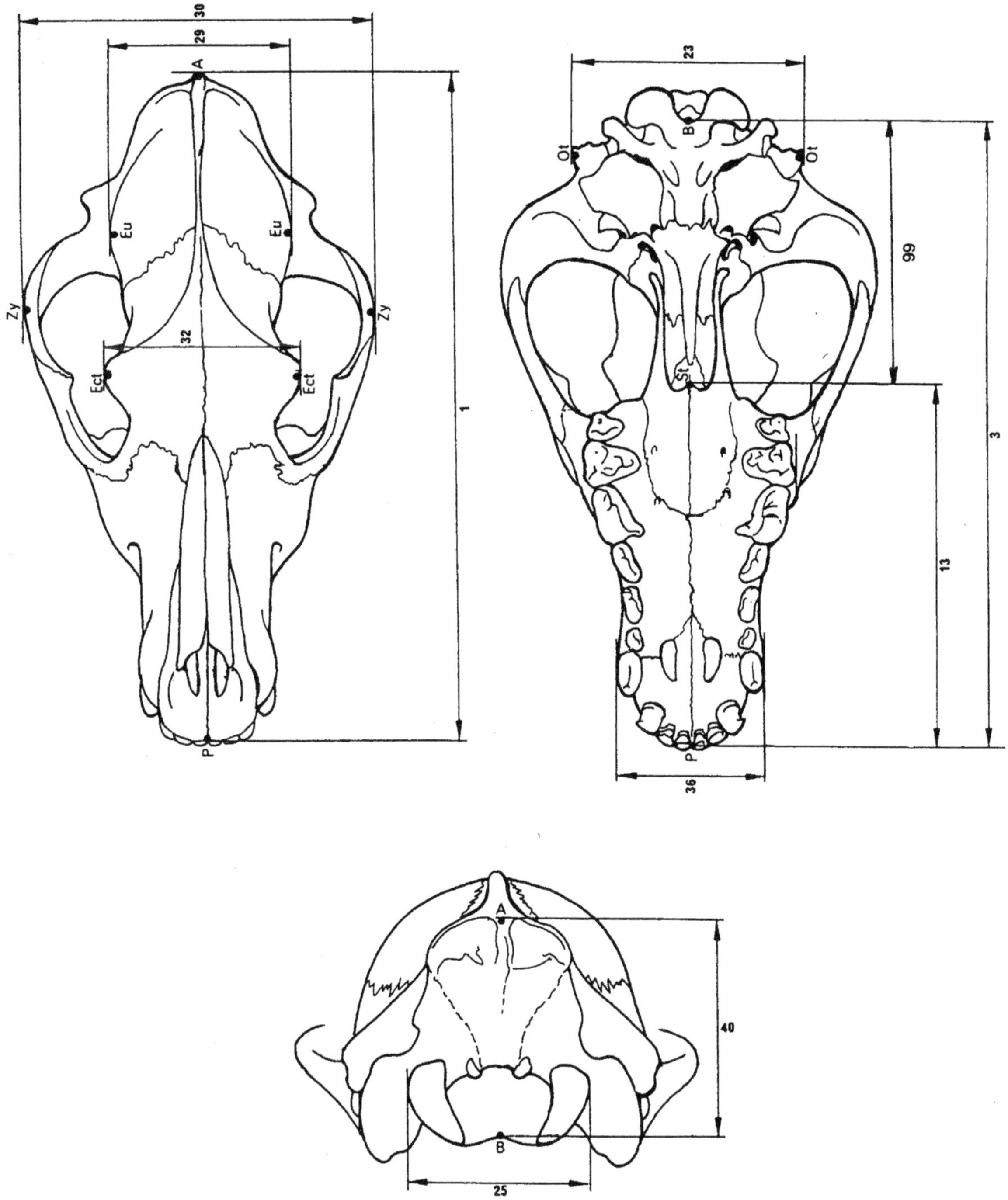


***Figure 2*** *Cranial measurements used in analyses (after von den Driesch, 1976)*

*limes* as marked by the Danube river. Separation is potentially increased by the Tisza, another major river flowing parallel to the Danube down through the Great Hungarian Plain (Fig.1).

Dog skulls are also available sporadically from numerous other Sarmatian and Roman provincial sites. This study, however, focused on these two major, spatially distinct settlements in order to minimize overlaps caused by dog skulls representing possibly more mixed dog populations from areas lying in-between.

Ten standard measurements for dog skulls, as listed by von den Dreisch (1976) have been selected for the purposes of analysis. Basion-staphylion length not listed for this species was used to complement median palatal length along the same axis (Table 1; Fig. 2).

These measurements were chosen on the basis of availability even on fragmented skulls and their potential for characterizing overall size as well as the robustness of the neuro- and visceroranium respectively.

Univariate statistical parameters were reviewed in the original material. Regional comparisons were made using Student's t-tests. The ratio between the number of cases was less than 2:1 (40-24), a proportion conventionally considered acceptable in such comparative studies. Discriminant analysis, a multivariate method that has been successfully used in testing *a priori* groupings in canid craniology (eg Clutton-Brock, Corbet & Hills, 1976; Benecke, 1987), yielded the same information in this study as simple univariate statistics and was thus abandoned.

The ratio between cases (64) and variables (11), on the other hand was greater than 2:1 which made the use of factor analysis realistic in appraising the structure of relationships between the cranial dimensions selected. Criteria for this calculation included latent roots $\lambda \geq 1$. These values were balanced between factors using simple Varimax rotation. Differences as well as factor loadings were considered statistically significant at the $p \leq 0.05$ level of probability. Owing to the methodological difficulties inherent in the parametric analysis of ration values (Atchley, Gaskins & Anderson, 1978), traditional index values were not included in the statistical evaluation.

## Results

Univariate statistics revealed only small differences between the mean values of the two groups. It is only frontal breadth measured between the ectorbitalia that was significantly wider in Sarmatian dogs, while the greatest breadth of the brain case and the breadth of the occipital condyle were more strongly developed in Roman provincial specimens (Table 2).

Although the majority of mean values was higher in the Sarmatian sample, these differences were not statistically significant on the required level of probability. A comparison of extreme values also reveals that this difference was actually caused by the presence of smaller specimens in the material from Tác-Gorsium (Fig. 3).

Meanwhile, the skulls of several large individuals occurred in the cranial assemblage from Gyoma 133. A comparison of coefficients of variation (Fig. 4) shows greater variability in the urban provincial material for almost all measurements. Deviations from this general pattern are also associated with significant differences between the two groups.

The differences noticed in individual measurements also influence relationships between cranial measurements. A factor analysis yielded two major factors that encompass almost three quarters (74.3%) of the variance represented by the 11 measurements.

Due to naturally high correlations, especially between measurements that represent the same dimensions (mediolateral breadths and lengths) the most characteristic factor may be labeled as "overall size". In spite of this, significantly different breadths of the brain case and the occipital condyle characterise a different group of measurements that are also closely connected to lengths, especially taken along the skull's base (basal length, median palatal length, basion-staphylion length). This seems to suggest that while overall size is prone to a certain variability, the functionally determined variability in the width of brain case and occipital condyle are also closely related to the basal length of skull. This factor, therefore, was named "basic size" which is associated with dimensions that are relatively stable even in the case of great overall size variability.

Factor scores of individual dog skulls were plotted against these two factors in order to help interpret differences in their morphometric character (Fig. 5). Distributions along the horizontal axis, representing ("overall Size"), show that data points representing the more variable Roman provincial dog skulls are more widely spread along Factor 1, while larger Sarmatian dogs tend to mostly show

| | **Roman** | | | | | **Sarmatian** | | | | | **Differences** | | |
|---|---|---|---|---|---|---|---|---|---|---|---|---|---|
| **Variable** | **n** | **Mean** | **Sd** | **min.-max.** | **cv** | **n** | **Mean** | **Sd** | **min.-max.** | **cv** | **t-value** | **df** | **p value** |
| A-P | 37 | 193.6 | 17.3 | 138.0-220.0 | 0.089 | 16 | 193.6 | 13.5 | 174.1-226.4 | 0.070 | -0.002 | 51 | 0.998 |
| B-P | 35 | 171.2 | 15.9 | 120.0-196.0 | 0.093 | 15 | 173.8 | 11.8 | 156.1-202.8 | 0.068 | -0.576 | 48 | 0.567 |
| B-A | 34 | 48.3 | 4.8 | 35.0-57.0 | 0.099 | 21 | 48.0 | 6.9 | 35.1-59.3 | 0.144 | 0.178 | 53 | 0.860 |
| B-St | 32 | 76.7 | 7.7 | 54.0-91.0 | 0.100 | 14 | 79.2 | 6.0 | 69.2-89.9 | 0.076 | -1.111 | 44 | 0.272 |
| St-P | 36 | 93.6 | 10.3 | 64.0-108.0 | 0.110 | 14 | 97.3 | 8.2 | 84.7-115.8 | 0.084 | -1.220 | 48 | 0.229 |
| C-C | 35 | 38.1 | 4.4 | 26.0-46.0 | 0.115 | 15 | 39.6 | 3.3 | 34.4-46.1 | 0.083 | -1.208 | 48 | 0.233 |
| Zyg-Zyg | 26 | 107.3 | 7.8 | 80.0-117.0 | 0.073 | 15 | 109.4 | 6.6 | 101.1-122.0 | 0.060 | -0.873 | 39 | 0.388 |
| **Ect-Ect** | **39** | **53.5** | **7.1** | **38.0-70.0** | **0.133** | **21** | **59.6** | **4.4** | **51.5-66.1** | **0.074** | **-3.565** | **58** | **0.001** |
| **Eu-Eu** | **37** | **61.9** | **4.2** | **47.0-70.5** | **0.068** | **19** | **59.9** | **1.8** | **56.1-62.5** | **0.030** | **2.008** | **54** | **0.050** |
| Ot-Ot | 35 | 62.2 | 5.3 | 44.5-72.0 | 0.085 | 22 | 64.5 | 3.7 | 57.3-71.0 | 0.057 | -1.805 | 55 | 0.077 |
| **Cond. occ.** | **34** | **38.6** | **3.6** | **28.0-46.0** | **0.093** | **21** | **35.6** | **6.3** | **26.1-49.2** | **0.177** | **2.281** | **53** | **0.027** |

***Table 2.*** *Univariate statistics used in the comparison of provincial Roman and Sarmatian dogs. Significant differences between mean values are marked by shading*

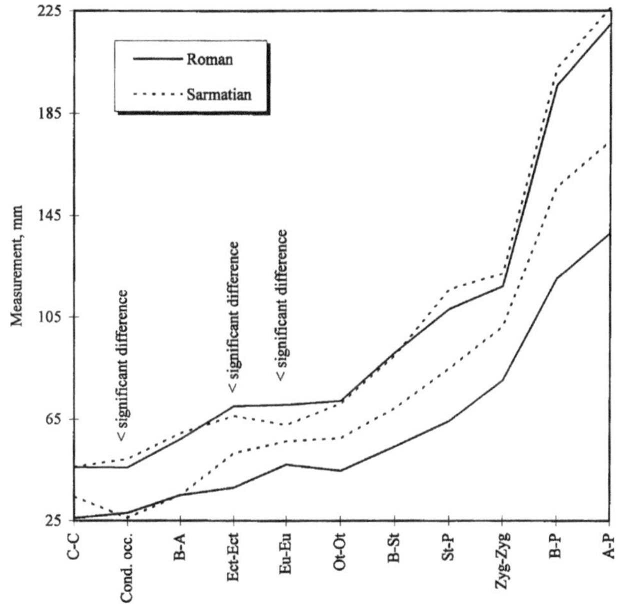


***Figure 3*** *The minima and maxima of cranial measurements arranged in an increasing order*

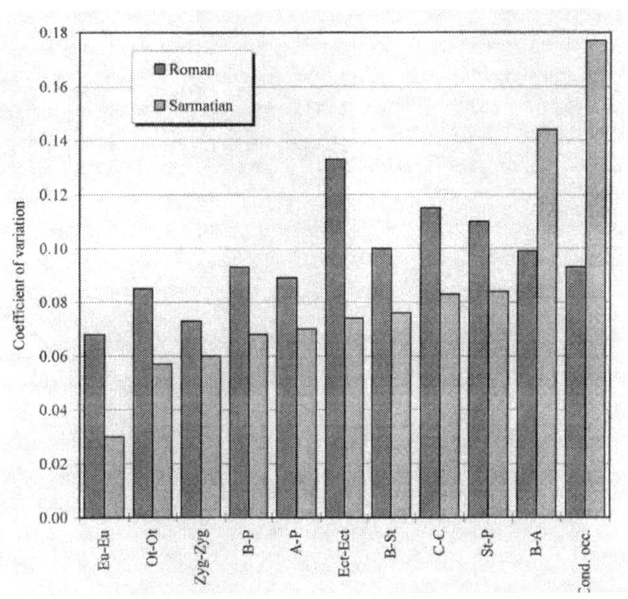


***Figure 4*** *Variability in cranial measurements as shown by coefficients of variation*

| Variable | Factor 1 Overall size | Factor 2 Basic size | Communalities |
|---|---|---|---|
| **Cond. occ.** | -0.029 | **0.920** | 0.847 |
| **Eu-Eu** | 0.543 | **0.677** | 0.754 |
| B-A | 0.637 | 0.395 | 0.562 |
| **B-St** | **0.662** | 0.554 | 0.744 |
| **A-P** | **0.720** | 0.611 | 0.891 |
| **Zyg-Zyg** | **0.734** | 0.256 | 0.605 |
| **B-P** | **0.742** | 0.593 | 0.903 |
| **St-P** | **0.751** | 0.410 | 0.731 |
| **C-C** | **0.751** | 0.242 | 0.623 |
| **Ot-Ot** | **0.807** | 0.265 | 0.721 |
| **Ect-Ect** | **0.884** | -0.091 | 0.789 |
| **Latent roots** | **5.307** | **2.863** | **8.170** |
| **% of total variance** | **48.2** | **26.1** | **74.3** |

***Table 3*** *Factor loading matrix of cranial measurements. Values greater than 0.66 are marked by shading*

positive factor scores in terms of overall size (18 of 24 individuals). The opposite of this phenomenon may be observed along the vertical axis ("basic Size"). In contrast to the skulls of Roman provincial dogs, only a few Sarmatian specimens (seven out of 24 individuals) are positively correlated with this latter group of basic measurements. Data points representing very small individuals (Basic size: factor score <-2) correspond to the specimens shown in Figure 6, plate 1.

**Discussion**

In this study, cranial measurements of dogs from two contemporary archaeological sites were compared. The settlements under discussion, originate from two major but contiguous geopolitical regions and represent different levels of social complexity. Only a few significant size differences were observed between the two groups of dog skulls. Larger skulls were relatively similar (Figure 7, plate 2). The presence of smaller individuals in the Roman provincial sample, however, resulted in greater variability,

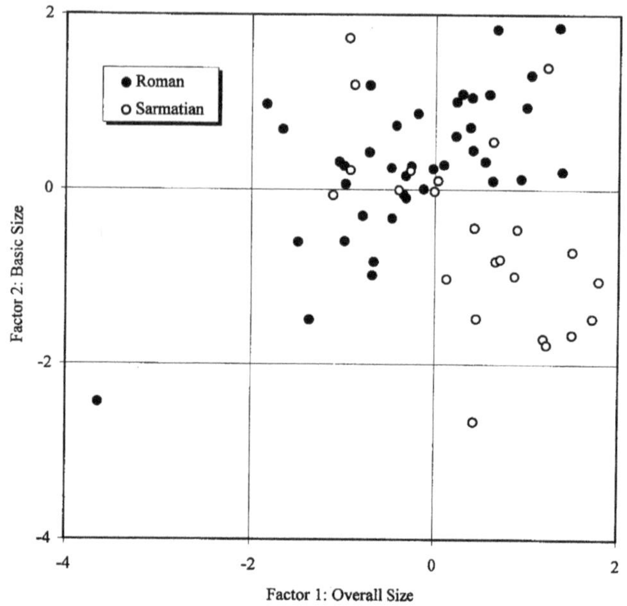


***Figure 5** The distribution of dog skulls in the plane defined by the two factors*

a phenomenon also observed in other parts of the Roman Empire. Hilzheimer (1932) distinguished three size groups of dog skulls on the basis of basal lengths in Roman Germania. He classified those skulls within size intervals (110-130; 150-160; 170-180 cm) of which the last was best represented in the materials under discussion here. Roman colonization brought about increasing variability in the size of dog skulls in Britannia and northern Gallia as well (Harcourt, 1974: 151; Lepetz, 1996: 68, Fig. 84). An even richer inventory of various types has been described from southern Gallia (Jourdan, 1976: 207).

In comparison to these examples in the literature, a special feature of the Gyoma 133 Sarmatian dogs is their relatively great overall size, ie the near complete absence of the small forms. Withers heights, not analyzed in this paper, indicated that most of these animals were taller than 50 cm (Bartosiewicz, 1996: 374), falling into Hornberger's (1970: 113) "d" category (largely corresponding to Alsatians or German Shephard dogs) and Bökönyi's group of "medium size" dogs (withers height, 48.2-68.1 cm). From the viewpoint of morphological variability, it is also interesting that while Bökönyi (1984: 66, Fig. 23) identified an entire series of tibiae showing a broad range of lengths and curvatures at Tác-Gorsium, only a single curved dog tibia was recovered at Gyoma 133 (Bartosiewicz, 1996: 432, Fig. 18).

In spite of the evidence for trade links between Roman and Sarmatian populations living on the two sides of the Danube, differences in the variability of Roman and Sarmatian dogs seem to be consistent with the research hypothesis. Given the great geographical distance between the two sites, it is unlikely that the two distant populations would have had an opportunity to regularly interbreed. If the possibility of target-oriented dog trading is excluded, the two groups of dogs must have been well separated from each other.

The usually tense relationships between the peoples of the Barbaricum and the Romans culminated in the Marcomannic wars during the AD170s, when the eastern parts of Transdanubia (including Tác-Gorsium) were overrun by armies from the Barbaricum. Finally, the Romans overcame their enemies and organized the province of *Sarmatia* in the broader region around Gyoma (Barkóczi, 1980: 98). It remains a question, however, whether such a major military incursion and the resulting backlash would have been accompanied by a noteworthy long-distance displacement of dogs across the natural barrier of the Danube.

More important than relative genetic isolation may have been the cultural and functional difference between the two settlements that influenced the cranial conformation in these dog populations. Bökönyi (1984: 92) has pointed out that, "with the exception of the luxury dogs (which even today are kept isolated from other breeds), all the watchdogs, herding dogs and pariah dogs had excellent opportunities to interbreed with each other at will" in the urban setting of Tác-Gorsium. Cranial similarity between late medieval urban dogs and pariah dogs was also explained by the near panmixis of such animals in the city of Vác (Bartosiewicz, 1995: 60).

It is possible that at the rural settlement of Gyoma 133, dogs may have been more explicitly watch dogs. Perhaps preference was given to larger animals, better suited to such uses. The larger dogs of eastern pastoralist groups may also have been related to different dog populations in the ancient Sarmatian homelands. Such unsubstantiated speculations, however, are impossible to reliably test using osteometric data alone.

## Conclusions

A comparison of cranial data from the Roman town of Tác-Gorsium and the Sarmatian rural settlement of Gyoma 133 displayed the greater variability of Roman provincial dog skulls, a phenomenon well known in the European literature. It is interesting, however, that most Sarmatian dog skulls fell within the upper range of the size variation of Roman dogs and also differed from them in the formation of basic measurements in the neurocranial region. Bones of small dogs were rare at the Sarmatian settlement of Gyoma 133.

It is suggested that differing selective pressures related to the dogs' function within the given society, rather than long standing genotypic differences, lie behind this phenomenon. In an urban environment less attention may have been paid to the strength and size of dogs, otherwise a selective advantage when animals are chosen for herding or guarding by rural people.

Sexual dimorphism is an often neglected aspect of dog studies, partly because sex-related craniometric differences are regarded as allometric, ie manifested primarily as a function of skull size. Phenotypic studies, however, could be fine-tuned by consistently recording sex and considering it during future analyses.

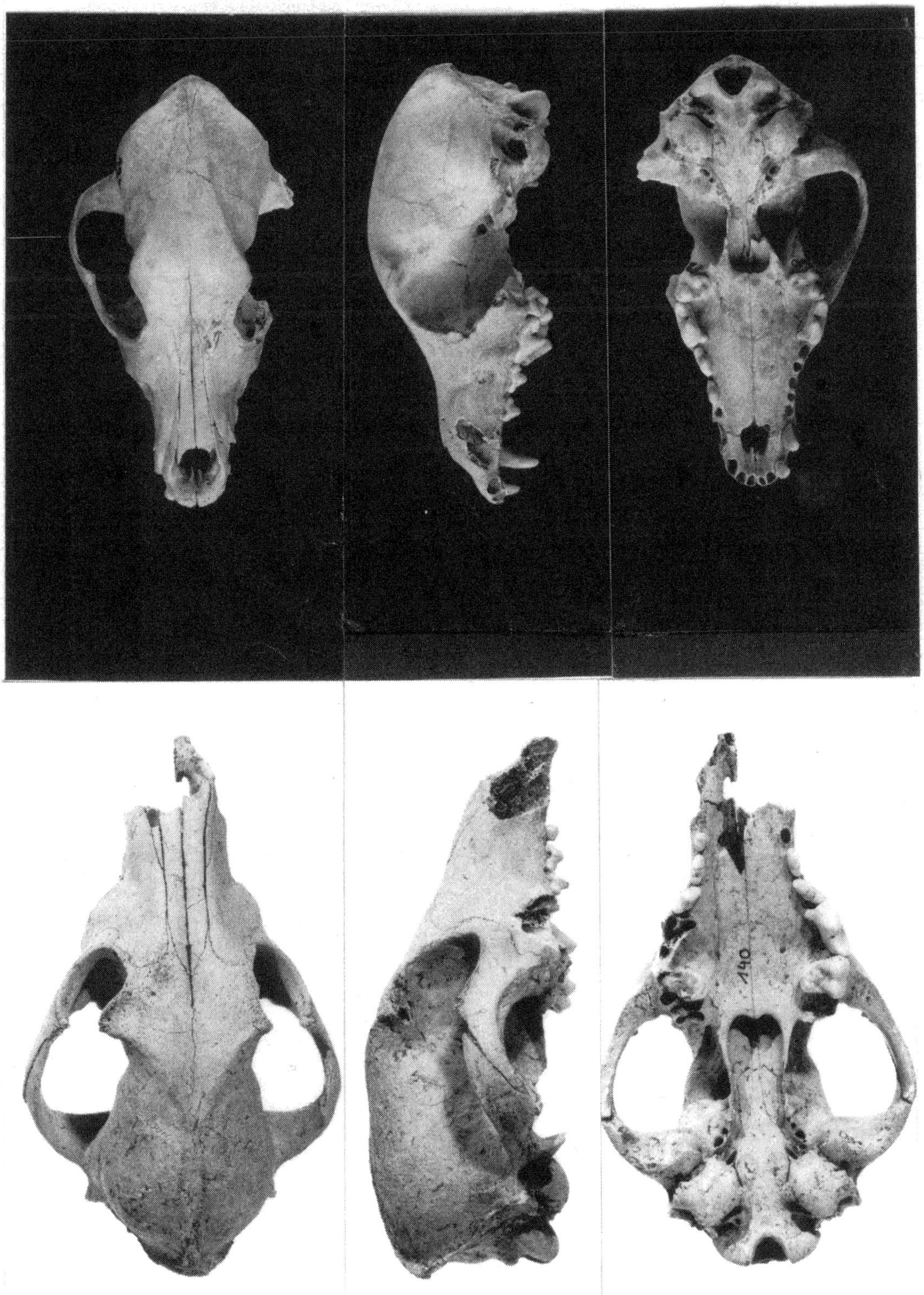


***Figure 6*** *Plate 1. One of the smallest dog skulls from Gyoma 133 in the Sarmatian Barbaricum (female, A-P=174.1 mm, left) and Tác-Gorsium in Roman Pannonia (female, A-P=138 mm; 1st type, after Bökönyi, 1984: 75). Scale ≈ 1:2*

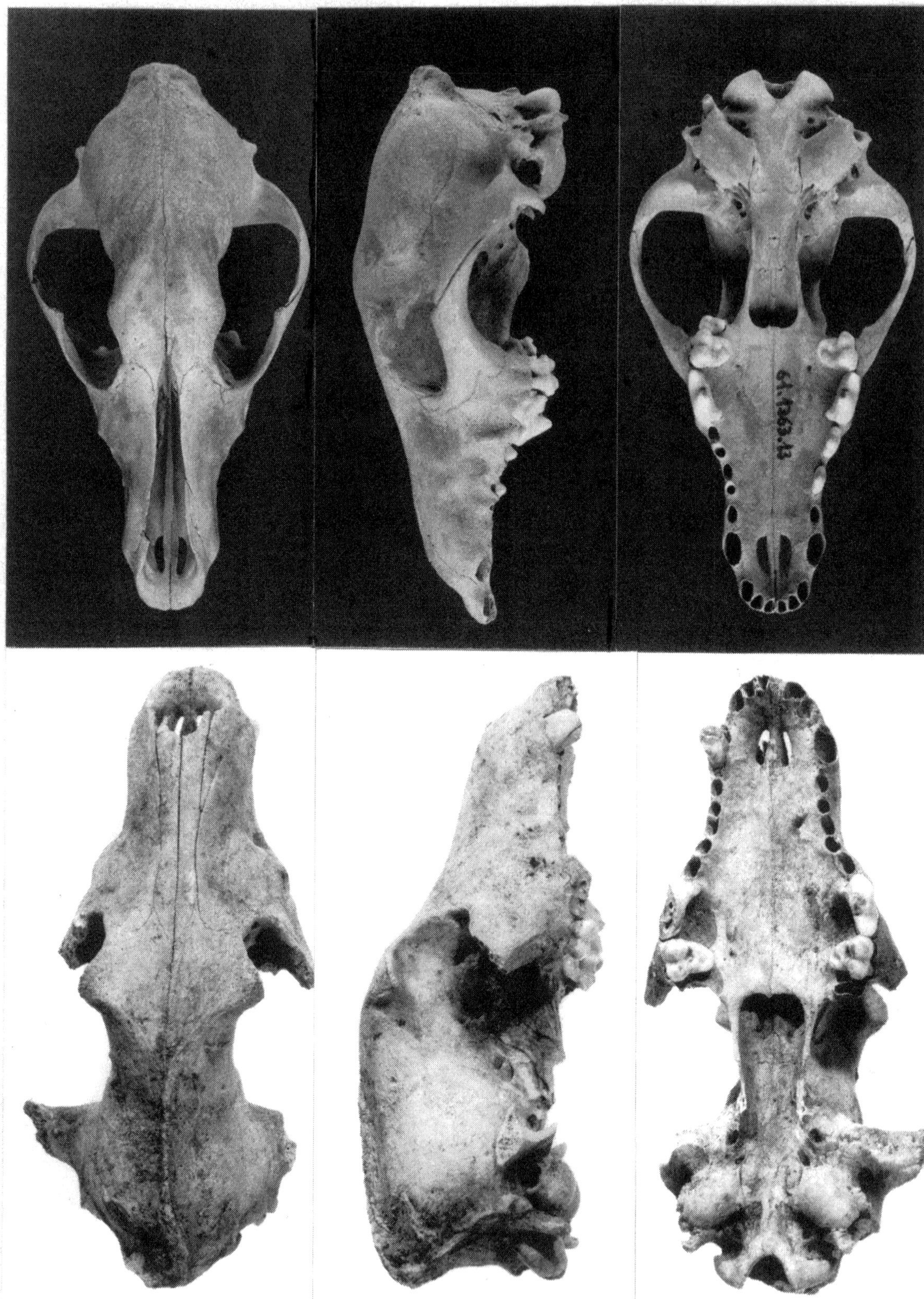


***Figure** 7 Plate 2. One of the largest dog skulls from Gyoma 133 in the Sarmatian Barbaricum male, A-P=197.2 mm, left) and Tác-Gorsium in Roman Pannonia (female, A-P=185.5 mm; 4th type, after Bökönyi, 1984: 75). Scale ≈ 1:2*

Terms such as "breed" and conscious vs. non-conscious "animal breeding" should be treated cautiously in archaeozoological research owing to the ethnocentric bias inherent to our modern concepts. Nevertheless, domestic animals always reflect the particular human (economic and cultural) environment they live in, and rapidly changing generations of dogs may have been easily affected by changing attitudes to animals. The attitudes toward domestic dogs living in a Romanised provincial town were certainly different from those kept at a rural settlement inhabited by Sarmatian pastoralists. Aside from settlement function, ethnic tradition (Bartosiewicz, 1998: 75) may also have influenced the way dogs were treated at these two settlements, resulting in phenotypic differences on the long run.

**References**


Atchley, A. H., Gaskins, C. T. & Anderson, D. (1976). Statistical properties of ratios I. Empirical results. *Systematic Zoology* **25,** 137-138.

Barkóczi, L. (1980). History of Pannonia. In (A. Lengyel & G. T. B. Radan. Eds) *The Archaeology of Roman Pannonia*. Budapest: The University Press of Kentucky–Akadémiai Kiadó, pp. 85-124.

Bartosiewicz, L. (1990). Species interferences and the interpretation of Neolithic animal exploitation. *Acta Archaeology of Hungary* **42,** 287-292.

Bartosiewicz, L. (1995). *Animals in the urban landscape in the wake of the Middle Ages:* Tempus Reparatum, Oxford: British Archaeological Reports International Series **609**.

Bartosiewicz, L. (1996). Animal exploitation at the Sarmatian site of Gyoma 133. In (S. Bökönyi, Ed) *Cultural and Landscape changes in South-East Hungary II.* Budapest: Archaeolingua, pp. 365-346.

Bartosiewicz, L. (1998). Attitudes to pets in the ethnolinguistic record. In (P. Anreiter, L. Bartosiewicz, E. Jerem & W. Meid, Eds) *Man and the Animal World: Studies in Memoriam Sándor Bökönyi.* Budapest: Archaeolingua Kiadó, pp. 65-78.

Benecke, N. (1987). Studies on early dog remains from Northern Europe. *Journal of Archaeological Science* **14,** 31-49.

Bökönyi, S. (1974). *History of Domestic Animals in Central and Eastern Europe.* Budapest: Akadémiai Kiadó.

Bökönyi, S. (1984). *Animal Husbandry and Hunting in Tác-Gorsium*. Budapest: Akadémiai Kiadó.

Clutton-Brock, J., Corbet, G. B. & Hills, M. (1976). A review of the family Canidae with a classification by numerical methods. *Bulletin of the British Museum (Natural History), Zoology* **29/3,** 117-199.

Driesch A., von den (1976). *Das Vermessen von Tierknochen aus vor- und frühgeschichtliche Siedlungen*, Institut für Paläoanatomie, Domestikationsforschung und Geschichte der Tiermedizin der Universität München.

Harcourt, R. A. (1974). The dog in prehistoric and early historic Britain. *Journal of Archaeological Science* **1,** 151-175.

Hilzheimer, M. (1932). Römische Hundeschädel aus Mainz, ein fränkischer Hundeschädel und ein Hundeschädel des 15. oder 16. Jahrhunderts ebendaher. *Biological Genetics* **8,** 91-126.

Hornberger, M. (1970). Gesamtbeurteilung der Tierknochenfunde aus der Stadt auf dem Magdalensberg in Kärnten (1948-1966). Klagenfurt: *Kärntn Museumsschr* **XLIX,** pp.1-144.

Jourdan, L. (1976). La faune du site gallo-romain et paléo-chrétien de la Bourse (Marseille). Paris.

Lepetz, S. (1996). *L'animal dans la société gallo-romaine de la France du Nord.* Revue Archéologique de Picardie Special **12**.

# Part 4

# Interpreting skeletal variation: non-Roman contexts

# PREHISPANIC DOG TYPES IN MIDDLE AMERICA

Raúl Valadez Azúa
Instituto de Investigaciones Antropológicas de la UNAM,
Circuito Exterior,
C.U. Delegación Coyoacan,
C.P. 04510, México D.F.

**Key words:** dogs, dog breeds, Mexico, Mexican hairless, Xoloitzcuintli, itzcuintli, tlalchichi, Sahagun, Prehispanic dogs, zooarchaeology

**Abstract**

Dogs were an important animal for Middle American people and therefore their remains are common in archaeological sites. Nevertheless, there is little information about their characteristics and diversity. The analysis of dog remains uncovered in Middle American settlements has identified four types of dogs. The first, and most common, has existed in the area since the Formative period: these animals were hairy and very similar to present-day Mexican or creole dogs. The second variety found was the hairless Mexican dog. The third was a short-legged type and the fourth was a small-sized brachycephalic dog.

This information, along with sixteenth century chronicles, confirms that the information given by Sahagun about dogs *(Códice Florentino)* is the most reliable because of the correspondence to types found in the analysis of archaeological remains. Therefore, the work presented by Francisco Hernández is considered less reliable because of the lack of corroboration. Accordingly, Middle American people appear to have had four types of dogs, the *itzcuintli* or Mesoamerican common dog, the *Xoloitzcuintli* or hairless dog, the *Tlalchichi* or "floor-dog" and the short-nosed dog.

## Introduction

Few subjects dealing with studies of Middle American ethnozoology have caused as much confusion as the issue of dog types in Mexico before the Spanish arrival. Five centuries of poor interpretations as well as a handful of formal studies on the subject have caused quite a research problem, until now that the field has been taken over by Mexican archaeologists.

Firstly we must consider the fact that even though the remains of these animals have been available, the information derived on this subject has been based only on historical records of Middle America compiled after the Spanish conquest of Mexico. Now there are enough zooachaeological studies so as to offer an interpretation based on these new data, which in turn can be complemented and clarified by Spanish chronicles.

## Ethnohistorical and ethnozoological background

The earliest references of dog types in Middle America date back to the sixteenth and seventeenth centuries, where in descriptions of the new territories, various Spanish explorers, naturalists and clergymen occassionally mentioned local dogs. Without a doubt, the most important documents in that collection are the *Códice Florentino* ("Florentine Codex") by Fray Bernandino de Sahagún (1979), finished in 1582 , and *Historia Natural de las Cosas de la Nueva España. Obras Completas* ("The Natural History of Things in New Spain. Complete Works"), by naturalist Francisco Hernández, which first appeared in 1651, in Latin. The first mentions three types of dogs and the second, four.

As of the nineteenth century, the subject of native dogs on the American continent became increasingly interesting to naturalists. This yielded an accumulation of data, which permitted the making of "Dogs of the American Aborigines" by Glover M. Allen, in 1920. The work describes 17 breeds of dogs, four of which belong to Mexican territory (Table 1). In order to list the descriptions according to different breeds, he used all the information available to him: published works, verbal descriptions, photographs, ancient chronicles, skeletons and zooarchaeological remains. The relevance of this study is such, that most North-American scholars continue to use it when trying to identify the breeds of newly discovered dog remains in their zooarchaeological research.

Generally speaking, Allen's work should be highly considered; however, it is important to point out that as far as dog types in Mexican territories are concerned, the lack of any bone remains maked it necessary for him to rely heavily on Spanish chroniclers, especially Francisco Hernández.

## Zooarchaeological background

Since dogs were the most important animal species in Middle American civilizations, there are many studies mentioning dog remains in archaeologicals excavations,

| Author | Name of dog | Description |
|---|---|---|
| Fray Bernadino de Sahagún (16th century) | *Itzcuintli, Chichi, Xochcocoyotl, Tetlami, Tehiutzol* | Common-looking dog, body with hair, fallen ears, medium-sized. |
| | *Xoloitzcuintle, Tehui* | Hairless dog, brown-colored, fallen or erect ears, medium-sized. |
| | *Tlalchichi* | Short, stout dog |
| Francisco Hernández (17th century) | *Techichi* | Dreary-looking common dog |
| | *Xoloitzcuintli* | Hairless, large-sized, dark color |
| | *Itzcuintepozotli* | Small-sized and small-headed dog, white a short neck and hump. |
| | *Tepeitzcuintli* | Small-sized, black-colored with white neck and head, untamed dog. |
| Glover M. Allen (1920) | Large of Common Indian Dog | Common American breed, with dolicocephalic skull, various colorings, very abundant, medium height, erect ears, lacking first inferior premolar. |
| | Mexican Hairless Dog | Hairless, dark-colored, medium-heigth dog with erect ears. |
| | Small Indian Dog or Techichi | Small-sized dog with long limbs, and a lean, fox-like face |
| | Short-nosed Indian Dog | Short-nosed, small sized dog, with abundant hair and fallen ears. |

***Table 1.*** *Dogs types described by Bernardino de Sahagún (1979), Francisco Hernández (1959), and Glover M. Allen (1920); the basic characteristics of each type are taken from each author.*

although as a rule few published papers go beyond the statement "in such a place dog bones were found," and even fewer still deal with reconstruction of organisms and identification of dog types found in archaaeological sites.

The first real instance of zooarchaeological research in which identified dog types found in archaeological deposits was given by Elizabeth Wing in 1970. Her work covered the study of 42 dogs discovered in Marismas Nacionales, Sinaloa (Fig. 1), and showed little variability in general characteristics, even though sizes varied widely. For this reason, she considered there were two dog types involved. The larger specimens are described as cases of the Common Indian dog (Table 1), and the smaller ones as *Techichis*, basing her classification on Allen's work.

Another interesting example is Nancy L.Hamblin's study (1984) in Cozumel (Fig. 1). Her work deals with the remains of 93 dogs, and she concludes, from her readings of Allen's paper, that they all were of the *Techichi* type, describing them as middle-sized, short-faced animals with complete dental formula.

My own experience is derived from the study of some 500 specimens found in 28 zooarchaeological collections, belonging to 25 archaeological projects in western, central and southeastern Mexico, spanning a chronological period that starts at the seventh century BC and extends to the present (Serra & Valadez, 1985; Valadez, 1992a, 1992b, 1993, 1994a, 1994b, 1995a,1996b, 1998; Valadez, Blanco & Rodríguez, 1998; Valadez, Paredes & Rodríguez, nd; Valadez & Mestre, nd; Valadez, Blanco & Rodríguez, nd) [Besides the published works, there are 38 unpublished technical reports on archaeozoological collections studied by the present author, in which dog remains were included among the material sampled as an important source of data. These reports and part of the collections examined are found in the Paleozoological Laboratory of the Instituto de Investigaciones Antropológicas of the Universidad Nacional Autónoma de México. At the time of writing this paper, there is an inter-institutional project aiming to gather all this new wealth of data in a single publication, in order to describe in detail all the different collections and establish guidelines for all those working on the identification of Middle American dog remains]. Such a wide data base has allowed me to examine in detail the types of dogs associated with each area and each time period. The dog collections found recently in Tula, Hidalgo (Tula, 80-82 project), in Teotihuacan, Mexico (Túneles y Cuevas project), in Punta Pájaros, Quintana Roo (Punta pájaros project), and in *Hunchavin*, Chiapas (*Hunchavin* project) (Fig. 1m), are worth mentioning, since they are particularly rich in remains, containing different types of specimens (Rodríguez, nd; Valadez, 1996b; Valadez, Blanco & Rodríguez, nd).

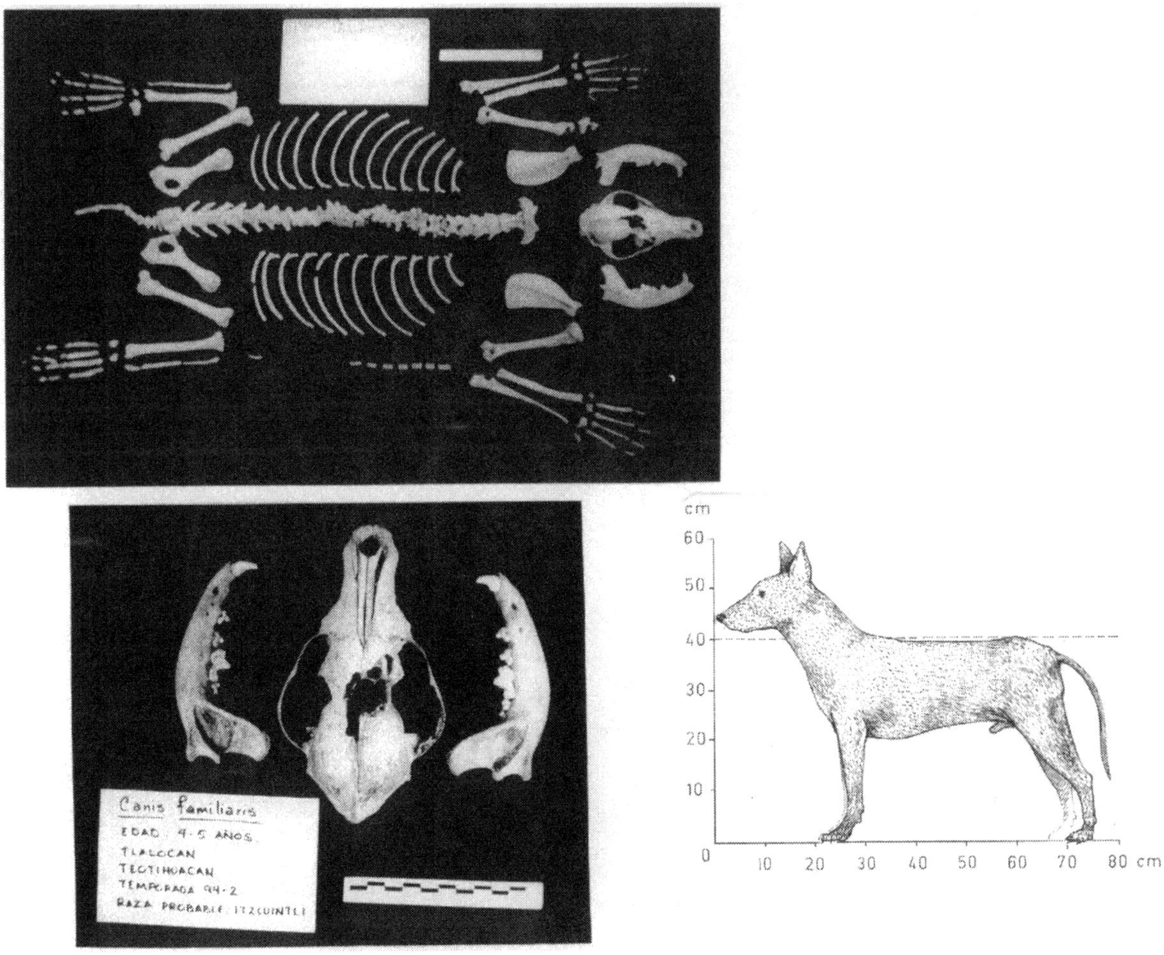


***Figure 1*** *Skeleton, skull and reconstruction of a Mesoamerican common dog. This was the most abundant type throughout the history of Middle America, as well as the forebearer of all other types. The specimen shown was found in a cave in Teotihuacan, and is 1,300 years old.*

Another source of data available is the skeletal collection of the hairless Mexican dog built in the Instituto de Investigaciones Antropológicas in 1989. Skeletal analyses of the *Xoloitzcuintle* in its two varieties (with and without hair), as well as of common dog bones, and of the tooth patterns of the specimens, has allowed me to determine the basic morphological patterns (skeletal and dental) of Mexican present-day dogs (Valadez, 1995a, 1995b) and to compare them to archaeological specimens.

It is important to highlight the relationship between dental formula and dog types in Middle America, especially referring to the Mexican hairless dog or *Xoloitzcuintle*. Since the middle of the twentieth century it has been known that a major feature of these animals is the absence of various teeth, mainly premolars and canines (Wright, 1965). Between 1960 and 1974, the existence of a close relationship between both traits was established (Schnaas, 1974), and it is now known that *Xoloitzcuintles* do not only have an incomplete dental formula, but also a significantly smaller and simpler one (Valadez, 1994a, 1995b). Nowdays this trait (lack of hair) is recognized as the result of a mutation, leading to some type of ectodermic displasia (Bekker, 1989) with repercussions on the development of various organs of ectodermic origin. Hence, lack of hair and an incomplete dental formula are genetically linked traits.

Keeping our zooarchaeological aims in sight, these data lead to the following principle: dog remains showing a complete dental formula (including samples showing loss of teeth due to age or circumstance) indicate specimens that had hair-covered bodies. Those remains showing dental formulas without premolars, and with smaller and simpler teeth (for example, first lower molar with three simple cusps) indicate that the dogs were hairless, that is to say, *Xoloitzcuintles*.

## Dog types in Prehispanic times

Mesoamerican common dog: *Itzcuintli* or Chichi (Sahagún); *Techichi* (Hernández); Common Indian dog (Allen).

This was the most common dog type in Mexican territory. They had dolicocephalic skulls and were of medium size

| TYPES OF DOGS (Meassurement in mm) | | | | | |
|---|---|---|---|---|---|
| | **Element** | **Mesoamerican Common Dog** | **Xoloitzcuintle (Female)** | **Tlalchichi (Male)** | **Short-nosed Indian Dog** |
| **CRANIUM** | Total length | 160 | | | 150 |
| | Facial length | 81 | | | 70.6 |
| | Median palatal lengt | 80.4 | | | 70.0 |
| | Length of entire tooth row | 82.3 | | | |
| | Length of cheeck tooth row | 60.3 | | | |
| | Length of molar row | 16.7 | | | |
| | Length of premolar row | 44.5 | | | |
| | Length of carnassial | 16.0 | | 16.0 | 16.1 |
| | Neurocranium breadth | 52.4 | | | |
| | Zygomatic breadth | 88.2 | | | |
| | Least breadth of cranium | 29.0 | | | 30.8 |
| | Greatest frontal breadth | 44.0 | | | 41.5 |
| | Greatest palatal breadth | 51.5 | | | 59.9 |
| | Maximum breadth of palatal at canine alveoli | 29.5 | | | 33.8 |
| **MANDIBLE** | Total length | 117.9 | 126.0 | | 114.0 |
| | Angular length | 118.2 | 125.7 | | 114.4 |
| | Indentation length | 113.2 | 120.8 | | 111.1 |
| | Indentation canine length | 104.2 | | 98.1 | 99.0 |
| | Tooth row length | 76.8 | | | 71.9 |
| | Molar tooth length | 29.2 | | 31.2 | 28.5 |
| | Premolar tooth length | 37.5 | | 31.6 | 31.3 |
| | Carnassial length | 18.7 | 18.2 | 18.5 | 18.7 |
| | Tickness of horizontal ramus | 22.0 | 21.8 | 17.1 | 21.9 |
| | Heigth of vertical ramus | 48.4 | 51.0 | 44.2 | |
| **LIMB BONES, VERTEBRAL COLUMN** | Humerus length | 128 | 138 | 97 | |
| | Ulna length | 140 | 159 | 112 | |
| | Femur length | 139 | 140 | 114 | |
| | Tibia length | 138 | 151 | 114 | 121 |
| | Atlas-sacrum length | 467.7 | 490 | 468 | 457.8 |

***Table 2*** *Measurements of long bones and vertebrae of specimens appearing in Figures 1-4.*

(*c.* 40 cm shoulder height and 70 cm head-trunk length), although there were specimens leaning to the smaller sizes (35 or 36 cm shoulder height and less than 65 cm long) (Valadez, 1995a, 1996b; Valadez, Paredes & Rodríguez, nd), while others were substantially larger (over 45 cm shoulder height and about 80 cm long) (Wing, 1970). Dental formalae were complete, showing that their bodies were covered with hair (Fig. 1, Table 2).

If one were to look for a distinctive feature of this dog, the clearest one would be its lack of specialization, which suggest it as the perfect common ancestor of all other types (Valadez, 1994b, 1995a). Indeed, it would be hard to find an archaeological site in Mexico in which this dog does not appear; hence, these findings are not only logical, but a must. Looking at the context in which they lie, it is seen that this dog was intensely used in all human activity, since it is found in burial sites as an offering, as well as in many faunistic motifs associated with food.

Such remains are to be found even in the earliest time frame (its existence is documented in sites over 3,000 years-old); therefore, this dog's origin would appear to be linked to that of the civilization of Middle America itself, or perhaps to even predate that event. It should perhaps be considered to be the same type of dog that arrived on this continent more than 8,000 years ago (Valadez, 1995a). Its wide-spread presence throughout Middle America shows records of a long life period for the type, as well as its success at survival.

It would be foolish to speak of archaeological sites where its remains have been found, for there are probably no exceptions to their showing up in sites. Let it suffice to say that this was the only type of dog existing in the Valley of Mexico up to the Posclassic Period (Valadez, 1992b, 1995a; Valadez, 1996b; Valadez, Paredes & Rodríguez, nd), and that all zooarchaeological dog remains found in Teotihuacan (Classic Period) belong to this type (Valadez, 1992a). As mentioned above, I have had the opportunity to analyze remains of some 500 specimens (Valadez, Blanco & Rodríguez, nd); from these, only 15 can be considered to come from other types.

Its proliferation and lack of specialization, which guaranteed its supremacy in the Prehispanic Period, were the reason for the disappearance of the common dog in Colonial times (Valadez & Mestre, nd). Both Sahagún (1979) and Hernández (1959) refer to a type of dog common to the area (*Itzcuintli* or *Chichi*, according to Sahagún; *Techichi*, according to Hernández), but without giving its characteristics. Doubtlessly, the features of this dog were so common that its description did not seem to require any specifics. Other chroniclers altogether ignore it (for the same reasons), and with the arrival of European types it became, instead of a common dog, a "non-entity" dog, until it was concluded that this was a creole type: of American birth but of European origin.

Whether this type of dog survives to this day in Mexico, let it suffice to say that upon leaving urban areas the most likely type of dog you can find is medium-sized, short-haired, dolicocephalic, lacking specialization, having erect or fallen ears, usually single-hued and, in general, with a morphology closely fitted to that of Prehispanic specimens (judging by zooarchaeological remains). All this makes it necesary to conclude that this type still exists and continues to be most prolific.

*Xoloitzcuintle* or Mexican hairless dog.

Although less frequent than the latter, there is no doubt of its Middle American origin, a fortunate fact since from the beginning of this century it has become a symbol of everything concerning dogs and Prehispanic cultures. The remains which have been identified show little change over time for this type: in the Prehispanic Period they were medium-sized (some 40 cm shoulder height, and 70 to 75 cm long), and had dolicocephalic skulls with a rounder cranium than common Middle American dogs, short parietals with prominent bony edges, and a wide post-orbital process. Dentition was limited to incisors, sometimes canines, and molars, all teeth being comparatively small and with a simple morphological structure (Valadez, 1995b). This detailed information has become invaluable to the zooarchaeological field in Mexico, for now it is possible to provide specimen identification through cranium and mandible fragments, and even single teeth. Both sixteenth century descriptions and present-day dogs of this type allow us to conclude that its coloring was rather dark, from dark brown to black.

I have recently conducted some research in iconography and ethnozoology (Valadez, 1996a) which leads to the impression that this type appeared in the western areas of Mexico about 2,000 years ago (Valadez & Mestre, nd), and remained there during half a millenium until the beginning of the *Chichimeca* migrations in the Epiclassic Period. Tula was one of the settlements where traditionally these migrating groups passed through (Fig.6, Map 1), and this city is the site where the first zooarchaeological specimens of the type were found (Tula, 80-82 project), which are also the oldest ones that have been identified to date (1,250-1,350 years old) (Valadez, Blanco & Rodríguez, 1998). Four specimens, two adults and two pups, were identified (Fig. 2).

After the Tula finds, the next record comes from an adult mandible found inside a cave located west of the Pyramid of the Sun in Teotihuacan (Túneles y Cuevas project, Dr. Linda Manzanilla), which was used as a dwelling during the Epiclassic Period (eighth and ninth centuries AD). Finally, the most recent finds belong to an individual discovered in the Punta Pajaros site in the State of Quintana Roo (Fig. 6, Map 1), which was a Mayan ceremonial center from the final phases of the Prehispanic Period (fifteenth century AD).

The number of records and their chronology suggest that after the fifth century AD, *Xoloitzcuintles* scattered constantly, even though slowly and within limits,

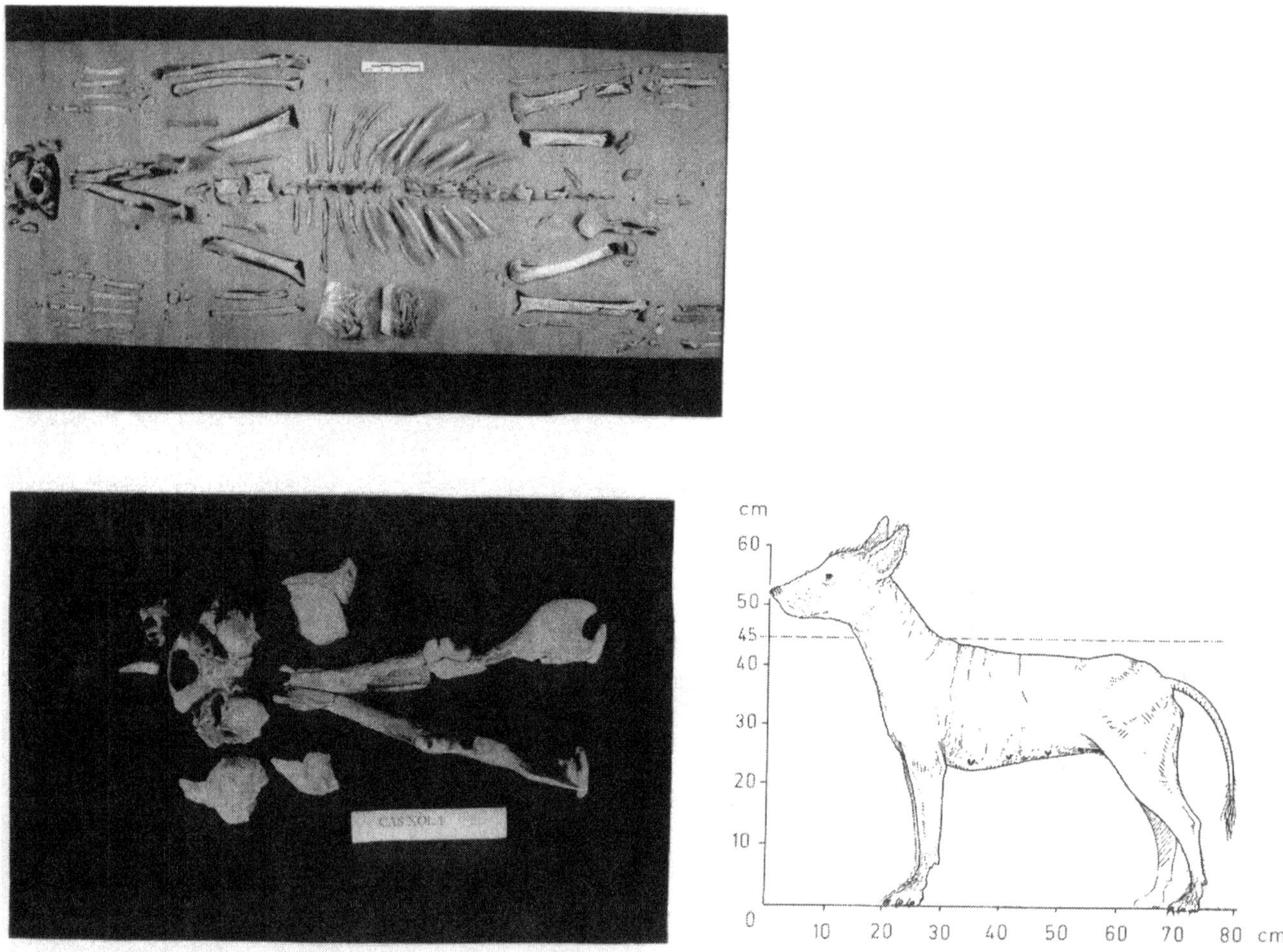


***Figure 2**. Skeleton, teeth, and reconstruction of Xoloitzcuintle found in the city of Tula. The specimen was an adult female of medium height and lived in the seventh century AD.*

throughout Middle American territory. By the end of the first millenium AD they were already known in Central Mexico, and three centuries later they had reached Mayan settlements (Fig. 7, Map 2). The remains show that this type was never very abundant, although probably their scarcity was offset by their impressive physique, which possibly explains the fact that they are the only dogs described by Spanish clergymen and explorers. This suggests, together with the ritual context in which remains have been found, that either these dogs were used exclusively by the elite, or that their use was restricted to special events.

*Tlalchichi*

Amongst the Tula 80-82 project dog collection, a medium-sized adult specimen was found (in a burial site) that has a dolicocephalic skull and complete dentition. It was a little under 70 cm long, not unlike common dogs and *Xoloitzcuintles* also found there, but with a shoulder height of barely 30 cm, because its legs were 30% shorter than the other specimens (Fig. 3). Zooarchaeologist Alicia Blanco and myself have recently begun to study another specimen of the same type found in a Prehispanic burial, which was uncovered at the beginning of this year by archaeologist Eduadro Perez Rivas, near the town of San Blas in the State of Nayarit (Fig. 6, Map 1).

In the *Códice Florentino* (Sahagún, 1979) a "shortish" type of dog named *Tlalchichi* (nahuatl for "floor-dog") is mentioned. If we understand the floor-dog concept as relating to a dog which is "near the floor" because of its short stature or for having short limbs, we may conclude that these remains show the necessary attributes needed to place them as evidence of a third type of Prehispanic dog. As mentioned above, trunk and skull sizes show no evidence of specialization; the short limbs are the only indication that this dog was different from the Mesoamerican common dog. Complete dentition shows that it had body hair.

Since some of these remains were found on the western coast of Mexico (San Blas) (Fig. 6, Map 1) and some in Tula (which as mentioned previously was used as a temporary stopping place by migrants coming from the West), this is clearly the case of a dog originating in the west of Middle America which then dispersed towards Central Mexico (Fig. 7, Map 2). A curious fact is that short-limbed dogs often appear in the iconography of the

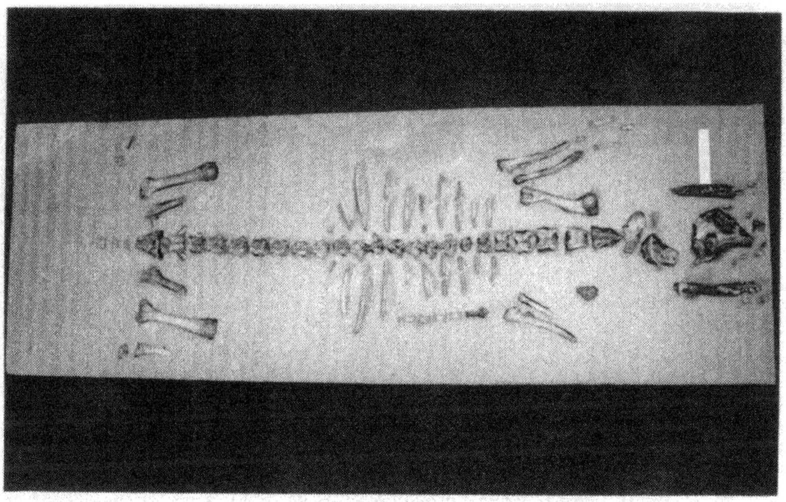

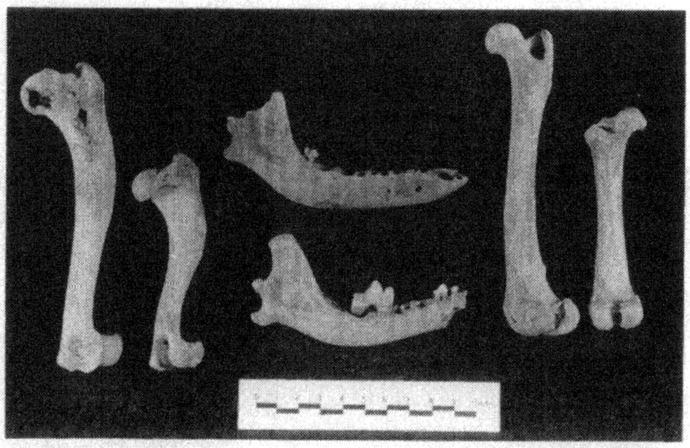

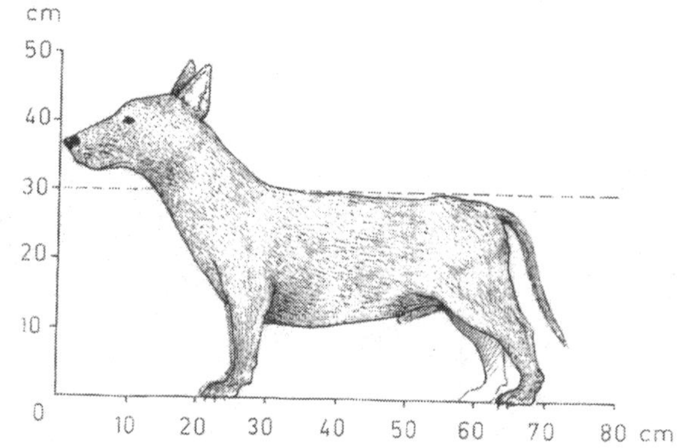


***Figure 3.*** *Skeleton and reconstruction of a short-legged dog (Tlalchichi) from the seventh century AD, found in Tula. The type is clearly identified upon comparison of the humerus, mandibles, and femurs belonging to this specimen with those of the Mesoamerican common dog.*

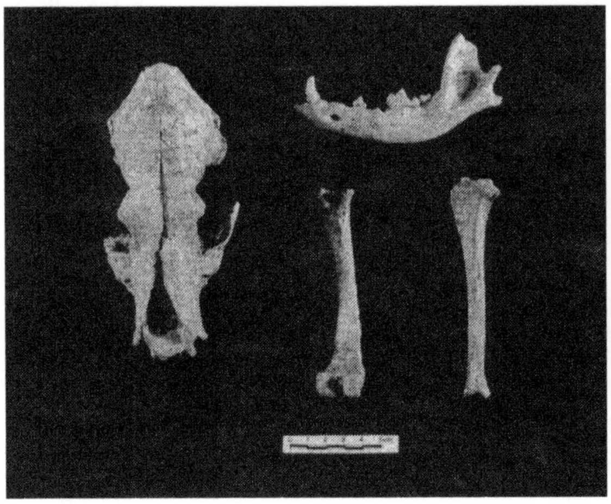

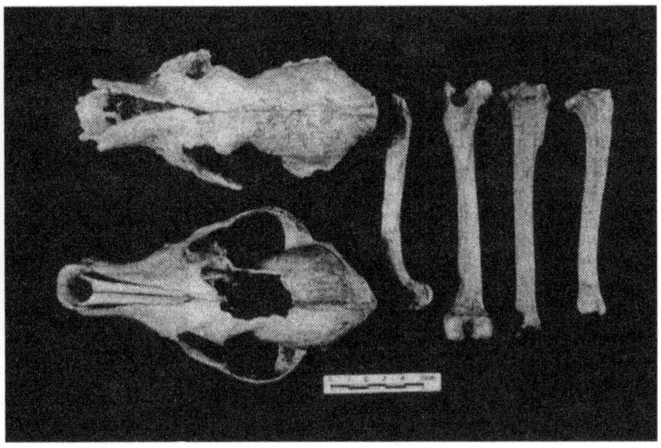

***Figure 4.*** *Left: skull, femur and tibia of a short-nosed Indian dog discovered in Punta Pajaros, Quintana Roo. Right: the same individual, shown compared to the Mesoamerican common dog from Figure 1. The finds of several specimens of this type in Southeastern Mexico suggest that it was restricted to the Mayan area.*

west (1,500 to 2,000 years ago), which reinforces this conclusion about their geographical origin. It must have gradually dispersed through the area after the seventh century AD, although in a moderate fashion, thus explaining its scarcity in archaeological contexts.

Short-nosed Indian dog.

Allen (1920) reports a small-sized dog breed, spread out from the United States to Peru, with the following features: short face, brachycephalic skull, hairy, and erect ears. Archaeologist Alicia Blanco and myself are currently working on two dog collections uncovered in the Mayan area. One of them consists of approximately 40 specimens found in burial sites within the ceremonial center of Chak-Mool in Punta Pajaros, Quintana Roo (Punta Pájaros project, Archaeologist Enrique Terrones), and the other one of ten specimens coming from *Hunchavin*, Chiapas (*Hunchavin* project, Archaeologist Akira Kaneco) (Fig. 6, Map 1). Interestingly enough, in both sites specimens were found (up to this moment, three of these have been identified) whose main features are complete dentition, comparatively short sizes (*c.* 35 cm shoulder height and 60 cm long), and brachycephalic skulls (Fig. 4). Even though these finds are still under study, preliminary data indicate differences in the morphological patterns that are usually found among Mesoamerican common dogs, enough to consider them as representatives of a fourth type of dog restricted to the Mayan areas (Fig. 7, Map 2).

In her study on dogs from Cozumel (Fig. 6, Map 1), Hamblin (1984) reports short-faced dogs. Unfortunately, in her work there are no photographs or drawings that could be compared with archaeological specimens. Nevertheless, the suggested correlation between her findings and ours remains an interesting possibility.

An unresolved case

The descriptions presented above may be considered the most advanced to date, for the data is based on clear and concrete evidence, including bone remains, iconograpic representations and written statements. It is more difficult to propose a new type of Middle American dog based only on a single data source, coming from only one find, even if it consists of bone remains. In cases such as these, it is necesary to conclude that this animal did exist, and maybe we're dealing with an unknown type. However, it is also feasible that this specimen is the result of a genetic anomaly, uncovered due to the fact that "peculiar organisms" (this applied also to human beings) were held in high esteem by these people, and often they were chosen for special ceremonies.

The case at hand refers to a specimen found in a cave close to the city of Cuernavaca, in the State of Morelos (*El Gallo* cave, Dr. Fernándo Sanchez) (Fig. 6, Map 1). Apparently, the site was used for a very special ritual some 2,500 years ago, and the offering partly consisted of a child and this dog. Environmental conditions were so favorable that the remains were very well conserved, and both individuals were completely recovered.

The dog was an adult animal that had body hair, brachycephalic skull, full dentition, 33 cm shoulder height and 42 cm long (Fig. 5), which identifies it as a short-faced small dog, definitely distinct from the afore mentioned cases (Valadez, 1998). Unfortunately, it is a unique find, quite isolated, and too distant in time (25 centuries) from presumable present-day descendants (such as the *Chihuahua* dog) to consider it anything other than a peculiar dog.

**Discussion**

Comparing zooarchaeological data and written sources.

Throughout this century it has been common practice for every author on the subject at hand to adjust his data to information coming from Colonial sources, and this has been detrimental for zooarchaeological research. Usually the researcher moves immediately from bone description to Colonial narrative, making no attempt whatsoever to derive conclusions from the remains themselves. As a consequence of relying on Spanish chronicles of scant scientific value in order to validate scientific data, such as those derived from bone remains, there has been considerable confusion. Only now are we able to revert the trend and do the right thing; that is, speak of Prehispanic types of dogs based on veryfiable data, and then use this information to evaluate the reliability of Colonial sources.

Hernández (1959) reports on four types of dogs: *Xoloitzcuintli* (nahuatl for "wrinkled dog"), a common looking dog called *Techichi* (nahuatl for "little dog"), a hunchback dog called *Itzcuintepozotli* (nahuatl for "hunchback in the guise of a dog"), and a small-sized dog called *tepeitzcuintle* (nahuatl for "mountain dog"). Of these four types, only two fit the zooarchaeological data (*Xoloitzcuintli* and *Techichi*); the others are probably a result of his informant's imagination (Valadez, 1994a).

Bernardino de Sahagún (1979) asserts there are three types of dogs in Central Mexico, and comparing his data to zooarchaeological information, the similarities between the two are very apparent:

- Sahagún describes and illustrates three types of dogs: a common-looking one, with hair, to which he assigns different names; the *Xoloitzcuintle;* and the *Tlalchichi.*

- it can be seen from his descriptions that whenever he speaks of dogs with hair he's speaking of dogs in general, as if they were one and the same, perhaps because they were so plentiful. The *Xoloitzcuintle* gets a detailed description and is treated as an interesting case, but the account is contradictory. It is as if his informants as well as himself knew of these dogs without being permitted direct access to them (either because of their rarity or because only certain people could have them). Finally, Sahagún declares that the *Tlalchichi* is a real and well-defined type,

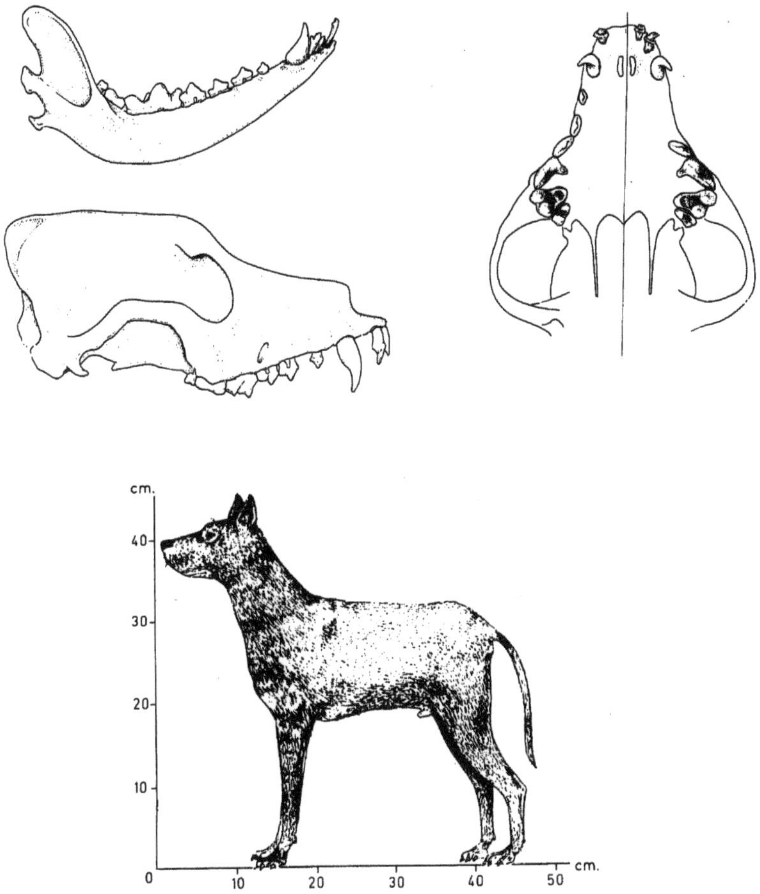


***Figure 5.** Skull, mandible and reconstruction of the specimen found in the "El Gallo" cave, near Cuernavaca, Morelos. Even though this a totally different dog from all of those know to date, this unique find suggests it might be a representative of an unknown type, if not a genetic anomaly.*

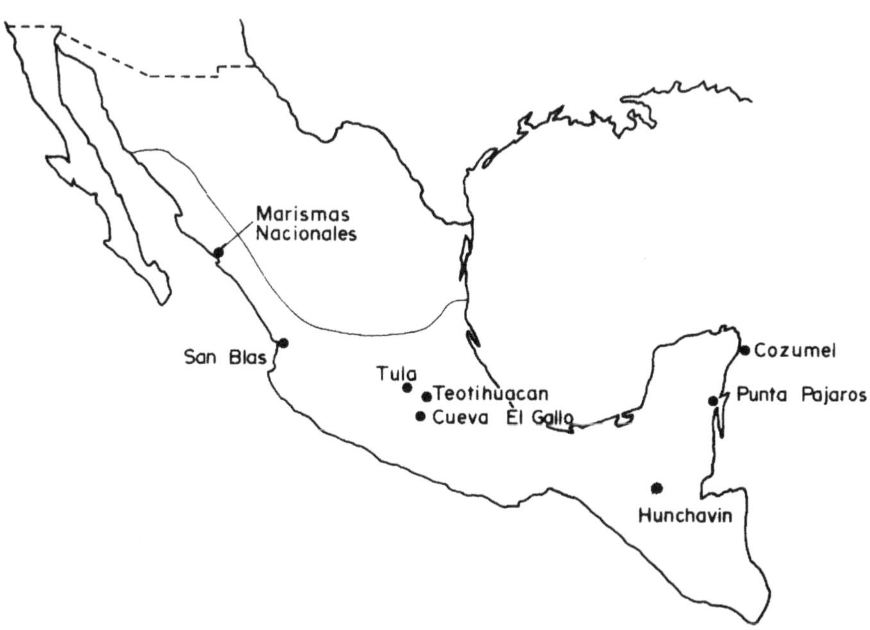


***Figure 6**. Map 1. Areas of maximum spread of Middle American civilization (eleventh to sixteenth centuries AD), and archaeological sites with dog remains mentioned in this paper.*

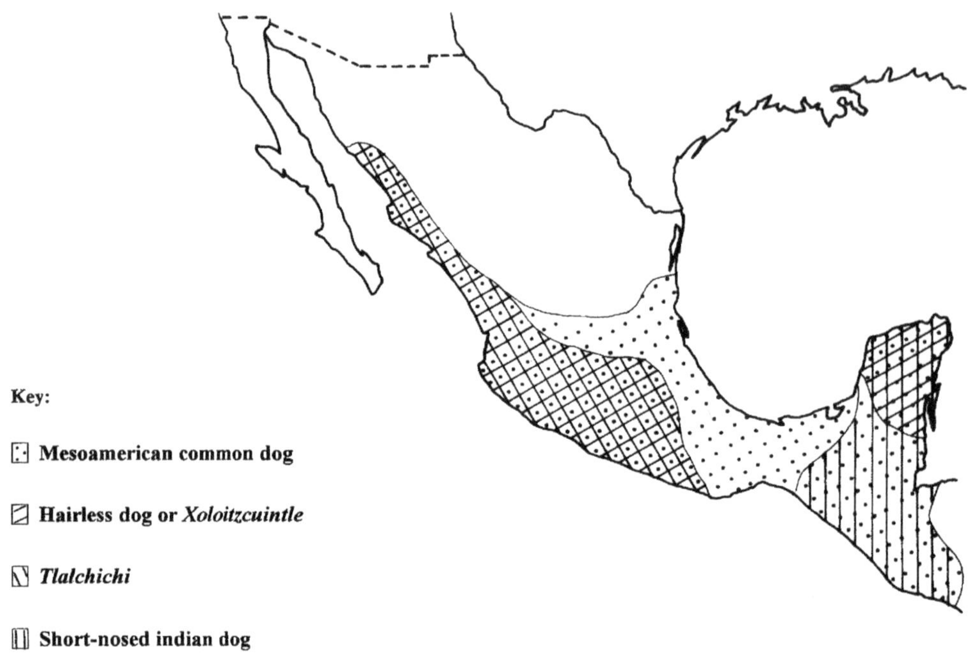


***Figure** 7. Map 2. Likely distribution area of the four types of dog described within Middle American territory at the end of the Prehispanic Period.*

but he only manages to say a couple of sentences about them.

- the three types described by the clergyman correspond to the three types in zooarchaeological records of Central Mexico. There is no mention of a short-nosed type; this type, however, belongs to the Mayan area.

Even if the obvious conclusion is that Sahagún's information is best fitted to reality, it is a tragic realization that all latter authors approaching the subject of Middle American dogs, including Allen, decided that Hernández was to be held as the best data source, thus encouraging the trend well into this century.

Finally, it is important to point out that in Allen's work (1920) the type of living organism denominated by the word *Techichi* is discussed at length. At the time, there was already a large list of possibilities, from the *Chihuahua* dog to those reporting this was the word for racoon (*Procyon lotor*). Since Hernández described this animal as a "dreary-looking dog, otherwise of common appearance," it all points to a general unawareness of the fact that he was referring to a dog common to *Nueva España*. The confusion derived from this reached Allen, who was then incapable of understanding that this *Techichi* and his common Indian dog were one and the same. Instead, he classified it as a long-limbed dog with a fox-like face, of which he had some reports coming from South America.

From that time on, the word *Techichi* had a two-fold meaning, according to which author one was reading. Allen's term is logically used by Wing and Hamblin in their papers, although it is rather peculiar that they use it to name small short-faced dogs.

Middle American dog genealogy and human interests.

The second relevant insight to be derived from this information is the manner in which the diversification process occurred, that is, the first appearance of the known different types of dogs throughout the history of Middle America.

Both due to its abundance and antiquity, the Mesoamerican common dog would seem to be the common ancestor of all latter forms, and the direct descendant of the dogs arriving to this continent more than 8,000 years ago (Valadez, 1995a).

We know that hairless Mexican dogs first appeared in the West about 20 centuries ago, and it might be that the *Tlalchichis* originate at that time. Short-nosed dogs seem to belong exclusively to the Mayan area, even though it is still not possible to date its origin. The fact that the three types are present at both ends of the Middle American territory shows that cultural isolation and geographical factors contributed to their apparition, while later they proliferated in the degree that they had dominant traits. Whether or not these people were breeding dogs we cannot know, but it does not seem likely in view of the very few existing samples of dog types different from the common dog. It is also possible that genotype manipulation was aimed at traits that do not appear in the zooarchaeological records, such as behavior, color or type of hair.

### Conclusions

During the last ten years, zooarchaeological research conducted in Mexico, backed by previous studies, allow us at the present time to recognize several forms of dogs existing in Middle America before the Spanish conquest. Although some cases are definitely better documented than others, we can speak now of four well-characterized forms, even though quite dissimilar in terms of their abundance and their area of distribution.

Mesoamerican common dogs and *Xoloitzcuintles* are doubtlessly the two best defined types, since there is supporting zooarchaeological, iconographic, ethnohistorical and biological evidence. The other two instances, the *Tlalchichi* and the short-nosed Indian dog, are backed by less evidence, but the main issue here is that there are bone remains which permit their characterization, enough to consider them a real type rather than unique events.

As progress is made in this field and new evidence is obtained, the need to clarify the following point must be more forcefully proposed: was the development of various types of dogs a random event, with minimal human intervention, or was it a guided process of which, as yet, we have no knowledge? Fortunately, Mexican archaeologists seem to have an increasing interest for zooarchaeological research. If this trend continues, it is possible that in a few years we shall be able to reconstruct more firmly the history of this species in this part of the American territory.

### Acknowledgements

I am thankful for the support of the biologist Bernardo Rodríguez Galicia in the writing of this article, and to archaeologist Alicia Blanco Padilla for allowing me to use materials presently under study. Drawings of dogs by César Fernández, photographs by José Saldaña.

### Appendix: Specimen accession data for figures 1-4.

**Figure 1**. Mesoamerican common dog.

<u>Project:</u> Tuneles y Cuevas (Instituto de Investigaciones Antropológicas de la Universidad Nacional Autónoma de México (IIA de la UNAM), Dr. Linda Manzanilla).

Ceremonial Burial in the "Pirul" cave, IX century AD.

<u>Archived</u> at Paleozoology Laboratory, IIA de la UNAM. México City.

**Figure 2**. Hairless Mexican dog.

<u>Project.</u> Tula 80-82 (Instituto Nacional de Antropología e Historia (INAH).

Domestic burial in "Plaza Charnay" area, VII-VIII century AD.

<u>Archived</u> at Paleozoology Laboratory, IIA de la UNAM. México City.

**Figure 3**. *Tlalchichi*.

<u>Project.</u> Tula 80-82 (Instituto Nacional de Antropología e Historia (INAH).

Domestic burial in the "Tunel Falso" area, VII-VIII century AD.

<u>Archived</u> at Paleozoology Laboratory, IIA de la UNAM. México City.

**Figure 4**. Short-nosed Indian dog.

<u>Project</u>. Punta Pájaros (Dirección de Salvamento Arqueológico del INAH, Enrique Terrones).

Ceremonial burial in ceremonial center of *Chaak-Mool,* XV century AD.

<u>Archived</u> at Sección de Biología de la Dirección de Salvamento Arqueológico del INAH. (Alicia Blanco Padilla). México City.

### References

Allen, G. M. (1920). *Dogs of the American Aborigines*. Cambridge: Harvard University Bulletin of the Museum of Comparative Zoology **63**.

Bekker, C (1989). *El xoloitzcuintle, ¿Un modelo de displasia ectodérmica autosómica dominante?* Biology Thesis, Guadalajara, Jalisco.

Hamblin, N. L. (1984). *Animal use by the Cozumel maya*. Tucson: University of Arizona Press.

Hernández, F. (1959). *Historia Natural de las Cosas de la Nueva España. Obras Completas, Tomo III, tratado quinto.* Mexico City: Universidad Nacional Autónoma de México.

Rodríguez, B. (nd). *Cánidos silvestres y domésticos descubiertos en cuevas teotihuacanas.* Master's Thesis, Universidad Nacional Autónoma de México.

Sahagún, B. (1979). *Códice Florentino.* Mexico City: Secretaría de Gobernación **III(libro 11)**.

Serra, M. C., & Valadez, R. (1985). Fauna de Terremote-Tlaltenco, D.F. *Universidad Nacional Autónoma de México Anales de Antropología* **XXII**, 159-213.

Schnaas, G. (1974). El perro pelón, mito, fantasía y biología. *Gaceta Médica de México*, **108**, 393-400.

Valadez, R. (1992a). *Impacto del recurso faunístico en la sociedad teotihuacana.* Ph..D. Thesis, Universidad Nacional Autónoma de México.

Valadez, R. (1992b). Fauna identificada en excavaciones de Temamatla, Estado de México. *Antropológicas* **6**, 69-84.

Valadez, R. (1993). Macrofósiles faunísticos. In (L.Manzanilla, Ed). *Anatomía de un conjunto residencial teotihuacano en Oztoyohualco*. Mexico City: Universidad Nacional Autónoma de México **II**, pp. 729-813.

Valadez, R. (1994a). ¿Cuántas razas de perros existieron en época prehispánica? *Veterinaria México* **25**, 1-12.

Valadez, R. (1994b). El perro prehispánico. *Universidad de México* **L (528-529)**, 15-20.

Valadez, R. (1995a). *El perro mexicano*. Mexico City: Instituto de Investigaciones Antropológicas, Universidad Nacional Autónoma de México.

Valadez, R. (1995b). Morfología dental del perro pelón mexicano. *Veterinaria México* **26**, 317-332.

Valadez, R. (1996a). ¿Qué es qué en las figurillas de perros de Colima? In (L. Manzanilla, Ed) *Memorias de la XXIV Mesa Redonda de la Sociedad Mexicana de Antropología* **2**. (Antropología e historia del occidente de México), Mexico City: Editorial Colofón, pp 379-390.

Valadez, R. (1996b). Perros descubiertos en entierros, Cueva del Pirul, Temporada 95. On file, Instituto de Investigaciones Antropológicas de la Universidad Nacional Autónoma de México, Mexico City.

Valadez, R. (1998). Un atípico caso de un típico ejemplo de un perro descubierto en un entierro prehispánico. *Asociación de Médicos Veterinarios Especialistas en Pequeñas Especies* **9**, 99-104.

Valadez, R., Blanco, A., & Rodríguez, B. (1998). Restos arqueozoológicos de xoloitzcuintles: 1994-1998. *Asociación de Médicos Veterinarios Especialistas en Pequeñas Especies* **9**, 205-212.

Valadez, R., & Mestre, G. (in press). *Historia del xoloitzcuintle en México*. Mexico City: Instituto de Investigaciones Antropológicas de la Universidad Nacional Autónoma de México, Museo Dolores Olmedo Patiño, Cámara de Diputados.

Valadez, R., Paredes, B., & Rodríguez, B. (in press). Entierros de perros descubiertos en la antigua ciudad de Tula, Hidalgo. *Latin American Antiquity*.

Valadez, R., Blanco, A., & Rodríguez, B. (nd). *El estudio del perro en la arqueología mesoamericana: Metodología, aplicaciones y reconstrucción de la historia del perro en México*. Mexico City: Instituto de Investigaciones Antropológicas de la Universidad Nacional Autónoma de México.

Wing, E. (1970). Dogs remains from the Marismas Nacionales. Paper presented at the Society of American Archaeology conference, Mexico City.

Wright, N. (1965). *El enigma del xoloitzcuintle*. Mexico City: Instituto Nacional de Antropología e Historia.

# PRELIMINARY RESULTS IN DETERMINING DOG TYPES FROM PREHISTORIC SITES IN THE NORTHEASTERN UNITED STATES

Brent M. Handley
University of Connecticut,
Department of Anthropology,
Storrs, Connecticut, USA
also
PAL, Inc. Pawtucket, RI USA

**Key words:** dogs, breeds, osteometry, archaeozoology, Maritime prehistory, North America

**Abstract**

This paper reports on the preliminary results of osteometric analysis of dog remains recovered from prehistoric sites in the northeastern United States. Two distinct dog types have been historically documented: one described as the size of a spaniel and the other reported as wolf-like in appearance. In order to assess the prehistoric existence of these dogs, and to be able to assign archaeological specimens to specific categories, 115 measurements were taken from the cranium, and post-cranial elements. Data from over forty specimens lend credence to speculation that more then one type of indigenous dog existed in northeastern North America before European contact.

## Introduction

This research reports preliminary findings from osteometric analysis of dog remains, *Canis familiaris,* from North American prehistoric sites located throughout New England and Long Island (Fig. 1). The goal was to distinguish the different types of dogs believed to have existed and interacted with Native Americans in this region (*cf,* Blick, 1988; Crockford, 1997; Howard, 1949; Krantz, 1959; Morey, 1986; 1992; Shigehara *et al.,* 1997). In this report the term "types" should not be confused with distinct breeds. Archaeologically breeds have been defined as the deliberate intervention by humans to procure certain traits (Crockford, this volume). Historic documents pertaining to dogs in the northeastern United States do not suggest different traits being exploited. In the data presented here distinctions between types of dogs will be based on size differences only.

*Canis familiaris* has been recovered in archaeological contexts throughout the region of the eastern United States and Canada known as the Maritime Peninsula. Many of the remains have been recovered from middens. These disarticulated bones are often interpreted as waste from the butchery process (Kerber *et al.*, 1989; Wyman, 1868). Burials are another context from which dog remains are recovered (Bourque, 1995; Kerber *et al,* 1989). The majority of documented burials are small pit features containing a single dog. Also recorded, but less frequently, are multiple dog burials and those accompanying human remains (Strong, 1985).

Both of these contexts have generated interest in the treatment and the role of dogs within indigenous cultures (Kerber *et al*, 1989; Kerber, 1997; Schwartz, 1997). Some of these uses include: pet, hunting aid, guard, and ceremonial median. In determining these roles, archaeologists use historic accounts as an aid in deciphering prehistoric cultural patterns. Ethnographic accounts have not only contributed understanding about the role and treatment of domestic dogs, they also provide insight into different types that were present.

Throughout the Maritime Peninsula, documentary evidence suggests at least two distinct dogs accompanied Native peoples. The first is described as a small dog, about the size of a spaniel (Roiser, 1906). This type of dog is often referred to as the "canoe dog". A specialty of the Wabanaki Tribe (Morrison, 1990), these small retrievers were excellent swimmers and thought to be used in hunting game birds. The second type was described as "wolf-like" in appearance (Rosier, 1906). This particular dog was often confused with *Canis lupus* due to its larger size and long hair (Wolley, 1860; Butler & Hadlock, 1948).

Archaeological research conducted along the coast of Maine in the early part of this century revealed a third type of dog. Loomis and Young (1912) recovered nearly sixty individuals they identified as *Canis familiaris.* Although the remains were fragmented, Loomis and Young proceeded to compare the various elements, coming up with three dog types; the Major Indian dog, the Common Indian dog, and lastly the smaller Minor Indian dog.

Loomis and Young did not disclose parameters for their research. It is unknown what demographic information was gathered on the specimens before osteometric

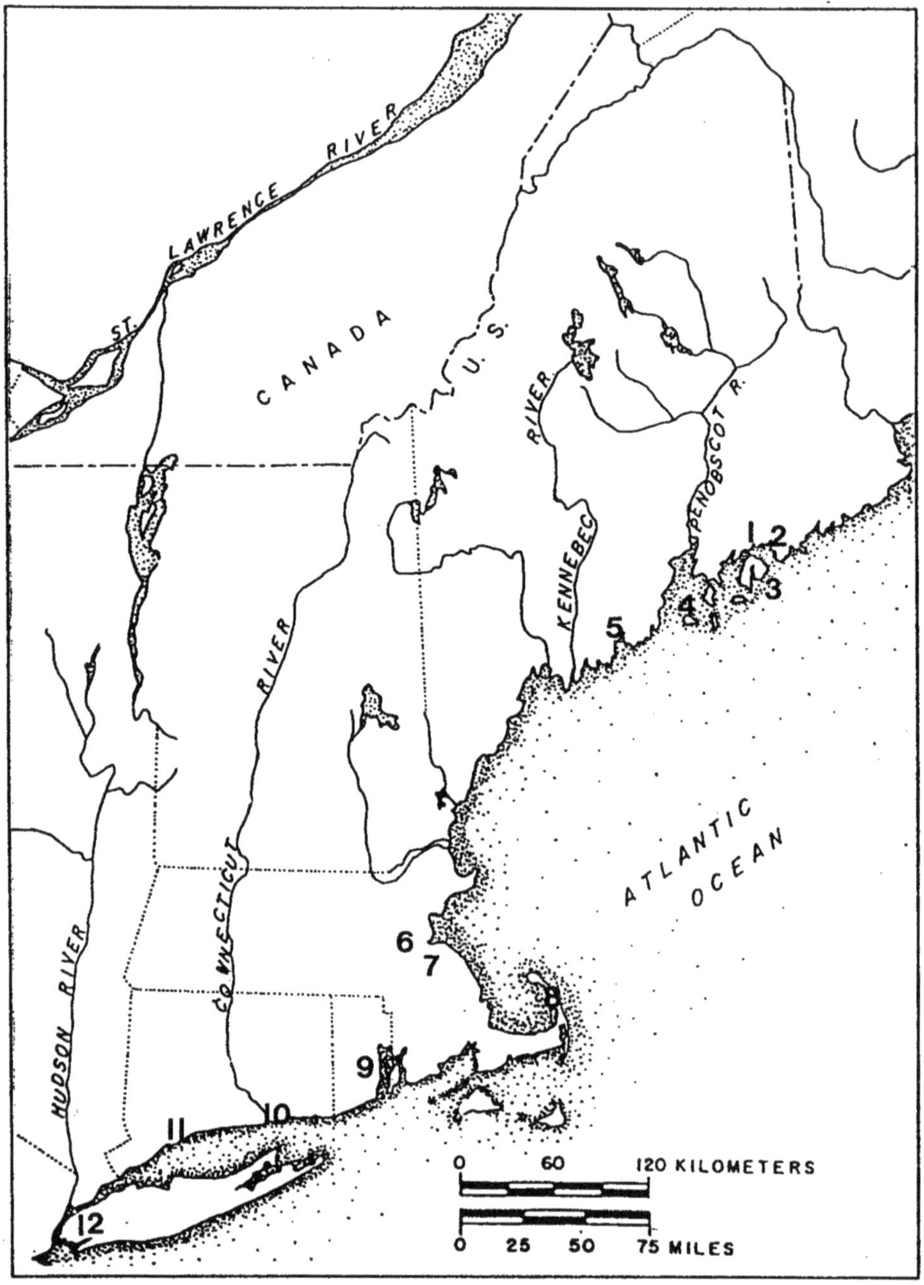


***Figure 1*** *Map of Northeastern United States with sites discussed in text. (1) Neven (2) Frenchman's Bay (3) Ruth Moore (4) Turner Farm (5) Whaleback Shell Heap (6) Watertown Arsenal (7) Squantum (8) Ryder Beach (9) Lambert Farm (10) Tubbs Shell Heap (11) Granniis Island (12) Port Washington*

analysis. Size comparisons of elements require certain criteria to be met in order to make inferences about one or more populations. Crockford (1997) concluded that without determining sex or age for the specimens, there will be a greater chance for inaccurate information. Determining these traits proved to be as important to this study as the metrics taken from each specimens.

## Materials

Over sixty dogs were made available for this analysis. The samples come from twelve sites located in Maine, Massachusetts, Rhode Island, Connecticut, and New York's Long Island (Fig. 1). All of the dogs represent burials recovered within the last 112 years of archaeological excavation. These were chosen over disarticulated remains because of the established criteria for selection, discussed below, which made it difficult to include fragmented bones.

The dog burials used in this analysis range in age from 4,000-6,00BP, corresponding with Northeastern cultural traditions known as the Late Archaic to Late Woodland periods (Tuck, 1978; Fitting, 1978). Within this 3000-year period, there seems to be no discernible pattern within mortuary practices involving dogs. Sites from both the Archaic and Woodland periods feature dogs buried in pit features, sometimes in the company of a human burial. There is a higher frequency of dogs associated with human graves in the Woodland traditions. This is also the same cultural period from which fragmented dog remains are often recovered from middens.

The sites with dog burials were all located along the coast. This does not necessarily suggest that *Canis familiaris* only occupied coastal regions. The features were all associated with shell, which has a neutralizing effect on the acidic nature of the soils in the Northeast. Some of the best bone preservation in this region is credited to its association with shell middens.

Despite burial in shell middens, the condition of many of the dog remains proved to be a problem. Many of the remains were recovered from shallow pits, and a large number of skeletons were fragmented. This limited the number of measurable elements.

| Elements | Epiphyseal union in weeks | |
|---|---|---|
| | Beginning | Closing |
| Humerus | | |
| proximal | 39 | 59 |
| distal | 21 | 33 |
| Radius | | |
| proximal | 35 | 47 |
| distal | 39 | 47 |
| Ulna | | |
| proximal | 29 | 37 |
| distal | 39 | 47 |
| Femur | | |
| proximal | 26 | 47 |
| distal | 37 | 52 |
| Tibia | | |
| proximal | 32 | 52 |
| distal | 37 | 47 |

***Table 1*** *Aging criteria for dog postcranial elements, from Gilbert (1993)*

| Tooth | Tooth eruption in weeks | |
|---|---|---|
| | Beginning | Closing |
| Incisor 1-3 | 12 | 22 |
| Canine | 21 | 24 |
| Premolar 1 | 16 | 20 |
| Premolar 2-4 | 21 | 24 |
| Molar 1 | 16 | 20 |
| Molar 2 | 21 | 24 |
| Molar 3 | 24 | 26 |

*Table 2 Aging criteria for dog tooth eruption, from Gilbert (1993)*

**Age of death**

Determining the level of maturity for an individual dog is an important procedure. Long bones and other elements with articulating surfaces have an epiphysial plate which fuses once maturity has been reached (Table 1). Teeth are another concern with regard to age. Although dogs replace their deciduous teeth at an early age, they do not grow to adult size until maturity (Table 2). Since juveniles to sub-adult dogs have not fully developed they were excluded from this analysis and only specimens that could be determined to be fully developed adults were measured. The methods used to make this determination include epiphysial union of the long bones and other articulating surfaces, in addition to the presence of permanent teeth as described by Klein and Cruz-Uribe (1984), Gilbert (1993), and Silver (1969).

The total sample of dog skeletal remains consisted of sixty-one individuals, forty-one of which were determined to be fully developed adults. The remaining specimens were assigned as sub-adult, young adult, or age indeterminate, and therefore not considered for this analysis.

**Sex Determination**

The next most crucial determination to be made was sex of individuals. Sexual dimorphism is quite common among most mammals and the domestic dog is no exception. Making this determination was important so as not to skew the data when attempting to ascertain distinct characteristics that might differentiate dog types based on size.

Determining sex from one or more populations is difficult without a living sample from which sexual dimorphism can be assessed. Studies have been conducted on a Japanese dog breed believed to represent descendants of a common prehistoric type (Shigehara *et al*, 1997). However, in the northeastern United States this was not possible, since no living samples are believed to represent descendants of the domestic dogs recovered from prehistoric contexts.

In order to minimize potential problems due to sexual dimorphism, several morphological traits were used in sex determination of prehistoric material. Besides a general size difference, only male dogs possess an *os penis* bone. Other sexual dimorphic traits include cranial shape and the form of the ascending ramus of the mandible (Shigehara *et al*, 1997). Male dogs tend to have a thicker sagitital crest than females. It has also been observed that the condyloid ridge, which forms the inferior margin of the masseter fossa of the mandible, is more acute in males. In females this ridge gently shifts downward.

Only one of the specimens used in this analysis possessed an *os penis* bone. However, using non-metric features twenty-seven dogs could be assigned to sex: eighteen males, and nine females (Table 3). Many of the dogs that

| Sample | Sex | Site | Cultural Affiliation | References |
|---|---|---|---|---|
| 001 | M | Tubbs | Late Woodland | Bellantoni, nd. |
| 002 | F | Granniss Is. (D36) | Archaic-Woodland | Thompson, nd. |
| 004 | M | Granniss Is. (R23) | Archaic-Woodland | " |
| 005 | F | Granniss Is. (B30) | Archaic-Woodland | " |
| 011 | M | Nevin Site:1 | Late Archaic | " |
| 012 | M | Nevin Site:2 | Late Archaic | " |
| 013 | F | Meeting House: 1 | Late Woodland | Yentsch *et al.*, 1979 |
| 015 | M | South Truro | Woodland | Moffet, 1949 |
| 016 | M | Squantum (Quincy) | Middle Woodland | Dincauze, 1973 |
| 017 | F | Watertown: 1 | Archaic-Woodland | Guernsey, nd |
| 018 | M | Watertown: 2 | Archaic-Woodland | Guernsey, nd |
| 019 | F | Frenchman's Bay: 1 | Archaic-Woodland | Hadlock, nd |
| 025 | M | Turner Farm: Fea. 26 | Late Archaic | Broque, 1995 |
| 027 | F | Turner Farm: Fea. 36 | Late Archaic | " |
| 028 | M | Turner Farm:W55-575 | Woodland | " |
| 030 | M | Port Washington: 1 | Woodland | Harrington, 1982 |
| 031 | M | Port Washington: 2 | Woodland | " |
| 032 | F | Port Washington: 3 | Woodland | " |
| 033 | M | Port Washington: 4 | Woodland | " |
| 034 | F | Port Washington: 5 | Woodland | " |
| 035 | M | Port Washington: 6 | Woodland | " |
| 036 | F | Port Washington: 7 | Woodland | " |
| 037 | M | Lambert Farm | Late Woodland | Kerber *et al.,* 1989 |
| 038 | M | Ruth Moore | Late Archaic | Lewis, nd. |
| 039 | M | Whaleback Shell Heap | Woodland | Gamage, 1886 |
| 040 | M | Whaleback Shell Heap | Woodland | Gamage, 1886 |
| 041 | M | Whaleback Shell Heap | Woodland | Gamage, 1886 |

*Table 3 List of all adult dog specimens discussed in text*

| Metric Codes | Number of Cases | Minimum | Maximum | Mean | Standard Deviation |
|---|---|---|---|---|---|
| 1 | 10 | 163.40 | 213.00 | 186.17 | 17.20 |
| 2 | 13 | 47.40 | 61.80 | 52.83 | 4.08 |
| 3 | 12 | 62.10 | 72.20 | 65.33 | 2.98 |
| 5 | 9 | 46.50 | 63.80 | 53.78 | 5.66 |
| 6 | 10 | 33.15 | 48.10 | 38.16 | 5.35 |
| 9 | 10 | 32.68 | 48.90 | 37.21 | 5.58 |
| 10 | 11 | 58.00 | 77.90 | 63.28 | 6.84 |
| 12 | 7 | 25.62 | 36.30 | 29.99 | 4.43 |
| 13 | 15 | 16.10 | 19.50 | 17.86 | 1.06 |
| 14 | 14 | 6.32 | 10.20 | 7.70 | 1.09 |
| 15 | 15 | 10.30 | 13.40 | 11.76 | 0.90 |
| 16 | 15 | 13.70 | 18.10 | 15.30 | 1.18 |
| 30 | 11 | 135.30 | 171.00 | 156.23 | 10.85 |
| 31 | 13 | 27.90 | 44.80 | 31.64 | 4.39 |
| 35 | 10 | 144.30 | 195.00 | 160.37 | 14.39 |
| 36 | 9 | 28.57 | 38.90 | 30.80 | 3.10 |
| 37 | 9 | 32.30 | 44.80 | 36.80 | 3.60 |

*Table 4 Measurement statistics for male dog specimens (mm)*

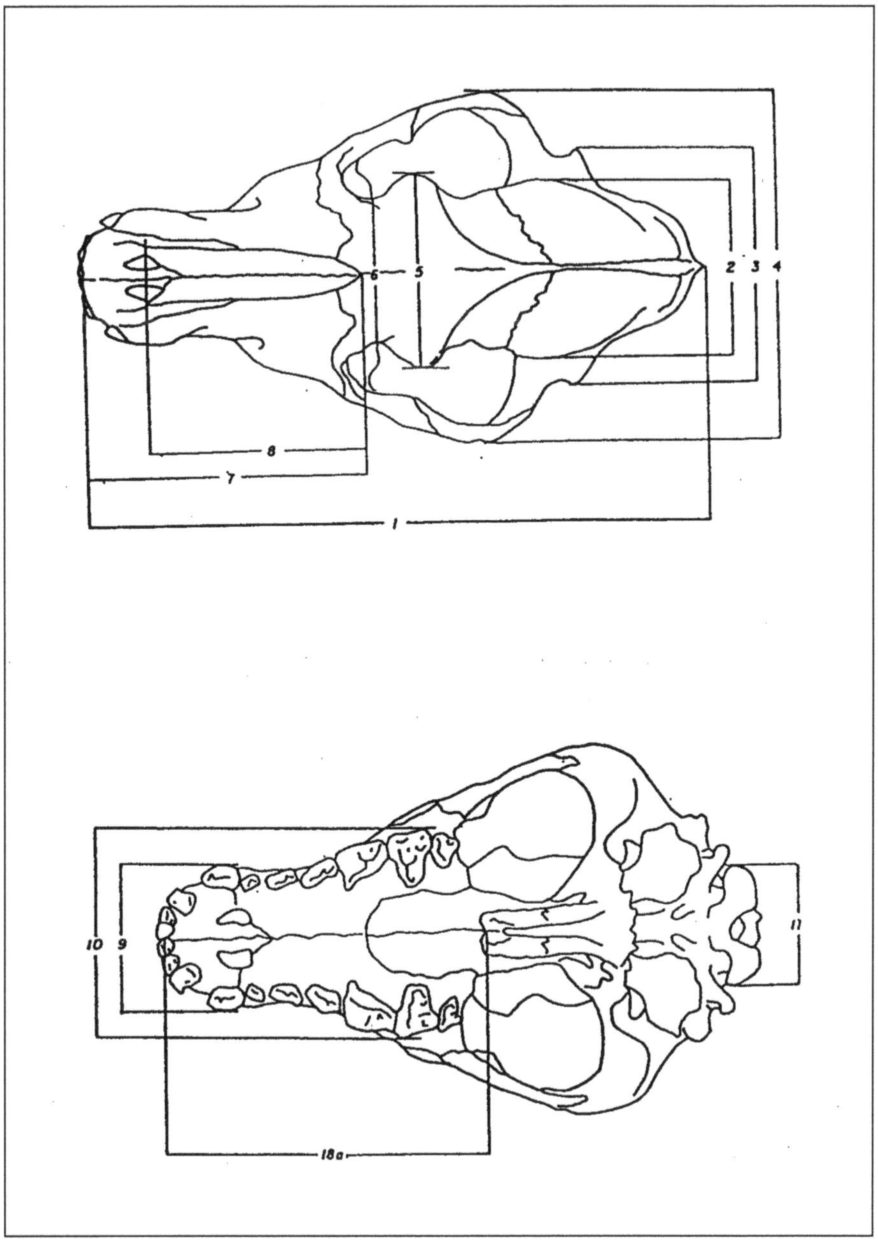


***Figure 2*** *Cranial diagram with metric codes indicated*

could not be assigned to one sex or the other were fragmented or had missing elements. Even though this lowered the sample size, it was deemed necessary to keep the sexes separated for the purpose of the study.

**Osteometry**

This analysis incorporates 115 measurements from cranial and post-cranial elements. The measurement methodology follows von den Driesch (1976), Blick (1988), Crockford (1997), and Morey (1986, 1992). For the purpose of this presentation only the measurements from the cranium, mandible, and selected long bone elements will be discussed (Fig. 2-4). The measurement descriptions used, with numeric codes assigned to them, are listed below.

1. Total cranial length
2. Greatest breadth of brain-case
3. Greatest breadth of mastoid
5. Frontal breadth
6. Least breadth between orbits
9. Breadth of canine alveoli
10. Greatest palatal breadth
12. Greatest inner height of orbit
13. Length of upper premolar 4
14. Width of upper premolar 4
15. Length of upper molar 1

16. Width of upper molar 1
30. Greatest length of humerus
31. Breadth of humerus distal end
35. Greatest length of femur
36. Breadth of femur distal end
37. Breadth of femur proximal end

Even with the small sample size, the large number of variables produced a sizable data set. To assess these variables, univariate analyses were applied (mean, maximum, minimum, and standard deviation). The assessment of the univariate variables was an initial procedure. Variables were chosen for further analyses based on number of specimens with significant statistics.

In order to interpret a data set of this nature several multivariate analytical methods could be applied. For this preliminary report, only scatter-plots of actual measurements have been produced. Statistically, plotting of variables is generally considered an initial step only (Drennan, 1996). Other methods used to interpret data pertaining to more than one population include discriminate function and cluster analyses. These methods require a larger sample than was available from this study in order to be significant. Until further specimens can be added to the data set univariate analyses and multivariate scatter-plots will be the methods utilized.

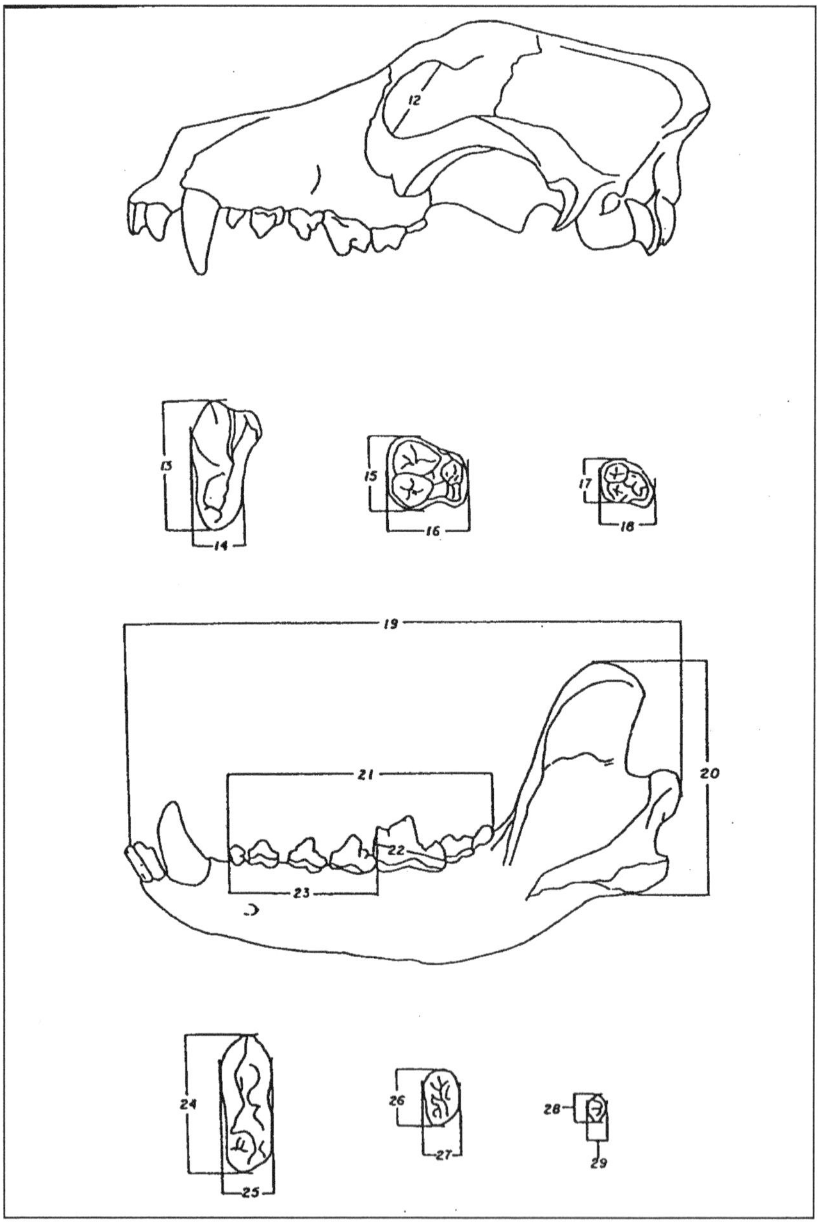


***Figure 3*** *Cranial and mandible diagram with metric codes indicated*

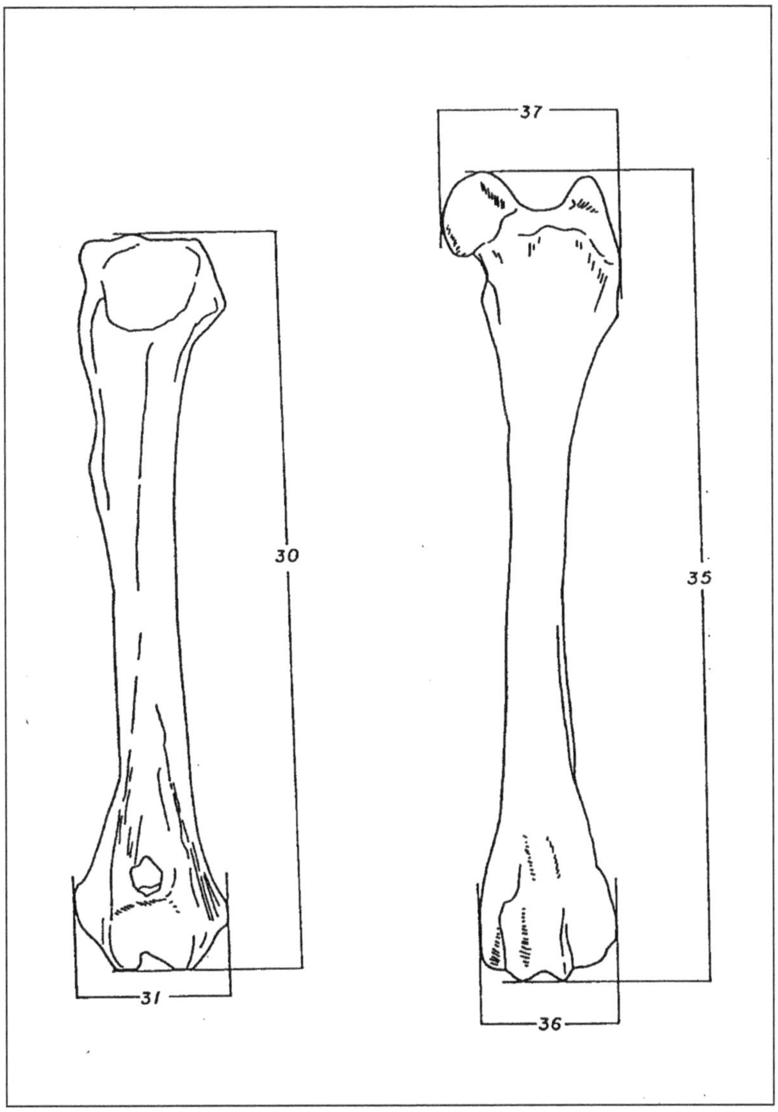


***Figure 4*** *Humerus (left) and femur (right) diagram with metric codes indicated*

**Results**

Although the number of individuals is low, results of the analytical methods used suggest that at least two distinct types of dogs are present. The larger number of adult male specimens proved to be the best in determining different types of dogs. To date twelve variables have been processed (Table 4), with number of cases ranging from seven to fifteen. By far the most representative elements with measurable dimensions, with regard to number of cases, were the teeth (#13-#16). These elements were almost always present. Other dimensions with high numbers of cases include maximum length of skull (#1), greatest neurocranium breadth (#2), and greatest mastoid breadth (#3).

Plot graphs using a combination of the dimensions revealed interesting results. Maximum cranial length (#1) plotted against greatest mastoid breadth (#3) suggests two distinct groupings (Fig. 5). The other plots for the male dog specimens, using upper tooth measurements, have a similar pattern (Fig. 6-7). The exception for this is maximum length (#13) and width (#14) of $P^4$. This graph has the largest number of cases, and shows a less obvious concentration. In fact a third grouping could be suggested from this plot.

What makes the male graphs interesting are the two points in the upper right-hand corner. In all of the graphs, these two points represents the same two dogs, specimens 016 and 018. These specimens were recovered from two sites on the periphery of the Greater Boston area (Fig. 1). Observations made while measuring indicated the unusual size of these dogs. Lawrence (1971) first analyzed these remains and compared them to the Eskimo dog as described by Allen (1920).

The nine female dog specimens did not produce the same pattern as the male specimens. At time of writing, nine measurements had been processed for this group, with case numbers ranging from seven to eight (Table 5). The dimensions with the highest number of individuals

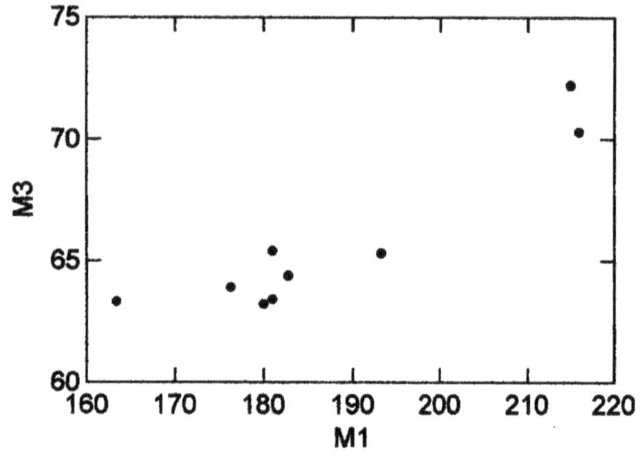


*Figure 5* Maximum cranial length (1) vs. greatest mastoid breadth (3) for adult male dogs

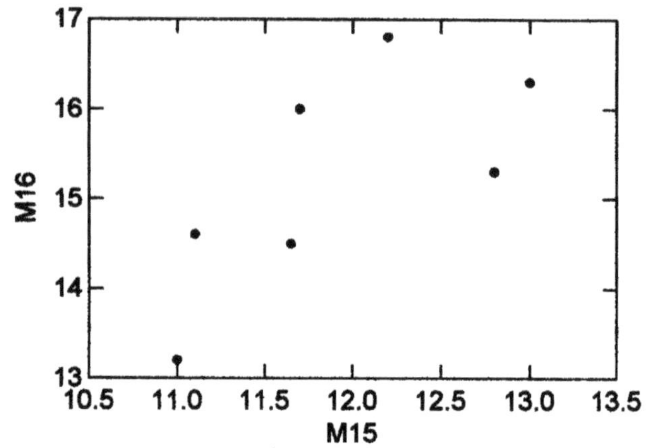


***Figure 8*** *Length of M¹ (15) vs. width of M¹ (16) for adults female dogs*

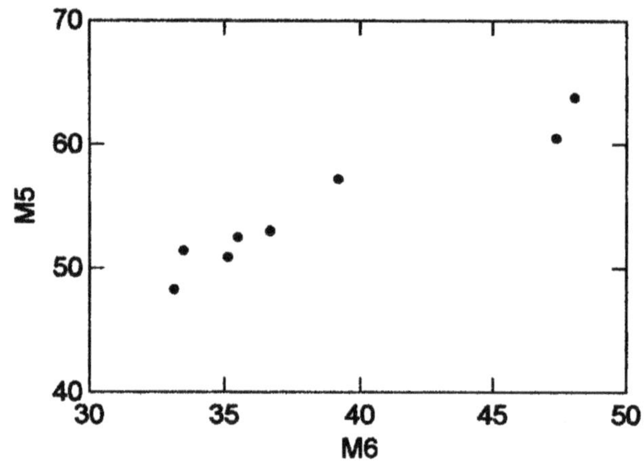


***Figure 6*** *Least breadth between orbits (6) vs. frontal breadth (5) for adult male dogs*

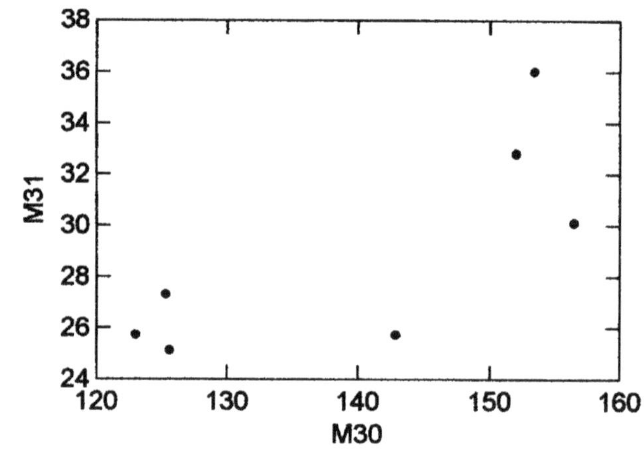


***Figure 9*** *Maximum length of humerus (30) vs. distal breadth of humerus (31) for adult female dogs*

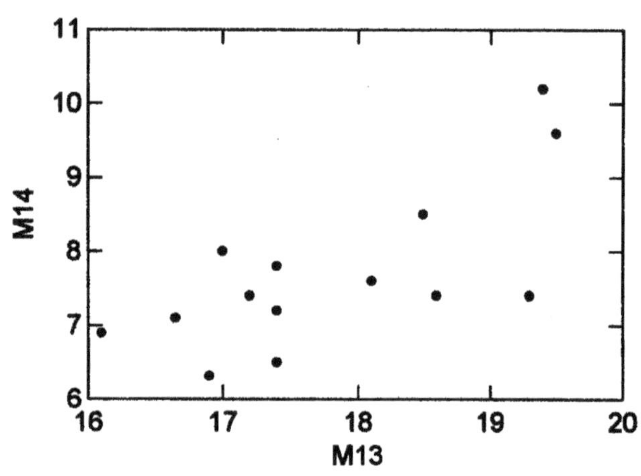


***Figure** 7 Length of P⁴ (13) vs. breadth of P⁴ (14) for adult male dogs*

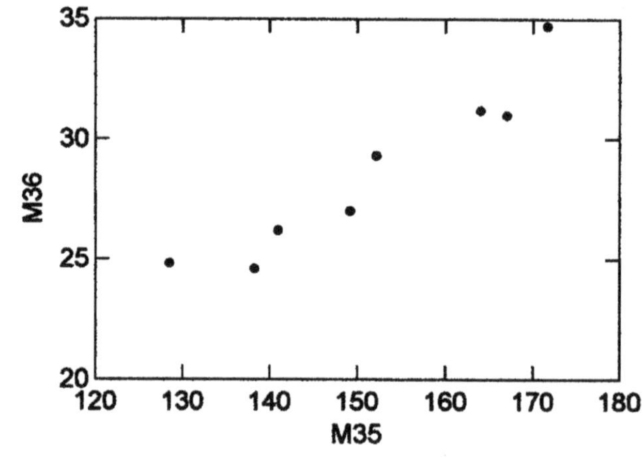


***Figure 10*** *Maximum length of femur (35) vs. distal breadth of femur (36) for adult femur dogs*

| Metric Codes | Number of Cases | Minimum | Maximum | Mean | Standard Deviation |
|---|---|---|---|---|---|
| 1 | 4 | 154.20 | 216.00 | 180.30 | 27.21 |
| 2 | 4 | 50.50 | 61.20 | 53.55 | 5.12 |
| 3 | 4 | 56.10 | 71.30 | 63.45 | 6.98 |
| 5 | 4 | 43.10 | 51.00 | 46.32 | 3.36 |
| 6 | 5 | 28.72 | 41.60 | 33.63 | 5.24 |
| 9 | 2 | 33.40 | 39.00 | 36.20 | 3.96 |
| 10 | 4 | 52.12 | 72.10 | 60.10 | 8.74 |
| 12 | 3 | 24.70 | 37.90 | 30.66 | 6.69 |
| 13 | 7 | 8.30 | 19.60 | 16.01 | 3.66 |
| 14 | 7 | 6.08 | 8.00 | 7.31 | 0.79 |
| 15 | 7 | 11.00 | 13.00 | 11.92 | 0.78 |
| 16 | 7 | 13.20 | 16.80 | 15.24 | 1.24 |
| 30 | 7 | 123.00 | 156.50 | 139.81 | 14.81 |
| 31 | 8 | 25.12 | 36.00 | 28.76 | 3.89 |
| 35 | 8 | 128.50 | 171.70 | 151.48 | 15.23 |
| 36 | 8 | 24.60 | 34.70 | 28.60 | 3.56 |
| 37 | 8 | 129.50 | 171.20 | 150.00 | 15.03 |

***Table 5*** *Measurement statistics for female dog specimens (mm)*

represented include; the maximum length and breadth of upper molar 1 (#15, 16), the greatest length and the greatest distal breadth of the humerus (# 30, 31), and the greatest length and the greatest breadth of the femur (#35, 36).

The unmodified plot graph of upper molar 1 (Fig. 8) shows a wide spread of points with no discernible pattern, although there does seem to be a break in the middle suggesting a larger and smaller dog. The humerus plot is more distinct: three groups are illustrated in the graph comparing length to distal breadth (Fig. 9). However, the middle-sized group is made up of a single point. Figure 10 illustrates the plot of the femur's length to distal breadth. Different from the others, the femur plot shows a linear regression with no obvious patterns in determining different types of dogs.

## Conclusion

Although the findings of this analysis are preliminary, they indicate a strong possibility that more than one type of dog existed in the Northeast during the prehistoric period. Through measuring specimens from many sites throughout the region, a pattern can be seen from several plot graphs. Male dogs, with a greater number of specimens, were the better sample and clearly show two types. Although the specimens that indicate the larger dog are only two in number, this is a good start in distinguishing the types encountered at archaeological sites. The possibility of a third type, as suggested by the female long bone plots, will need further study.

The evidence of two types of dogs suggested by this osteometric analysis seems to agree with the historic documents. A large dog, large enough to be confused by European explorers as a wolf, and a smaller one. As for Loomis and Young's third dog, there were only two graphs produced by this study that suggested another type. Unfortunately, the specimens they used were not available for re-examination.

## Future Research

Future research should include the addition of other specimens to the collected data, and a more complete analysis using other dimensions. A comprehensive look at the cultural traditions represented by the dogs with regard to type is also needed. With the information presented here, it appears that Archaic and Woodland period dogs from Maine are similar in size and shape to those in southern New England and Long Island. The exceptions are specimens 016, 017, and 018, three large dogs from the Boston basin area. Although specimens 017 and 018 have no assigned cultural affiliation, 016 had collagen dating producing a $C^{13}$ corrective date of 1710 ± 70BP (Nelson, 1989). This places the dog in the Middle to Late Woodland period. This type of information is of great importance to understanding migration by different human cultural groups, and the possible introduction of different types of dog into this region.

## Acknowledgments

The specimens used in this analysis were made available by Peabody Museum of Ethnology and Archaeology, at Harvard University, the R. S. Peabody Foundation, at Andover Academy, the Maine State Museum, the American Museum of Natural History, the New York State Museum, the Newport Historic Society, the Offices of the Connecticut State Archaeologist, and David Thompson, of the New Haven Archaeological Society. Robert Lewis provided the Ruth Moore dog metrics, Skeletal diagrams were drawn by Dana Ricardi. I would like to thank Claudia Milne for editing an earlier draft of this report. Special thanks to Susan Crockford for including this research at the 1998 ICAZ dog symposium, and to Jennifer Kirby for all her help and moral support.

**References**


Allen, G. M. (1920). *Dogs of the American Aborigines*. Bulletin of the Museum of Comparative Zoology at Harvard College **63**.

Blick, J. P. (1988). A preliminary report on the osteometric analysis of some aboriginal dogs (*Canis familiaris*) from Weyanoke Old Town, 44 PG 51, Prince George County, Virginia. *Quarterly Bulletin of the Archaeological Society of Virginia* **43**.

Bourque, B. J. (1995). *Diversity and Complexity in Prehistoric Maritime Societies: A Gulf of Maine Perspective*. New York: Plenum Press.

Butler, E. M. & Hadlock, W. S. (1949). Dogs of the northeastern woodland Indians. *Bulletin of the Massachusetts Archaeological Society* **10**, 17-35.

Crockford, S. J. (1997). *Osteometry of Makah and Coast Salish Dogs*. Simon Fraser University, Burnaby: Archaeology Press **22**.

Driesch, A., von den (1976). *A Guide to the Measurement of Animal Bones from Archaeological Sites*. Peabody Museum of Archaeology and Ethnology, Harvard University: Peabody Museum Bulletin **1**.

Dincauze, D. F. (1973). *Archaeological Reconnaissance in the Greater Boston Area: 1969-1972*. Harvard Peabody Museum Library: National Science Foundation Grant Report **GS-2706**.

Drennan, R. D. (1996). *Statistics for Archaeologists*. New York: Plenum Press.

Fitting, J. (1978). Regional cultural development, 300 BC to AD 1000. In *Handbook of North American Indians* **15**, 44-57.

Gamage, A. (1886). Personal notes and correspondence. *Accession files, Peabody Museum of Archaeology and Ethnology*.

Gilbert, B. M. (1993). *Mammalian Osteology*. Missouri Archaeological Society.

Harrington, M. (1982). Exploration of an ancient burial ground and village site near Port Washington, Long Island. In (J. Truex, Ed) *The Second Coastal Archaeology Reader*. New York: Suffolk County Archaeological Association, pp. 83-89.

Howard, W. E. (1949). A means to distinguish skulls of coyotes and domestic dogs. *Journal of Mammalogy* **30**,169-171.

Kerber, J. E., Leveillee, A. D. & Greenspan, R. L. (1989). An unusual dog burial feature at the Lambert Farm site, Warwick, Rhode Island: preliminary observations. *Archaeology of Eastern North America* **17**, 165-174.

Kerber, J. E. (1997). Native American treatment of dogs in northeastern North America: archaeological and ethnohistorical prespective. *Archaeology of Eastern North America* **25**, 81-96.

Klein, R. G., & Cruz-Uribe, K. (1984). *The Analysis of Animal Bones from Archaeological Sites*. Chicago: University of Chicago Press.

Krantz, G. S. (1959). Distinctions between the skulls of coyotes and dogs. *Kroeber Anthropological Society* **21**.

Lawrence, B. (1971). Accession files on site 19-NF-81, Peabody Museum of Ethnology and Archaeology, Harvard University.

Loomis, F. B., & Young, D. B. (1912). On the shell heaps of Maine. *American Journal of Science* **4**.

Moffet, R. (1953). Matters disclosed by erosion at Ryder Beach site. *Bulletin of the Massachusetts Archaeological Society* **XIV**, 69-72.

Morey, D. F. (1992). Size, shape and development in the evolution of the domestic dog. *Journal of Archaeological Science* **19**, 181-204.

Morey , D. F. (1986). Studies on Amerindian dogs: taxonomic analysis of canid crania from the northern Plains. *Journal of Archaeological Science* **13**, 119-145.

Morrison, A. H. (1990). Dawnland dog-feast: Wabanaki warfare, c. 1600-1760. In (W. Cowan, Ed) *Papers of the Twenty-First Algonquian Conference*. Ottawa: Carleton University.

Nelson, C. M. (1989). Radiocarbon age of the dog burial from Squantum, Massachusetts. *Bulletin of the Massachusetts Archaeological Society* **50**, 29.

Rosier, J. (1906). True relation of Captain Waymouth's voyage, 1605. In (H. S. Surrage, Ed) *Early English and French Voyages, 1534-1608*. New York: Original Narratives of Early American History.

Schwartz, M. (1997). *A History of Dogs in the Early Americas*. New Haven: Yale University Press.

Shigehara, N., Onodera, S. & Eto, M. (1997). Appendix A: Sex determination by discriminant analysis and evaluation of non-metric traits in the dog skeleton. In (S. J. Crockford) *Osteometry of Makah and Coast Salish Dogs*. Simon Fraser University, Burnaby: Archaeology Press **22**.

Silver, I. A. (1969). The ageing of domestic animals. In (D. Brothwell & E. S. Higgs, Eds) *Science in Archaeology*. London: Thames & Hudson, pp. 283-302.

Strong, J. A. (1985). Late Woodland dog ceremonialism on Long Island in comparative and temporal perspective. *The Bulletin and Journal of Archaeological for New York State* **91,** 32-38.

Tuck, J. A. (1978). Regional cultural development, 3000 to 300 BC. In *Handbook of North American Indians* **15**, 28-43.

Willoughby, C. C. (1935). *Antiquities of the New England Indians*. Peabody Museum American Archaeology and Ethnology, Harvard University.

Wolley, C. (1860). *A Two Years Journal in New York*. (E.B. O'Callaghan, Ed) New York.

Wyman, J. (1868). An account of some kjockkenmoeddings, or shell heaps, in Maine and

Massachusetts. *American Naturalist* **I**, 561-584.

Yentsch, A., Brown, M. & Deetz, J. (1979). *Review of Excavations during 1969-1970 at the Old Quaker Meeting House Site in Newport, Rhode Island.* On file, Plymouth: Archaeology Laboratory, Plymouth Plantation.

# THE FORM AND MEANING OF MAYA AND MISSISSIPPIAN DOG REPRESENTATIONS

Marion Schwartz
Anthropology Department,
Yale University, P.O. Box 208277,
New Haven, CT 06520-8277 USA

**Key words:** dogs, Mayan dogs, Mayan art, Mississippian art, prehistoric art

**Abstract**

Artistic representations of dogs can contribute to a greater understanding of the prehistory of canines in two ways. First, depictions of dogs may shed light on the type of dogs present and secondly, they have the potential to provide a sense of the way dogs were integrated into their human society. Two cases in point are: the 'dog' effigy pots frequently recovered from Mississippian burials (AD1000-1500) in the Southeastern region of the United States; the Underworld dog monsters illustrated on the Late Classic (AD550-850) polychrome vases unearthed in the Maya area of Mesoamerica. Both types of depictions may have been inspired by races of dogs that existed at the time of contact and disappeared shortly thereafter.

## Introduction

Broadly speaking, the patterning of dog remains in the archaeological record of southeastern North America stands in contrast to that found in the Maya area of Mesoamerica. More buried dogs have been recovered from the Southeast than from lowland Mesoamerica, although the practice of burying dogs drops off substantially by Mississippian time. On the other hand, dog remains are much more frequent in midden contexts in lowland Mesoamerica, although, here again, the frequency declines through time. Moreover, artistic representations of dogs from the Classic period of the Maya (AD550-850) and the Late Mississippian Period (AD1100-1500) suggest a different symbolic focus at least as far as canines are concerned among elite populations in these two areas. See Figure 1 for the rough boundaries of the two areas under discussion.

Above and beyond undoubted cultural explanations for these distinctions, what, if anything, can be concluded about the actual dogs involved, ritually and practically, in the everyday lives of pre-Columbian Americans? Does evidence exist, either from faunal remains or ethnohistoric accounts, that artists left behind an accurate record of what pre-Columbian dogs looked like? Were there, in the Southeast, short-legged, small dogs with tails that curved over their backs as depicted in funerary ware. Were there, in lowland Mesoamerica, the loose-skinned, scalloped eared dogs that appear on the drinking vessels of the Maya elite?

As elsewhere around the world, dog remains are found in various contexts in Mesoamerica and the Southeast. Often the only indication that dogs lived in a prehistoric community is the recovery of the occasional unassociated skeletal element, either in residential or ceremonial areas. Dog bones also occur in refuse, sometimes with cut or burn marks suggesting that dog was being exploited as a food source. Dogs were deliberately buried, as whole or partial skeletons, either with humans or alone, with burial goods or without, or in some type of ritual sacrifice. In Mesoamerica it is not unusual to find that a high proportion of the mid-sized mammals in middens turns out to be dog (Fig. 2) (van der Schalie & Parmalee, 1960; Wing, 1977; Olsen, 1978; Wing, 1978; Hamblin, 1984; Carr, 1985; Pohl, 1985; Wing & Scudder, 1991; Pohl, 1991; Pohl, 1994). At the Olmec site of San Lorenzo (1200-900BC) in coastal Vera Cruz, over 50% of the mammalian remains were dog (Wing, 1978). In Central Mexico during the Santa Maria phase (900-200BC) at Teohuacán dogs had become an increasingly important part of the diet (Flannery, 1967). At Preclassic Cuello (1200BC-AD250) Clutton-Brock and Hammond (1994) concluded that dogs, based on abundant remains and mortality pattern, were raised for food and killed at the end of their first year of life. Fauna collected from the island of Cozumel, off the Yucatan Peninsula, dates mostly to the Post Classic (AD800-1200). Deer, due in part to their size, were generally the most common faunal remains at Maya sites, but did not live on the island. Therefore, all the mammalian remains were of mid-sized animals with dogs, opossums, and rabbits leading the list.

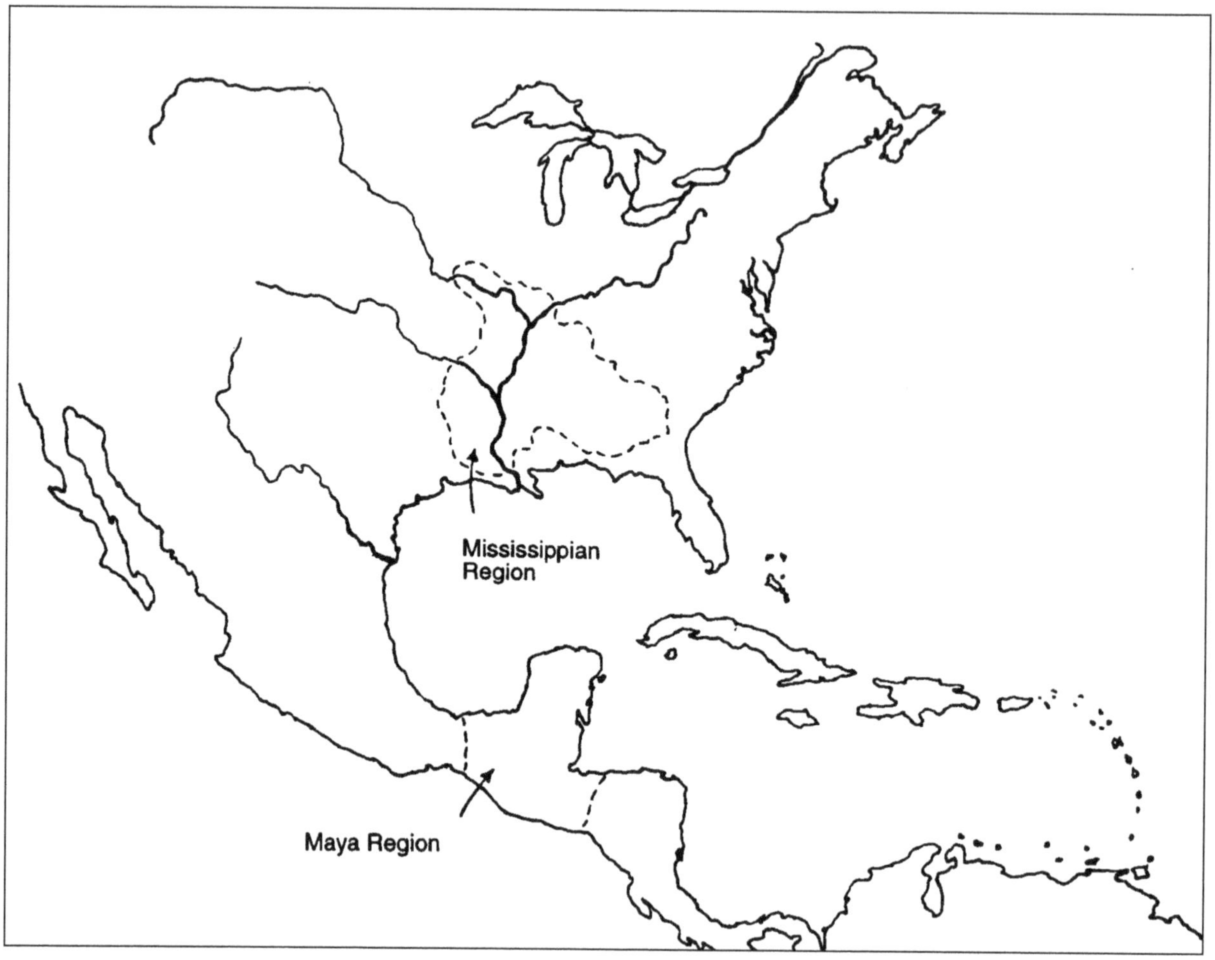


*Figure 1* Map delineating the boundaries of the Mississippian and May Regions.

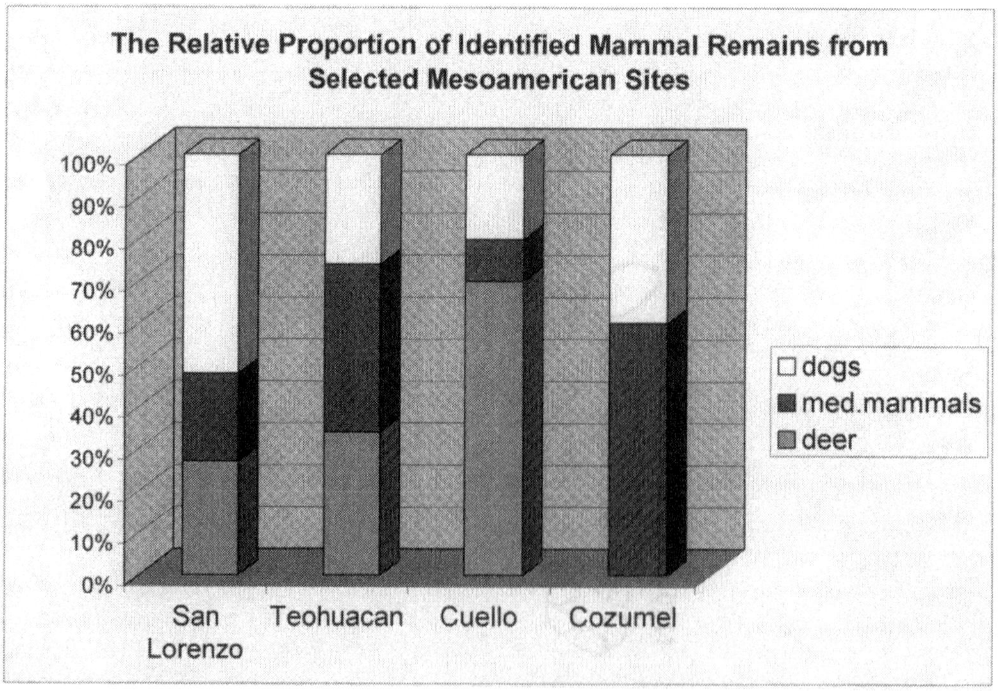


*Figure 2* The distribution of mammalian fauna in midden contexts. Data sources from left to right; Wing, 1978; Flannery, 1967, Wing & Scudder, 1991; Hamblin, 1984.

In contrast to the "smoking gun" evidence that dogs were regularly consumed in Mesoamerica, in the Southeast the best evidence we have that dogs were raised with consumption in mind comes from the DeSoto chronicles, in which Rodrigo Rangel stated that "The Indians came forth in peace and gave them....a few little dogs, which are good food. These are little dogs that do not bark, and they rear them in houses in order to eat them" (Clayton *et al.*, 1993:263). Dog feasts in preparation for warfare among the Natchez were reported by Le Page du Pratz in 1758. According to the Frenchman, the meat of dog was chosen "by way of preference in their feasts and ceremonies." (Le Page du Pratz, 1975: 368). Ethnohistoric accounts concur that the occasional consumption of dog was widespread throughout eastern America and out onto the Plains (*eg*, Gyles, 1869; Thwaites, 1898; Sagard, 1939; Radin, 1990). And yet, faunal studies usually do not show dog remains in any substantial numbers, at least in eastern North America (Parmalee, 1957; Neitzel, 1965; Smith, 1975; Schnell, Knight & Schnell, 1981; Welch, 1991; but see Houck, 1951; Milanich, 1994; Wheeler & McGee, 1994 - sites in Florida seem to be the exception to the rule). DeSoto's men mentioned that the dogs they were given were "little" and perhaps archaeological evidence of dogs of different sizes might help establish whether, indeed, a small breed existed that quickly became extinct after the arrival of the hungry Spanish. Webb (1959) reports dogs of two sizes were recovered from Belcher Mound in Louisiana, as does Wing (1965) in Florida.

On the other hand, in Southeastern North America, buried dogs are much more common than anywhere in Mesoamerica, with the exception of a few related sites in the Marismas Nacionales in West Mexico (Wing, nd). However, the vast majority of Southeastern dog burials date to Archaic (5000-3000BC) sites such as Eva, and Indian Knoll and other shell mounds in Kentucky, Tennessee, and Alabama (Table 1).

| Site | Date | Description | Reference |
|---|---|---|---|
| Eva (Tennessee) | 5000BC | 18 dogs buried with or near humans | Lewis & Lewis, 1961 |
| Hiwassee Island (Tennessee) | 3000BC | 8 buried dogs with or near humans | Harrington, 1922 |
| Read Shell Midden (Kentucky) | 3000BC | 63 dogs buried with humans in shell heap | Webb, 1950b |
| Indian Knoll (Kentucky) | 3000BC | 21 dogs buried with humans in shell heap | Webb, 1946 |
| Carlson Annis (Kentucky) | 3000BC | 28 dogs buried with humans in shell heap | Webb, 1950a |
| Perry Site, Pickwick Basin (Alabama) | 3000BC | 36 dogs buried with humans in shell heap | Webb & DeJarnette, 1942 |
| Hiwassee Island (Tennessee) | AD1100-1500 | no buried dogs | Lewis & Kneberg, 1946 |
| Pickwick Basin Sites (Alabama) | AD1100-1500 | no buried dogs | Webb & DeJarnette, 1942 |
| Moundville (Alabama) | AD1100-1500 | no buried dogs | Welch, 1991 |
| Etowah (Georgia) | AD1100-1500 | 2 dogs with tailed curled over back buried in temple mound | van der Schalie & Parmalee, 1960 |
| Gorden Site (Mississippi) | Late Prehistoric | 1 dog buried with arrowhead | Cotter, 1952 |
| Fatherland Site (Louisiana) | Protohistoric | no buried dogs | Neitzel, 1965 |

*Table 1 The patterning of dog burials in the US Southeast*

Of course, dog burials do occur in succeeding periods but their incidence is much reduced. As there is no reason to assume that the numbers of dogs decreased, changes in the ritual behavior of humans, noted in other aspects of the material culture, must also have taken place regarding people's relationship to dogs. It appears that the valued canine hunter of the Arachic Period became less common than the mangy village cur. Perhaps with the advent of agriculture, the status of dogs declined as their number increased with the expanding Mississippian human population. Now they could be consumed instead of buried because they could be fed on agricultural surpluses. With the emergence of an aristocratic class, mortuary practices changed and dogs were, for the most part, left out of the richest burial mounds (Moore, 1897; but see also Milner, 1984).

During the Woodland Period (AD200-800), animal effigy figures tended to be "allegorical" and naturalistic symbols that were referable to some human characteristic. For example, dogs and vultures were paired on a pottery vessel from the Weeden Island McKeithern Site I (AD500), in northern Florida. Dogs and vultures form a logical pair because of their habit of eating carrion. They may have also been regarded by this early Florida society as being symbols of impropriety and incest (Milanich, 1994). But, with the emergence of the Mississippian elite, artists started to incorporate deliberately obscure animal motifs, what Vernon Knight (1989) refers to as 'Mississippian Monsters.' Suddenly, the thunderbird and the antlered flying rattle snake join other composite characters with esoteric meanings. As Mary Helms has held for Panamanian chiefdom (1991), political power may be gained if knowledge, which in this case is the meaning of ritual symbols, is monopolized. If the common people feel mystified and lack understanding of how to participate in ritual, they can be relegated to watchers, only. Dogs, however, do not appear to have ever entered the realm of the deliberate obscurism that was the product of a newly developed class system. I have found only one example of an animal that may possibly be a dog that is part of the Southeastern ceremonial complex art produced by Mississippian mound building peoples. This ritual art, sometimes called "Southern Cult" art, possesses characteristic motifs and artifact forms but in regionally variable styles. Engraved shell and painted and effigy pottery, as well as other typical artifacts, have been recovered in profusion from the large mound sites of Moundville, Etowah, and Spiro, located in Alabama, Georgia, and Oklahoma respectively.

Spiro, a "Caddoan" site on the flood plain of the Arkansas River in eastern Oklahoma has produced some of the most spectacular mortuary grave goods anywhere in the Southeast. Most intriguing are the hundreds of intricately engraved shell cups and gorgets. Many of these feature avian themes, ranging from men wearing bird costumes, to half men/half birds, to birds with no human characters. In some of these scenes, it is clear that actual ceremonies are being portrayed as in this fragmentary gorget from the Craig Mound, phase A (Fig. 3). An animal is being held in the grasp of man with a bird (headdress?) on its head. Although the animal is identified as a racoon by Phillips and Brown (1984) I do not see the goggle eyes that other unmistakable raccoon figures have. It is possible that this animal is, instead, a dog about to be sacrificed in some unknown ceremony by men in highly symbolic garb. Certainly similar sacrifice scenes have been illustrated in Mesoamerica. For example, in the Mixtec Codex Nutall, a lady and lord are offering a sacrifice of a dog to two priests (Fig. 4). In light of the reported ceremonial use of dog by the Natchez, a highly stratified Southeastern group, the Spiro dog sacrifice hypothesis has some credence.

***Figure 3*** *Shell gorget, Craig school, phase A Spiro (Phillips and Brown 1984, plate 128), copyright 1978 & 1984 (President & Fellows of Harvard College). Courtesy of Peabody Museum Press.*

***Figure 4*** *Blood spurts from the chest of a dog whose heart has been removed as an offering being made by the prince and his wife to two priests. A turkey has also been sacrificed. Codex Nuttall, Mixtec.*

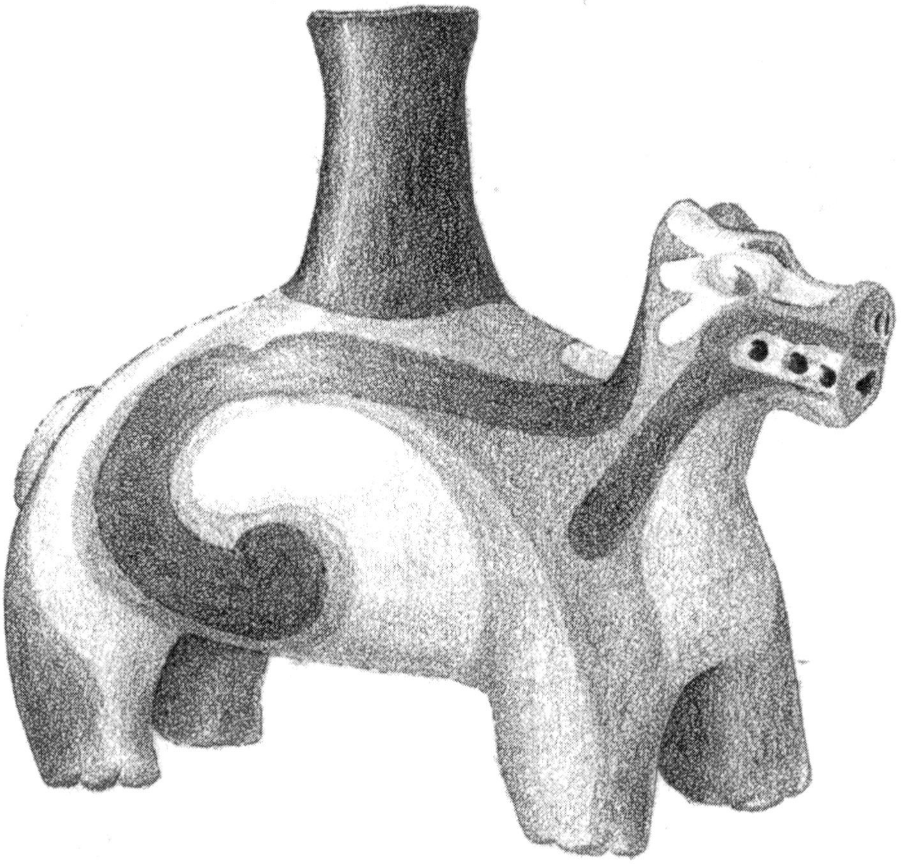

***Figure 5*** *Dog effigy pot with squat body form. Effigies such as these may represent the small dogs given to DeSoto during his march through the south. Tennessee. Mississippian.*

In general, however, Mississippian depictions of dogs are familiar and domestic. Symbols, yes, but an obscure part of the ritual of Southern cult priests, no. Consider the squat dog pots (Fig. 5) collected from numerous human grave sites in Missouri, Tennessee, Georgia, and Alabama (see Heye, Hodge & Pepper, 1919; Kelly, 1938; Webb & DeJarnette, 1942; Chapman, 1985; Dye & Wharey, 1989 for reports on these pots). Their very comforting familiarity suggests a close relationship between dogs and people. In burial 4 at Cemochechobee, a Mississippian ceremonial center in Georgia, excavators uncovered a secondary internment of a skull, possibly female, with two clay vessels placed near it (Schnell, Knight & Schnell, 1981). The dog pot (Fig. 6) recovered from the grave had ten oval spots painted on her belly indicating she was a pregnant or lactating bitch. Her tail curls up in a flat loop. Interestingly, the dog was decorated with two southern cult motifs, the "weeping eye" and the "ogee."

A curious ceramic dog-effigy pipe (Fig. 7) came from Kogers Island burial 23, the most elaborate grave site reported in Alabama outside of Moundville. Recovered from the burial of an extended male with sixty-two other artifacts (Webb & DeJernette, 1942), this effigy is of a short-legged dog with its tail tightly curled over its back. Curled tails carried over the back characterize northern spitz breeds of dog, and it is intriguing that some Mississipian dog depictions have curled and even cork screw tails. This is a trait not found in dog effigies in Meso and South America (Schwartz, 1997). Further evidence that effigy dogs may depict real dogs comes from the van der Schalie & Parmalee (1960) report that, at Etowah, the two dogs buried in Mound C had their tails curled over their backs.

The best example of using effigy dogs to infer what real dogs looked like is, of course, in depicting dogs with wrinkles to indicate a hairless condition, as was done in South America and Mesoamerica. Lines on faces or bodies of dog effigies has long been thought to be an artistic convention used to represent the wrinkled skin so visible on hairless dogs. Aboriginal hairless dogs still exist in Mexico and Peru and archaeologists have used the first appearance of wrinkled dog effigies to date the appearance of the hairless dog in different parts of the Americas (Weiss, 1970; Cordy-Collins, 1994). The Mexican hairless, best known by its Aztec name, *Xoloitzcuintli*, was certainly present in the Maya region and was called *ah bil* by the Yucatec Maya. These nude Maya dogs were used to hunt birds and deer as well as being eaten (Roys, 1976).

The mutant gene that produces hairlessness in dogs can also cause abnormal dentition, so that a dog with such a gene has a high probability of lacking a full set of teeth. If mandibles with missing premolars are preserved in the archaeological record it would serve as an independent

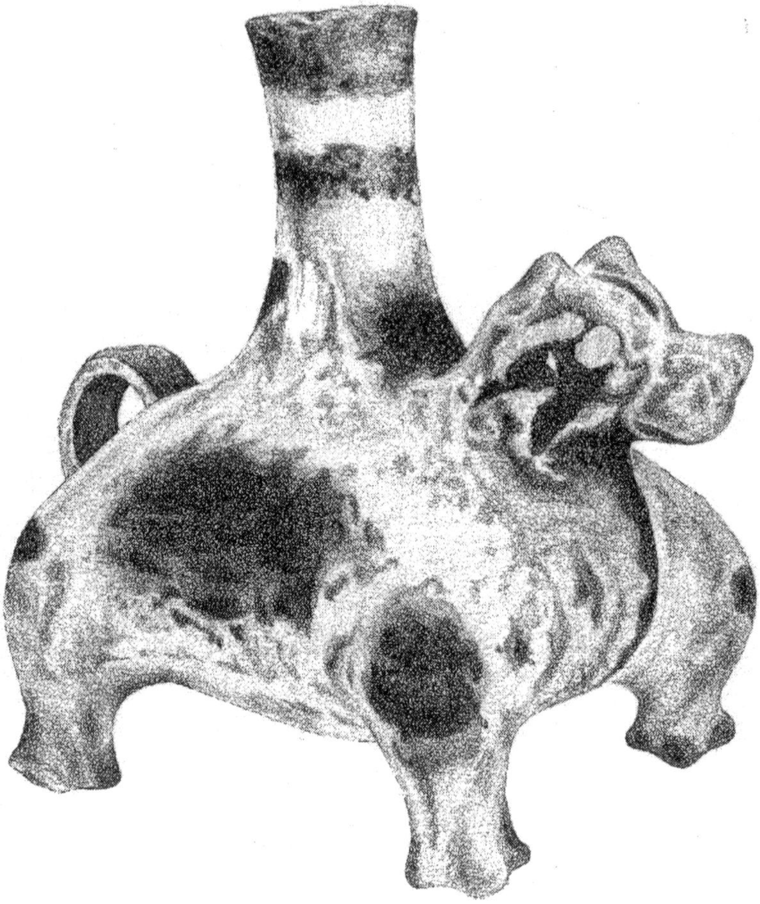

***Figure 6*** *Dog effigy pot from burial 4 at Cemochechobee, Georgia. The underside of the pot has ten oval spots in rows to signify nipples. Mississippian.*

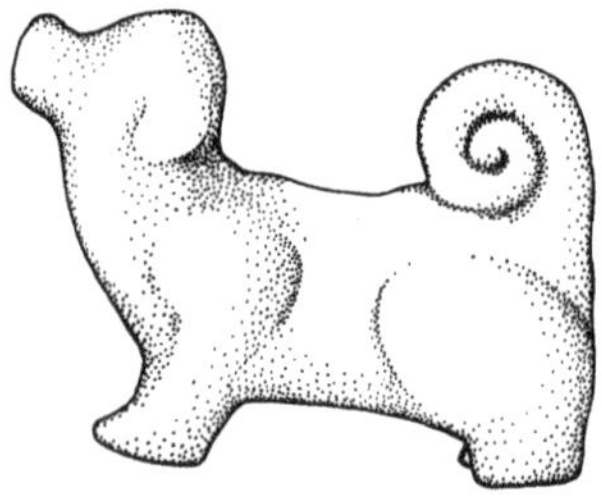

***Figure 7*** *Dog effigy pipe with curled tail. 4 cm. Ceramic. Burial 23, Kogers Island, Alabama. Mississippian.*

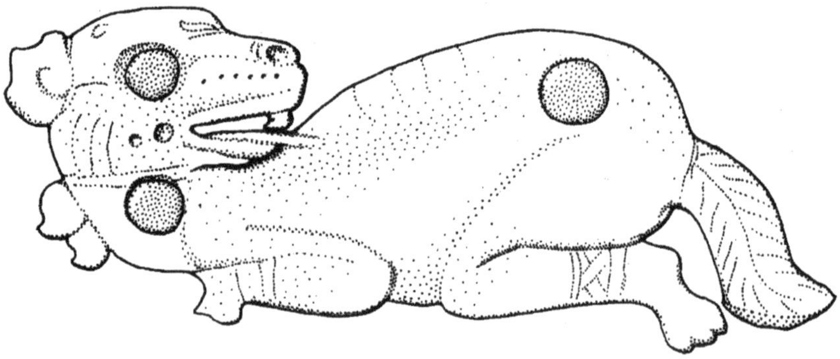

***Figure 8*** *Crouching dog with scalloped ears. Shell with Jade inlay. 14 cm. Maya.*

means of identifying hairlessness in dogs. Mary Pohl (1994), in studying the fauna from a Late Classic Maya site in the Copan Valley, Honduras, reported a partial dog mandible which lacked premolars and perhaps a canine. On the other hand, Elizabeth Wing (nd) has examined a number of buried dogs from West Mexican Post-Classic (AD700-1300) sites in the Marismas Nacionles and did not detect any abnormal dentition. Raul Valadez has found quite a few examples of mandibles from hairless dogs at Tula, in Central Mexico, and has been able to distinguish isolated teeth from hairless dogs due to their distinctive morphology (Valadez, 1995). Continued work on canid remains from Mesoamerican sites may help give a better idea about the numbers of hairless dogs relative to their certainly more common hirsute relatives in the pre-Columbian past.

Many Maya representations of dogs also show a ragged or scalloped ear, whereas effigies from West Mexico as well as those from North and South America show no such deformities. According to J.E.S. Thompson (1961) the sore and ripped dog ears shown in Maya art are probably the result of Chiclero's ulcer *(Leishmaniasis),* a disease caused by a biting fly. Chiclero's ulcer must have had a substantial impact on Maya dog populations since mangled ears show up so often on dog effigies (Fig. 8). At the Late Classic Maya site of Tonina, located in the southwestern part of the Maya region, a stone sculpture of dog 360 centimeters long was discovered lying above a human burial (Graham & Von Euw, 1996). This resting dog sculpture has very small ears and conspicuous wrinkles on its face. On its back a text has been engraved which reads "(it is) the dog of..." followed by the personal name of a Maya lord who presumably owned it. This is the only known pre-Columbian example of a portrait of a specific animal owned by an identified individual (David Stuart, pers. comm.). Because it is a portrait of a particular animal, this sculpture can be considered further evidence that hairless and small-eared dogs were part of the Maya landscape.

Because the Maya possessed a written language, the elite classes, particularly scribes and priests, were able to hide ritual meanings from the mostly illiterate population in ways far surpassing the pictorial attempts made by Mississippian political leaders. By the Late Classic Period (AD550-850), the surface of Maya pottery had become the canvas for highly complex fine line paintings which served the elite in ceremonial, economic and political ways. Best known are the polychrome vases whose inscriptions tell us that each was the drinking vessel of a particular individual for the consumption of a favorite chocolate drink. Here, as in Mississippian areas, regional craft shops developed their own recognizable styles, but the level of sophistication and layers of meaning are unsurpassed throughout the Americas. Although themes relating to *Xibalba*, the Maya Underworld, frequenting the surface of these vases indicate an ultimate funerary function, wear patterns suggest the vessels were used before death (Reents-Budet, 1994).

The depicted dogs of the Maya are domestic and familiar, but they also enter the supernatural sphere. Maya dogs are special companions of the gods of the Underworld. This is not surprising. Because of their ability to guide and to guard, as well as their ability to bridge the gap between nature and culture, dogs have found their way to the Next World in many of the world's early cultures (see Schwartz, 1997: Chap.4). Maya dogs are also associated with the Young Maize God. Also known as the First Father, the Young Maize God can be recognized by his elongate head, shaped like an ear of corn (Fig. 9). This association of dogs with the Maize God may have Underworld connotations but there may be fertility symbolism as well.

***Figure 9*** *A dog with necklace and shawl kneels before the Young Maize God. Late Classic Maya Vase.*

Although some dogs illustrated on Maya polychrome pots are white or spotted, others are shown with scalloped ears and wrinkled faces. These loose skinned dogs are accompanied by the *way* glyph; meaning that they are animal spirit companions to people of noble birth. These dog *ways* occupy *Xibalba* and wear fringed sacrificial shawls and carry an eyeball and a long bone in a ceremonial bowl. The lower half of their bodies have jaguar spots and their tails are feathered (Fig. 10). Inscriptions refer to their tails as a 'fire tails.'(Grube & Nahm, 1994). Their companion *ways* are often spider monkeys, usually sporting rabbit ears, and toads with jaguar spots. This extraordinary supernatural dog has a combination of features that join a host of concepts into a single creation. This creature is the embodiment of death, fire, disease, and sacrifice and is the most abstract, yet most inclusive, metaphorical dog that pre-Columbian America ever produced.

***Figure 10*** *The scallop-eared dog is one of an Underworld Triad which includes a toad, also with jaguar spots, and a spider monkey with deer ears. Both the dog and the toad carry sacrificial bowls containing body parts. The inscriptions tells us that these creatures are wayob, animal spirit companions. Late Classic Maya Vase. Rollout photograph K1181© Justin Kerr 1980.*

Peter Roe (1995) has put forth the argument that dogs, on the predator-poor Caribbean Islands, functioned as mythic substitutions for jaguars of a distantly remembered past. In a similar way, the urban Maya, living in an almost completely deforested habitat (Rice, Rice & Deevey, 1985), may have chosen dogs for stand-ins for wild carnivores. The jaguar was, of course, a powerful symbol for Maya people even though real felines must have been an exceedingly rare sight. The people living in the eastern woodlands accepted large carnivores such as bears, pumas, and wolves as a very real part of their landscape, which may have given rise to a different conceptualization of the dog's place in the universe. However, Mississippian people did place dog pots in their graves so it is possible that dogs were regarded as guides to the Next World here as well in the Maya area.

The agricultural base had a greater time depth by about two thousand years in the Maya region than in the Southeast, and dogs, fed on maize (Gerry, 1993) could be husbanded (Pohl, 1982). In the Southeast, the evidence is more shaky regarding the presence a small barkless dog raised especially to be eaten. Still, we have every reason to believe that a short-legged, curly-tailed dog existed because so many Mississippian artists depicted it. Moreover, effigy figures provide evidence that the small-eared, hairless dog of the Maya was also a real animal. Art can function as a primary source on pre-Columbian animals as conceived by pre-Columbian people.

**Acknowledgements**

I am grateful to Susan Hochgraf who provided all the drawings and to David Stuart for bringing to my attention the stone dog from Tonina.

**References**

Carr, H.S. (1985). Subsistence and ceremony: faunal utilization in a Late Preclassic community at Cerros, Belize. In (M. Pohl, Ed) *Prehistoric Lowland Maya Environment and Subsistence Economy.* Cambridge: Peabody Museum Press.

Chapman, J. (1985). *Tellico Archaeology: 12,000 Years of Native American History.* Tennessee Valley Authority Publications in Anthropology **40**..

Clayton, L.A., Knight, V. J. & Moore, E.C. (1993). *The DeSoto Chronicles: the Expedition of Hernando de Soto to North America in 1539-1543.* Tuscaloosa: University of Alabama Press.

Clutton-Brock, J. & Hammond, N. (1994). Hot dogs: comestible canids in Preclassic Maya culture at Cuello, Belize. *Journal of Archaeological Science* **21**, 819-826.

Cordy-Collins, A. (1994). An unshaggy dog story. *Natural History* **103**, 34-41.

Cotter, J.L. (1952). The Gorden site in southern Mississippi. *American Antiquity* **18**, 110-126.

Dye, D.H. & Wharey, C. (1989). Exhibition catalog. In (Galloway, P., Ed) *The Southern Ceremonial Complex: Artifacts and Analysis.* Lincoln: University of Nebraska Press. pp.321-379.

Flannery, K.V. (1967). Vertebrate Fauna and Hunting Patterns. In (D. S. Byers, Ed) *The Prehistory of the Tehuacan Valley,* Vol. 1. Austin: University of Texas Press, pp. 132-177.

Gerry, J. (1993). *Diet and Status among the Classic Maya: An Isotopic Perspective.* Ph.D. Thesis, Harvard University.

Graham, I. & Euw, E., von (1996). *Corpus of Maya Hieroglyphic Inscriptions, Vol. 6/2.* Cambridge: Harvard University Press.

Grube, N. & Nahm, W. (1994). A census of Xibalba: a complete inventory of *way* characters on Maya ceramics. In (J. Kerr, Ed) *The Maya Vase Book, Vol. 4.* New York: Kerr Associates, pp. 686-712.

Gyles, J. (1869). *Memoirs of Odd Adventures, Strange Deliverances, etc. in the Captivity of John Gyles, Esq.* Cincinnati: Spiller and Gates.

Hamblin, N.L. (1984). *Animal Use by the Cozumel Maya.* Tucson: University of Arizona Press.

Harrington, M.R. (1922). Cherokee and earlier remains on upper Tennessee River. In (F. W. Hodge, Ed) *Indian Notes and Monographs.* New York: Museum of the American Indian Heye Foundation.

Helms, M. (1991). Esoteric knowledge, geographical distance, and the elaboration of leadership status. In (Rambo & Gillogly, Eds) *Profiles in Cultural Evolution.* University of Michigan Anthropological Papers **85,** pp.330-350.

Heye, G.G., Hodge, F. W. & Pepper, G. W. (1918). *The Nacoochee Mound in Georgia: Contributions from the Museum of the American Indian Heye Foundation,* Volume IV/3. New York: Museum of the American Indian Heye Foundation.

Houck, M. (1951). Animal remains from south Indian Field. *Yale University Publications in Anthropology* **45**, pp. 51-60.

Le Page du Pratz, A. (1975). *The History of Louisiana.* Baton Rouge: Louisiana State University Press.

Kelly, A.R. (1938). A preliminary report on archaeological explorations at Macon, Ga. Washington: *Smithsonian Bureau of American Ethnology Bulletin* **119**.

Knight, V.J. (1989). Some speculations on Mississippian monsters. In (P. Galloway, Ed) *The Southern Ceremonial Complex: Artifacts and Analysis.* Lincoln: University of Nebraska Press, pp. 205-210.

Lewis, T.M. & Kneberg, M. (1946). *Hiwassee Island: An Archaeological Account of Four Tennessee Indian Peoples.* Knoxville: University of Tennessee Press.

Lewis, T.M. & Lewis, M. K. (1961). *Eva: An Archaic Site.* Knoxville: University of Tennessee Press.

Milanich, J.T. (1994). *Archaeology of Precolumbian Florida.* Gainsville: University Press of Florida.

Milner, G.R. (1984). Social and temporal implications of variation among American Bottom Mississippian cemeteries. *American Antiquity* **49,** 468-488.

Moore, C. B. (1897). *Certain aboriginal mounds of the Georgia coast.* Journal of the Academy of Natural Sciences of Philadelphia **XI**.

Neitzel, R.S. (1965). *Archaeology of the Fatherland Site: the Grand Village of the Natchez.* Anthropological Papers of the American Museum of Natural History **51**.

Olsen, S.J. (1978). Vertebrate faunal remains. In (G. Willey, Ed) *Excavations at Seibel., Vol 1.* Cambridge: Papers of the Peabody Museum of Archaeology and Ethnology **70**, pp. 172-176.

Parmalee, P.W. (1957). Vertebrate remains from Cahokia site, Illinois. *Transactions of the Illinois State Academy of Science* **50,** 235-242.

Phillips, P. & Brown, J. A. (1984). *Pre-Columbian Shell Engravings from the Craig Mound at Spiro, Oklahoma, Part 2.* Cambridge: Peabody Museum Press.

Pohl, M. (1985). The privileges of Maya elites: prehistoric vertebrate fauna from Seibel. In (M. Pohl, Ed) *Prehistoric Lowland Maya Environment and Subsistence Economy.* Cambridge: Peabody Museum Press.

Pohl, M. (1991). Women, animal rearing, and social status: the case of the formative period Maya of Central America. In *The Archaeology of Gender (Proceedings of the twenty-second annual conference of the Archaeological Association of the University of Calgary).* Calgary: University of Calgary Press, pp. 295-312.

Pohl, M. (1994). Appendix D, Late Classic Maya fauna from settlements in the Copan Valley, Honduras: assertion of social status through animal consumption. In *Ceramics and Artifacts from Excavation in the Copan Residential Zone.* Cambridge: Papers of the Peabody Museum of Archeology and Ethnology **80**, pp. 459-476.

Pohl, M. & Feldman, L. H. (1982). The traditional role of women and animals in lowland Maya economy. In (K. Flannery, Ed) *Maya Subsistence.* New York: Academic Press, pp. 353-367.

Radin, P. (1990). *The Winnebago Tribe.* Lincoln: University of Nebraska Press.

Reents-Budet, D. (1994). *Painting the Maya Universe: Royal Ceramics of the Classic Period.* Durham: Duke University Press.

Rice, D.S., Rice, P. & Deevey, E. S. (1985). Paradise lost: Classic Maya impact on a lacustrine environment. In (M. Pohl, Ed) *Prehistoric Lowland Maya Environment and Subsistence Economy.* Cambridge: Peabody Museum Press, pp.91-103.

Roe, P. G. (1995). Eternal companions: Amerindian dogs from Tierra Firma to the Antilles. In (R. Alegria & M. Rodriguez, Eds) *Proceeding of the XVth International Congress on Caribbean Archaeology.* San Juan, Puerto Rico, pp. 155-172.

Roys, R.L. (1976). *The Ethno-Botany of the Maya.* Philadelphia: Institute for the Study of Human Issues.

Sagard, G. (1939). *The Long Journey to the Country of the Hurons.* (G.M. Wrong, Ed) Toronto: The Champlain Society.

Schalie, H., van der & Parmalee, P. W. (1960). Animal remains from the Etowah Site, Mound C, Bartow County, Georgia. *The Florida Anthropologist* **13,** 37-54.

Schnell, F.T., Knight, V. J. & Schnell, G. S. (1981). *Cemochechobee: Archaeology of a Mississippian ceremonial Center on the Chattahoochee River.* Gainsville: University Presses of Florida.

Schwartz, M. (1997). *A History of Dogs in the Early Americas*. New Haven: Yale University Press.

Smith, B.D. (1975). *Middle Mississippi Exploitation of Animal Populations*. Anthropological Papers of the University of Michigan Museum of Anthropology **57**, 102-180.

Thompson, J.E.S. (1961). Notes on a plumbate vessel with shell inlay and on Chiclero's ulcer. In (S. K. Lothrop *et al.*, Eds) *Essays in Pre-Columbian Art and Archaeology*. Cambridge: Harvard University Press.

Thwaites, R.G. (1898). *The Jesuit Relations and Related Documents*. Cleveland: Borrows Brothers.

Valadez, R. (1995). Morfologia dental del perro pel6n Mexicano. *Veterinaria México UNAM* **26**, 317-332.

Webb, C.H. (1959). *The Belcher Mound: A Stratified Caddoan Site in Caddo Parish*. Memoirs of the Society for American Archaeology **16**.

Webb, W.S. (1946). *Indian Knoll*. University of Kentucky Reports in Anthropology **4**, 115-365.

Webb, W.S. (1950a). *The Carlson Annis Mound, Site 5*. University of Kentucky Reports in Anthropology **7**, 267-354.

Webb, W.S. (1950b). *The Read Shell Midden*. University of Kentucky Reports in Anthropology **7**, 358-367.

Webb, W.S. & W. DeJarnette. (1942). *An Archeological Survey of Pickwick Basin in the Adjacent Portions of the States of Alabama, Mississippi and Tennessee*. Washington: Bureau of American Ethnology Bulletin **129**.

Weiss, P. (1970). El Perro peruano sin pelo. *Acta Herediana .Lima* **3**, 33-45

Welch, P.D. (1991). *Moundville's Economy*. Tuscaloosa: University of Alabama Press.

Wheeler, R.J. & McGee, R. (1994). Report of preliminary zooarchaeological analysis: Groves' Orange Midden. *The Florida Anthropologist* **47**, 393-403.

Wing, E.S. (1965). Animal bones associated with two Indian sites on Marco Island, Florida. *Florida Anthropologist* **18**, 21-27.

Wing, E.S. (1977). Vertebrates. In (B. Stark, Ed) *Prehistoric Ecology at Patarata 52, Veracruz, Mexico*. Vanderbilt University, Nashville: Publications in Anthropology, pp. 205-212.

Wing, E.S. (1978). Use of dogs for food: an adaptation to the coastal environment. In (B. Stark & B. Voorhies, Eds) *Prehistoric Coastal Adaptations: the Economy and Ecology of Maritime Middle America*. New York: Academic Press, pp. 29-42.

Wing, E.S. & Scudder, S.J. (1991). The exploitation of animals. In ( N. Hammond, Ed.) *Cuello: An Early Maya Community in Belize*. Cambridge: Cambridge University Press, pp. 84-97.

Wing, E.S. (nd). Vertebrate remains from the archeological sites in the Marismas Nacionales. Unpublished manuscript.

# DOGS OF ARMENIA

N. H. Manaserian
&
L. Antonian
Institute of Zoology, (ANAS)
P. Sevak st. 7., Yerevan 375014, Armenia

**Key words:** dog, Armenia, rock art, jackel, archaeozoology, Bronze Age, wolf

**Abstract**

In the archaeological collections of Armenia, bone remains of dogs occur regularly, although in small numbers. They are generally poorly preserved and present few opportunities for osteometric description. However, of the material recovered from 45 sites, 12 dog skulls were well preserved. Associated ceramics indicate that the skulls concerned can be traced back to the Bronze Age (fourth-second millennia BC). Analysis results showed that the skulls represent two different types: a medium-sized dog and a larger one. Artistic representations of dogs suggest their uses in the ancient Armenian culture.

**Introduction**

The dog is the only animal whose bones are rather well-preserved in the excavated archaeological material of Armenia. Although no definitely identified dog remains have been found in the faunal material of the Early Palaeolithic, bone fragments of dogs (humerus, metapodials, mandibles) as well as skulls have been found in "kitchen middens" dating from the Neolithic (eg Mokhrablur, fourth millennium BC) to the Middle Ages. Of 45 sites containing osteological material, only 12 sites contained intact dog skulls (Fig.1). The condition of the complete skull shown in Figure 2 (specimen N19) suggests that dogs were not used as food, since in this era animal brains were a favourite food which required smashing the cranium to extract.

**Osteometric analysis**

The cranial characteristics of the well-preserved dog skulls found in various sites of Armenia (fourth millennia BC to the Middle Ages) suggest that their general dimensions oscillate within constrained limits, evident also in specimens taken from a single excavated site (Fig. 3). In other words, there is little variation between skulls. The *processus postorbitalis* is slightly developed and the "stop" is small. Differences do occur in the sagittal crest (*crista sagittalis*): in some of the specimens it is poorly developed (Tsamakaberd 11, Sevan 15, Beniamin 1); in others, moderately developed (Keteel, Mokhrablur 19); in still others, well developed (Mokhrablur 18, Arteek 3, Gilli 5, Shirakavan 1). The nasal bones are generally shortened, comprising 34-40% of basal length. The muzzle is long, comprising 34-38% of total length, with a breadth of 20-22%.

The condition of the cranial sutures, as well as tooth eruption, indicate that all specimens were adult individuals. The basal skull length ranges from 176-205 mm, and the total length varies from 190-224 mm. As shown in Table 1, most of the specimens vary only within small limits: the skull excavated from the site at Gilli, however, is noticably larger than the others (Fig. 3).

According to Dal (1952), the dog of ancient Armenia was a rather archaic animal, close to a "peat spitz dog" in conformation. However, the dog recovered from the town site of Teinshaben (Iron Age) is very close to a jackal in size and shape, leading Mejlumian (1972) to suggest that descendents of both jackals and wolves are found in deposits from the Lake Sevan basin (Middle to Late Bronze Age). Therefore, some osteometric comparisons were undertaken to determine if there might be any basis for this conclusion. According to Gromov (1963), the temporal index in wolf-dogs is 22-28% and in jackal-dogs, 11-20%. Given that osteometric analysis of the archaeological material indicated 9 out of 13 dogs with a temporal index of 27-41% and 4 with a temporal index of only 15-21%, further comparisons were undertaken. Analysis of other canid skulls produced the following results for the temporal index: wolf (n=10), 22-43%; jackal (n=1), 15%; modern dog (n=3), 23-32%.

It was also noted that the *cingulum* on the first upper molar appeared to differ in shape between wolf and jackal. The

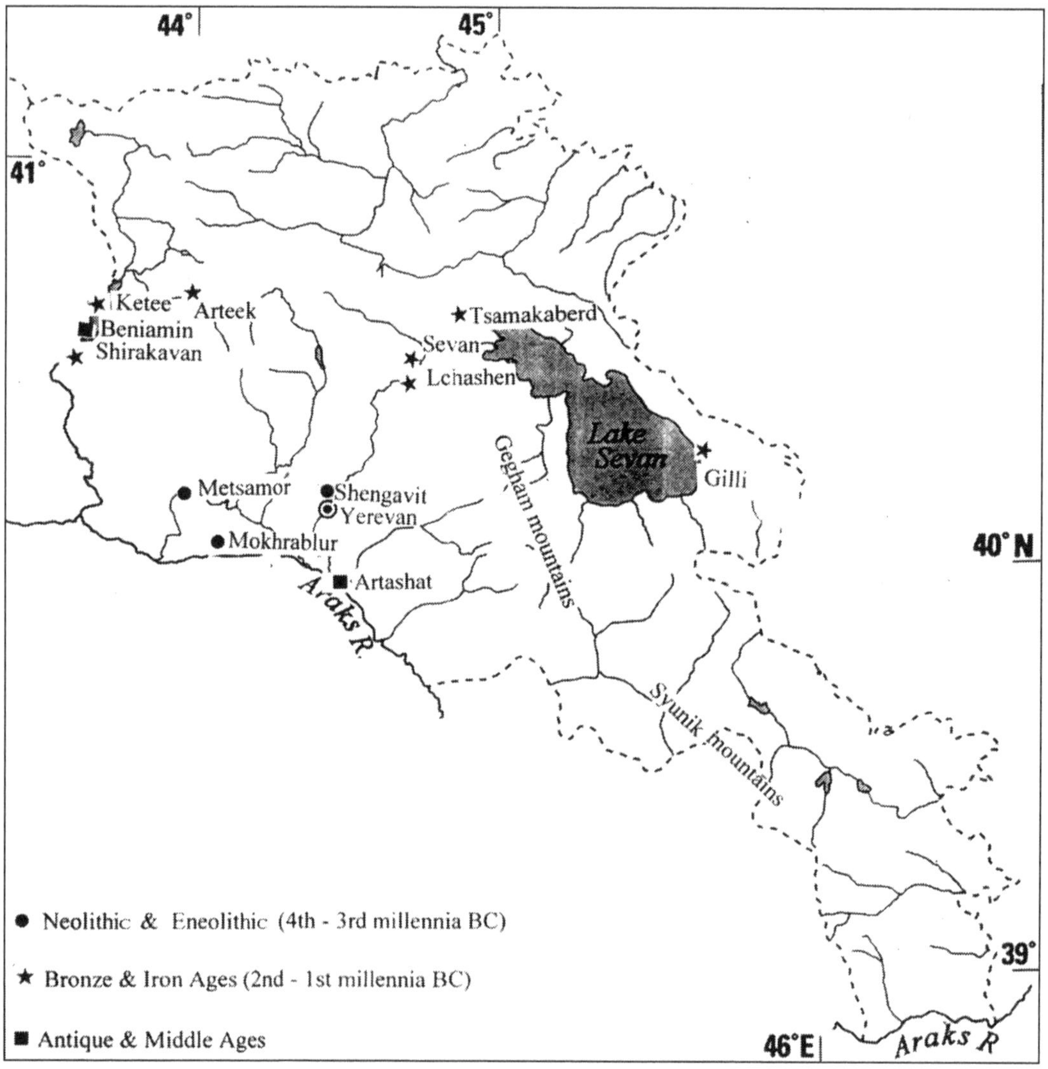


***Figure 1*** *Map of the archaeological sites in Armenia where dog remains have been found*

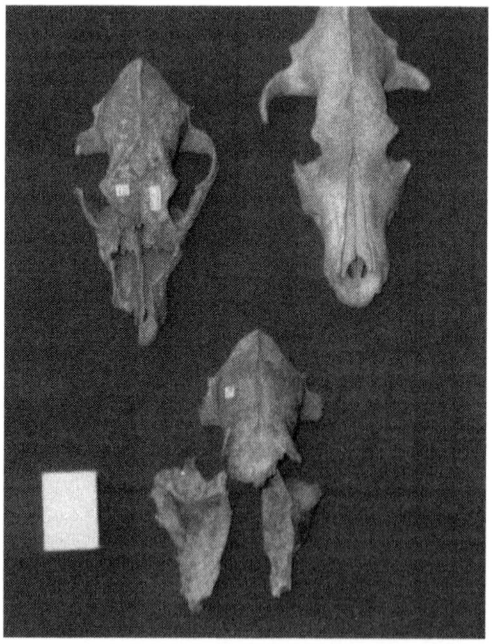

***Figure 2*** *Three Neolithic dog skulls from Mokhrablur, fourth millennium BC. Specimen N19 is upper R*

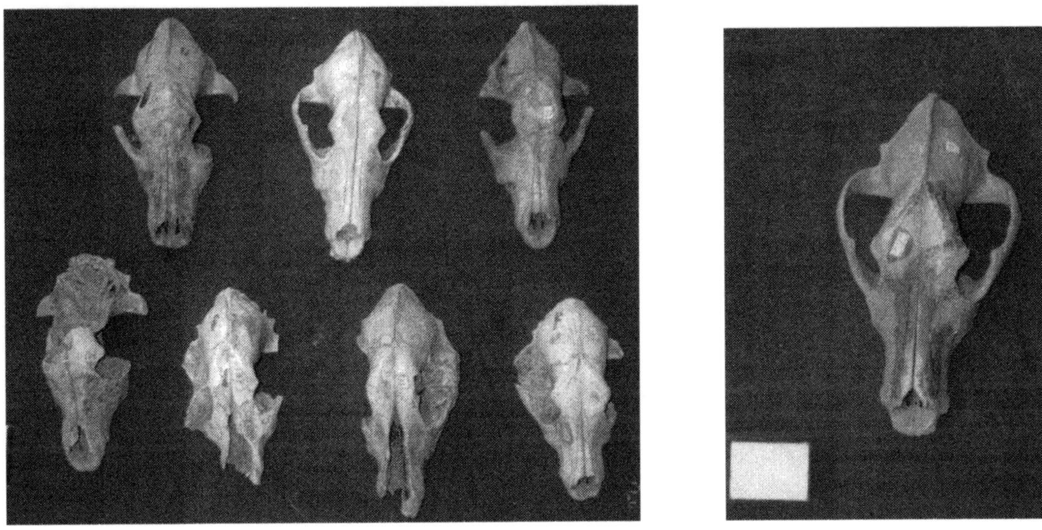

***Figure 3*** *A variety of Bronze & Iron Age dog skulls from Armenia, second-first millennium BC; right photo is the skull from Gilli (N28), see Figure 1 for location*

| Codes[a] | M#19[b] | M#21 | M#18 | K#1 | G#5 | B#1 | Ak#3 | L#27 | S#6 | At#3 | Sh#1 | T#2 | T#11 |
|---|---|---|---|---|---|---|---|---|---|---|---|---|---|
| 1 | 193 | 190 | 213 | | 224 | | 207 | | | | 193 | 211 | 198 |
| 2 | 176 | 178 | 198 | | 205 | 182 | 190 | | | | 182 | 194 | 175 |
| 3 | 183 | 180 | 198 | | 205 | 184 | | | | 190 | 189 | 203 | 186 |
| 4 | | | | | 120 | 117 | | | | | | | |
| 5 | | 36 | 41 | 50 | 47 | 47 | 41 | 42 | | 40 | 39 | | |
| 6 | 50 | 49 | 58 | | 55 | 55 | 55 | 57 | 56 | | 65 | | |
| 7 | 35 | 37 | 36 | | 35 | 37 | 37 | 41 | 40 | | 39 | 36 | 35 |
| 8 | | | 43 | 47 | 47 | 44 | | | 42 | 41 | 41 | 44 | 39 |
| 9 | | 57 | 66 | 68 | 68 | 62 | | 62 | 66 | 63 | 62 | | |
| 10 | | 43 | 45 | 47 | 48 | 43 | | 42 | 45 | 44 | 47 | | |
| 11 | | | 68 | 76 | 82 | | | 57 | | 68 | 63 | | |
| 12 | | 58 | 65 | 70 | 71 | 63 | 61 | | | 66 | 58 | | |
| 13 | | 84 | 94 | 94 | 97 | 79 | 91 | | | 89 | 90 | | |
| 14 | | 11 | 12 | 10 | 11 | 11 | 10 | | | | 37 | | |
| 15 | 14 | 14 | | | 15 | | 14 | | | | | | |
| 16 | | 6.0 | | | | | | | | | 13 | | |
| 17 | 20 | 18 | 20 | 20 | 18 | 19 | 20 | | 19 | | 22 | | |
| 18 | | | 13 | 12 | | | 12 | | 12 | | | | |
| 19 | | 78 | | | 81 | 80 | 83 | 75 | 76 | | 80 | | |
| | | | | | | | | | | | | | |
| | | | | | | | | | | | | | |

a) Measurement codes for Table 1: 1 Total length; 2 Basal length; 3 Condylobasal length; 4 Breadth between *os zygomaticum;* 5 Smallest length between orbits; 6 Greatest breadth of occiput; 7 Smallest breadth of occiput; 8 Breadth between canines; 9 Greatest breadth of occiput at the anterior root of $M^1$; 10 Breadth between *foramen infraoribitale;* 11 Length of nasal suture; 12 Length from *premaxilre* to *foramen infraorbitale;* 13 Length from *os palatinum* to *foramen magnum;* 14 Length between *incisivi* to *foramen palatinum;* 15 Length of *foramen palatinum;* 16 Smallest breadth of malar arc (*arcus zygomaticum*); 17 Greatest diameter of $P^1$ ; 18 Length of canine; 19 Breadth of canine

b) Prefix codes for site designations: M, Mokhrablur; K, Ketee; G, Gilli; B, Benjamin; Ak, Arteek; L, Lehashen; S, Sevan; At, Artashat; Sh, Shirakava; T, Tsamakaberd. See figure 1 for locations

***Table 1*** *Measurements (mm) of the skulls of dogs from Armenia*

*cingulum* shape on two archaeological specimens, those recovered from Mokhrablur (N21) and Benjamin (N3), showed the jackal type while in five other prehistoric skulls it is of wolf type. In addition, on the occiput of the wolf a cylindrical shaped swelling occurs next to the inner, anterior margin of the *bullae tympani*. This swelling is lacking in the jackal.

We have concluded that two types of dog, differing in size and skull conformation, definitely existed in ancient Armenia. Although preliminary analysis indicates these dogs could be descendants of different ancestral species, additional archaeological specimens from this and other regions are needed for a more comprehensive analysis. It would also be useful to assess the influence of sexual dimorphism on the variation in skull shape and size seen in ancient dogs. However, skeletal remains of dogs are not numerous, despite the fact that considerable osteological material has been recovered from deposits ranging from the Neolithic to the Middle Ages in Armenia, perhaps because so far only "kitchen middens" have been excavated. The fact that dogs do not appear to have been eaten means that their remains are seldom recovered from these types of deposits.

**Artistic representations**

Evidence for the role of dogs in prehistoric Armenia comes from two types of artistic representations. A number of dog and wolf figures have been found among zoomorphic statuettes from Early Bronze Age (third millenium BC), Bronze Age (nineteenth century BC) and Iron Age (ninth–eighth century BC) deposits. The earthenware and bronze statuettes shown in Figure 4 are of Early Bronze Age (Navur, Nor Bayazet). A cluster of statuettes and pendants (13 items) shown in Figure 5 are from the Iron Age (Ayrum, see Yesaian, 1980). All of the animals depicted have large heads and open mouths with frightful, bared teeth. The animal figures are strained, the front paws stretched forward, the back ones slightly tucked in and the ears upright: one has a feeling that the dogs are growling or barking at someone or something. These statuettes are very similar, the only distinction is in their tails: some have twisted tails, others have straight ones (this can be seen in rock carvings as well). In deposits from Paravakar (first millennium BC), a bronze statuette of a dog with a long body and flop-ears on a large head has been found among the statuettes of men and women. The dog with a "leash" suggests its use as a house guard (Fig. 6, #7).

The statuettes may have had protective significance linked with the existence of a cult of ancient Armenians called "araleses", which means idols with a human body and a dog head (Khorensky, 1893). The legends say they lick the wounds of injured men and revive them.

Diverse rock carvings depicting dogs are well represented in the Gegham (Martirosian, 1981) and Syunik (Karakhanian & Saphian, 1970) mountains: a herdsman with dogs, third-first millennia BC (Fig. 7 & 8); a hunter with dogs (Fig. 9); hunting for goats, third millennium BC (Fig. 10); a hunting scene with leopard, third-second millennia BC (Fig. 11); hunting for lion, second millennium BC (Fig. 12); a child with dogs (Fig. 13). Three dog shapes are repeated: one with a short straight tail (as in the statuettes), resembling a Caucasian sheepdog; one with a curled tail and cocked ears, strongly resembling a husky; another type, perhaps resembling the so-called Armenian greyhound ("Tazi").

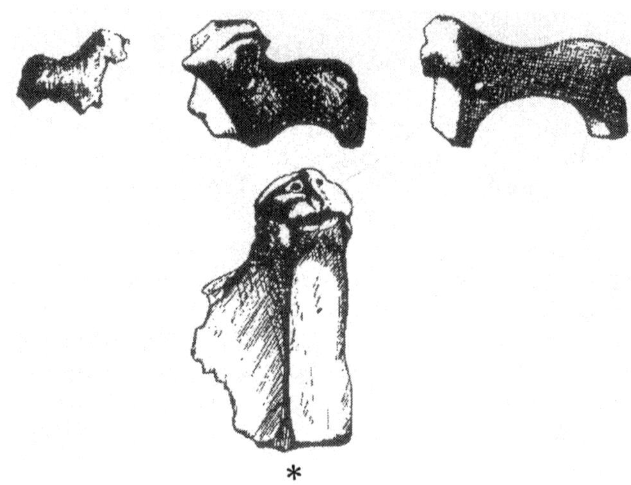


***Figure 4*** *Earthenware and bronze statuettes of Early Bronze Age (Navur & Nor Bayazet)*

***Figure 5*** *Statuettes and pendants from Ayrum (Iron Age)*

The possibility that the rock art figures represent actual dog races, rather than casual dog types, is suggested by their frequent occurrence in various pictures. It is likely that people living during this time were aware of a number of dog races used for various purposes (hunting, house guarding, flock guarding, etc). The origin of such diversity is the question that remains to be answered.

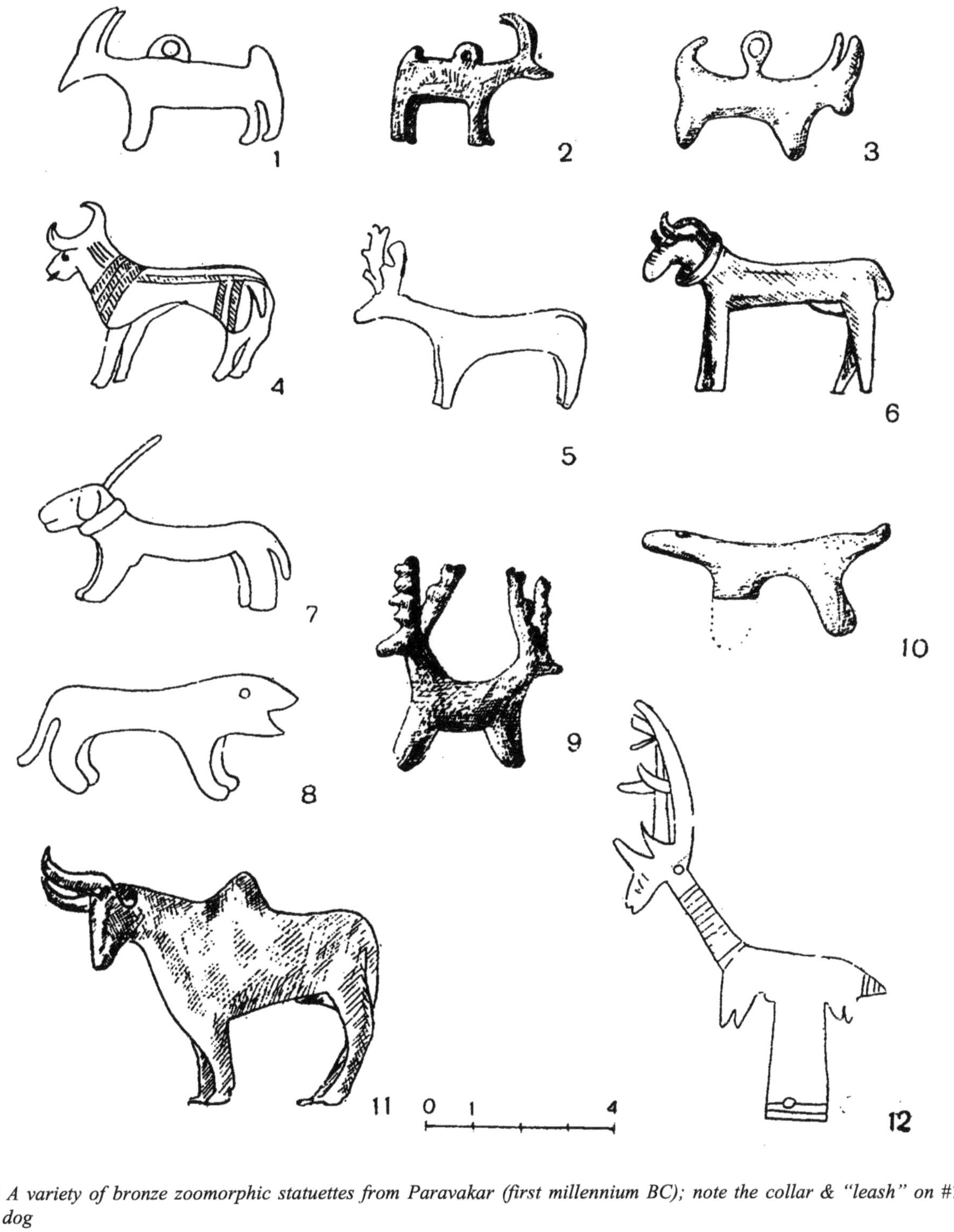


***Figure 6*** *A variety of bronze zoomorphic statuettes from Paravakar (first millennium BC); note the collar & "leash" on #7, which depicts a dog*

***Figure** 7 Pictographs of herdsman with dogs (*), third to first millennium BC*

***Figure 8** Pictograph of herdsman with dogs (*), third to first millennium BC, in another style*

***Figure 9*** *Pictograph of a hunter with a dog (*)*

***Figure 10*** *Pictograph of hunting goats with dogs (*), third millennium BC*

***Figure 11*** *Pictograph of a hunting scene, with dogs (*) and leopards*

***Figure 12*** *Pictograph of a lion hunt, with a dog (*) and goats, second millennium BC*

***Figure 13*** *Pictograph of a child with dogs (*) and a goat*

**References**


Dal, S. (1952). Research results of mammals from archaeological Teinshaben town of Urarthu. *News of the Armenian National Academy of Sciences* **1**.

Gromov, E. M. & Gureev, A. A. (1963). Mammals. In, *Fauna in the USSR* **II**, pp. 752.

Karakhanian, G. & Saphian, P. (1970). *Rock Carvings of Syunik*. Yerevan.

Khorensky, M. (1893). *History of Armenia*. Yerevan.

Martirosian, H. (1981). *Rock Carvings on the Ghegham Mountains*. Yerevan.

Mejlumian, S. (1972). *Palaeofauna of Eneolith, Bronze and Iron Ages in the Territory of Armenia*. Yerevan.

Yesaian, S. (1980). *The Sculpture of Ancient Armenia*. Yerevan.

# Part 5

# Modern primitive dogs

# THE NEW GUINEA SINGING DOG: A LIVING PRIMITIVE DOG

Janice Koler-Matznick
The New Guinea Singing Dog Conservation Society
5265 Old Stage Road, Central Point, OR 97502, USA

I. Lehr Brisbin, Jr.
University of Georgia, Savannah River Ecology Laboratory
P. O. Drawer E, Aiken, SC 29802, USA

James K. McIntyre
South West Pacific Research Foundation
631 Tarpon Avenue, #6391, Fernandina Beach, FL 32034

**Key words:** dog, primitive dog, New Guinea, Hallstrom's dog, NGSD, singing dog, dingo

**Abstract**

The New Guinea singing dog (NGSD), also known as Hallstrom's dog, was originally described in 1957 by E. Troughton as the new species, *Canis hallstromi*. Most authorities now favor one of the following designations: *Canis dingo*, *Canis lupus dingo*, *Canis familiaris dingo*, or *Canis familiaris hallstromi*. NGSD taxonomic affiliations need to be clarified by morphological and DNA studies. This canid originally occurred as free ranging populations in the remote mountain highlands of Papua New Guinea and Irian Jaya and some such populations might still survive in the most remote areas. Tamed specimens have been commensal with the local people. Recent anecdotal evidence indicates wild NGSD populations are declining or becoming rapidly hybridized with domestic dogs. The documented captive population in 1998 is about 100 animals. The NGSD has never been carefully studied in the wild. The few studies that have included NGSDs have demonstrated their uniqueness in relation to other canids. Several of their behaviors and vocalizations are unique and captive studies suggest that the NGSD is not a pack animal. NGSDs have 6-PGD and G-6-PD blood isoenzymes identical to the coyote and different from that of domestic dogs, Australian dingoes and wolves. NGSDs have an annual breeding season, with estrus normally beginning in August. However, if not impregnated, NGSDs frequently recycle into estrus again in eight to twelve weeks, a pattern never before recorded for any other canid. The above facts suggest that NGSDs should be preserved and studied both in a genetically pure and free ranging state and in captivity.

## Introduction

Hallstrom's dog, or as it is now more commonly known, the New Guinea singing dog (NGSD), was discovered by the scientific community in 1957 when the first pair was brought out of the Huri-Duna country of the Southern Highland District of Papua New Guinea to the Taronga Zoo, Sydney, Australia (Troughton, 1957). It was initially described as a new species, *Canis hallstromi*, by E. Troughton (1957). However, this designation has been challenged (Schultz, 1969), and most authorities (Bino, 1996; Brisbin *et. al.*, 1994; Corbett, 1995) now favor one of the following designations: *Canis dingo, Canis lupus dingo*, *Canis familiaris dingo*, or *Canis familiaris hallstromi*.

The authors know of no fossil NGSDs from the Papua New Guinea highlands. There are fossil dogs from the coastal area dated at 2,400BP but their cranial morphology does not match that of the NGSD (Gollan, 1982). However, the highland environment, with high humidity/annual rainfall and a plethora of endemic small scavengers, may not in general be conducive to bone preservation. Canids are in any case usually poorly represented in most fossil fauna, especially from non-anthropogenic sites, because they are very agile and rarely die in places, such as lake shores and sink holes, that facilitate fossilization (Olsen, 1985). In addition, as comparatively few scientific excavations have taken place in the Papuan Highlands, the sampling of the area is hardly comprehensive. Thus, the lack of "hard" evidence does not negate the possibility that the NGSD may have been on the island prior to, or coincident with, the arrival of the dingo in Australia. Current fossil evidence indicates that the dingo was present in Australia from at least 3,500BP, which is prior to the spread of domestic dogs through the Pacific islands (Gollan, 1982).

Although early explorers reported free-ranging *Canis familiaris* (domestic dogs) occurring from coastal to mountain habitats in New Guinea, free-ranging dogs resembling the original species description of *C. hallstromi* are now found only in a few remote locations of the Central Highlands. NGSDs occur in alpine, sub-alpine and lower to mid-montane mossy cloud forest habitats. There are other superficially similar types of *C. familiaris* in the lowland and coastal areas of Papua, New Guinea, but they are morphologically distinct from the NGSD and appear to be more closely related to Indonesian and Pacific Island dogs (Gollan, 1982).

The NGSD has never been directly studied in the wild and few scientific studies have included them as subjects. However, observations of captive NGSDs suggest some hypotheses about their behavior, anatomy and physiology. In this paper, we will update the findings of Brisbin *et al.* (1994). Specifically, we will review some observations of captive animals from the North American population, and suggest conservation measures that are now urgently needed for the few remaining genetically pure, free-living populations.

**Physical description**

According to a survey taken by the senior author in 1996 (n = 26 living NGSDs), the average adult in North America is about 40.5 cm tall (range 35.6-45.7) at the withers and weighs about 11.4 kg (range 9.5-14.1). These measurements compare favorably to those of Schultz (1971) whose NGSD heights ranged from 34 to 42 cm and weights from 8.2 to 12 kg. There is sexual dimorphism, with females being approximately 10% smaller overall than males. The NGSD body is slender and longer than tall due to disproportionately short legs. Despite the fact that NGSD heads are proportionately much wider than those of any fox, their general vulpine resemblance is frequently commented upon by casual observers today and was often remarked upon in earlier reports (Titcomb, 1969; Troughton, 1957). This superficial similarity is accentuated in red phase specimens with thick double coats, and by the NGSD's fluid, cat-like movements. Figure 1 shows a typical adult female NGSD.

The coat has three layers of hair (except for the ventral area, which consists mainly of finely textured medium length hair). They have a soft downy undercoat that becomes extremely dense in cold seasons, a short (0.5-1cm on the sides) to moderately long (2.5-3cm at the hackles) middle layer of guard hair, and an outer layer of sparse, bristle-like hairs slightly longer than the middle layer. This coarse outer layer is most prominent on the back, with especially stiff hairs over the rump. The pale cream-colored hair on the underside of the tail is about 5.5 to 6 cm long.

NGSDs have two color phases: red, and black and tan. The red coat varies from a brown-gold to a deep auburn and normally has a few black-tipped hairs admixed, especially on the backs of the ears and the tail. In a minority of specimens the black-tipped hairs are frequent enough to create a "sabled" effect. The black and tan variety may be merely an extension of the black already present in the red coat as the pattern does not have the tan spots over the eyes seen in most black and tan *C. familiaris*. NGSDs are born very dark sable, providing a basis for this adult color variation. There are no black and tans in the North American population, but they were fairly common in Germany (Schultz, 1969). They have also been reported from free-ranging and village populations in Papua New Guinea (McIntyre, personal observation & unpublished anecdotal reports by New Guinea highland residents) and one is shown in a photo in Flannery (1995: 58). In red NGSDs, the coat is counter-shaded to pale cream on the ventral areas. Young red specimens have a black muzzle, which fades to gray/tan by about five years of age. The coats of older adult red-phase (black and tan phase have not been observed by the authors) NGSDs usually become pale and grizzled with white. By an age of ten to fifteen years, the entire coat may begin to appear almost completely white with only faint shadings of red pigment and the black-tipped guard hairs remaining.

***Figure 1*** *A typical adult female New Guinea singing dog from the North American captive population.*

The guard hairs of the dorsal areas of the red color phase are banded from distal to proximate ends: red-cream-dark gray-cream (K.-M., personal observation). The presence of two distinct bands (defined as a color bordered by two other colors) is seen in *Canis lupus* (gray wolf) and *Canis latrans* (coyote) but C. *familiaris* normally has only one band (Kennedy, 1982). The hair on the skull between the ears has an especially wide gray band, and as the hair there is short and stands-off, the gray shows through the red giving that area an umber cast. Nearly all NGSDs have white on the toes, underside of the chin and tail tip. The possible adaptive significance of these markings is discussed below in the section on Social Signaling.

The NGSD head is distinctive. The eyes are almond to triangular in shape, obliquely set, and the mouth, nose and eye rims have black pigment. The ears are rather small, highly mobile, and capable of independent movement. They have a unique "conched" or "hooded" shape similar to that of a tulip petal and are set rather forward on the skull. The muzzle is equal to or just slightly shorter than the length of the skull, but the muzzle in live specimens appears small in proportion due to the extreme width of the zygomata.

Schultz (1969) found NGSD zygomata to be nearly as broad as those of Australian dingoes that were almost twice NGSD body size. Zygomatic widths in Schultz's NGSD specimens ranged from 84 to 96.5 mm in females (vs.102-115 in female dingoes) and 96.5 to 103 mm in males (vs.107-114.4 in male dingoes). Schultz's NGSD condylobasal lengths averaged 146.7 mm (range 142-151, n = 5) in males and 142 mm (range 124.5-142, n = 9) in females, while his dingo skulls averaged 182.3 mm in males (range 176-187, n = 4) and 172.8 mm in females (range 169-179, n = 6). Wood-Jones (1929) measured zygomatic widths on two un-sexed *Canis* specimens collected on Mt. Scratchley, which probably represented non-hybridized NGSD. These specimens measured 88 mm in zygomatic width for a 147 mm long skull and 90 mm for one 149 mm long. Two sexed skulls from NGSDs believed to be second-generation San Diego Zoo stock (third generation captive-born), in the collection of the University of California at Berkeley Museum of Vertebrate Zoology, Berkeley, California, U.S.A., were measured by the senior author (M.-K.). The male skull was 158.4 mm in length with a zygomatic width of 106.1 mm. The female's measurements were 141.4 mm and 88.3 mm respectively. Measurements of two contemporary skulls of North American captives in the senior author's collection, were 173.3 mm length and 105.5 mm zygomatic width for a large male that weighed 14 kilos in life and 155.2 and 97.2 for an average-sized 11 kilo female. Troughton (1971) gave the holotype specimen's condylobasal length as 151 mm and the allotype's as 149 mm, but did not give zygomatic width measurements.

As noted by Schultz (1969), a distinctive feature of the NGSD skull is the angle of the processus julularis, which are nearly perpendicular. Because of this, the crista saggitalis externa extends beyond the occipital condyli and the foramen magnum opens almost directly downward. Also, the upper carnassial tooth is relatively large, being about 10% of the condylobasal length (Wood-Jones, 1929), although NGSDs do not exhibit tooth crowding.

In addition to the skull measurements given above, Schultz supplies measurements for height at withers, head-body length, and length of the femur, humerus, tibia and radius (Schultz 1969: 71) and of the palatal length, molar row length, length of carnassial, frontal width, muzzle width, distance between canines, and length of canine (Schultz 1969: 70). The only other NGSD measurements located in the literature are in Troughton (1957), and are of the live holotype's ear (length 75 mm, width 40 mm), head and body length (650mm), tail (245 mm), fore paw (dew claw to ground: 25 mm) and hind paw length (heel to end of toe:145 mm).

Gollan (1982) compared NGSD skull measurements to those of the Australian dingo and South East Asian island domestic dogs. His analysis indicated that the NGSD's skull and tooth measurements appeared to be more similar to the South East Asian dogs (although NGSDs are more robust and generally larger), than to the Australian dingo. However, recent preliminary mtDNA sequencing (Matisoo-Smith *et al.*, 1998), has revealed that the NGSD and the Australian dingo are related to other Austronesian dogs, but that they share a mtDNA point mutation. These mtDNA sequences (T. Glenn & I. L. Brisbin, Jr., pers. comm.) were not found in the Asian Pacific dogs and other purebred domestic dogs tested. Preliminary nuclear DNA sequencing (Gergits & Brisbin, 1995) revealed the NGSD has 2-3 genetic markers in the size range of 7-10 Kb that were not found in the Alaskan gray wolves or seven domestic dog breeds tested. Thus, molecular techniques suggest that the dingo and the NGSD are indeed closely related.

## NGSD traits

NGSDs appear to have several unique characteristics when compared to *C. lupus* and *C. familiaris*. Little scientific study has been done on their anatomical or physiological characteristics, but casual observations and one blood enzyme study indicate they have attributes that are unusual.

Their joints, especially those of the spinal column, seem to be extremely flexible. They rotate the radius, femur and humerus and flex the digits to a greater degree than most *C. familiaris*. Like a cat, they can pass their bodies through any space wide enough to admit their zygomata. This flexibility may be an adaptation to hunting in thick forest and rugged mountain habitats. Certainly it facilitates their impressive jumping ability (a NGSD 41 cm at the shoulder can effortlessly spring 1.5 m up or out from a standing position).

The tapetums of NGSD eyes are highly reflective, glowing bright green when the animal looks out from a low light area into a brighter light source. The effect is quite striking, as if the eyes are lit from within. No anatomical studies have been done on NGSDs, so the structural basis of this peculiar degree of reflectance is not known. It may be that NGSD pupils dilate to a greater diameter than most canids and allow more of the reflected light to escape the eye. Alternatively, their tapetums, like those of the cat, may actually reflect a greater proportion of the ambient light. Either way, this would seem to be an adaptation that would facilitate hunting in low light. Anecdotal reports by New Guinea highland people gathered by McIntyre in 1998 indicate NGSDs are most active (most frequently sighted) during the crepuscular hours and at night.

Three other NGSD characteristics are unique compared to the *C. lupus* species group. Two were ascertained in the laboratory tests of V. Simonsen (1976), who reported that NGSDs have two blood isoenzymes that differ electrophoretically from those of the tested domestic dogs, dingoes and wolves. The NGSD 6-phosphogluconate dehydrogenase enzyme matched that of *C. latrans* and their Glucose-6-phosphate dehydrogenase matched that of *C. latrans* and *Vulpes vulpes* (red fox).

Although no tests have been done to determine the underlying physiological mechanisms, NGSDs also appear to exhibit a unique estrus pattern (Brisbin *et al.*, 1994). Their breeding season, July through September, coincides with that of *C. l. pallipes* (Indian wolf) (G. Das, pers. comm.), *C. dingo* (Corbett, 1995), and pariah dogs from India (Oppenheimer & Oppenheimer, 1975), and Africa (Coe, 1994). However, about 76% of NGSD females that are not impregnated during their annual cycle will recycle into estrus a second time 8 – 12 weeks later and, more rarely, even for a third time. One female kept by the senior author (K.-M.), when bred during an abbreviated cycle, whelped a litter of six puppies, the maximum recorded litter size for captive NGSDs. Thus, the short cycles are physiologically complete. This pattern, shown in Figure 2, has not been reported for any other canids and could represent an adaptation of this non-pack canid to low population numbers.

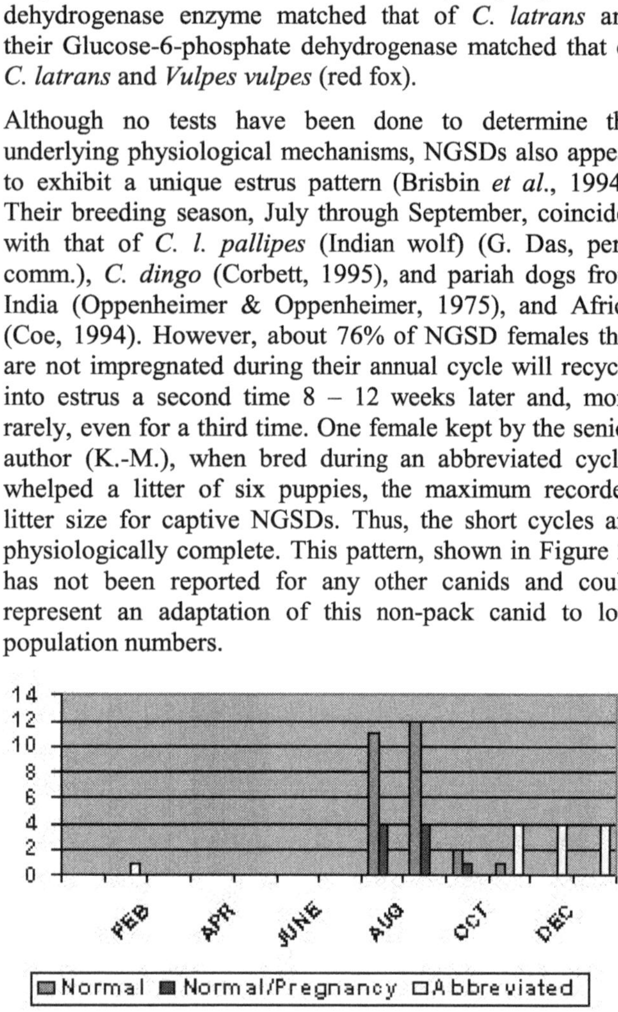


***Figure 2*** *Frequencies of estrus cycles of New Guinea singing dogs. Data are based on a total of 41 cycles recorded for six captive specimens in the authors' collections. In about 76% of the cycles which did not result in pregnancy, the females cycled again in two to three months.*

## Captive propagation

The North American NGSD population has had only four founders. Two of these were offspring of the first captive Papuan pair at the Taronga Zoo, Australia. The progeny of this original Taronga Zoo pair were distributed to zoos around the world, including a pair sent to the San Diego Zoo, California, in 1958 (Brisbin *et al.*, 1994; Titcomb, 1969). This San Diego Zoo brother/sister pair were the original founders of the North American population and no unrelated specimens were added to the gene pool until a female was imported from Germany in 1987 by the Sedgwick County Zoo, Wichita, Kansas (Brisbin *et al.*, 1994). This German female descended from five NGSDs obtained in Irian Jaya, collected by W. Nelke in 1976 (Voth, 1988) from the Eipo tribe in the Famek and Eipomek river valley districts of Irian Jaya while on an expedition for the Museum of Ethnography, State Museums, Berlin, Germany. In 1994, Brisbin imported the fourth founder, a 14-year-old male, bred by the Baiyer River Sanctuary, Papua, New Guinea. This male's sire was from the original Taronga Zoo/Papuan line (ie the same as the pair imported to the San Diego Zoo), but his dam was reported to have been wild caught in the vicinity of Wapamanda in the Papua New Guinea highlands (R. McKay, pers. comm.). Descendants from the Papuan/Irian Jaya crosses and the offspring of one Papuan/Irian Jaya cross female and the Baiyer River male now form the North American breeding population.

## Behavior

In general, NGSDs show all of the common behaviors of *C. lupus* and *C. familiaris*. However, of the 246 behaviors recently described for NGSDs (Koler-Matznick & Brisbin, unpublished ms.), 25 (about 10%) are unique to them or are expressed differently (lower or higher threshold, different frequency, etc.) compared to the same behaviors as described for *C. lupus* and *C. familiaris* (Table 1). A review of the literature (eg Bekoff, 1974; Bekoff 1977; Bradshaw & Nott, 1995; Cohen & Fox,1976; Fagen, 1981; Fox, 1971; Fox, 1975; Gittleman, 1996; Goodman & Klinghammer, 1990; Kleiman, 1967; Schenkel, 1967; Scott & Fuller,1965; Serpell,1995) suggests that 17 of these behaviors (about 7%) have not yet been described for any other *Canis* species.

The NGSD "head toss" attention-soliciting motion is especially distinctive. During the full expression of the "head toss" the neck is flexed far to one side and the head and neck are rotated through a 90-degree arc and brought forward to the starting position. This complete behavior is executed in about 1.0 second. In an abbreviated "head toss," the head may be merely flicked quickly to one side and back. Anecdotal reports indicate that an apparently similar "head toss," also performed as a solicitation behavior, may be present in certain aboriginal breeds of *C. familiaris*, namely the Karelian Bear Dog (D.Deeley, pers.comm.) and the African Basenji (J.Dalgleish, pers. comm.).

Another behavior not found previously described is the "cheek rub" greeting/marking behavior in which, after an initial visual and olfactory inspection, the NGSD rapidly wipes one or both cheeks in one direction on an object. The objects "cheek rubbed" are diverse, and include toys, conspecifics and unfamiliar humans. Except for the musk of their interdigital glands, NGSDs are, in general, odorless to a human. However, the cheek area has a faint, sweet musk odor that may indicate the presence of a scent gland and this "cheek rub" may be a marking behavior

| Behavior Name | Behavior Function | Unique | Different |
|---|---|---|---|
| Barrier Lick | displacement (?) | X | |
| Cheek Rub | greeting/marking | X | |
| Chorus Howl | not determined | | X (form) |
| Copulation Contraction | not determined | X | |
| Copulation Scream | not determined | X | |
| Counterpoint howl | not determined | X | |
| Crow | greeting | X | |
| Genital Bite | play/aggression | X | |
| Grin | greeting/submission | | X (form) |
| Head Toss | solicitation of attention | X | |
| Hind Paw Groom | self care | X | |
| Howl | variable/contact seeking | | X (form) |
| Mouth Wipe | self care | X | |
| Observation Stand | predation | X | |
| Open Mouth Bite | play | | X (as in C. latrans) |
| Open Mouth Whine | attention soliciting | X | |
| Perching | resting/observing | X | |
| Place Food | facilitates feeding | X | |
| Scent Rub | not determined | | X (lower threshold) |
| Sheath Cleaning | self care | X | |
| Sneeze | reaction to volatiles | | X (cause) |
| Tail Flagging | social communication | | X (form) |
| Tail Grab | play | | X (frequency) |
| Tooth Gnashing | controlled excitement | X | |
| Trill Vocalization | not determined | X | |
| Wide Sit | resting posture | | X (form) |

***Table 1** Behaviors that are unique or expressed differently in NGSDs compared to previously describe behaviors for C. lupus and C. familiaris. The reason the behavior is considered different is indicated - see text for a partial list of literature citations surveyed*

comparable to a similar behavior in *Felis domesticus*, the domestic cat. The cheek scent is very noticeable in dams with young puppies and the puppies are "perfumed" with it.

NGSDs appear to lack one behavior common in all other *Canis* species: the stereotyped "play bow" invitation to engage in play, in which the forehand is lowered and rump elevated, usually with tail wagging, and the position held for an average of 0.34 second (Bekoff, 1977). This specific behavior has not been seen by the authors in more than 2,500 combined hours of NGSD observation of scores of individual specimens including all age groups. Why the NGSD should lack this behavior is so far a mystery.

**Social organization**

The behavior of captive NGSDs suggests that this canid must have a monogamous, non-pack social organization in the wild. Family groups of parents and offspring can be maintained together until the juveniles begin to reach sexual maturity. However, with only a few exceptions, in all North American collections, adult same-sex pairs, even siblings raised together, show extreme aggression toward each other during estrus season when a potential mate is present. If the same-sex individuals are not separated, or the potential mate removed, this aggression has in most cases resulted in severe, even fatal, injuries. In this respect, they differ from nearly all other forms of *Canis*.

Given enough space, as in the wild or living as free-ranging commensals with the indigenous people of New Guinea, perhaps NGSDs can maintain casual associations with familiar same-sex individuals, and avoid them during the reproductive season by dispersing into separate territories. The possibility that this social capacity exists has been demonstrated at the Neumunster Zoo, Germany, which was able to keep a colony of up to 17 individuals of the Irian Jaya line in a large exhibit (Voth,1988). However, there were adult fatalities in this group, and most of the pups were killed before weaning age.

Anecdotal reports from several NGSD keepers indicate that, unlike domestic dog males, male NGSDs will regularly provide parental care and regurgitate for puppies if they are allowed to remain with the dam and offspring.

**Social signaling**

Two of the arguments for the domesticated status of the NGSD that proved instrumental in prompting the scientific community to re-classify this canid as a feral domestic dog were: (1) they have white markings and (2) they are submissive and friendly toward familiar humans (Schultz, 1969). However, white markings are not rare in the Canidae (Ortolani & Caro, 1996) and there appear to be sound adaptive functions for the NGSD's markings.

Like many primitive feral/pariah forms of *C. familiaris*, NGSDs have fish-hook shaped or slightly curled tails with

***Figure 3*** *New Guinea singing dogs "flagging" their tails in excitement.*

pale fluffy undersides. The tails are held above the level of the back when the animal is at the alert (Fig. 3). Anecdotal evidence from captive animals suggests that the prominent display ("flagging") of the underside of the tail may have social significance and help coordinate the efforts of individuals in hunting or enhance signaling during social behaviors.

The white tail tip accentuates tail movements and positions (ie the "friendly" wag or the stiffly held dominant tail position), especially in dim light such as moon light. The specific differences of NGSD tail shape, carriage and markings allow identification of individuals by humans from a hundred meters or more, and presumably by NGSDs also. The white stripe on the NGSD under-jaw is flashed during their "head toss" solicitation motion and during passive submissive behaviors such as "rolling over." In young animals the contrast of this ventral white stripe with the black muzzle is striking and would facilitate communication of non-antagonistic intent while the animals are establishing their relationships. Thus, NGSD markings and tails seem to provide adaptive communication functions for this aggressive, non-pack canid.

It should be noted that the white markings on the captive NGSDs are more prominent today than when the Schultz article was published in 1969. The Irian Jaya strain was not imported into Germany until 1976 and was introduced into the USA in 1987 and Australia in 1989. These Irian Jaya village NGSDs reportedly had been artificially selected by the Eipo people for large white markings (Voth, 1988). If this is true, to our knowledge this is the only directed artificial selection to which NGSDs have been subjected. Thus, today, in addition to the original Papuan strain's white chin/throat, toes, tail tip, and an occasional small patch on the nape, many captive NGSDs also have white blazes on their faces, partial to full white collars and white stockings occasionally extending as much as half way up the legs. However, they are never piebald spotted, a commonly accepted marker for domestication.

The ease with which NGSDs are tamed and their friendliness toward people are probably not reliable indicators of domestication. Many captive-raised wild canids become "friendly" toward humans. Captive NGSDs are extremely shy of humans if not well socialized when young, so their tameness is conditional upon their early rearing regime. Although captive-born NGSDs may form social bonds with familiar humans, actively seeking attention from them by greeting them and submitting to them as they would to an older, familiar conspecific, wild-born, free-ranging individuals in the Papuan highlands are extremely difficult to observe (Bino,1996; McIntyre, personal observation,1996). Unlike most *C. familiaris*, even socialized captive NGSDs retain a fundamental psychological and emotional independence and autonomy from humans (ie they do not become "devoted"), a trait previously noted for captive raised wild canids (Fox, 1971).

## Vocalizations

One of the most striking features of NGSDs is their unique vocalizations. The howling which gave rise to their popular name is marked by an extraordinary degree of frequency modulation (Brisbin *et al.*, 1994) compared to that of *C. lupus* and *C. familiaris*. NGSD group howls also show a division of "roles" within the group with lead "soloists" and supporting "harmonizers" that howl in apparent response to the leader. These dogs have many calls in their vocal repertoire that are different from those reported for other canids, including some that, in sonogram form, resemble bird calls (Brisbin *et al.*, 1994). However, the details of the structural complexity and functional significance of NGSD vocal patterns are not yet well understood. Mark Feinstein, Hampshire College, Amherst, MA, USA, is presently studying NGSD vocalizations.

**Population status and conservation**

Although an accurate NGSD world population census has proven difficult due to language barriers and the diverse record-keeping habits of zoos, recent inquiries to zoos in Australia and Germany, and a review of the current NGSD listings with the International Species Inventory System (ISIS) (Earnhardt *et al.*, 1995), indicate a striking decline in documented breeding NGSDs over the past twenty years. Even the Taronga Zoo in Australia, long one of the principle institutions to hold and breed NGSDs, no longer has any in their collection. The zoo population is reduced because once their species designation was changed to feral *C. familiaris*, their perceived value as zoo exhibits declined and zoos either divested themselves of their NGSDs or stopped breeding them. We believe there are less than 100 documented, reproductive NGSDs in captivity today.

In North America there also exists an undocumented NGSD population, begun about 10 years ago when zoos started to sell their NGSDs to private exotic animal breeders. These breeders frequently fail to keep pedigree and origin information on their NGSDs. In some cases, moreover, some of these undocumented animals may have been hybridized, either inadvertently or deliberately, with domestic dogs. Thus, until such time as a molecular marker can be devised to identify genetically pure specimens, these NGSDs, which perhaps number as many as the documented population, are unfortunately lost to any conservation breeding program.

Practically nothing is known of the status of the wild NGSD population, except that near human habitations it is rapidly being hybridized with village and imported domestic dogs. As early as 1969, the locals in many areas of the Central Highlands reported that they no longer heard NGSD howls in areas where they previously were heard frequently (Titcomb,1969). In addition, virtually nothing is known of NGSD natural history under free-ranging conditions, aside from two investigations, both based on indirect evidence (Bino,1996; McIntyre, unpublished data). Thus, even rough estimates of minimum population levels are unavailable at this time. Brisbin *et al.* (1994) name the few locations where anecdotal evidence indicated there might still be pure NGSDs, including Mt. Giluwe ("Walmilako" in the local language , meaning "the dog" ), Mt. Wilhelm (both the former are national parks where hunting is forbidden), and the mountain range in Sanduan Province, western Papua New Guinea, that includes Mt. Stolle. Mt. Stolle was visited in 1996 by one of the authors (McIntyre). During this visit McIntyre collected casts of NGSD tracks, scats, and reports from local informants in the area of the remote Mekil (another local name for Mt. Stolle) Research Station, maintained by the OK Tedi mining company, which is recovering gold from the area. His findings suggested free-ranging NGSDs were still present in this area although no estimate of possible numbers could be determined. The scats have not, as of this date, been analyzed. In the same area, from 1994 to 1996, Bino (1996) observed NGSD prey remains and resting areas, collected scats for analysis, and obtained verbal reports of NGSD habits from local villagers. Again, no estimate of population numbers could be made.

Neither investigator was able to directly observe wild NGSDs. Even the local hunters interviewed during the above studies commented they rarely caught glimpses of these elusive canids, so studying free-ranging NGSDs would undoubtedly be an extraordinary but extremely interesting challenge.

McIntyre found that the local people admire the wild NGSDs and the Highland Wild Dog 'mystique' has a strong foothold in the lives of the people living at the base of the mountain. Native people differentiate between the "wild" NGSDs and domestic dogs, and have separate names for them. NGSDs figure prominently in many of the native people's folklore and myths, including several "origin" stories, which relate how the NGSD brought language, fire, taro root (a staple crop), or even that the natives themselves descended from the NGSD (McIntyre, unpublished data; Titcomb, 1969; Voth, 1988). This strong "spiritual" connection to the NGSD will perhaps facilitate the cooperation of local people in preserving remnant wild populations, a necessary prerequisite to any conservation effort.

In this regard, it should be noted that McIntyre spoke to the villagers nearest Mekil about the importance of keeping the wild NGSDs free from cross-breeding, and asked if they could refrain from taking their domestic dogs with them when they went to Mekil (a two hour walk) to hunt. They agreed, but went one step further and gave their dogs to more distant relatives. Since that time no fresh NGSD tracks or droppings have been reported by researchers working on other studies in the Mekil area. Bino (1996) found that NGSDs moved around a great deal, as resting sites were not used for more than a day or two consecutively. He also discovered that they travel to areas where the *Elaeocarpus* sp. and *Panandus* sp. fruits are ripe (whether to eat the fruits or the potential prey attracted by the fruits is unknown but seeds of these plants were found in NGSD scats). Thus, the Mekil NGSDs may merely have moved on to another portion of their territory and may return in the future.

We believe these unique dogs should be preserved and studied in both free-ranging and captive populations. Due to their long-term isolation and the general absence of artificial selection, the NGSD may offer one of the few remaining opportunities to learn important information about the origin of the domestic dog, and to gain new insights into canid adaptation and evolution. Below are our conservation recommendations and the urgent questions that we feel need to be answered.

**Conservation priorities**

1 Determine the population status of free-living NGSDs in Papua New Guinea and Irian Jaya.

2 Develop molecular genetic markers to assist in detecting possible hybrids in both free-ranging and captive populations, and to assist in the management of the captive population.

3 Determine the genetic integrity and threat of hybridization from domestic dogs for all remaining populations.

4 Obtain new animals and/or semen from free-living populations to increase the founder base of the current captive population.

5 Seek support from appropriate governmental agencies in Papua, New Guinea and Irian Jaya, to limit the spread of free-ranging domestic dogs in habitats where genetically pure NGSDs may still occur.

**Unanswered questions**

1 Is the NGSD a form of primitive feral domesticated dog, or is it a unique wild species, a type of "proto-domesticate" or commensal, that has maintained its ancestral identity due to isolation from other canids and the absence of directed artificial selection?

2 What is the basic social organization, ecology, and natural history of free-ranging NGSDs?

3 In light of their suggested monogamous social structure and unique vocalizations, what is the relationship of these dogs to other primitive *C. familiaris*, including the Australian dingo?

4 How do NGSD physiological characteristics and internal anatomy compare to those of other *Canis* species?

5 What ethical considerations are associated with permitting the captive NGSD population to progress to a state of full domestication (eg through acceptance into the registry and exhibition activities of organizations such as, in the USA, the United Kennel Club or the American Kennel Club)? A number are already living as house pets in the USA - what are the possible ramifications of the long-term selection of such NGSDs for "pet" traits that will inevitably result from this situation?

**Acknowledgements**

Manuscript preparation was supported in part under financial assistance award no. DE-SC09-96R18546 from the U.S. Department of Energy to the University of Georgia. We acknowledge Darwin Matznick for his support and assistance with the senior author's study, and Nicolaus C. Rhoden for stimulating conversations and for sharing information from her research on the NGSD. We also thank Juliet Clutton-Brock and Marc Bekoff for their encouragement and insights, and the two anonymous reviewers for their helpful comments and suggestions.

**References**

Bekoff, M. (1974). Social play in coyotes, wolves and dogs. *BioScience* **24,** 225-230.

Bekoff, M. (1977). Social communication in canids: evidence for the existence of a stereotyped mammalian display. *Science* **197**, 1097-1099.

Bino, R. (1996). Notes on the behaviour of New Guinea singing dogs (*Canis lupus dingo*). *Science in New Guinea* **22**, 43-47.

Bradshaw, J. W. & Nott, H. M. R. (1995). Social and communication behavior of companion dogs. In (J. Serpell, Ed.) *The Domestic Dog*. Cambridge, MA: Cambridge University Press, pp. 115-130.

Brisbin, I. L., Jr., Coppinger, R. P., Feinstein, M. H., Austad, S., & Mayer, J. J. (1994). The New Guinea singing dog: taxonomy, captive studies and conservation priorities. *Science in New Guinea* **20**, 27-38.

Coe, S. (1994). *The Basenji: Out of Africa to You.* Wilsonville, OR: Doral Publishers.

Cohen, J. A. & Fox, M. W. (1976). Vocalizations in wild canids and possible effects of domestication. *Behavioural Processes* **1,** 77-92.

Corbett, L. (1995). *The Dingo in Australia and Asia.* Ithaca, NY: Cornell University Press. Earnhardt, J. M., Thompson, S. D., & Willis, K. (1995). ISIS database: an evaluation of records essential for captive management. *Zoo Biology* **14**, 493-508.

Fagen, R. (1981). *Animal Play Behavior.* New York: Oxford University Press.

Flannery, T. (1995). *Mammals of New Guinea*. Ithaca, NY: Cornell University Press.

Fox, M. W. (1971). *Behavior of Wolves, Dogs and Related Canids*. Malabar, FL: R. E.. Krieger.

Fox, M. W. (1975). *The Wild Canids: Their Systematics, Behavioral Ecology and Evolution.* Malabar, FL: R.E. Krieger.

Gergits, W. F. & I. L. Brisbin, Jr. (1995). A taxonomic reassessment of the New Guinea singing dog. Paper presented at the American Society of Mammalogists Conference, Burlington, IL.

Gittleman, J. L. (1996). *Carnivore Behavior, Ecology and Evolution Vol. 2.* Ithaca, NY: Cornell University Press.

Gollan, K. (1982). *The prehistoric dingo.* Ph.D. Thesis. Australian National University, Canberra.

Goodman, P. A. & Klinghammer, E. (1990). *Wolf Ethogram.* Battleground, IN: North American Wildlife Park Foundation.

Kennedy, A. J. (1982). Distinguishing characteristics of the hairs of wild and domestic canids from Alberta. *Canadian Journal of Zoology* **60**, 536-541.

Kleiman, D. (1967). Some aspects of social behavior in the Canidae. *American Zoologist* **7**, 365-372.

Matisoo-Smith, E., *et al.* (1998). The Austronesian origins of the dingo and other Pacific dogs: evidence from mitochondrial DNA. Paper presented at the 8th International Congress of Archaeozoology, August, 1998, Victoria, B.C., Canada.

Olsen, S. (1985). *Origins of the Domestic Dog: The Fossil Record.* Tucson: University of Arizona Press.

Ortolani, A. & Caro, T. M. (1996). The adaptive significance of color patterns in carnivores: phylogenetic tests of classic hypotheses. In (J. L. Gittleman, Ed.) *Carnivore Behavior, Ecology and Evolution Vol. 2*. Ithaca, NY: Cornell University Press, pp. 132-188.

Oppenheimer, E. C. & Oppenheimer, J. R. (1975). Certain behavioral features in the pariah dog (*Canis familiaris*) in West Bengal. *Applied Animal Ethology* **2**, 81-92.

Schultz, W. (1969). On Hallstrom's dog (*Canis hallstromi*, Troughton 1957). *Zoologischer Anzeiger* **183**, 47-72.

Scott, J. P. & Fuller, J. L. (1965). *Genetics and the Social Behavior of the Dog*. Chicago: University of Chicago Press.

Schenkel. R. (1967). Submission: its features and function in the wolf and dog. *American Zoologist* **7**, 319-329.

Serpell, J. (1995). *The Domestic Dog: Its Evolution, Behavior and Interactions with People.* NY: Cambridge University Press.

Simonsen, V. (1976). Electrophoretic studies on the blood proteins of domestic dogs and other Canidae. *Hereditas* **82,** 7-28.

Titcomb, M. (1969). *Dog and Man in the Ancient Pacific.* Honolulu: Bernice P. Bishop Museum Special Publication **59**.

Troughton, E. (1957). A new native dog from the Papuan Highlands. *Proceedings of the Royal Zoological Society of New South Wales* **1955-56**, 93-94.

Troughton, E. (1971). The early history and relationships of the New Guinea highland dog (*Canis hallstromi*). *Proceedings of the Linnean Society of New South Wales* **96**, 93-97.

Voth, I. (1988). Social behavior of New Guinea dingoes (*Canis lupus f. familiaris*): expressive behavior, social organization and rank relationships. D.V.M. Thesis, Ludwig Maxmillian University, Munich.

Wood-Jones, F. (1929). The cranial characters of the Papuan dog. *Journal of Mammalogy* **10**, 329-33.

# Part 6

## Archaeozoological analysis methods & results

# BREED DIFFERENCES IN DOG HAIR

Michael L. Ryder
4 Osprey Close, Lord's Wood
Southampton, SO16 8EX, UK

**Key words:** dog, hair, breeds, hair analysis, archaeozoology

**Abstract**

Preliminary coat length and hair diameter data are presented for ten breeds covering five Kennel Club groups of dog. In general larger breeds had greater hair diameters than smaller breeds, and males had greater hair diameters than females. Such coat variation in archaeological remains could throw light on dog breed origins as it has done in sheep. Most breeds had a skewed hair diameter distribution in which the bulk of the hairs were fine (the undercoat) with fewer coarser fibres (the outer hair). In some animals there was a seasonal narrowing of the hair in autumn, which indicates a seasonal cycle of growth with a spring moult.

## Introduction

Hair and wool fibre diameter measurements made on archaeological remains have not only thrown light on the coat type of livestock - cattle and goats, and notably the fleece types of sheep (Ryder, 1983) - but also on husbandry practices (Ryder, 1998). There have been few reports of remains of dog hair, but as more are found, and hair shaft diameter measurements made, similar light is likely to be thrown on breed origins in domestic dogs (*Canis familiaris*), which could have had hair differences even before, like sheep, they were bred for coat type. I here put on record similar measurements of dog hair to start a data base for archaeozoologists.

There are over 400 breeds of dog varying greatly in size and shape from the tiny Yorkshire terrier to the Great Dane as a result of human selective breeding. This wide variety of breeds and the fact that they tend to be classified by function (as in the British Kennel Club's used here), makes it difficult to discern their genetic affinities.

## Material and methods

Hair from domestic pets was measured as part of a trial of dietary additives aimed at improving coat condition in dogs. This was organised by A.van den Broek of the Royal (Dick) School of Veterinary Studies, Edinburgh. The ten breeds investigated are listed in Table 1 under the British Kennel Club's classification, which does not necessarily have any biological or historical relevance. The letters L and S indicate the designations as either large or small breeds for the purposes of the trial. The (popular) coat descriptions given come from Breed Society specifications. The number of animals investigated in each breed is indicated in brackets in Table 2.

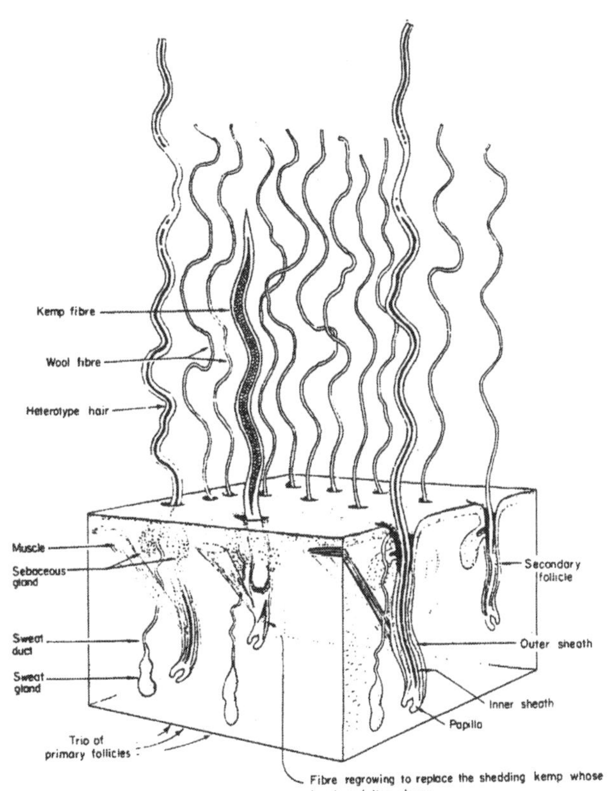


***Figure 1** A typical hair follicle group illustrated by an idealised group in sheep. The (three) larger primary follicles lie on one side of the group and the smaller and more numerous secondaries (growing the underwool) lie on the other. The outer (guard) hairs grown by the primary follicles comprise bristles (kemps in sheep) produced by the central primaries, and awns (heterotypes in sheep) grown by the two lateral primary follicles of the trio. Note the latticed medulla of the bristles and the non-latticed medulla of the awns (from Ryder, 1973).*

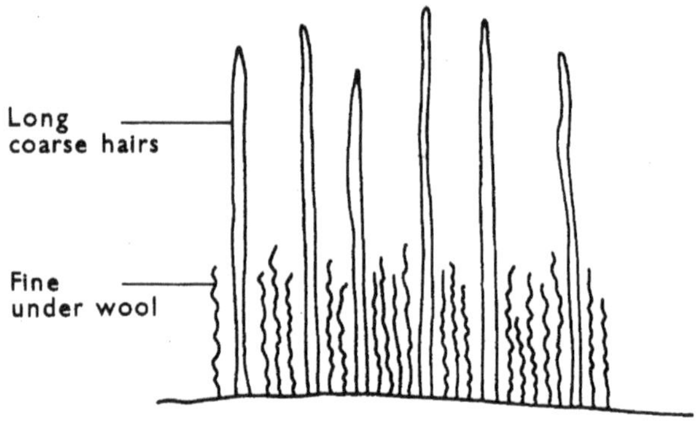


***Figure** 2 Generalised mammalian double coat (from Ryder, 1973).*

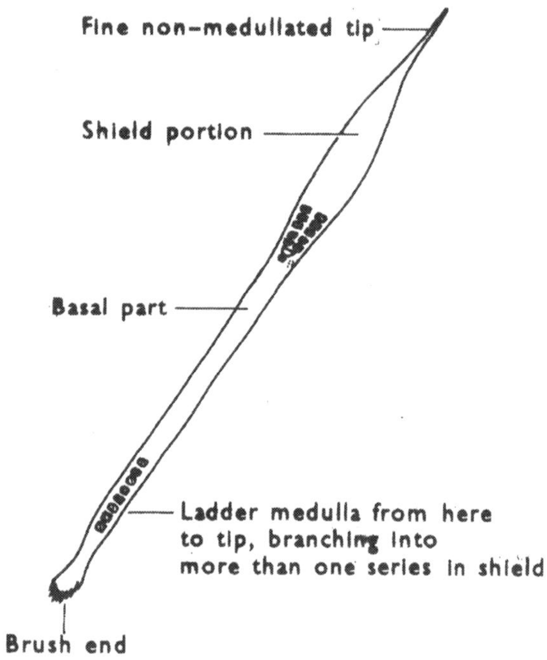


***Figure** 3 Idealised fur-type guard hair with ladder medulla (from Ryder, 1973).*

Hair tuft samples were cut off with scissors by a veterinarian at the level of the skin surface and taken from the same mid-side area at monthly intervals for up to four times from October to January as indicated in Table 2. Dogs dropping out of the trial were the reason why some animals were sampled fewer than four times. On receipt, the length of the tuft was measured against a mm scale in order to ascertain rate of growth. A subsample 1mm long was cut from the base (skin end) of the tuft, and a whole mount microscopic preparation of these short lengths of hair was made. The shaft diameter of 100 hairs was then measured in each sample using a projection microscope at a magnification of 500X. As with the wool method, the measurements were made in micrometres (µm) to the nearest half micrometre. The results are expressed as histograms indicating diameter distributions, and the range and shape of these defines coat type and in particular fleece type in sheep.

The mode is used in conjunction with the mean to indicate skewness in the diameter distribution and compensates for the fact that different ranges of diameter can have the same mean. The hair diameter would be expected to increase when the coat is growing well. The percentages of naturally-pigmented and medullated hairs, features associated with vigorous growth, were also recorded.

## Results and discussion

There was no increase in hair diameter from the first sample, when the dietary additives began, to the last sample. The sleekness of a coat in "good condition", therefore, appears to owe more to abundant secretion of the sebaceous glands in the skin than to the dimensions of the hair. The lack of a clear change in hair dimensions as a result of the dietary additives means that the present data can be regarded as a normal sample. For the present paper the results from up to four hair samples have been pooled for each dog to give mean or otherwise typical values. In Table 2 therefore, the diameter range is the overall range of individual hair diameter over the successive samples, whereas the coat length is the maximum observed. Where two values are given, "w" is the length of the undercoat and "h" of the outer hair. The mean hair diameter is a mean of means as is the standard deviation (s.d.) and the mode. Where a seasonal narrowing of the hairs was obvious, more than one value is given.

## Coat structure and hair types

As with other domestic animals such as cattle, one would not expect a wide variation in coat type in dogs like that seen in domestic sheep (and to a lesser extent in goats and camelids) as a result of human selective breeding of different fleece types for different uses (Ryder, 1983; 1998). However, in common with many mammals, the hair follicles of dogs are of two types: primaries, which are arranged in trios and grow the outer "guard" hairs, and secondaries, which grow the shorter and finer undercoat (Baker, 1966). Therefore, one would expect dogs to have a double coat, and this was apparent in the hair length measurements of at least one dog in each of the groups (Table 2). In only two breeds, however, was this obvious from a bimodal distribution in the hair diameter measurements - the Yorkshire "Terrier" and the West Highland, and in the latter, as well as the finer hair grown by the secondary follicles, the two kinds of guard hair could be seen by eye. These were the chalky white bristles (equivalent to kemps in sheep) growing in the central follicles of the primary trios, and the awns growing in the lateral primaries (Fig. 1).

Most breeds had a modified double coat as in many domestic sheep. This was shown by a skewed hair diameter distribution, in which the bulk of the hairs were fine (the undercoat), with a "tail" of coarser fibres (the outer hair). A double coat is also indicated by the modes being less than the means (Table 2). No dog had a double

| BCK Groups[a] | BCK Breeds[b] | Coat description |
|---|---|---|
| **Gundogs** | Munsterlander, lg (l) | long & dense, not curly or coarse |
| | Golden retriever (l) | flat or wavy, with dense undercoat |
| | Labrador retriever (l) | short, straight, dense (hard); with undercoat |
| **Terriers** | West Highland White (s) | 5cm outer hair; short, soft undercoat |
| | Jack Russel (s) | short and dense (hard) |
| **Utility** | Bulldog (l) | short, close and smooth with fine texture |
| | Poodle, toy (s) | long: dense & harse; short: close, thick & curly |
| **Working** | Doberman (l) | short, smooth and dense (hard) |
| | German shepherd (l) | straight, hard, dense outer; thick undercoat |
| **Toy** | Yorkshire terrier (s) | long, straight & glossy, fine silky texture |

a) British Kennel Club designations; b) s, small; l, large

***Table 1** Breeds of dogs investigated in hair study*

| Breeds (individuals) | Coat length,mm | Diametre, μm | mean | s.d. | mode[e] | % pigment | % medullation |
|---|---|---|---|---|---|---|---|
| **Munsterlander (2)** | | | | | | | |
| (male,3 yrs;3 samples) | 30 | 10-108 | 45.2 | 17.5 | 30 | 37 | 60 |
| (female,5 yrs;3 samples) | w5 h35[a,b] | 12-86 | 48.3 | 16.0 | 52 | 9 | 72 |
| | 40[c] | 10-90 | 40.0 | 19.3 | 28 | 44 | 65 |
| **Golden retriever (3)** | | | | | | | |
| (male,5 yrs;3 samples) | 35 | 10-94 | 36.8 | 12.9 | 32 | 21 | 96 |
| (female,2 yrs;4 samples) | 15[b] | 10-84 | 34.2 | 19.1 | 30 | 38 | 78 |
| (female,?? yrs;4 samples) | w5 h15[a] | 10-98 | 33.8 | 17.3 | 29 | 1 | 86 |
| **Labrador retriever (3)** | | | | | | | |
| (female,10 yrs;3 samples) | 30 | 10-100 | 38.5 | 16.5 | 30 | 99 | 66 |
| (male,?? yrs;4 samples) | w5 h30 | 10-120 | 37.1 | 19.7 | 27 | 2 | 90 |
| (X,male,13 yrs;2 samples) | 25 | 10-100 | 31.4 | 14.5 | 29 | 26 | 90 |
| **W. Highland White (2)** | | | | | | | |
| (female,10 yrs;4 samples) | 60 | 10-76 | 31.2 | 10.1 | 28 | 0 | 17 |
| (male,6.5 yrs;2 samples) | 75 | 10-90 | 34.0 | 15.0 | 25 | 0 | 30 |
| **Jack Russel (2)** | | | | | | | |
| (male,8 yrs;4 samples) | w5 h20 | 10-52 | 26.6 | 8.3 | 22 | 92 | 95 |
| (male,3 yrs;2 samples) | 30 | 10-66 (90, 112)[d] | 33.2 | 14.4 | 32 | 59 | 58 |
| **Bulldog (1)** | | | | | | | |
| (male,1.5 yrs;2 samples) | 10 | 8-116 | 44.8 | 25.9 | 27 | 47 | 90 |
| **Toy Poodle (1)** | | | | | | | |
| (female,7.5 yrs;4 samples) | w5 h20 | 10-46 (60)[d] | 21.9 | 8.0 | 18 | 18 | 16 |
| **Doberman (3)** | | | | | | | |
| (male,1 yr;3 samples) | 25 | 8-126 | 45.0 | 24.6 | 30 | 100 | 50 |
| (female,7mnths;4 samples) | w5 h15 | 10-118 | 43.5 | 24.8 | 30 | 98 | 80 |
| (female,2 yrs;3 samples) | w5 h15 | 10-108 | 46.9 | 22.7 | 30 | 100 | 80 |
| **German shephard (2)** | | | | | | | |
| (female,?? ;4 samples) | w5 h45 | 10-124 | 47.9 | 19.4 | 32 | 37 | 92 |
| (male,?? ;2 samples) | w10 h50 | 10-74 (90)[d] | 27.4 | 10.6 | 24 | 63 | 96 |
| **Yorkshire terrier (1)** | | | | | | | |
| (male,1.5 yrs;4 samples) | w20 h180 | 8-56 (82)[d] | 18.5 | 10.6 | 10 | 30 | 1 |

a) h = outer hair (total coat length), w = underwool; b) autumn coat; c) winter coat; d) diameters outside the main range listed separately in brackets; e) the mode is the most frequent diameter.

***Table 2** Hair measurements of dog breeds*

coat as primitive as that of wild ruminants (Fig 2) in which the diameter distributions of the outer hair and underwool are completely separate. A few dogs, notably in the larger breeds, had a continuous or symmetrical hair diameter distribution. Domestic cattle lack an undercoat and have a similar symmetrical hair diameter distribution.

Many hairs had a medulla, a largely hollow core in general found in coarser hairs. The high proportion of medullated fibres (Table 2) accords with the observation that the finer hairs, too, had a medulla. This was the ladder type of medulla found in fur fibres (Wildman, 1954) (Fig. 3). The coarser hairs had a continuous, non-latticed, medulla; no hairs were observed with a wide latticed medulla like that seen in the outer hairs of ruminants such as deer (Fig. 1).

### Coat length

According to Robinson (1989) the main coat variation in dogs is length of coats, long hair being inherited as a recessive to short hair, as in Angora rabbits. In the present observations, the only breed with a distinctly long coat was the Yorkshire Terrier. The overall coat length ranged from 10mm in the Bulldog to 180mm in the Yorkshire Terrier (Table 2). Most coats, however, fell within the range 15mm to 50mm, the mode being 30mm. Since the last sample was taken in January, most of these measurements are likely to represent the maximum length (Ryder, 1973), and so they are relatively short compared with values of 30mm in the cat (Ryder, 1976) and 50mm in the wild sheep (Ryder, 1983). This accords with the view of Robinson (1989) that the coat length of most domestic dogs - the smooth, sleek, "tight" coat of the Labrador for instance - results from human selective breeding and is shorter than "normal" ie presumably the ancestral coat length.

### Hair diameter

The hair diameter measurements were recorded initially as histograms, which show the range and mode in micrometres: a mean diameter, plus a standard deviation, was calculated from the 100 measurements. The full data are listed in Table 2. The typical mean hair diameter in the larger dogs was around 40μm, ie within the medium range. This value compares with cattle hair (Ryder, 1998) and is coarser than most clothing wool, which is less than 30μm in diameter and cashmere from goats, which is about 15μm (Ryder, 1998). The maximum hair diameters were around 100μm, which is the same as in domestic sheep, whereas the maximum in goats is about 200μm and in red deer, 300μm (Ryder, 1977). The modes were about 30μm, which contrasts with 20μm in the fleeces of sheep, with skewed fibre diameter distribution.

In general the hair of males was coarser than that of females (eg 37μm compared with 27 to 34μm in the Golden Retriever), such a sex difference being found in sheep (Ryder & Stephenson, 1968). The other major difference was associated with body size, the mean diameter range in small dogs being from 19 to 34μm, and that in the larger dogs from 27 to 48μm. Species differences in hair shaft diameter linked to body size are obvious: for instance, the hairs of mice are only a few micrometres thick (Ryder, unpublished). However, I am not aware of any other intra-specific variations in hair fineness associated with body size. The great variations in the fibre diameter of wool between breeds of sheep are not associated with differences in body size (Ryder & Stephenson, 1968). The explanation in dogs may be that there is much greater breed variation in body size than in other species. The low mean diameter of 19μm in the Yorkshire Terrier is comparable with fine wool in sheep and particularly remarkable, the more so since this breed also had the longest coat. In sheep in contrast the longer the fleece, the coarser the wool fibres (Ryder & Stephenson, 1968).

### Seasonal change

As indicated above in the discussion on hair diameter distributions and in Table 2, some dogs appeared to show seasonal narrowing of hair diameter during the autumn-winter sampling period. This implies a synchronous rest period followed by a spring regrowth, which stimulates the moult of the old coat. This type of coat replacement was not marked in the dogs investigated and indeed Robinson (1989) considered that hair replacement in dogs is "diffuse" (as in man), ie there is a continuous replacement of a proportion of the hairs. In reality, as in the cat in which winter follicle inactivity never reached 100% (Ryder, 1976), gradual hair replacement is probably imposed on a basic seasonal cycle of growth. There is also likely to be breed variation in the extent of the moult as in sheep (Ryder and Stephenson, 1968).

In the past when most dogs were kept entirely outdoors, unshielded from the changes in day-length that control the seasonal hair growth cycle, moults are likely to have been more marked. This is important in an archaeological context since goats, for instance, have differences in the hair diameter distribution between summer and winter, and so hair remains can indicate the season of death (Ryder, 1998).

### Conclusions

The breed differences in the coat revealed by this investigation could in an archaeological context throw light on dog-breed origins as they have done in sheep.

### References


Baker, K. P. (1966). Postnatal development of the dog's skin. *British Veterinary Journal* 122: 344-347.

Robinson, Roy (1989). An insight into genetics. *Our Dogs*.10 November, p. 10; 8 December, p. 7.

Ryder, M. L. (1973). Hair. In *Studies in Biology* 41. London: Arnold.

Ryder, M. L. (1976). Seasonal changes in the coat of the cat. *Research in Veterinary Science* 21: 280-283.

Ryder, M. L. (1977). Seasonal coat changes in red deer. *Journal of Zoology, London* **181**, 137-143

Ryder, M. L. (1983). *Sheep and Man.* London: Duckworth.

Ryder, M. L. (1998). Animal hair in medieval ship caulking throws light on livestock types. *Environmental Archaeology - the Journal of Human Palaeoecology* **2**, 61-66.

Ryder, M. L. & Stephenson, S. K. (1968). *Wool Growth.* London & New York: Academic Press.

Wildman, A. B. (1954). Wool. In *The Microscopy of Animal Textile Fibres*. Leeds, UK: Industries Research Association.

# AN ANOMALOUS WOLF, *Canis lupus arctos,* FROM ELLESMERE ISLAND AND THE PROBLEM OF HYBRIDISATION BETWEEN WILD AND DOMESTIC CANIDS

J. Clutton-Brock
Department of Zoology,
The Natural History Museum,
Cromwell Road, London SW7 5BD

A. C. Kitchener
Department of Geology & Zoology,
National Museums of Scotland,
Chambers Street, Edinburgh EH1 1JF, UK

**Key words**: dog, wolf, archaeozoology, hybridisation, Arctic

**Abstract**

The discovery of an anomalous Arctic wolf, *Canis lupus arctos*, on Ellesmere Island in 1986 prompted an investigation into the possibility of introgressive hybridisation with Eskimo huskies, *C. familiaris*. A ratio diagram of six cranial measurements with cranial volume suggested that the Ellesmere canid was just an anomalous wolf, but also indicated that Arctic wolves have become smaller during the twentieth century. A principal components analysis of 28 skull measurements revealed that there has been a change in shape and size of Arctic wolf skulls during this period. Skulls from the period 1930-50 are more similar to a known F1 wolf/husky hybrid, but revert to being more wolf-like subsequently. Similar changes in skull morphology have been recorded for the European wildcat, *Felis silvestris*, in Scotland in the twentieth century, due to introgressive hybridisation with domestic cats, *F. catus*.

Low population density of the wild species, an imbalance in sex ratio, rapid recolonisation of former range and an abundance of alternative potential mates are common factors that may help to explain the more or less extensive introgression observed in Arctic wolves as well as in endangered canids, such as the red wolf, *C. rufus*, and the Ethiopian wolf, *C. simensis*. Historical records show that hybridisation between wolves and Huskies was frequent and even encouraged

The problems in conservation biology caused by hybridisation between wild and domestic species, and opportunities for research into its process and progress are discussed.

## Introduction

Recent work on mitochondrial DNA has shown that the dog is genetically indistinguishable from the grey wolf (Wayne, 1993), although dogs and wolves do not, in general, share the same mitochondrial DNA haplotypes, and where they do, there are distinct differences in their sequences (Vilà & Wayne, 1999). However, in view of their many phenetic and behavioural differences, we still consider that the dog and wolf should be classified as separate species, *Canis familiaris* and *Canis lupus* (Gentry *et al.*, 1996; Clutton-Brock, 1999). Similarly the domestic cat, *Felis catus*, is significantly different from the wildcat, *Felis silvestris*, in terms of morphology and behaviour, and we believe they also should be classified as separate species, although their genetic differences are very small (Randi & Ragni, 1991).

However, despite disagreements between biologists about how to classify them, the fact remains that under certain circumstances domestic or feral domestic mammals will readily hybridise with their wild counterparts and will product fertile offspring. This has been well recorded for dogs and wolves in northern Italy at the present day (Boitani, 1982), and it is estimated that somewhere between 10,000 and 100,000 dog/wolf hybrids currently inhabit the USA (Garcia-Moreno *et al.*, 1996; Steinhart 1995). This has obvious implications for conservation biology, where the introduction of a limited range of the original gene pool via the domestic population may disrupt co-adapted gene complexes, which may compromise the survival of the indigenous species. At the very least, it may adversely affect the natural gene frequencies in small wild populations, resulting in morphologically variable hybrid swarms. This was described in the 1960s for the expanding population of wild canids in New England, which Lawrence & Bossert (1969) believed were predominantly coyote, *Canis latrans*, but with some introgression from dogs and wolves. Molecular analysis of the wolf population over a wide area of Minnesota and south east Canada shows that individual wolves often carry coyote mitochondrial DNA, indicating hybridisation between male wolves and female coyotes (Wayne *et al.*, 1996). More recently Wayne & Jenks (1991; see also Wayne *et al.*, 1996) have shown from analysis of mitochondrial DNA that the existing population of red wolves (*Canis rufus*) can be considered as a 'new species', resulting from hybridisation between coyotes and grey wolves. However, this view has been strongly refuted by Nowak (1996) and Nowak *et al.* (1996) on the basis of the fossil record of the red wolf (Nowak, 1979), morphological discontinuities between the red wolf and both the coyote and grey wolf (Nowak, 1979), and the problems with inferring a hybridised population from mitochondrial DNA data. Clearly, accepting the red wolf or other wild canid populations as comprising mainly of hybrids has important implications for the validity of the recovery programmes for these species, and their subsequent legal protection (O'Brien *et al.*, 1990; O'Brien & Mayr, 1991; Wayne *et al.*, 1996; Pennock & Dimmick, 1997; Sillero-Zubiri & Macdonald, 1998). Similarly there are many different colour forms of cats in Scotland as a result of hybridising wildcats and domestic cats (French *et al.*, 1988; Kitchener & Easterbee, 1992; Daniels *et al.*, 1998; Kitchener, 1998).

On the islands of Melville and Ellesmere in the Canadian Arctic there is a distinct population of white wolves that has been placed in the subspecies *Canis lupus arctos* Pocock, 1935 (the Arctic wolf). In August 1986 Dr Russell Marris from near Lake Hazen collected the carcasses of two of these wolves on Ellesmere Island, North West Territories, Canada. Marris presented one carcass to the Royal Ontario Museum (ROM), Toronto, and the other to the Natural History Museum (NHM), London. One of us (JC-B) was struck by the anomalous condition of the NHM skull, which has many dog-like features. This raised the question of whether the Arctic wolves are, or have been in the past, in serious danger from introgression from domestic or feral dogs and so we decided to investigate the status of the wolves by measuring a series of skulls collected at different times over the last 150 years.

Accordingly a principal components analysis was carried out on measurements of a large series of Arctic wolf skulls, the earliest of which was collected in 1834-5 (Clutton-Brock *et al.*, 1994). The two skulls collected in 1986 were the most recent in our time series.

**Description of the anomalous wolf**

The two wolves appeared to have died together and were part of a population being studied in the wild by David Mech of the United States Fish and Wildlife Service (Brandenburg, 1988; Mech, 1987, 1988). Mech knew of the two carcasses and wrote to the Natural History Museum (13 September 1986) to enquire whether there was evidence of wounds on the body of the wolf, because he was interested to know whether the wolves could have been killed by a musk-ox (*Ovibos moschatus*), the remains of which had also lain near by (Fig. 1). There were, however, no signs of a struggle and no fresh broken bones, although there were old breaks on the right zygomatic arch of the skull and on the ribs of the wolf presented to the Royal Ontario Museum (Pasitschniak-Arts *et al.*, 1988). It was evident that the wolves had most probably died from natural causes, perhaps during the previous winter.

The Arctic wolf, *Canis lupus arctos*, is the most northern of the five subspecies of *Canis lupus* that are currently recognised by Nowak (1983; 1996) from North America, on the basis of a multivariate statistical analysis of measurements of skulls from 580 male wolves from throughout the world. Although recent analysis of mitochondrial DNA in different grey wolf populations does not support the recognition of genetically distinct separate subspecies (Wayne, 1993), this does not invalidate

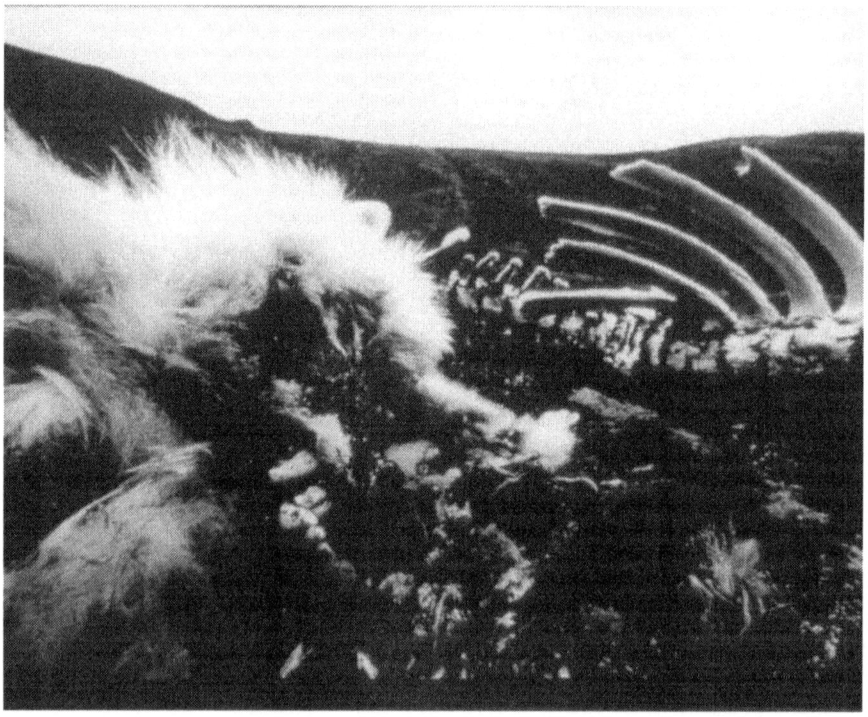

***Figure 1:*** *The carcass of an Arctic wolf on Ellesmere Island, Canada, close to the carcass of a Musk ox. It was this wolf that was presented to The Natural History Museum, London, in 1986. The grossly undershot mandible can be seen in the photo, which is reproduced by kind permission of Jim Brandenburg (see also Brandenburg, 1988).*

the phenetic separation of a form of wolf that is adapted to arctic conditions.

Pocock (1935) originally described the Arctic wolf on the basis of a skull from Melville Island, but also stated that wolves from Ellesmere Island belong to this subspecies. Goldman (1964: 430), referred to the descriptions of Pocock (1935) and defined *Canis lupus arctos* Pocock, 1935 as:

> a nearly white subspecies of medium size. Differs from *occidentalis* of Mackenzie in smaller size, more predominantly white coloration, and narrower braincase. Closely resembles *orion* of Greenland in coloration, but may be larger; skull with more elevated frontal region.

Goldman had examined two skulls from Ellesmere Island in the American Museum of Natural History, which he claimed closely agreed with Pocock's description of the type specimen from Melville Island and a second skull from Ellesmere Island. On the basis of his morphometric analysis, Nowak (1983; 1996) characterised *arctos* as having relatively large carnassials and extended its range to Greenland (formerly *orion*), Banks (formerly *bernardi*) and Victoria Islands as well as all the Queen Elizabeth Islands. Nowak claimed (pers. comm., 1987) that, 'Anterior shortening and crowding of the upper premolars is not unusual in *arctos*, and can even be considered characteristic of the subspecies'.

The white pelt of the wolf received at The Natural History Museum was in too poor a condition to be tanned and had to be destroyed. The wolf was an adult male, which from the fusion of the skull sutures, could have been more than four years old. After one look at the skull it appeared only too probable that the wolf had starved to death, indeed it was remarkable that the animal had survived so long since it suffered from extreme malocclusion of the jaws. The cranium, tympanic bullae, and mandible are within the size range and proportions of a small Arctic wolf, but the facial region of the skull is extremely short and wide and is undershot by the mandible to an extraordinary degree (Fig. 2). The upper premolars are compacted and the upper third premolars are turned through 90°. The lower canines are completely worn down from occluding with the upper incisors while the lower incisors are much displaced and compacted. There is evidence for extensive periodontal disease on both the right maxilla and the right mandibular ramus.

At first sight this anomalous skull appears to have the cranium and mandible of a wolf with the facial region of a domestic or feral dog. As is well known, reciprocal crosses between grey wolves and dogs are viable and fully fertile in both sexes (Gray, 1972). Dogs have been present wherever and whenever there have been humans in the Arctic Circle and their remains are found on all archaeological sites. It is therefore quite possible that, for many hundreds of years, there has been some crossbreeding between wolves and dogs in the Arctic Circle. A description of the traditional Eskimo (husky) dog was given by Allen in his classic review of 'Dogs of the American Aborigines' (1920).

**The status of the anomalous wolf**

The investigation into the status of the NHM wolf was begun by comparing the measurements of the skull with those of all the Arctic wolves in The Natural History Museum collections together with those of a pure-bred husky dog ('Farthest North', died 1900), a F1 hybrid wolf/husky dog, and a dog from an archaeological deposit at Brooman Point, Bathurst island, dated to around AD 850 (McGhee, 1981). Six linear dimensions and the measurement of cranial capacity were selected from this data set and are presented as a ratio diagram in Figure 4. This ratio diagram supports the visual impression that the anomalous skull is not from a domestic dog or hybrid but from a small wolf with abnormally short upper carnassial teeth and an excessively short maxillary region (Fig. 4). In two key characteristics that can be used to separate the skulls of dogs from wolves, these being cranial capacity and the length of the tympanic bulla, the skull is strongly wolf.

It became evident from this preliminary analysis that the skulls of Arctic wolves that were dated to the last century and the beginning of this century were considerably larger overall than the more recently collected skulls. This can be seen in Figure 4 where the dimensions of the type specimen (Fig. 3), collected in 1834-5, fall at the larger end of the range for each measurement.

Direct comparison of the measurements (Table 1) of the two new skulls show that the NHM skull (male) is much shorter than the ROM skull (sex unknown) but the mandible is about the same length. Apart from the extraordinarily short palatal region of the NHM specimen, both skulls look like those of small wolves. This raised the broader question of whether there have been quantifiable changes in the size and morphology of the skulls of Arctic wolves over the last hundred years, as have been recorded for the European wildcat in Scotland by French *et al.* (1988). It was therefore decided to carry out a statistical analysis of skull measurements from 33 Arctic wolves (from various collections, see Clutton-Brock *et al.*, 1994), 7 husky dogs, and a known F1 hybrid to investigate the changes and to assess the status of the anomalous canid from Ellesmere Island.

A principal components analysis of 28 skull measurements revealed two changes in the size and shape of Arctic wolf skulls during the present century (Clutton-Brock *et al.*, 1994). The skulls of wolves from the period 1930-50 became more similar in shape and size to the known F1 hybrid between a wolf and a husky. However, since 1950 there has been a reversion in skull morphology, so that recent Arctic wolves resemble more closely those collected before 1930. The analysis also revealed that the skull of the NHM canid was not an introgressive hybrid, but was indeed just an anomalous, small wolf.

This view has recently been confirmed by Federoff (1996) who studied the inheritance of a similar skull

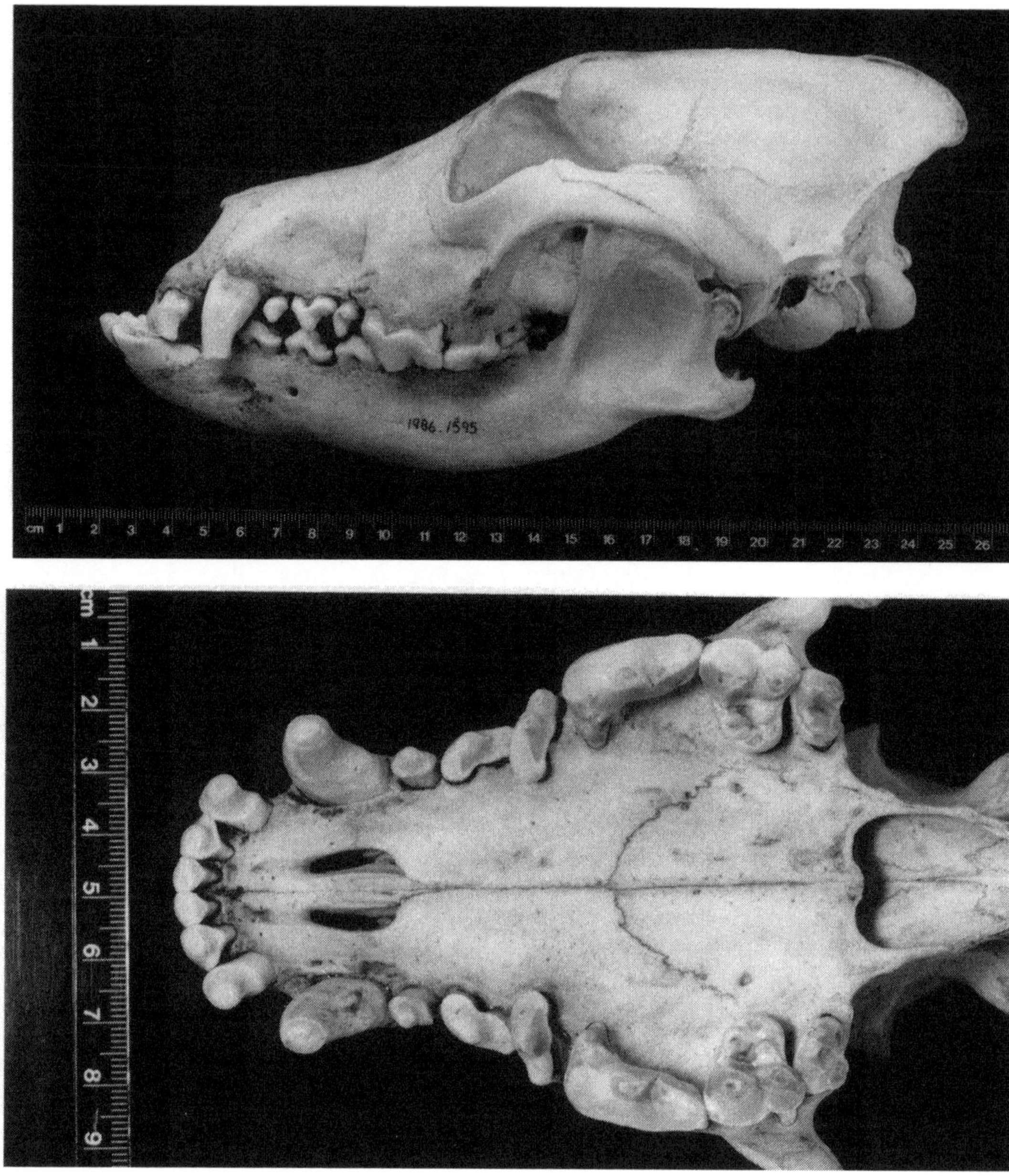


***Figure 2:*** *a. Left side of the skull of the 1986 male wolf (Canis lupus arctos, NHM 1986.1595) to show the undershot jaw. b. Palatal view of the same skull to show the compacted teeth. Photos by Phil Hurst, Natural History Museum, London.*

deformity in a captive population of Arctic wolves derived from a single pair from Axel Heiburg Island to the west of Ellesmere. Therefore this skull malformation occurs naturally in Arctic wolves, and is probably expressed phenotypically when severe selection pressures limit the development of pack size, resulting in inbreeding, as had occurred with the captive population. This congenital defect resulting in muzzle shortening is different and more extreme than that proposed by Jolicoeur (1959) (see below).

**Possible causes of morphological change**

The results of statistical analysis clearly showed that there had been changes in both size and shape of the skulls of the Arctic wolves during the twentieth century, but the problem is, what has caused these changes? Firstly the diet of the wolves may have changed over this period. Jolicoeur (1959) suggested that Arctic wolves have relatively shorter snouts and wider skulls than other wolves because the poor availability of food in their first northern winter (at 4-5 months of age) would not allow for the full growth of the facial region of the skull relative to the cranium. The reduction in size and shortening of the skulls of Arctic wolves during this century suggests that this situation may have worsened and could have been caused by a decline in the populations of large prey animals, such as musk-oxen. However, the balance between the population dynamics of herd animals and wolves on mainland Canada has been greatly affected by wolf control programmes while we have no figures for the number of wolves and their prey on the Arctic islands (Carbyn, 1983).

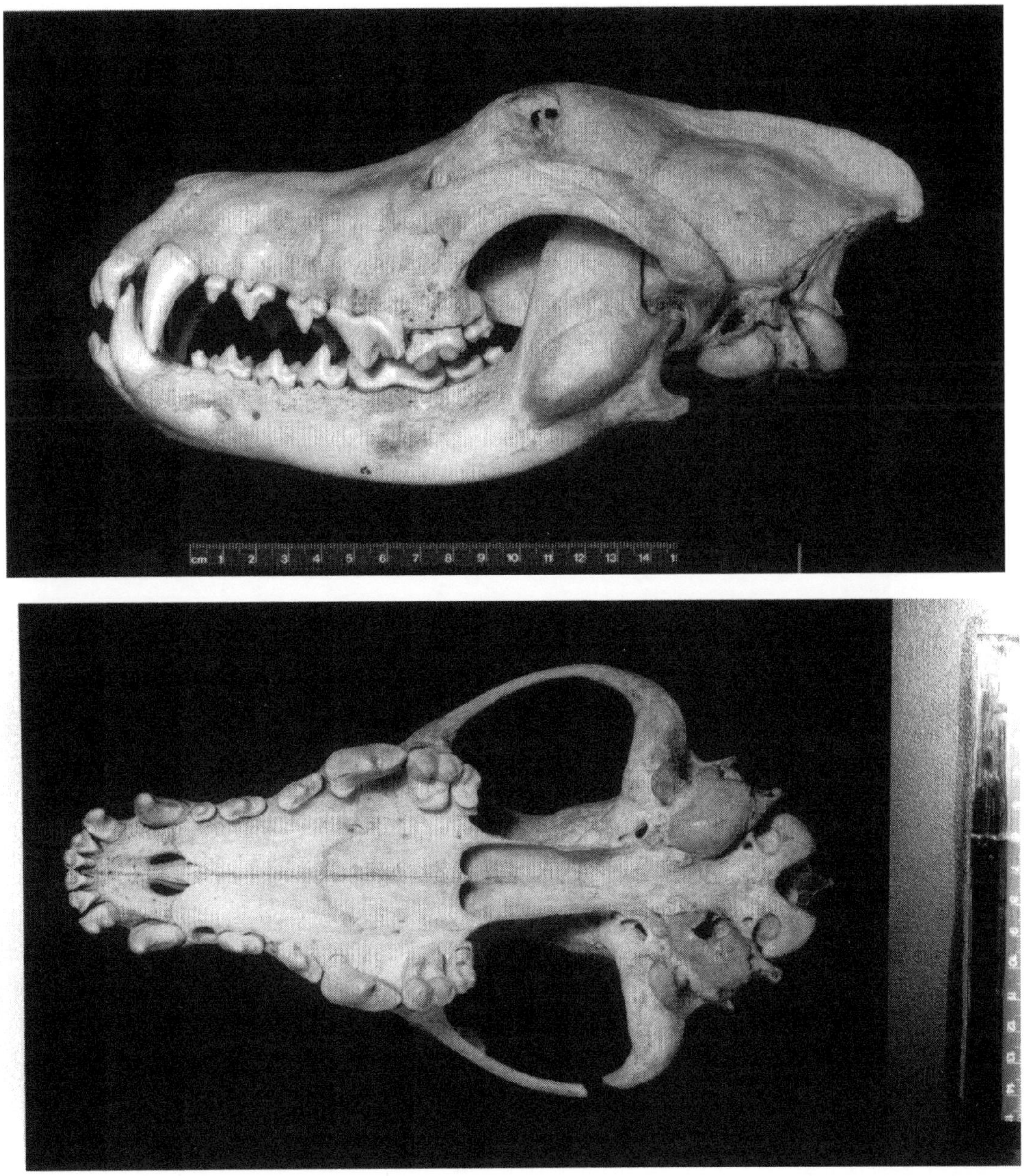

***Figure 3**: a. Side view of the skull of the type specimen of Canis lupus arctos Pocock, 1935. Note the bullet hole in the frontal sinus. (BM (NH) 1855.11.26.74). Collected by Sir E. Belcher from Melville Island in 1853-4. b. Palatal view of the same skull to show the large, well-spaced teeth. Photos by Phil Hurst, Natural History Museum, London.*

In recent years the Arctic wolves of Ellesmere Island have been feeding on scraps from the research station there (Mech, 1988; R. Burton, pers. comm., 1992). Perhaps a major change in obtaining food from hunting prey to scavenging may have caused a reduction in skull size and shape, ie. the beginning of a process of domestication, as first observed by Wolfgramm (1894) for wolves born in captivity. This might explain the cramped teeth seen in modern Arctic wolves, and in particular the 1986 male wolf. Although European settlements have been established on Ellesmere Island since 1901, scavenging from humans is a very recent and local event and does not explain the particular pattern of changes that have occurred over the last 70 years, especially in the skulls from the 1930-50 period. If skull shape had shifted toward F1 dog/wolf hybrid shape because of such a change in diet, we would expect a very high proportion of the wolves' diet to be scavenged, but we have no evidence for this.

Oscillations in climate could also have an effect on skull size. Geist (1987) has shown that the skulls of wolves increase with latitude until about 65° north, when the duration of the annual productivity pulse begins to decline, leading to reduction in body size of both predators and prey because of a decline in nutrients and energy that are available during periods of growth. This is, of course, related to the availability of food, as discussed above. Therefore, if climate was affecting skull size of Arctic wolves, we would expect to find evidence that the climate had deteriorated during the middle of this century. However, such evidence is decidedly lacking.

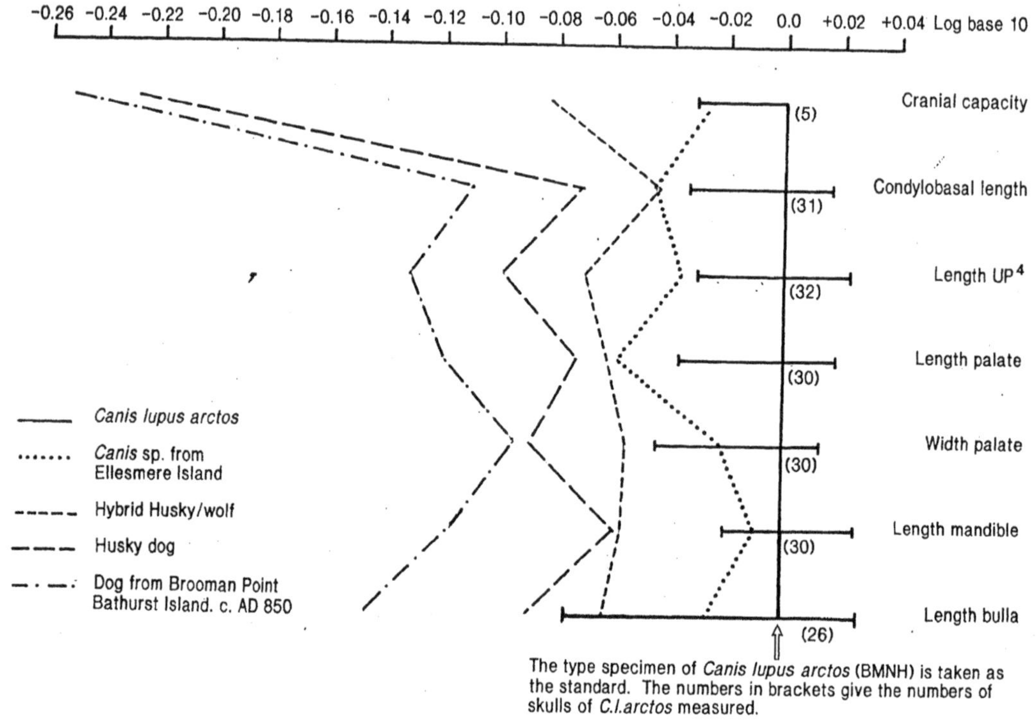


***Figure 4:*** *Ratio diagram of seven measurements of the skull of the 1986 male to compare with skulls of Arctic wolves, two dog skulls and a hybrid husky/wolf (after Simpson. 1941). The type specimen of Canis lupus arctos Pocock, 1935 is taken as the standard. The numbers in brackets give the numbers of skulls of* C. l. arctos *measured.*

***Table 1****: Measurements of the specimens of Arctic wolf collected by Russell Marris in 1986. The numbers and points of measurement correspond to those described and illustrated in von den Driesch (1976). Skull and mandible 1 and the bones of the skeleton are from the specimen in The Natural History Museum, London, no 1986.1595. Skull and mandible 2 are in the Royal Ontario Museum, Toronto, no 94176. Note that limb bone measurements are all from the right side.*

| | | **NHM ♂ Skull** | **ROM Skull** |
|---|---|---|---|
| 1. | Total length | 229.5 | 232.0 |
| 2. | Condylobasal length | 210.1 | 223 |
| 3. | Basal length | 197.6 | 211 |
| 13. | Palatal length | 108.5 | 119.4 |
| 15. | Length of cheek teeth row | 73.5 | 83.4 |
| 16. | Length of molar row | 24.9 | 25.8 |
| 17. | Length of premolar row | 54.9 | 63.6 |
| 18. | Length of the carnassial | 24.7 | 25.0 |
| 22. | Greatest diameter of the auditory bulla | 30.05 | 28.9 |
| 30. | Zygomatic breadth | 135.0 | 134.9 |
| 32. | Frontal breadth | 60.0 | 58.7 |
| 34. | Greatest palatal breadth | 80.0 | 78.3 |
| 36. | Breadth at the canine alveoli | 45.0 | 45.9 |
| 42. | Neurocranium capacity | 153 cc | - |

| | | **NHM ♂ Mandible** | **ROM Mandible** |
|---|---|---|---|
| 1. | Total length (from condyle) | 176.0 | 174.3 |
| 2. | Length (from angular process) | 178.8 | 174.2 |
| 4. | Length: condyle process to the aboral border of canine alveolus | 152.7 | 153.7 |
| 7. | Length: aboral border of alveolus of $M^3$ to aboral border of canine alveolus | 96.4 | 99.2 |
| 10. | Length of molar row | 43.1 | - |
| 11. | Length of premolar row | 50.5 | 48.4 |
| 13. | Length of carnassial | 28.3 | 28.8 |
| 18. | Height of vertical ramus | 69.4 | - |
| 19. | Height of mandible behind $M^1$ | 28.9 | 31.0 |
| 20. | Height of mandible between $P^2$ and $P^3$ | 24.1 | 24.5 |

**Atlas**

| | |
|---|---|
| GB | Greatest breadth over wings - 94.7 |
| BFcr | Greatest breadth of cranial articular surface - 47.0 |
| BFcd | Greatest breadth of caudal articular surface - 38.5 |
| Lad | Length of dorsal arch - 20.8 |

**Axis**

| | |
|---|---|
| LCDe | Greatest length in body - 59.9 |
| LAPa | Greatest length of arch - 63.7 |
| BFcr | 36.1 |
| BFcd | 21.8 |
| H | Height - 44.4 |

**Scapula**

| | |
|---|---|
| HS | Height of scapula - 166.6 |
| SLC | Smallest length of neck - 33.7 |
| GLP | Greatest length of glenoid process - 40.1 |
| BG | Breadth of glenoid cavity - 24.3 |

**Humerus**

| | |
|---|---|
| GL | Greatest length - 222.4 |
| Dp | Depth of proximal end - 53.8 |
| SD | Smallest breadth of shaft - 15.4 |
| Bd | Greatest breadth of distal end - 41.7 |

**Radius**

| | |
|---|---|
| GL | 216.3 |
| BP | Greatest breadth of proximal end - 23.7 |
| SD | 17.0 |
| Bd | 32.2 |

**Ulna**

| | |
|---|---|
| GL | 250.0 |
| LO | Length of olecranon - 39.6 |
| SDO | Smallest depth of olecranon - 26.8 |
| BPC | Greatest breadth across coronoid process - 23.4 |

**Pelvis**

| | |
|---|---|
| GL | 183.5 |
| LAR | Length of acetabulum on the rim - 27.7 |
| GBTi | Greatest breadth across ischial tuberosities - 128.2 |

**Baculum**

| | |
|---|---|
| GL | 1103.0 |

**Femur**

| | |
|---|---|
| GL | 231.5 |
| Bp | 49.0 |
| DC | Greatest depth of femur head - 24.6 |
| SD | 15.5 |
| Bd | 43.1 |

**Tibia**

| | |
|---|---|
| GL | 248.1 |
| Bp | 46.7 |
| SD | 15.1 |
| Bd | 28.4 |

**Calcaneum**

| | |
|---|---|
| GL | 57.9 |
| GB | Greatest breadth - 22.45 |

**Talus**

| | |
|---|---|
| GL | 34.6 |

**Metacarpal 4**

| | |
|---|---|
| GL | 86.45 |
| Bd | 12.0 |

**Ist Phalanx 4 fore**

| | |
|---|---|
| GL | 35.6 |
| Bp | 12.15 |

**Metatarsal 4**

| | |
|---|---|
| GL | 85.9 |
| Bd | 11.25 |

**Ist Phalanx 4 hind**

| | |
|---|---|
| GL | 30.9 |
| Bp | 11.1 |

A third possible cause of reduction in skull size and a change in skull shape is offered by hybridisation with domestic dogs. In other parts of the world the genetic integrity of wolf populations is thought to be threatened in this way, for example in Italy (Boitani, 1982: 165). However, Vilà & Wayne (1999) have recently questioned whether any significant introgression has occurred anywhere in the world, based on the lack of dog mtDNA in any extant wolf population, including small and fragmented populations in Italy, Spain and Israel. While introgression between domestic dogs and the highly endangered Ethiopian wolf is a serious cause for concern (Wayne & Gottelli, 1998; Vilà & Wayne, 1999), unlike the grey wolf, female Ethiopian wolves leave their natal packs to mate with males from other packs, but because domestic dogs massively outnumber them, hybridisation is frequent. The mated females return to their natal groups to give birth and hence, the hybrids are socialised as Ethiopian wolves. As a result 8-17% of Ethiopian wolves from one of a handful of surviving populations have a domestic dog ancestry (Vilà & Wayne, 1999). Vilà & Wayne (1999) suggest that grey wolf/dog hybrids are unlikely to survive because male domestic dogs do not assist in rearing the young (although this has been observed in Alaska (D.Weir, pers. comm.)), and they also suggest that hybridisation is relatively rare because of a mismatch in the breeding seasons for wolves and domestic dogs (see below).

### Hybridisation between wolves and husky dogs

> "Farewell, poor Carlo hapless dog, adieu
> We mourn thy fate unknown*; but fate too sure;
> For fled is ev'ry hope that Fancy drew,
> Nor could thy frame the piercing frost endure.

*Poor Carlo continued his daily rambles with the she-wolf until, at length he went and returned no more. His fate was never ascertained; but there can be little doubt that he fell a victim in an encounter with the male, as Botswain, another of our dogs, who was decoyed away in the same manner a few days afterwards, returned to us after a combat, in which it is evident he had hardly escaped with his life."

The seven stanzas (of which the one quoted above is the first) on the death of a favourite setter at Winter Harbour, Melville Island, were published in the *Northern Gazette* on St Valentine's Day, 1820, during Parry's first voyage to the N. W. Passage. The poem and footnote, which were published by Sabine (1821), can probably serve as a memorial to encounters that happened, not just this once but many times in the past, between male domestic dogs and wild wolves. It is doubtful, however, whether a highly bred European dog, such as a setter, would have mated successfully with a female wolf in competition with male wolves.

On the other hand, the mating of male Arctic wolves with husky bitches has probably always occurred, even though at a low level. In the past it was encouraged by the Inuit in order to produce dogs which were stronger and more robust than normal. Degerbøl & Freuchen (1935: 145) described how:

> "From times immemorial the Eskimoes have wanted to get wolf blood into their dogs... But as is so often the case, so much has been said and fabled about it that people believe it has happened more frequently than actually can be borne out by facts."

See also Allen (1920). Degerbøl & Freuchen (1935: 146) described the life of a crossbred wolf/dog female that was taken to Cape York from Stockholm in 1916. This hybrid could never be trained as a sledge dog but survived well by killing birds and scavenging on the scraps left by hunters until it joined a pack of three wolves and was eventually shot. In the same report Degerbøl & Freuchen (1935: 163) reported that husky bitches in north Greenland sometimes ran far away when on heat and were followed by male dogs which could be lost in a snow storm or drift out to sea on an ice flow. Such dogs had been known to run up to a sledge after being a year in the wild. Hybridisation between these feral dogs and wolves may have become more frequent as the wolves decreased in numbers and their social and reproductive behaviour was disrupted.

Hybridisation between Arctic wolves and husky dogs would explain the reduction in size of the wolves, because it is clear that the F1 hybrid is intermediate in size (Clutton-Brock *et al.*, 1994). It is interesting to note that the skull characters used by Nowak (1979 & pers. comm., 1987, see above) to distinguish Arctic wolves from other wolves may also be interpreted as characters associated with wolf/dog hybrids. Skulls of Arctic wolves in museum collections in North America mostly date from 1930 onwards.

The history of wolf populations in North America may offer clues to why hybridisation with huskies has apparently occurred. As a result of the 1930s depression and World War Two, persecution of wolves decreased allowing populations to increase in numbers and partially recolonise their former ranges (Nowak, 1983; Carbyn, 1987). For example, anthropogenic habitat change in Minnesota and south east Canada has resulted in colonisation by coyotes, which have introgressed extensively with wolves (Wayne *et al.*, 1996). We had originally believed that the numbers of working huskies decreased with the introduction of motorised sleds (MacRury, 1991), so that unwanted dogs became feral and began to hybridise with the wolves. However, motor sleds did not begin to be used by the Inuit until the 1960s, because they were expensive to buy and maintain, and this was not possible for most Inuit until economic development began at that time (eg. Southampton Island; Moyer, 1972/73).

On Ellesmere Island it is unlikely that Eskimo dogs were responsible for hybridisation, because there were no Inuit settlements there during the early 20th century (D. Moyer, pers. comm.). Moreover, the Inuit were much more likely to kill surplus dogs rather than to turn them loose, since their skins and meat could be put to practical use (D. Moyer, pers. comm.). Instead, it is more likely that dogs were lost by fur trappers, who were forced into these areas to hunt more intensively for declining numbers of Arctic foxes, *Alopex lagopus,* at this time. Longer travel distances required larger teams of husky dogs, from which losses were more likely. In addition, we note that six Royal Canadian Mounted Police posts were established on Ellesmere Island between 1920 and 1956 (Dept. of Energy, Mines and Resources, 1971), which could be an alternative source of husky dogs, although we have been unable to check whether policemen and/or dog teams went missing during this time.

Anderson (1943) described a new subspecies of wolf, *Canis lupus bernardi*, from Banks Island (south of the range of *arctos, sensu stricto* on Melville Island) from specimens collected in 1914-16. However, Manning & McPherson (1958) found that a second series of wolves collected from the same locality in 1953-5 had a different skull shape. Nowak (1979; 1983) considered the idea that *arctos* had replaced *bernardi*, but rejected this on the grounds that all samples from Banks Island were still separable from *arctos*. More recently, Nowak (1996) has regarded *bernardi* as a synonym of *arctos*, and suggested that differences in the availability of food during development may be responsible for the differing skull shapes (see Jolicoeur, 1959). However, perhaps the change in skull shape of Banks Island Arctic wolves was also caused by hybridisation with released or escaped husky dogs.

The reversion in the skulls of Arctic wolves collected after 1950 to the shape of those collected before 1930 suggests that hybridisation between wolves and husky dogs no longer occurs and that there has been rapid natural selection for the wolf skull morphology. Certainly, we assume that husky dogs have now been virtually replaced by motor sleds throughout the Canadian Arctic, so that further possibilities for hybridisation must be very limited.

## Discussion

The degree of introgression between canid species is very variable. For example complete introgression has been apparently recorded between the coyote and the grey wolf, to produce the red wolf, in the S.E. USA (but see Nowak (1996) and Nowak *et al.* (1996)) and a widespread and highly introgressed population of canids in Minnesota and southeastern Canada (Wayne *et al.*, 1996). Significant introgression with domestic dogs threatens the highly endangered and distinctive Ethiopian wolf (Wayne &

***Figure 5**: Map of Canadian Arctic showing distribution of* C.l. arctos.

Gottelli, 1998), while grey wolf populations are apparently unaffected, owing to behavioural and physiological differences with domestic dogs (Vilà & Wayne, 1999). However, we have presented strong circumstantial evidence that changes in the skull morphology of Arctic wolves has been the result of hybridisation with escaped or released husky dogs during the first half of the 20th century (Clutton-Brock *et al.*, 1994). Moreover, because the period of hybridisation has been limited and the selection pressures on Arctic wolves has apparently been very strong, there has been a reversion in skull morphology to a more wolf-like form, which parallels a similar apparent morphological change in the skulls of Scottish wildcats during the same period, as a result of hybridisation with domestic cats during recolonisation of its former range, followed by consolidation of the wildcat population (French *et al.*, 1988). As yet we have a poor understanding of the factors that promote introgression (see above). Hybridisation can even be very difficult to achieve in captive breeding, owing to factors such as a mismatch in seasonality of oestrus and male fertility between wild and domestic species. Mengel (1971), who reviewed this question in his description of dog/coyote hybrids, claimed that male coyotes (and he believed male wolves also) are only fertile for about two months in the year; wild female canids of all species come into oestrus only once a year, while domestic male dogs are always in breeding condition, and domestic female dogs normally have two oestrus seasons each year. Mengel (1971) found that dog/wolf hybrids and coy/dogs retained the annual breeding cycle of their wild parent.

As well as the mismatches in breeding season between wild and domestic pairs, there may also be an agonistic interaction between the larger and stronger wild forms and the weaker domestic, and this may occur most of the time in the wild, as exemplified in the case of 'poor Carlo', quoted above.

Given our lack of knowledge of the factors that cause introgression, every care should be taken to avoid the meeting of wild animals with their domestic counterparts. This is especially important when the wild species is endangered, but it is also necessary to prevent the mating of so-called 'primitive' indigenous breeds of domestic animals with modern introductions. For example Allen (1920) described how the original Eskimo dogs in North America were already in his day almost extinct through interbreeding with introduced European dogs. Today, the dingo-like pariah dogs have almost disappeared from South East Asia through introgression with domestic dogs (Corbett, 1995). The dingo in Australia and the New Guinea singing dog are also in equal danger from hybridisation with feral mongrels of European descent (Corbett, 1995).

From what we have seen in our own studies of Arctic wolves and other studies of canids where introgression occurs, a number of key factors seems to be necessary, including a low population density of the wild or endangered species, rapid (re)colonisation of an area by the abundant species, and an imbalance of sex ratios. Despite the evidence from Vilà & Wayne's (1999) study of grey wolves, which proposes that introgression is not a significant threat to any population in the world, we have shown that significant changes in skull morphology can take place due to introgression with domestic dogs in populations of wolves living at very low population densities.

Walker and Frison (1982) present compelling evidence that hybridisation occurred in some areas of the US northern plains during a prehistoric period that spanned almost 8,000 years. The lack of mitochondrial DNA evidence for introgression in modern grey wolves is perplexing: it suggests that the young from matings of male wolf and female dogs are not recruited to the wolf population (since mtDNA is maternally inherited). However, despite the known occurrence of hybrid animals (eg. Randi *et al.*, 1993), mtDNA will not pick up the existence of young from female wolf and male dog matings. Vilà & Wayne (1999) present only scant evidence for biparentally inherited markers that would indicate matings between male dogs and female wolves. While these authors cite studies that show there may be 10,000 wolf/dog hybrids (Garcia-Moreno *et al.*, 1996) in the USA, with a history of hybridisation that stretches back at least 10,000 years (Schwartz, 1997), they still maintain that wolf populations have not been affected to any significant degree by hybridisation with dogs. We suggest it is possible that subtle human-induced changes in wolf environments may occasionally tip the balance in encouraging high levels of hybridisation, including increased mortality due to disease or road-kills, deliberate extermination, declines in natural prey populations, and habitat loss or modification. While hybridisation may normally occur irregularly and infrequently, further research is urgently required on the long-term repercussions of this phenomenon.

Hybridisation has important implications for the conservation of endangered species. Clearly the disruption of coadapted gene complexes that promote the survival of a species locally may compromise the future survival of that species. Secondly, hybrids are rarely recognised legally for protection of endangered species (Mayr & O'Brien, 1990; Pennock & Dimmick, 1997), so that if a population is regarded as highly introgressed or the identification of individuals as belonging to an endangered species becomes difficult, the species may inadvertently lose its legal protection. We recommend that endangered populations of canids, which are thought to be threatened by hybridisation with other canids, especially domestic dogs, should be carefully monitored for the changes in genetics and morphology, as well as the incidence of diseases acquired from the introduced species. Behavioural and field studies of introgressing populations and non-introgressing sympatric populations could shed important new light on the factors that promote or prevent hybridisation in the wild.

**References**


Allen, G. M. (1920). Dogs of the American Aborigines. *Bulletin of the Museum of Comparative Zoology Harvard College* **63:** 431-517.

Anderson, R. M. (1943). Summary of the large wolves of Canada, with description of three new Arctic races. *Journal of Mammalogy* **24:** 386-393.

Boitani, L. (1982). Wolf management in intensively used areas of Italy. In (R. E. Harrington & P. C. Paquet, Eds) *Wolves of the World*. New Jersey: Noyes Publications, pp. 158-172..

Brandenburg, J. (1988). *White Wolf: Living with an Arctic Legend*. In (J. S Thornton, Ed) Minocqua, WI: NorthWord Press.

Carbyn, L. N. (1983). *Wolves in Canada and Alaska*. Ottawa: Canadian Wildlife Service Report Series **45**.

Carbyn, L. N. (1987). Grey wolf and red wolf. In (M. Novak, J. A. Baker, M. E. Obbard & B. Malloch (Eds) *Wild Furbearer Management and Conservation in North America*. Ontario: Ministry of Natural Resources, pp. 358-377.

Clutton-Brock, J. (1999). *The Natural History of Domesticated Mammals*. 2nd edition. Cambridge: Cambridge University Press.

Clutton-Brock, J., Kitchener, A. C. & Lynch, J. (1994). Changes in the skull morphology of the Arctic wolf, *Canis lupus arctos*, during the twentieth century. *Journal of Zoology London* **223:** 19-36.

Corbett, L. K. (1995). *The Dingo in Australia and Asia*. Ithaca & London: Constock/Cornell.

Daniels, M. J., Balharry, D., Hirst, D., Kitchener, A. C., Aspinall, R. J. (1998). Morphological and pelage characteristics of wild living cats in Scotland: Implications for defining the "wildcat". *Journal of Zoology London* **244:** 231-248.

Degerbøl, M. & Freuchen, P. (1935). Mammals. *Report of the Fifth Thule Expedition 1921-24*. vol 2. Copenhagen.

Department of Energy, Mines and Resources. (1971). *The National Atlas of Canada*. Ottawa: Department of Energy, Mines and Resources.

Driesch, A., von den (1976). *A Guide to the Measurement of Animal Bones from Archaeological Sites*. Cambridge, MA: Harvard University Peabody Museum Bulletin **1**

Federoff, N.E. (1996). Malocclusion in the jaws of captive bred Arctic wolves. *Canadian Field-Naturalist* **110:** 683-687.

French, D. D., Corbett, L. K. & Easterbee, N. (1988). Morphological discriminants of Scottish wildcats (*Felis silvestris*), domestic cats (*Felis catus*) and their hybrids. *Journal of Zoology London* **214:** 235-260.

Garcia-Moreno, J., Matocq, M. D., Roy, M. S., Geffen, E. & Wayne, R. K. (1996). Relationships and genetic purity of the endangered Mexican wolf based on analysis of microsatellite loci. *Conservation Biology* **10:** 376-389.

Geist, V. (1987). Bergmann's Rule is invalid. *Canadian Journal of Zoology* **65**: 1035-1038.

Gentry, A., Clutton-Brock, J. & Groves, C. P. (1996). Case 3010. Proposed conservation of usage of 15 mammal specific names based on wild species which are antedated by or contemporary with those based on domestic animals. *Bulletin of Zoological Nomenclature* **53:** 28-37.

Goldman, E. A. (1964). Classification of wolves. In (S. P. Young & A. E. Goldman Eds) *The Wolves of North America Part 2*. New York: Dover.

Gray, A. P. (1972). *Mammalian Hybrids*. Famham Royal: Commonwealth Agricultural Bureaux.

Jolicoeur, P. (1959). Multivariate geographical variation in the wolf *Canis lupus* L. *Evolution* **13:** 282-299.

Kitchener, A.C. (1998). The Scottish wildcat: The cat with an identity crisis? *British Wildlife*. **9:** 232-242.

Kitchener, A. C. & Easterbee, N. (1992). The taxonomic status of black wild felids in Scotland. *Journal of Zoology London* **227:** 342-346.

Lawrence, B. & Bossert, W. H. (1969). *The Cranial Evidence for Hybridization in New England* Canis. Cambridge, MA: Breviora Museum of Comparative Zoology **330**.

MacRury, I. K. (1991). *The Inuit dog: its provenance, environment and history*. M.Phil. Thesis, University of Cambridge.

Manning, T. H. and MacPherson, A. H. (1958). *The Mammals of Banks Island*. Ottawa: Arctic Institute of North America Technical Paper **2**.

McGhee, R. (1981). A tale of two cultures: a prehistoric village in the Canadian Arctic. *Archaeology* **34**: 44-51.

Mech, L. D. (1987). At home with the Arctic wolf. *National Geographic* **171:** 562-594.

Mech, L. D. (1988). *The Arctic Wolf*. Stillwater: Voyageur.

Mengel, R. M. (1971). A study of dog-coyote hybrids and implications concerning hybridization in *Canis*. *Journal of Mammalogy* **52:** 316-336.

Moyer, D. S. (1972/3). Wage income and skidoos: The non-productive significance of an Eskimo consumer item. *Folk* **14-15:** 145-162.

Nowak, R. M. (1979). *North American Quaternary* Canis. University of Kansas: Museum Natural History Monograph **6**.

Nowak, R M. (1983). A perspective on the taxonomy of wolves in North America. In (L. N. Carbyn, Ed) *Wolves in Canada and Alaska*. Ottawa: Canadian Wildlife Service Report Series **45**, pp. 10-19, 127-33.

Nowak, R. M. (1996). Another look at wolf taxonomy. In (L. N. Carbyn, S. H. Fritts, & D. R. Seip, Eds) *Ecology and Conservation of Wolves in a Changing World*. Edmonton: Canadian Circumpolar Institute, pp. 375-397.

Nowak, R. M., Phillips, M. K., Henry, V. G., Hunter, W. C. & Smith, R. (1996). The origin and fate of the red wolf. In (L. N. Carbyn, S. H. Fritts, & D. R. Seip, Eds) *Ecology and Conservation of Wolves in a Changing World*. Edmonton: Canadian Circumpolar Institute, pp. 409-415.

O'Brien, S. J. & Mayr, E. (1991). Bureaucratic mischief: Recognizing endangered species and subspecies. *Science* **251:** 1187-1188.

O'Brien, S. J., Roelke, M. E., Yuhki, N., Richards, K. W., Johnson, W. E., Franklin, W. L., Anderson, A. E., Bass, O. L., Belden, R. C. & Martenson, J. S. (1990). Genetic introgression within the Florida panther (*Felis concolor coryi*). *National Geographic Research* **6:** 485-494.

Pasitschniak-Arts, M., Taylor, M. E. & Mech, L. D. (1988). Note on skeletal injuries in an adult Arctic wolf. *Arctic and Alpine Research* **20:** 360-65.

Pennock, D. S. & Dimmick, W. W. (1997). Critique of the Evolutionary Significant Unit as a definition for "distinct population segments" under the US Endangered Species Act. *Conservation Biology* **11:** 611-619.

Pocock, R. I. (1935). The races of *Canis lupus*. *Proceedings Zoological Society London* **III**: 647-86.

Randi, E. & Ragni, B. (1991). Genetic variability and biochemical systematics of domestic and wild cat populations (*Felis silvestris*: Felidae). *Journal of Mammalogy* **72:** 79-88.

Randi, E., Lucchini, V. & Fransisci, F. (1993). Allozyme variability in the Italian wolf (*Canis lupus*) population. *Heredity* **71:** 516-522.

Sabine, E. (1821). Stanzas on the death of a favourite setter dog. *The North Georgia Gazette and Winter Chronicle*. London: John Murray.

Schwartz, M. (1997). *A History of Dogs in the Early Americas*. New Haven: Yale University Press.

Sillero-Zubiri, C. & Macdonald, D. (1998) *The Ethiopian wolf. Status Survey and Conservation Action Plan*. Gland: IUCN.

Steinhart, P. (1995). *The Company of Wolves*. New York: Vintage Books.

Vilà, C. & Wayne, R. K. (1999). Hybridization between wolves and dogs. *Conservation Biology* **13:** 195-198.

Walker, D. N. & Frison, G. C. (1982). Studies on Amerindian dogs, 3: prehistoric wolf/dog hybrids from the Northwestern plains. *Journal of Archaeological Science* **9**: 125-172.

Wayne, R. K. (1993). Molecular evolution of the dog family. *Trends in Genetics* **9:** 187-225.

Wayne, R. K. & Gottelli, D. (1998). Systematics, population genetics and genetic management of the Ethiopian wolf. In (C. Sillero-Zubiri & D. Macdonald, Eds) *The Ethiopian Wolf. Status Survey and Conservation Action Plan*. Gland: IUCN, pp. 43-50.

Wayne, R. K. & Jenks, S. M. (1991). Mitochondrial DNA analysis implying extensive hybridization of the endangered red wolf *Canis rufus*. *Nature* **351:** 565-568.

Wayne, R. K., Lehman, N. & Fuller, T. K. (1996). Conservation genetics of the gray wolf. In (L. N. Carbyn, S. H. Fritts, & D. R. Seip, Eds) *Ecology and Conservation of Wolves in a Changing*

*World.* Edmonton: Canadian Circumpolar Institute, pp. 399-407.
Wolfgramm, A. (1894). Die Einwirkung der Gefangenschaft auf die Gestaltung des Wolfschädels. *Zoologisches Jahrbucher Systematik* **7:** 773-822.

# USE OF THE MASTOID REGION OF THE CRANIA OF CANIDS TO DISTINGUISH WOLVES, DOGS, AND WOLF/DOG HYBRIDS

Bonnie C. Yates
National Fish & Wildlife Forensics Laboratory,
U.S. Fish & Wildlife Service
Ashland, Oregon,
USA

**Key words:** dog, wolf, wolf/dog hybrid, mastoid, hybridisation, archaeozoology, osteology

**Abstract**

This brief paper describes a method devised for distinguishing wolves, dogs and wolf/dog hybrids, based on the shape and development of the mastoid region of the skull. This research found that the extent of the elevation of the insertion for *m. sternomastoideus* and *m. cleidomastoideus* in the mastoid region clearly separates *Canis lupus* from *C. lupus* X *C. familiaris* with an accuracy rate of 88%. These muscles are responsible for neck and gait movements, and the sculpting of this cranial region in the living animal appears to be related to different gaits exhibited by wild and domestic canids, and perhaps also to dominance behaviours. However, these results are still preliminary and indications at this point suggest that for wildlife forensic purposes, additional morphological features must also be considered in distinguishing wolf/dog hybrids from wolves.

## Introduction

The osteological differences between closely related canids (Carnivora: Canidae) become slight and crucial when inferences are sought that have bearing on site interpretation, domestication timing, and more recently, endangered species issues. Traditional morphological examinations include cranial measurements and non-metric observations of cranial features (eg bulla inflation, orbital angle) to evaluate where on the continuum between wolf and domestic dog the questioned animal lies.

The shape of the mastoid region, for example, is "more deeply sculpted" in wolves than in large breeds and mixed breeds of domestic dogs.

## Mastoid shape differences

This research found that the extent of the elevation of the insertion for *m. sternomastoideus* and *m. cleidomastoideus* in the mastoid region clearly separates *Canis lupus* from *Canis lupus* X *C. familiaris* with an accuracy rate of 88%, and *C. lupus* from *C. familiaris* at < 95%. These muscles are responsible for neck and gait movements (Evans, 1993), and the sculpting of this cranial region in the living animal appears to be related to different gaits exhibited by wild and domestic canids, and perhaps also to dominance behaviours. Results of this preliminary study indicate that for wildlife forensic purposes, however, additional morphological features must also be considered in distinguishing wolf/dog hybrids from wolves.

The photo provided (Fig. 1) is meant to guide you as you examine skulls in hand, to assist in orienting the skulls and finding the process. Note that the white box merely draws your attention to the general location of the mastoid process. The labelled image is a scanned skull, tilted so that the mastoid lay directly on the scanning surface. This accounts for the strange angle of view, but it allowed consistency in producing the scanned image.

The swale or channel that separates the two insertion areas is wider and deeper in adult wolves (*Canis lupus*) with a greater lateral extension of the cleidomastoideus process; while in the domestic dog (*C. familiaris*), this channel is shallow, narrow, and the lateral projection of the cleidomastoidues process is less pronounced. In wolf/dog hybrids, this process is generally intermediate in separation and projection, but this feature is undoubtedly affected by degree of hybridisation in the individual animal: F1 or F2 animals should approximate the wolf by virtue of size, but this may be modified by the behaviour of the living animal.

It may eventually be demonstrated that if an individual demonstrates wild tendencies and/or alpha-type behaviour, then the feature will be more wolf-like; and conversely, if the individual demonstrates submissiveness, the feature will be more dog-like. Low-ranking female wolves appear to exhibit less separation and projection. Immature animals show deep channels, lots of separation, and usually a dimple or small depression in the channel; the bone is less ossified, as expected in juveniles, and this condition may change during maturation depending on behaviour and pack rank.

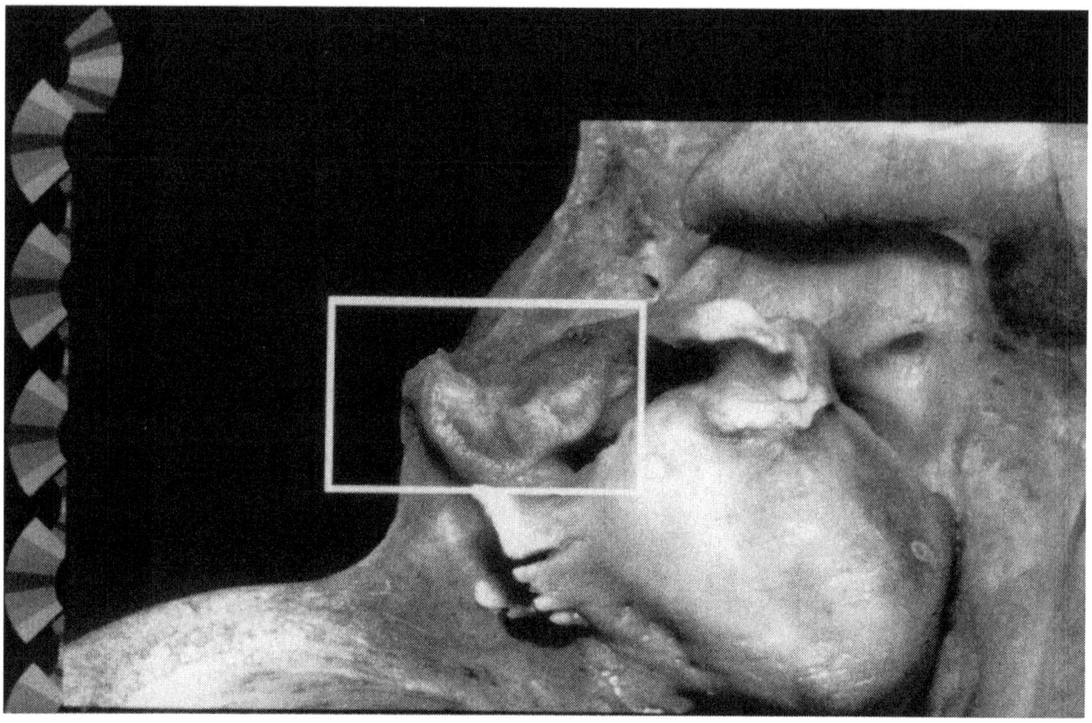

*The mastoid process includes two muscle insertion areas, outlined by the boxes in the ventral views: (1) the cleidomastoideus, which attaches to the lateral-most projection, and is involved in forward motion of the front leg, and (2) the sternomastoideus, which inserts on the small, circular mound situated above the bulla, and is involved with side to side motion of the neck (Evans, 1993).*

**Accuracy of the method**

A preliminary test was designed in which a suite of observations describing the mastoid process in skulls of wolves, dogs, and known wolf/dog hybrids was given a score. Fifty-six skulls of known identity were scored by six people with varying levels of experience and no prior or visible knowledge of the identity of the skulls. The results indicated that the scoring method had 88% or better effectiveness in assigning the correct morphotype to each skull.

**Future work**

Of course, more research is needed. I am presently collaborating with a researcher who will study various skulls in collections along the West Coast: the Burke Museum, Seattle, Washington; the LA County Museum, California; Cal-Academy of Science, San Francisco, California; Oregon State University, Corvallis; University of California (Davis and Berkeley) in hopes of expanding the data base. I will continue my work with the Smithsonian, Washington, DC collection. We also want to study living animals from a behavioral perspective and then collect the skull when the animal dies to see if its living behavior is indeed reflected in this process.

**Reference**

Evans, H. E. (1993). *Miller's Anatomy of the Dog,* third edition. Philadelphia: W. B. Saunders Co.

# ANCIENT DNA EVIDENCE OF A SEPARATE ORIGIN FOR NORTH AMERICAN INDIGENOUS DOGS

Ben F. Koop
Maryann Burbidge
Ashley Byun
Ute Rink
Centre for Environmental Health,
University of Victoria,
Victoria B.C., Canada
V8W 3N5

Susan J. Crockford
Pacific Identifications Inc.,
6011 Oldfield Rd.,
Victoria B.C., Canada
V9E 2J4

**Key words:** dogs, genetics, mtDNA, ancient DNA, dog evolution, hybridisation, canine genetics

**Abstract**

To determine whether indigenous North American dogs evolved independently from European canids after a separate domestication event, we have expanded the existing data set of previous reported canid mitochondrial DNA sequences. Our study adds 718bp D-loop sequences of extant fox, coyote, wolf (four subspecies) and dog (eleven breeds), and is the first to include several 300bp sequences of extinct indigenous North American dog (three varieties), extracted from hide and bone. From a limited number of representatives we have found evidence for independent evolution of North American indigenous dogs. We also identified a unique North American mtDNA lineage that contained two wolves and four of the five indigenous dogs, providing preliminary support for an independent origin of North American dogs. Thus at least two domestication events of dogs from wolves must have occurred.

## Introduction

The dog has the distinction of being the oldest domesticated animal. The earliest well-documented archaeological remains of fully domestic dogs (*Canis familiaris*) date to *c.* 10,000BP in North America, *c.* 14,000BP in northern Europe and *c.* 12,000BP in the Near East (Clutton-Brock, 1995). There are disputed finds dating further back, to the late Pleistocene or earlier, from the Near East (Dayan, 1994) and Siberia (Olsen, 1985).

Although all *Canis* species are reproductively compatible (Iljin, 1941; Mengel, 1971; Roy *et al.,* 1996), molecular methods have found no support for the suggestion that some dogs are descendants of jackals or coyotes and uphold the ancestral relationship of the grey wolf (*C. lupus*) to the domestic dog (Roy *et al.,* 1994; Wayne, 1993). However, the issue of monophyletic versus polyphyletic origins of dogs has remained contentious, despite repeated attempts to settle it using conventional systematic approaches (Epstein, 1971; Morey, 1990, 1992; Olsen & Olsen, 1977; Tanabe *et al.,* 1991; Wayne, Nash & O'Brien, 1987; Wayne & O'Brien, 1987). This debate centers on explaining the extreme range of morphological variation found among modern dog breeds (Wayne, 1986; Wilcox & Walkowicz, 1989).

Two hypotheses have been advanced to explain morphological variation in the dog. The prevailing theory (Clutton-Brock, 1995; Morey, 1994) suggests that a polyphyletic origin for dogs is probable and that as many as four domestication events (from European, Indian, Chinese and North American wolf subspecies) gave rise to four distinct dog morphotypes (mastiffs, sight hounds, oriental spaniels and Arctic spitzs, respectively). This theory suggests that these ancestral types may also have been subsequently modified through interbreeding with local wolves as humans migrated throughout the world. A second theory embraces the concept of a single domestication event, suggesting that local human preferences for certain physical and behavioural types modified a "basic" primitive dog through controlled breeding to give us such disparate groups as sight hounds, spitzs, spaniels and mastiffs (Scott, 1967).

Three recent studies using mitochondrial DNA sequences

from the D-loop of extant (living) canids have attempted to distinguish between wolf and dog species, and to determine the origins of the domestic dog (Okumura *et al.*, 1996; Vilà *et al.*, 1997; Tsuda *et al.*, 1997). Okumura *et al.* (1996) looked at 750 base pairs (bp) from 94 individual dogs representing eight indigenous Japanese breeds plus 16 other breeds and found 40 different haplotypes, but none that were breed specific. Vilà *et al.* (1997) looked at 261bp from 162 wolves from 27 localities and 140 dogs (67 breeds) and identified 27 wolf haplotypes and 26 dog haplotypes that grouped into four clades of dogs of which three were intermixed with wolves (a 1030bp extension of 24 selected canids confirmed these overall results). Tsuda *et al.* (1997) looked at 672bp from 34 modern dogs (24 breeds) and 19 wolves representing three subspecies, and found two major groups in which both dogs and wolves were represented. In this study we combine the data of these previous studies and contribute 718bp from 29 extant dogs from 11 different breeds, 13 wolves representing four subspecies, three coyotes and one red fox, in order to provide a robust framework in which to assess the relationships of extinct indigenous dogs from the Pacific Northwest region of North America (*cf.* DeSalle, 1994; Eernisse & Kluge, 1993;. Hagelberg, 1994; Horai *et al.*, 1991; Kringer *et al.*,1997; Richards *et al.*,1993).

We have incorporated "ancient" genetic material from museum skins and archaeological bones of three extinct varieties of North American dogs into this study of extant canids. The Tahltan bear dog (Fig.1) was used by the Tahltan First Nation to hunt bear and other game in northwestern British Columbia, Canada. The Tahltan was officially registered by the Canadian Kennel Club in 1942, but is now considered extinct. It was a relatively small, fox-sized dog, 12 to 15 inches at the shoulder, with a variously-patterned black and white coat (Crisp, 1956; Wilcox & Walkowicz, 1989). The Tahltan samples used in this study were taken from hides collected in 1938 and 1940. Samples were also extracted from several skeletal specimens of prehistoric dogs excavated from archaeological sites on the south coast of British Columbia (Fig. 2). Two distinct morphotypes, the "village" dog and the "wool" dog (Fig.3), had been reported by early (*c.* 1800) European visitors to this area (Crockford, 1997; Schulting, 1994). The dingo-like village dog, known also as the "Plains Indian dog" (Allen, 1920), was considered common throughout western North America from British Columbia south to California and eastward to the Great Plains. The distribution of the smaller, spitz-type wool dog or "Clallam-Indian dog" (Allen, 1920), was restricted to southwestern British Columbia in Canada and to the Puget Sound and Olympic Peninsula regions of Washington State, USA. The wool dog was said to have been kept exclusively for its thick fur, which was used in blanket weaving (Amoss, 1993; Howay, 1918), a role for dogs which is unique to this area. The wool dog as a distinct type was considered extinct by 1858 (Howay, 1918) due to interbreeding with both village dogs and imported European breeds. No specimens of either Coast Salish indigenous variety were ever collected (Haag, 1948), leaving skeletal remains recovered from archaeological sites as the only source of genetic material available (Crockford, 1997).

## Materials and Methods

Twenty-nine blood samples were collected from 11 contemporary dog breeds (Great Pyrenees, Maremma, Alaskan malamute, Siberian husky, Samoyed, Norwegian elkhound, Akita, Shar-Pei, Greyhound, Saluki and Whippet; Table 1). All individuals are registered purebreds. Blood or muscle tissue was also collected from 13 wolves (two *C. l. tundrarum* ; three *C. l. crassodon*; six *C. l. lupus*; two *C. l. columbianus*), three coyotes (*C. latrans* ) and one fox (*Vulpes vulpes*.). Three of the "ancient" samples were taken from indigenous dog bone recovered from previously excavated Coast Salish archaeological sites. Ribs from an intact burial (specimen

***Figure 1*** *Tahltan Bear Dog. Photo courtesy Winnie Acheson, Atlin B.C. (taken 1969).*

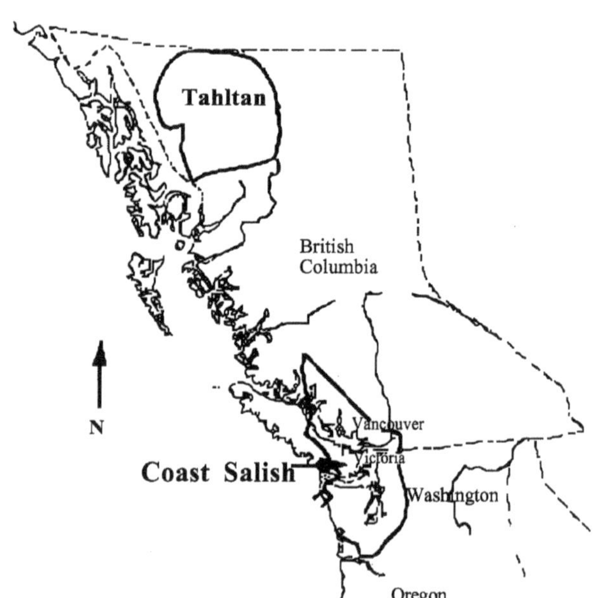


***Figure 2*** *Map of British Columbia, showing First Nations territories associated with indigenous dogs sampled.*

***Figure 3*** *Village Dog (left) and Wool Dog (right). Forensic reconstructions by Cpl. Cameron Pye, Royal Canadian Mounted Police, Vancouver (Crockford & Pye, 1997)*

SM89:0400), recovered from the 1989 salvage excavation at the St. Mungo Cannery site (DgRr2) Vancouver, were sampled for wool dog 64 (Crockford, 1997). An isolated femur (specimen WH88:1081), from the 1988 salvage excavations at the Beach Grove golf course (DgRs30) on the Fraser Delta, was sampled for village dog 74 (Crockford, 1997). Samples of Village dog 41 were obtained from ribs, radius and ulna of an intact burial recovered during the 1992 salvage excavation at DgRr1 site on the Fraser Delta (Stryde, pers. comm.) . The two other "ancient" samples were taken from museum skins of Tahltan bear dogs collected from the Liard River area, archived at the Royal British Columbia Museum, Victoria.

DNA from all contemporary tissues was extracted following Gustincich *et al.*'s (1991) DTAB/CTAB extraction method. DNA from ancient skins was extracted using the CTAB method (Doyle & Doyle, 1987; Byun, Koop & Reimchen, 1997). The bone samples were sterilized, drilled and powder collected in disposable sterile 15 mL tubes (Purdue & Patton, 1992; Sykes, 1993). DNA was extracted from the powdered bone using the methods described by Thomas *et al.* (1990). Stringent sterilization measures were used with ancient DNA to prevent contamination and multiple positive and negative controls were done (*cf.* Hagelberg & Clegg, 1991; Handt *et al.*, 1994; Höss *et al.*, 1996; Pääbo, 1993; Richards *et al.*, 1995). Only those samples in which two independent DNA extractions resulted in identical results are presented in this paper (Burbidge, 1997).

The primers H15980 5'-GCTGGTACCACC ATCAGCACCCAAAGCT-3' and L650 5'-RYTGGTACCAAGGCTRGGACCAAACCT-3' were used to amplify a 1.2 Kb PCR product of the D-loop of the mitochondrial genome in contemporary samples. The primers for ancient DNA samples were 5'-ACTATTCCCTGACACCCCTACATT-3' (H15942) and 5'-GGCCCGGAGCGAGAAGA-3' (L15682), which yielded a 300bp product.
PCR reactions (50ul) contained PCR buffer (for contemporary samples,1X PCR buffer of 500 mM KCl, 100 mm $MgCl_2$ and 15 mM Tris-HCl pH 8.4; for ancient samples, either 10X PCR buffer of 500mM KCL, 25mM $MgCl_2$, 100mM Tris-HCL pH 8.3, plus 0.01% gelatin or 1X Promega PCR buffer of 500mM KCL, 100mM Tris-HCL pH 8.8, 1% Trixton X-100 plus $MgCl_2$ added to a final concentration of 2.5uM), 0.2 mM of each dNTP, 0.5 uM of each primer, 0.1 units/ul *Taq* polymerase (Gibco, Pharmacia) and 10.0 ul DNA template (regardless of concentration). One ul of sample, as well as 10, 100 and 1000 fold template dilutions containing 0.4 mg/mL BSA, were used in ancient PCR reactions. An initial denaturation of 5 minutes (4 for ancient samples) preceded the 30 PCR cycles (35 for ancient samples), which consisted of DNA melting (45 sec. at 94°C, primer annealing (45 sec. at 50°C) and primer extension (2 min. at 72°C). These cycles were followed by a final 7 minute elongation at 72°C.

Contemporary dog PCR products were cut with Kpn I (45 units, Pharmacia) and PCR product bands were excised from a 2% Nusieve agarose gel and purified with subsequent phenol/ chloroform extractions. Inserts were ligated into a Kpn I cut, dephosphorylated M13mp18 vector, and transformed into MAX Efficiency DH□F'IQ™ Competent Cells (GibcoBRL). Isolation of recombinant M13 clone DNA and other routine DNA manipulations followed standard procedures (Sambrock, Fritsch & Maniatis, 1989). Ancient PCR products were purified from 2% NuSieve agarose using Wizard PCR Preps (Promega), and cloned directly using a TA cloning kit (Invitrogen). Plasmid preparations of TA clones were performed according to the Promega procedures (Titus, 1991).

Clones were sequenced using ABI 373 and 377 automated sequencers. The final sequence for each individual was determined from a consensus of a minimum of three sequence reads.

A multiple alignment of all of the sequences was determined using both MAP (Huang, 1994) and Megalign (DNASTAR, Madison WI). The final alignment was checked by eye and all differences were rechecked back to the original sequence runs. The muliple alignment was imported into programs PAUPv3.1 (Swofford, 1993; Swofford *et al.,* 1996) and PHYLIP for further phylogenetic analyses. In addition to our own sequences, the recently published D-loop sequences from other canids were obtained from GenBank and included in phylogenetic analyses (Table 1).

DNA sequence divergence between pairs of sequences was calculated using Kimura 2 parameter corrections (Kimura, 1980). Neighbor-Joining distance analysis (Saitou & Nei, 1987) was performed using equal probability for site variation as well as gamma corrections of 0.25 and 0.5. Similar results were obtained with the different correction methods. Support for the various branches was determined from over 500 bootstrap replicates.

The parsimony analysis yielded a very large number of equally most parsimonious trees. Therefore, to avoid the problem of local minima trees biasing the overall branching results, over 5000 equally parsimonious trees were determined from 500 random sequence addition starting points (in which up to 50 equally parsimonious trees were kept for branch swapping). A strict consensus of these results was used to generate the final tree. The strength of the data supporting the various groupings was determined from 500 bootstrap replicates, in which each replicate consisted of the most parsimonious tree derived from the iteration of 40 random addition starting points and branch swapping on the 40 most-parsimonious trees in each iteration.

**Results**

The 1.2kb PCR products spanning the D-loop from several individual dogs were initially sequenced and examined for variability and suitability for phylogenetic studies. The 3' portion of the 1.2kb sequence contained a 10bp repeat sequence (GTACGTACAC) in which the repeat number varied (avg.=15) within individuals. As the variation in repeat number could not be assigned solely to simple PCR amplification artifacts nor to the presence of heteroplasmic mitochondrial DNAs (Hoelzel *et al.,* 1994; Rothuizen *et al.,* 1996), only the 5' 718bp region was used for subsequent analysis. This 718bp region of the D-loop was sequenced from 48 contemporary canids: 29 domestic dogs (11 different breeds), 13 wolves (four subspecies), three coyotes, and one red fox. In addition, a 300bp segment through the most variable section of the 718bp was amplified and sequenced for five ancient specimens from extinct indigenous dogs. All of the new sequences are shown in Figure 4. Included in our analysis (Table 1) are 22 of the extended sequences (14 dog; eight wolf) from Vilà *et al.* (1997), 35 750bp sequences (35 dog) from Okumura *et al.* (1996) and 32 672bp sequences (24 dog; eight wolf) from Tsuda *et al.* (1997).

Within the data set of 129 sequences (> 650bp) there were 247 variable positions with 157 of these containing bases shared by two or more individuals. The distribution of these variable sites skewed slightly to the 3' 300bp segment, which contained 31 of the 57 informative sites. An analysis of pairwise sequence similarities showed a very high level of similarity among dogs and wolves. For the 718bp fragment of the D-loop, dogs and wolves are 0-3% different from one another, coyotes and dogs/wolves differ by 5.8% and the fox differs from coyotes, wolves and dogs by 16.4%. The average difference between pairs of dog sequences was 1.4%, although the variance was quite large (4.1%) and a bimodal distribution of pairwise difference values was evident. Approximately one half of the individuals differed by an average of 0.6% and the other half differed by 2.0%, suggesting some structuring within dogs. Similarly, wolves differed from each other by 1.6% but the variance was smaller (2.5%). A UPGMA analysis showed five major clusters of domestic dogs and wolves (tree not shown, but similar to Figure 5). Within each cluster, divergence averaged 0.6% and divergence between clusters averaged 2%, in agreement with the distribution of pairwise sequence divergence values. The "Wolf" cluster contained 15 of the 25 wolves and 17 of the 97 dogs, the "Dog" cluster contained 68 dogs and four wolves, the "Elkhound" cluster contained three Norwegian elkhounds and three wolves, the "Indigenous Dog" cluster contained two wolves and four of the five indigenous dog samples, while the "Dog2" cluster contained eight modern dogs and one indigenous dog. Individual dog breeds and wolf subspecies showed no evidence of clustering.

Five major dog/wolf groupings were also identified in maximum parsimony and Neighbor-Joining phylogenetic analyses, which gave essentially identical results (Fig. 5): a "Wolf" group, a "Dog" group, a second "Dog2" group, an "Indigenous dog" group, and an "Elkhound" group. Well supported groups (as indicated by high bootstrap values of 99-100%) included the fox group, the coyote group and the combined dog/wolf group. Bootstrap values over 50% support the Dog2 (78%) and Indigenous dog (61%) groups but the Wolf, Dog and Elkhound groups are very weakly supported. Low bootstrap values (less than 50%) for separate "wolf" and "dog" branches indicate there is little support for distinct wolf or dog mtDNA genotypes. However, in support of the five major groupings, all but three wolf sequences (AB007376, W12, and W15) consistently fell within the five dog/wolf groups in independent UPGMA, parsimony and Neighbor-Joining analyses.

The "Wolf" group contained 15 of the 25 wolf sequences and 17 of the 103 extant dog sequences. Representatives of 6 different wolf subspecies from Europe, North America, and Asia are found in this group (*C. l. lupus, pallipes, chanco, tundrarum, crassodon, columbianus*). While wolves are represented in four of the five major dog/wolf groups, the "Wolf" group contains the largest number and diversity of wolves. All of the dogs that fell within this

| Accession# | Name | Species/subsp. | (bp) | Locality | Reference |
|---|---|---|---|---|---|
| D83639 | Fox | Vulpes vulpes / red | 666 | Unknown | Okumura et al., 1996 |
| AF098155 | Fox | Vulpes vulpes / red | 678 | Yukon | this study |
| AF098153 | Coyote14 | C. latrans | 702 | Yukon | this study |
| AF098154 | Coyote16 | C. latrans | 701 | Yukon | this study |
| AF008158 | coyotel | C. latrans | 708 | Yukon | Vila et al. 1997 |
| AF008137 | W1 | C. lupes | 712 | Spain | Vila et al. 1997 |
| AF008138 | W4 | C. lupes | 713 | Italy | Vila et al. 1997 |
| AF008139 | W5 | C. lupes | 712 | Greece, Romania | Vila et al. 1997 |
| AF008136 | W6 | C. lupes | 712 | Romainia, Russia | Vila et al. 1997 |
| AF008140 | W12 | C. lupes | 712 | Saudi Arabia | Vila et al. 1997 |
| AF008142 | W14 | C. lupes | 712 | Saudi Arabia | Vila et al. 1997 |
| AF008141 | W15 | C. lupes | 712 | Saudi Arabia | Vila et al. 1997 |
| AF008135 | W27 | C. lupes | 713 | China | Vila et al. 1997 |
| AF098115 | Arctic Wolf1 | C. lupes tundrarum | 705 | Arctic | this study |
| AF098116 | European Wolf1 | C. lupes lupus | 706 | Europe | this study |
| AF098117 | European Wolf2 | C. lupes lupus | 706 | Europe | this study |
| AF098118 | Yukon Wolf2 | C. lupes columbianus | 706 | Yukon | this study |
| AF098119 | Yukon Wolf4 | C. lupes columbianus | 708 | Yukon | this study |
| AF098120 | Yukon Wolf5 | C. lupes columbianus | 706 | Yukon | this study |
| AF098121 | Yukon Wolf6 | C. lupes columbianus | 706 | Yukon | this study |
| AF098122 | Yukon Wolf22 | C. lupes columbianus | 708 | Yukon | this study |
| AF098123 | VanIsld Wolf1, Whippet | C. lupes crassodon | 707 | Vanc. Isl. | this study |
| AF098124 | VanIsld Wolf2 | C. lupes crassodon | 707 | Vanc. Isl. | this study |
| AF098125 | VanIsld Wolf3, Arctic Wolf, Yukon Wolf | C. lupes crassodon/col./tun. | 707 | VI, Arctic, Yukon | this study |
| AB007372 | European Wolf a, c, d | C. lupus lupus | 672 | Yugoslavia | Tsuda et al., 1997 |
| AB007376 | Chinese Wolf c, g | C. lupus chanco | 672 | Mongolia | Tsuda et al., 1997 |
| AB007378 | Chinese Wolf e | C. lupus chanco | 672 | Mongolia | Tsuda et al., 1997 |
| AB007377 | Chinese Wolf d | C. lupus chanco | 672 | Mongolia | Tsuda et al., 1997 |
| AB007379 | Chinese Wolf f | C. lupus chanco | 672 | Mongolia | Tsuda et al., 1997 |
| AB007375 | Chinese Wolf a,b | C. lupus chanco | 672 | Mongolia | Tsuda et al., 1997 |
| AB007374 | Indian Wolf a,b,c,d,e | C. lupus pallipes | 672 | Afghanistan | Tsuda et al., 1997 |
| AB007373 | European Wolf b,e,f,g | C. lupus lupus | 672 | Yugoslavia | Tsuda et al., 1997 |
| AF098148 | Tahltan1 | C. familiaris | 300 | Liard River BC | this study |
| AF098149 | Tahltan2 | C. familiaris | 300 | Liard River BC | this study |
| AF098150 | VillageDog41 | C. familiaris | 300 | Crescent Beach BC | this study |
| AF098151 | VillageDog74 | C. familiaris | 300 | Beach Grove BC | this study |
| AF098152 | WoolDog64 | C. familiaris | 299 | Vancouver B.C. | this study |
| AF098126 | Akita1 | C. familiaris | 706 | Victoria BC* | this study |
| AF098127 | Akita23 | C. familiaris | 707 | Victoria BC* | this study |
| U03575 | Unknown | C. familiaris | 407 | Unknown | Slade et al., 1994 |
| AF098128 | Greyhound2(2), Malamute, Saluki, Siberian Husky | C. familiaris | 707 | Victoria BC* | this study |
| AF098129 | Great Pyrenees17, Greyhound | C. familiaris | 707 | Victoria BC* | this study |
| AF098130 | Great Pyrenees27 | C. familiaris | 707 | Victoria BC* | this study |
| AF098131 | Malamute20 | C. familiaris | 707 | Victoria BC* | this study |
| AF098132 | Malamute22 | C. familiaris | 707 | Victoria BC* | this study |
| AF098133 | Malamute24 | C. familiaris | 707 | Victoria BC* | this study |
| AF098134 | Maremma28 | C. familiaris | 707 | Victoria BC* | this study |
| AF098135 | Maremma(2) | C. familiaris | 707 | Victoria BC* | this study |
| AF098136 | Maremma33 | C. familiaris | 707 | Victoria BC* | this study |
| AF098137 | Norwegian Elkhound5 | C. familiaris | 707 | Victoria BC* | this study |
| AF098138 | Norwegian Elkhound19 | C. familiaris | 707 | Victoria BC* | this study |
| AF098139 | Samoyed7 | C. familiaris | 707 | Victoria BC* | this study |
| AF098140 | Samoyed8 | C. familiaris | 708 | Victoria BC* | this study |
| AF098141 | SharPei25 | C. familiaris | 707 | Victoria BC* | this study |
| AF098142 | Siberian Husky10 | C. familiaris | 708 | Victoria BC* | this study |
| AF098143 | Siberian Husky11 | C. familiaris | 706 | Victoria BC* | this study |
| X97343 | Unknown | C. familiaris | ### | Unknown | Rothuizen,J et al. |
| AF098145 | Whippet13 | C. familiaris | 708 | Victoria BC* | this study |
| AF098146 | Whippet15 | C. familiaris | 708 | Victoria BC* | this study |
| AF098147 | Whippet16 | C. familiaris | 706 | Victoria BC* | this study |
| AF098148 | Whippet18 | C. familiaris | 708 | Victoria BC* | this study |
| AF008146 | D3, Chow, Norwegian Elkhound, Mexican , hairless, Siberian Husky, Papillon, English Setter, Icelandic Sheepdog, Springer Spaniel, Spitz, Border terrier, Fox terrier, Whippet, Poodle, Rottwiler | C. familiaris | 713 | | Vila et al. 1997 |
| AF008148 | D4, boxer, German shepherd, Alaskan husky, Labrador retriever, Samoyed, Schipperke, Giant Schnauzer, Norfolk terrier | C. familiaris | 713 | | Vila et al. 1997 |
| AF008144 | D6a, German shepherd, Groenendael, Mexican hairless, Afghan, Maremma, Otter hound, Toy poodle, Golden retriever, Irish water spaniel, Tibetan spaniel | C. familiaris | 713 | | Vila et al. 1997 |
| AF008150 | D6b, as above | C. familiaris | 713 | | Vila et al. 1997 |

| | | | | | |
|---|---|---|---|---|---|
| AF008152 | D7a, German shepherd, Alaskan husky, Siberian Husky, Swedish elkhounds, Giant schnauzer, Airedale terrier, Siberian Husky | C. familiaris | 713 | | Vila et al. 1997 |
| AF008156 | D7b, as above | C. familiaris | 713 | | Vila et al. 1997 |
| AF008157 | D7c, as above | C. familiaris | 684 | | Vila et al. 1997 |
| AF008155 | D8b, Norwegian elkhound, Swedish elkhound | C. familiaris | 713 | | Vila et al. 1997 |
| AF008145 | D18a, Dingo, Golden retriever | C. familiaris | 713 | Austrialia, New Guinae | Vila et al. 1997 |
| AF008154 | D18b, as above | C. familiaris | 713 | Austrailia, New Guinae | Vila et al. 1997 |
| AF008151 | D21, Mexican hairless | C. familiaris | 713 | Mexico | Vila et al. 1997 |
| AF008143 | D24, Golden retriever | C. familiaris | 713 | | Vila et al. 1997 |
| AF008147 | D26a, Mexican hairless, Pug, Rhodesian ridgeback, Shar | C. familiaris | 713 | | Vila et al. 1997 |
| AF008149 | D26b, as above | C. familiaris | 713 | | Vila et al. 1997 |
| D83599 | Afgan | C. familiaris | 984 | Japan* | Okumura et al., 1996 |
| D83600 | Akita (2) | C. familiaris | 984 | Japan | Okumura et al., 1996 |
| D83601 | Akita (5) | C. familiaris | 984 | Japan | Okumura et al., 1996 |
| D83602 | Cavelier King Charles Spaniel | C. familiaris | 984 | Japan* | Okumura et al., 1996 |
| D83604 | Doberman | C. familiaris | 984 | Japan* | Okumura et al., 1996 |
| D83605 | Eskimo | C. familiaris | 984 | Japan* | Okumura et al., 1996 |
| D83606 | Great Pyrenees | C. familiaris | 984 | Japan* | Okumura et al., 1996 |
| D83607 | Kokkaido (3), Akita (3), Iki (2) | C. familiaris | 984 | Japan | Okumura et al., 1996 |
| D83608 | Kokkaido | C. familiaris | 984 | Japan | Okumura et al., 1996 |
| D83609 | Hokkaido (3), Shikoku, Kishu, Shiba, Akita, Pug | C. familiaris | 984 | Japan, Japan* | Okumura et al., 1996 |
| D83610 | Iki (3) | C. familiaris | 984 | Japan | Okumura et al., 1996 |
| D83611 | Kishu | C. familiaris | 984 | Japan | Okumura et al., 1996 |
| D83612 | Kishu | C. familiaris | 984 | Japan | Okumura et al., 1996 |
| D83613 | Kishu | C. familiaris | 984 | Japan | Okumura et al., 1996 |
| D83616 | Pointer | C. familiaris | 984 | Japan* | Okumura et al., 1996 |
| D83618 | Pug | C. familiaris | 984 | Japan* | Okumura et al., 1996 |
| D83619 | Ryukyu(4) | C. familiaris | 984 | Japan | Okumura et al., 1996 |
| D83620 | Ryukyu, Kishu | C. familiaris | 984 | Japan | Okumura et al., 1996 |
| D83621 | Ryukyu | C. familiaris | 984 | Japan | Okumura et al., 1996 |
| D83622 | Ryukyu | C. familiaris | 984 | Japan | Okumura et al., 1996 |
| D83623 | Ryukyu | C. familiaris | 984 | Japan | Okumura et al., 1996 |
| D83624 | Ryukyu | C. familiaris | 984 | Japan | Okumura et al., 1996 |
| D83625 | Saluki | C. familiaris | 984 | Japan | Okumura et al., 1996 |
| D83626 | Sheltie(2) | C. familiaris | 984 | Japan* | Okumura et al., 1996 |
| D83627 | Shiba(2), Akita | C. familiaris | 984 | Japan | Okumura et al., 1996 |
| D83628 | Shiba (9), Iki | C. familiaris | 984 | Japan | Okumura et al., 1996 |
| D83629 | Shiba | C. familiaris | 984 | Japan | Okumura et al., 1996 |
| D83631 | Shiba | C. familiaris | 984 | Japan | Okumura et al., 1996 |
| D83632 | Shiba | C. familiaris | 984 | Japan | Okumura et al., 1996 |
| D83633 | Shiba(4), Akita, Hokkaido, Iki(2), Kai, Dalmation, Beagle, Poodle, Eskimo | C. familiaris | 984 | Japan, Japan* | Okumura et al., 1996 |
| D83634 | Shi Tzu | C. familiaris | 984 | Japan | Okumura et al., 1996 |
| D83635 | Shi Tzu | C. familiaris | 984 | Japan | Okumura et al., 1996 |
| D83636 | Shikoky | C. familiaris | 984 | Japan | Okumura et al., 1996 |
| D83637 | Siberian Husky | C. familiaris | 984 | Japan* | Okumura et al., 1996 |
| D83638 | Siberian Husky | C. familiaris | 984 | Japan* | Okumura et al., 1996 |
| AB007403 | Ryukyu a, b | C. familiaris | 672 | Japan | Tsuda et al., 1997 |
| AB007388 | Mongolian native b | C. familiaris | 672 | Mongolia | Tsuda et al., 1997 |
| AB007385 | Mikawa a,b, Shikoku b, Siberian Husky | C. familiaris | 672 | Japan | Tsuda et al., 1997 |
| AB007394 | English Setter | C. familiaris | 672 | Japan* | Tsuda et al., 1997 |
| AB007390 | Beagle | C. familiaris | 672 | Japan* | Tsuda et al., 1997 |
| AB007389 | Shiba (Shinshu a,b) | C. familiaris | 672 | Japan | Tsuda et al., 1997 |
| AB007392 | Maltese | C. familiaris | 672 | Japan* | Tsuda et al., 1997 |
| AB007382 | Shetland Sheepdog | C. familiaris | 672 | Japan* | Tsuda et al., 1997 |
| AB007395 | Hokkaido a, Kai b | C. familiaris | 672 | Japan | Tsuda et al., 1997 |
| AB007386 | Shiba (Mino a) | C. familiaris | 672 | Japan | Tsuda et al., 1997 |
| AB007384 | Kishu a,b | C. familiaris | 672 | Japan | Tsuda et al., 1997 |
| AB007383 | Hokkaido b, Shikoku a | C. familiaris | 672 | Japan | Tsuda et al., 1997 |
| AB007393 | Pointer | C. familiaris | 672 | Japan* | Tsuda et al., 1997 |
| AB007396 | German Shephard a | C. familiaris | 672 | Japan* | Tsuda et al., 1997 |
| AB007397 | German Shephard b | C. familiaris | 672 | Japan* | Tsuda et al., 1997 |
| AB007400 | Korean native a | C. familiaris | 672 | Japan | Tsuda et al., 1997 |
| AB007398 | Indonesian native a | C. familiaris | 672 | Kalimantan | Tsuda et al., 1997 |
| AB007399 | Indonesian native b | C. familiaris | 672 | Kalimantan | Tsuda et al., 1997 |
| AB007401 | Korean native b | C. familiaris | 672 | Korea | Tsuda et al., 1997 |
| AB007380 | Akita a | C. familiaris | 672 | Japan | Tsuda et al., 1997 |
| AB007391 | Doberman Pinscher | C. familiaris | 672 | Japan* | Tsuda et al., 1997 |
| AB007387 | Shiba (Mino b), Kai a, Dackshund | C. familiaris | 672 | Japan, Japan* | Tsuda et al., 1997 |
| AB007402 | Mongolian native a | C. familiaris | 672 | Mongolia | Tsuda et al., 1997 |
| AB007381 | Akita b | C. familiaris | 672 | Japan | Tsuda et al., 1997 |

***Table 1*** *A listing of all canids used in phylogenetic analyses, including: GenBank accession numbers, species, subspecies, breeds, sequences, locations (if known) and references.*

```
                                          1111111111111111111111222222222222222222222222222
                  1223444445556666677788889992222355566788888999990000001111222222223344558999
Taxon             390789234690252467807802895693678978967512568034568025679236703456794506585057
---------------------------------------------------------------------------------------------------
Fox               ATTCGTACCATGTGACCGCCTCC---TGCCCTAATACAAAACTGTTTCTTCAATC-TATGGCACATGCCGTCATTGAG
Coyote14          ATCTGTTATTAACCGGTACTCTCATCC--AGCTAATACTGATTGCTCCTTCGCTTGT-TACTGGTTAATACCCATGTA
Coyote16          ATCTGTTATTAACCGGTACTCTCATCC--GGCTAATACTGACTGCTCCTTCGTCTGT-TACTGGTTAATACCCATGTA
ArcticWolf        ATCCATTATTAACCGGTATTCTCGCCTCTAGTTAATATCGATCACCTCTCTATTTGTRTACTAATCAATATCCACGTA
EuropeanWo1       ATCCATTATTAACCGGTATTCTCGCCTCTAGTTAATATCGATCACCTCTCTATTTGTGTACTAATCAATATCCACGTA
EuropeanWo2       ATCCATTATTAACCGGTATTCTCGCCTCTAGTTAATATCGATCACCTCTCCATTTGTATACTAATCAATATCCACGTA
YukonWolf2        ATCCATTATTAACCGGTATTCTCGCCTCTAGTTAATATCGATTGCCTCCCCATTTGTATACTAATCAACATCCACGTA
YukonWolf4        ATCCATTATTAACCGGTATTCTCGCCTCTAGTTAATATCGATTACCTCTCTATCTGTATGCTAATCAATATCCACGTA
YukonWolf5        ATCCATTATTAACCGGTATTCTCGCCTCTAGTTAATATCGATCACCTCTCCATTTGTATGCTAATCAATATCCACGTA
YukonWolf6        ATCCATTATTAACCGGTATTCTYGCCTCTAGTTRATATCGATCACCTCTCCATTTGTATGCTAATCAATATCCACGTA
YukonWolf22       ATCCATTATTAACCGGTATTCTCGCCTCTAGTTAATATCGATTACCTTTCTATCTGTATACTAATCAATATCCACGTA
VanIsldWol1       GTCCATTATTAACCGGTATTCTCGCCCCTAGTTAATACCGATTACCTCTCTATCTGTTTACTAATCGATATCCACGTA
VanIsldWol2       GTCCATTATTAACCGGTATTCTCGCCCCTAGTTAATACCGATTACCTCTCTATCTGTTTACTAATCGATATCCACGTA
VanIsldWol3       ATCCATTATTAACCGGTATTCTCGCCTCTAGTTAATATCGATCACCTCTCCATTTGTATACTAATCAATATCCACGTA
Tahltan1             CAYTATTAACCGGTATTCTCGCCTCTAGTTAATATCGRTTACCTTTCTATCTGTATACTAATCAATATCCACGTA
Tahltan2             CATTATTAACCGGTATTCTCGCCTCTAGTTAATATCGATTACCTTTCTATCTGTATACTAATCRATATCCACGTA
VillageDog41         CATTATTAACCGGTATTCTCGCCTCTAGTTAATATCGATTACCTTTCTATCTGTATACTAATCGATATCCACCTA
VillageDog74         CATTATTAACCGGTATTCTCGCCTCTAGTTAATATCGATTACCTTTCTATCTGTATACTAATCAACATTCACGTA
WoolDog64            CATTATTAACCGGTATTTTCGCCTCTAGTTAATACCGACTACCTCTCTATCTGTGTACTAACCGATATCCACGTA
Akita1            ATCCATTATTAACCGGTATTCTCGCCCCTAGTTAATACCGATTACCTCTCTGTCTGTATACTAATCGATATCCACGTA
Akita23           RTCCATTATTAACCGGTATTCTCGCCCCTAGTTAATACCGATTACCYCTCTGTCTGTATACTAATCGATAYCCACGTA
Greyhound2        GTCCATTATTAACCGGTATTCTCGCCCCTAGTTAATACCGATTACCTCTCTATCTGTATACTAATCGATATCCACGTA
GtPyrenees1       GTCCATTATTAACCGGTATTCTCGCCCCTAGTTAATACCGATTACCTCTCTGTCTGTGTACTAATCGATATCCACGTA
GtPyrenees1       GYCCATTATTAACCGGTATTCTCGCCCCTAGTTAATACCGATTACCTCTCTGTTTGTATACTAATCGATATCCACGTA
Malamute20        GTCCATTATTAACCGGTATTCTCGCYCCTAGCTAATACCGATTACCTCTCTATCTGTATACTAATCGATATCCACGTA
Malamute22        GTCCATTATTAACCGGTATTCTCGCCCCTAGCTAATACCGATTACCTCTCTATCTGTATACTAATCGATATCCACGTA
Malamute24        GYCCATTATTAACCGGTATTCTCGCCCCTAGTTAATACCGATTACCCCTCTGTCTGTATACTAATCGATATCCACGTA
Maremma28         GTCCATTATTAACCGGTATTCTCGCCCCTAGTTAATACCGATTACCTCTCTATCTGTATACTAATCGATATCCACGTA
Maremma31         GYCCATTATTAACCGGTATTCTCGCCCCTAGTTAATACCGATTACCTCTCTATCTGTATACTAATCGATATCCACGTA
Maremma33         GYCCATTATTAACCGGTATTCTCGCCCCTAGTTAATACCGATTACCTCTCTATCTGTWTACTAATCGATATCCACGTA
NrwElkhoun5       ATCCATTATTAACCGGTATTCTCGCCCCTAGTTAATACCGATTACCTCTCCATTTACGTGCTAATCGATATCCACGTA
NrwElkhoun19      ATCCATTATTAACCGGTATTCTCGCCCCTAGTTAATACCGATTACCTCTCCATTTACGTGCTAATCGATATCCACGTA
Samoyed7          GYCCATTATTAACCGGTATTCTCGCCCCTAGTTAATACCGATTACCCCTCTGTCTGTATACTAATCGATATCCACGTA
Samoyed8          GTCCATTATTAACCGGTATTCTCGCCCCTAGTTAATACCGATTACCCCTCTGTCTGTATACTAATCGATATCCACGTA
SharPei25         GYCCATTATTAACCGGTATTCTCGCCCCTAGTTAATACCGATTACCTCTCTGTCTGTATACTAATCAATACCCACGTA
SHusky10          GYCCATTATTAACCGGTATTCTCGCCCCTAGTTAATACCGATTACCTCTCTATCTGTATACTAATCGATATCCACGTA
SHusky11          GYCCATTATTAACCGGTATTCTCGCCCCTAGTTAATACCGATTACCTCTCTGTCTGTACACTAATCGATATCCACGTA
Whippet13         GYCCATTATTAACCGGTATTCTCGCCCCTAGTTAATACCGATTACCTCTCTATCTGTTTACTAATCGATATCCACGTA
Whippet15         GTCCATTATTAACCGGTATTCTCGCCCCTAGTTAATACCGATTACCTCTCTATCTGTTTACTAATCGATATCCACGTA
Whippet16         GTCCATTATTAACCGGTATTCTCGCCCCYAGTTAATACCGATTACCTCTCTATCTGTTTACTAATCGATATCCACGTA
Whippet18         GTCCATTATTAACCGGTATTCTCGCCCCTAGTTAATACCGATTACCTCTCTATCTGTTTACTAATCGATATCCACGTA

                  23333333333333344444455555555555555666666666666666666666666666677777777777777
                  90223447888999911129900000123366789000111111223444555666666666789000000111111111
Taxon/Node        935694867891247589509238990372731712412357829845718901234567881623467901234789
---------------------------------------------------------------------------------------------------
Fox               ACTCTCGCTCTCTGCCCTGTCTTGGCTGTCRAAGCGCAGAGTCAATTTAATGCACAACTCATGAAG---GAATTTACA
Coyote14          GCTCTTGCGATCTATTTGGTTAAC-TCTCTTTAGTATAAAACTAATTTAACATAAAATCAATGAAATCTGGACACGTA
Coyote16          GCTCTCGCAATCTATTTGGTTAAC-TCTTTTTAGTATAAAACTAATTTAACATAAAATCAATGAAATCTGGACACGTA
ArcticWolf        GCCTACACAACTCATTCGGTTAATATCTCTTTACTATAAAACTAATTTAACATAAGATCAATGAAGTTTNGGTACGTA
EuropeanWo1       GCCTACACAACTCATTCGGTTAATATCTCTTTAGTATAAAACTAATTTAACATAAGATCAATGAAGTTTAGGTACGTA
EuropeanWo2       GCCTACACAACTCATTCGGTTAATATCTCTTTGGTATAAAACTAATTTAACATAAGATCAATGAAGTT-GGGTACGTA
YukonWolf2        GCCTACGCAACTCATTCGGTTAATATCTCTTTGGTATAAAACTAATTTAACATAAGATCAATGAAGTTTGGGTACGTA
YukonWolf4        GCCTACACAACTTATTCGGTTAATATCTCTTTGGTATAAAACTAATTTAACATAAAATCAATGAAGTTTGGGTACGTA
YukonWolf5        GMCTACACAACTCATTCGGTTAATATCTCTTTGGTATAAAACTAATTTAACATAAGATCAATGAAGTTTGGGTACGTA
YukonWolf6        GCCTACACAACTCATTCGGTTAATATCTCTTTGGTATAAAACTAATTTAACATAAGATCAATGAAGTTTGGGTACGTA
YukonWolf22       GCCTACACAACTTATTCGGTTAATATCTCTTTGGTATAAAACTAATTTAACATAAAATCAATGGAGTTNGGGTACGTG
VanIsldWol1       GCCTACATAACCTATTCGGCTAATATCYCTTTGGTATAAAACTAATTTAACATAARATCAATGAAGTTTGGGTACGTA
VanIsldWol2       GCCTACATAACCTATTCGGCTAATATCCCTTTAGTATAAAACTAATTTAACATAAAATCAATGAAGTTTGGGTACGTA
VanIsldWol3       GCCTACACAACTCATTCGGTTAATATCTCTTTGGTATAAAACTAATTTAACATAAGATCAATGAAGTTTGGGTACGTA
Tahltan1          GC
Tahltan2          GC
VillageDog41      GC
VillageDog72      GC
WoolDog64         GC
Akita1            GCCCACATAACTTATTCGGCTAATATCCCTTTAGTACAAAACTAATTTAACATAAAATCAATGAAGTTTGGGTACGTA
Akita23           GCCYACATAACTTATTCGGCTAATATCCCTTTAGTACAAAACTAATTTAACATAAAATCAATGAAGTTTGGGTACGTA
Greyhound2        GCCTACATAACTTATTCGGCTAATATCCCTTTAGTACAAAACTAATTTAACATAAAATCAATGAAGTTTGGGTACGTA
GtPyrenees        GCCTACATAACTTATTCGGCTAATATCCCTTTAGTACAAAACTAATTTAACATAAAATCAATGAAGTTTGGGTACGTA
GtPyrenees        GCCTACATAACTTATTCGGCTAATATCCCTTTAGTACAAAACTAATTTAACATAAAATCAATGAAGTTTGGGTACGTA
Malamute20        GCCTACATAACTTATTCGGCTAATATCCCTTTAGTATAAAACTAATTTAACATAAAATCAATGAAGTTTGGGTACGTA
Malamute22        GCCTACATAACTTATTCGGCTAATATCCCTTTAGTATAANACTAATTTAACATAAAATCAATGANGTTTGGGTACGTA
Malamute24        GCCTACATAACTTATTCGGCTAATATCTCTTTAGTATAAAGCTGTNCANCCATNAARTCARYGNAGNTTGAGTACGTA
Maremma28         GCCTACATAACTTATTCGGCTATTATCCCTTTAGTACAAAACTAATTTAACATAAAATCAATNAAGTTTGGGTACGTG
Maremma31         GCCTACATAACTTATTCGGCTATTATCCCTTTAGTACAAAACTAATTTAACATAAAATCAATGAAGTTTGGGTACGTA
Maremma33         GCCTACATAACTTATTCGGCTAATATCCCTTTAGTATAAAACTAATTTAACATAAAATCAATGAAGTTTGGGTACGTA
NrwElkhoun5       GCCTACACAACTCATTCGATTAATATCTTTTTGGTATAAAACTAATTTAACATAAAATCAATGAAGTTTGGGTACGTA
NrwElkhoun19      GCCTACACAACTCATTCGATTAATATCTTTTTGGTATAAAACTAATTTAANATAAAATCAATGAAGTTTGGGTACGTA
Samoyed7          GCCTACATAACTTATTCGGCTAATATCCCTTTAGTATAAAACTAATTTAACATAAAATCAATGAANTTTGGGTACGTA
Samoyed8          GCCTACATAACTTATTCGGCTAATATCCCTTTAGTATAAAACTAATTTAACATAAAATCAATGAAGTTTGGGTACGTG
SharPei25         GCCTACATAACTTATTCGGCTAATATCCCTTTAGTATAAAACTAATTTAACATAAAATCAATGAAGTTTGGGTACGTA
SHusky10          GCCTACATAACTTATTCGGCTAATATCCCTTTAGTACAAAACTAATTTAACATAAAATCAATGAAGTTTGGGTACGTA
SHusky11          GCCTACATAACTTATTCGGCTAAT-TCTCTTTAGTACGAAACTAATTTAACATAAAATCAATGAAGTTTGGGTACGTA
Whippet13         GCCTACATAACTTATTCGGCTAATATCCCTTTAGTATAAAACTAATTTAACATAAAATCNATGAAGTTTGGGTACGTA
Whippet15         GCCTACATAACTTATTCGGCTADTATCCCTTTAGTATAAAACTAATTTAACATAAAATCAATGAAGTTTGGGTACGTA
Whippet16         GCCTACATAACTTATTCGGCTAATATCCCTTTAGTATAAAACTAATTTAACATAAAATCAATGAAGTTTGGGTACGTA
Whippet18         GCCTACATAACTTATTCGGCTAATATCCCTTTAGTATAAAACTAATTTAACATAAAATCAATGATGTTTGGGTACGTA
```

***Figure 4**. Alignment of novel canid DNA sequences. Only variable sites and their locations are indicated, the complete sequences are available from GenBank. The position numbering structure refers to the alignment of DNA sequences.*

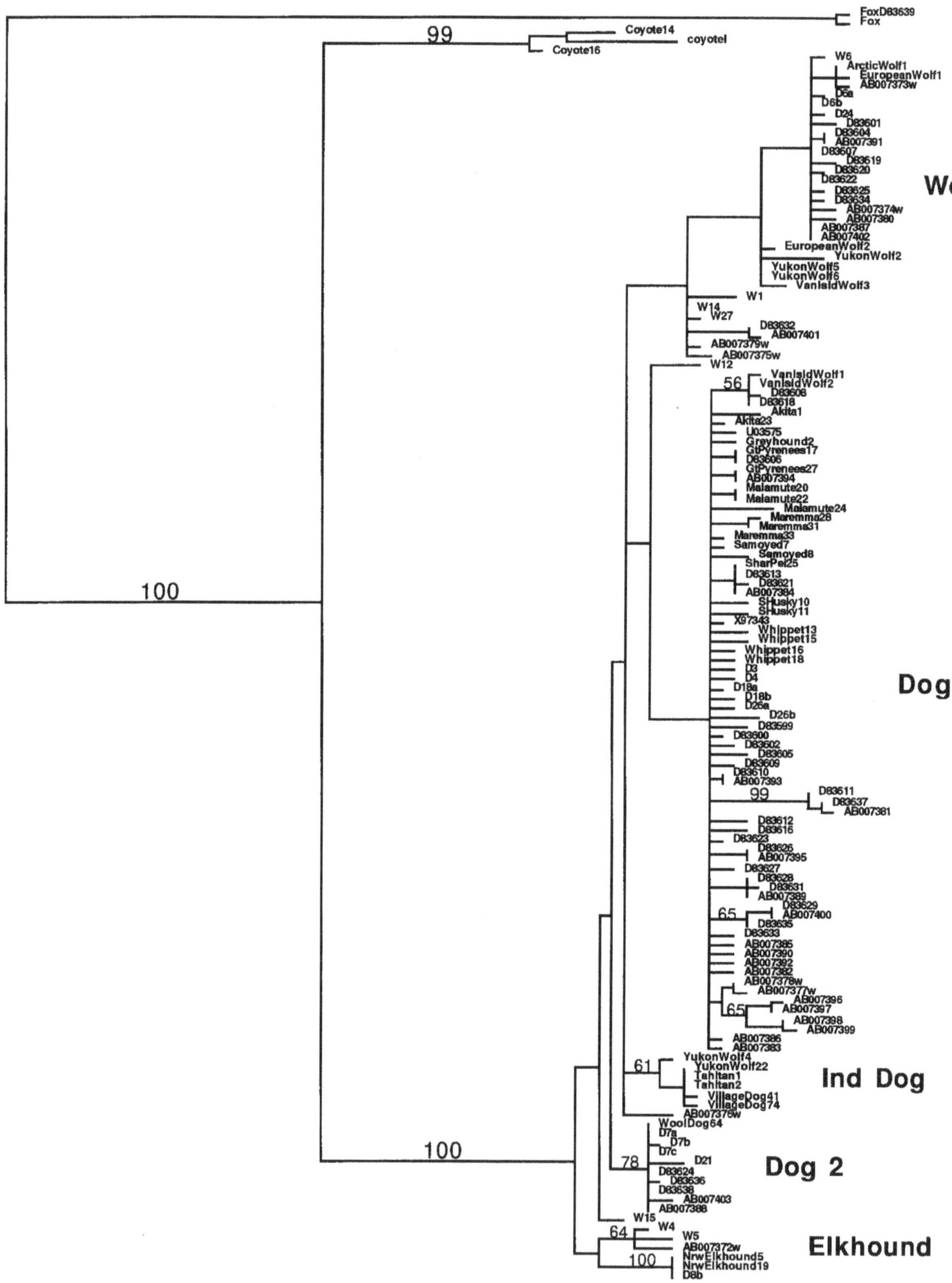


***Figure 5*** *Canid phylogenetic tree. This phylogenetic tree depicts a phylogram obtained by inputting into PAUP a single strict consensus tree obtained from the best 5000 equally parsimonious trees. In this way, estimates of amounts of evolution occurring along the different branch lengths are shown. All of the mtDNA sequences used in this study are presented. Written out names indicate sequences from the current study. Short number designations with a "W" or "D" prefix indicate wolf & dog sequences from Vilà et al. (1997). Long number designations (GenBank accession numbers) with a "D" prefex indicate dog sequences from Okumura et al. (1996) and those with an "AB" prefix only indicate dog sequences from Tsuda et al. (1997); Tsuda et al. sequences for wolves begin with "AB" and end with "w". Bootstrap values over 50% are presented along the branch lengths.*

group (Akita, German shepard, Groenendael, Mexican hairless, Afghan, Maremma, Otter hound, Toy poodle, Golden retriever, Irish water spaniel, Tibetan spaniel, Doberman, Hokkaido, Iki, Ryukyu, Kishu, Saluki, Shi Tzu, Mongolian Native dog, Shiba, Kai dog, Dackshund, Korean Native dog) have representatives in other groups. The "Wolf" clade can be defined by a "G" at position 663, with two exceptions found within the group and two exceptions from other groups. In addition, a "T" at position 166 is found in 24 of the 31 "Wolf" group sequences (but also occurs in all of the "Indigenous Dog" group sequences).

The "Dog" group contains 69 of the 103 dog sequences and contains representatives of all of the extant dog breeds assessed in this study. Many of the dog breed representatives are found only in this group. Of the 69 sequences in this group all but three have a "C" at position 490 (all other dog/wolves have, with 1 exception, a "T") and all but five have a "T" at position 376 (all other dog/wolves have a "C"). In addition, position 3 may be diagnostic of dogs (28 of 29 have a "G") but sample size is currently limited. At position 95, all of the "Dog" group and all of the Elkhound" group have a "C", as opposed to a "T" in all but one of the other dog/wolf sequences. Among the sequences of different breeds there is no evidence of clustering. The "Dog" group also contains two of the three wolf sequences from Vancouver Island and two of eight wolf sequences from China. For the wolves from Vancouver Island, there is morphological evidence for very recent hybridization with extant dogs (Friis, 1985) which would explain their phylogenetic placement (ie indicating offspring of female dogs and male wolves living in the wild and identified morphologically as wolves).

The "Indigenous Dog" group consists of sequences from four of the five extinct indigenous dog samples, two of which are from the same geographic region as two Canadian wolves represented in this group. Yukon Wolf4 and 22 are full length sequences and the indigenous dog sequences are 299bp (positions 15-320) in length. The two Tahltan sequences are identical to the Wolf22 sequence except for one and two ambiguities. The two village Dog sequences each differ by 2bp from the Wolf22 sequence. Wolf 4 differs from Wolf 22 by 4bp over 718bp. The "Indigenous Dog" group is defined by a "C" insertion at position 104 and by a "T" at position 166 that is shared with the "Wolf" group. Interestingly, four of these canids come from the southern Yukon territory and the northern portion of the adjacent province of British Columbia, while the other two are from southern British Columbia (Fig. 2).

The "Dog2" group contains ten nearly identical sequences from modern dogs (Alaskan husky, Airedale terrier, German shepherd, Siberian husky, Mexican hairless, Ryukyu, Shikoku, Mongolian Native Dog) and one "ancient" sequence (indigenous wool dog). Diagnostic of this group is a "T" at position 77 and a "C" at position 225. Although all of the modern dog breeds from "Dog2" are also represented in the "Dog" group, this is the only group that does not contain a wolf representative.

The "Elkhound" group contains three European wolf sequences and three nearly identical Norwegian elkhound sequences (each differing by one ambiguous base). The three wolf sequences differ from each other by 0.5% and from the elkhounds by 1.5%. This group is defined by a "C" at position 209 and an "A" at position 425.

In addition to the five major dog/wolf groups, there are three wolf sequences (W12 , W15 (Middle East and India) and AB007376w (China)) that are basally positioned and inconsistently placed within the dog/wolf phylogenies obtained from parsimony, UPGMA and Neighbor Joining analyses. With regard to affinities between the different dog/wolf groups, base positions 95, 166, 227 and 581 have the most direct relevance to phylogenetic relationships but provide no consistent support for the relationships among the five dog/wolf groups and the three ancestral wolf lineages.

The fox, coyote and dog/wolf sequences showed an average divergence between fox and coyotes of 15.5% +/- 0.7, an average divergence between fox and dog/wolf of 17.4% +/- 0.5 and a divergence between coyotes and dog/wolf of 5.8% +/- 0.4 (Kimura, 1987; Thorpe, 1996). Using a simple rate test, it appears that since their common ancestor, the dog/wolf lineage has accumulated significantly more ($P > 0.05$) changes than the coyote lineage (3.8% in the dog/wolf lineage and 2.0% in the coyote lineage).

**Discussion**

Three separate studies have recently examined the relationship of dog and wolf mitochondrial D-loop sequences based on samples from extant animals. Okumura *et al.* (1996) found four dog lineages but no breed-specific haplotypes, while Vilà *et al.* (1997) found four clades of dogs (three of which were intermixed with wolves) and very little structure among wolves. Vilà *et al.* (1997) interpreted the presence of a large clade exclusive to dogs as evidence for a separate and older origin of domestic dogs than was previously thought (>100,000 years). The conclusions of Tsuda *et al.* (1997) however, were that dogs and wolves are very similar and found two major clades, neither of which were specific to dogs or wolves. They concluded, in agreement with Vilà *et al.* (1997) that extant breeds of dogs have maintained a large degree of polymorphism introduced from their ancestral wolf populations and that extensive interbreeding had occurred among multiple matriarchal origins. However, in contrast to Vilà *et al.* (1997) they concluded that there was no evidence for a distinct dog lineage and thus could not confirm a >100,000 year origin for domestic dogs. In all of these studies, the particular groupings had very low levels of confidence, as indicated by bootstrap values of less than 50%. In the present study we have combined the data from the three previous studies and added 718bp from 29 modern dogs (11 breeds), 13 wolves (four subspecies), three coyotes and one red fox. In addition, we have obtained five indigenous North American dog sequences from three extinct varieties and attempted to assess their relationship to modern dogs and wolves.

It has been suggested that the extreme morphological diversity of dogs may be the result of a varied genetic diversity caused by multiple founding events and occasional interbreeding with wolves (Vilà *et al.*, 1997). However, we have found that the overall level of genetic diversity in dogs is similar to that of wolves, coyotes and foxes. The level of genetic diversity within the dog/wolf cluster varies between 0 and 3% and appears to have a bimodal distribution with one group averaging 0.6% and another group averaging 2.0%, with the overall average being 1.5%±0.8%. Genetic diversity values obtained from six foxes range from 0-3.1%, with an average of 1.8% (Tsuda *et al.*, 1997; this study) and values obtained from 3 coyotes average 1.5% (this study). Thus the level of genetic variation within the dog/wolf group appears to be at least comparable to that of foxes and coyotes but provides no clear evidence for differentiation of two genetic groups within the dog/wolf clade.

While the evidence from phylogenetic analyses are weak, within the dog/wolf group appear to be at least eight ancestral mitochondrial DNA lineages (Fig.5). These lineages (three of which are represented by a single sequence) have diverged from each other by at least 2% and 0.5-1 million years (assuming a molecular clock and a rate of 2-4% per million years (Brown *et al.*, 1986; Kimura, 1987; Vilà *et al.*, 1997; Tsuda *et al.*, 1997). If, as suggested by the rate test, mtDNA from wolves and dogs are changing at a faster rate than mtDNA from coyotes then these time estimates may be overestimated by a factor of two. Still, even allowing for an increased rate of evolution in dogs and wolves, the presence of both dogs and wolves in several ancient mtDNA lineages indicates that dog and wolf taxonomic groupings are independent of the ancient mtDNA lineages and thus the age of the different mtDNA lineages can not be used to estimate when dogs were derived from wolves.

The most striking feature of the dog and wolf sequence phylogeny is the lack of any clear resolution despite the presence of a large number of variable sites (Fig.5 & 6). There was no evidence of two distinct mitochondrial sequence clades which would have supported the presence of two genetically isolated populations over an extended period of time. Instead, the data indicate at least eight, more-or-less equally distinct mtDNA haplotypes of which seven are found in animals classified as wolves and five in dogs (Fig.6). The broader representation of mitochondrial lineages in the wolves argues to a limited extent for the ancestral origins of modern dogs from wolves and the sharing of so many haplotypes by so many different individuals indicates that the dogs and wolves sampled share a very recent common ancestry. The limited degree of structure that is apparent (Fig.6) suggests five major dog/wolf groups; a large predominantly dog "Dog" group (69 dog and four wolf sequences), a second, small "Dog2" group (ten dog sequences), a small "Elkhound" group (three elkhounds and three wolves), a small North American "Indigenous Dog" group (four of the five indigenous North American dogs and two North American

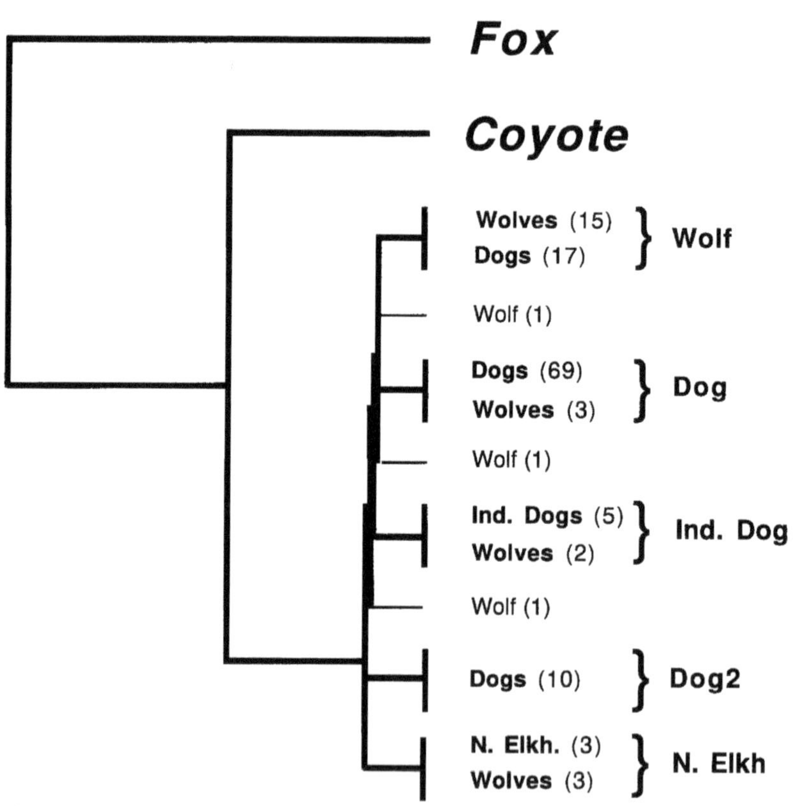


***Figure 6*** *Summary of canid relationships. The five major groups of dog mtDNA sequences are shown together with the three unique wolf sequences. Fox and coyote serve to root the tree. "N.Elkh" indicates Norwegian Elkhound; "Ind. Dog" indicates Indigenous dogs. Numbers in brackets are counts of distinct haplotypes represented.*

wolves) and a large "Wolf" group (15 of the 25 wolves and 17 dogs). Both dog and wolf sequences were found in all but the Dog2 group. The presence of three unique wolf sequences that did not consistently fall within any these groups suggests that additional, rare ancestral lineages may also exist.

There are two lines of reasoning that can be followed in interpreting these data; one that considers a single domestication event and another that proposes multiple founding events. The primary genetic evidence for the hypothesis that dogs could have evolved from wolves in a single domestication event comes from the observation that the majority of dogs have mtDNA haplotypes that belong to the "Dog" group and from the paucity of wolf representatives in this group. That nearly all of the modern breeds of dogs are represented in this "Dog" group argues for a distinct common ancestry and a more limited ancestral gene pool from which domestic dogs arose. In this argument, at least seven highly divergent mitochondrial haplotypes existed in the wolf population and the presence of these wolf haplotypes in dogs would be the result of introgression in a substantially unidirectional manner (offspring of male dogs and female wolves identified as dogs). The small, exclusively dog "Dog2" group could be the result of drift or small sample sizes and the four wolf sequences within the "Dog" group could be explained by rare introgression of the dog haplotypes into wolves (the offspring of male wolves and female dogs identified as wolves). Thus, the presence of a large, predominantly dog group supports the idea of a single domestication of dogs from wolves, with introgression between distinct genetic lineages explaining the multiple haplotypes found in dogs and wolves. A prediction of this model is that some indigenous North American dog sequences would come from an ancestral "Dog" mtDNA haplotype. As shown in figure 5, this prediction is not observed; none of the indigenous North American dogs samples have the "Dog" haplotype.

The alternative to a single domestication origin of dogs is a multiple or plural domestication process in which dogs were derived independently several times from different wolf ancestors (polyphyletic origins). In this model, different dog lineages would likely share several different mtDNA haplotypes with wolves because they both draw genes from a common gene pool. The overwhelming majority of dogs in the "Dog" group could be explained by the rapid proliferation of predominantly European dog breeds over the past few hundred years and the influence of Kennel Clubs that rigorously monitor and control breeding programs. Under a polyphyletic origins model and given the data in figure 5, the majority of European dog breeds came from a single mtDNA lineage that was cultivated and propagated by human breeders. In this model, we would not expect ancient dog sequences from outside Europe to resemble the main "Dog" group but rather any one of the eight haplotypes, and very likely a haplotype found in the same geographic region. This later observation is observed in indigenous North American dog samples. The two Tahltan and two Village dog sequences are nearly identical to a rare wolf haplotype that is also found in northern British Columbia. In addition, the single ancient Wool dog sequence is similar to contemporary dogs in a second, small, uniquely dog "Dog2" group. That the five ancient indigenous dog sequences clearly did not come from the common "Dog" lineage supports a plural domestication model. Partial support for the polyphyletic model may also come from the distinct "Elkhound" group in which the majority of the Elkhound sequences group together with three European wolf sequences. However, the evidence for polyphyletic origins does not support the hypothesis that different dog morphotypes originated from different wolf subspecies.

The confounding factor in both multiple and single domestication models is the facile interbreeding between dogs and wolves (Iljin, 1941; Mech, 1970; Steinhart, 1996). Furthermore it is likely that this interbreeding is not just a recent phenomena but has persisted throughout dog and wolf history (Clutton-Brock, Kitchener & Lynch, 1994; Walker & Frison, 1982). The use of maternally inherited mtDNA to study dog origins limits our observations to only the female side of any hybridization event (*cf.* Cronin *et al.,* 1995; Polzeihn *et al.,* 1995; Wayne & Jenks, 1991), but from this analysis it is clear that morphologically distinct female dogs or wolves can contain any one of several different mtDNA lineages. Examples of nearly identical mtDNA sequences in both wolf and dog representatives in four of the eight major dog/wolf mtDNA lineages suggests that there is reciprocal hybridization between wolves and dogs (*cf.* Carr & Hughes, 1993). Thus, in contrast to the study of Loftus *et al.* (1994) on origins of domestic cattle, it becomes very difficult to sort out multiple versus single domestication models because hybridization can nearly always explain the same phylogenetic patterns.

Indigenous dog samples from North America, however, offer a unique opportunity to examine an independent dog history dating back to the late Pleistocene. Humans (and perhaps their dogs) crossed into Alaska from Asia and moved south into North, Central and South America, presumably in several waves of immigration (Adovasio & Pedler, 1997; Fladmark, 1979; Meltzer, 1989). Indigenous dogs in the Americas have been relatively isolated, up until 300-400 years ago (Allen, 1920; Schwartz, 1997), from many of the events shaping dog evolution over the past 10-15,000 years in Europe and Africa. The presence of two distinct (bootstrap values greater than 50%; fig.5) and rare mtDNA lineages in North American indigenous dogs, one of which is found only in North America, strongly supports the independent evolution of indigenous North American dogs.

The origin of the indigenous North American dogs however is less clear. North American indigenous dogs may have come from a source population in Asia and followed human migrations (Olsen & Olsen, 1977; Olsen, 1985; Clutton-Brock, 1995). If this were the case we would expect to see genetic evidence for a close affinity of North American indigenous dogs to native Asian dogs or wolves. For one of the five indigenous samples collected (Wool dog) this appears to be the case. The wool dog sample shares a rare mtDNA lineage (Dog2) with a few predominantly Asian dog breeds as well as a Mexican hairless. However, if we operate under the multiple domestication model explained above, we might expect a distinctly American dog haplotype to resemble that of local wolves. Our finding that four of the five indigenous dogs possess a uniquely North American mtDNA lineage that is shared with at least two wolves from North America is consistent with an independent domestication in North America. While hybridisation (female dog mating with male wolf to give wolf offspring) could explain one of the wolf haplotypes (Yukon Wolf22), the presence of a distinct Yukon Wolf4 mtDNA haplotype in this group supports this independent domestication scenario. It appears then that based on a small number of rare mtDNA haplotypes, some North American indigenous dogs may have come from Asian origins and others may have originated independently in the Americas. Clearly however, additional sampling of North American and Asian wolves, and prehistoric dogs, will be required to verify these preliminary results, especially given the absence of any evidence for the cultural antecedents of primary domestication in early North American archaeological sites (Crabtree, 1993; Meltzer, 1989; Tchernov, 1993).

The use of mtDNA from extant animals to understand dog evolution has provided evidence for the very close relationship between dogs and wolves as a distinct group from coyotes (Fig.6). It is also clear that dogs and wolves share many mtDNA sequences because of recent domestication and that reciprocal hybridisation continues to result in the exchange of genetic information between them. A significant conclusion of the analysis is that this exchange of genetic information precludes associating the age of different mtDNA haplotypes with the timing of domestication events. Although the problem of exchange of genetic information by hybridisation will always be present, some of the difficulties inherent in using modern samples may be ameliorated through the use of ancient samples, particularly those that predate the following: trans-continental transport of dogs by humans into areas with indigenous varieties (Allen, 1920; Schwartz, 1997); population declines of wolves that facilitate hybridisation with dogs (Carbyn, Tritts, and Seip, 1995; Clutton-Brock *et al.,* 1994; Maagaard & Graugaard 1994; Mech, 1970, 1995; Wayne *et al.,* 1994); deliberate dog/wolf hybridisation (Steinhart, 1996). The use of ancient samples with a shorter independent evolutionary history, however, may have less variation upon which to draw conclusions. Nevertheless, while the present study incorporates only five ancient samples, it has contributed invaluable information about dog evolution. Additional samples of both prehistoric dogs and late Pleistocene wolves have the potential for expanding our understanding of wolf/dog relationships, and relationships between regional populations of dogs, without the "background noise" introduced by recent perturbations in the demographics of both species.

**Acknowledgments**

Contemporary dog samples were collected by Dr. Layne Bixby of the Quadra Animal Clinic in Victoria. Contemporary wild canid samples were donated by: Mary Gamberg, Watson Lake (Yukon Territory); Philip Merchant, Yukon Fish and Wildlife (Yukon Territory); William A. Rapley, Metro Toronto Zoo (Ontario); Drs. Coates and Schwantje, Wildlife Branch (British Columbia); and D. Heinrich, Christian-Albrechts Universität zu Kiel, Institut für Haustierkunde (Germany). We thank Richard Meadow (Harvard University, USA), Daniel Bradley (Trinity College, Ireland), Juliet Clutton-Brock (Natural History Museum, London) and Geoffrey Chambers (Victoria University, New Zealand) for reviewing the manuscript. Ancient skins were donated by the Royal British Columbia Museum, Victoria. This project was funded jointly by the National Science and Engineering Research Council of Canada (Ottawa), Pacific Identifications Inc. (Victoria), and the British Columbia Heritage Trust (Victoria).

**References**

Adovasio, J.M. & Pedler, D.R. (1997). Monte Verde and the antiquity of humankind in the Americas. *Antiquity* **71**, 573-80.

Allen GM. (1920). Dogs of the American aborigines. Harvard University, *Museum of Comparative Zoology Bulletin* **63**, 431-517.

Amoss, P.T. (1993). Hair of the dog: unraveling pre-contact Coast Salish social stratification. In (A. Mattina & T. Montler, Eds) *American Indian Linguistics and Ethnography in Honor of Laurence C. Thompson.* Missoula: University of Montana, pp. 3-35.

Brown, G.G., Gadaleta, G., Pepe, G., Saccone, C. & Sbisà, E. (1986). Structural conservation and variation in the d-loop-containing region of vertebrate mitochondrial DNA. *Journal of Molecular Biology* **192**, 503-511.

Burbidge, M. J. (1998). Polyphyletic origins of extinct and extant domestic dogs. Masters thesis, University of Victoria, British Columbia.

Byun, S.A., Koop, B.F. & Reimchen, T.E. (1997). North American black bear mtDNA phylogeography: implications for morphology and the Haida Gwaii glacial refugium controversy. *Evolution* **51**, 1647-1653.

Carbyn, L.N., Fritts, S.H. & Seip, D.R. (1995). *Ecology and Conservation of Wolves in a Changing World.* Canadian Circumpolar Institute, Occasional Publication **35**, Edmonton.

Carr, S.M. & Hughes, G.A. (1993). Direction of introgressive hybridization between species of North American deer (*Odocoileus*) as inferred from mitochondrial *cytochrome-b* sequences. *Journal of Mammalogy* **74,** 331-342.

Clutton-Brock, J. (1995). Origins of the dog: domestication and early history. In (J. Serpell, Ed) *The Domestic Dog: Its Evolution, Behaviour, and Interactions with People*. Cambridge: Cambridge University Press, pp. 8-20.

Clutton-Brock, J., Kitchener, A.C. & Lynch, J.M. (1994). Changes in the skull morphology of the Arctic wolf, *Canis lupus arctos*, during the twentieth century. *Journal of Zoology, London* **233**, 19-36.

Crabtree, P.J. (1993). Early animal domestication in the middle East and Europe. In (M.B. Schiffer, Ed) *Archaeological Method and Theory*. Tuscon: University of Arizona Press, pp. 201-245.

Crockford, S.J. (1997). *Osteometry of Makah and Coast Salish dogs*. Burnaby: Simon Fraser University, Archaeology Press **22.**

Crockford, S.J. & Pye, C.J. (1997). Forensic reconstruction of prehistoric dogs from the Northwest Coast. *Canadian Journal of Archaeology* **21**, 149-153.

Cronin, M.A., Renecker, L, Pierson, B.J. & Patton, J.C. (1995). Genetic variation in domestic reindeer and wild caribou in Alaska. *Animal Genetics* **26**, 427-434.

Crisp, W.G. (1956). Tahltan bear dog. *The Beaver* **287**, 38-41.

Dayan, T. (1994). Early domestic dogs of the Near East. *Journal of Archaeological Science* **21**, 633-640.

DeSalle, R. (1994). Implications of ancient DNA for phylogenetic studies. *Experientia* **50**, 543-550.

Doyle, J.J. & Doyle, J.L. (1987). A rapid DNA isolation procedure for small quantities of fresh leaf tissue. *Phytochemical Bulletin* **19**, 11-15.

Eernisse, D.J. & Kluge, A.G. (1993). Taxonomic congruence versus total evidence, and amniote phylogeny inferred from fossils, molecules, and morphology. *Molecular Biology and Evolution* **10**, 1170-1195.

Epstein, H. (1971). *The Origin of the Domestic Animals of Africa*. New York: Africana Publishing Corp.Friis, L.K. (1985). *An investigation of subspecies relationships of the grey wolf, Canis lupus, in British Columbia*. Masters thesis. University of Victoria.

Fladmark, K.R. (1979). Routes: alternative migration corridors for early man in North America. *American Antiquity* **44,** 55-69.

Gustincich, S., Manfioletti, G., Del Sal, G., Schneider, C. & Carninci, P. (1991). A fast method for high-quality genomic DNA extraction from whole human blood. *Biotechniques* **11**, 298.

Haag W.G. (1948). An osteometric analysis of some aboriginal dogs. Lexington: University of Kentucky, *Reports in Anthropology* **7**.

Hagelberg, E. (1994). Mitochondrial DNA from ancient bones. In (B. Herrmann & S. Hummel, Eds) *Ancient DNA*. New York: Springer–Verlag.

Hagelberg, E. & Clegg, J.B. (1991). Isolation and characterization of DNA from archaeological bone. *Proceedings of the Royal Society of London, B series* **244**, 45–50.

Handt, O., Höss, M., Krings, M. & Pääbo, S. (1994). Ancient DNA: methodological challenges. *Experientia* **50**, 524-529.

Hoelzel, A.R., Lopez, J.V., Dover, G.A. & O'Brien, S.J. (1994). Rapid evolution of a heteroplasmic repetitive sequence in the mitochondrial control region of carnivores. *Journal of Molecular Evolution* **39**, 191-199.

Horai, S., Kondo, R., Murayama, K., Hayashi, S., Koike, H., & Nakai, N. (1991). Phylogenetic affiliation of ancient and contemporary humans inferred from mtDNA. *Philosophical Transactions of the Royal Society, London, series B* **333**, 409-417.

Höss, M., Jaruga, P., Zastawny, T.Z., Dizdaroglu, M. & Pääbo, S. (1996). DNA damage and DNA sequence retrieval from ancient tissues. *Nucleic Acids Research* **24**, 1304-1307.

Howay, F.W. (1918). The dog's hair blankets of the Coast Salish. *Washington Historical Quarterly* **9**, 83-92.

Huang, X. (1994). On global sequence alignment. *Computer Applications in the Biosciences* **10,** 227-235.

Iljin, N.A. (1941). Wolf-dog genetics. *Journal of Genetics* **42**, 359-414.

Kimura, M. (1980). A simple method for estimating evolutionary rates of base substitutions through comparative studies of nucleotide sequences. *Journal of Molecular Evolution* **16**, 111-120.

Kimura, M. (1987). Molecular evolutionary clock and the neutral theory. *Journal of Molecular Evolution* **26**, 24–33.

Kringer, M., Stone, A., Schmitz, R.W., Krainitzki, H., Stoneking, M. & Pääbo, S. (1997). Neanderthal DNA sequences and the origin of modern humans. *Cell* **90**, 19-30.

Loftus, R.T., MacHugh, D.E., Bradley, D.G, Sharp, P.M & Cunningham, P. (1994). Evidence for two independent domestications of cattle. *Proceedings of the National Academy of Science USA* **91**, 2757-2761.

Maagaard, L. & Graugaard, J. (1994). Female Arctic wolf, *Canis lupus arctos*, mating with domestic dogs, *Canis familiaris*, in Northeast Greenland. *Canadian Field-Naturalist* **108**, 374-375.

Mech, L.D. (1970). *The Wolf*. Garden City: Natural History Press.

Mech, L.D. (1995). The challenge and opportunity of recovering wolf populations. *Conservation Biology* **9**, 270-278.

Meltzer, D.J. (1989). Why don't we know when the first people came to North America? *American Antiquity* **54,** 471-490.

Mengel, R.M. (1971). A study of dog-coyote hybrids and implications concerning hybridization in *Canis*. *Journal of Mammalogy* **52**, 316-336.

Morey, D.F. (1990). *Cranial allometry and the evolution of the domestic dog*. Ph.D. thesis, University of Tennessee.

Morey, D.F. (1992). Size, shape and development in the evolution of the domestic dog. *Journal of Archaeological Science* **19**, 181–204.

Morey, D.F. (1994). The early evolution of the domestic

dog. *American Scientist* **82**, 336–347.

Okumura, N., Ishiguro, N., Nakano, M., Matsui, A. & Sahara, M. (1996). Intra- and interbreed genetic variations of mitochondrial DNA major non-coding regions in Japanese native dog breeds (*Canis familiaris*). *Animal Genetics* **27**, 397-405.

Olsen, S.J. (1985). *Origins of the Domestic Dog*. Tucson: University of Arizona Press.

Olsen, S.J. & Olsen, J.W. (1977). The Chinese wolf, ancestor of New World dogs. *Science* **197**, 433-35.

Pääbo, S. (1993). Ancient DNA. *Scientific American* **269**, 86-92.

Polziehn, R.O., Strobeck C., Sheraton J. & Beech, R. (1995). Bovine mtDNA discovered in North American bison populations. *Conservation Biology* **9,** 1638-1643.

Richards, M., Smalley, K., Sykes, B. & Hedges, R. (1993). Archaeology and genetics: analyzing DNA from skeletal remains. *World Archaeology* **25**, 18-28.

Richards, M.B., Sykes, B.C. & Hedges, R.E.M. (1995). Authenticating DNA extracted from ancient skeletal remains. *Journal of Archaeological Science* **22**, 291-299.

Rothuizen, J., de Gouw, H., van der Vugt, H.H.J., Plas, M.A., Dorresyein, G.M., Hoelzel, A.R. & Lenstra, J.A. (1996). Variation in the mitochondrial tandem repeat in dogs and bears. Unpublished manuscript.

Roy, M.S., Geffen, E., Smith, D., Ostrander, E.A. & Wayne, R.K. (1994). Patterns of differentiation and hybridization in North American wolflike canids, revealed by analysis of microsatellite loci. *Molecular Biology and Evolution* **11**, 553-570.

Roy, M.S., Geffen, E., Smith, D. & Wayne, R.K. (1996). Molecular genetics of pre-1940 red wolves. *Conservation Biology* **10**, 1413-1424.

Sambrock, J., Fritsch, E.F. & Maniatis, T. (1989). *Molecular Cloning: A Laboratory Manual.* Cold Spring Harbor, NY: Cold Spring Harbor Laboratory.

Saitou, N. & Nei, M. (1987). The neighbor-joining method: a new method for reconstructing phylogenetic trees. *Molecular Biology and Evolution* **4**, 406-425.

Schulting, R. (1994). The hair of the dog: the identification of a coast Salish dog–hair blanket from Yale, British Columbia. *Canadian Journal of Archaeology* **18**, 57–76.

Schwartz, M. (1997). *A History of Dogs in the Early Americas*. New Haven: Yale University Press.

Scott, J.P. (1967). The evolution of social behaviour in dogs and wolves. *American Zoologist* **7**, 373-81.

Steinhart, P. (1996). *The Company of Wolves*. New York: Vintage Books.

Swofford, D.L. (1993). *PAUP: Phylogenetic Analysis Using Parsimony*. Champaign: Illinois Natural History Survey.

Swofford, D.L., Olsen, G.J. , Waddell, P.J. & Hillis, D.M. (1996). Phylogenetic inference. In (D.M. Hillis, C. Moritz & B.K. Mable, Eds) *Molecular Systematics*, 2nd ed.. Sunderland: Sinauer.

Sykes, B. (1993). Contamination and chemical modification. Paper presented at the Ancient DNA Conference, Washington, D.C., May 1993.

Tanabe, Y., Ota, K, Ito, S., Hashimoto, Y., Sung, Y.Y., Ryu, J.K. & Faruque, M.O. (1991). Biochemical-genetic relationships among Asian and European dogs and the ancestry of the Japanese native dog. *Journal of Animal Breeding and Genetics* **108,** 455-478.

Tchernov, E. (1993). From sedentism to domestication-a preliminary review for the southern Levant. In (A.Clasen, S. Payne & H.P. Uerpmann, Eds) *Skeletons in her Cupboard: Festschrift for Juliet Clutton-Brock.* Oxford: Oxbow Books. Monograph **34**, pp.189-233.

Thomas, W.K., Pääbo, S.. Villablanca, F.X. & Wilson, A.C. (1990). Spatial and temporal continuity of Kangaroo rat populations shown by sequencing mitochondrial DNA from museum specimens. *Journal of Molecular Evolution* **31**, 101-112.

Thorpe, R.S. (1996). The use of DNA divergence to help determine the correlates of evolution of morphological characters. *Evolution* **50**, 524-531.

Tsuda, K. Kikkawa, Y., Yonekawa, H. & Tanabe, Y. (1997). Extensive interbreeding occurred among multiple matriarchal ancestors during the domestication of dogs: evidence from inter- and intraspecies polymorphisms in the D-loop regions of mitochondrial DNA between dogs and wolves. *Genes and Genetic Systems* **72**, 229-238.

Titus, D.E. (1991). *Promega Protocols and Applications Guide*. 2nd ed., Madison.

Vilà, C., Savolainen, P., Maldonado, J.E., Amorim, I.R., Rice, J.E., Honeycutt, R.L., Crandall, K.A., Lundeberg, J. & Wayne, R.K. (1997). Multiple and ancient origins of the domestic dog. *Science* **276**, 1687-1689.

Walker, D.N. & Frison, G.C. (1982). Studies on Amerindian dogs, 3: prehistoric wolf/dog hybrids from the Northwest Plains. *Journal of Archaeological Science* **9**, 125-172.

Wayne, R.K. (1986). Cranial morphology of domestic and wild canids: the influence of development on morphological change. *Evolution* **40**, 243-61.

Wayne, R.K. (1993). Molecular evolution of the dog family. *Trends in Genetics* **9**, 218-225.

Wayne, R.K. & Jenks, S.M. (1991). Mitochondrial DNA analysis implying extensive hybridization of the endangered red wolf *Canis rufus*. *Nature* **351**, 565–568.

Wayne, R.K., Lehman, N.M., Allard, M.W. & Honeycutt, R.L. (1994). Mitochondrial DNA variability of the gray wolf: genetic consequences of population decline and habitat fragmentation. *Conservation Biology* **6**, 559-569.

Wayne, R.K., Nash, W.G. & O'Brien, S.J. (1987). Chromosomal evolution of the Canidae. I. Species with high diploid numbers. *Cytogenetics and Cell Genetics* **44**, 134-41.

Wayne, R.K. & O'Brien, S.J. (1987). Allozyme divergence

within the Canidae. *Systematic Zoology* **36**, 339-55.

Wilcox B.W. & Walkowicz, C. (1989). *The Atlas of Dog Breeds of the World*. Neptune City: T.F.H Publications.

# MOLECULAR GENETIC ANALYSIS OF ANCIENT JAPANESE DOGS

Naotaka Ishiguro
Department of Veterinary Public Health
Obihiro University of Agriculture and Veterinary Medicine,
Obihiro City, Hokkaido, 080-8555, Japan

Naohiko Okumura
Department of Bioresource Chemistry,
Obihiro University of Agriculture and Veterinary Medicine,
Obihiro City, Hokkaido, 080-8555, Japan

Akira Matsui
Nara National Cultural Properties Research Institute,
Nijyo-cho, Nara City,
Nara 630-8577, Japan

Nobuo Shigehara
Primate Research Institute,
Kyoto University,
Section of Phylogeny and Classification,
Inuyama City, Aichi 484-8548, Japan

**Key words:** mtDNA, Jomon, dog, ancient DNA, archaeozoology, canine genetics

**Abstract**

Japanese native dogs are classified into six breeds (Hokkaido dog, Akita dog, Kai dog, Shiba dog, Kishu dog and Shikoku dog) from their locality and physical characteristics. These dogs are thought to be descendants of the Japanese indigenous dogs of ancient times. To estimate the genetic relationship between ancient dogs and modern Japanese dogs, sequences of ancient mitochondrial DNA (mtDNA) from archaeological remains of Japanese dogs were determined and compared with those of modern indigenous dog breeds.A total of 120 ancient dog samples (mainly bones) were derived from the Jomon (thirtyith to third century BC), Yayoi (third century BC to third century AD), Kofun (third to sixth century AD), Okhotsk (seventh to twelfth century AD) and Kamakura (twelfth to fourteenth century AD) periods. Highly variable sequences of the mtDNA D-loop region (198 base pairs) from the DNA extracts were amplified by the polymerase chain reaction (PCR), and were compared with those of 178 modern dogs. A total of 55 out of 120 ancient samples were amplified by PCR. Fifteen haplotypes (M2, M5 , M10, M11 and A1 to A11) were detected in the ancient D-loop sequences, among which eleven haplotypes (A1 to A11) were unique sequences not observed in modern dogs. Haplotype M5 was widely distributed in the northern part of Japan, while haplotype M2 was widely distributed in southern Japan. All the ancient dogs belonged to the C1 cluster of four clusters (C1 to C4), which were established by phylogenetic analysis (UPGMA) of 178 modern dogs. These results indicate that the dogs of the C1 cluster were widely distributed in the Japanese archipelago from the Jomon Period.

## Introduction

The dog (*Canis familiaris*) is the oldest domesticated animal, and its ancestor is thought to have evolved from the wolf, *Canis lupus* (Hemmer, 1990; Scott & Fuller, 1965). More than 400 breeds of domestic dog occur today in the world. Because dogs have a close relationship with humans, they would always have migrated with them, and also have been evolved by artificial selection and cross breeding. In Japan, six Japanese dog breeds (Hokkaido dog, Akita dog, Kai dog, Shiba dog, Kishu dog and Shikoku dog) or varieties are now preserved as indigenous breeds (Tanabe, 1991), and they are claimed to be descendants of the Japanese indigenous dogs of ancient times. For a long time, most Japanese native dog breeds were raised as hunting dogs to catch wild boars, deer and bears in several mountainous districts.

Mitochondrial DNA (mtDNA) polymorphism has been widely used to study the relationships between closely related species and the genetic structure within populations (Higuchi *et al.*, 1987; Thomas *et al.*, 1989; Horai *et al.*, 1991; Hagelberg & Clegg, 1993). Okumura *et al.* (1996) showed that the mitochondrial DNA control region of 94 modern dogs has 38 different haplotypes and classified these dogs into four phylogenetic clusters whose lineages did not directly agree with distinct dog breeds. Tsuda *et al.* (1997) and Vilà *et al.* (1997) reported evidence of genetic exchange between the dog and wolf populations. These abundant genetic backgrounds are thought to reflect the complexity produced by genetic exchanges between dogs and wolves or among dog breeds.

To date, many ancient dog samples have been discovered from archaeological sites in Japan. Shigehara (1994) investigated the morphology of ancient dog bones, but little is known of their genetic background. To estimate the genetic relationship with modern Japanese dog breeds, ancient mtDNA sequences were amplified from ancient dog bones by the polymerase chain reaction (PCR) and were compared with modern dog mtDNA lineages by phylogenetic analysis.

## Materials and methods

### mtDNA data base from modern dogs

The mtDNA data base was constructed from 124 modern Japanese native dogs and 54 non-Japanese dogs (Okumura *et al.*, 1996). The modern Japanese dog samples comprised six dog breeds native to the main island (Shiba dog, Hokkaido dog, Akita dog, Kai dog, Kishu dog and Shikoku dog) and two breeds indigenous to the Iki and Ryukyu islands (Iki dog and Ryukyu dog). A 198bp segment of the mtDNA control region was used to construct the modern dog mtDNA data base and was compared with the corresponding region from ancient dog samples.

### Archaeological bone samples of ancient Japanese dogs

A total of 120 well-preserved ancient dog specimens (48 mandibles, 16 femurs, four ulnas, five radii, nine humeri, 11 tibiae and 27 tooth roots) were collected from 27 archaeological sites. The ancient dog specimens were obtained from archaeological sites within the following chronological periods: 20 sites of the Jomon Period (*c.* thirtyith to third century BC, 54 specimens); 2 sites of the Yayoi Period (*c.* third century BC to third century AD, 34 specimens); 1 site of the Kofun Period (*c.* third to sixth century AD, 1 specimen); 1 site of the Kamakura Period (1192-1333 AD, 21 specimens); and 3 sites of the Okhotsk Period (seventh to twelfth century AD, 10 specimen).

### DNA extraction and mtDNA amplification by PCR

Bone powder (0.2 to 0.5g) was obtained from the archaeological specimens using an electric drill and was suspended in 0.5M ethylenediamine tetraacetic acid (EDTA). The supernatant was removed by centrifugation, and the pellet of bone powder was repeatedly decalcified with 0.5M EDTA. Then, the bone powder sample was treated overnight in 0.5M EDTA with proteinase K (300 μg/ml) and N-lauryl sarcosine (0.5%) (Hardy *et al.*, 1995, with slight modifications). The sample was extracted with phenol and again with chloroform:phenol (1:1), was concentrated using a Centricon 30 micro-concentrator (Amicon, Beverly, MA) and was washed with distilled water. The extracted DNA samples were directly used for PCR. The DNA extracts were independently prepared two or three times to verify the results.

To prevent contamination of the DNA from modern dogs, all reagents and implements were steriliged under ultraviolet irradiation and they were handled in a biohazard hood with a ventilation system. Reagents without bone powder were always used as controls to check for contamination in the ancient bone samples.

### DNA amplification by PCR and DNA sequencing

Two primers mitL57: 5'-TTAAACTATTCCCTGACACCCC and mitH52:5'-TCGAGGCATGGTGATTAAG were used to amplify the 258bp D-loop region. The first PCR test was done with primers mitL57 and mitH52 in the following PCR conditions: one cycle of DNA denaturation and activation of ampliTaq Gold (Perkin Elmer, Norwalk, CT) at 95C for 10 min, annealing at 55C for 1 min, extension at 72C for 1 min, followed by 50 cycles of denaturation at 94C for 30 sec, annealing at 55C for 1 min, and extension at 72C for 1 min. For a seminested PCR, the second PCR was made with primers mitH52 and mitL63: 5'-ACCCCTACATTCATATATTGAAT(26), and 1μl of the first PCR products, for 30 cycles under the same PCR conditions. After PCR amplification, the product was checked by electrophoresis, the primers were removed by Centricon 100 and the DNA sequence was determined by direct sequencing using a 373S DNA Sequencer (Applied Biosystems, Inc., Foster City, CA).

### Phylogenetic analysis

The DNA sequence data were analyzed using GENETYX-MAC software (Software Development Co., LTD., Tokyo) for multiple sequence alignment. Phylogenetic analysis was made using the PHYLIP program package, version 3.572 (Felsenstein, 1995). The mean numbers of the nucleotide substitutions per nucleotide site between the sequences were calculated by the two-parameter method. From the estimated distance matrices, a dendrogram was made by the UPGMA method. To determine the confidence intervals of the phylogenies, the bootstrap method was used from 1,000 replications.

## Results

### Amplification of mtDNA from ancient dog samples

Figure 1 shows an outline of this study. Twenty-eight haplotypes (M1 to M28) were differentiated in the 198-bp of the mtDNA D-loop region from 178 modern dogs. The 28 haplotypes were used as an mtDNA data base to compare with those from ancient dogs (Figure 2).

The DNA was extracted from 120 ancient dog bones and was used to amplify the ancient mtDNA. To confirm the accuracy of the DNA sequences, the DNA was extracted at

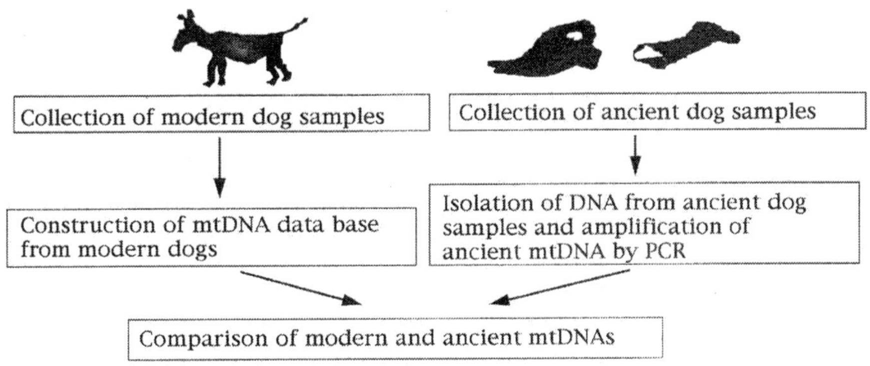


1) Origin of modern Japanese dogs
2) Evolutional sequence diversity of Japanese dogs
3) Genetic relationship between dog breeds

***Figure 1*** *An outline of the ancient dog study.*

```
                 Base substitutions and gaps in 198-bp        Archaeological Periods
                                       11111111111111111112   of ancient dogs
                 223334445566666777777904556677778888889 91   (No. of specimens)
                 792450492825789012345800675602357013489570
Modern      M1   CCCACCGTCGACTCTCCCCT-AGCTTTCTATCCATTAACTGT ┐
dog         M2   ....................................T..... │  Jomon(9),Yayoi(3),Kamakura(5)
haplotype   M3   ...................................CT..... │
            M4   .................................G..T..... │
            M5   .............................G............ │  Jomon(5),Okhotsk(6)
            M6   ..........................C..G............ │
            M7   .............................G....C.....AC │
            M8   ...........................T.....G..T..... │C1
            M9   ...........................T.............C │
            M10  .............................G..........A. │  Okhotsk(1)
            M11  .............................G..........AC │  Kamakura(4)
            M12  T............................G............ │
            M13  .............................G......G..... │
            M14  ..........................C..G...........C │
            M15  ..........................C.CG............ │
            M23  ...............T.....G..........T.......A. │
            M24  ............................C...T.C....... │
            M25  .....................G.................... │
            M28  ..........C..................G...G..T..... ┘
            M16  ........T......T........C...........G..C.. ┐C2
            M27  ........T......T........C..............C.. ┘
            M17  ...GAT......C.......C...C....GC.........A. ┐C3
            M18  ...G........--------C...C....GC.........A. ┘
            M19  ...............T.........C......T...GG..A. ┐
            M20  ...............T.......T.C......T...GG..A. │C4
            M21  ...............T.....G.T.C......T...GG..A. │
            M22  ..........C....T.......T.C......T...GG..A. │
            M26  ..G............T.......T.C......T...GG..A. ┘
Ancient     A1   .T...........................G............ ┐  Okhotsk(1)
dog         A2   ......A......................G............ │  Jomon(5)
haplotype   A3   ......AC.................................. │  Jomon(4)
            A4   .........A.................T.....G..T..... │  Yayoi(2)
            A5   ...........TC................G............ │C1 Okhotsk(1)
            A6   ............C................G............ │  Jomon(4)
            A7   ............C.......C........G...G........ │  Okhotsk(1)
            A8   .....................G.......G..........AC │  Kofun(1)
            A9   ......................A..........G..T..... │  Yayoi(1)
            A10  .............................G.T......T... │  Yayoi(1)
            A11  ........................................A. ┘  Jomon(1)
```

***Figure 2*** *Thirty-nine different haplotypes of 198bp of the D-loop region in modern and ancient Japanese dogs. Their identities are shown by dots. Nucleotide position 1 corresponds to the base position 26 described by Okumura et al. (1996) of the whole dog mtDNA control region. This figure shows the correlation of the clusters (C1 to C4) and the haplotype of ancient dogs* least twice from the same specimens, and was amplified at least twice per DNA extract. Amplification of mtDNA by the PCR was successful in 55 (46%) out of 120 samples. In general, the samples from the shell mounds were good to amplify the DNA because their compact bones were hard, they had little colour and they had little PCR inhibitor (Paabo, 1989; Tuross, 1994). The amplification of ancient mtDNA appeared to be markedly correlated with the preservation times and conditions of the archaeological specimens. The amplification of mtDNA was successful in all ten samples from the latest Okhotsk Period. In contrast to the samples from the Okhotsk Period, the PCR reactions were less successful in 28 (52%) of 54 samples from the oldest Jomon Period. Fifty-five ancient DNA sequences were classified into 15 haplotypes, five modern haplotypes (M2, M5, M10 and M11) and eleven ancient types (A1 to A11), that were not observed in the mtDNA data base from 178 modern dogs.

Figure 3 shows a map of the mtDNA haplotypes amplified from the bone samples from archaeological sites. Haplotype M5 was widely distributed in the northern part of Japan including Hokkaido Island, while haplotype M2 was widely distributed in the southern part of Japan. Modern haplotypes M2 and M5 were simultaneously found in archaeological sites in the northern half of Honshu Island. Haplotype M11 was specifically obtained from one archaeological site (Kusadosengen) of the Yayoi Period. Haplotypes A2 and A3 were each found from two sites near to each other, Tagara and Hosoura for A2 and Obora and Tagara for A3, suggesting that dogs migrated frequently. Haplotypes A3 and A6 were found in Ikawazu and Kasori archaeological sites, respectively. Haplotypes A1, A4, A5, A7, A8, A9, A10 and A11 were from dogs found at one site, indicating that dogs with a unique genetic marker were distributed in a restricted region.

Among the haplotypes from the archaeological specimens, two modern haplotypes (M2 and M5) and four ancient haplotypes (A2, A3, A6 and A11) were obtained from the archaeological sites of the Jomon Period (Figure 2). From this result, the native dogs in the Jomon Period (the oldest archaeological period studied) had already some genetic diversity and the dog population with haplotype M2 was well established in Japan since the Jomon Period.

Phylogenetic analysis of ancient mtDNA and origin of anceient Japanese dog

A total 178 DNA modern samples of the 198-bp D-loop region were classifeid into the C1 to C4 lineages by the unweighted pair-group method using an arithmetic mean (UPGMA) as described by Okumura et al. (1996). Figure 4 shows a UPGMA tree constructed from 28 modern haplotypes and 11 ancient haplotypes detected in ancient DNA samples. All haplotypes from 55 ancient dogs were classified into the C1 lineage (Fig. 4).

Haplotype M2 found in dogs isolated from archaeological sites of the Jomon, Yayoi and Kamakura Periods was widely detected in six modern Japanese native dogs (Shiba dog, Akita dog, Hokkaido dog, Iki dog, Kai dog and Ryukyu dog) showing that the M2 haplotype is the most common haplotype (Okumura et al.,1996). Haplotype M5

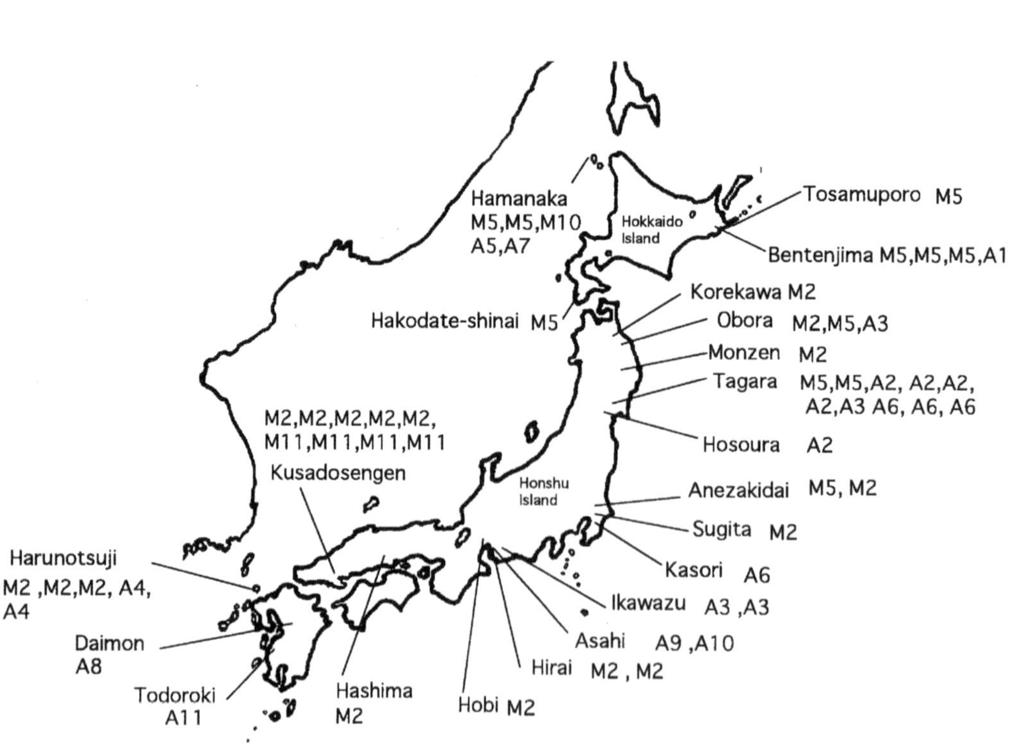


***Figure 3*** *Haplotypes and isolation sites of ancient dog amplified mtDNA sequences. Archaeological sites of isolated ancient dog samples in each Period:* ***Jomon Period****: Hakodate-shinai, Korekawa, Obora, Monzen, Tagara, Hosoura, Anezakidai, Kasori, Sugita, Hirai, Ikawazu, Hobi, Hashima, Todoroki;* ***Yayoi Period:*** *Asahi, Harunotsuji;* ***Kofun Period:*** *Daimon;* ***Kamakura Period:*** *Kusadosengen;* ***Okhotsk Period:*** *Tosamuporo, Bentenjima, Hamanaka.*

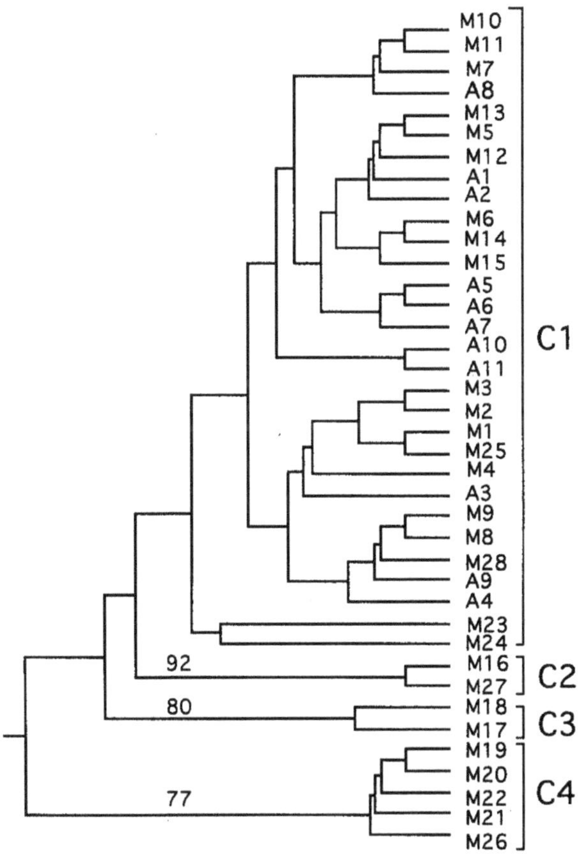


*Figure 4 The UPGMA tree of 28 modern and 11 ancient dog D-loop sequences. Designation of each D-loop haplotype follows Figure 2. Numbers on the branch indicate bootstrap values (%, 1000 replicates). The tree topology shows four distinct clusters (Cl to C4).*

from the northern sites of Honshu Island and Hokkaido Island in both Jomon and Okhotsk Periods was detected only in the modern Ryukyu dog and the Shetland Sheep dog, indicating that the M5 haplotype is rare in modern dog breeds. Because haplotype M2 was found in most of the Japanese native dog breeds, identifying ancestral dog breed of the ancient Japanese dog seems to be difficult from its haplotype.

## Discussion

We amplified ancient mtDNA from ancient dog specimens from archaeological sites of the Jomon, Yoyoi, Kofun, Kamakura and Okhotsk Periods, compared it with the mtDNA data base constructed from modern dogs and then speculated the origin of the Japanese dog breeds. A 198bp ancient mtDNA was amplified from 55(46%) of 120 dog specimens by semi-nested PCR. From phylogenetic analysis of the haplotypes, dog breeds belonging to the C1 lineage appears to have been widely distributed in the Japanese archipelago in the Jomon Period. Among them, haplotypes M2 and M5 were frequently isolated from ancient dogs in Jomon Period. Haplotypes M2 and M5 have a unique geographic distribution. The distribution pattern of haplotype M2 resembles the distribution pattern of the mtDNA haplotype of the mouse *Mus musculus musculus* found in the southern two-thirds of Japan, whereas the distribution pattern of haplotype M5 resembles that of the mtDNA haplotype of *M. m. castaneus* found in the north of Japan (Yonekawa *et al.*, 1988). The geographic distribution of haplotypes of ancient dogs may suggest useful anthropological information of the distribution or migration of the human population in Japanese history, because dogs would always have migrated with people.

The morphological investigation of skulls indicates that ancient dogs in the Jomon Period had slightly smaller bodies than those found in the Yayoi and Kofun Periods. Archeological evidence of the skull morphology (Nishimoto, 1989; Shigehara, 1989), Tanabe (1991) suggests that the origin of modern Japanese dogs was established by crossbreeding of two main lineages: an older dog population brought from southeast Asia in the Jomon Period, and a newcomer population that came through the Korean peninsula in the Yayoi and Kofun Periods. Haplotype M11 dogs from the Kusadosengen archaeological site may be a new haplotype brought to Japan after the Yayoi Period. The appearance of new haplotypes found in the archaeological sites later than the Jomon or Yayoi Periods suggests that a dog population showing a different genetic background was brought to Japan from neighbour countries. Dogs belonging to the C2 to C4 lineages would have been brought to Japan after the Kamakura Period (Figure 4), and they may subsequently have participated in the formation of the gene composition of native Japanese dogs.

Vilà *et al.*(1997) recently classified 26 haplotypes of 261bp mtDNA from 140 modern domestic dogs into four clades by phylogenetic analysis, and reported that some dog haplotypes are closely related with wolf lineages. The modern dog haplotype clusters C1 and C4 constructed using the UPGMA method in this study correspond to the two clades, I and IV, of Vilà *et al.* (1997). Ancient Japanese dogs classified in the C1 cluster obviously belong to a divergent monophyletic clade I sharing no sequence with wolves. Because Japanese wolf (*Canis lupus sp.*) populations became extinct as early as about 70 years ago on Honshu Island, no sequence for its mtDNA is available. Therefore, unfortunately, because no evidence exists, genetic exchange between ancient dog and Japanese wolf populations could not be analyzed.

## References

Felsenstein, J. (1993). PHYLIP version 3.5C. *Executables for Macintosh.*

Hagelberg, E. & Clegg, J. B.(1993). Genetic polymorphisms in prehistoric Pacific islands determined by analysis of ancient bone DNA. *Philosophical Transactions of the Royal Society of London Series B* **252**, 163-170.

Hardy, C., Callou, C., Vigne, J. D., Casane, D., Dennebouy, N., Mounolou, J. C. & Monnerot, M. 1995). Rabbit mitochondrial DNA diversity

from prehistoric to modern times. *Journal of Molecular Evolution* **40**, 227-237.
Hemmer, H. (1990). *Domestication: The Decline of Environmental Appreciation*. Cambridge: Cambridge University Press.
Higuchi, R., Wrischnik, L.A., Oakes, E., George, M., Tong, B. & Wilson, A.C. (1987). Mitochondrial DNA of the extinct quagga: relatedness and extent of postmortem change. *Journal of Molecular Evolution* **25**, 283-287.
Horai, S., Kondo, R., Murayama, K., Hayashi, S., Koike, H. & Nakai, N. (1991). Phylogenetic affiliation of ancient and contemporary humans inferred from mitochondrial DNA. *Philosophical Transactions of the Royal Society of London Series B* **233**, 409-417.
Nishimoto, T. (1989). Animal remains in Shimoguri Kuwanae Iseki. *Reports on Culture & Assets Oita Prefecture* **80**, 48-61 (In Japanese).
Okumura, N., Ishiguro, N., Nakano, M., Matsui, A. & Sahara, M. (1996). Intra- and interbreed genetic variations of mitochondrial DNA major noncoding regions in Japanese native dog breeds (*Canis familiaris*). *Animal Genetics* **27**, 397-405.
Paabo, S. (1989). Ancient DNA: extraction, characterization, molecular cloning, and enzymatic amplification. *Proceedings of the National Academy of Science, USA* **86**, 1939-1943.
Scott, J.P. & Fuller, J. L. (1965). *Dog Behavior: The Genetic Basis*. Chicago: University of Chicago Press.
Shigehara, N. (1989). Morphological changes in ancient Japanese dogs. *Archeology of Japan* **303**, 22-27 (In Japanese)
Shigehara, N. (1994). Morphological changes in Japanese ancient dogs. *Archaeozoologia* **7**, 79-94.
Tanabe, Y. (1991). The origin of Japanese dogs and their association with Japanese people. *Zoological Science* **8**, 639-651.
Thomas, R.H. Schaffner, W., Wilson, A.C. & Paabo, S. (1989). DNA phylogeny of the extinct marsupial wolf. *Nature* **340**, 465-467.
Tsuda, K., Kikkawa, Y., Yonekawa, H. & Tanabe, Y. (1997). Extensive interbreeding occurred among multiple matriarchal ancestors during the domestication of dog: evidence from inter- and intraspecies polymorphisms in the D-loop region of mitochondrial DNA between dogs and wolves. *Genes & Genetic Systems* **72**, 229-238.
Tuross, N. (1994). The biochemistry of ancient DNA in bone. *Experientia* **50**, 530-535.
Vilà, C., Savolainen, P., Maldonado, J. E., Amorim, I. R., Rice, J. E., Honeycutt, R. L., Crandall, K. A., Lundeberg, J. & Wayne, R. K. (1997). Multiple and ancient origins of the domestic dog. *Science* **276**, 1687-1689.
Yonekawa, H., Moriwaki, K., Gotoh, O., Miyashita, N., Matsushima, Y., Shi, L., Cho, W.S., Zhen, X.L. & Tagashira, Y. (1988). Hybrid origin of Japanese mice "*Mus musculus molossinus*". Evidence from restriction analysis of mitochondrial DNA. *Molecular Biology & Evolution* **5**, 63-78.

# Part 7

# Discussion & additional references

# A COMMENTARY ON DOG EVOLUTION: REGIONAL VARIATION, BREED DEVELOPMENT AND HYBRIDISATION WITH WOLVES

Susan J. Crockford
Pacific Identifications Inc.,
6011 Oldfield Rd., R.R. #3,
Victoria, B.C. Canada
V8X 3X1

**Key words:** dog, evolution, wolf, wolf/dog hybrid, archaeozoology, breed, *Canis,* mtDNA, sympatry

**Abstract**

In this concluding chapter, some of the issues associated with dog evolution raised by other authors are examined from a broader perspective. In particular, the potential significance of certain morphological similarities among regional populations of prehistoric dogs is discussed, along with the development of distinct breeds. In addition, I address the unique situation resulting from the early prehistoric expansion of the dog (*Canis familiaris*) into territories occupied by its ancestral species, the grey wolf (*Canis lupus*). The extraordinary extent of sympatric distribution of dog and wolf is an unprecedented situation among domesticates, resulting in a myriad of potential hybrid zones that have existed for thousands of years. Indeed, both morphological and genetic evidence suggest that occasional hybridisation must have occurred between these two species and that wolves may have been more affected genetically by these events than dogs. A plea is made for more detailed and consistent reporting of archaeological canid remains and for additional ancient DNA studies on both species.

## Introduction

The unqualified success of the domestic dog, based on absolute numbers of animals and their current global distribution, results from a number of factors. Compared to other domesticates, dogs possess certain biological and historical characteristics that make them unique (Fig.1). Specific examples have been presented by many of the contributors to this volume, some of which I discuss in a broader, evolutionary context. In the first section of this paper, the origin of dogs and potential significance of certain morphological similarities among prehistoric regional populations is discussed, along with the development of distinct breeds. In the second part, I address the biological repercussions of the sympatric distribution which developed as a result of the early prehistoric expansion of the dog (*Canis familiaris*) into territory occupied by its ancestral species, the grey wolf (*Canis lupus*). The long-neglected topic of wolf/dog hybridisation is discussed in some detail, particularly its potential impact on the interpretation of mtDNA evidence for the evolutionary relationships between wolves and dogs.

## The significance of regional variation

### Origins: morphological and genetic evidence

There are several important questions that remain unresolved regarding dog origins. We still do not know for certain what populations or subspecies of wolf are ancestral to living, modern dogs and to extinct varieties – or precisely when those lineages arose. Also unresolved is the identification of the particular ecological, biological and cultural parameters responsible for initiating the transformation of wolf into dog (Clutton-Brock, this volume). Although I have presented what I believe is a plausible, testable hypothesis for how and why the speciation process of *protodomestication* occurred, at present no real consensus has been reached (Crockford, in press; Crockford, this volume). Indeed, some researchers still argue that the dog does not qualify for formal species designation. Many agree, however, as I do, that as *Canis familiaris* retains a suite of morphological, behavioural and physiological traits that are distinct from *Canis lupus,* it should be regarded as a separate species (Clutton-Brock, this volume; Crockford, this volume; Gentry, Clutton-Brock & Groves, 1996).

Despite this uncertainty, genetic evidence presented by Koop *et al.* (this volume) suggests that at least two separate domestication events from genetically distinct ancestors must have occurred, with the possibility that as many as five were actually involved. This evidence, combined with morphological evidence provided by Musil (this volume), suggests that we must finally abandon the notion that all dogs must have descended from small wolf subspecies and therefore, that all early dogs were of similar size (see also Dayan, 1994; Clutton-Brock, this volume). Specimens of early Israeli dogs, for example, although described as "small," are actually about the size of a modern Dalmatian (Crockford, 1997a). Early Japanese (Jomon Period) dogs were somewhat smaller, about the size of modern shiba dogs. Surprisingly (Fig. 2), the 10,000 year old dog from the French Alps (Chaix, this volume) is about the same size as early Jomon dogs. Early northern European dogs,

## The unique biological and historical characteristics of dogs, compared to other domestic taxa

- dogs were the first animal to enter into a domestic relationship – dogs thus have the longest shared history with humans

- dogs expanded their geographic range very early, moving with various human populations as they migrated throughout the world - by 7,000BP dogs were well established in all of Eurasia and the Americas

- dogs were once a common inclusion in human burials (or were buried in graves of their own), phenomena that started very early and persisted for thousands of years across cultural and geographic boundaries – this suggests that one of the first "roles" for dogs may have been ritual rather than practical in nature

- although dogs were eaten regularly in some areas, the practice was not widespread geographically or chronologically – consumption thus appears to have developed relatively late and to have been only one of several roles dogs filled

- for thousands of years, the dog has shared all of the Northern Hemisphere with populations of its closest living relative, the grey wolf – the extent and duration of this sympatric distribution is an unprecedented situation among domesticates that has made hybridisation between them far more likely to occur

***Figure 1*** *The unique historical characteristics of dog evolution, compared to other domestic taxa*

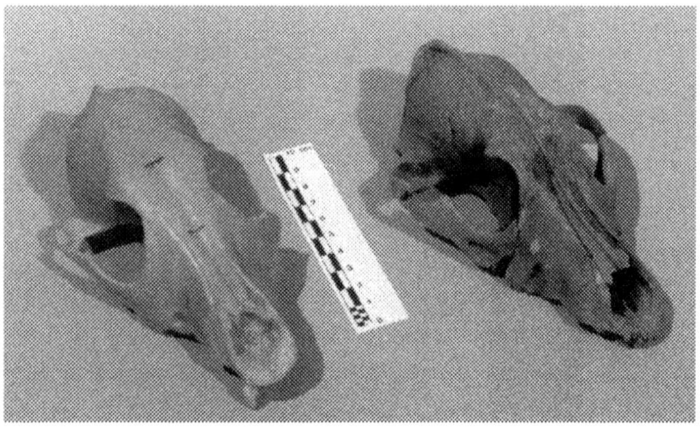
a

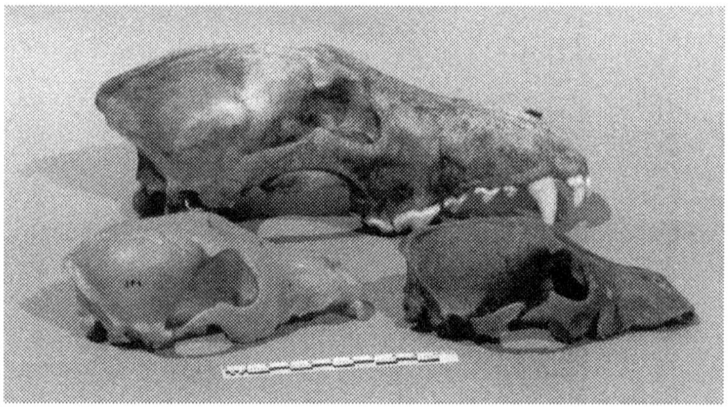
b

***Figure 2*** *Early dogs (L→R):* ***a)*** *cast of Preboreal Age (♀) cranium from Saint-Thibaud-de-Couz (Savoie, France) dated 10, 050 ± 100BP (specimen # JP1, courtesy L. Chaix); early Jomon Age (♀) cranium from the Torihama site (Fukui, Japan) dated c. 7,000-4,500BP (courtesy N. Shigehara);* ***b)*** *same dog specimens as in "a", with modern cranium of NA grey wolf (♂) at rear (courtesy B. Yates).*

described as "large," are actually dingo-sized (Gollan 1980), with crania of *c.* 200mm total length (similar in size to those in Fig. 3 & 4).

Dogs clearly do not display "extreme phenotypic diversity, even during the early stages of domestication," as Vilà *et al.* (1997) have suggested. In fact, the truly remarkable feature of early dogs is that despite some overall size differences between regional populations, early dogs show an astonishing similarity in conformation that persists in most regions for thousands of years (Dayan & Galili, this volume; Shigehara & Hongo, this volume). Extreme diversity in size and shape is a relatively late and localised development, as I discuss in more detail later. Homogeneity in cranial size and conformation of Pacific Rim dogs, for example, is apparent in Figures 5 and 6. The relatively small size difference in postcranial elements from other regional populations is illustrated in Figure 7.

Modest size variations between populations of early dogs, however, suggest that this trait may be useful for identifying locations of protodomestication and routes of subsequent dispersal, if indeed this trait reflects significant size differences in ancestral wolf subspecies. Thus, initial domestication events appear to have occurred in northern Germany at *c.* 13,000 BP, Israel at *c.* 15,000 (or somewhat before) and somewhere in Asia at *c.* 12,000 BP - producing relatively large early dogs in northern Europe (Clutton-Brock, 1995; Crockford, this volume; Musil, this volume); a somewhat smaller form in the Middle East (Dayan &

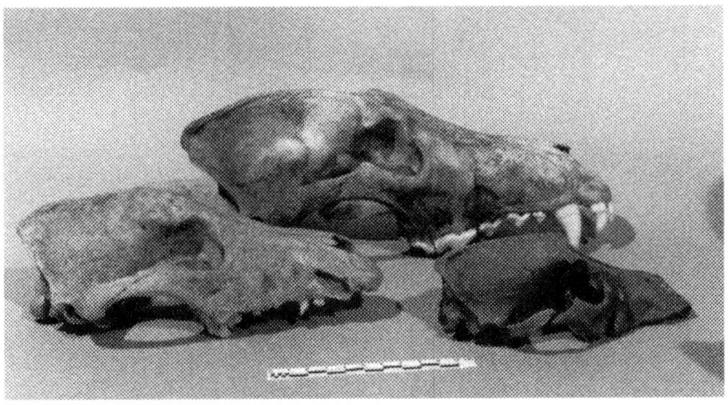

***Figure 3*** *Early dogs (L→R): Modern cranium of NA grey wolf (♂) (courtesy B. Yates) vs. Neolithic Age cranium (♂ from Tsamakaberd (the Caucuses, Armenia) dated c. second millennium BC (specimen #2, courtesy N. Maneserian); and early Jomon Age (♀) cranium from the Torihama site (Fukui, Japan) dated c. 7,000-4,500BP (courtesy N. Shigehara).*

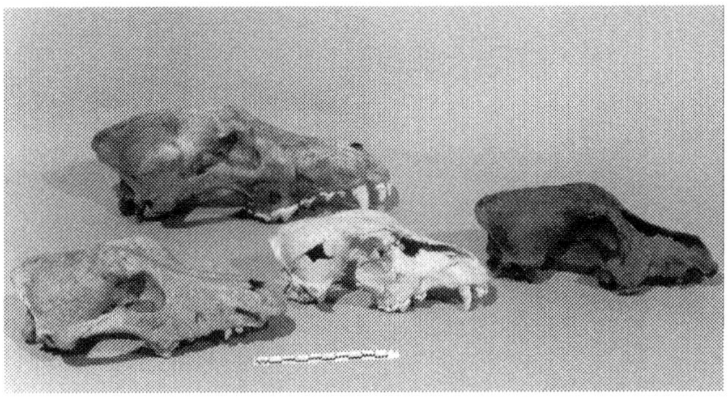

***Figure 4*** *Pre-Roman Age European dogs: Modern cranium of NA grey (♂) wolf (courtesy B. Yates) at rear vs. (L→R) Neolithic Age cranium (♂) from Tsamakaberd (the Caucuses, Armenia) dated c. second millennium BC (specimen #2, courtesy N. Maneserian); Iron Age (♂) cranium from Berkshire (UK) dated c. 500-0BC (specimen # BLEWB, courtesy K. Clark); Bronze Age (♂) cranium from the Thames floodplain (UK) (specimen #YFP92-100, courtesy K. Clark)*

***Figure 5*** *European vs. Pacific Rim dogs (L→R): Roman Age (♂) cranium from Bavenstedt (Lower Saxony, Germany) dated c. third-fifth century AD (specimen # 90/16, courtesy S. Hanik, Potsdam); Maori dog (♀) cranium from New Zealand dated c. 800-200BP (courtesy R. McGovern-Wilson, Rotorua, NZ); Gulf of Georgia Age NW Coast Wool dog (♂) cranium from Vancouver Island (BC, Canada) dated c. 1,400-250BP (specimen #2400, courtesy S. Crockford); Okhotsk Culture Age (♂) cranium from the Hamanada No. 2 site (Hokkaido, Japan) dated c. 1,200BP (courtesy N. Shigehara)*

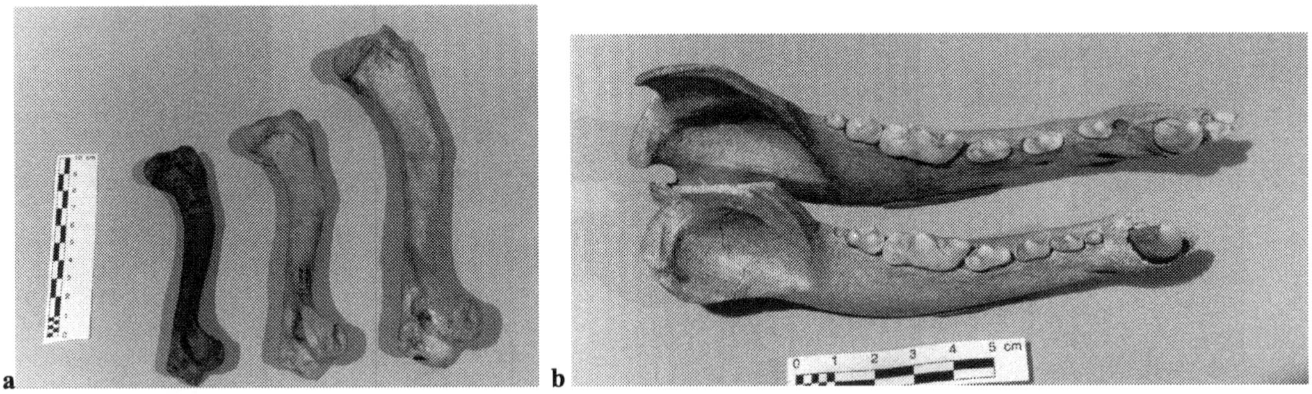


***Figure 6*** *European vs. Pacific Rim dogs.* ***a)*** *humerus (L→R): Maori dog (♀) from New Zealand dated c. 800-200BP (courtesy R. McGovern-Wilson, Rotorua, NZ); Gulf of Georgia Age NW Coast Wool dog (sex unknown) from Vancouver Island (BC, Canada) dated c. 1,400-250BP (specimen #2407, courtesy S. Crockford); Roman Age (♂) from Bavenstedt (Lower Saxony, Germany) dated c. third-fifth century AD (specimen # 90/16, courtesy S. Hanik, Potsdam).* ***b)*** *mandibles, illustrating tooth crowding (top→bottom): Roman Age (♂) from Bavenstedt (Lower Saxony, Germany) dated c. third-fifth century AD (specimen # 90/16, courtesy S. Hanik, Potsdam); Okhotsk Culture Age (sex unknown) from the Hamanada No. 2 site (Hokkaido, Japan) dated c. 800BP (courtesy N. Shigehara).*

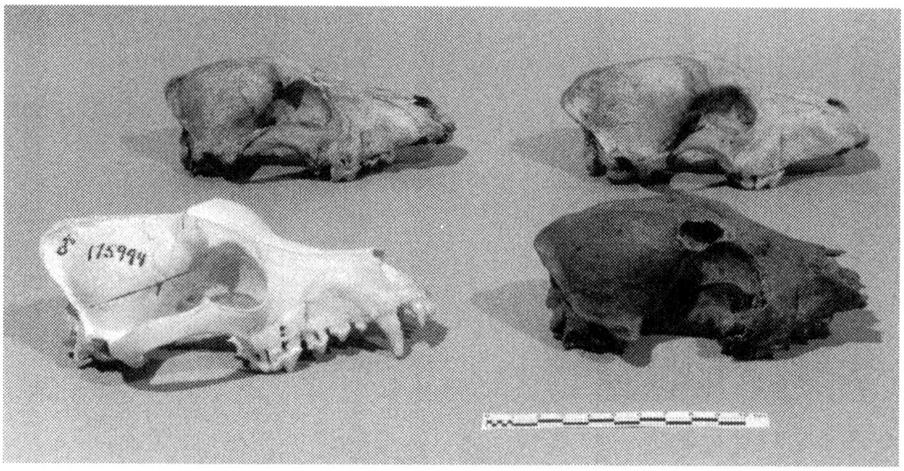

***Figure** 7 Pacific Rim dogs (L→R, back row first): Okhotsk Culture Age (♂) cranium from the Hamanada No. 2 site (Hokkaido, Japan) dated c. 1,200BP (courtesy N. Shigehara); Gulf of Georgia Age NW Coast Wool dog (♂) from Vancouver Island (BC, Canada) dated c. 1,400-250BP (specimen #2400, courtesy S. Crockford); recent New Guinea singing dog (♂) cranium (specimen 175994, collected 1964, courtesy J. Koler-Matznick & UCLA, Berkeley Museum); Maori dog (♀) cranium from New Zealand dated c. 800-200BP (courtesy R. McGovern-Wilson, Rotorua, NZ);*

Galili, this volume; Meadow, this volume); and a still smaller type in ancient Japan and China (Shigehara & Hongo, this volume). Early North American dogs (*c.* 8,000-10,000BP) were smaller than early northern European forms but larger than early Jomon dogs (Crockford, 1997a; Morey & Wiant, 1992; Yohe & Pavesic, this volume), suggesting possible affinities with yet another Asian ancestor.

Regardless of where and when they occurred, it appears that multiple speciation events are the rule rather than the exception for domesticates. For example, genetic evidence supports the hypothesis of multiple speciation events in cattle (Bradley *et al.*, 1998) and sheep (Hiendleder *et al.*, 1998). Data from genetic analysis of extinct as well as extant populations now provide similar evidence for dogs (Koop *et al.,* this volume). As expected, these data suggest a common geographic origin for indigenous North American (NA) dogs, from a wolf lineage distinct from ancestors of most modern purebreds.

In contrast, when the data for prehistoric dogs based on 200bp (Ishiguro *et al.*, this volume) was compared to Koop *et al.*'s sequences of greater length (300bp), ancient Japanese dogs placed within the same clade that contained most modern dog sequences (unpublished data, 1998). On close examination, diagnostic sites for ancient North American dogs were found to lay on both sides of the 200bp region used by Ishiguro *et al.* (Ben Koop, pers. comm.), suggesting that ancient mtDNA segments short enough to be easily extracted and amplified (eg, Matisoo-Smith *et al.*, 1998; based on *c.* 200bp of South Pacific dogs) may not provide enough genetic variation for phylogenetic analysis in *Canis*. Although more difficult to obtain from ancient tissues, mtDNA segments of at least 300bp appear to be essential for meaningful results.

Genetic information has thus provided corroboration of evidence based on morphology that multiple protodomestication events occurred. The diversity of the "Dog" clade (Koop *et al.,* this volume: fig. 5) suggests it may have derived from a clade or subspecies of genetically distinct wolf, probably now extinct (Morell, 1997). A recent study suggests polar bears were similarly derived from a distinct clade of brown bear: only 1.1% sequence difference in mtDNA was found between the polar bear (*Ursus maritimus*) and one geographically isolated lineage of brown bear (*U. arctos*), leading Talbot and Shields (1996: 574) to conclude that "morphological features distinguishing polar bears from brown bears have evolved rapidly in response to selective pressures of adapting to a new environment, prior to the emergence of distinguishing molecular features." Such a pattern is also seen in *Bos*, where phylogenetic studies suggest the four regional subspecies of aurochsen which gave rise to genetically distinct lineages of domestic cattle had been separate for thousands of years prior to protodomestication events (Bradley *et al.*, 1998).

Molecular data has also provided important insights into breed development. For example, all studies agree that there are no breed-specific, or morphotype-specific, mtDNA haplotypes. This means that genetically distinct lineages did not generate specific dog morphotypes, such as spitzes and mastiffs: apparently, any dog lineage may generate these morphotypes. This supports the contention of Coppinger and Schneider (1995) that so-called "wolf-like" breeds are no more closely related to wolves than other breeds or necessarily a result of repeated hybridisation: their somewhat less specialised characteristics are merely one of many breed-specific patterns of development that many be generated out of any dog lineage (see also Crockford, this volume).

## Development of regional variation

Although a pattern of multiple protodomestication events similar to that suggested for dogs has also been shown for

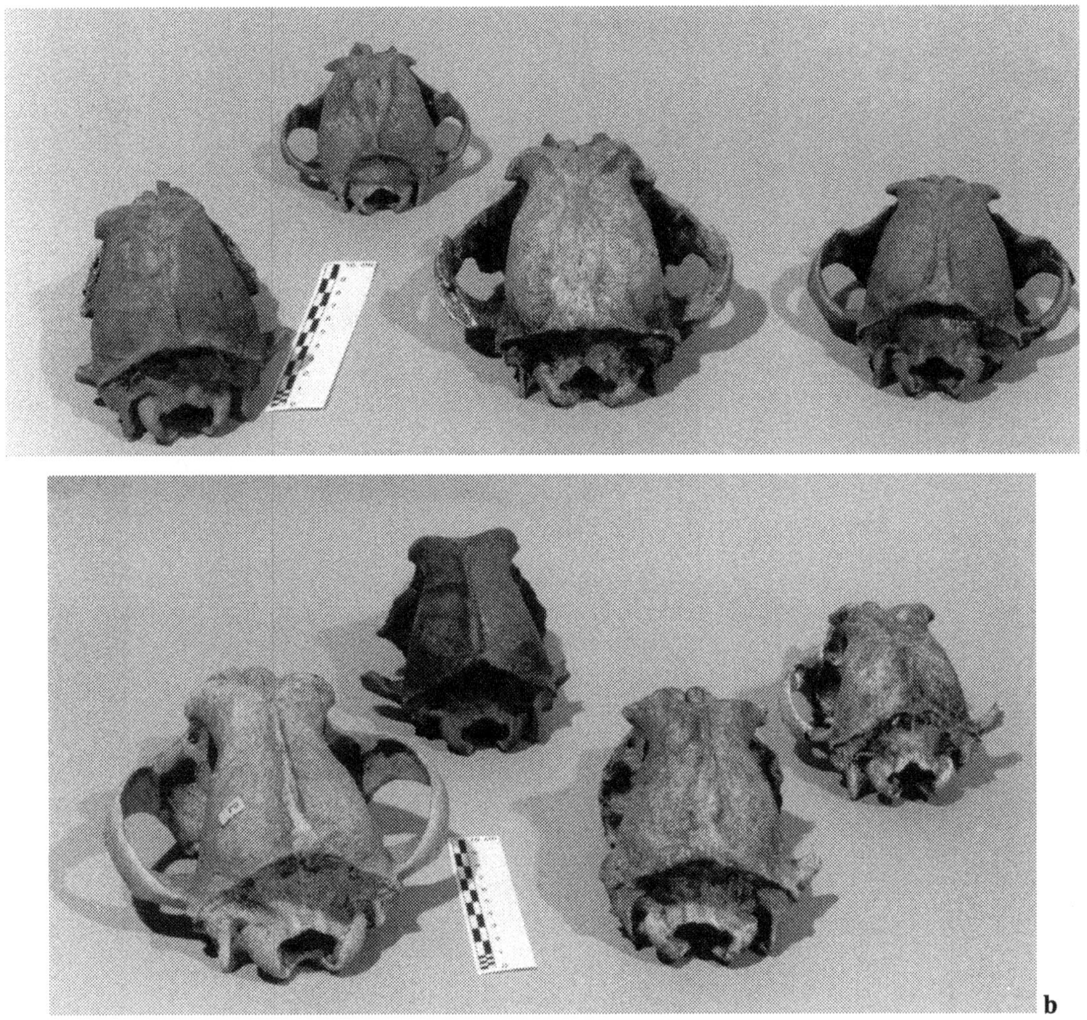


***Figure 8*** *Examples of sexual dimorphism in dog crania.* ***a)*** *females (L→R): Maori dog from New Zealand dated c. 800-200BP (courtesy R. McGovern-Wilson, Rotorua, NZ); Gulf of Georgia Age NW Coast Wool dog from Vancouver Island (BC, Canada) dated c. 1,400-250BP (specimen #0801, courtesy S. Crockford); early Jomon Age dog from the Torihama site (Fukui, Japan) dated c. 7,000-4,500BP (courtesy N. Shigehara); Roman Age dog from Bavenstedt (Lower Saxony, Germany) dated c. third-fifth century AD (unnumbered specimen, courtesy S. Hanik, Potsdam) - note the flat, broad sagittal crest and hooked ectorbitales.* ***b)*** *males (L→R): Neolithic Age dog (♂) from Tsamakaberd (the Caucuses, Armenia) dated c. second millennium BC (specimen #2, courtesy N. Maneserian); Roman Age dog from Silchester (UK) dated first-fourth century AD (specimen #1653, courtesy L. Cram); Gulf of Georgia Age NW Coast Village dog from the Straight of Georgia region (BC, Canada) dated c. 1,400-250BP (specimen #0925, courtesy S. Crockford); Final Jomon Age dog from the Hobi site (Aichi, Japan) dated c. 3,000-2,3000BP (courtesy N. Shigehara) - note the narrow, elevated sagittal crest and bulbous swelling behind the ectorbitales.*

cattle and sheep, as previously discussed, here the similarity in their history ends. Dogs were relatively more mobile and their earlier presence gave them opportunities to invade areas of the world that cattle and sheep didn't reach until much later. By *c.* 7,000BP, dogs had become established in most major continents (Fig. 10), appearing in Australia and Arctic regions somewhat later. After *c.* 500AD, dogs are found in archaeological deposits in sub-Sahara regions of Africa, and on remote Pacific Islands such as New Zealand and Hawaii. Such an early dispersal of dogs allowed regional varieties to develop in many areas, well before breed development began.

Regional variation in skeletal morphology can be quite marked. For example, although all Pacific Rim dogs except the dingo are relatively small, Polynesian dogs found on the islands of Oceania are also distinctly shortlegged (Clark, 1997a & b). In contrast, some populations share much more subtle anomalies, such as the congenital loss of the first premolar common to NA dogs (Crockford, 1997a; Handley, this volume). Surprisingly, at the Dog Symposium "show & tell" session, it was noticed that all European dogs present (from Britain, Turkey, Armenia, Germany & the Republic of Georgia, at all time periods represented) displayed a distinct pattern of tooth wear. In all dogs where there was significant wear, the post-canine teeth were worn level, parallel to the gum line (Fig. 11a). In contrast, all Pacific Rim dogs with worn teeth (from Japan, the Northwest coast of NA & New Zealand) shared a similar, concave pattern of tooth wear. In these dogs, the post-canine teeth were worn much more heavily on the inside (lingual) surface (Fig. 11b), giving the molar teeth in particular a pronounced "dished" appearance. Morphological differences in occlusion due to different

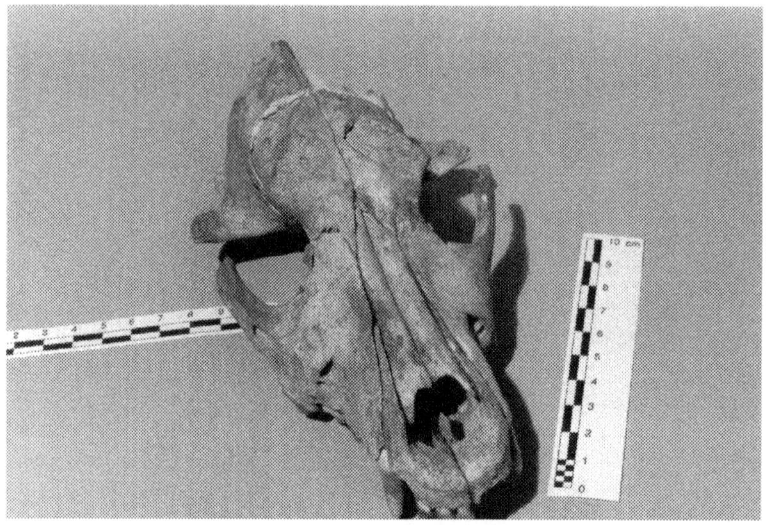

a

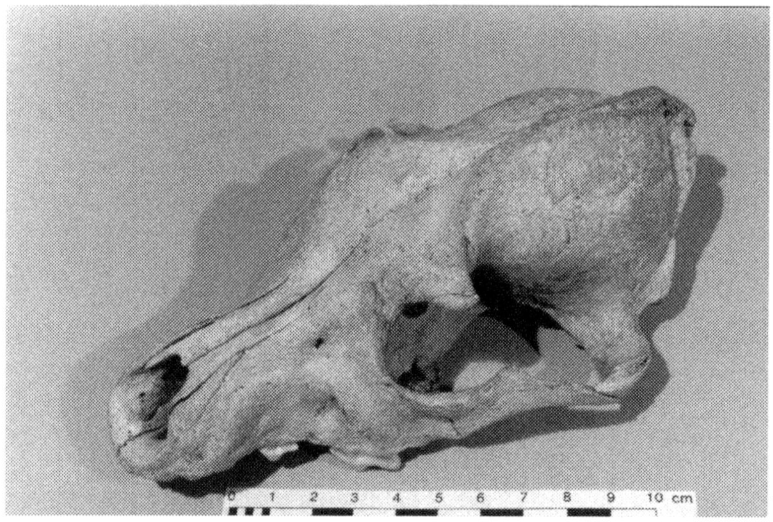

b

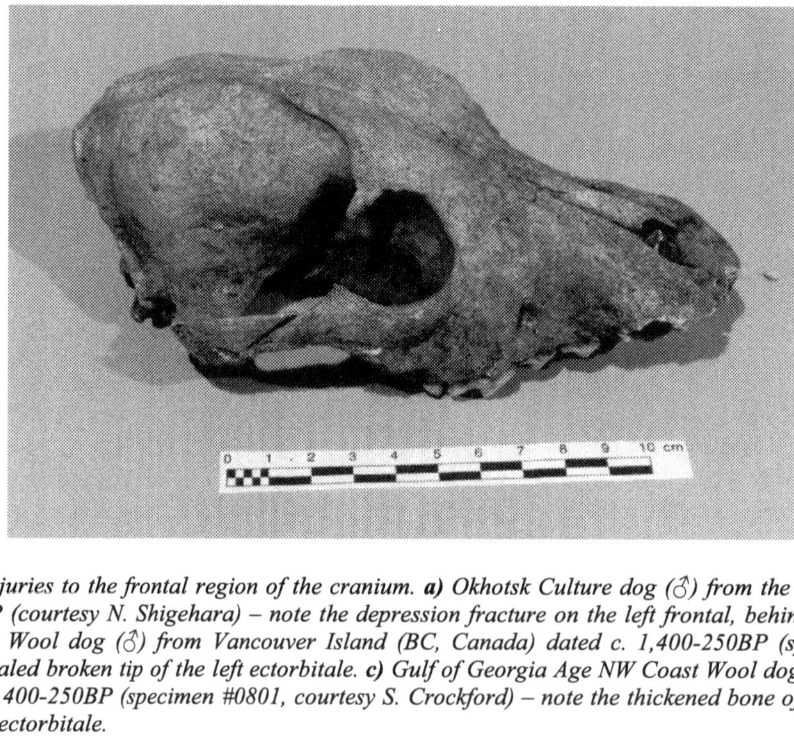

c

***Figure 9*** *Examples of injuries to the frontal region of the cranium.* ***a)*** *Okhotsk Culture dog (♂) from the Benten-jima site (Hokkaido, Japan) dated c. 1,200BP (courtesy N. Shigehara) – note the depression fracture on the left frontal, behind the ectorbitale.* ***b)*** *Gulf of Georgia Age NW Coast Wool dog (♂) from Vancouver Island (BC, Canada) dated c. 1,400-250BP (specimen #2400, courtesy S. Crockford) – note the healed broken tip of the left ectorbitale.* ***c)*** *Gulf of Georgia Age NW Coast Wool dog (♂) from Vancouver Island (BC, Canada) dated c. 1,400-250BP (specimen #0801, courtesy S. Crockford) – note the thickened bone of both frontals and excessive development of the right ectorbitale.*

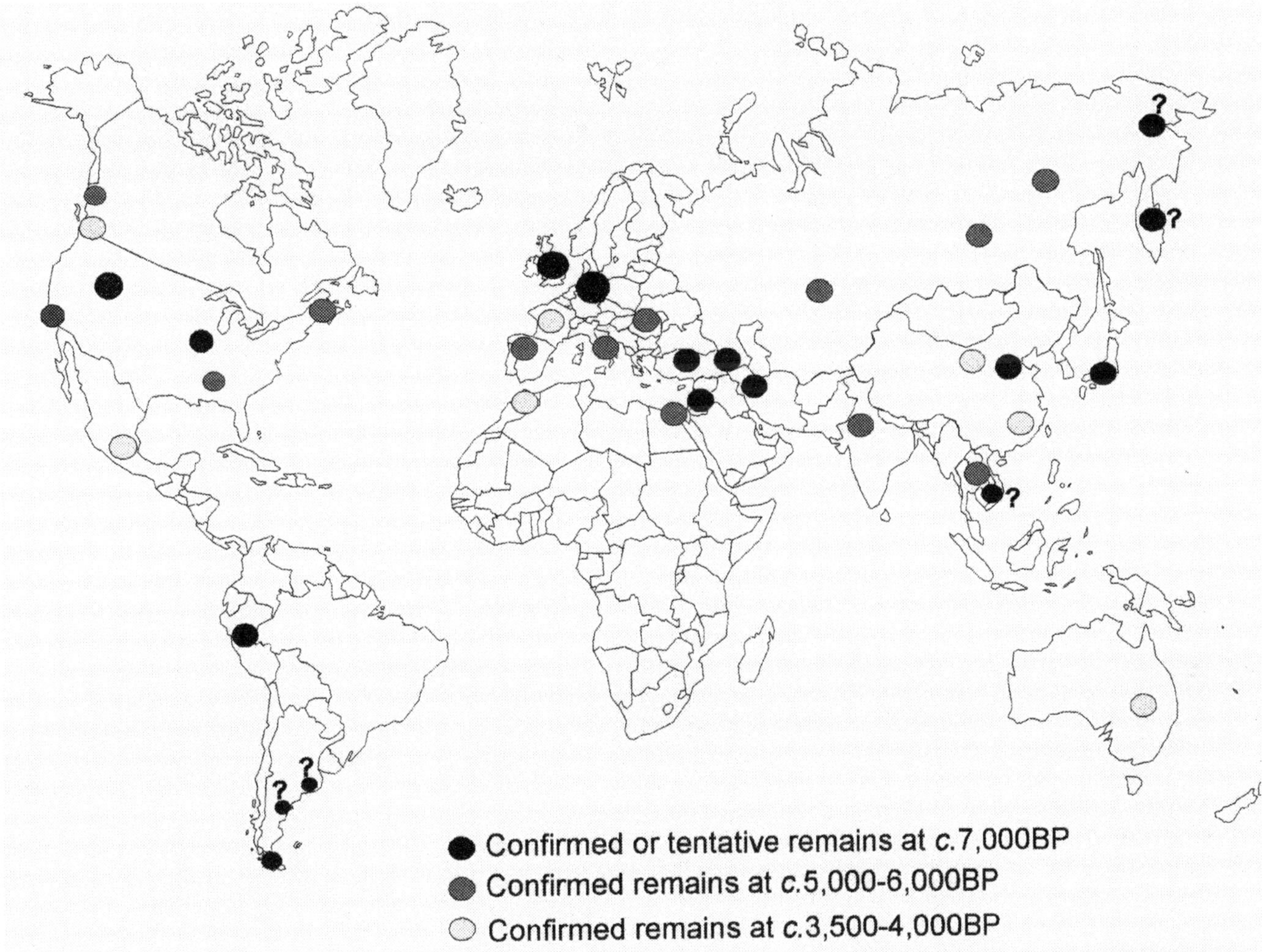


***Figure 10*** *Global distribution of significant archaeological remains of Canis familiaris at c.7,000BP, c. 5,000-6,000BP, and c. 3,500-4,000BP (tentative finds in eastern Siberia (?) associated with disputed dates; tentative finds in South America (?) are disputed on taxonomic grounds). Many more finds exist in NA and Europe, but no remains have been found in sub-Sahara Africa before 500AD (Brothwell et al. 1979; Cannon, Schwarcz & Knyf, 1999; Clutton-Brock, 1981, 1995; Erlandson, 1994; Gollan, 1980; Handley, this volume; Olsen, this volume; Olsen, 1985).*

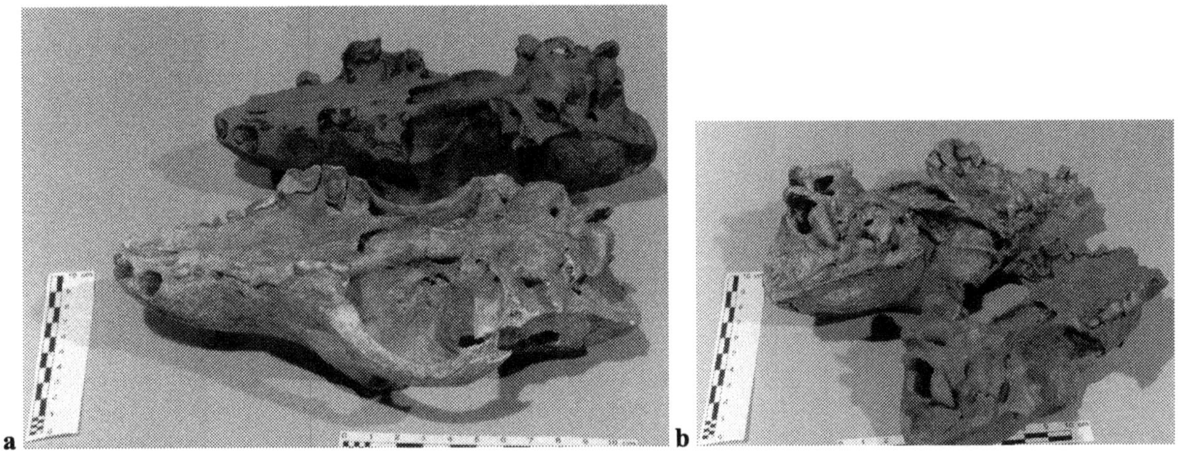


***Figure 11*** *Tooth wear: **a)** pattern typical of some European dogs – flattened wear, parallel to the gum line on all teeth: rear, Bronze Age (♂) cranium from the Thames floodplain (UK) (specimen #YFP92-100, courtesy K. Clark); front, Iron Age (♂) cranium from Berkshire (UK) dated c. 500-0BC (specimen # BLEWB, courtesy K. Clark). **b)** pattern typical of some Pacific Rim dogs – heavy concave wear on lingual surfaces of molar teeth especially: rear, Gulf of Georgia Age NW Coast Village dog (♂) from Vancouver region (BC, Canada) dated c. 1,400-250BP (specimen #925, courtesy S. Crockford); front, Maori dog (♀) cranium from New Zealand dated c. 800-200BP (courtesy R. McGovern-Wilson, Rotorua, NZ).*

origins was one suggestion for such a broad geographic pattern of tooth wear, although the small size of this particular sample obviously precludes drawing any real conclusions. It does, however, point out the need for reporting traits like tooth wear in a universally consistent manner (eg. Herbert-Horard, this volume) so that regional comparisons can eventually be made.

Another morphological trait that shows some broad variation is the shape of the coronoid process of the mandible, the upright portion (ascending ramus) of the lower jaw. Olsen & Olsen (1977) had previously proposed that the consistently hooked shape of the coronoid process in all NA prehistoric dogs suggested a Chinese wolf ancestor, since *C. l. chanco* is the only wolf subspecies that shows a similarly hooked coronoid process. Gollan (1980), however, points out that some specimens of *C. l. pallipes* and *C. l. arabs* also possess a curved coronoid process. Figure 12 compares mandibles made available during the Dog Symposium of several canids with a *straight* coronoid process: an ancient dog from Iran (Hotu Cave, *c.* 6,500BC); a modern, purebred Alaskan malamute; a modern NA wolf, and a modern wolf from the Republic of Georgia. Figure 13 compares mandibles of prehistoric dogs that share a *hooked* coronoid process: specimens from both early and late Jomon deposits in Japan, the Northwest (NW) Coast of NA and the *c.* 10,000 year old dog from the French Alps. Again, this small sample of dogs from diverse regions suggests that the coronoid process may be an important non-metric trait that, with consistent reporting of its shape for all dogs, may shed light on dog origins, between-lineage crosses, and wolf/dog hybridisation.

Roles of early dogs

One of the earliest dog remains recovered is that of a young animal buried with a human female *c.* 12,000 years ago in Israel (Davis & Valla, 1978). Thus, the practice of including dogs with human burials existed very early in the relationship between dogs and humans, a phenomenon that suggests an early "role" of dogs may have been sacred or ritual rather than functional in nature (see also Yohe & Pavesic, this volume). Deliberate burial of dogs on their own is also an old and geographically widespread phenomenon that transcended cultural boundaries (Morey & Wiant, 1992; Olsen, 1985; Olsen, this volume; Schwartz, this volume; Shigehara & Hongo, this volume) and may be yet another manifestation of this early sacred role. Could an initial sacred role for dogs have been a natural human response to protodomestication occurring so rapidly - ie, within one or two human generations (see Crockford, this volume) - that people in all cultures were simply awed by the sudden transformation of wolves into dogs? Although the suggestion is purely speculative, it illustrates why understanding the initial process of protodomestication is so important to understanding subsequent dog/human relationships and for interpreting the archaeological record.

When were other, more practical roles for dogs established? While we assume that dogs must have assisted in hunting at an early stage, there is no archaeological evidence of that practice. Only much later do we have evidence from artistic images that some dogs assisted in hunting (eg, Manaserian & Antonian, this volume; Schwartz, 1997 & this volume). Skeletal evidence that

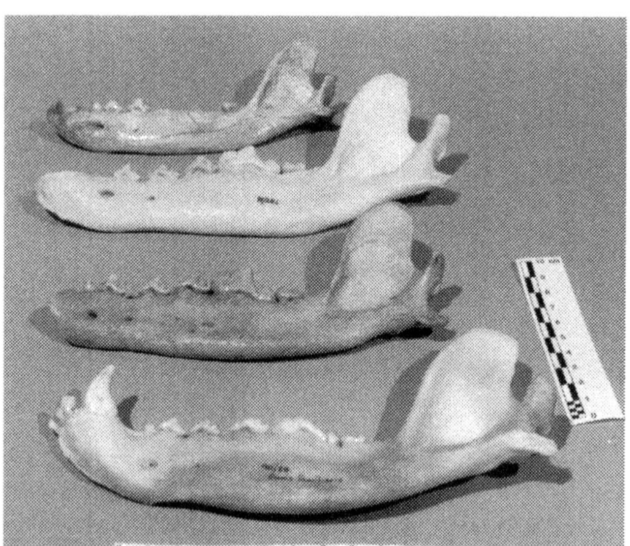

***Figure 12*** *Coronoid process, straight posterior margin (top→bottom): early Middle Eastern dog mandible (Hotu Cave, Iran) dated c. 6500BC (courtesy R. Meadow); recent NA wolf (♀) mandible, Alaska (courtesy University of Victoria, Anthropology Dept.); recent European wolf mandible, Republic of Georgia (courtesy O. Bendukidze, Tbilisi, Georgia); recent CKC registered Alaskan malamute (♂) mandible (courtesy University of Victoria, Anthropology Dept.) - note the tooth crowding of $P_4$ & $M_1$ in the two wolf specimens – although identified as "wolf," the Alaska specimen is suspected of being a hybrid based on shape of the mastoid process described by Yates (this volume).*

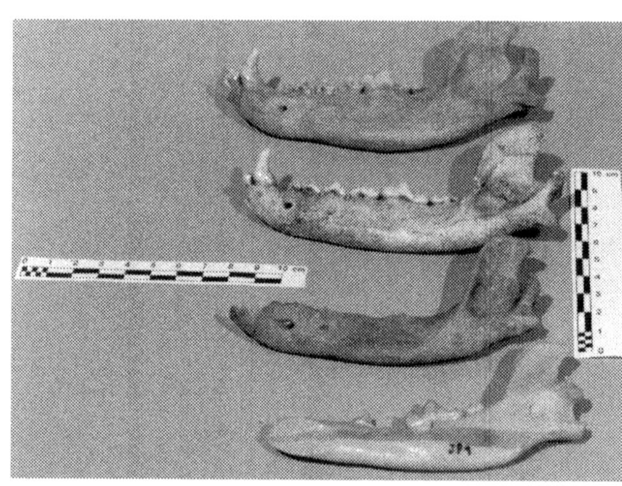

***Figure 13*** *Coronoid process, curved posterior margin (top→bottom): Okhotsk Culture Age (sex unknown) mandible from the Hamanada No. 2 site (Hokkaido, Japan) dated c. 800BP (courtesy N. Shigehara); Final Jomon (♂) mandible from the Hobi site (Aichi, Japan) dated c. 3,000-2,300BP (courtesy N. Shigehara); Gulf of Georgia Age NW Coast Wool dog (sex unknown) mandible from Vancouver Island (BC, Canada) dated c. 1,400-250BP (specimen #805, courtesy S. Crockford); cast of Preboreal Age (♀) mandible from Saint-Thibaud-de-Couz (Savoie, France) dated 10, 050 ± 100BP (specimen # JP1, courtesy L.Chaix).*

dogs were used prehistorically as sledge haulers or pack animals may come from skeletal pathologies (eg, Warren, this volume), but there are few data of this kind anywhere. Ethnographic reports in North America document, however, that at the time of European contact (and probably before), dogs were used to pull sledges or simple wooden travois, or to carry loaded packs on their backs (Schwartz, 1997). While dogs were certainly eaten in some areas regularly (Clark, 1995; Clutton-Brock & Hammond, 1994; Horard-Herbin, this volume), and in others on occasion, chronologically this evidence appears to come from the last 5,000 years of their history rather than the first 5,000-10,000 years. Whether these patterns are a result of excavation practices, reporting biases, limited dispersal of literature, or poor preservation of older material, is hard to say. Perhaps with more consistent and detailed reporting of dog remains, it will be possible to fill in some of these gaps in our understanding of early dog/human relationships.

It is obvious that few of the answers to the questions that remain regarding the history of our relationships with dogs will be forthcoming without more consistent and rigorous data collection by both field and laboratory archaeozoologists. I have suggested previously (Crockford, 1997a) that treating dog remains with the same care and attention to detail as we do human remains should solve many of the problems: *careful field collection*, with photos or drawings of *in situ* remains; *complete reporting of isolated and associated remains, with photos* (including: lists of elements recovered; age and sex of individuals (Fig. 8); tooth wear and tooth anomalies; injuries and pathologies (Fig. 9); shape of the coronoid process; osteometric data; *chronological dating* of actual material; *reporting of data separate and apart from subsistence fauna* (perhaps included in both parts of a report if dogs were clearly being eaten). Interpretation will always be problematic, but we can certainly improve on our recording practices for essential raw data of archaeological remains (see also Olsen, this volume).

## Breed development

It is apparent from the above discussion that the variation seen in the skeletal remains of early dogs is either the result of their origination from geographically different wolf populations, the effect of fixation of genetic mutations within regional populations, or both. Some anomalous morphotypes may also be preserved archaeologically, such as individual dogs affected by chance mutations (eg, Valadez, this volume) and obviously, archaeozoologists must be wary of how they interpret such idiosyncratic remains. Nevertheless, archaeological evidence of true dog husbandry - the repeated occurrence of similar, distinctive morphotypes - does not appear until the Roman period, *c.* 3,000BP in Europe. While there are both written records and artistic evidence of distinct breeds from this period in both Egypt and China (Clutton-Brock, 1981; Olsen, 1985), little in the way of archaeological skeletal evidence of breeds in these regions has been reported. In contrast, many archaeological sites from the first few centuries AD onwards in Italy and much of the rest of Europe (Part 3, this volume), contain the remains of distinctly different dog morphotypes: small terriers, pug-nosed lap-dogs, dwarfed dogs, and even truly "large" dogs (Figs.14 & 15). This explosion of variation certainly appears unprecedented, although as Clark (this volume) has pointed out, the archaeological record in Britain and other areas may be biased by both excavation and reporting practices, and by poor preservation of older material. Nevertheless, the Roman contribution to the creation of numerous physically distinct dog breeds is undeniable.

The Romans, however, had the benefit of writing to keep track of breeding strategies. Could there have been true husbandry for specific breeds before there were written records to keep track of individual relationships and lineages? Our modern definition of a breed describes a "group of animals that has been selected by man to possess a uniform appearance that is inheritable and distinguishes it from other groups of animals within the same species" (Clutton-Brock 1979: 37). I suggest that the NW Coast of North America provides a good example of such prehistoric breeding, summarised below from Crockford (1997a & b) and several other comprehensive accounts.

Early historic and ethnographic references (Howay, 1918; Schulting, 1994) report the presence of two types of indigenous dogs kept by aboriginal peoples on the southern portion of the NW Coast at the time of European contact in the late 1700s (see Koop *et al.,* this volume: figs. 1-3). In general, these accounts describe a medium sized, dingo-like animal sometimes used for hunting (which I call the "Village dog") and a smaller, longhaired dog kept almost exclusively for its thick soft fur which was woven into blankets (traditionally called the "Wool dog"). Measures were reportedly taken to keep the two types from interbreeding. One of these methods was to sequester Wool dogs together, with buried dried fish provided for food, on a small off-shore island when they could not be closely monitored, such as during the summer fishing and hunting season. Both dog types were well-proportioned, robust animals; archaeological evidence (Crockford, 1997a) suggests the Wool dog averaged 44 cm at the shoulder and the Village dog 52 cm (Table 1).

The larger Village dog was said to have had short fur in various shades of brown (perhaps with white markings) and appears to have been a widely distributed type across western North America. The Wool dog, however, was reported as noticeably smaller than the Village dog, with most being white or white with black, and possessing upright ears and a curled tail. The Wool dogs, which had a very restricted distribution on the NW Coast, were reported to have had their long thick fur sheared like sheep several times a year. Woven blankets (of both dog hair and mountain goat wool) were not utilitarian items but used instead to indicate wealth in the highly stratified societies of this region (Amoss, 1993). The dogs kept for their wool appear to have been owned by women, who also undertook the weaving and presumably oversaw the details of husbandry. In the early historic period (*c.* 1800), dog-hair blankets were replaced as a preferred status item by

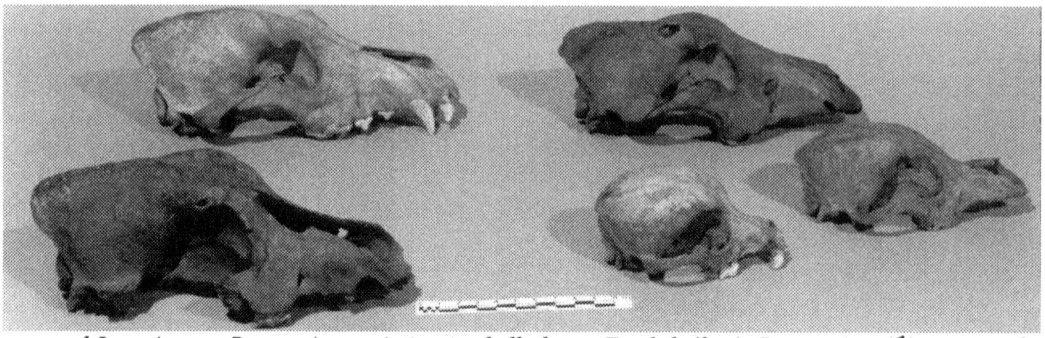

***Figure 14*** *Bronze and Iron Age vs. Roman Age variation in skull shape. Far left (fore), Bronze Age (♂) cranium from the Thames floodplain (UK) (specimen #YFP92-100, courtesy K. Clark); far left (rear), Iron Age (♂) cranium from Berkshire (UK) dated c. 500-0BC (specimen # BLEWB, courtesy K. Clark); far right (rear), large Roman Age (♂) cranium from Bavenstedt (Lower Saxony, Germany) dated c. third-fifth century AD (specimen # 90/16, courtesy S. Hanik, Potsdam); far right (fore, L→R), Pomeranian-type Roman Age (♂?) cranium from Bavenstedt (Lower Saxony, Germany) dated c. third-fifth century AD (specimen #53/3/005, courtesy S. Hanik, Potsdam) – note that despite the diminution in size, all teeth are present (including $PM^1$) and show extreme, flattened tooth wear (parallel to the gum line); terrier-type Roman Age (♀?) cranium from Bavenstedt (Lower Saxony, Germany) dated c. third-fifth century AD (unnumbered specimen, courtesy S. Hanik, Potsdam) – note this specimen also shows extreme, flattened tooth wear on its molars.*

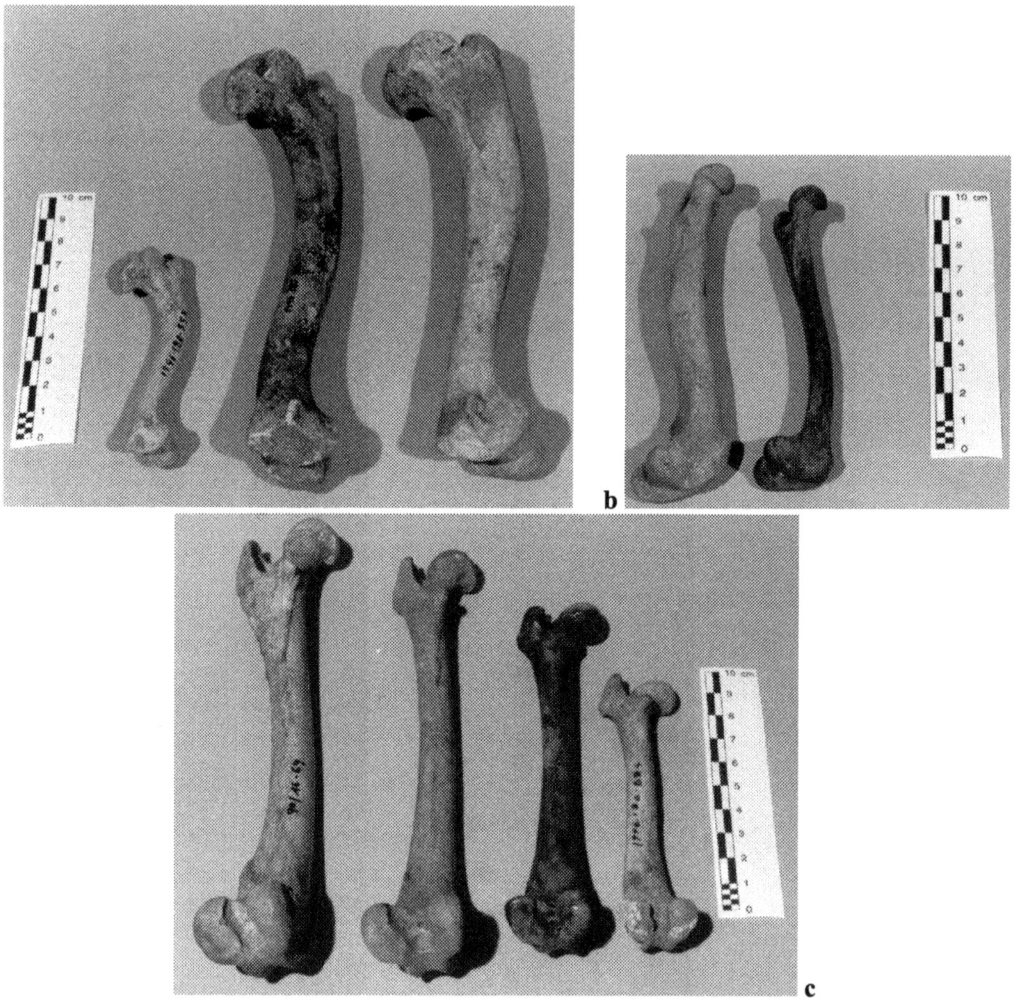

***Figure 15*** *Variation in post-cranial elements.* ***a)*** *humerus (L→R): small Roman Age dog from Silchester (UK) dated first-fourth century AD (specimen #550, courtesy L. Cram); large Roman Age dog from Silchester (UK) dated first-fourth century AD (specimen #80, courtesy L. Cram); large Roman Age dog from Bavenstedt (Lower Saxony, Germany) dated c. third-fifth century AD (specimen #90/16/64, courtesy S. Hanik, Potsdam).* ***b)*** *femur (L→R): small Roman Age dog from Silchester (UK) dated first-fourth century AD (specimen #109, courtesy L. Cram), bowed to the inside; small Roman Age dog from Silchester (UK) dated first-fourth century AD (specimen #1862, courtesy L. Cram), bowed to the outside.* ***c)*** *femur (L→R): large Roman Age dog from Bavenstedt (Lower Saxony, Germany) dated c. third-fifth century AD (specimen #90/16/64, courtesy S. Hanik, Potsdam); large Roman Age dog from Silchester (UK) dated first-fourth century AD (specimen #658, courtesy L. Cram); Charles Age NW Coast Wool dog from the Fraser Delta (BC, Canada) dated c. 4,400-3,000BP (specimen #0400, courtesy S. Crockford); small Roman Age dog from Silchester (UK) dated first-fourth century AD (specimen #684, courtesy L. Cram).*

sheep's wool and Hudson's Bay trade blankets and thus, the weaving of dog-hair blankets was rapidly abandoned. The impetus for keeping Wool dogs reproductively isolated from Village dogs was soon lost and the Wool dog as a separate type rapidly became extinct.

No other aboriginal group has this recorded pattern of dog use. In contrast to other NA archaeological samples where size variation in skeletal remains has been reported, or where ethnographic reports suggest different types of dogs existed (eg, Handley, this volume; Walker & Frison, 1982), on the NW Coast there is strong ethnographic evidence for actual dog husbandry. The accounts here describe two types of dogs deliberately maintained as separate populations, based on distinct heritable morphological traits, with an explicitly stated economic rationale. The economic motivation for lineage segregation of Wool dogs from Village dogs must have been compelling indeed, in order to overcome the practical difficulties of dog husbandry without benefit of non-chewable restraints and fencing materials. The Wool dog certainly appears to fit the definition of a modern breed, but how could it have been developed and maintained under such conditions?

We now know that long hair is a recessive trait to short hair (see also Ryder, this volume), and that first-generation hybrids (F1) between these two extremes generally possess hair of intermediate length. This suggests that even without a knowledge of genetics and inheritance, the effect of interbreeding the two types would have been readily apparent, since hybrid offspring would resemble the shorthaired Village dog more than the longhaired Wool dog in coat type. Such hybrids between the two types would be immediately and continuously identifiable throughout the dog's life, even without written records. Without the desired long woolly coat it is doubtful these animals would have been used for future deliberate breeding. The recessive inheritance pattern of the desired trait must have been the key factor, making it obvious that keeping individuals with long hair reproductively isolated from others was necessary for the continued production of longhaired dogs. A recessive trait also explains why the breed disappeared so quickly once deliberate husbandry efforts ceased.

Therefore, the NW Coast Wool dog may constitute a rare example of a true indigenous dog "breed," developed

| **Cranium measurements** | **Wool dog small type range (mm)** | **Village dog large type range (mm)** | **Vertebrae measurements (con't)** | **Wool dog small type range (mm)** | **Village dog large type range (mm)** |
|---|---|---|---|---|---|
| #1 (10/8) | 146-173 | 176-203 | Lumbar 2 (PL) (10/6) | 22-24 | 25-28 |
| #2 (9/8) | 140-162 | 164-188 | Lumbar 3 (PL) (8/9) | 22.0-24.5 | 25.5-28.0 |
| #12 (10/9) | 59-73 | 76-87 | Lumbar 4 (PL) (6/7) | 22-25 | 26.5-29.0 |
| #13 (10/5) | 73-86 | 89-98 | Lumbar 5 (PL) (5/8) | 21-25 | 26.5-28.0 |
| #15B (10/9) | 76-85 | 90-99 | Lumbar 6 (PL) (7/7) | 22.0-24.5 | 25.5-28.0 |
| #23 (9/8) | 56-60 | 64-71 | Lumbar 7 (PL) (11/9) | 18.0-19.5 | 20.5-23.0 |
| #34 (9/7) | 56-58 | 65-69 | Sacrum (PL) (8/7) | 30.0-32.5 | 33.5-38.0 |
| **Mandible measurements** | | | **Limb elements** | | |
| #1 (22/14) | 103-128 | 135-151 | Scapula (HS) (8/8) | 101-126 | 129-142 |
| #4 (21/14) | 85-112 | 113-132 | Humerus (GL) (14/15) | 137-151 | 153-179 |
| #6 (11/8) | 87-113 | 118-131 | Ulna (GL) (11/10) | 140-167 | 170-203 |
| #7 (20/14) | 64-74 | 76-84 | Radius (GL) (11/10) | 123-141 | 145-156 |
| #17 (22/14) | 17-20 | 25-27 | Metacarpal II (GL) (19/12) | 38-49 | 50-58 |
| #19 (21/14) | 18-21 | 26-27 | Metacarpal III (GL) (17/13) | 44-58 | 59-69 |
| **Vertebrae measurements** | | | Metacarpal IV (GL) (16/16) | 45-58 | 59.5-70.0 |
| Cervical 1 (LAd) (17/13) | 12-14 | 15-17 | Metacarpal V (GL) (12/14) | 41.0-49.5 | 50.5-60.0 |
| Cervical 2 (LCDd) (14/13) | 37-44 | 45-52 | | | |
| Cervical 3 (PL) (10/14) | 21-24 | 25-30 | Femur (GL) (13/12) | 142-164 | 167-182 |
| Cervical 4 (PL) (15/13) | 20-22 | 24-27 | Tibia (GL) (11/13) | 139-158 | 159-177 |
| Cervical 5 (PL) (9/7) | 18-20 | 21-25 | Fibula (GL) (6/4) | 135-148 | 154-157 |
| Cervical 6 (PL) (7/8) | 16-18 | 19-21 | Calcaneus (GL) (26/24 | 35.0-40.5 | 41.5-51.0 |
| Cervical 7 (PL) (6/5) | 16-18 | 19-20 | Talus (GL) (10/7) | 21-58 | 25-27 |
| Thoracic 3 (PL) (4/6) | 15.0-15.5 | 16.5-17.0 | Metatarsal II (GL) (17/15) | 49-58 | 59-68 |
| Thoracic 12 (PL) (7/8) | 17-19 | 20.5-21.0 | Metatarsal III (GL) (20/21) | 55.0-66.5 | 68-75 |
| Thoracic 13 (PL) (8/8) | 19.0-21.5 | 22.5-24.0 | Metatarsal IV (GL) (16/13) | 58-68 | 69-78 |
| Lumbar 1 (PL) (9/8) | 20-22 | 23.5-26.0 | Metatarsal V (GL) (17/16) | 48-59 | 60-67 |
| | | | | | |
| | | | **Shoulder height estimate (cm) (91/119)** | **35-50 (avg. 44)** | **47-59 (avg. 52)** |

***Table 1*** *Selected osteometric characteristics of Northwest Coast North American dogs, c. 4,000-500 BP. Measurement codes from von den Driesch (1976), figures from Crockford (1997a). Note that sample sizes (n) are different for each dimension and for each type (small/large)*

without benefit of written records. Can the same be said of the hairless Mexican *xoloitzcuintle* (Valadez, this volume)? Hairlessness inherits as a *SEMI-lethal dominant* trait (Wilcox & Walkowicz, 1989: 75), as does tailessness in Manx cats (Weaver & Hedrick, 1992). Semi-lethal dominant means that affected individuals carry only one dominant gene (eg, "Hh"). Foetal offspring that carry two dominant genes (eg, "HH") almost always fail to develop, while offspring without the dominant gene for the trait (eg, "hh") are unaffected – in this case, they have hair. Since offspring with two dominant genes rarely survive, lineages of hairless dogs cannot "breed true". Only one hairless parent per mating need be used to perpetuate the trait: the other parent may be any type. This explains why all modern "Xolos" tested by Vilà, Maldonado & Wayne (1999) possessed only "modern" dog mtDNA sequences – the hairless trait by itself is not an indicator of a pure lineage. However, human protection was probably necessary for the survival of the relatively vulnerable hairless dogs. If hairless dogs were valued for their hairless quality alone, then the "Xolo" could perhaps also be considered a rare indigenous breed (since there may well have been attempts made at deliberate propagation, even if they were not often successful).

In contrast, the gene governing shortleggedness in the other Mexican variety (Fig. 16), the *Tlalchichi* (Valadez, this volume), probably inherits as a simple dominant trait, as achondroplastic dwarfism does in humans (Weaver & Hedrick, 1992). This mutation probably did not require deliberate breeding in order to become regularly expressed in the population, a reminder that not all skeletal variants represent true dog husbandry, even if they occur often (see also Clark, 1997b).

**The complex evolutionary history of dogs and wolves**

The significance of extensive sympatry and hybridisation

The evolutionary relationship between wolf ancestor and dog descendant is not nearly as simple as has recently been portrayed by authors of genetic analyses (Vilà *et al.*, 1997; Vilà, Maldonado & Wayne, 1999; Wayne, 1993). Wolves and humans shared territories in many regions of the world for thousands of years before dogs evolved (Olsen, 1985; Hall & Sharp, 1978). The relationship between humans and dogs that began when dogs became a unique entity must have changed the historical relationship that had previously existed between humans and wolves. In addition, the human/dog relationship altered the course of the natural evolutionary ancestor/descendant relationship that would normally have developed between wolves and dogs. Genetic data, to be meaningful, must be interpreted within a framework of this unique and complex evolutionary history. Not surprisingly, the story for dogs has proven to be as complicated and hotly debated as the quest for genetic evidence of human origins and migrations (Brown *et al.*, 1998; Krings *et al*, 1998).

Genetic analyses have so far provided disappointingly little clarification on relationships within and between dog and wolf populations. Extant wolves overall show remarkably similar levels of within-species genetic diversity as dogs, and some haplotypes are found in both species. There are neither breed-specific haplotypes for dogs nor subspecies-specific haplotypes for wolves, despite a very large number of animals tested from different geographic areas. Koop *et al.'s* study (this volume: fig.6) identifies eight more-or-less equally distinct mtDNA haplotypes, shared by most dogs and wolves, that differ from each other by at least 2%

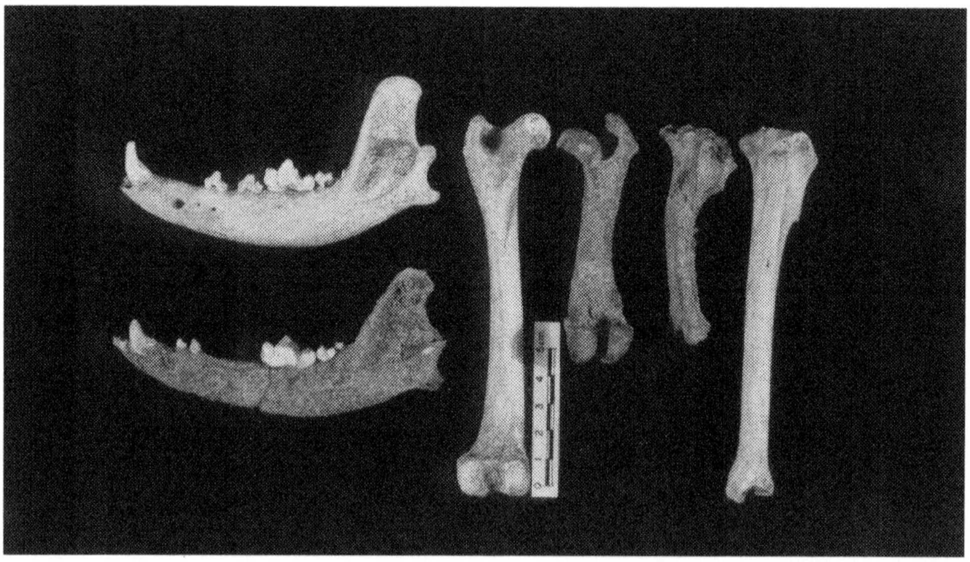

***Figure 16*** *Skeletal elements of Meso-American Common dog vs. short-legged Tlalchichi (courtesy R. A. Valadez, Mexico City) – note that mandibles of both types are about equal in size, although the femur and tibia of the Tlalchichi are only about ½ the length of the Common dog.*

(compare this to the 1.1% sequence difference between one lineage of brown bear and the polar bear (Talbot & Shields, 1996)). In some respects, this analysis has provided more questions than answers. Although some of the difficulty may result from the use of genetic analysis methods more appropriate for comparing wild taxa than for examining diversity within and between closely related species (cf, the *median network* approach used for analysis of cattle and human mtDNA samples (Bailey *et al.*, 1996; Brown *et al.*, 1998)), I suggest that most of the problem comes from a reliance on samples from living animals.

I believe that there is a good reason for the confusing results generated in canid phylogenetic studies. In contrast to extant cattle and sheep populations, where genetic analysis shows evidence of geographic and phylogenetic structure (Bradley *et al.*, 1998; Hiendleder *et al.*, 1998), using tissue samples from living animals is an inappropriate strategy for studying phylogenetic relationships of dogs and wolves because of the long-standing potential for hybridisation between them. While Koop *et al.* (this volume) discuss briefly the effect that hybridisation might have had on the interpretation of their mtDNA analysis results, I believe that further comment on this phenomenon is warranted.

The wolf is the only ancestral species of a domestic animal that has survived with an extensive distribution and relatively high population numbers into the 20th century (Mech, 1970). The last aurochsen, for example, disappeared about 500 years ago and were probably rare long before that time; while wild pigs and sheep still exist, their distributions are scattered and population numbers low (Reid, 1984). As mentioned previously (Fig. 10), by *c.* 7,000BP dogs - along with their human companions - had dispersed throughout the entire Northern Hemisphere, invading territory already occupied by various subspecies of wolf. Regions that are now sparsely inhabited, or even uninhabited (such as the Brooks Peninsula on the west coast of Vancouver Island, British Columbia), often supported substantial prehistoric human populations (Haggarty & Inglis, 1997). Almost everywhere there were people prehistorically, there were dogs – lots of dogs. Dog remains often comprise a large proportion of the vertebrate fauna recovered from prehistoric archaeological sites and there is some evidence to suggest that population control measures for dogs were practised in some areas (Crockford, 1997a; Schwartz, this volume).

In contrast, we know very little about population numbers of wolves prehistorically; dogs may have outnumbered local wolf populations in some areas. These populations of dogs were also largely "uncontrolled" for almost all of their history. Little evidence exists for confining dogs in ancient times, except for noble classes in later eras (Clutton-Brock, 1981). Dogs and wolves must have had opportunities for interspecies encounters wherever local populations overlapped. Therefore, the potential for hybridisation with their ancestral species has been substantially higher for dogs, over a much longer period of time, than for any other domesticate. Has this hybridisation, however, been reciprocal (affecting both species) or largely unidirectional (affecting one species predominantly)?

Offspring inherit the mtDNA of their mother, regardless of the genotype of the father, and females in turn pass their mother's mtDNA to their own offspring. Although this maternal inheritance pattern gives a one-sided picture of genetic relationships, it is often very useful for identifying unidirectional hybridisation events (Arnold, 1997; Carr & Hughes, 1993; Dowling & DeMarais, 1993). For example, in several studies of hybridisation of domestic stock with wild species (eg, reindeer & caribou; cattle & bison), the domestic species is almost always the female partner because domestic males are seldom large enough, or aggressive enough, to interest wild females (Cronin *et al.*, 1995; Polziehn *et al.*, 1995). Such essentially unidirectional hybridisation also occurs between closely related wild species pairs (Arnold, 1997; Harrison, 1993). One species - usually the smaller or less aggressive one - is most often the female partner, as between coyotes and wolves (Lehman *et al.*, 1991). This pattern of hybridisation can produce descendant populations of morphologically and behaviourally distinct taxa that possess mtDNA haplotypes of a closely related species (Dowling & DeMarais, 1993).

If recent hybridisation of male wolves with female dogs explains a few wolf mtDNA haplotypes that group with the major modern dog clade (eg, Vancouver Island wolves, Koop *et al.*, this volume), how would ancient hybridisation events involving prehistoric male wolves and prehistoric female dogs from various distinct ancestral lineages affect this picture? In addition to the possible explanations provided by Koop *et al.* for the phylogenetic results of their study, another that must be considered is this: that wolf mtDNA haplotypes are similar to dog haplotypes because essentially unidirectional, introgressive hybridisation has gone on for so long that extant wolf populations now have predominantly dog mtDNA.

I suggest that *occasional* hybridisation of dog females with wolf males must have occurred, over thousands of years and throughout all wolf territory. Severe declines in population sizes of wolves worldwide (Mech, 1970; Boitani, 1995) and subsequent population rebounds (Clutton-Brock & Kitchener, this volume; Mech, 1995), could potentially have left most existing wolves with a confusing mixture of mtDNA haplotypes derived from distinct ancestral dog lineages - even if they still look and act like wolves. While such a situation is unprecedented, there is good reason for this to be so: the long, human-mediated history of extensive sympatry between these species is unique in the animal kingdom (*cf*, domestic and wild cats, Daniels *et al.*, 1998).

Consider first of all the potential for recent hybridisation events in the living populations of canids sampled for mtDNA. For dogs, registered purebreds (or crosses between them) have been used exclusively in all recent molecular studies (Koop *et al.*, this volume). These dogs are, by definition, animals whose breeding has been tightly controlled for many generations. Such dogs would not be allowed to produce wolf hybrids that were subsequently used for breed production. Therefore, we would not expect to find genes of modern wolves in this group. Indeed, this is one good reason for using purebred dogs to represent "dogs". Not surprisingly, one very distinct mtDNA haplotype is common to most of this group (Koop *et al.*, this volume), including most "wolf-like" breeds.

The integrity of samples used to represent "wolves" in recent genetic studies, however, is less clearly defined. The example presented by Clutton-Brock and Kitchener (this volume) - that some wolves in museum collections may actually be wolf/dog hybrids - is a reminder that biologists and archaeozoologists at museums often assume, as do molecular biologists, that animals identified as wolves by field biologists are indeed wolves suitable for taxonomic and systematic research. However, it is a well recognised fact that even first generation hybrids may not be distinguishable from pure wolves by experienced personnel (Steinhart, 1995; Yates, this volume), whether they are living animals, carcasses or skeletons. Hybrids are more likely to be shot as nuisance animals, or to be hit by vehicles, because their reduced fear of people often puts them in close proximity to human populations (Steinhart, 1995). These animals thus have a much higher likelihood of becoming museum specimens and sources of genetic samples than animals living in remote regions.

For example, during the dog symposium "show and tell" session, Bonnie Yates (US Fish & Wildlife Forensic Laboratory) identified an Alaskan "wolf" skeletal specimen in the University of Victoria collection as a potential hybrid animal (Yates, this volume; Fig. 12). She suspects this animal was part of a pack culled by government wildlife officers, who distributed the carcasses to various collections (including the one at her institution). Skeletons from this donation "event" that now reside in her collection also possess hybrid characteristics. In another example (Koop *et al.*, this volume), three wolf tissue specimens donated by government wildlife personnel were taken from animals shot in a region where hybridisation with dogs had been previously documented (Vancouver Island, BC). On analysis, two of these three samples were found to possess purebred dog mtDNA haplotypes and were clearly of recent hybrid origin, apparently without anyone suspecting the animals of being hybrids. Vilà *et al.* (1997) identified a putative hybrid in their study as well (haplotype W6).

The Vancouver Island example given above suggests that since most modern purebred dogs are genetically discernible from wolves, very recent hybridisation events of purebred dog females with wolves are often readily distinguished by their "purebred" mtDNA haplotypes. However, modern purebreds are not the only representatives of their species: feral dogs and unconfined dogs of indigenous heritage in all countries that roam and breed at will are just as valid members. These uncontrolled animals are far more likely to interbreed with wolves than closely monitored purebreds, yet except for the data provided by Koop *et al* (this volume) for NA indigenous dogs, we have no record of their mtDNA haplotypes – with one exception, which I discuss later.

Vilà & Wayne (1999) have recently argued that hybridisation between dogs and wolves is minimal and has had no appreciable effect on wolf populations worldwide, a conclusion that I contend is premature. Although anecdotal information abounds, true scientific study on wolf/dog hybridisation is conspicuously absent. Only one systematic study on the nature and frequency of hybridisation has been undertaken (Boitani, 1983; D. Mech, pers. comm., 1998), even though for decades substantial numbers of wolf/dog hybrids have been deliberately generated in both North America and Europe. Although no accurate figures are available, Randall Lockwood of the US Humane Society (pers. comm., 1999) estimates that the number of wolf/dog hybrids deliberately produced each year in the US alone is between 10,000 and 100,000 animals. Unfortunately, one fact is clear from all accounts of hybrid behaviour: these animals hate to be confined and many either escape, or are released, into the wild (Lockwood, pers. comm., 1999; Steinhart, 1995). Many of these wolf hybrids will possess dog mtDNA haplotypes. Despite this, no studies at all seem to have been done on the fate of known hybrids living a wild existence (although, see the Russian literature listed in Walker, this volume).

While Vilà and Wayne (1999) suggest that reproductive timing differences between dogs and wolves preclude mating of female dogs and male wolves, this conclusion does not fit the facts they provide. They state that in dog females of most breeds, oestrus occurs in spring and fall, although it may occur in any month, while grey wolf males produce sperm only during the late winter/early spring. Rather than indicating "timing incompatibility," these patterns point to considerable potential overlap in reproductive receptivity for these species during the late winter/early spring period. By suggesting that physiological factors preclude mating of dog females with wolf males, Vilà and Wayne ignore a whole body of literature that suggests this may be the easiest way to produce a hybrid animal (see literature listed in Walker, this volume). Iljin (1941: 363), for example, reported no difficulties mating a female dog with a male wolf. He also cites Pliny (AD 23-79) as stating that the Gauls staked bitches to trees in order to have them mated with wolves, a practice also described for Inuit and North American Plains Indians (Allen, 1920; Clutton-Brock & Kitchener, this volume).While an assumption is clearly being made by many reading these reports that the purpose of staking bitches was to produce hybrid offspring, it is perhaps more likely that it was used to lure wolves in for an easy kill during copulation, as described in Steinhart (1995: 300).

In reality, therefore, the practice of staking bitches out for wolves may only rarely have resulted in the production of hybrid offspring, as Haag (1948) has already suggested. Nevertheless, given the unpredictable behaviour of most modern hybrids (Hope, 1993; Steinhart, 1995), such offspring would surely have been just as difficult to manage in prehistoric and historic times as they are today, making it doubtful that hybridisation was routinely encouraged. It is perhaps more likely that unconfined dog bitches were occasionally impregnated by dispersing male wolves, but would return home to give birth. Offspring produced from such ancient hybridisation events would almost certainly crave freedom as strongly as modern hybrids do and would chose a wild existence at their first opportunity. It is the fate of such hybrid offspring that is of real concern.

Carr and Harris (1993: 337), in their discussion of introgressive hybridisation between North American deer species (genus *Odocoileus*), state that in order for mtDNA to move from one species to another "female hybrids must be fertile and must change species, that is, the hybrids must leave their natal maternal group and be absorbed into the paternal species." The issue thus becomes a question of whether female hybrid offspring of female dogs and male wolves, raised as dogs, are later able to join wild wolf populations and successfully produce backcrossed offspring that are raised as wolves. The argument against this, as presented by Vilà & Wayne (1999), depends on the premise that even if dispersing hybrid females were accepted into a pack, they would not be allowed to reproduce. They cite data based on a single study of zoo animals that only the most dominant or "alpha" female of a grey wolf pack produces pups (Mech, 1970). This "alpha female-exclusive" pattern of reproduction, however, was never a proven rule for wolves (Mech, 1970) and has since been shown to be an individual phenomenon at best. Evidence from a study of many packs of wolves in Alaska demonstrates that production of multiple litters (two-three) is relatively common (Meier *et al.*, 1995).

Given this perspective on grey wolf reproduction, it is possible to suggest that a hybrid female who lived long enough to find a wolf pack that would accept her, or a dispersing wolf male to join forces with, could very well have the necessary assistance to successfully raise backcrossed offspring in the wild. I'm not suggesting that this necessarily is, or ever was, a common phenomenon. Introgressive hybridisation, however, need occur only very occasionally to move mtDNA from dogs into wolves. Once a female wolf hybrid with dog mtDNA became successfully established as a breeding animal amongst wolves, whether 50 years ago or 5,000 years ago, dog mtDNA haplotypes would have increased dramatically within that local wolf population over subsequent generations, as seen in several populations of NA deer (Carr & Harris, 1993).

For example, in the only study on the genetic composition of modern free-ranging dogs available, Vilà & Wayne (1999) report unpublished data on Spanish feral dogs discovered to have mtDNA haplotypes found otherwise only in Iberian grey wolves. The authors interpret these results as indicating mating of female wolves with male feral dogs (in other words, the dogs have wolf mtDNA). I presume this conclusion was reached because the authors expected feral dogs to have the same mtDNA haplotypes as modern purebreds and because the possibility of mating female dogs with male wolves had already been discounted on physiological grounds. However, I suggest it is more likely that the Iberian wolves have feral dog mtDNA, although a haplotype distinct from modern purebreds. As previously discussed, female dogs are much more likely to mate successfully with male wolves on the rare occasions when crosses do occur and their hybrid offspring are more likely to leave their maternal group. Also, modern purebred mtDNA haplotypes are clearly only one pattern of several found in dogs.

Therefore, I contend the possibility exists that rare but recurring prehistoric hybridisation events involving female dogs could have moved distinctive dog mtDNA haplotypes into local wolf populations in many locations throughout the Northern Hemisphere. Thus, recent hybridisation involving dogs of modern purebred ancestry would account for only a small proportion of the possible effects of introgression of dog genes into wolf populations. Although this is not the only possible explanation for recent genetic results, or a conclusive one, it is a possibility that has not been considered before now. I suggest, however, that we cannot construct accurate phylogenetic trees of dogs and wolves without taking the historical factors of their long and extensive sympatric relationship into account.

We *must* know what the genetic composition of wolf populations was before dogs existed if we hope to untangle the mess of dog genetic history: multiple origins and widespread dispersal, accidental and deliberate lineage crosses, and interspecific hybridisation with wolves. Subfossil remains with the potential to fill in essential gaps in our knowledge of wolf mtDNA include samples of small Pleistocene wolves that lived in Israel during the period 20,000 to 30,000 BP (Dayan 1994), large northern European wolves pre-dating the Magdalenean period (Musil, this volume), prehistoric dingo (Golan 1980), pre-Jomon period Japanese wolves (Ishiguro *et al.*, this volume), and pre-domestication period Chinese wolves (Olsen, 1985). Subfossil specimens of NA wolves that predate the bulk of human migration into the New World, such as those reported from Natural Trap Cave, Wyoming (*c.* 14,000 to 20,000 BP), are also prime candidates for ancient DNA testing (Walker & Frison, 1982).

We need to be wary, however, of using archaeological wolf remains as sources of mtDNA. Remains of definitive hybrids are rare in archaeological sites worldwide (Olsen, 1985), although this record may be biased by analysis methods and reporting practices. Walker and Frison (1982), for example, present a thorough analysis of rare examples of hybridisation from sites in Wyoming (the US northern Plains region). A few canid skulls from many sites in this area appear to be wolf/dog hybrids, although they are distinctly wolf-like rather than dog-like: all had been previously identified as wolves. The authors suggest that other "wolf" remains recovered archaeologically, in this and other regions, may also represent hybrids and suggest that cranial material of all "wolves" should be re-examined.

With a few rare exceptions, therefore, it seems the archaeological record of canids worldwide consists of many remains of morphologically typical dogs (even in arctic regions, where extensive natural hybridisation with wolves has been suggested to occur), and a few "wolves" (which may or may not be hybrids). The few "wolf" remains recovered (eg, Olsen, this volume) may represent wild animals hunted for ritual purposes or furs, hybrid nuisance animals dispatched for safety, or hybrid animals accepted and utilised as dogs. Obviously, wolf-sized canids recovered from archaeological sites must be examined more closely for their potential as ancient hybrids and the context of their remains analysed carefully.

If Walker and Frison (1982) are correct, however - that further analysis of "wolves" recovered from sites on the North American Plains (New Mexico to Alberta, Canada) will reveal them to be hybrids - then some particular aspect of prehistoric life in these areas must have substantially increased the incidence of natural hybridisation. Clearly more "wolf" remains are recovered from these sites than is usual. Mass hunting strategies for large, herd animals (ie, bison) may be the answer. Bison "kills" of several animals may have generated enough surplus meat to have drawn local wolves closer to habitations that usual, placing them in close proximity to indigenous dogs. Occasional hybridisation would have been a natural consequence of this close association between indigenous dogs and wolves. It is doubtful if humans could have prevented interbreeding; they could only have "managed" the hybrid population by killing truly dangerous or threatening individuals and perhaps, taking in behaviourally acceptable animals and treating them as dogs. Compare this situation to that described by Musil (this volume) for Magdelanean horse hunters: since dogs did not yet exist, a similar situation appears to have set up the conditions necessary for protodomestication to occur.

One last point that perhaps deserves mentioning is the cultural implications of long-standing hybridisation between dogs and wolves. Reports of unpredictable, often inappropriately aggressive behaviour demonstrated by wolf/dog hybrids (Hope, 1993; Steinhart, 1995; literature listed in Walker, this volume) suggest that these animals are potentially dangerous to humans and their livestock. If this is true now, it must also have been true for hybrids generated hundreds or thousands of years ago and suggests the possibility that wolf hybrids have been a problem for human populations for a very long time. Unfortunately, then as now, the hybrid ancestry of attacking animals would rarely have been apparent. As Steinhart (1995: 313-314) has pointed out, however, certain details provided on several incidents of both recent and historically documented "wolf" attacks on humans indicate they may well have involved wolf/dog hybrids. It is possible, therefore, that there is a legitimate basis for the entrenched fear and condemnation of "wolves" traditionally held by Europeans (*cf*, Boitani, 1995).

I suggest we must consider the possibility that hybridisation with dogs could have affected the mtDNA composition of wolf populations significantly over the last 10,000 years. This does not mean that all wolves are "hybrids" or that they are any less deserving of conservation efforts. As for other species, when introgressive hybridisation is so rare that the recipient population remains morphologically, physiologically and behaviourally distinct, it is the height of genetic chauvinism to call them hybrids on mtDNA evidence alone (Brownlow, 1996; Dowling & DeMarais, 1993; Nowak, 1995). Hybridisation as an adaptive evolutionary strategy has been shown to be neither rare nor "unnatural" in animals (Arnold, 1997; Endler, 1998; Harrison, 1993), and as more genetic studies are done, additional evidence of this phenomenon is sure to accumulate (Brownlow, 1996).

The possibility of long-standing introgressive hybridisation with dogs does mean, however, that we cannot use mtDNA samples from living wolves, or those from archaeological contexts, in phylogenetic studies. Dogs have a unique evolutionary history and wolves, as a consequence, do also. Additional genetic and morphological investigations are needed for both species, in order to give due consideration to the impact that humans, and human activities, have had on their evolution.

**Acknowledgments**

I extend my sincere thanks to the colleagues who contributed to this volume, who provided me with such fascinating examples of important issues in dog history that I could discuss in more detail. I especially thank Ashley Byun, Juliet Clutton-Brock, Andrew Kitchener, Sandra Olsen, Danny Walker, and Bonnie Yates for reviewing this essay and offering their comments and suggestions.

**References**

Allen, G. M. (1920). Dogs of the American aborigines. *Harvard University Museum of Comparative Zoology Bulletin* **63**: 431-517.

Amoss, P. T. (1993). Hair of the dog: unraveling pre-contact Coast Salish social stratification. In (A. Mattina & T. Montler, Eds) *American Indian Linguistics and Ethnography in Honor of Laurence C. Thompson*. Missoula: University of Montana, pp. 3-35.

Arnold, M. L. (1997). *Natural Hybridization and Evolution*. Oxford University Press.

Bailey, J. F., Richards, M. B., Macaulay, V. A., Colson, I. B., James, I. T., Bradley, D. G., Hedges, R. E. M. & Sykes, B.C. (1996). Ancient DNA suggests a recent expansion of European cattle from diverse wild progenitor species. *Proceedings of the Royal Society of London* B **263**, 1467-1473.

Botaini, L. (1983).Wolf and dog competition in Italy. *Acta Zoologica Fennica* **174**: 259-264.

Botaini, L. (1995). Ecological and cultural diversities in the evolution of wolf-human relationships. In (L. N. Carbyn, S. H. Fritts & D. R. Seip, Eds) *Ecology and Conservation of Wolves in a Changing World.* Edmonton: Canadian Circumpolar Institute Occasional Publication **35**, pp. 3-11.

Bradley, D. G., Loftus, R. T., Cunningham, P. & MacHugh, D. E. (1998). Genetics and domestic cattle origins. *Evolutionary Anthropology* **6,** 79-86.

Brothwell, D., Malega, A. & Burleigh, R. (1979). Studies on Amerindian dogs, 2: variation in early

Peruvian dogs. *Journal of Archaeological Science* **6**: 139-161.

Brown, M. D., Hosseini, S. H., Torroni, A., Bandelt, H.-J., Allen, J. C., Schurr, T. G., Scozzari, R., Cruciani, F. & Wallace, D. C. (1998). mtDNA haplogroup X: an ancient link between Europe/Western Asia and North America? *American Journal of Human Genetics* **63**: 1852-1861.

Brownlow, C. A. (1996). Molecular taxonomy and the conservation of the red wolf and other endangered carnivores. *Conservation Biology* **10**: 390-396.

Cannon, A., Schwarcz, H. P. & Knyf, M. (1999). Marine-based subsistence trends and the stable isotope analysis of dog bones from Namu, British Columbia. *Journal of Archaeological Science* **26**: 399-407.

Carr, S. M. & Hughes, G. A. (1993). Direction of introgressive hybridization between species of North American deer (*Odocoileus*) as inferred from mitochondrial cytochrome b sequences. *Journal of Mammalogy* **74**: 331-342.

Clark, G. R. (1995).*The kuri in prehistory: a skeletal analysis of the extinct Maori dog*. MA thesis, University of Otago, New Zealand.

Clark, G. R. (1997a). Osteology of the Kuri Maori: the prehistoric dog of New Zealand. *Journal of Archaeological Science* **24**: 113-126.

Clark, G. R. (1997b). Anthropogenic factors and prehistoric dog morphology: a case study from Polynesia. *Archaeology of Oceania* **32**: 124-130.

Clutton-Brock, J. (1979). The definition of a breed. In (M. Kubasiewicz, Ed) *Archaeozoology I*. Szczecin, Poland: Agricultural Academy.

Clutton-Brock, J. (1981). *Domesticated Animals From Early Times*. London: British Museum (Natural History)/ Heinemann.

Clutton-Brock, J. (1995). Origins of the dog: domestication and early history. In (J. Serpell, Ed) *The Domestic Dog: Its Evolution, Behaviour and Interactions with People.* Cambridge: Cambridge University Press, pp.8-20.

Clutton-Brock, J. & Hammond, N. (1994). Hot dogs: comestible canids in Preclassic Maya culture at Cuello, Belize. *Journal of Archaeological Science* **21**: 819-826.

Coppinger, R. & Schneider, R. (1995). Evolution of working dogs. In (J. Serpell, Ed) *The Domestic Dog: Its Evolution, Behaviour and Interactions with People.* Cambridge: Cambridge University Press, pp. 21-47.

Crockford, S. J. (1997a). *Osteometry of Makah and Coast Salish Dogs*. Burnaby, BC: Simon Fraser University Archaeology Press **22**.

Crockford, S. J. & Pye, C. J. (1997b). Forensic reconstruction of prehistoric dogs from the Northwest Coast of North America. *Canadian Journal of Archaeology* **21**: 149-153.

Crockford, S. J. (in press). Animal domestication and heterochronic speciation: the role of thyroid hormone. In (K. McNamara & N. Minugh-Purvis, Eds) *Human Evolution Through Developmental Change.* Baltimore: Johns Hopkins University Press.

Cronin, M.A., Renecker, L., Pierson, B. J. & Patton, J. C. (1995). Genetic variation in domestic reindeer and wild caribou in Alaska. *Animal Genetics* **26**: 427-434.

Daniels, M. J., Balharry, D., Hirst, D., Kitchener, A. C. & Aspinall, R. J. (1998). Morphological and pelage characteristics of wild living cats in Scotland: implications for defining the "wildcat". *Journal of the Zoological Society of London* **244**: 231-247.

Davis, S. J. M. & Valla, F. R. (1978). Evidence for domestication of the dog 12,000 years ago in the Natufian of Israel. *Nature* **276**, 608-610.

Dayan, T. (1994). Early domesticated dogs of the Near East. *Journal of Archaeological Science* **21**, 633-640.

Dowling, T. E. & DeMarais, B. D. (1993). Evolutionary significance of introgressive hybridization in cyprinid fishes. *Nature* **362**: 444-446.

Driesch, A., von den (1976). *A Guide to the Measurement of Animal Bones from Archaeological Sites.* Peabody Museum of Archaeology and Ethnology, Harvard University: Peabody Museum Bulletin **1**.

Endler, J. A. (1998). The place of hybridization in evolution – a review of *Natural Hybridization and Evolution* (1997) by M. L. Arnold. *Evolution* **52**: 640-644.

Erlandson, J. M. (1994). *Early Hunter-Gatherers of the California Coast*. New York: Plenum Press.

Gentry, A., Clutton-Brock, J. & Groves, C. P. (1996). Case 3010. Proposed conservation of usage of 15 mammal specific names based on wild species which are antedated by or contemporary with those based on domestic animals. *Bulletin of Zoological Nomenclature* **53**: 28-37.

Gollan, K. (1980). *Prehistoric dingo*. Ph.D. Thesis. Australian National University, Canberra.

Haag, W. G. (1948). *An Osteometric Analysis of Some Aboriginal Dogs*. Lexington: University of Kentucky Reports in Anthropology **7**.

Haggarty, J. C. & Inglis, R. I. (1997). Archaeology of Brooks Peninsula. In (R. J. Hebda & J. C. Haggarty, Eds) *Brooks Peninsula: An Ice Age Refugium on Vancouver Island.* Victoria, BC: Ministry of Environment, Lands and Parks Occasional Paper **5**, pp. 14.1-14.27.

Hall, R. L. & Sharp, H. S. (1978). *Wolf and Man: Evolution in Parallel.* New York: Academic Press.

Harrison, R. G. (1993). *Hybrid Zones and the Evolutionary Process.* New York: Oxford University Press.

Hiendleder, S., Mainz, K., Plante, Y. & Lewalski, H. (1998). Analysis of mitochondrial DNA indicates that domestic sheep are derived from two different ancestral sources: no evidence for contributions from urial and argali sheep. *Journal of Heredity* **89**: 113-120.

Hope, J. (1994).Wolves and wolf hybrids as pets are big business – but a bad idea. *Smithsonian* (June): 34-40.

Howay, F.W. (1918). The dog's hair blanket of the Coast Salish. *Washington Historical Quarterly* **9:** 83-92.

Iljin, N.A. (1941). Wolf-dog genetics. *Journal of Genetics* **42**: 359-414.

Krings, M., Stone, A., Schmitz, R. W., Krainitzki, H., Stoneking, M. & Pääbo, S. (1997). Neanderthal DNA sequences and the origin of modern humans. *Cell* **90**, 19-30.

Lehman, N., Eisenhawer, A., Hansen, K., Mech, L. D., Peterson, R. O., Gogan, P. J. & Wayne, R. K. (1991). Introgression of coyote mitochondrial DNA into sympatric North American gray wolf populations. *Evolution* **45**: 104-119.

Matisoo-Smith, E., Allen, J. S., Wilton, A. N., Laycock, J. L., Ladefoged, T. N. & M. A. Allen (1998). The Austronesian origins of the dingo and other Pacific dogs? Evidence from mitochondrial DNA. Paper delivered at the 8th Congress of the International Council for Archaeozoology, Victoria.

Mech, L. D. (1970). *The Wolf.* Garden City, NY: Natural History Press.

Mech, L. D. (1995). The challenge and opportunity of recovering wolf populations. *Conservation Biology* **9**, 270-278.

Meier, T. J., Burch, J. W., Mech, L. D. & Adams, L. G. (1995). Pack structure and genetic relatedness among wolf packs in a naturally-regulated population. In (L. N. Carbyn, S. H. Fritts & D. R. Seip, Eds). *Ecology and Conservation of Wolves in a Changing World.* Edmonton: Canadian Circumpolar Institute Occasional Publication **35**, pp. 293-302

Morell, V. (1997). The origin of dogs: running with the wolves. *Science* **276**, 1647-1648.

Morey, D. F. & Wiant, D. M. (1992). Early Holocene domestic dog burials from the North American midwest. *Current Anthropology* **33**: 224-229.

Nowak, R. M., Phillips, M. K., Henry, V. G., Hunter, W. C. & Smith, R. (1995).The origin and fate of the red wolf. In (L. N. Carbyn, S. H. Fritts & D. R. Seip, Eds). *Ecology and Conservation of Wolves in a Changing World.* Edmonton: Canadian Circumpolar Institute Occasional Publication **35**, pp. 409-415.

Olsen, S. J. (1985). *Origins of the Domestic Dog: The Fossil Record.* Tucson: University of Arizona Press.

Olsen, S.J & Olsen, J. W. (1977). The Chinese wolf, ancestor of New World dogs. *Science* **197**: 433-435.

Polziehn, R.O., Strobeck, C., Sheraton, J. & Beech, R. (1995). Bovine mtDNA discovered in North American bison populations. *Conservation Biology* **9**: 1638-1643.

Reed, C. A. (1984). The beginnings of animal domestication. In (I. L. Mason, Ed) *Evolution of Domesticated Animals.* London: Longman Co., pp. 1-6.

Schulting, R. (1994). The hair of the dog: the identification of a coast Salish dog-hair blanket from Yale, B.C. *Canadian Journal of Archaeology* **18**: 57-76.

Schwartz, M. (1997). *A History of Dogs in the Early Americas*. New Haven: Yale University Press.

Steinhart, P. (1995). *The Company of Wolves*. New York: Vintage Books.

Talbot, S.L. & Shields, G. F. (1996). A phylogeny of the bears (Ursidae) inferred from complete sequences of three mitochondrial genes. *Molecular Phylogenetics and Evolution* **5**: 567-575.

Vilà, C., Maldonado, J. E. & Wayne, R. K. (1999). Phylogenetic relationships, evolution, and genetic diversity of the domestic dog. *Journal of Heredity* **90**, 71-77.

Vilà, C., Savolainen, P., Maldonado, J. E., Amorim, I. R., Rice, J. E., Honeycutt, R. L., Crandall, K. A., Lundeberg, J. & Wayne, R. K. (1997). Multiple and ancient origins of the domestic dog. *Science* **276**, 1687-1689.

Vilà, C. & Wayne, R. K. (1999). Hybridization between wolves and dogs. *Conservation Biology* **13**: 195-198.

Walker, D. N. & Frison, G. C. (1982). Studies on Amerindian dogs, 3: prehistoric wolf/dog hybrids from the Northwestern plains. *Journal of Archaeological Science* **9**: 125-172.

Wayne, R. K. (1993). Molecular evolution of the dog family. *Trends in Genetics* **9,** 218-224.

Weaver, R. F. & Hedrick, P. W. (1992). *Genetics.* Dubuque, IA: Wm. C. Brown, Publishers.

Wilcox, B. W. & Walkowicz, C. (1989). *The Atlas of Dog Breeds of the World.* Neptune City: T.F.H. Publications.

# PRELIMINARY BIBLIOGRAPHY ON DOGS AND WOLVES, STRESSING THEIR PREHISTORIC AND HISTORIC OCCURRENCE, HYBRIDISATION, AND DOMESTICATION

Danny N. Walker
Wyoming State Archaeologist Office
Department of Anthropology
University of Wyoming
P.O. Box 3431
Laramie, WY 82071-3431
USA

**Key words:** dog, wolf, hybridisation, domestication, evolution, paleontology, archaeozoology

**Abstract**

The following working bibliography has been compiled to document published references on prehistoric dogs, wolf-dog hybrids, and North American wolves and other wild canids, stressing the morphology, identification, and behavior relating to domestication. It should not, however, be considered a complete, exhaustive listing, particularly since references to North American prehistoric domestic dogs are biased by availability. Also, only wof bibliographic references that are generally, or specifically, addressed to the topics of morphology, identification, and behavior relating to domestication are listed. Missing also will be many references in a language other than English: although many European and Near Eastern literature citations discussing domestic dogs are included, these should not be considered a comprehensive listing. The working bibliography has been arbitrarily divided into 10 categories: prehistoric and early historic domestic dogs; prehistoric and early historic wolves; canid hybridization; domestication theory; folklore and general readings; osteology; morphology; behavior and ecology; biochemical identification studies; and wild canids.

**Introduction**

The following working bibliography has been compiled to document published references on prehistoric dogs, wolf-dog hybrids, and North American wolves and other wild canids, stressing the morphology, identification, and behavior relating to domestication. It should not be considered a complete, exhaustive listing of references on North American prehistoric domestic dogs, for many localities and sites are not represented (especially those in Canada). It also does not include those references where a discussion of canid remains are a minor part of a site report, or unpublished conference papers. I have attempted, however, to be fairly complete for primary research into prehistoric canids and modern wolf-dog hybrids of the United States. Many additional references may be available on these specific topics, such as site excavation reports (some listed here) that discuss prehistoric domestic canid remains from the site fauna. If there is a fault with this bibliography, it is this, because many such site reports are published in local or regional journals or exist only in the "grey literature" of cultural resource management and thus are not readily available. The prehistoric dog data available in FAUNMAP (FAUNMAP Working Group, 1994) and the Bibliography of Fossil Vertebrates (1998) list many references not included in this bibliography. I have compiled those references into a more exhaustive bibliography which will eventually be made available through the Wyoming State Archaeologist's Office web site.

Several wolf [both grey (*Canis lupus*) and red (*Canis rufus*)] bibliographies are more complete (see Aylsworth, 1997; Knee, 1995; Searching Wolf, 1998a, 1998b, 1998c; Smithsonian Institute, 1993; University of Wyoming, 1995) and should be consulted for details about wolves. Only those wolf bibliographic references that are generally, or specifically, addressed to the topics of morphology, identification, and behavior relating to domestication are listed. Additional references will be cited within these publications of a more general nature. Also missing will be many references in a language other than English. For example, there has been much study on domestic dogs from Central America and the Caribbean, but these are normally published by local journals or publishers in Spanish. Some Spanish dog references discussed by other authors are listed here, but again, have not been directly examined for additional citations. Many European and Near Eastern literature citations discussing domestic dogs are listed here, but based on availability of literature, these should not be considered a comprehensive listing.

The working bibliography has been arbitrarily divided into 10 categories: prehistoric and early historic domestic dogs; prehistoric and early historic wolves; canid hybridization; domestication theory; folklore and general readings; osteology; morphology; behavior and ecology; biochemical identification studies; and wild canids. Many of the references listed could fall into more than one of these categories, if a comprehensive indexing of these references were desired. However, for brevity purposes, I have placed the listings into the major topical category. Likewise, if both prehistoric domestic dog and wolf have been recorded from the same locality, the reference for that locality appears under the domestic dog section.

## The Working Bibliography

### Prehistoric and early historic dogs

Alaska Geographic. (1987). Dogs of the North. *Alaska Geographic* **14**.

Allen, G. M. (1920). Dogs of the American aborigines. *Bulletin of theMuseum of Comparative Zoology, Harvard University* **63,** 431-517.

Allen, G. M. (1939). Dog skulls from Uyak Bay, Kodiak Island. *Journal of Mammalogy* **20**, 336-340.

Allen, G. M. (1954). Canid remains from Pueblo Bonito and Pueblo del Arroyo. *Smithsonian Miscellaneous Collections* **124**, 385-389.

Allen, J. A. (1876). Geographical variation among North American mammals. *United States Geological and Geographical Survey of the Territories* **2**, 309-344.

Allison, M. J., Focacci, G. & Santoro, C. (1982). The pre-Columbian dog from Arica, Chile. *American Journal of Physical Anthropology* **59**, 299-304.

Altuna, J. & Mariezkurrena, K. (1992). Perros enanos in yacimentos romanos de la penninsula iberica. *Archaeofauna* **1**, 83-86. [dwarf dogs from two Roman sites in Spain; Spanish, English summary].

Amoss, P.T. (1993). Hair of the dog: unraveling pre-contact Coast Salish social stratification. In (A. Mattina & T. Montler (Eds) *American Indian Linguistics and Ethnography in Honor of Laurence C. Thompson*. Missoula: University of Montana, pp. 3-35..

Anderson, A. J. (1981). Pre-European hunting dogs in the South Island, New Zealand. *New Zealand Journal of Archaeology* **3**, 143-160.

Anderson, A. J. (1990). Kuri. In (C. M. King, Ed) *The Handbook of New Zealand Mammals*. Auckland: Oxford University Press, pp. 280-287.

Anderson, D. C. (1966). The Gordon Creek burial. *Southwestern Lore* **32**, 1-9.

Anell, B. (1969). Running down and driving of game in North America. *Studia Ethnographica Upsaliensia* **30**.

Arnesson-Westerdahl, A. (1983). A discovery of a mesolithic dog near Lake Hornborg. *Fauna och Flora* **78**, 71-72. [Swedish].

Baker, F. C. (1941). A study in ethnozoology of the prehistoric Indians in Illinois. *Transactions of the American Philosophical Society* **32**, 51-78.

Bannister, H. M. (1869). The Esquimaux dog. *American Naturalist* **3**, 522-530.

Barber, M. B., & Baroody, J. C. (1977). Analysis of vertebrate faunal remains from the Shannon site, Montgomery County, Virginia. *Quarterly Bulletin of the Archeological Society Virginia* **31**, 101-113.

Bardwell, J. (1981). *The paleoecological and social significance of the Zooarchaeological remains from Central Plains tradition earthlodges of the Glenwood locality, Mills County, Iowa*. Master's Thesis, University of Iowa.

Barnett, H. G. (1955). *The Coast Salish of British Columbia*. Eugene: University of Oregon Press.

Barton, B.S. (1803). On Indian dogs. *Alexander Tillock's Philosophical Magazine* **15**, 1-9; 136-143.

Barton, B.S. (1805). Some accounts of the different species and varieties of native American or Indian dogs. *Philadelphia Medical and Physical Journal* **1**, 3-31.

Bartosiewicz, L. (1994). Late neolithic dog exploitation: chronology and function. *Acta Archaeologica* **46**, 59-71.

Bayham, F. E. (1982). *A diachronic analysis of prehistoric animal exploitation at Ventana Cave*. Ph.D. Thesis, Arizona State University.

Beebe, B. F. (1978). Two new Pleistocene mammals from Beringia. *American Quaternary Association Abstracts, 5th Biennial meeting*, p. 159.

Beebe, B. F. (1980). A domestic dog (*Canis familiaris* L.) of probable Pleistocene age from Old Crow, Yukon Territory, Canada. *Canadian Journal of Archaeology* **4**, 161-168.

Benecke, N. (1987). Studies on early dog remains from northern Europe. *Journal of Archaeological Science* **14**, 31-49.

Benecke, N. (1994). Archaozoologische Studien zur Entwicklung der Haustierhaltung in Mitteleuropa und Skandinavien von den Anfangen bis zum ausgehenden Mittel-alter. *Schriften zur Ur- und Fruhgeschichte* 46 [more male skeletons than female in dog burials].

Berta, A. (1981). Fossil wolves, coyotes, and dogs of Florida. *Plaster Jacket* **36**, 8-24.

Billat, S. E. (1985). *A study of Fremont subsistence at the Smoking Pipe site*. Master's Thesis, Brigham Young University.

Blick, J. P. (1988). A preliminary report on the osteometric analysis of some aboriginal dogs (*Canis familiaris*) from Weyanoke Old Town, 44PG51, Prince George County, Virginia. *Quarterly Bulletin of the Archaeological Society of Virginia* **43**, 1-13.

Blumbergs, Z., Gejvall, N. G., & Verenius, C. (1974). Subboreal faunal remains from a peat-bog in Sandakra Village, in the parish of Skurup, Scania, S. Sweden. *Ossa* **1**, 24-37.

Boessneck, J. & Driesch, A. von den. (1978). Preliminary analysis of the animal bones from Tell Hesban. *Andrews University Seminary Studies* **16**, 259-304.

Bokonyi, S. (1975). Vlasac: An early site of dog domestication. In (A. I. Clason, Ed*) Archaeozoological Studies*. New York: American Elsevier, pp. 167-178.

Bokonyi, S. (1984). Animal husbandry and hunting in TacGorsium: the vertebrate fauna of a Roman town in Pannonia. *Studia Archaeologica* **VIII**. Budapest: Akademiai Kiado.

Bozell, J.R. (1988). Changes in the role of the dog in Protohistoric-Historic Pawnee culture. *Plains Anthropologist* **33**, 95-111.

Bozell, J. R. (1991). Fauna from the Hulme site and comments on central plains tradition subsistence variability. *Plains Anthropologist* **36**, 229-253.

Bozell, J. R. & Rogers, M.K. (1989). Great Oasis fauna from central Nebraska. *Central Plains Archaeology* **1**, 3-36.

Bozell, J. R., Snyder, L. M. & Falk, C.R. (1982). Descriptive analysis of unmodified vertebrate remains recovered from the Schuyler site (25CX1), Colfax County, Nebraska. *University of Nebraska, Department of Anthropology Technical Report* **82-06**, 1-110.

Brackenridge, H. M. (1904). Journal of a voyage up the river Missouri performed in eighteen hundred and eleven. In (R. G. Thwaites, Ed) *Early Western Travels, Volume 6*. Cleveland: Arthur H. Clark, pp. 22-166.

Bradbury, J. (1904). Travels in the interior of America in the years 1809, 1810, and 1811. In (R. G. Thwaites, Ed) *Early Western Travels*, *Volume 5*. Cleveland: Arthur H. Clark, pp. 22-320.

Breitburg, E. (1982). Analysis of Area A fauna. *Southern Illinois University, Center for Archeological Investigations Research Paper* **33**, 861-957.

Breitburg, E. (1983). An analysis of faunal remains from Wynnewood State Historical site, Summer County, Tennessee, and its implications to Tennessee plantation site archaeology in the central basin. *Tennessee Anthropologist* **8**, 182-198.

Breitburg, E. (1992). Faunal assemblage. *Illinois University, Center for Archeological Investigations Research Paper* **58**, 182-194.

Brightman, R. A. (1993). *Grateful prey: Rock Cree human-animal relationships*. Berkeley: University of California Press. [webbed feet in Peruvian dogs].

Brizinski, M. & Savage, H. (1983). Dog sacrifices among the Algonkian Indians: An example from the Frank Bay Site. *Ontario Archaeology* **39**, 33-40.

Brothwell, D., Malaga, A. & Burleigh, R. (1979). Studies on Amerindian dogs, 2: Variation in early Peruvian dogs. *Journal of Archaeological Science* **6**, 139-161.

Brown, C., Kelley, J., Penman, J. & Sparling, J. (1973). Faunal analysis of the Cole Quarry Archaic site. *Bulletin of the New York State Archeological Association* **58**, 25-40.

Brown, O. H. (1974). An Indian's wealth was his dogs. *The Denver Post* November 10, 45-47.

Bullen, R. P. & Benson, C. A. (1967). Cut wolf jaws from Tick Island, Florida. *Florida Anthropologist* **20**, 175-177.

Burleigh, R., Clutton-Brock, J., Felder, P. J. & Sieveking, G. de G. (1977). A further consideration of Neolithic dogs with special reference to a skeleton from Grime's Graves (Norfolk), England. *Journal of Archaeological Science* **4**, 353-366.

Burleigh, R. & Brothwell, D. (1978). Studies on Amerindian dogs, 1: Carbon isotopes in relation to maize in the diet of domestic dogs from early Peru and Ecuador. *Journal of Archaeological Science* **5**, 355-362.

Butler, E. M. & Hadlock, W. S. 1949. Dogs of the Northeastern Woodland Indians. *Bulletin of the Massachusetts Archaeological Society* **10**, 17-35.

Byrd, K. M. (1974). *Tchefuncte subsistence patterns: Morton Shell Mound, Iberia Parish, Louisiana*. Master's Thesis, Louisiana State University.

Cardinal, E. A. (1975). Faunal remains from the Zimmerman site, 1970. *Illinois State Museum Report Investigations* **32**, 73-79.

Churcher, C. S. (1993). Dogs from Ein Tirghi cemetery, Balat, Dakhleh Oasis, western desert of Egypt. In (A.Clason, S.Payne & H.-P.Uerpmann, Eds) *Skeletons in her Cupboard: Festschrift for Juliet Clutton-Brock.* Oxford: Oxbow Books, Monograph 34, pp. 39-60.

Clark, C. P. (1990). A dog burial from Isle Royale, Lake Superior: an example of household ritual sacrifice in the terminal Woodland period. *Midcontinent Journal of Archaeology* **15**, 265-78

Clark, C., Canaday, T. W. & Petersen, K. L. (1987). Domestic dog in the Dolores Archaeological Program faunal assemblage. In (K. L. Peterson & J. D. Orcutt, Eds) *Dolores Archaeological Program: Supporting Studies: Settlement and Environment.* Denver: United States Bureau of Reclamation, pp. 275-288.

Clark, G. R. (1995). *The kuri in prehistory: a skeletal analysis of the extinct Maori dog*. Masters Thesis, University of Otago, New Zealand.

Clark, G. R. (1997). Osteology of the kuri Maori: the prehistoric dog of New Zealand. *Journal of Archaeological Science* **24**, 113-126.

Clark, G. R. (1997). Anthropogenic factors and prehistoric dog morphology: a case study from Polynesia. *Archaeology Oceania* **32**, 124-130.

Clark, K. M. (1995). The later prehistoric and protohistoric dog: the emergence of canine diversity. *Archaeozoologia* **7**, 9-32.

Clark, K. M. (1996). Neolithic dogs: a reappraisal based on evidence from the remains of a large canid deposited in a ritual feature. *International Journal of Osteoarchaeology* **6**, 211-219.

Cleland, C. E. (1965a). Appendix 2: Analysis of the faunal remains of the Fatherland site. *Anthropology Papers American Museum Natural History* **51**, 96-101.

Cleland, C. E. (1965b). Faunal remains from bluff shelters in northwest Arkansas. *Arkansas Archeologist* **6**, 39-63.

Cleland, C. E. (1966). The prehistoric animal ecology and ethnozoology of the Upper Great Lakes Region. *University of Michigan Museum of Anthropology Papers* **29**.

Cleland, C. E. (1970). Comparison of the faunal remains from French and British refuse pits at Fort Michilimackinac: A study in changing subsistence pattern. *Canadian Historical Site OccasionalPapers in Archeology and History* **3**, 3-23.

Cleland, C.E. & Kearney, J. (1967). The vertebrate fauna of the Chesser Cave site, Athens County, Ohio. In (O. H. Prufer & D. H. McKenzie, Eds) *Studies in Ohio Archaeology.* Cleveland: Press of Western Reserve University, pp. 43-48.

Clutton-Brock, J. (1959). Niah's Neolithic dog. *The Sarawak Museum Journal* **9**, 13-14.

Clutton-Brock, J. (1961). Near Eastern canids and the affinities of the Natufiean dogs. *Zeitschrift fürTierzuchtung und Zuchtungsbiologis* **76**, 326-333.

Clutton-Brock, J. (1979). Mammalian remains from the Jerico Tell. *Proceedings of the PrehistoricSociety* **45**, 135-157.

Clutton-Brock, J. (1983). The skeleton of a dog from Machrins, Colonsay. *Proceedings of the Society of Antiquaries of Scotland* **111**, 276-278.

Clutton-Brock, J. (1989). A dog and a donkey excavated at Tell Brak. *Iraq* **51**, 217-224.

Clutton-Brock, J. (1996). The legacy of Iron Age dogs and livestock in southern Africa. *Azania* **XXXIX-XXX**, 161-167.

Clutton-Brock, J. & Burleigh, R. (1992). The osteology and dating of the skeleton of a large hunting dog. In (V. K. Blockley, Ed) *Archaeology of Canterbury.* HBMC & Canterbury Archaeology Trust.

Clutton-Brock, J. & Hammond, N. (1994). Hot dogs: comestible canids in Preclassic Maya culture at Cuello, Belize. *Journal of Archaeological Science* **21**, 819-826.

Clutton-Brock, J. & Noe-Nygaard, N. (1990). New osteological and C-isotope evidence on Mesolithic dogs: comparisons to hunters and fishers at Star Carr, Seamer Carr and Kongemose. *Journal of Archaeological Science* **17**, 643-653.

Colbert, E. H. (1950). The fossil vertebrates. In (E. W. Haury, Ed) *The Stratigraphy and Archaeology of Ventana Cave.* Tucson: University of Arizona Press, pp. 126-148.

Colburn, M. L. & Martin, T. J. (1985). P-Flat site, 47AS47, Manitou Island, Wisconsin: Faunal remains from the 1984 excavations. Ms. on file. Springfield: Illinois State Museum.

Colton, H. S. (1970). The aboriginal southwestern Indian dog. *American Antiquity* **35**, 153-159.
Cope, E. D. (1883). On the extinct dogs of North America. *The American Naturalist* **17**, 235-249.
Coues, E. (1893). *History of the Expedition Under the Command of Lewis and Clark.* 3 volumes. New York: Francis P. Harper.
Crellin, D. (1994). *Is there a dog in the house?: The cultural significance of prehistoric domesticated dogs in the mid Fraser region of British Columbia.* Masters Thesis, Simon Fraser University, Burnaby.
Crockford, S. J. (1997). *Osteometry of Makah and Coast Salish Dogs.* Burnaby: Simon Fraser University Archaeology Press **22.**
Crockford, S. J. & Pye, C. J. (1997). Forensic reconstruction of prehistoric dogs from the Northwest Coast of North America. *Canadian Journal of Archaeology* **21**, 149-153.
Cumbaa, S. L. (1971). A comparison of animal bones from six Indian sites on Marco Island, Florida. Ms.on file. Gainesville: Florida Museum Natural History.
Curran, C. B. (1976). Prehistoric and early historic occupation of the Mobile Bay and Mobile Delta area of Alabama with an emphasis on subsistence. *Journal of Alabama Archaeology* **22**, 61-85.
Curran, C. B., Reitz, B. & Walden, J. (1977). Chapter VI faunal remains, bone and shell artifacts. *University of Alabama Office of Archaeological Research Series* **2,** 173-192.
Cybulski, J. S. (1992). *A Greenville Burial Ground: Human Remains and Mortuary Elements in British Columbia Coast Prehistory.* Archaeological Survey of Canada Mercury Series **146.**
Dalquest, W. W., Roth, E. & Judd, F. (1969). The mammal fauna from Schultze Cave, Edwards County, Texas. *Florida State Museum Bulletin* **13**, 206-276.
Davis, L. C. (1987). Late Pleistocene/ Holocene environmental changes in the central plains of the United States: The mammalian record. *Illinois State Museum Scientific Papers* **22**, 88-143.
Davis, S. J. M. (1981). The effects of temperature change and domestication on the body size of Late Pleistocene to Holocene mammals of Israel. *Paleobiology* **7**, 101-114.
Davis, S. J. M. & Valla, F. R. (1978). Evidence for domestication of the dog 12,000 years ago in the Natufian of Israel. *Nature* **276**, 608-610.
Day, L. P. (1984). Dog burials in the Greek world. *American Journal of Archaeology* **88**, 21-32.
Day, S. P. (1996). Dogs, deer and diet at Star Carr: A reconsideration of C-isotope evidence from Early Mesolithic dog remains from the Vale of Pickering, Yorkshire, England. *Journal of Archaeological Science* **23**, 783-787.
Dayan, T. (1994). Early domesticated dogs of the Near East. *Journal of Archaeological Science* **21**, 633-640.
De Bonis, L. & Lebeaux, M. O. (1974). Discovery of a canid skull in the quaternary of Cantal: Multi-dimensional study of the discriminant features between dogs and wolves. *Mammalia* **38**, 717-728. [French, English summary].
De Grossi Mazzorin, J. & Tagliacozzo, A. (1998). Dog remains in Italy from the Neolithic to the Roman period. *Anthropozoologica* **25/26**, 429-440.
De Smet, Father P., S. J. (1905). *Life, Letters and Travels of Father Pierre-Jean De Smet, S.J. 1801-1873.* 4 volumes. (H. M. Chittenden & A. Talbot, Eds) New York: Francis P. Harper.
Degerbl, M. (1961). On a find of a Preboreal domestic dog (*Canis familiaris* L.) from Star Carr, Yorkshire, with remarks on other Mesolithic dogs. *Proceedings of the Prehistoric Society* **27**, 35-55.
Dixon, E. J. & Smith, G. S. (1986). Broken canines from Alaskan cave deposits: Re-evaluating evidence for domesticated dog and early humans in Alaska. *American Antiquity* **51**, 341-351.
Douglas, C. & Leslie, Jr., D. M. (1982). Two kinds of small Indian dogs from Black Mesa, Arizona. *Journal of the Arizona-Nevada Academy of Science* **16**, 88-90.
Driver, J. C. (1985). Zooarchaeology of six prehistoric sites in the Sierra Blanca Region, New Mexico. *University of Michigan Museum of Anthropology Technical Reports* **17.**
Duffield, L. F. (1969). The vertebrate faunal remains from the School Land I and School Land II sites, Delaware County, Oklahoma. *Bulletin of the Oklahoma Anthropology Society* **18**, 47-66.
Duffield, L. F. (1970). *Some Panhandle aspect sites: their vertebrates and paleoecology.* Ph.D. Thesis.University of Wisconsin, Madison.
Dunbar, Rev. J. B. (1918). Letters concerning the Presbyterian mission in the Pawnee country, near Bellevue, Nebraska, 1831-1849. *Kansas State Historical Society Collections* **14**, 570-689.
Dyck, I. & Morlan, R. E. (1995). The Sjovold site: A river crossing campsite in the Northern Plains. *Archaeological Survey of Canada Mercury Series Paper* **151**, 1-615.
Eakin, D. (1979). Faunal remains from the Garrett Allen site, Wyoming. Ms. on file. Laramie: University of Wyoming.
Emslie, S. D. (1978). Dog burials from Mancos Canyon, Colorado. *The Kiva* **43**, 167-182.
Emslie, S. D. (1986a). Faunal remains from Dos Casas hamlet. In (D. Breternitz, Ed) *Dolores Archaeological Orogram: Anasazi Communities at Dolores: Early Anasazi Sites in the Sagehen Flats Area.* Denver: Department of the Interior, Bureau of Reclamation, pp. 631-635.
Emslie, S. D. (1986b). Faunal report for Prairie Dog hamlet. In (D. Breternitz, Ed) *Dolores Archaeological Program: Anasazi Communities at Dolores: Early Anasazi Sites in the Sagehen Flats Area.* Denver: Department of the Interior, Bureau of Reclamation, pp. 481-493.
Emslie, S. D. (1986c). Faunal remains from Casa Bodega hamlet. In (D. Breternitz, Ed) *Dolores Archaeological Program: Anasazi Communities at Dolores: Early Anasazi Sites in the Sagehen Flats Area.* Denver: Department of the Interior, Bureau of Reclamation, pp *537-538.*
Eshelman, R. & Grady, F. (1986). Quaternary vertebrate localities of Virginia and their avian and mammalian fauna. *Virginia Division Mines, Minerals, Resources Publication* **75**, 43-70.
Ewers, J. C. (1960). A Blood Indian's conception of tribal life in Dog days. *The Blue Jay* **18**, 44-48.
Falk, C. R. (1969a). Faunal remains; faunal evidence. *Plains Anthropologist Memoir* **6**, 44-51.
Farnham, T. J. (1904). Travels in the great western prairies, the Anahuac and Rocky mountains, and the Oregon country. In (R. G. Thwaites, Ed) *Early Western Travels*, Vol. 28. Cleveland: Arthur H. Clark, pp. 25-380.

Faunmap Working Group. (1994). FAUNMAP: A database documenting Late Quaternary distributions of mammal species in the United States. *Illinois State Museum Scientific Papers* **25**.

Fradkin, A. (1979). Faunal remains from the Alderman site, Volusia County, Florida. *Florida Anthropologist* **32**, 75-83.

Fraser, F. C. & King, J. E. (1949). A preliminary report on the Star Carr bone remains. *Proceedings of the Prehistoric Society* **15**, 67.

Frison, G. C., Wilson, M. & Walker, D. N. (1978). The Big Goose Creek site: bison procurement and faunal analysis. *Occasional Papers on Wyoming Archeology* **1**, 1-50.

Funk, R. E., Fisher, D. W. & Reilly, Jr., E. M. (1970). Caribou and Paleo-indian in New York state: a presumed association. *American Journal Science* **268**, 181-186.

Futch, R. S. (1979). *Prehistoric human ecology at the Morton Shell Mound (16IB3), Iberia Parish, Louisiana.* Masters Thesis, Louisiana State University.

Gilmore, M. R. (1933). The Arikara method of preparing a dog for a feast. *Papers of the Michigan Academy of Science, Arts and Letters* **19**, 37-38.

Gilmore, R. M. (1946). Mammals in archeological collections from southwestern Pennsylvania. *Journal of Mammalogy* **27**, 227-234.

Gilmore, R. M. (1950). Fauna and ethnozoology of South America. *Bulletin of the Bureau of American Ethnology* **143**, 345-364.

Gleeson, P. F. (1970). *Dog remains from the Ozette village archaeological site.* Masters Thesis, Washington State University.

Gollan, K. (1980). *Prehistoric dingo.* Ph.D. Thesis, Australian National University.

Graham, M. A., Wilson, M. C. & Graham, R. W. (1987). Paleoenvironments and mammalian faunas of Montana, southern Alberta, and southern Saskatchewan. *Illinois State Museum Scientific Papers* **22**, 410-459.

Graham, R. W. (1987). Late Quaternary mammalian faunas and paleoenvironments of the southwestern Plains of the United States. *Illinois State Museum Scientific Papers* **22**, 24-86.

Grayson, D. K. (1973). *The avian and mammalian remains from Nightfire Island.* Ph.D. Thesis, University of Oregon.

Grayson, D. K. (1974). The Riverhaven No. 2 vertebrate fauna: comments on methods in faunal analysis and on aspects of the subsistence potential of prehistoric New York. *Man in the Northeast* **8**, 23-39.

Grayson, D. K. (1977). Paleoclimatic implications of the Dirty Shame Rockshelter mammalian fauna. *Tebiwa* **9**, 1-26.

Grayson, D. K. (1988). Danger Cave, Last Supper Cave, and Hanging Rock Shelter: the faunas. *Anthropological Papers of the American Museum of Natural History* **66**.

Green, M. (1992). *Animals in Celtic Life and Myth.* London: Routledge.

Greene, J. L. & Matthews, T. W. (1976). Faunal study of unworked mammal bones. In (E. W. Haury, Ed) *The Hohokam: Desert Farmers and Craftsmen: Excavations at Snaketown, 1964-1965.* Tucson: University of Arizona Press, pp. 367-373.

Guernsey, S. J. & Kidder, A. V. (1921). Basket Maker caves of northeastern Arizona. *Papers of the Peabody Museum of American Archaeology and Ethnology Harvard University* **7**.

Guilday, J. E. (1955). Animal remains from an Indian village site, Indiana County, Pennsylvania. *Pennsylvania Archaeologist* **25**, 142-147.

Guilday, J. E. (1961). Vertebrate remains from the Varner site. *Pennsylvania Archaeologist* **31**, 119-23.

Guilday, J. E. (1963). Bone refuse from the Oakfield site, Genesee County, New York. *Pennsylvania Archaeologist* **33**, 12-15.

Guilday, J. E. (1965). Bone refuse from the Lamoka Lake site. In (W. A. Ritchie, Ed) *The Archeology of New York State.* Garden City: Natural History Press, pp. 56-58.

Guilday, J. E. (1969). Faunal remains from Dutchess Quarry Cave No. 1. *Pennsylvania Archaeologist* **39**, 7-22.

Guilday, J. E. (1971). Biological and archeological analysis of bone from a 17th century Indian village (46PU31), Putnam County, West Virginia. *West Virginia Geological Economy Survey, Report on Archeological Investigations* **4**, 1-64.

Guilday, J. E. (1973a). Faunal remains from the Nahrwold No. 1 site. *New York State Museum Memoir* **20**, 288-290.

Guilday, J. E. (1973b). Vertebrate faunal remains from the Garoga site, Fulton County, New York. *New York State Museum Memoir* **20**, 329-330.

Guilday, J. E. & Adam, E. K. (1967). Small mammal remains from Jaguar Cave, Lemhi County, Idaho. *Tebiwa* **10**, 26-36.

Guilday, J. E. & Parmalee, P. W. (1965). Animal remains from the Sheep Rock Shelter (36 HU 1), Huntington County, Pennsylvania. *Pennsylvania Archaeologist* **35**, 34-49.

Guilday, J. E. & Parmalee, P. W. (1975). Faunal remains from the Zebree site. *Archeological Survey Research Report* **4**, 228-234.

Guilday, J. E., Parmalee, P. W. & Tanner, D. P. (1962). Aboriginal butchering techniques at the Eschelman site (36LA12), Lancaster County, Pennsylvania. *Pennsylvania Archaeologist* **32**, 59-83.

Guilday, J. E. & Tanner, D. P. (1965). Vertebrate remains from the Mount Carbon site, (46-Fa-7), Fayette County, West Virginia. *West Virginia Archeologist* **18**, 1-14.

Guilday, J. E. & Tanner, D. P. (1966). Animal remains from the Westmoreland-Barber site (40Mi-11), Marion County, Tennessee. In (C. Faulkner & J. B. Graham, Eds) *Westmoreland-Barber Site (40Mi-11) Nickajack Reservoir season II*, Knoxvile: University of Tennessee, pp. 138-140.

Guilday, J. E. & Tanner, D. P. (1966). The vertebrate fauna from Mound City. In *Mound City Revisted.* Columbus: Report to the National Park Service by the Ohio Historical Society.

Guilday, J. E. & Tanner, D. P. (1969). Vertebrate remains from the Fairchance Mound (46Mr13), Marshall County, West Virginia. *West Virginia Archeologist* **21**, 41-54.

Gustafson, P. (1980). *Salish Weaving.* Vancouver: Douglas & McIntyre Ltd.

Haag, W. G. (1947). Dog remains in archaeological sites. *Plains Archaeology Conference Newsletter* **1**, 27-28.

Haag, W. G. (1948). An osteometric analysis of some aboriginal dogs. *University of Kentucky Department of Anthropology Reports in Anthropology* **7**, 107-264.

Haag, W. G. (1948). Dog remains in archeological sites. *Plains Archeological Conference News Letter* **1**, 27-28.

Haag, W. G. (1954). A mummified dog from the Lighthouse Site, Supe. In (G. R. Willey & J. M. Corbett, Eds) *Early Ancon and Early Supe Culture..* New York: Columbia University Press, pp. 138-140.

Haag, W. G. (1966). Two dog skeletons from Caldwell Village. *University of Utah Anthropological Papers* **84**, 97-101.

Haag, W. G. (1970). Dog remains from Hogup Cave. *University of Utah Anthropological Papers* **93**, 273-274.

Haag, W. G. (1983). Appendix G: Dog remains from the Wildcat Canyon site. *University of Oregon Anthropological Papers* **30**.

Haag, W. G. & Heizer, R. F. (1953). A dog burial from the Sacramento Valley. *American Antiquity* **18**, 263-264.

Hale, H. S. (1989). *Prehistoric subsistence strategies and settlement patterns in the Lake Okeechobee Basin of the South Florida Peninsula*. Ph.D. Thesis, University of Florida.

Hamblin, N. L. (1984). *Animal Use by the Cozumel Maya*. Tucson: University of Arizona Press.

Hammond, G. P. & Rey, A. (1940). Narratives of the Coronado Expedition, 1540-1542. *University of New Mexico, Coronado Historical Series* **II**.

Harcourt, R. A. (1967). Osteoarthritis in a Romano-British dog. *Journal of Small Animal Practice* **8,** 521-522.

Harcourt, R. A. (1974). The dog in prehistoric and Early Historic Britain. *Journal of Archaeological Science* **1**, 151-175.

Harris, A. H. (1963). *Vertebrate Remains and Past Environmental Reconstruction in the Navajo Reservoir District*. Museum of New Mexico Papers in Anthropology **11**.

Harrison, T. (1965). Three 'secret' communication systems among Borneo nomads (and their dogs). *Journal of the Malaysian Branch of the Royal Asiatic Society* **38**, 67-86.

Harten, L. B. (1975). *The osteology of the human skeletal material from the Braden site, 10WN117, in western Idaho*. Masters Thesis, Idaho State University.

Henderson, N. (1994). Replicating dog travois travel on the Northern Plains. *Plains Anthropologist* **39**, 145-159.

Higham, C. (1968). Size trends in prehistoric European domestic fauna and the problem of local domestication. *Acta Zoologica Fennica* **120**, 1-21.

Higham, C. F. W., Kijngam, A. & Manly, B. F. J. (1980). An analysis of prehistoric canid remains from Thailand. *Journal of Archaeological Science* **7,** 149-165.

Hill, F. C. (1972). *A Middle Archaic Dog Burial in Illinois*. Evanston: Foundation for Illinois Archaeology.

Hirshfeld, S. E. (1968). Vertebrate fauna of Nichol's Hammock, a natural trap. *Quarterly Journal Florida Academy Science* **31**, 177-189.

Howay, F. W. (1918). The dog's hair blankets of the Coast Salish. *Washington Historical Quarterly* **9**, 83-92.

Hudson, J. (1993). The impacts of domestic dogs on bone in forager camps. *Southern Illinois University at Carbondale Center for Archaeological Investigations Occasional Paper* **21**, 301-323.

Irving, J. T., Jr. (1835). *Indian Sketches Taken During an Expedition to the Pawnee Tribes (1833)*. Philadelphia: Carey, Lea & Blanchard.

Jackson, J. W. (1935). The animal remains from the Stonehenge Excavations of 1920-1926. *Antiquaries Journal* **15**, 434. .

Jackson, J. W. (1935). Report on the skeleton of the dog from ash pit C. *Wiltshire Archaeological and Natural History Magazine* **47**, 76-78.

James, E. (1904). An account of an expedition from Pittsburgh to the Rocky Mountains performed in the years 1819 & 1820. In (R. G. Thwaites, Ed) *Early Western Travels*, Volume 2. Cleveland: Arthur H. Clark, pp.1-321.

Jelinek, A. J. (1969). Identification of mammalian fauna. *University of Michigan, Museum of Anthropology Papers* **40**, 127-133.

Jenkins, J. T. & Semken, Jr., H. A. (1972). Faunal analysis of the Lane Enclosure, Allamakee County, Iowa. *Proceedings of the Iowa Academy of Science* **78**, 76-78.

Jennings, J. D. (1957). *Danger Cave*. University of Utah Anthropology Papers **27**.

Johnson, E. M. (1975). Faunal and floral material from a Kansas City Hopewell site: analysis and interpretation. *Museum Texas Tech University Occasional Papers* **36**, 1-37.

Johnson, E. M. (1987). Vertebrate remains. In (E. Johnson, Ed) *Lubbock Lake: Late Quaternary Studies on the Southern High Plains*. College Station: Texas A&M University Press, pp. 49-89.

Jones, R. (1970). Tasmanian aborigines and dogs. *Mankind* **7**, 256-271.

Kay, M. (1975). *Central Missouri Hopewell subsistence-settlement system*. Ph.D. Thesis, University of Colorado.

Kay, M. (1980). The central Missouri Hopewell subsistence-settlement system. *Missouri Archaeological Society* **15**.

Keber, E. Q. (1992). Xolotl: Dogs, death, and deities in Aztec myth. *Latin American Indian Literatures Journal* **8**, 229-239.

Keddie, G. (1993). Prehistoric dogs of BC: wolves in sheep's clothing. *The Midden* **25**, 3-5.

Kelley, J. E. (1975). Zooarchaeological analysis at Antelope House: Behavioral inferences from distribution data. *The Kiva* **41**, 81-85.

Kent, S. (1981). The dog: an archaeologist's best friend or worst enemy. The spatial distribution of faunal remains. *Journal of Field Archaeology* **8**, 367-372.

Kent, S. (1993). Variability in faunal assemblages: the influence of hunting skill, sharing, dogs, and mode of cooking on faunal remains at a sedentary Kalahari community. *Journal of Anthropological Archaeology* **12**, 323-385.

Kerber, J. E. (1997a). *Lambert Farm: Public Archaeology and Canine Burials Along Narragansett Bay*. New York: Harcourt Brace.

Kerber, J. E. (1997b). Native American treatment of dogs in northeastern North America: archaeological and ethnohistorical perspective. *Archaeology of Eastern North America* **25**, 81-96.

Kerber, J. E., Leveille, A. & Greenspan, R. (1989). An unusual dog burial feature at the Lambert Farm site, Warwick, Rhode Island: preliminary observations. *Archaeology of Eastern North America* **17,** 165-174.

Kroeber, A. L. (1941). Culture element distributions: XV. Salt, dogs, tobacco. *University of California Anthropological Records* **6,** 1-20.

Krozser, K. (1991). Canid remains at kill sites: a case study from the Oldman River Dam Project. *Alberta Archaeological Survey Occasional Paper* **33**, 81-100.

Kurtn, B. & Anderson, E. (1972). The sediments and fauna of Jaguar Cave, II: the fauna. *Tebiwa* **15**, 21-45.

Lang, R. W. (1968). The natural environment and subsistence economy of the McKees Rocks Village site. *Pennsylvania Archaeologist* **38,** 50-80.

Lang, R. W. & Harris, A. H. (1984). The faunal remains from Arroyo Hondo Pueblo, New Mexico: A study in short-term subsistence change. *School of American Research Press Arroyo Hondo Archaeological Series* **5**.

Lawrence, B. (1944). Bones from the Governador Area. In (E. T. Hall, Jr.) *Early Stockaded Settlements in the Governador New Mexico*. New York: Columbia University Press, pp. 73-78.

Lawrence, B. (1951). Mammals found at the Awatovi site. *Peabody Museum, Harvard University* **4**.

Lawrence, B. (1951). Discussion of canid remains from the Southwest. *Papers of the Peabody Museum of American Archaeology and Ethnology* **35**, 5-43.

Lawrence, B. (1953). Appendix E: Cave fauna. *Papers of the Peabody Museum of American Archaeology and Ethnology* **48**, 80-81.

Lawrence, B. (1966). Early domestic dogs. *Zeitschrift für Saugetierkunde* **32**, 44-59.

Lawrence, B. (1968). Antiquity of large dogs in North America. *Tebiwa* **11**, 43-49.

Lawrence, B. (1977). Dogs from the Dominican Republic. *Cuadernos del Cendia. Centro Dominicano de Investigaciones Anthropologicas (Cendia)* **268**, 3-19.

Lawrence, B. & Reed, C. A. (1983): The dogs from Jarmo. In (L. S. Braidwood, R. J. Braidwood, B. Howe, C. A. Reed, & P. J. Watson, Eds) *Prehistoric Archaeology along the Zagros Flanks*. Chicago: University of Chicago Oriental Institute, pp. 485-489.

Leonard, R. D. W. (1989). *Anasazi Faunal Exploitation: Prehistoric Aubsistence on Northern Black Mesa, Arizona*. Southern Illinois University Center for Archaeological Investigations Occasional Paper **13**.

Leveillee, A. (1993). Eastern Woodland mortuary practices as reflected in canine burial features at the Lambert Farm site, Warwick, Rhode Island. *Bulletin of the Massachusetts Archaeological Society* **54**, 19-24.

Lewis, R. B. (1974). Mississippian exploitative strategies: a southeast Missouri example. *Missouri Archaeological Society* **11**, 1-63.

Lewis, R. B. (1979). *Hunter-gatherer foraging: some theoretical explanations and archeological tests*. Ph.D. Thesis, University of Illinois.

Lippold, L. K. (1973a). Animal resource utilization at the Cooper's Shore site (47RO2) Rock County, Wisconsin. *Wisconsin Archeologist* **54**, 36-62.

Lippold, L. K. (1973b). Animal resource utilization at Wisconsin Effigy mound sites. *Wisconsin Archeologist* **54**, 135-152.

Livingston, S. (1984). Faunal analysis. In (E. Lohse, Ed) *Archeological Investigations at Site 45-OK-11 Chief Joseph Dam project, Washington*. Seattle: University of Washington (Institute of Environmental Studies), pp. 181-192.

Livingston, S. (1985). Faunal analysis. In (M. Jaehnig, Ed) *Archeological Investigations at Site 45-OK-258, Chief Joseph Dam Project, Washington*. Seattle: University of Washington (Institute of Environmental Studies), pp. 117-133.

Livingston, S. (1988). *The avian and mammalian faunas from Lovelock Cave and the Humboldt Lakebed site*. Ph.D. dissertation, University of Washington, Seattle.

Livingston, S. & Lyman, R. L. (1984). Faunal analysis. In (E. Lohse, Ed) *Archeological Investigations at Site 45-DO-242 and 45-DO-243. Chief Joseph Dam Project, Washington*. Seattle: University of Washington (Institute of Environmental Studies), pp. 131-142.

Long, J. (1904). Voyages and travels of an Indian interpreter and trader, describing the manners and customs of the North American Indians. In (R. G. Thwaites, Ed) *Early Western Travels*, Vol. 2. Cleveland: Arthur H. Clark, pp.22-329.

Lopex, J. & Wisniewski, S. (1956). Discovery of a possible ceremonial dog burial in the City of Greater New York. *Bulletin of the Archeological Society of Connecticut* **29**, 14-19.

Lukens, P. W. , Jr. (1963). *Some ethnozoological implications of mammalian faunas from Minnesota archeological sites*. Ph.D. Thesis, University of Minnesota.

Lundelius, E. L., Jr. (1967). Late Pleistocene and Holocene faunal history of central Texas. In (P. S. Martin & H. E. Wright, Jr., Eds) *Pleistocene Extinctions: A Search for a Cause*. New Haven: Yale University Press, pp. 287-319.

Lundelius, E. L., Jr., Graham, R. W., Anderson, E., Guilday, J., Holman, J. A., Steadman, D. W. & Webb, S. D. (1983). Terrestrial vertebrate faunas. In (S.C. Porter, Ed) *Late-Quaternary Environments of the United States: Vol. 1, the Late Pleistocene*. Minneapolis: University of Minnesota Press, pp.311-353.

Lupo, K. D. & Janetski, J. C. (1994). Evidence of domesticated dogs and some related canids in the eastern Great Basin. *Journal of California and Great Basin Anthropology* **16**, 199-220.

Lyman, R. L. (1976). *A cultural analysis of faunal remains from the Alpowa locality*. Master's Thesis, Washington State University.

Lyman, R. L. (1987). Elk Creek zooarchaeology. In (R. M. Pettigrew & C. G. Lebow, Eds) *Data Recovery at Sites 35JA27, 35JA59, and 35JA100, Elk Creek Lake Project, Jackson County, Oregon*. Portland: U.S. Army Corps of Engineers, pp. D.1-D.37.

Lyon P. J. (1970). Differential bone destruction: An ethnographic example. *American Antiquity* **35**, 213-215.

Mack, J. M. (1983). Archaeological investigations in the Salt Cave locality: Subsistence uniformity and cultural diversity on the Klamath River, Oregon. *University of Oregon Anthropology Papers* **29**.

Manson, C., MacCord, H. A. & Griffin, J.B. (1943). The culture of the Keyser Farm site. *Papers of the Michigan Academy Science, Arts and Letters* **29**, 375-418.

Marcus, L. F. (1960). A census of the abundant large Pleistocene mammals from Rancho La Brea. *Los Angeles County Museum Contributions to Science*

Marrinan, R. A. (1980). Analysis of faunal remains from Sliding Slab Rock Shelter, Sebastian County, Arkansas. Ms. on file. Pauls Valley, OK: Environmental Assessments, Inc.

Martin, R. A. & Webb, S. D. (1974). Late Pleistocene mammals from the Devil's Den fauna, Levi County. In (S. D. Webb, Ed) *Pleistocene Mammals of Florida*. Gainesville: University of Florida Presses, pp. 114-145.

Martin, T. J. & Colburn, M. L. (1989). Faunal remains. *Michigan Archaeologist* **35**, 135-152.

McKee, D. F. (1988). *A faunal analysis of the River Bend site (48NO202): evidence of Protohistoric subsistence on the Northwest Plains*. Master's Thesis, University of Wyoming.

McKern, S. T. (1988). Archaeological investigations at the Bessie Bottom site (48UT1186): late Prehistoric animal procurement in the Upper Bear River Drainage. *Western Wyoming College Archaeological Services, Cultural Resource Management Report* **45**.

McMillan, R. B. (1970). Early canid burial from the Western Ozark Highland. *Science* **167**, 1246-1247.

Michalik, L. K. (1982). An ecological perspective on the Huber phase subsistence-settlement system. *Minnesota Publications in Anthropology* **1**, 29-53.
Mick, L. S. (1983). *An ecological evaluation of faunal diversity in the Central Plains tradition.* Master's Thesis, University of Nebraska.
Misner, E. J. & Reitz, E. J. (1992). Vertebrate fauna from Pump Canal (16SC27), Louisiana. Ms. on file. Athens: University of Georgia (Museum Natural History).
Monk, S. (1984). *A Summary of Identified Vertebrate Remains Recovered from Mound City Group National Monument, Ross County, Ohio.* Lincoln, NB: Midwest Archeological Center.
Mohl, J. (1986). Dog remains from a paleoeskimo settlement in West Greenland. *Arctic Anthropology* **23**, 81-89.
Molloy, P. (1993). Hunting practices at an Historic Plains Indian village: Kansa ethnoarchaeology and faunal analysis. *Plains Anthropologist* **38**, 187-197.
Montgomery, J. A. (1979). *Prehistoric subsistence at Semiahmoo Spit, 45WH17.* Master's Thesis, Western Washington University.
Morey , D. F. (1986). Studies on Amerindian dogs: taxonomic analysis of canid crania from the Northern Plains. *Journal of Archaeological Science* **13**, 119-145.
Morey, D. F. & Klippel, W. F. (1991). Canid scavenging and deer bone survivorship at an Archaic Period site in Tennessee. *Archaeozoologica* **4**, 11-28.
Morey, D. F. & Wiant, M. D. (1992). Early Holocene domestic dog burials from the North American Midwest. *Current Anthropology* **33**, 224-229.
Morrison, D. (1984). A note on Thule culture dogs from Coronation Gulf, N.W.T. *Canadian Journal of Archaeology* **8**, 149-157.
Muehl, C. (1991). Dogs of the old Southwest. *Arizona Highways* **May**, 34-37.
Murie, O. J. (1936). Dog skulls from St. Lawrence Island, Alaska. *Miscellaneous Publications of the University of Alaska* **2**, 347-357.
Murie, O. J. 1948. Dog skulls from Ipiutak. *Anthropological Papers of the American Museum of Natural History* **42**, 255-264.
Murphy, D. A. (1996). Dogs of destiny, hounds from Hell. American soldiers and canines in the Mexican War. *Military History of the West* **26**, 35-48.
Musil, R. (1970). Domestication of the dog already in the Magdalenian? *Anthropologie (Brno): International Journal of the Science of Man* **8**, 87-88.
Neumann, G. K. (1937). Faunal remains from Fulton County sites. In (F. C. Cole & T. Deuel, Eds) *Rediscovering Illinois.* Chicago: University of Chicago Press, pp. 265-268.
Neusius, S. W. (1986a). Faunal remains from Windy Wheat Hamlet. In (D. Breternitz, Ed) *Anasazi Communities at Dolores: Early Anasazi Sites in the Sagehen Flats Area.* Denver, CO: Department Interior Bureau Reclamation, pp. 855-862.
Neusius, S. W. (1986b). Faunal remains from Prince Hamlet. In (D. Breternitz, Ed) *Anasazi Communities at Dolores: Early Anasazi Sites in the Sagehen Flats Area.* Denver, CO: Department Interior Bureau Reclamation, pp.483-498.
Neusius, S. W. (1986c). Faunal remains from Periman Hamlet Area 1. In (D. Breternitz, Ed) *Anasazi Communities at Dolores: Early Anasazi Sites in the Sagehen Flats Area.* Denver, CO: Department Interior Bureau Reclamation, pp.187-198.
Neusius, S. W. (1986d). Faunal remains from Lemoc Shelter. In (D. Breternitz, Ed) *Anasazi Communities at Dolores: Early Anasazi Sites in the Sagehen Flats Area.* Denver, CO: Department Interior Bureau Reclamation, pp.279-296.
Neusius, S. W. (1986e). Faunal remains from Aldea Sierritas. In (D. Breternitz, Ed) *Anasazi Communities at Dolores: Early Anasazi Sites in the Sagehen Flats Area.* Denver, CO: Department Interior Bureau Reclamation, pp.381-398.
Neusius, S. W. & Gould, M. (1988). Faunal remains: implications for Dolores Anasazi adaptations. In (D. Breternitz, Ed) *Anasazi Communities at Dolores: Grass Mesa Village* Denver, CO: Department Interior Bureau Reclamation, pp.1049-1136.
Noe-Nygaard, N. (1988). C13 values of dog bones reveal the nature of changes in man's food resources at the Mesolithic-Neolithic transition, Denmark. *Chemical Geology* **73**, 87-96.
Olsen, J. W. (1982). Prehistoric environmental reconstruction by vertebrate faunal analysis, Grasshopper Pueblo. *Anthropology Papers of the University of Arizona* **40**, 63-72.
Olsen, J. W. (1990). Vertebrate faunal remains from Grasshopper Pueblo, Arizona. *University of Michigan Museum of Anthropology Papers* **83**.
Olsen, S. J. (1968). Canid remains from Grasshopper Ruin. *The Kiva* **34**, 33-40.
Olsen, S. J. (1970). Two pre-Columbian dog burials from Georgia. *Bulletin of the Georgia Academy of Science* **28**, 69-72.
Olsen, S. J. (1972). Animal remains from Altar de Sacrificios. *Papers of the Peabody Museum of Archaeology and Ethnology* **64**, 243-246.
Olsen, S. J. (1972). The small Indian dogs of Black Mesa, Arizona. *Plateau* **45**, 47-54.
Olsen, S. J. (1974). Early domestic dogs in North America and their origins. *Journal of Field Archaeology* **1**, 343-345.
Olsen, S. J. (1976). The dogs of Awatovi. *American Antiquity* **41**, 102-106.
Olsen, S. J. (1978). Vertebrate faunal remains. *Memoirs of the Peabody Museum of Archaeology and Ethnology* **14**, 172-176.
Olsen, S. J., Olsen, J. W. & Guo-qin, Q. (1980). Domestic dogs from the Neolithic of China. *Explorers Journal* **Dec.**, 165-167.
Parmalee, P. W. (1959). Animal remains from the Raddatz Rockshelter, Sk5, Wisconsin. *Wisconsin Archeologist* **40**, 83-90.
Parmalee, P. W. (1960a). Animal remains from the Aztalan site, Jefferson County, Wisconsin. *Wisconsin Archeologist* **41**, 1-10.
Parmalee, P. W. 1960b. Animal remains from the Durst Rockshelter, Sauk County, Wisconsin. *Wisconsin Archeologist* **41**, 11-17.
Parmalee, P. W. (1962a). Identification of the faunal remains from the Lawhorn site. *Missouri Archaeologist* **24**, 95-96.
Parmalee, P. W. (1962b). The faunal complex of the Fisher site, Illinois. *American Midland Naturalist* **68**, 399-408.
Parmalee, P. W. (1963). Vertebrate remains from the Bell site, Winnebago County, Wisconsin. *Wisconsin Archeologist* **44**, 58-69.
Parmalee, P. W. (1964). Vertebrate remains from an historic archaeological site in Rock Island County, Illinois.

*Transactions of the Illinois Academy of Science* **57**, 167-174.

Parmalee, P. W. (1963). Vertebrate remains from the Bell site, Winnebago County, Wisconsin. *Wisconsin Archeology* **44**, 58-69.

Parmalee, P. W. (1965). The food economy of Archaic and Woodland peoples at the Tick Creek Cave site, Missouri. *Missouri Archaeologist* **27**, 1-34.

Parmalee, P. W. (1965). The vertebrate fauna. *Cleveland Museum of Natural History Scientific Publications New Series* **4**, 115-118.

Parmalee, P. W. (1967). A recent cave bone deposit in southwestern Illinois. *Bulletin of the National Speleological Society* **29**, 119-147.

Parmalee, P. W. (1968). Vertebrate remains from the Mason site (40Fr8), Franklin County, Tennessee. *University of Tennessee, Department Anthropology Report Investigations* **6**, 256-262,.

Parmalee, P. W. (1971). Fisher and porcupine remains from cave deposits in Missouri. *Transactions of the Illinois Academy of Science* **64**, 225-229.

Parmalee, P. W. (1972). Vertebrate remains from the Fifield site, Porter County, Indiana. *Indiana Historical Society Prehistoric Research Series* **5**, 202-205.

Parmalee, P. W. (1973). Faunal remains from the Kane Village site (Ms-194), Madison County, Illinois. *Bulletin of the Illinois Archaeological Survey* **9**, 49-52.

Parmalee, P. W. (1979). Inferred Arikara subsistence patterns based on a selected faunal assemblage from the Mobridge site, South Dakota. *Kiva* **44**, 191-218.

Parmalee, P. W. & Bogan, A. E. (1978). Cherokee and Dallas dog burials from the Little Tennessee River Valley. *Tennessee Anthropologist* **3**, 100-112.

Parmalee, P. W. & Bogan, A. E. (1980a). Vertebrate remains from early European and historic Indian occupations at the Waterman site, Randolph County, Illinois. *Transactions of the Illinois State Academy of Science* **73**, 49-54.

Parmalee, P. W. & Bogan, A. E. (1980b). A summary of the animal remains from the Noble-Wieting site (11ML28), McLean County, Illinois. *Transactions of the Illinois Academy of Science* **73**, 1-6.

Parmalee, P. W. & Klippel, W. E. (1983). The role of native animals in the food economy of the historic Kickapoo in central Illinois. *University of Michigan, Museum of Anthropology Papers* **72**, 253-324.

Parmalee, P. W., McMillan, R. B. & King, F. B. (1976). Changing subsistence patterns at Rodgers Shelter. In (W. R. Wood & R. B. McMillan, Eds) *Prehistoric Man and his Environments: A Case Study in the Ozark Highland.* New York: Academic Press, pp. 141-161.

Parmalee, P. W. & Oesch, R. D. (1972). Pleistocene and recent faunas from the Brynjulfson caves, Missouri. *Illinois State Museum Report of Investigations* **25**.

Parmalee, P. W., Oesch, R. D. & Guilday, J. E. (1969). Pleistocene and recent vertebrate faunas from Crankshaft Cave, Missouri. *Illinois State Museum Report of Investigations* **14**.

Parmalee, P. W., Paloumpis, A. A. & Wilson, N. (1972). Animals utilized by Woodland peoples occupying the Apple Creek site, Illinois. *Illinois State Museum Reports of Investigations* **23**, 1-62.

Parmalee, P. W. & Reyman, J. E. (1971). Faunal remains from Peters Cave, Jackson County, Illinois. *Southern Illinois University Museum Research Records* **7**, 61-64.

Parmalee, P. W. & Shane, O. C. (1970). The Blain site vertebrate fauna. In (O. H. Prufer & O. C. Shane, Eds) *Blaine Village and the Fort Ancient Traditions in Ohio.* Kent, OH: Kent State University Press, pp. 185-206.

Pavesic, M. G., Miller, S. J., Gamel, P. A. & Green, T. J. (1993). The DeMoss site: a material culture and faunal update. *Idaho Archaeologist* **16**, 3-16.

Pearson, C. E. (1984). Red Bud Creek: Late Prehistoric material culture and subsistence in coastal Georgia. *Early Georgia* **12**, 1-39.

Pferd, W., III. (1987). *Dogs of the American Indians.* Denlinger's Publishers, Fairfax, Virginia.

Pillaert, E. E. (1969). Faunal remains from the Millville site (47GT53), Grant County, Wisconsin. *Wisconsin Archeologist* **50**, 93-108.

Pippin, L. C. (1987). Prehistory and paleoecology of Guadalupe Ruin, New Mexico. *University of Utah Anthropology Papers* **112**, 129-173.

Pohl, M. E. D. (1976). *Ethnozoology of the Maya: an analysis of fauna from five sites in the Peten, Guatemala.* Ph.D. Thesis, Harvard University.

Pollack, H. E. D. & Ray, C. E. (1957). Notes on vertebrate animal remains from Mayapan. *Carnegie Institution Report* **41**.

Post, R. W. (1987). Dog remains from Devon Island, N.W.T.: Archaeological and osteological evidence for domestic dog use in Thule Culture. *Arctic* **40**, 184-190.

Potter, M. A. & Baby, R. S. (1964). Hopewellian dogs. *Ohio Journal of Science* **64**, 36-40.

Prufer, O. H. (1965). The McGraw siteùa study in Hopewellian dynamics. *Scientific Publications of the Cleveland Museum of Natural History* **3**, 1-144.

Prummel, W. (1992). Early Medieval dog burials among the Germanic tribes. *Helinium* **32**, 132-194.

Purdue, J. R. (1982). The environmental implications of the fauna recovered from Rodgers Shelter. In (M. Kay, Ed) *Holocene Adaptations within the Lower Pomme de Terre River Valley, Missouri, Vol. 1.* Kansas City: U. S. Army Corps of Engineers, pp 199-261.

Putnam, F. W. (1899). [On finding of skulls and bones of Indian dogs.]. *Report of Harvard College* 1897-98, **269**.

Quinones, K. E. (1991). Xolotl: dogs, death, and deities in Aztec myth. *Journal of Latin America Indian Literature* **7**, 229-39.

Rasmussen, D. L. (1974). New Quaternary mammal localities in the upper Clark Fork River valley, Western Montana. *Northwest Geology* **3**, 62-70.

Ray, C. E., Wills, D. L. & Palmquist, J. C. (1968). Fossil musk oxen of Illinois. *Illinois State Academy of Science* **61**, 282-292.

Reitz, E. J. (1985a). Comparison of Spanish and aboriginal subsistence on the Atlantic coastal plain. *Southeastern Archaeology* **4**, 41-50.

Reitz, E. J. (1985b). Faunal evidence for sixteenth century Spanish subsistence at St. Augustine, Florida. *Florida Anthropologist* **38**, 54-69.

Reitz, E. J. (1987a). Vertebrate and invertebrate fauna. In (E. M. Futato, Ed) *Archaeological Investigations at Shell Bluff and White Springs, Two Late Woodland Sites in the Tombigbee River Multi-resource District.* Mobile: US Army Corps of Engineers, pp. 193-199.

Reitz, E. J. (1987a). Report on the faunal material excavated by West Georgia College from Colonel's Island. *West Georgia College Studies in the Social Sciences* **26**, 63-79.

Reitz, E. J. & Cumbaa, S. L. (1983). Diet and foodways of eighteenth-century Spanish St. Augustine. In (K. Deagan, Ed) *Spanish St. Augustine: The Archaeology of a Colonial Creole Community*. New York: Academic Press, pp.151-186.

Reitz, E. J., Marrinan, R. A. & Scott, S. L. (1987). Survey of vertebrate remains from prehistoric sites in the Savannah River valley. *Journal of Ethnobiology* **7**, 195-221.

Reynolds, R .L. (1985). Domestic dog associated with human remains at Rancho La Brea. *Bulletin of the Southern California Academy of Science* **84**, 76-85.

Rock, C. (1980). A preliminary analysis of faunal remains at the Abercrombie site, Alabama. Ms. on file, Athens, GA: University of Georgia.

Roe, F. G. (1939). From dogs to horses among the western Indian tribes. *Transactions of the Royal Society of Canada*, Sec. II/ Ser. 3, **33**, 209-275.

Roe, F. G. (1955). *The Indian and the Horse*. Norman: University of Oklahoma Press.

Roe, P. G. (1995). Eternal companions: Amerindian dogs from Tierra firma to the Antilles. In (R. Alegra & M. Rodriguez, Eds) *Proceedings of the XV International Congress for Caribbean Archaeology*. San Juan, Puerto Rico, pp. 155-172.

Rogers, K. (1975). Faunal remains from the Zimmerman site, 1972-Grid B. *Illinois State Museum Report of Investigations* **32**, 86-91.

Roll, T. E. & Deaver, K. (1980). *Bootlegger Trail Site: A Spring Bison Kill*. Washington, DC: Heritage Conservation and Recreation Service, Interagency Archeological Service.

Roth, J. (1979). Analysis of faunal remains. In (D. J. Hally, Ed) Archaeological Investigations of the Little Egypt Site (9Mu102), Murray County, Georgia. Ms.on file. Athens, GA: University of Georgia, pp. 570-591.

Roys, R. L. (1931). *The Ethnobotany of the Maya*. New Orleans, LA: Tulane University, Middle American Research Institute Publication **2**.

Sadek-Kooros, H. (1972). The sediments and fauna of Jaguar Cave. I :the sediments. *Tebiwa* **15**, 1-20.

Sammis, C. G. (1978). *Mammalian osteo-archaeology of the Mitchell site, (39DV2), Mitchell, South Dakota*. Master's Thesis, University of Iowa.

Schmitt, D. (1986). *Zooarchaeology and taphonomic investigations of site 35JA42, upper Applegate River, southwestern Oregon*. Master's Thesis, Oregon State University.

Schulting, R. (1994). The hair of the dog: the identification of a coast Salish dog-hair blanket from Yale, B.C. *Canadian Journal of Archaeology* **18**, 57-76.

Schwartz, M. (1997). *A History of Dogs in the Early Americas*. New Haven: Yale University Press.

Scott, E. M. (1985). French subsistence at Fort Michilimackinac 1715-1781: The clergy and the traders. *Mackinac Island State Park Commission, Archeological Completion Report Series* **9**.

Scott, K. W. & Wilson, M. (1984). Dead Indian Creek local fauna. *The Wyoming Archaeologist* **27**, 51- 62.

Seler, E. (1890). The dog in ancient Mexico. *Congress des Amricanistes*, Berlin 1888, Session **7**, 321-324.

Semken, H. A., Jr. (1973). Faunal identifications. *University of Iowa Archeologist Report* **4**, 66-67.

Semken, H. A., Jr. (1980). Holocene climatic reconstructions derived from the three micromammal cultural horizons of the Cherokee Sewer site, Northwestern Iowa. In (D. C. Anderson & H. A. Semken, Jr., Eds) *The Cherokee Excavations: Holocene Ecology and Human Adaptations in Northwestern Iowa*. New York: Academic Press, pp. 67-99.

Semken, H. A., Jr., & Falk, C. R. (1987). Late Pleistocene/Holocene mammalian faunas and environmental changes on the northern plains of the United States. *Illinois State Museum Scientific Papers* **22**, 176-313.

Semken, H. A., Jr., Miller, C. & Cordell, J. (1990). Inventory of identified vertebrates recovered from the Ft. Randall Historic site, South Dakota (39GR15). Ms. on file. Omaha, NE: US Army Corps of Engineers.

Shigehara, N. (1994). Morphological changes in Japanese ancient dogs. *Archaeozoologica* **6 (2)**, 79-94.

Shigehara, N., Guoqin, Q., Komiya, H. & Jing, Y. (1998). Morphological study of the ancient dogs from three Neolithic sites in China. *International Journal of Osteoarchaeology* **8**, 11-22.

Shigehara, N. & Onodera, S. (1984). Skeletal remains of the domestic dogs from Tagara shellmound. *Journal of the Anthropological Society of Nippon* **92**, 187-210. [Japanese, English tables & abstract].

Simpson, G. G. (1928). Pleistocene mammalian fauna of the Seminole Field, Pinellas County, florida. *Bulletin of the American Museum of Natural History* **56**, 561-599.

Simoons, F. J. (1994). *Eat Not This flesh: Food Avoidances From Prehistory to the Present*. Madison: University of Wisconsin Press.

Singer, M. (1968). Pygmies and their dogs: A note on culturally constituted defence mechanisms. *Ethos* **6**, 270-279.

Skaggs, Opal. (1946). A study of the dog skeletons from Indian Knoll with special reference to the coyote as progenitor. *University of Kentucky Reports in Anthropology* **4**, 341-355.

Smith, B. A. (1985). The use of animals at the 17th century mission of St. Ignace. *Michigan Archaeologist* **31**, 97-122.

Smith, B. D. (1975). Middle Mississippi exploitation of animal populations. *University of Michigan Museum of Anthropology Papers* **57**.

Smith, J. H. (1972). Faunal remains from the Evans Mound site. In (M. S. Berry, Ed) *The Evans Site*. Salt Lake City: University of Utah, Anthropology Special Report, pp. 220-233.

Snyder, Lynn M. (1991). Barking mutton: ethnohistoric, archaeological and nutritional evidence pertaining to the domestic dog as a Native American food resource on the Plains. *Illinois State Museum Scientific Papers* **23**, 359-378.

Soper, J. D. (1940). Eskimo dogs of the Canadian Arctic. *Canadian Geographical Journal* **20**, 96-108.

Speck, F. G. (1925). Dogs of the Labrador Indians. *Natural History* **25**, 58-64.

Sprott, J. E. (1997). Christmas, basketball, and sled dog races: common and uncommon themes in the new seasonal round in an Inupiaq village. *Arctic Anthropology* **34**, 68-85.

Stallibrass, S. (1990). Canid damage to animal bones: two current lines of research. In (D. E. Robinson, Ed) *Experimentation and Reconstruction in Environmental Archaeology*. Oxford: Oxbow Books, pp. 151-166.

Stewart, F. L. (1974). *Faunal Remains from the Nodwell Site (BcHi - 3) and from Four Other Sites in Bruce County, Ontario*. Archaeological Survey of Canada Mercury Series **16**.

Stewart, M. C. (1979). Subsistence in the St. Johns regions in the Alderman site. *Florida Anthropologist* **32**, 52-74.
Stewart, T. D. (1963). Skeletal remains of aboriginal dogs. *The Archeolog* **15**, 54-58.
Strong, J. A. (1985). Late Woodland dog ceremonialism on Long Island in comparative and temporal perspective. *The Bulletin and Journal of Archaeology for New York State* **91**, 32-38.
Styles, B. W. (nd). Archaic faunal exploitation at the Napoleon Hollow site. Ms. on file. Springfield: Illinois State Museum.
Styles, B. W. (1981). Faunal exploitation and resource selection. *Northwestern University Archeological Program Scientific Papers* **3**.
Styles, B. W. & Purdue, J. R. (1984). Faunal exploitation at the Cedar Grove site. *Arkansas Archeological Survey Research Series* **23**, 211-226.
Styles, B. W. & Purdue, J. R. (1986). Middle Woodland faunal exploitation. *Kampsville Archeology Center Research Series* **6**, 513-526.
Styles, B. W., Purdue, J. R. & Colburn, M. L. (1985). Analysis of faunal remains. *Arkansas Archeological Survey Research Series* **24**, 58-72.
Swanson, E. H. (1972). *Birch Creek: Human Ecology in the Cool Desert of the Northern Rocky Mountains 9,000 B.C. - A.D. 1950.* Pocatello: Idaho State University Press.
Swigart, E. K. (1987). The Woodruff Rockshelter site - 61FL26: An interim report: Faunal analysis as a means to evaluate environment and culture. *Archeological Society of Connecticut* **50**, 43-75.
Switzer, L. L. (1992). *An osteometric comparison of Northwest aboriginal dogs*. Master's Thesis, Washington State University.
Teichert, M. (1993). Size and utilization of the most important domesticated animals in Central Europe from the beginning of domestication until the late Middle Ages. In (A. Clason, S. Payne & H.-P. Uerpmann, Eds) *Skeletons in Her Cupboard: Festschrift for Juliet Clutton-Brock.* Oxford: Oxbow Books Monograph **34**, pp. 235-238.
Tchernov, E. (1993). From sedentism to domestication-a preliminary review for the southern Levant. In (A. Clason, S. Payne & H.-P. Uerpmann, Eds) *Skeletons in Her Cupboard: Festschrift for Juliet Clutton-Brock.* Oxford: Oxbow Books Monograph **34**, pp. 189-233.
Tchernov, E. (1993). The impact of sedentism on animal exploitation in the southern Levant. In (H. Buitenhuis & A. T. Clason, Eds) *Archaeozoology of the Near East: Proceedings of the First International Symposium on the Archaeozoology of South-western Asia and Adjacent Areas,*. Universal Book Services, Leiden, pp. 10-26.
Tchernov, E. & Horwitz, L. K. (1991). Body size diminution under domestication: unconscious selection in primeval domesticates. *Journal of Anthropological Archaeology* **10**, 54-75.
Tchernov, E. & Valla, F. F. (1997). Two new dogs, and other Natufian dogs, from the Southern Levant. *Journal of Archaeological Science* **24**, 65-95.
Theler, J. L. (1981). Faunal remains (47-Bf-45). *Wisconsin Archeologist* **62**, 34-44.
Theler, J. L. (1987). Woodland tradition economic strategies: animal resource utilization in southwestern Wisconsin and northeastern Iowa. *University of Iowa Archeologist Report* **17**.
Theler, J. L. (1989). The Pammel Creek site faunal remains. *Wisconsin Archeologist* **70**, 157-242.
Theler, J. L. (1992). Faunal remains from the Tremaine site (47LC95). *Mississippi Valley Archaeological Center, University of Wisconsin-La Crosse, Report Investigations* **133**, 50-58.
Thurman, M. D. (1988). On the identity of the Chariticas [Sarh Rikka]: Dog eating and pre-horse adaptation on the High Plains. *Plains Anthropologist* **33**, 159-170.
Titcomb, M. (1969). *Dog and Man in the Ancient Pacific*. Bishop Museum Special Publication **59**.
Tonni, E. P. & Politis, G. G. (1981). Un gran canido del Holoceno de la Provincia de Buenos Aires y el registro prehispanico de *Canis (Canis) familiaris* en las areas pampeana y patagonica. *Ameghiniana. Revista de la Asociacin Paleontologica Argentina* **18**, 251-265. [A large canid from the Holocene of Buenos Aires Province and the pre-Hispanic record of *Canis (Canis) familiaris* in the Pampean and Patagonian areas.] [Spanish, English summary].
Torre, D. (1981). The Ruscinian and Villafranchian dogs of Europe. In (M. V. A. Sastry et al., Eds) *Field Conference of the Neogene-Quaternary Boundary. India, 1979.* Calcutta: Proceedings of the Geological Survey of India, pp. 199-201.
Tozzer, A. M. & Allen, G. M. (1910). Animal figures in the Maya Codices. *Papers of the Peabody Museum of American Archaeology and Ethnology* **4**.
Turnbull, P. F. & Charles A. Reed. (1974). The fauna from the terminal Pleistocene of Palegawra Cave. *Fieldiana Anthropology* **63**, 81-164.
Varner, J. G. & Varner, J. J. (1983). *Dogs of the Conquest.* Norman: University of Oklahoma Press.
Walker, D. N. (1975). Non-bison remains from the Vore Bison Jump, Crook County, Wyoming. *Plains Anthropologist* **20**, 217-224.
Walker, D. N. (1980a). The Garnsey site canid. *University of Michigan, Museum of Anthropology Technical Reports* **12**, 344-365.
Walker, D. N. (1980b). The Vore site local fauna. *Plains Anthropologists Memoir* **16**, 154-169.
Walker, D. N. (1982). Early Holocene vertebrate fauna. In (G. C. Frison & D. J. Stanford, Eds) *The Agate Basin Site: A Record of the Paleoindian Occupation of the Northwestern High Plains.* New York: Academic Press, pp. 274-308.
Walker, D. N. (1983). Faunal remains and subsistence practices at 5MF605, a fur trading post in northwest Colorado. *Southwestern Lore* **49**, 6-29.
Walker, D. N. (1985). REVIEW: Origins of the domestic dog, by Stanley J. Olsen. *Zooarchaeological Research News* **4**, 8-10.
Walker, D. N. (1987a). Late Pleistocene/Holocene environmental changes in Wyoming: the mammalian record. *Illinois State Museum Scientific Papers* **22**, 334-393.
Walker, D. N. (1987b). Horner site local fauna: vertebrates. In (G. C. Frison & L. C. Todd, Eds) *The Horner Site: Type Site of the Cody Cultural Complex.* New York: Academic Press, pp. 327-345.
Walker, D. N. (1990a). Canid remains from two sites (48CR4815 & 48CR4816) at Seminoe Reservoir, Carbon County, Wyoming. Ms. on file. Laramie, WY: Office of the Wyoming State Archaeologist.
Walker, D. N. (1990b). Preliminary report on faunal remains from Sand Canyon Pueblo (5MT765), Colorado. Ms. on file. Cortez, CO: Crow Canyon Archeological Center.

Walker, D. N. (1991). Comments on canid skulls from 48YE1 (Nebraska). Ms. on file. Laramie, WY: University of Wyoming.
Walker, D. N. (1992a). Appendix I: A canid skull from 48CA582, Campbell County, Wyoming. In: Results of the archaeological investigations at 48CA582, Campbell County, Wyoming. Ms. on file. Laramie, WY: Larson-Tibesar Associates, Inc.
Walker, D. N. (1992b). Zooarchaeology of Camp Grafton (32ED85). Ms. on file. Eagle, CO: Metcalf Archaeological Consultants.
Walker, D. N. (1993). Faunal remains. *Crow Canyon Archaeological Center Occasional Paper* **3**, 239-252.
Walker, D. N. (1995a). Faunal remains recovered from 1994 archaeological monitoring, Upper Sanger site (32OL12), North Dakota. Ms on file. Eagle, CO: Metcalf Archaeological Consultants.
Walker, D. N. (1995b). Canids from the Alkali Creek site. In (M. D. Metcalfe, Ed) Alkali Creek: A stratified record of prehistoric flint mining in North Dakota. Ms on file. Eagle, CO: Metcalf Archaeological Consultants, pp. 405-437.
Walker, D. N. (1995c). Faunal remains from 1994 excavations at the Rock Springs site, southeast Idaho. Ms. on file. Ogden, Utah: Weber State University.
Walker, D. N. (1996). Canid remains from the Powder River Supply Depot, Prairie County, Montana (24PE231). *Archaeology in Montana* **36**, 63-82.
Walker, D. N. & Frison, G. C. (1982). Studies on Amerindian dogs, 3: Prehistoric wolf/dog hybrids from the Northwestern Plains. *Journal of Archaeological Science* **9**, 125-172.
Walters, I. (1984). Gone to the dogs: A study of bone attrition at a central Australian campsite. *Mankind* **14**, 389-400.
Warren, R. E. (1986). Ice Glider faunal analysis and Yanktonai ethnohistory. *South Dakota Archaeological Society Special Publication* **10**, 146.
Waters, J. H. (1965). Animal remains from some New England Woodland sites. *Bulletin of the Archeological Society of Connecticut* **33**, 5-11.
Webb, S. D. (1974). *Pleistocene Mammals of Florida*. Gainesville: University Presses of Florida.
Webb, S. D. (1992). Surprise Cave local fauna. Ms. on file. Gainsville: Florida State Museum.
Wheeler, C. (1980). Mammalian remains from the burial at 5EA128. *Southwestern Lore* **46**, 43.
White, J. R. (1981). Faunal remains from the Eaton (Hopewell) Furnace site. *Ohio Journal of Science* **81**, 138-141.
White, T. E. (1955). The dog bones from the Buffalo Pasture site. *American Antiquity* **21**, 170-178.
White, T. E. (1955). Observations on the butchering technique of some aboriginal peoples: 7, 8, & 9. *American Antiquity* **21**, 170-178.
Wilson, G. L. (1924). The horse and the dog in Hidatsa culture. *Anthropological Papers of the American Museum of Natural History* **15**, 123-311.
Wilson, I. R., Crockford, S. J., Dahlstrom, B. & Twohig, K. (1995). EdQx43: Non-subsistence fauna. 1994 excavations at EdQx43 & EdQx44, Monte Creek, B.C. Ms. on file. Victoria, BC: Archaeology Branch **1994-160**, pp. 66-74.
Wilson, M. (1975). The Cactus site local fauna. *Archaeological Survey of Canada Mercury Series* **46**, 201-234.
Wing, E. S. (1965). Animal bones associated with two Indian sites on Marco Island, Florida. *Florida Anthropologist* **18**, 21-28.
Wing, E. S. (1978). Use of dogs for food: an adaptation to the coastal environment. In (B. L. Stark & B. Voorhies, Eds) *Prehistoric Coastal Adaptations: the Economy and Ecology of Maritime Middle America*. New York: Academic Press, pp 29-41.
Wing, E. S. (1984). Use and abuse of dogs. *Carnegie Museum Special Publication* **8**, 228-232.
Wing, E. S. (1989). Human use of canids in the Central Andes. *Advances in Neotropical Mammalogy* **1989**, 265-278.
Wing, E. S. (1991). Dog remains from the Sorc site on Vieques Island, Puerto Rico. *Illinois State Museum Scientific Papers* **23**, 379-386.
Wing, E. S. & Reitz, E. J. (1982). Prehistoric fishing communities of the Caribbean. *Journal of New World Archeology* **5**, 13-32.
Witty, T. A. & Frantz, W. P. (1964). The historic uses of dogs among the Plains Indians. *Kansas Anthropological Association Newsletter* **9**.
Wright, G. A., Proulx, R. & Koenig, T. (1982). A Native American burial from 48YE1, Fishing Bridge Peninsula, Yellowstone National Park. *The Wyoming Archaeologist* **25**, 2-26.
Young, E. R. (1890). *By Canoe and Dog Train Among the Cree and Salteaux Indians*. London: Charles E. Kelley.
Young, S. P. (1943). What was the early Indian dog? *American Forests* **49**, 571-573, 594, 603.
Young, S. P. (1944). What was the early Indian dog? *American Forests* **50**, 26-28, 32, 45.
Zeuner, F. E. (1963). *A History of Domesticated Animals*. London: Hutchinson.
Zeuner, F. E. (1958). Dog and cat in the Neolithic of Jericho. *Palestine Exploration Quarterly* **52-55**.

**Prehistoric and early historic wolves**

Aaris-Sorensen, K. (1977). The subfossil wolf, *Canis lupus* L., in Denmark. *Videnskabelige Meddelelser fra Dansk naturhistorisk Forening* **140**, 129-146.
Agenbroad, L. D. (1978). Excavation at the Hot Springs Mammoth site: a late Pleistocene animal trap. *Transactions of the Nebraska Academy of Science* **6**, 127-130.
Agenbroad, L. D. & Mead, J. I. (1986). Large carnivores from Hot Springs Mammoth Site, South Dakota. *National Geographic Research* **2**, 508-516.
Allen, J. A. (1876). Descriptions of some remains of an extinct species of wolf and an extinct species of deer from the Lead Region of the Upper Mississippi. *American Journal of Science* **3**, 47-51.
Allen, J. A. (1885). On an extinct type of dog from Ely Cave, Lee Co., Va. *Memoirs of the Museum of Comparative Zoology* **10**, 3-13.
Allison, I. S. (1966). *Fossil Lake, Oregon: Its Geology and Fossil Faunas*. Oregon State Monograph Studies Series, Earth Science **9**.
Anderson, E. (1968). *Fauna of the Little Box Elder Cave, Converse County, Wyoming: the Carnivora*. University of Colorado Studies Series, Earth Science **6**.

Aybes, C. & Yalden, D. W. (1995). Place-name evidence for the former distribution and status of wolves and beaver in Britain. *Mammal Review* **25**, 201-227.

Bibliography of Fossil Vertebrates. (1998). Bibliography of fossil vertebrates. On the Internet: http://eteweb.lscf.ucsb.edu/bfv/

Bonnichsen, R. (1985). High-altitude adaptation at False Cougar Cave. *National Geographic Research Report* **20**, 27-40.

Bonnichsen, R., Graham, R. W., Geppert, T., Jacobson, G. L., Konrad, V., Mead, J. I., Oliver, J. S., Oliver, S. G., Purdue, J. R., Schnurrenberger, D., Stuckenrath, R., Tratebas, A. & Young, D. E. (1985). False Cougar and Shield Trap: two high altitude caves, Pryor Mountains, Montana. *Center for the Study of Early Man, Pryor Mountain Paper* **22**, 1-18.

Burns, J. A. (1984). *Late Quaternary palaeoecology and zoogeography of southwestern Alberta: vertebrate and palynological evidence from two Rocky Mountain caves*. Ph.D. Thesis, University of Toronto.

Butzeck, S., Stubbe, M. & Piechocki, R. (1988). Contribution to the history of the mammal fauna of East Germany: 3. The wolf *Canis lupus* L., 1758. *Hercynia* **25**, 278-317. [German].

Cannon, K. P. (1992). A review of archeological and paleontological evidence for prehistoric presence of wolf and related prey species in the Northern and Central Rockies Physiographic Provinces. In (J. D. Varley & W. G. Brewster, Eds) *Wolves for Yellowstone? Volume VI: Research and Analysis.* Yellowstone National Park, WY: US National Park Service, pp 175-265.

Chomko, S. A. (1976). Faunal exploitation in the Initial Middle Missouri variant. *Missouri Archaeological Society Research Series* **13**, 35-41.

Chomko, S. A. & Gilbert, B. M. (1987). The late Pleistocene/Holocene faunal record in the northern Bighorn Mountains, Wyoming. *Illinois State Museum Scientific Papers* **22**, 394-409.

Chorn, J. (1976). A subrecent mammalian fauna from the Big Snowy Mountains. Ms. on file. Springfield: Illinois State Museum.

Churcher, C. S. (1959). Fossil *Canis* from the tar pits of La Brea, Peru. *Science* **130**, 564-565.

Cope, E. D. (1883). On the extinct dogs of North America. *American Naturalist* **17**, 235-249.

Corner, R. G.. (1977). A late Pleistocene-Holocene vertebrate fauna from Red Willow county, Nebraska. *Transactions of the Nebraska Academy of Science* **4**, 77-93.

Corner, R. G. & Myers, T. P. (1976). Paleontological and archaeological remains from gravel pits in Red Willow County, Nebraska. *Proceedings of the Nebraska Academy of Sciences* **86**, 45.

Cowan, I. McT. (1954). The occurrence of the Pleistocene wolf *Canis dirus* in the Rocky Mountains of Central Alberta. *Canadian Field Naturalist* **68**, 44.

Curnow, E. (1969). *The history of the eradication of the wolf in Montana*. Master's Thesis, University of Montana.

Czaplewski, N. J., & Mead, J. I. (1990). Small mammals from Hot Springs Mammoth site, South Dakota. *Mammoth Site of Hot Springs, South Dakota Scientific Papers* **1**, 40-41.

Dallman, J. E. (1983). Choice of diet: response to climate change. *University of Iowa Archaeologist Report* **16**, 1-133.

Dalquest, W. W., Roth, E., & Judd, F. (1969). The mammal fauna of Schulze Cave, Edwards County, Texas. *Bulletin of the Florida State Museum* **13**, 205-276.

Dalquest, W. W. & Schultz, G. E. (1992). *Ice Age Mammals of Northwestern Texas*. Wichita Falls, TX: Midwestern State University Press.

Davis, L. C. (1975). *Late Pleistocene geology and paleoecology of the Spring Valley Basin, Mead County, Kansas*. Ph.D. Thesis, University of Iowa.

Davis, L. C. (1987). Late Pleistocene/Holocene environmental changes in the central plains of the United States: the mammalian record. *Illinois State Museum Scientific Papers* **22**, 88-143.

Downs, T., Howard, H., Clements, T. A. & Smith, G. A. (1959). Quaternary animals from Schuiling Cave in the Mojave Desert, California. *Contributions in Science of the Los Angeles County Museum* **29**, 1-21.

Driver, J. C. (1988). Late Pleistocene and Holocene vertebrates and paleoenvironments from Charlie Lake Cave, northeast British Columbia. *Canadian Journal of Earth Science* **25**, 1545-1553.

Easterla, D. & Jackson, G. (1961). Discovery of *Canis dirus* in Missouri. *The Bluebird* **28**, 3-4.

Eckles, D., Wedel, D. & Walker, D. N. (1993). An early historic period horse skeleton from southwestern Wyoming. *The Wyoming Archaeologist* **38**, 55-68.

Fleharty, E. D. (1995). *Wild Animals and Settlers on the Great Plains*. Norman: University of Oklahoma Press.

Fonda, S. S. & Czebieniak, A. P. (1986). Late Pleistocene fauna of the Hollidaysburg Fissure. *Virginia Division of Mines Publication* **75**, 128-129.

Frankforter, W.D. (1969). Faunal study of large ruminants from Mill Creek culture sites in northwestern Iowa. *Journal of the Iowa Archeaological Society* **16**, 286-301.

Frison, G. C., Wilson, M. & Wilson, D. J. (1976). Fossil bison and artifacts from an early altithermal period arroyo trap in Wyoming. *American Antiquity* **41**, 28-57.

Furlong, E. L. (1906). The exploration of Samwell Cave. *American Journal of Science* **22**, 235-247.

Galbreath, E. C. (1938). Post-glacial fossil vertebrates from east central Illinois. *Field Museum of Natural History Geological Series* **VI**, 303-396.

Galbreath, E. C. (1964). A dire wolf skeleton and Powder Mill Creek Cave, Missouri. *Transactions of the Illinois State Academy of Science* **57**, 224-242.

Goetze, J. R. (1989). Mammalian faunas of a Late Pleistocene-Holocene terrace of the Red River, Tillman County, Oklahoma. *Texas Journal of Science* **41**, 205-209.

Goslin, R. M. (1955). Animal remains from Ohio rockshelters. *Ohio Journal of Science* **55**, 358-362.

Grau, J. M. T., Puig, R. & Ruiz-Olma, J. (1990). Hunting of the wolf (*Canis lupus* L., 1978) in Giron (N.E. Iberia) during the 18th and 19th centuries: example of the use of archive data. *Miscellania Zoologica* **14**, 217-223.

Grayson, D. K. (1985). The paleontology of Hidden Cave: birds and mammals. *American Museum of Natural History Anthropological Papers* **61**, 125-161.

Guilday, J. E. (1962). The Pleistocene local fauna of the Natural Chimneys, Augusta County, Virginia. *Annals of the Carnegie Museum of Natural History* **36**, 87-122.

Guilday, J. E. & Bender, M.S. (1958). A recent fissure deposit in Bedford County, Pennsylvania. *Annals of the Carnegie Museum of Natural History* **35**, 127-138.

Guilday, J. E. & M. S. Bender, M. S. (1960). Late Pleistocene records of the yellow-cheeked vole, Microtus xanthognathus (Leach). *Annals of the Carnegie Museum of Natural History* **35**, 315-330.

Guilday, J. E. Martin, P. S. & McCrady, A. D. (1964). New Paris No. 4: a Pleistocene cave deposit in Bedford County, Pennsylvania. *Bulletin of the National Speleological Society* **26**, 121-194.

Guilday, J. E. & Tanner, D. P. (1962). Animal remains from the Quaker State Rockshelter (36Ve27), Venango County, Pennsylvania. *Pennsylvania Archaeologist* **32**, 131-137.

Gustafson, C. E. (1972). *Faunal remains from the Marmes Rockshelter and related archeological sites in the Columbia Basin.* Ph.D. Thesis, Washington State University.

Hager, M. W. (1972). A late Wisconsinan-Recent vertebrate fauna from Chimney Rock Animal Trap, Larimer County, Colorado. *University of Wyoming Contributions in Geology* **11**, 63-71.

Harris, A. H. (1970). The Dry Cave mammalian fauna and late pluvial conditions in southeastern New Mexico. *Texas Journal of Science* **22**, 3-27.

Harris, A. H. (1977). Wisconsin age environments in the northern Chihuahuan desert evidence from the higher vertebrates. *National Park Service Transactions and Proceedings Series* **3**, 23-52.

Harris, A. H. (1980). The paleocology of Dry Cave, New Mexico. *National Geographic Society Research Report* **12**, 331-338.

Harris, A. H. (1985). *Late Pleistocene Vertebrate Paleoecology of the West.* Austin: University of Texas Press.

Harris, A.. H. & Findley, J. S. (1964). Pleistocene-Recent fauna of the Isleta Caves, Bernalillo County, New Mexico. *American Journal of Science* **262**, 114-120.

Harting, J. E. (1994). *A Short History of the Wolf in Britain.* Whitestable, UK: Pryor Publications.

Hawksley, O. (1986). Remains of Quaternary vertebrates from Ozark caves and other miscellaneous sites. *Missouri Speleologist* **26**, 1-67.

Hawksley, O., Reynolds, J. & McGowan. J. (1963). The dire wolf in Missouri. *Missouri Speleology* **5**, 63-72.

Hay, O. P. (1914). The Pleistocene mammals of Iowa. *Iowa Geolgoical Survey* **23**, 1-653.

Heaton, T. H. (1985). Quaternary paleontology and paleoecology of Crystal Ball Cave, Millard County, Utah: with emphasis on mammals and description of a new species of fossil skunk. *Great Basin Naturalist* **45**, 337-390.

Heaton, T. H. (1987). Initial investigation of vertebrate remains from Snake Creek Burial Cave, White Pine County, Nevada. *Current Research in the Pleistocene* **4**, 107-109.

Heizer, R. F. & Napton, L. K. (1970). Archaeology and the prehistoric Great Basin lacustrine subsistence regime as seen from Lovelock Cave, Nevada. *University of California, Contributions in Archaeological Research* **10**.

Indeck, J. (1987). *Sediment analysis and mammal faunal remains from Little Box Elder Cave, Wyoming.* Ph.D. Thesis, University of Colorado.

Janis, C. M. & Wilhelmm P. B. (1993). Were there mammalian pursuit predators in the Tertiary? Dances with wolf avatars. *Journal of Mammalian Evolution* **1,** 103-125.

Jefferson, G. T. (1991). Rancholabrean age vertebrates from the Southeastern Mojave Desert, California. In (R. E. Reynolds, Ed) *Crossing the Borders: Quaternary Studies in Eastern California and Southwestern Nevada.* Redlands: San Bernardino County Museum Association Special Publication, pp. 163-175.

Jefferson, G. T. (1991). *A Catalogue of Late Quaternary Vertebrates from California: Two, Mammals.* Museum of Natural History of Los Angeles County Technical Report **7**.

Johnson, E. (1987). *Lubbock Lake, Late Quaternary Studies on the Southern High Plains.* College Station: Texas A&M University Press.

Kelly, L. S. (1979). Animal resource exploitation by early Cahokia populations on the Merrell Tract. *Illinois Archaeological. Survey Circular* **4**, 1-28.

Kurtén, B. (1974). A history of coyote-like dogs (Canidae, Mammalia). *Acta Zoologica Fennica* **140**.

Kurtén, B. (1984). Geographic differentiation in the Rancholabrean dire wolf (*Canis dirus* Leidy) in North America. *Special Publication of the Carnegie Museum of Natural History* **8**, 219-227.

Kurtén, B. (1988). *On Evolution and Fossil Mammals.* New York: Columbia University Press.

Kurtén, B. & Anderson, E. (1980). *Pleistocene Mammals of North America.* New York: Columbia University Press.

Kurtén, B. & Kaye, J. M. (1982). Late Quaternary Carnivora from the Black Belt, Mississippi. *Boreas* **11**, 47-52.

Leidy, J. (1889). Notes and description of fossils in caves and crevices of the limestone rocks of Pennsylvania. *Geological Survey of Pennsylvania Annual Report* **1887**, 1-20.

Lindsay, E. H. (1978). Late Cenozoic vertebrate faunas, southeastern Arizona. *New Mexico Geological Society Guidebook, 29th Field Conference* **b**, 269-275,

Lipps, L. & Ray, C. E. (1967). The Pleistocene fossiliferous deposit at Ladds, Bartow County, Georgia. *Bulletin of the Georgia Academy of Science* **25**, 113-119.

Lundelius, E. L., Jr. (1972). Vertebrate remains from the Gray Sand. In (J. J. Hester, Ed) *Blackwater Locality No. 1: A Stratified Early Man Site in Eastern New Mexico.* Dallas, TX: Fort Burgwin Research Center, Southern Methodist University, pp. 148-163.

Lundelius, E. L., Jr. (1972). Fossil vertebrates from the Late Pleistocene Ingleside Fauna, San Patricio County, Texas. *Economic Geology, University of Texas at Austin Report of Investigations* **77**.

Lyman, R. L. (1985). The paleozoology at the Avey's Orchard site. *Eastern Washington University, Report Archaeology and History* **100-42**, 243-320.

Lyman, R. L. & Livingston, S. D. (1983). Late Quaternary mammalian zoogeography of eastern Washington. *Quaternary Research* **20**, 360-373.

Martin, T. J. (1986). *A faunal analysis of Fort Ouiatenon, an eighteenth century trading post in the Wabash Valley of Indiana.* Ph.D. Thesis, Michigan State University.

Martin, T. J. (1988). Animal remains from the Cahokia Wedge site. *Illinois Historical Preservation Agency Studies in Illinois Archaeology* **3**, 221-234.

Martin, T. J. & Masulis, M. C. (1988). Preliminary report on animal remains from Fort De Chartres Appendix D, in (D. Keene, Ed.) *Archeological Excavations at Fort De Chartres, 1985-1987.* Springfield, IL: Historical Preservation Agency.

Martin, T. J. & Masulis, M. C. (1992). Faunal exploitation at the Little Muddy Rockshelter. *American Resources Group Ltd., Cultural Resource Management Report* **186**, 427-472.

Matthew, W. D. (1916). The grim wolf of the tar pits. *American Museum Journal* **16**, 45-47.

Mead, E. M. & Mead, J. I. (1989). Snake Creek Burial Cave and a review of the Quaternary mustelids of the Great Basin. *Great Basin Naturalist* **49**, 143-154.

Melton, W. G. (1985). Late Pleistocene mammals and artifacts recovered from Blacktail Cave, Lewis and Clark County, Montana. *National Geographic Society Research Report* **19**, 415-421.

Merriam, J. C. (1903). The Pliocene and Quaternary Canidae of the Great Valley of California. *University of California Bulletin of the Department of Geology* **3**, 277-290.

Merriam, J. C. (1912). The fauna of Rancho La Brea. Part II. Canidae. *Memoirs of the University of California* **1**, 215-272.

Miller, S. J. (1979). The archeological fauna of four sites in Smith Creek Canyon. *Nevada State Museum Anthropology Papers* **17**, 273-329.

Moodie, R. L. (1928). Studies in Paleonathology, XXII -- Pyorrhea Alveolaris in a Pleistocene wolf. *Annals of Medical History* **10**.

Moodie, R. L. (1928). The evidence of pyorrhea, dead teeth and gingival infections in the mandibles of the Pleistocene giant wolf (*Aenocyon dirus*) from the Rancho La Brea. *The Pacific Dental Gazette* **136**, 414-419.

Moodie, R. L. (1929). Studies in Paleonathology, XIX -- An alveolar abscess in a fossil mammal. *The Pacific Dental Gazette* **37**, 428-433.

Moodie, R. L. (1930). A suggestion of rickets in the Pleistocene. *American Journal of Surgery* **10**, 162-163.

Nelson, M. E. & Madsen, J. H., Jr. (1980). A summary of Pleistocene, fossil vertebrate localities in the northern Bonneville Basin of Utah. *Utah Geology and Mining Survey, Bulletin* **116**, 97-113.

Nelson, M. E. & Madsen, J. H., Jr. (1986). Canids from the Late Pleistocene of Utah. *Great Basin Naturalist* **46**, 415-420.

Nelson, M. E. & Madsen, J. H., Jr. (1987). A review of Lake Bonneville shoreline faunas (late Pleistocene) of northern Utah. *Utah Geological Association Publication* **16**, 320-333.

Nigra, J. O. & Lance, J. F. (1947). A statistical study of the metapodials of the Dire wolf group from the Pleistocene of Rancho La Brea. *Bulletin of the Southern California Academy of Science* **46**, 26-34.

Nowak, R. M. (1979). North American Quaternary *Canis*. *University of Kansas Museum of Natural History Monograph* **6**, 1-154.

Oliver, J. S. (1989). Analogues and site context: bone damages from Shield Trap Cave (24CB91), Carbon County, Montana. In (R. Bonnichsen & M. Sorg, Eds.) *Bone Modification*. Orono: Center for the Study of Early Man, pp. 73-98.

Olsen, E. C. (1940). A late Pleistocene fauna from Herculaneum, Missouri. *Journal of Geology* **48**, 32-57.

Packard, E. L. (1950). A large wolf from the Pleistocene of Willamette Valley, Oregon. *Geology News Letter of the Geological Society of Oregon* **16**, 89-90.

Parmalee, P. W. (1962). Additional faunal records from the Kingston Lake site, Illinois. *Transactions of the Illinois Academy of Science* **55**, 6-12.

Parmalee, P. W. (1959). Use of mammalian skulls and mandibles by prehistoric Indians of Illinois. *Transactions of the Illinois Academy of Science* **52**, 85-95.

Parmalee, P. W. (1973). Identification of Late Woodland faunal material from the Schild sites, Illinois. *Illinois Archaeological Survey Bulletin* **9**, 138-140.

Parmalee, P. W. (1975). A general summary of the vertebrate fauna from Cahokia. *Illinois Archaeological Survey Bulletin* **10**, 137-155.

Parmalee, P. W. & Stephens, D. (1972). A wolf mask and other carnivore skull artifacts from the Palestine site, Illinois. *Pennsylvania Archaeologist* **42**, 71-75.

Purdue, J. R. & Styles, B. W. (1987). Changes in the mammalian fauna of Illinois and Missouri during the late Pleistocene and Holocene. *Illinois State Museum Scientific Papers* **22**, 144-175.

Ray, C. E. (1958). Additions to the Pleistocene mammalian fauna from Melbourne, Florida. *Bulletin of the Museum Comparative Zoology Harvard* **119**, 421-449.

Ray, C. E. (1967). Pleistocene mammals from Ladds, Bartow County, Georgia. *Bulletin of the Georgia Academy of Science* **25**, 120-150.

Ray, C. E. & Lipps, L. (1968). Additional notes on the Pleistocene mammals from Ladds, Georgia. *Bulletin of the Georgia Academy of Science* **26**, 63.

Reynolds, J. F. (1962). *A preliminary study of the Pleistocene vertebrate fauna from (Pul-103) Cave, Missouri.* Master's Thesis, Central Missouri State College.

Schultz, C. B. & Howard, E. B. (1935). The fauna of Burnet Cave, Guadalupe Mountains, New Mexico. *Proceedings of the Academy of Natural Science of Philadelphia* **87**, 273-298.

Semken, H. A., Jr. (nd). Vertebrate remains from Coldwater Cave, northeastern Iowa. Ms. on file. Iowa City: University of Iowa.

Shakespear, S. (1975). An osteometric description of the Pleistocene dire wolf, *Canis dirus*, from the Maricopa Brea of California (abstr.). *Proceedings of the Utah Academy of Sciences, Arts, and Letters* **52**, 77-78.

Sharp, H. S. (1978). Comparative ethnology of the wolf and the Chipewyan. In (R.Hall & H. S. Sharp, Eds) *Wolf and Man: Evolution in Parallel*. New York: Academic Press, pp 55-80.

Skinner, M. F. (1942). The fauna of Papago Springs Cave, Arizona. *Bulletin of the American Museum of Natural History* **80**, 143-220.

Slaughter, B. H. (1975). Ecological interpretation of Brown Sand Wedge local fauna. In (F. Wendorf & J. J. Hester, Eds) *Late Pleistocene Environments on the Southern High Plains*. Dallas, TX: Fort Burgwin Research Center, Southern Methodist University, pp. 179-192.

Slaughter, B. H., Crook, W. W., Jr., Harris, R. K., Allen, D. C. & Seifert, M. (1962). The Hill-Shuler local faunas of the Upper Trinity River, Dallas and Denton Counties, Texas. *University of Texas Bureau of Economic Geology Report of Investigations* **48**.

Stevenson, M.. (1978). Dire wolf systematics and behavior. In (R.Hall & H. S. Sharp, Eds) *Wolf and Man: Evolution in Parallel*. New York: Academic Press, pp. 179-196.

Stock, C. (1918). The Pleistocene fauna of Hawver Cave. *University of California Bulletin of the Department of Geology* **10**, 462-515.

Stock, C. (1929). A census of the Pleistocene mammals of Rancho La Brea, based on the collection of the Los Angeles Museum. *Journal of Mammalogy* **10**, 281-289.

Stock, C. (1938). A coyote-like wolf jaw from the Rancho La Brea Pleistocene. *Bulletin of the Southern California Academy of Sciences* **37**, 49-51.

Stock, C. & Lance, J. F. (1948). The relative lengths of limb elements in *Canis dirus*. *Bulletin of the Southern California Academy of Science* **47**, 79-84.

Stock, C., Lance, J. F. & Nigra, J. O. (1946). A newly mounted skeleton of the extinct dire wolf from the Pleistocene of Rancho La Brea. *Bulletin of the Southern California Academy of Science* **45**, 108-110.

Townsend, J. K. (1850). On the giant wolf of North America. *Journal of the Academy of Natural Science of Philadelphia* **2**, 1-334.

Van Valkenburgh, B. & Hertel, F. (1993). Tough times at La Brea: Tooth breakage in large carnivores of the late Pleistocene. *Science* **261**, 456-459.

Voorhies, M. R. (1990). Vertebrate paleontology of the proposed Norden Reservoir area: Brown, Cherry, & Keya Paha counties, Nebraska. *University of Nebraska, Archeological Research Technical Report* **82-09**.

Walker, D. N. (1975). *A cultural and ecological analysis of the vertebrate fauna from the Medicine Lodge Creek site (48BH499).* Master's Thesis, University of Wyoming.

Walker, D. N. (1996). Preliminary results of initial paleontological testing at Wolf Den Cave, Bighorn County, Wyoming. Ms. on file. Cody, Wyoming: Bureau of Land Management.

Walker, D. N. (1984). The wolves of Natural Trap Cave, Wyoming. In (L.Martin & B. M.Gilbert, Eds) *Natural Trap Cave, Wyoming*. University of Kansas Museum of Natural History Publications.

Wang, Xiaoming. (1990). Pleistocene dire wolf remains from the Kansas river with notes on dire wolves in Kansas. *Occasional Papers of the Kansas Museum of Natural History* **137**, 1-8.

Weber, F. R., Hamilton, T. D., Hopkins, D. M., Repenning, C. A. & Haas, H. (1981). Canyon Creek: a Late Pleistocene vertebrate locality in interior Alaska. *Quaternary Research* **16**, 167-180.

Weems, R. E. & Higgins, B. B. (1977). Post-Wisconsinan vertebrate remains from a fissure deposit near Ripplemead, Virginia. *National Speleological Society Bulletin* **39**, 106-108.

Weinand, D. C. & Reitz, E. J. (1992). Vertebrate fauna from the 1970 excavation at Stallings Island (9Co1), Georgia. Ms.on file. Athens: University of Georgia.

White, J. A., McDonald, H. G., Anderson, E. & Soiset, J. M. (1984). Lava blisters as carnivore traps. *Carnegie Museum of Natural History Special Publication* **8**, 241-256.

Wineman, E. (1930). White wolf, foe of cattlemen, is dead. *Montana Wild Life* **2**, 6-7.

Wyman, W. D. (1984). *Wisconsin and North Country Wolf and Bear Stories: Seventy-five Stories Told by Pioneer Settlers, Lumberjacks, and Other Yarn Spinners*. River Falls, WI: University of Wisconsin-River Falls Press.

Young, S. P. (1925). The old wolf "Three Toes" of Harding County, South Dakota. *Daily Science News Bulletin*. August 13.

Young, S. P. (1946). *The Wolf in North American History*. Caldwell, ID: The Caxton Printers, Ltd.

Zeimans, G. & Walker, D. N. (1974). Bell Cave, Wyoming: preliminary archaeological and paleontological investigations. *Wyoming Geological Survey Report of Investigations* **10**, 88-90.

**Canid hybridisation**

Bee, J. W. & Hall, E. R. (1951). An instance of coyote-dog hybridization. *Transactions of the Kansas Academy of Science* **54**, 73-77.

Belshaw, M. (1990). *All the Loving Wolves: Living and Learning with Wolf Hybrids*. Sante Fe, MN: Red Crane Books.

Bezdek, H. (1944). A red-gray fox hybrid. *Journal of Mammalogy* **25**, 90.

Cole, L. J. & Shackelford, R. M. (1946). Fox Hybrids. *Transactions of the Wisconsin Academy Science Arts and Letters* **38**, 315-332.

Cook, R. (1952). The coy-dog: hybrid with a future? *Journal of Heredity* **43**, 71-73.

Danilkin, A. (1979). Hunting of roe deer by wolf-dog hybrids. *Okhota i okhotnich'e khozyzystov* **3**, 18-19.

Dice, L. R. (1942). A family of dog-coyote hybrids. *Journal of Mammalogy* **23**, 186-192.

Fox, M. W., Halperin, S., Wise, A. & Kohn, E. (1976). Species and hybrid differences in frequency of play and agonistic actions in canids. *Zeitschrift fuer Tierpsychologie* **40**, 194-209.

Freeman, R. C. (1976). *Coyote x dog hybridization and red wolf influence in the wild Canis of Oklahoma*. Master's Thesis, University of Oklahoma.

Freeman, R. C. & Shaw, J. H. (1979). Hybridization in *Canis* (Canidae) in Oklahoma. *Southwestern Naturalist* **24**, 485-500.

Freeman, R. S., Adorjan, A. & Pimlott, D. H. (1961). Cestodes of wolves, coyotes, and coyote-dog hybrids in Ontario. *Canadian Journal of Zoology* **39**, 527-532.

Fuller, A. R. (1885). Dog-wolf hybrid. *Forest and Stream* **25**, 507.

Galaka, B. A. (1969). Hybrid wolves in Vinnitsa region. *Zbirnyk Prats' Zoolohichnoho Muzeyu* **33**, 113-114. [Russian; English summary].

Gipson, P. S. & Sealander, J. A. (1976). Changing food habits of wild *Canis* in Arkansas with emphasis on coyotes hybrids and feral dogs. *American Midland Naturalist* **95**, 249-253.

Gloyd, J. S. (1992). Wolf hybrids: A biological time bomb? *Journal of the American Veterinary Medical Association* **201**, 381-382.

Gray, A. P. (1972). Mammalian hybrids: A checklist with bibliography. *Commonwealth Bureau Animal Breeders Genetics Technical Communication* **10** (revised).

Gurskii, I. G. (1975). Hybridization between wolves and dogs in nature. *Byulleten' Moskovskogo Obshchestva Ispytatelei Prirody Otdel Biologicheskii* **80**, 131-136. [Russian, English summary].

Hall, E. R. (1943). Cranial characters of a dog-coyote hybrid. *American Midland Naturalist* **29**, 371-74.

Hinske, G. (1974). Studies of coyote skulls and dubious hybrid populations. *Zoologischer Anzeiger* **192**, 98-137. [German].

Hoffmeister, D. F. (1952). "Coydog" in Illinois? *Outdoors in Illinois* **Spring**, 34.

Hollis, Burr. (1885). Hybrid wolves. *Forest and Stream* **25**, 467.

Hope, Jack. (1994). Wolves and wolf hybrids as pets are big business - but a bad idea. *Smithsonian* **25**, 34-45.

Iljin, N. A. (1934). Segregation in crosses between a wolf and a dog, and material on the genetics of the dog. *Trudy Po Dinamike Razvitiia* **8**, 105-165. [Russian, English summary]

Iljin, N. A. (1941). Wolf-dog genetics. *Journal of Genetics* **42**, 359-414.

Jacob, M. (1996). The Humane Society's attack on the wolfdog. *The Wildlife Guardian,* Autumn 1996. [on the internet: http://www/hooked.net/~michaelj/gow/wild_002.htm ].

Jay, M. T., Reilly, K. F., Debess, E. E., Haynes, E. H., Bader, D. R. & Barrett, L. R. (1994). Rabies in a vaccinated wolf-dog hybrid. *Journal of the American Veterinary Medical Association* **205**, 1729-1732.

Kennelly, J. J. & Roberts, J. D. (1969). Fertility of coyote-dog-hybrids. *Journal Mammalogy* **50**, 830-831.

Keyes, B. (1992). Is that a pet wolf next door? *USA Weekend* Feb. 28-Mar. 1, 5.

Kolenosky, G. B. (1971). Hybridization between wolf and coyote. *Journal of Mammalogy* **52**, 446-449.

Kramek, B. J. (1992). The hybrids howl: legislators listen – these animals aren't crying wolf. *Rutgers Law Journal* **23**, 633-656.

Lehman, N., Eisenhawer, A., Hansen, K., Mech, L. D., Peterson, R. O., Gogan, P. J. P. & Wayne, R. K. (1991). Introgression of coyote mitochondrial DNA into sympatric North American gray wolf populations. *Evolution* **45**, 104-119.

Linzey, D. W. (1971). Animal harvested in south Alabama; probably a coyote-red wolf hybrid. *Alabama Conservation* **41**, 6-7.

Maargaard, L. & Graugaard, J. (1994). Female Arctic wolf, *Canis lupus arctos*, mating with domestic dogs, *Canis familiaris*, in northeast Greenland. *Canadian Field-Naturalist* **108**, 374-375.

Mahan, B. R. & Gipson, P. S. (1978). Osteoarthrosis in a coyote x dog hybrid from Nebraska. *Journal of Wildlife Disease* **9**, 395-398.

Mahan, B. R., Gipson, P. S. & Caw, R. M. (1978). Characteristics and distribution of coyote x dog hybrids collected in Nebraska. *American Midland Naturalist* **100**, 408-415.

Masuch, S. (nd). *The Wolfdog; A factual overview.* [on the internet: http://www.kc.net/~wolf2dog/wolfdog.htm ].

Marx, D. (1994). Wolf hybrids: To own or not to own? *Dog World* **79**, 42-45.

Mengel, R. M. (1971). A study of dog-coyote hybrids and implications concerning hybridization in *Canis*. *Journal of Mammalogy* **52**, 316-336.

Manners-Smith, J. (1901). Wolf hybrids in Gilgit. *Journal of the Asiatic Society of Bengal* **70**, 143-146.

Moore, Tom D., Walker, D. N. & Fisher, R. A. (1983). Wolves, dogs and hybrids: Law enforcement problems in Wyoming. *Proceedings of the Western Association of Fish and Wildlife Agencies* **63**, 69-81.

Newsome, A. E. & Corbett. L. (1982). The identity of the dingo, II. Hybridization with domestic dogs in captivity and in the wild. *Australian Journal of Zoology* **30**, 365-374.

Newsome, A. E. & Corbett, L. K. (1985). The identity of the dingo, III. The incidence of dingos, dogs and hybrids and their coat colours in remote and settled regions of Australia. *Australian Journal of Zoology* **33**, 363-375.

Noack, T. (1887). Wolfbastarde. *Zoologischer Garter* **28**, 106-111.

Pastoral Revue. (1934). Alsatian-Coyote cross killing stock in California. *Pastoral Review and Grazier's Record* **4**, 1045.

Paul, J. R. (1969). Coyotes and coy-dogs. *Living Museum* (Illinois State Museum) **31**, 114-115.

Pierce, M. P. (1885). Hybrid wolves. *Forest and Stream* **24**, 426-427.

Pilgrim, K.L., Boyd, D. K. & Forbes, S. H. (1998). Testing for wolf-coyote hybridization in the Rocky Mountains using mitochondrial DNA. *Journal of Wildlife Management* **62**, 683-689.

Polsky, R. H. (1995). Wolf hybrids: Are they suitable as pets? *Veterinary Medicine* **90**, 1122-1124.

Prendergast, D. (1989). *The Wolf Hybrid*. Gallup, NM: Rudelhaus Enterprises.

Prendergast, D. (1995). *Above Reproach: A Guide for Wolf Hybrid Owners*. Gallup, NM: Rudelhaus Enterprises.

Riabov, L. (1963). Wolf-dog hybrids. *Okhota I Okhotnich'e Khoziaistvo* **11**, 29-30. [Russian]

Riabov, L. S. (1973). Wolf-dog hybrids in the Voronezh Oblast. *Byulleten' Moskovskogo Obshchestva Ispytatelei Prirody Otdel Biologicheskii* **78**, 25-39. [Russian, English summary]

Riabov, L. S. (1978). New data on wolves and wolf-dog hybrids in the Voronezh Oblast *Byulleten' Moskovskogo Obshchestva Ispytatelei Prirody Otdel Biologicheskii* **83**, 39-45. [Russian, English summary]

Riabov, L. S. (1980). O povedenii brodyachikh i odichavshikh sobak i volko-sobachikh gibridov v prirode [On behavior of strayed and feral dogs and wolf-dog hybrids in the wild]. in (D. I. Bibikov, Ed) *Povedenie volka*. Moscow: AN SSSR, pp. 136-146.

Roy, M. S., Geffen, E., Smith, D., Ostrander, E. & Wayne, R. K.. (1995). Patterns of differentiation and hybridisation in North American wolf-like canids revealed by analysis of microsatellite loci. *Molecular Biology and Evolution* **11**, 553-570.

Schleifenbaum, C. (1976). Fur marks and coat color in wolves and in the poodle-wolf bastards, raised in Kiel. *Zeitschrift fuer Saugetierkunde* **41**, 147-167.

Schmitz, O. J. & Kolenosky, G. B. (1985). Hybridization between wolf and coyote in captivity. *Journal of Mammalogy* **66**, 402-405.

Seitz, A. (1984). Canid hybrids. *Zool. Garten* **54**, 330-332.

Seligmann, J. (1991). Caution: wolf-dogs can be hazardous. *Newsweek* **118**, 50.

Silver, H. & Silver, W. T. (1969). Growth and behavior of the coyote-like canid of Northern New England with observations on canid hybrids. *Wildlife Monographs* **17**.

Slaughter, B. H. (1977). Wolves, coyotes, ducks, and hybridism. *Texas Journal of Science* **28**, 351-354.

Stanfill, J. (1994). Wolf-dogs: the pets from hell? *U.S. News and World Report* **117**, 78.

Stebbins, G. L. (1959). The role of hybridization in evolution. *Proceedings of the American Philosophical Society* **103**, 231-251.

Stockard, C. R. (1941). *The Genetic and Endrocrinic Basis for Differences in Form and Behavior (as elucidated by studies of contrasted pure-line dog breeds and their hybrids).* Philadelphia: Wistar Institute, American Anatomical Memoirs **19**.

Tek, C. (1972). Breeding wolf x dog hybrids; *Canis lupus* x *Canis familiaris* at Coucy-les-Eppes Zoo. *International Zoo Yearbook* **12**, 115.

Thornton, W. A., Creel, G. C. & Trimble, R. E. (1971). Hybridization in the fox genus *Vulpes* in west Texas. *Southwestern Naturalist* **15**, 473-484.

Van Gelder, R. G. (1977). Mammalian hybrids and generic limits. *American Museum Novitates* **2635**, 1-25.

Wayne, R. K. & Jenks, S. M. (1991). Mitochondrial DNA analysis implying extensive hybridization of the endangered red wolf *Canis rufus*. *Nature* **351**, 565-568.

Weidemann, W. (1970). Body weight and body weight correlation in wolves and poodles and in their cross breeding generation N-1 and N-2. *Zeitschrift fuer Saugetierkunde* **35**, 238-247. [German; English summary].

Wilson, P. (1992). On inferring hybridity from morphological intermediacy. *Taxon* **41**, 11-23.

Zimen, E. (1987). Ontogeny of approach and flight behavior towards humans in wolves, poodles and wolf-poodle hybrids. In (H. Frank, Ed) *Wolf and Man: Advances, Issues and.Problems in Captive Wolf Research*. Dordrecht, The Netherlands: Dr W. Junk Publishers, pp. 75-292.

**Domestication theory**

Ash, E. C. (1927). *Dogs: Their History and Development*. 2 vols. London.

Beck, A. M. (1973). *The Ecology of Stray Dogs: A Study of Free-ranging Urban Animals*. Baltimore: York Press.

Beck, A. M. (1975). The ecology of "feral" and free-roving dogs in Baltimore. In (M. W. Fox, Ed) *The Wild Canids: Their Systematics, Behavioral Ecology and Evolution*. Malabar, FL: Krieger Publishing Co., pp. 380-390.

Belayev, D.K. (1979). Destabilizing selection as a factor in domestication. *Journal of Heredity* **70**, 301-308.

Belyaev, D. K., Ruvinsky, A. O. & Trut, L. N. (1981). Inherited activation-inactivation of the star gene in foxes: its bearing on the problem of domestication. *Journal of Heredity* **72**, 267-274.

Bjarvall, A. & Sahlin, H. (1977). The wolf that became a dog: how to distinguish between the two species. *Fauna och Flora* **72**, 234-238. [Swedish].

Bokonyi, S. (1969). Archaeological problems and methods of recognizing animal domestication. In (P. J. Ucko & G. Dimbleby, Eds) *The Domestication and Exploitation of Plants and Animals*. London: Duckworth, pp. 219-229.

Bokonyi, S. (1974). *History of Domestic Mammals in Central and Eastern Europe*. Budapest: Akadmaiai Kiad.

Borchelt, P. L., Lockwood, R., Beck, A. M. & Voith, V. L. (1983). Attacks by packs of dogs involving predation on human beings. *Public Health Report* **98**, 57-66.

Brisbin, I. L., Jr. (1976). The domestication of the dog. *American Kennel Gazette* **93,** 22-27.

Brisbin, I. L., Jr. (1977). The pariah: Its ecology and importance to the origin, development and study of purebred dogs. *American Kennel Gazette* **94**, 22-29.

Brisbin, I. L., Jr. & Risch, T. S. (1997). Primitive dogs, their ecology and behavior: unique opportunities to study the early development of the human-canine bond. *Journal of the American Veterinary Medical Association* **210**, 1122-1126.

Budiansky, S. (1992). *The Covenant of the Wild: Why Animals Chose Domestication*. London: Weidenfeld & Nicolson.

Burns, M. & Fraser, M. N. (1966). *Genetics of the Dog*. 2nd edition. Edinburgh: Oliver & Boyd.

Clutton-Brock, J. (1970). The origins of the dog. In (D. Brothwell & E. Higgs, Eds) *Science in Archaeology*. New York: Praeger Publishers, 303-309.

Clutton-Brock, J. (1976). The historical background to the domestication of animals. *International Zoo Yearbook* **16**, 240-246.

Clutton-Brock, J. (1977). Man-made dogs. *Science* **197**, 1340-1342.

Clutton-Brock, J. (1980). The domestication of the dog with special reference to social attitudes to the wolf. *Carnivore* **3**, 27-33.

Clutton-Brock, J. (1981). *Domesticated Animals from Early Times*. Austin: University of Texas Press.

Clutton-Brock, J. (1984). Dog. In (I. L. Mason, Ed.) *Evolution of Domesticated Animals*. London: Longman, pp. 198-211.

Clutton-Brock, J. (1991). *The Dog*. New York: A.A. Knopf.

Clutton-Brock, J. (1992a). Domestication of animals. In (S. Jones, R. Martin & D. Pilbeam, Eds) *The Cambridge Encyclopedia of Human Evolution,* Cambridge: Cambridge University Press, pp. 380-385.

Clutton-Brock, J. (1992b). The process of domestication. *Mammal Review* **22**, 79-85.

Clutton-Brock, J. (1994). The unnatural world: Behavioral aspects of humans and animals in the process of domestication. In (A.Manning & J. Serpell, Eds) *Animals and Human Society: Changing Perspectives*. London & New York: Routledge, pp. 23-25..

Clutton-Brock, J. (1995). Early origins of the Saluqi. In (G. Goodman, Ed) *The Saluqi Cursing Hound of the East*. Arizona: Midbar, pp. 75-82.

Clutton-Brock, J. (1995). Origins of the dog: domestication and early history. In (J. Serpell, Ed) *The Domestic Dog: Its Evolution, Behaviour and Interactions With People.* Cambridge: Cambridge University Press, pp 7-20.

Clutton-Brock, J. & Jewell, P. (1993). Origin and domestication of the dog. In (H. E. Evans, Ed) *Miller's Anatomy of the Dog*,. 3rd edition. Philadelphia: W. B. Saunders, pp. 21-31.

Cohn, J. (1977). How wild wolves became domestic dogs. *Bioscience* **47**, 725-728.

Cohen, J. A. & Fox, M. W. (1976). Vocalizations in wild canids and possible effects of domestication. *Behaviourial Processes* **1**, 77-92.

Coppinger, R. & Schneider, R. (1995). Evolution of working dogs. In (J. Serpell, Ed) *The Domestic Dog: Its Evolution, Behaviour and Interactions With People.* Cambridge: Cambridge University Press, pp 21-47.

Crabtree, P. J. (1993). Early animal domestication in the middle East and Europe. In (M. B. Schiffer, Ed) *Archaeological Method and Theory.* Tucson: University of Arizona Press, pp. 201-245.

Crockford, S. J. (in press). Animal domestication and heterochronic speciation: the role of thyroid hormone. In (K. McNamara & N. Minugh-Purvis, Eds) *Human Evolution Through Developmental Change*. Baltimore: Johns Hopkins University Press.

Ducos, P. (1978). "Domestication" defined and methodological approaches to its recognition in faunal assemblages. *Harvard University Peabody Museum Bulletin* **2**, 53-56.

Eaton, R. L. (1969). Cooperative hunting by cheetahs and jackals and a theory of domestication of the dog. *Mammalia* **33**, 87-92.

Epstein, H. (1971). *The Origin of the Domestic Animals of Africa* (Vol. 1). New York: Africana Publishing.

Fox, M. W. (1978). Man, wolf and dog. In (R.Hall & H. S. Sharp, Eds) *Wolf and Man: Evolution in Parallel.* New York: Academic Press, pp. 9-30.

Fox, M. W. (1987). *The Dog: Its Domestication and Behavior.* Malabar, FL: Krieger Publishing Co. [1978 edition reprinted]

Frank, H. & Frank, M. G. (1982). On the effects of domestication on canine social development and behavior. *Applied Animal Ethology* **8**, 507-525.

Hall, R. L. (1977). Paleobiology and systematics of canids and hominids. *Journal of Human Evolution* 6, 519-531.

Hall, R. L. (1978). Variability and speciation in canids and hominids. In (R.Hall & H. S. Sharp, Eds) *Wolf and Man: Evolution in Parallel.* New York: Academic Press, pp. 153-178.

Hall, R. L. & Sharp, H. S. (1978). *Wolf and Man: Evolution in Parallel.* New York: Academic Press.

Hecker, J. M. (1982). Domestication revisited: Its implications for faunal analysis. *Journal of Field Archaeology* **9**, 217-236.

Hemmer, H. (1975). The origin of the domestic dog and changes in relative brain size due to domestication. *Zoologische Beitraege* **21**, 97-104. [German, English summary].

Hemmer, H. (1990). *Domestication: The Decline of Environmental Appreciation.* Cambridge: Cambridge University Press.

Hunter, J. (1787). Observations tending to show that the wolf, jackal and dog, are all of the same species. *Philosophical Transactions of the Royal Society of London* **77**, 253-266.

Hunter, J. (1789). A supplementary letter on the identity of the species of the dog, wolf, and jackal. *Philosophical Transactions of the Royal Society of London* **79**, 160-161.

Jesse, G. (1886). *Researches into the History of the British dog.* London: Hardwicke.

Lumer, H. (1940). Evolutionary allometry in the skeleton of the domesticated dog. *American Naturalist* **74**, 439-467.

Macintosh, N. W. G. (1975). The origin of the dingo: An enigma. In (M. W. Fox, Ed) *The Wild Canids: Their Systematics, Behavioral Ecology and Evolution.* Malabar, FL: Krieger Publishing Co., pp. 87-106.

Manwell, C. & Baker, C. M. A. (1983). Origin of the dog: from wolf or wild *Canis familiaris*? *Speculations in Science and Technology* **6**, 213-224.

Matthew, W. D. (1930). The phylogeny of dogs. *Journal of Mammalogy* **11**, 117-138.

Mizuno, A. (1995). Japanese monkeys, dogs and wolves. *Honyurui Kagaku* **35**, 53-55. [Japanese].

Morell, V. (1997). The origin of dogs: running with the wolves. *Science* **276**, 1647-1648.

Morey, D. F. (1990). *Cranial allometry and the evolution of the domestic dog.* Ph.D. Thesis, University of Tennessee.

Morey, D. F. (1992). Size, shape and development in the evolution of the domestic dog. *Journal of Archaeological Science* **13**, 119-145.

Morey, D. F. (1994). The early evolution of the domestic dog. *American Scientist* **82**, 336-347.

Morton, S. G. (1851). Comments on dogs. *Proceedings of the Philadelphia Academy of Natural Science* **5**, 139-140.

Morton, S. G. (1852). Observations on the antiquity of some races of dog. *Proceedings of the Philadelphia Academy of Natural Science* **5**, 85-89.

Morton, S. G. (1852). Communications on the races of dogs. *Proceedings of the Philadelphia Academy of Natural Science* **5**, 139-140.

Musil, R. (1984). The first known domestication of wolves in Central Europe. *BAR International Series* **227**, 23-25.

Olsen, S. J. (1979). Archaeologically, what constitutes an early domestic animal? *Advances in Archaeological Method and Theory* **2**, 175-197.

Olsen, S. J. (1985). *Origins of the Domestic Dog: The Fossil Record.* Tucson: University of Arizona Press.

Olsen, S. J. & Olsen, J. W. (1977). The Chinese wolf, ancestor of New World Dogs. *Science* **197**, 533-535.

Olsen, S. J., Olsen, J. W. & Qi, G. (1982). The position of *Canis lupus variabilis*, from Zhoukoudian, in the ancestral lineage of the domestic dog, *Canis familiaris*. *Vertebrata Palasiatica* **20**, 264-267. [Chinese, English summary].

Ota, K. (1980). The dog: Its domestication and development, with special reference to the origin and the history of the Japanese native dog (a review). *Report of the Society for Researches on Native Livestock* **9**, 53-94. [Japanese, English summary].

Packard, A. S. (1885). Origin of the American varieties of the dog. *American Naturalist* **19**, 896-901.

Radinsky, L. (1973). Evolution of the canid brain. *Brain, Behavior and Evolution* **7**, 169-202.

Saito, H. (1936). On the discrimination between the ancient dogs and the Japanese wolves. *Nihonken* **5**. [Japanese].

Scott, J. P. (1954). The effects of selection and domestication upon the behavior of the dog. *Journal National Cancer Institute* **15**, 739-758.

Scott, J. P. (1967). The evolution of social behavior in dogs and wolves. *American Zoologist* **7**, 373-381.

Scott, J. P. (1968). Evolution and domestication of the dog. *Evolutionary Biology* **2**, 243-275.

Serpell, J. (1995). *The Domestic Dog: Its Evolution, Behaviour and Interactions With People.* Cambridge: Cambridge University Press.

Szinak, J. & Veress, I. (1976). How did the wolf become a dog? *Buvar* **31**, 173-176.

Tanabe, Y. (1985). *An Approach to the Mystery of Ancient Japanese by Tracing the Modern Japanese Dogs.* Tokyo: PHP Kenkyusho. [Japanese].

Tchernov, E. & Horwitz, L. K. (1991). Body size dimunution under domestication: Unconscious selection in primeval domesticates. *Journal of Anthropological Archaeology* **10**, 54-75.

Troughton, E. Le G. (1971). The early history and relationships of the New Guinea highland dog (*Canis hallstromi*). *Proceedings Linnean Society New South Wales* **96**, 93-98.

Trut, L. N. (1999). Early canid domestication: the fox-farm experiment. *American Scientist* **87**, 160-169.

Trut, L.N. & Osadchuk, L.V. (1997). *Evolutionary-genetic and Genetic-physiol;ogical Aspects of Fur Animal Domestication: A Collection of Reports.* Oslo: IFASA/Scientifur.

Van Gelder, R. G. (1978). A review of canid classification. *American Museum Novitates* **2646**, 1-10.

Wayne, R. K. (1993). Molecular evolution of the dog family. *Trends in Genetics* **9**, 218-224.

Willis, M. B. (1989). *Genetics of the Dog.* New York: Howell Book House.

**Folklore and general reading**


Allen, D. L. (1979, 1993). *Wolves of Minong: Their Vital Role in a Wild Community*. Ann Arbor: University of Michigan Press..

Anonymous. (1852). Children suckled by wolves. *Chamber's Journal* (n.s.) **17**, 122-123.

Anonymous. (1853). The wolf as an object of superstition. *Chamber's Edinburgh Journal* **20**, 27-30.

Anonymous. (1854). Wolf nurses in India. *Harper's New Monthly Magazine* **9**, 199.

Anonymous. (1869). Wolf-hunting in Chicago. *Appleton's Journal* **2**, 230-231.

Anonymous. (1882). Wolf children. *Chamber's Journal* **59**, 597-599.

Anonymous. (1927). Wolf children of India. *Living Age* **332**, 1021-1022.

Anonymous. (1927). India's wolf children found in caves. *Literary Digest* **95**, 54-56.

Anonymous. (1939). Trained timber wolves. *Outdoor Life* **83**, 65.

Ball, V. (1873). Notes on children found living with wolves in the northwestern provinces and Oudh. *Proceedings Asiatic Society Bengal* **8**, 128-129.

Bauer, P. (1995). *Wild Puppies*. San Francisco: Chronicle Books.

Baynes, E. H. (1923). Timber wolves - wild and tame. *Nature Magazine* **2**, 333-338.

Beck, A. M. & Jones, B. A. (1985). Unreported dog bites in children. *Public Health Reports* **100**, 315- 321.

Bettelheim, B. (1959). Feral children and autistic children. *American Journal of Sociology* **64**, 455-467.

Bomford, L. (1993). *The Complete Wolf*. New York: St. Martin's Press.

Bradshaw, J. (1992). *The Wolf: The World's Wild Dogs*. Ada, OK: Garrett Educational Corporation.

Brandenburg, J. (1990). *White Wolf: Living with an Arctic Legend*. Minocqua, WI: NorthWord Press.

Braund, K. (1975). *The Uncommon Dog Breeds*. New York: Arco Publishing Company.

Brisbin, I. L, Jr. (1992). Carolina dog. *Wolves and Related Canids* **(Fall)**, 41-44.

Carson, L. (1939). Black wolves of Revillagigedo. *The Alaska Sportsman* **5**, 8-9, 25-29.

Casey, D. & Clark,T. W. (1996). *Tales of the Wolf: Fifty-one Stories of Wolf Encounters in the Wild*. Moose, WY: Homestead Publishing.

Clarke, C. H. D. (1971). The Beast of Gevaudan. *Natural History* **80**, 44-51.

Clutton-Brock, J. (1995). Aristotle, the scale of nature, and modern attitudes to animals. *Social Research* **62**, 415-838.

Clutton-Brock, J. (1996). Competitors, companions, status symbols, or pests: A review of human associations with other carnivores. In (J. L. Gittleman, Ed) *Carnivore Behaviour, Ecology, and Evolution, Vol. 2*. Ithica: Comstock Publications, pp. 375-392.

Colbert, E. H. (1939). Wild dogs and tame - past and present. *Natural History* **43**, 90-95.

Coppinger, R. & Feinstein, M. (1991). Hark! Hark! The dogs do bark and bark and bark. *Smithsonian* **21**, 119-129.

Cordy-Collins, A. (1994). An unshaggy dog story. *Natural History* **103**, 34-41.

Crane, C. (1991). Wolves in dogs' clothing. *Animals* **124**, 28-31.

Crisp, W. G. (1956). Tahltan bear dog. *The Beaver* **287**, 38-41.

Dalrymple, B. (1919). *The Gray Wolf of South Dakota*. Altoona, PA: Altoona Tribune.

Doogan, M. (1991). A dog in wolf's clothing. *Alaska* **57**, 16-18.

Drummond, L. (1977). Structure and process in the interpretation of South American myth: The Arawak Dog Spirit People. *American Anthropologist* **79**, 842-868.

Eckels, R. P. (1937). *Greek wolf Lore*. Philadelphia: University of Pennsylvania.

Editor. (1938). Trains wolves for sled team. *The Outdoorsman* **77**, 6.

Ekblaw, W. E. (1936). Eskimo dogs - forgotten heros. *Natural History* **37**, 173-184.

Fiennes, R. & Fiennes, A. (1968). *The Natural History of the Dog*. London: Weidenfeld & Nicholson.

Fogle, B. (1995). *The Encyclopedia of the Dog*. Willowdale, Ontario: Firefly Books.

Fox, M. W. (1980). *The Soul of the Wolf*. Boston: Little Brown.

Fox, M. W. (1980, 1992). *The Soul of the Wolf: A Meditation on Wolves and Man*. New York: Lyons & Burford.

Gittleman, J. L. & Pimm, S. L. (1991). Crying wolf in North America. *Nature* **351**, 524-525.

Gordon, W. R. (1942). I tamed a wolf for my dog team. *The Alaska Sportsman* **8**, 10-11, 29-31.

Gotch, A. F. (1979). Introduction, and Dogs, weasels, lions and their kin. In *Mammals - Their Latin Names Explained*. Poole, UK: Blandford Press, pp. 11-21, 157-162.

Gould, S. J. (1994). A dog's life in Galton's polyhedron. In *Eight Little Piggies*. New York: W. W. Norton & Co., pp. 382-395.

Gregg, J. (1943). The strange "dog" of Fort Selkirk. *The Beaver* **273,** 31.

Gregory, W. K. (1933). Nature's wild dog show. *Bulletin New York Zoological Society* **36**, 83-96.

Grinnell, G. B. (1896). About wolves and coyotes. *Forest and Stream* **47**, 511-512.

Grinnell, G. B. (1907). Dog-wolf partnership. *Forest and Stream* **69**, 772.

Hampton, B. (1997). *The Great American Wolf*. New York: Henry Holt & Co.

Harry, J. E. (1936). *Dog and Dogs*. New York & London.

Hellmuth, J. (1964). *A Wolf in the Family*. New York: The New American Library.

Hochmeyer, E. (1885). Were-wolves and wolf-children. *American Field* **24**, 8.

Hormell, R. S. (1940). Notes on the history of rheumatism and gout. *The New England Journal of Medicine* **223**, 754-760. [use by ancient Greeks of hot wolf oil to cure arthritis].

Johnson, S. A. & Aamodt, A. (1985). *Wolf Pack: Tracking Wolves in the Wild*. Minneapolis: Lerner Publications Co.

Kappeler, M. (1991). *Dogs: Wild and Domestic (Animal Families)*. Milwaukee: Gareth Stevens Children's Books.

Kneafsey, B. & Condon, K. C. (1995). Severe dog-bite injuries, introducing the concept of the pack attack: A literature review of seven case reports. *Injury* **26**, 37-41.

Laing, H. M. (1939). Leave it to lobo. *Field and Stream* **44**, 36-38, 63.

Lane, B. (1941). Wolves are queer devils. *The Alaska Sportsman* **7**, 10-11, 32-34.

Lantier-Sampon, P. (1992). *The Wonder of Wolves*. Milwaukee: Gareth Stevens Publishing.

Lascelles, T. (1939). The nature album: why don't huskies bark like domestic dogs? *Forest and Outdoors* **35**, 354-356.

Lawrence, R. D. (1986, 1988). *In Praise of Wolves*. New York: Ballantine.

Lawrence, R. D. (1990). *Wolves*. San Francisco: Sierra Club Books/ Boston: Little, Brown & Co.

Lawrence, R. D. (1993). *Trail of the Wolf.* Emmaus, PA: Rodale Press.

Leach, M. (1961). *God had a Dog*. New Brunswick, NJ: Rutgers University Press

Lepthien, E. U. (1991). *Wolves*. Chicago: Childrens Press.

Leslie, R. F. (1974). *In the Shadow of a Rainbow: The True Story of a Friendship Between Man and Wolf.* New York: W.W. Norton & Co.

Lever, C. (1985). *Naturalized Mammals of the World.* London: Longman.

Ling, M. (1991). *Amazing Wolves, Dogs, and Foxes*. New York: Alfred A. Knopf.

Lockwood, R. & Rindy, K. (1987). Are 'pit bulls' different? An analysis of the pit bull terrier controversy. *Anthrozoos* **1**, 2-8.

Long, K. (1996). *Wolves: A Wildlife Handbook.* Boulder, CO: Johnson Books.

Lopez, B. H. (1978). *Of Wolves and Men*. New York: Charles Scribner's Sons.

McLoughlin, J. C. (1983). *The Canine Clan: A New Look at Man's Best Friend*. New York: The Viking Press.

Mech, L. D. (1977). Where can the wolf survive? *National Geographic Magazine* **152**, 518-537.

Mech, L. D. (1987). At home with the Arctic wolf. *National Geographic Magazine* **171**, 562-594.

Mech, L. D. (1988). *The Arctic Wolf: Living with the Pack.* Minnesota: Voyageur Press.

Meintaefel, P. A. (1937). Wolf myth exploded. *Christian Science Monitor*. **July 7**, 5.

Merriam, C. H. (1923). Wolf attack on humans denied. *Outdoor Life* **51**, 493.

Mestel, R. (1994). Ascent of the Dog. *Discover: The World of Science* **15**, 90-98.

Morbidity and Mortality Weekly Reports. (1997). Dog-bite related fatalities – United States, 1995-1996. *Morbidity and Mortality Weekly Reports* **46**, 467.

Ogburn, W. F. & Bose, N. K. (1959). On the trail of the wolf-children. *Genetic Psychology Monthly* **60**, 117-193.

Podberscek, A. (1994). Dog on a tightrope: The position of the dog in British society as influenced by press reports on dog attacks (1988-1992). *Anthrozoos* **7**, 232-241.

Pollard, J. (1964, 1991). *Wolves and Werewolves*. London: R. Hale.

Powers, W. K. & Powers, M. N. (1986). Putting on the dog. *Natural History* **95**, 6-17.

Peterson, R. L. (1947). A record of a timber wolf attacking a man. *Journal of Mammalogy* **28**, 294-295.

Rapoport, R. (1994). *The Wolf.* Berkeley, CA: RDR Books.

Ray, Mrs. G. R. (1940). The mad wolf at Fort Churchill. *The Beaver* **271**, 30.

Rine, J. Z. (1965). *The World of Dogs*. New York: Doubleday.

Rosenberger, J. (1995). Wolves, dogs, and danger. *Animals* **128**, 13-17.

Rue, L. L., III. (1994). *Wolves: A Portrait of the Animal World.* New York: Smithmark Publishers, Inc.

Rutter, R. J. & Pimlott, D. H. (1968). *The World of the Wolf.* Philadelphia & New York: J. B. Lippincott Co.

Sacks, J. J., Sattin, R. W. & Bonzo, S. E. (1989). Dog bite-related fataliaties from 1979 through 1988. *Journal of the American Medical Association* **262**, 1489-1492.

Savage, C. (1989). *Wolves.* San Francisco: Sierra Club Books.

Savage, C. (1996). *The World of the Wolf.* San Francisco: Sierra Club Books.

Senn, H. A. (1982). *Were-wolf and Vampire in Romania.* New York: Columbia University Press.

Serpell, J. A. (1983). The personality of the dog and its influence on the pet-owner bond. In (A. H. Katcher & A. M. Beck, Eds) *New Perspectives on our Lives with Companion Animals*. Philadelphia: University of Pennsylvania Press, pp. 57-65.

Shaw, J. H. & Jordan, P. A. (1977). The wolf that lost its genes. *Natural History* **86**, 80-88.

Sheldon, J. W. (1992). *Wild Dogs: The Natural History of the Nondomestic Canidae.* San Diego: Academic Press.

Shchepanskaia, T. B. (1993). The dog as a guide on the brink of worlds. *Etnograficheskoe obozrenie* **1**, 71-79. [Russian, English summary].

Shockwill, G. A. (1898). Wolf children. *Lippincott Monthly Magazine* **61**, 115-125.

Singh, J. A. L. & Zingg, R. M. (1942). *Wolf Children and Feral Man*. New York: Harper & Brothers.

Sleeman, Sir W. H. (1888). Wolves nurturing children in their dens. *The Zoologist* **12**, 87-98.

Smart, G. F. (1931). A wolf pack howls in his back yard. *American Magazine* **111**, 81, 129.

Smith, C. H. (1846). Dogs. *The Naturalist's Library*, vols. 18-19. Edinburgh

Speelan, S. R. & Williams, J. O. (1926). Breeds of dogs. *United States Department Agriculture Farmers' Bulletin* **1491**, 1-46.

Tedford, R. (1978). *History of Dogs and Cats*. St. Louis: Ralston Purina Co.Texas Department of Health. (1997). *Texas 1996 severe attack surveillance summary*. [on the internet: http://www.kc.net/~wolf2dog/texasbite.htm ].

Thomas, E. M. (1993). *The Hidden Life of Dogs*. Boston: Houghton Mifflin Co.

Traup, N. E. (1888). Wolves nurturing children. *The Zoologist* **12**, 221.

Voelker, R. (1997). Dog bites recognized as public health problem. *Journal American Medical Association* **277**, 278-280.

Walsh, J. H. (1879). *The Dogs of Great Britain, America, and Other Countries*. New York: Orange Judd.

White, D. (1944). Man-sized wolf. *Field and Stream* **48**, 75.

White, D. G. (1991). *Myths of the Dog-Man*. Chicago: University of Chicago Press.

Wilcox, B. W. & Walkowicz, C. (1989). *The Atlas of Dog Breeds of the World*. Neptune City, NJ: T. F. H. Publications.

Wright, J. C. (1991). Canine aggression toward people; bite scenarios and prevention. *Veterinary Clinics of North America Small Animal Practice* **21**, 299-314.

Young, S. P. (1930). Conquering wolfdom and catdom. *Southwest Wilds and Waters* **2**, 6-7, 47.

Young, S. P. (1941). The evolution of the steel wolf trap in North America. *Western Sportsman* **6**, 10- 11, 30-31.

Zimen, E. (1981). *The Wolf: A Species in Danger*. New York: Delacorte Press.

Zingg, R. M. (1941). India's wolf children - two human infants reared by wolves. *Scientific American* **164**, 135-137.

**Taxonomy**

Clutton-Brock, J., Corbet, G. B. & Hills, M. (1976). A review of the family Canidae with a classification by numerical methods. *Bulletin of the British Museum (Natural History), Zoology* **29**, 119-199.

Corbet, G. B. & Clutton-Brock, J. (1984). Appendix: Taxonomy and nomenclature. In (I. L. Mason, Ed) *Evolution of Domesticated Animals*. London: Longman, pp. 434-438.

Corbet, G. B. & Hill, J. E. (1991). Introduction and Family Canidae. In *A World List of Mammalian Species*, 3rd edition. Oxford: Oxford University Press, pp. 1-8, 101-102.

Gentry, A., Clutton-Brock, J. & Groves, C. P. (1996). Case 3010: Proposed conservation of usage of 15 mammal specific names based on wild species which are antedated by or contemporary with those based on domestic animals. *Bulletin of Zoological Nomenclature* **53**, 28-37.

Groves, C. P. (1971). Request for a declaration modifying Article 1 so as to exclude names proposed for domestic animals from zoological nomenclature. *Bulletin of Zoological Nomenclature* **27**, 269-272.

International Commission on Zoological Nomenclature. (1926). Opinion 91. Thirty-five generic names of mammals placed in the official list of generic names. *Smithsonian Miscellaneous Collections* **73**, 1-2.

International Commission on Zoological Nomenclature. (1957). Opinion 451. Use of the plenary powers to secure that the specific name dingo Meyer, 1793, as published in the combination *Canis dingo*, shall be the oldest available name of the dingo of Australia (Class Mammalia). *Opinions International Commission Zoological Nomenclature* **15**, 329-338.

Kennedy, M. L. & Smith, R. A. (nd). Field identification of coyotes and domestic dogs. In *Systematics, Food Habits, and Parasites of the Coyote (*Canis latrans*) in Tennessee*. Nashville: Tennessee Wildlife Resources Agency.

Longman, H. A. (1928). Note on the dingo, the Indian wild dog and a Papuan dog. *Memoirs of the Queensland Museum* **9**, 151-157.

Lull, R. S. (1922). Variations or specific distinctions: which? *Bulletin of the Geological Society of America* **33**, 599, 604.

May, R. (1990). Taxonomy as destiny. *Nature* **347**, 129-130.

Melville, R.V. & Smith, J. D. D. (1987). *Official Lists and Indexes of Names and Works in Zoology*. London: International Trust for Zoological Nomenclature.

O'Brien, S. J. & Mayr, E. (1991). Bureaucratic mischief: recognizing endangered species and subspecies. *Science* **251**, 1187-1188.

Palmeirim, J. M. (1998). Analysis of skull measurements and measurers: Can we use data obtained by various observers? *Journal of Mammalogy* **79**, 1021-1028.

Pocock, R. I. (1935). The races of *Canis lupus*. *Proceedings of the Zoological Society of London* **3**, 647-686.

Whitlock, S. C. (1948). Identification of wolf, coyote, dog. *Michigan Conservationist* **17**, 6-15.

**Osteology**

Adams, L. A. (1935). Dental variations of the dog, *Canis domesticus*. *Transactions of the Illinois Academy of Science* **28**, 243-244.

Anderson, R. J. (1901). Abnormal dentition in the dog. *Irish Naturalist* **10**, 89-90.

Baumann, F. & Huber, W. (1946). On the development and juvenile skull shape of different canids. *Archiv Fuer Julius Klaus-Stiftung Fuer Ver. Soz. Rass.* **21**, 352-361.

Bradley, O. C. (1908). Note on the interparietal region of the skull of the dog and horse. *Veterinary Journal* **15**, 485-490.

Bradley, O. C. (1948). *Topographic Anatomy of the Dog*. 5th ed. Edinburgh: Oliver & Boyd.

Buchalczyk, T., Dynowski, J. & Szteyn, S. (1981). Variations in number of teeth and asymmetry of the skull in the wolf. *Acta Theriologica* **26**, 23-30.

Bueckner, H. J. (1971). Allometric studies on the front legs of adult canids. *Zoologischer Anzeiger* **186**, 11-46. [in German].

Cahill, D. R. & Marks, S. C., Jr. (1982) Chronology and histology of exfoliation and eruption of mandibular premolars in dogs. *Journal of Morphology* **171**, 213-218.

Campbell, A. (1842). Note on the skeletons of the Buansu, the Pariah dog, and Jackal. *Calcutta Journal of Natural History* **2**, 209-211.

Carson, H. S. (1962). Coyote, coy-dog, or dog. *Maine Fish and Game*, Spring, 4-7.

Churcher, C. S. (1960). Cranial variation in the North American Red fox. *Journal of Mammalogy* **41**, 349-360.

Clutton-Brock, J., Kitchener, A. C. & Lynch, J. M. (1994). Changes in the skull morphology of the Arctic wolf, *Canis lupus arctos*, during the twentieth century. *Journal of Zoology* **233**, 19-36.

Cross, E. C. (1940). Arthritis among wolves. *Canadian Field-Naturalist* **54**, 2-4.

Daigo, M. (1956). Comparative studies on the osteology of dogs by breeds I (cranium). *Bulletin of the Nippon Veterinary And Zootechical College* **6,** 43-50. [Japanese, English summary].

Daigo, M. (1957). Comparative studies on the osteology of dogs by breeds II (mandibulae). *Bulletin of the Nippon Veterinary And Zootechnical College* **6**, 70-79. [Japanese, English summary].

Daigo, M. (1957). Comparative studies on the osteology of dogs by breeds III (pelvis). *Jyu-i Chikusan Shinpo* **306**, 725-729. [Japanese].

Dayan, T., Simberloff, D., Tchernov, E. & Yom-Tov, Y. (1992). Canine carnassials: Character displacement in the wolves, jackals and foxes of Israel. *Biological Journal of the Linnean Society* **45**, 315-331.

de Serres, M. M. (1835). On the distinctive characters of the dog, the wolf, and the fox, as supplied by the skeleton. *Edinburgh New Philosophical Journal* **19**, 244-253.

Dolgov, V. A. & Rossolimo, O. L. (1964). Dental abnormalities in *Canis lupus* Linnaeus, 1758. *Acta Theriologica* **8**, 237-244. [Russian, English summary].

Endo, H., Obara, I., Yoshida, T., Kurohmaru, M., Hayashi, Y. & Suzuki, N. (1997). Osteometrical and CT examination of the Japanese wolf skull. *Journal of Veterinary Medical Science* **59**, 531-538.

Federoff, N. E. (1996). Malocclusion in the jaws of captive bred Arctic wolves, *Canis lupus arctos*. *Canadian Field-Naturalist* **110**, 683-687.

Fine, M. D. (1964). An abnormal P/2 in *Canis* cf. *C. latrans* from the Hagerman Fauna of Idaho. *Journal of Mammalogy* **45**, 483-485.

Flower, W. H. (1869). On the value of the characters of the base of the cranium in the classification of the order Carnivora, and on the systematic position of *Bassaris* and other disputed forms. *Proceedings of the Zoological Society of London* **1868**, 4-37.

Flower, W. H. (1870). *An Intorduction to the Osteology of the Mammalia*. London: Macmillan.

Flower, W. H. (1871). On the composition of the carpus of the dog. *Journal of Anatomy and Physiology* **6**, 62-64.

Fritts, S. H. & Caywood, D. D. (1980). Osteoarthrosis in a wolf (*Canis lupus*) radio-tracked in Minnesota. *Journal of Wildlife Diseases* **16**, 413-417.

Gilbert, B. M. (1980). Cranial traits separating canids. In *Mammalian Osteology*. Laramie, Wyoming: B. Miles Gilbert.

Giles, E. (1960). Multivariate analysis of Pleistocene and recent coyotes (*Canis latrans*) from California. *University of California Publications in Geological Sciences* **36**, 369-390.

Gimmelreikh, G. A., Manzyi, S. F., Klikov, V. J., Melnik, K. P. & Pilipchuk, O. Y. (1985). Morphology and Skeleton. In (D. 1. Bibikov, Ed.) *The Wolf: History, Systematics, Morphology,Ecology*. Moscow: Nauka, pp. 125-175. [Russian].

Goodwin, E. A. & Ballard, W. B. (1985). Use of tooth cementum for age determination of gray wolves. *Journal of Wildlife Management* **49**, 313-316.

Gray, J. E. (1868). Notes on the skulls of the species of dogs, wolves and foxes (Canidae) in the collection of the British Museum. *Proceedings of the Zoological Society of London* **1868**, 492-525.

Guilday, J. E. (1962). Supernumerary molars of *Otocyon*. *Journal of Mammalogy* **43**, 455-462.

Gunson, J. R. & Nowak, R. M. (1979). Largest gray wolf skulls found in Alberta. *Canadian Field Naturalist* **93**, 308-309.

Hall, E. R. (1940). Supernumerary and missing teeth in wild mammals of the orders Insectivora and Carnivora, with some notes on disease. *Journal of Dental Research* **19**, 103-143.

Harrison, D. L. (1973). Some comparative features of the skulls of wolves and pariah dogs. *Bonner Zoologische Beitraege* **24**, 186-191.

Hare, W. C. D. (1961). Radiographic anatomy of the cervical region of the canine vertebral column. Part II: Developing vertebrae. *Journal of the American Veterinary Medical Association* **139**, 217-220.

Hell, P. (1972). Craniometric study performed on wolves (*Canis lupus* L.). *Lynx* **13**, 5-14. [Czech, German summary].

Hell, P. (1990). Dental anomalies among West Carpathian wolves and their significance in dog breeding. *Zeitschrift fuer Jagdwissenschaft* **36**, 266-269. [German, English summary].

Hell, P. & Duricka, J. (1989). Anomalie chrupu zapadokarpatskeho vlka (*Canis lupus*) [Anomalies of the dentition of the West Carpathian wolf]. *Folia venatoria* **19**, 339-347. [Slovak, English & German summaries].

Hildebrand, M. (1952). An analysis of body proportions in the Canidae. *American Journal of Anatomy* **90**, 217-256.

Hildebrand, M. (1954). Comparative morphology of the body skeleton in Recent Canidae. *University of California Publications in Zoology* **52**, 399-470.

Holler, W. (1978). Vergleichende Untersuchung an den obersten ersten Molaren bei Wolf (*Canis lupus lupus*), Haushund (*Canis familiaris*-Dobermann) und Goldschakal (*Canis aureus*) mit Hilfe der Moire-Topographie. [Comparative study of the first upper molar of the wolf (*Canis lupus lupus*), house dog (*Canis familiaris*) (dobermann pinscher), and the oriental jackal (*Canis aureus*) via moire-topography]. *Saugetierkundliche Mitteilungen* **26**, 131-139.

Hough, J. R. (1948). The auditory region in some members of the Procyonidae, Canidae, and Ursidae. *Bulletin of the American Museum of Natural History* **92**, 67-118.

Howard, W. E. (1949). A means to distinguish skulls of coyotes and domestic dogs. *Journal of Mammalogy* **30**, 169-171.

Huber, W. & Lueps, P. (1987). Problems of the skull of the domestic dog from the recent point of view. *Lynx* **11**, 67-73.

Humphries, J. M., Bookstein, F. L., Chernoff, B., Smith, G. R., Elder, R. L. & Poss, S.G. (1981). Multivariate discrimination by shape in relation to size. *Systematic Zoology* **30**, 291-308.

Huxley, T. H. (1880). On the cranial and dental characters of the Canidae. *Proceedings of the Zoological Society London* **16**, 238-287.

Huxley, T. H. (1880). On the epipubis in the dog and fox. *Proceedings of the Royal Society of London* **30**, 162-163.

Jolicoeur, P. (1959). Multivariate geographic variation in the wolf. *Evolution* **13**, 283-301.

Jolicoeur, P. (1975). Sexual dimorphism and geographical distance as factors of skull variation in the wolf *Canis lupus* L. In (M. W. Fox, Ed) *The Wild Canids: Their Systematics, Behavioral Ecology and Evolution*. Malabar, FL: Krieger Publishing Co., pp. 54-61.

Kato, K. (1957). Morphological study of the macerated dog pelvis bones. *Report from the Department of Anatomy, Jikei University School of Medicine* **17**, 1-36 [Japanese].

Kaudelka, F. (1885). *Die Tierknochenfunde aus den spatromischen Siedlungsschichten von Lauriacum*. Ph.D. Thesis. Munich University [discusses canid shoulder height based on long bone length ratios].

Krantz, G. S. (1959). Distinctions between the skulls of coyotes and dogs. *Kroeber Anthropological Society Papers* **21**, 40-45.

Lokhotop, R. I. (1994). Anomalii cherepa i zubnoi sistemy volka s territorii Ukrainy [Cranial and dentition anomalies in wolf of the Ukranian territory]. *Vestnik Zoologii* **94**, 45-50. [Russian, English summary].

Lueps, P. & Huber, W. (1969). Attempt at a differentiated characterization of the skull base of the wolf and the domestic dog. *Mitteilugen der Naturforschenden Gesellshaft in Bern* **26**, 1-29 [German].

Mallory, F. F., Hillis, T. L., Blomme, C. G. & Hurst, W. G. (1994). Skeletal injuries of an adult timber wolf, *Canis lupus*, in northern Ontario. *Canadian Field-Naturalist* **108**, 230-232.

Marrett Tims, H. W. (1896). On the tooth genesis in the Canidae. *Journal Linnean Society (Zoology)* **25**, 445-480.
Meyer, P. (1974). Teeth of female wolves. *Deutsche Tieraerzliche Wochenschrift* **81**, 171 [German].
Mivart, St. G.. (1890). Note on canine dental abnormalities. *Proceedings of the Zoological Society of London* **1890**, 376-378.
Morris, E. L. (1971). *Detection of bone disease in dogs by radioisotope scanning.* Master's Thesis, Texas A&M University.
Nenkoff, G. I. (1935). Variations in the ribs and vertebrae of the dogs. *Sofia Selsko-Stapanska Akademiia* **10**, 461-497. [Russian]
Newsome, A. E., Corbett, L. K. & Carpenter, S. M. (1980). The identity of the dingo. I. Morphological discriminants of dingo and dog skulls. *Australian Journal of Zoology* **28**, 615-625.
Nussbaumer, M. (1982). On the variability of dorsal-basal curvatures in skulls of domestic dogs. *Zoologischer Anzeiger* **209**, 1-32. (German, English summary).
Obara, I. (1980). Cranial and dental characters of the Japanese native dogs. *Report of the Society for Researches on Native Livestock* **9**, 139-154. [Japanese, English summary].
Ogata, M., Wakuri, H., Suzuki,T. & Sugiura, K. (1979). Statistic study on dental numerical anomalies in the Japanese mongrel dog *(Canis familiaris). Journal of the Mammalian Society of Nippon* **8**, 33-39. [Japanese, English summary].
Okarma, H. & Buchalczyk, T. (1993). Craniometrical characteristics of wolves *Canis lupus* from Poland. *Acta Theriologica* **38**, 253-262.
Olsson, S. (1971). Degenerative joint disease (osteoarthrosis): A review with special reference to the dog. *Journal of Small Animal Practice* **12**, 333-342.
Onodera, S., Shigehara, N. & Eto, M. (1987). Discriminant analysis of the sexual differences of the skeletons in shiba dogs (*Canis familiaris*). *Acta Anatomica Nipponica* **62** [Japanese, English summary].
Ostergard, M. (1980). X-ray diffraction investigations of bones from domestic and wild animals. *American Antiquity* **45**, 59-63.
Parker, G. R. & Maxwell, J. W. (1986). Identification of pups and yearling wolves, *Canis lupus*, by dentine width in the canine. *Arctic* **39**, 180-181.
Pasitschniak-Arts, M., Taylor, M. E. & Mech, D. L. (1988). Note on skeletal injuries in an adult Arctic wolf. *Arctic and Alpine Research* **20**, 360-365.
Pichette, C. & Voight, D. R. (1971). A multivariate analysis of some Ontario and Quebec wolf (*Canis lupus*) skulls. *Bulletin Service de la faune du Quebec* **13**.
Riser, W. H., Parkes, L. J. & Shirer, J. F. (1967). Canine craniomandibular osteopathy. *American Veterinarian Radiologic Society* **8**, 23-31.
Rossolino, O. L. & Dolgov, V. A. (1965). Variability of the skull in *Canis lupus* Linnaeus, 1758 from the U.S.S.R. *Acta Theriologica* **10**, 195-207.
Schebitz, H. C. H. & Wilkens, H. (1986). *Atlas of Radiographic Anatomy of the Dog and Cat.* Toronto: W. B. Saunders.
Serres, M. de. (1835). On the distinctive characters of the dog, the wolf and the fox, as supplied by the skeleton. *Edinburgh New Philosophical Journal* **19**, 244-253.
Shigehara, N., Onodera, S. & Eto, M. (1997). Sex determination by discriminant analysis and evaluation of non-metric traits in the dog skeleton. In (S. J. Crockford) *Osteometry of Makah and Coast Salish Dogs.* Burnaby, BC: Simon Fraser University Archaeology Press **22**, pp. 113-126.
Shipp, A. D. (1991). Crown reduction - disarming of biting pets. *Journal of Veterinary Dentistry* **8**, 4-6.
Shufeldt, R. W. (1897-1901). The osteology of *Vulpes macrotis. Journal of the Philadelphia Academy of Natural Science*, **2/11**, 395-414.
Silver, I. A. (1969). The ageing of domestic animals. In (D. Brothwell & E. Higgs, Eds) *Science in Archaeology.* London: Thames & Hudson, pp. 283-302.
Smith, R. N. & Allcock, J. (1960). Epiphysial fusion in the greyhound. *The Veterinary Record* **72**, 75-79.
Stains, H. J. (1975). Calcanea of members of the Canidae. *Bulletin of the Southern California Academy of Sciences* **74**, 143-155.
Suminski, P. & Kobryn, K. (1980). Assessment of the usefulness of a discrimination analysis method of sex determination of wolf skulls (*Canis lupus* L. 1758). *Zeitschrift fuer Jagdwissenshaft* **26**, 121-126. [German, Russian & English summaries].
Sumner-Smith, G. (1966). Observations on epiphyseal fusion of the canine appendicular skeleton. *Journal of Small Animal Practice* **7**, 303-311.
Teichert, M. (1987). Brachymel dogs. *Archaeozoologia* **I**, 69-75.
The, T. L. & Trouth, C. O. (1976). Sexual dimorphism in the basilar part of the occipital bone of the dog (*Canis familiaris*). *Acta Anatomica* **95**, 565-571.
Thomson, P. C. & Rose, K. (1992). Age determination of dingoes from characteristics of canine teeth. *Wildlife Research* **19**, 597-599.
Valverde, J. A. & Hidalgo, A. (1974). On the wolf, *Canis lupus*, in Iberia: I. Sexual dimorphism in skulls. *Donana Acta Vertebrata* **1**, 233-244. [Spanish, English summary].
Van Valkenburgh, B. (1988). Incidence of tooth breakage among large predatory mammals. *American Naturalist* **131**, 291-302.
Vaughan, L. C. (1987). Disorders of the tarsus in the dog. I. *British Veterinary Journal* **143**, 388-401.
Vilà, C., Urios, V. & Castroviejo, J. (1993). Tooth losses and anomalies in the wolf (*Canis lupus*). *Canadian Journal of Zoology* **71**, 968-971.
Vitums, A. (1952). Hyperphalangism of an extremely well developed first digit accompanied by an extra dewclaw in the hindpaws of a dog. *American Veterinary Medical Association Journal* **121**, 93-95.
Voros, I. & Szathmary, L. (1977). The results of osteometric examinations on the wolf, *Canis lupus* Linne, 1758 of Barabas (northeast Hungary, Szabolcs-Szatmar County). *Acta Biologica Debrecina* **14**, 159-164. [Hungarian, English & German summaries].
Watkins, J.D. & Bradley, R. Craniomandibular osteopathy in a Labrador puppy. *Veterinarian Record* **79**, 262-264.
Wayne, R. K. (1984). *A comparative study of skeletal growth and morphology in domestic and wild canids.* Master's Thesis, Johns Hopkins University.
Wayne, R. K. (1986a). Cranial morphology of domestic and wild canids: the influence of development on morphological change. *Evolution* **40**, 243-261.
Wayne, R. K. (1986b). Developmental constraints on limb bone growth in domestic and some wild canids. *Journal of Zoology (London) A* **210**, 381-399.
Wayne, R. K. (1986c). Limb morphology of domestic and wild canids: the influence of development on morphological change. *Journal of Morphology* **187**, 301-319.

Wilder, B. G. (1874). On the composition of the carpus in dogs. *Proceedings of the American Association of the Advancement of Science Portland* **1873**, 301-302.

Windle, B. C. A. (1880). On some cranial and dental characters of the domestic dog. *Proceedings of the Zoological Society London* **1890**, 5-29.

Wood-Jones, F. (1929). The cranial characters of the Papuan dog. *American Journal of Mammalogy* **10**, 329-333.

Zoolitsch, H. (1969). Metrische untersuchungen an schadeln adulter Wildwolfe und Goldschakale. [Metric investigations of the wild wolf and jackal.] *Zoologischer Anzeiger* **182**, 153-182. [German]

## Morphology

Adam, W. S. (1970). *Microscopic Anatomy of the Dog: A Photographic Atlas*. Springfield, IL: C. C. Thomas.

Alberch, P. (1985). Developmental constraints: why St. Bernards often have an extra digit and poodles never do. *American Naturalist* **126**, 430-433.

Atkins, D. L. & Dillon, L. S. (1971). Evolution of the cerebellum in the genus *Canis*. *Journal of Mammalogy* **52**, 96-107.

Bekoff, M. & Jamieson, R. (1975). Physical development in coyotes (*Canis latrans*) with a comparison to other canids. *Journal of Mammalogy* **56**, 685-692.

Bookstein, F. L. (1989). "Size and Shape;" a comment on semantics. *Systematic Zoology* **38**, 173-180.

Corbett, L. K. (1985). Morphological comparisons of Australian and Thai dingos: A reappraisal of dingo status, distribution and ancestry. *Proceedings of the Ecological Society of Australia* **13**, 277-291.

Dawson, W. M. (1937). Heredity in the dog. *United States Department of Agriculture Yearbook*. **1937**, 1314-1349.

Dennis-Bryan, K. & Clutton-Brock, J. (1988). *Dogs of the Last Hundred Years at the British Museum (Natural History)*. London: British Museum (Natural History).

Ebinger, P. (1980). Relationship of brain weight and body weight in wolves and domestic dogs. *Zeitschrift fuer Saeugetierkunde* **45**, 148-153. [German, English summary].

Eilam, D. (1994). Influence of body morphology on turning behavior in carnivores. *Journal of Motor Behavior* **26**, 3-12.

Elton, C. (1931). Epidemics among sledge dogs in the Canadian Arctic and their relation to disease in the Arctic fox. *Canadian Journal of Research* **5**, 673-692.

Evans, H. E. (1993). *Miller's Anatomy of the Dog*. Philadelphia: W.B. Saunders.

Evans, H. E. & Christensen, G. C. (1979). *Miller's Anatomy of the Dog*. Philadelphia: W.B. Saunders.

Harris, R. B. & Ream, R. R. (1983). A method to aid in discrimination of tracks from wolves and dogs. *Canadian Wildlife Service Report* **45**, 120-124.

Hildebrand, M. (1952). The integument in Canidae. *Journal of Mammalogy* **33**, 419-428.

Hillis, T. L. & Mallory, F. F. (1996). Fetal development in wolves, *Canis lupus*, of the Keewatin District, Northwest Territories, Canada. *Canadian Journal of Zoology* **74**, 2211-2218.

Hillis, T. L. & Mallory, F. F. (1996). Sexual dimorphism in wolves (*Canis lupus*) of the Keewatin District, Northwest Territories, Canada. *Canadian Journal of Zoology* **74**, 721-725.

Howell, W. H. (1889). *Dissection of the Dog as a Basis for the Study of Physiology*. New York: H. Holt & Co.

Iljin, N. A. (1926). Principal laws of inheritance and cynology. *Dog Breeding and Training URSS* **1**, pp. 4-6. [Russian].

Iljin, N. A. (1928). Ruby eye in dogs. *Dog Breeding and Training, U.R.S.S.* **1**, 1-3. [Russian].

Iljin, N. A. (1931). Coat colour inheritance in Dobermann-pincher. *Trud. Dinam. Razvit.* **6**, 241-151. [Russian, English summary].

Kazmierczak, J. J., Burgess, E. C. & Mundson, T. E. (1988). Susceptibility of the gray wolf (*Canis lupus*) to infection with the Lyme disease agent, *Borrelia burgdorferi*. *Journal of Wildlife Diseases* **24**, 522-527.

Kennedy, A. T. (1982). Distinguishing characteristics of the hairs of wild and domestic canids from Alberta. *Canadian Journal of Zoology* **60**, 536-549.

Kennedy, P. K., Kennedy, M. L., Clarkson, P. L. & Liepins, I. S. (1991). Genetic variability in natural populations of the gray wolf, *Canis lupus*. *Canadian Journal of Zoology* **69**, 1183-1188.

Klinghammer, E. (1979). *The Behavior and Ecology of Wolves*. New York: Garland STPM Press.

Kolenosky, G. B. & Standfield. R. O. (1975). Morphological and ecological variation among gray wolves (*Canis lupus*) of Ontario, Canada. In (M. W. Fox, Ed) *The Wild Canids: Their Systematics, Behavioral Ecology and Evolution*. Malabar, FL: Krieger Publishing Co., pp. 62-72.

LaJeunesse, T. A. & Peterson, R. O. (1993). Marrow and kidney fat as condition indices in gray wolves. *Wildlife Society Bulletin* **21**, 87-90.

Landon, D. B., Waite, C. A., Peterson, R. O. & Mech, L. D. (1998). Evaluation of age determination techniques for gray wolves. *Journal of Wildlife Management* **62**, 674-682.

Mech, L. D. (1988). Longevity in wild wolves. *Journal of Mammalogy* **69**, 197-198.

Mech, L. D. (1997). An example of endurance in an old wolf, *Canis lupus*. *Canadian Field-Naturalist* **111**, 654-655.

Mech, L. D. & Goyal, S. M. (1993). Canine parvovirus effect on wolf population change and pup survival. *Journal of Wildlife Diseases* **29**, 330-333.

Mech, L. D. & Goyal, S. M. (1995). Effects of canine parvovirus on gray wolves in Minnesota. *Journal of Wildlife Management* **59**, 565-570.

Mech, L. D., Meier, T. J. & Seal, U. S. (1993). Wolf nipple measurements as indices of age and breeding status. *American Midland Naturalist* **129**, 266-271.

Miller, M. E. (1965). *Anatomy of the Dog*. Philadelphia: W. B. Saunders Co.

Moneer, M. A., Farah, I. O., Pomeroy, K. A., Goyal, S. M. & Mech, L. D. (1988). Detection of parvoviruses in wolf feces by electron microscopy. *Journal of Wildlife Diseases* **24**, 170-172.

Okarma, H. (1991). Marrow fat content, sex and age of red deer killed by wolves in winter in the Carpathian Mountains. *Holarctic Ecology* **14**, 169-172.
Ozoga, J.J. & Harger, E. M. (1966). Occurrence of albino and melanistic coyotes in Michigan. *Journal of Mammalogy* **47**, 339-340.
Parker, R. R. (1926). The susceptibility of the coyote (*Canis lestes*) to tularemia. *Reports of the United States Public Health Service* **41**, 1407-1410.
Pederson, S. (1983). Geographic variation in Alaskan wolves. *Canadian Wildlife Service Report Series* **45**, 345-361.
Peichl, L. (1992). Morphological types of ganglion cells in the dog and wolf retina. *Journal of Comparative Neurology* **324**, 590-602.
Peichl, L. (1992). Topography of ganglion cells in the dog and wolf retina. *Journal of Comparative Neurology* **324**, 603-620.
Pocock, R. I. (1914). On the feet of domestic dogs. *Proceedings of the Zoological Society of London* **1914**, 478-484.
Pocock, R. I. (1914). On the feet and other external features of the Canidae and Ursidae. *Proceedings of the Zoological Society of London* **1914**, 913-941.
Ryder, M. L. (1997). A note on the hair measurements of domestic dogs. *The Linnean* **13**, 39-44.
Schmitz, O. J. & Lavigne, D. M. (1987). Factors affecting body size in sympatric Ontario *Canis*. *Journal of Mammalogy* **68**, 92-99.
Schultz, W. (1969). On Hallstrom's Dog (*Canis hallstromi*, Troughton 1957). *Zoologischer Anzeiger* **183**, 47-72.
Scott, H. H. (1928). Carcinoma on the tonsil in a common wolf (*Canis lupus*). *Proceedings of the Zoological Society of London* **1928**, 43-47.
Scott, J. P., Fuller, J. L. & King, J. A. (1959). The inheritance of annual breeding cycles in hybrid basenji-cocker spaniel dogs. *Journal of Heredity* **50**, 254-261.
Skeel, M. A. & Carbyn, L. N. (1977). The morphological relationship of gray wolves. *Canadian Journal of Zoology* **55**, 737-747.
Suminski, P. (1966). Comparative studies on the *os penis* and *os glandis* in domestic predatory animals. *Folia Morphologica* **25**, 415-429.
Theberge, J. B., Forbes, G. J., Barker, I. K. & Bollinger, T. (1994). Rabies in wolves of the Great Lakes region. *Journal of Wildlife Diseases* **30**, 563-566.
Thieking, A., Goyal, S. M., Bey, R. F., Loken, K. I., Mech, L. D., Thiel, R. P. & O'Connor, T. P. (1992). Seroprevalence of Lyme disease in gray wolves from Minnesota and Wisconsin. *Journal of Wildlife Diseases* **28**, 177-182.
Thurber, J. M. & Peterson, R. O. (1991). Changes in body size associated with the range expansion in the coyote (*Canis latrans*). *Journal of Mammalogy* **72**, 750-755.
Weiler, G. J., Garner, G. W. & Ritter, D. G. (1995). Occurrence of rabies in a wolf population in northeastern Alaska. *Journal of Wildlife Diseases* **31**, 79-82.
Wheler, C. L. & Pocknell. A. M. (1996). Ectopic perianal gland tumor in a timber wolf. *Canadian Veterinary Journal* **37**, 41-42.

**Behaviour & ecology**

Baridze, Y. K. & Todu, S. G. (1988). Some findings on the ontogenesis of wolf parental behavior. *Soobshcheniya Akademii Nauk Gruzinskoi S.S.R.* **132**, 385-388 [Russian, English summary].
Bekoff, M. (1975). Social behavior and ecology of the African Canidae: a review. In (M. W. Fox, Ed) *The Wild Canids: Their Systematics, Behavioral Ecology and Evolution*. Malabar, FL: Krieger Publishing Co., pp. 120-142.
Bekoff, M. (1977). Social communication in canids. *Science* **197**, 1097-1099.
Bekoff, M., Daniels, T. J. & Gittleman, J. L. (1984). Life history patterns and the comparative social ecology of carnivores. *Annual Review of Ecology and Systematics* **15**, 191-232.
Bekoff, M., Hill, H. L. & Mitton, J. B. (1975). Behavioral taxonomy in canids by discriminant function analysis. *Science* **190**, 1223-1225.
Bekoff, M. & Wells, M. C. (1980). The social ecology of coyotes. *Scientific American* **242**, 130-148.
Bekoff, M. & Wells, M. C. (1982). Behavioral ecology of coyotes: Social organization, rearing patterns, space use, and resource defense. *Zeitschrift fur Tierpsychologie* **60**, 281-305.
Bekoff, M. & Wells, M. C. (1986). Social ecology and behavior of coyotes. *Advances in the Study of Behavior* **16**, 251-338.
Berman, M. & Dunbar, I. (1983). The social behaviour of free-ranging suburban dogs. *Applied Animal Ethology* **10**, 5-17.
Boitani, L. (1983). Wolf and dog competition in Italy. *Acta Zoologica Fennica* **174**, 259-264.
Boitani, L. & Cuicci, P. (1995). Comparative social ecology of feral dogs and wolves. *Ethology Ecology and Evolution* **7**, 49-72.
Boitani, L., Francisci, F., Ciucci, P. & Andreoli, G. (1995). Population biology and ecology of feral dogs in central Italy. In (J. Serpell, Ed) *The Domestic Dog: Its Evolution, Behaviour and Interactions With People*. Cambridge: Cambridge University Press, pp. 217-244.
Boyd, D. K., Pletscher, D. H. & Brewster, W. G. (1993). Evidence of wolves, *Canis lupus*, burying dead wolf pups. *Canadian Field-Naturalist* **107**, 230-231.
Brisbin, I. L., Jr., Coppinger, R. P., Feinstein, M. H., Austad, S. & Mayer, J. J. (1989). The New Guinea singing dog: taxonomy, captive studies and conservation priorities. *Science in New Guinea* **20**, 27-38.
Corbett, L. K. (1988). Social dynamics of a captive dingo pack: Population regulation by dominant female infanticide. *Ethology* **78**, 177-198.
Corbett, L. (1995). *The Dingo in Australia and Asia*. New York: Cornell University Press.
Corbett, L. & Newsome, A. (1975). Dingo society and its maintenance: a preliminary analysis. In (M. W. Fox, Ed) *The Wild Canids: Their Systematics, Behavioral Ecology and Evolution*. Malabar, FL: Krieger Publishing Co., pp. 369-379.
Coren, S. (1994). *The Intelligence of Dogs: Canine Consciousness and Capabilities*. New York: The Free Press.
Daniels, T. J. (1983). The social organization of free-ranging urban dogs. I. Non-estrous social behavior. *Applied Animal Ethology* **10**, 341-363.

Daniels, T. J. (1983). The social organization of free-ranging urban dogs. II. Estrous groups and the mating system. *Applied Animal Ethology* **10,** 365-373.

Dietz, J. M. (1984). Ecology and social organization of the Maned Wolf (*Chrysocyon brachyurus*). *Smithsonian Contributions to Zoology* **392,** 1-51.

Feddersen-Petersen, D. (1994). Comparative studies of behavioural development of wolves (*Canis lupus* L.) and domestic dogs (*Canis lupus* f. *fam.*): Domestication traits and selective breeding. *Tierarztliche Umschau* **49,** 527.

Fentress, J. C. (1967). Observations on the behavioral development of a hand reared male timber wolf. *American Zoologist* **7,** 339-351.

Fox, M. W. (1970). A comparative study of the development of facial expressions in canids; wolf, coyote and foxes. *Behavior (Leiden)* **36,** 49-73.

Fox, M. W. (1987). *Behaviour of Wolves, Dogs and Related Canids*. Malabar, FL: Krieger Publishing Co. [reprint].

Fox, M., Beck, A. M. & Blackman, E.. (1975). Behavior and ecology of a small group of urban dogs (*Canis familiaris*). *Applied Animal Ethology* **1,** 119-137.

Frank, H. & Frank, M. G. (1982). Comparison of problem-solving performance in six-week-old wolves and dogs. *Animal Behavior* **30,** 95-98.

Frank, H. & Frank, M. G. (1983). Inhibition training in wolves and dogs. *Behavioural Processes* **8,** 363-377.

Frank, H. & Frank, M. G. (1984). Information processing in wolves and dogs. *Acta Zoologica Fennica* **171,** 225-228.

Frank, H. & Frank, M. G. (1985). Comparative manipulation-test performance in ten-week-old wolves *Canis lupus* and Alaskan malamutes *Canis familiaris*: a Piagetian interpretation. *Journal of Comparative Psychology* **99,** 266-274.

Frank, H., Frank, M. G., Hasselbach, L. M. & Littleton, D. M. (1989). Motivation and insight in wolf (*Canis lupus*) and Alaskan malamute (*Canis familiaris*) visual discrimination learning. *Bulletin of the Psychonomic Society* **27,** 455-458.

Fritts, S .H. & Paul, W. J. (1989). Interactions of wolves and dogs in Minnesota. *Wildlife Society Bulletin* **17,** 121-123.

Fritts, S. H. & Paul, W. J. (1993). Interactions of wolves and dogs in Minnesota. *International Wolf* **3,** 26-27.

Gittleman, J. L. (1989). *Carnivore Behavior, Ecology, and Evolution*. Ithaca, NY: Comstock Publishing Association, Cornell University Press.

Gollan, K. (1984). The Australian dingo: in the shadow of man. In (M. Archer & Clayton, Eds) *Vertebrate Zoogeography and Evolution in Australasia*. Carlisle, Australia: Hesperian Press, pp. 921-927.

Goodwin, D., Bradshaw, J. W. S. & Wickets, S. M. (1997). Paedomorphosis affects agonistic visual signals of domestic dogs. *Animal Behavior* **53,** 297-304.

Kleiman, D. (1967). Some aspects of social behaviour in the Canidae. *American Zoologist* **7,** 365-372.

Laikre, L. & Ryman, N. (1991). Inbreeding depression in a captive wolf (*Canis lupus*) population. *Conservation Biology* **5,** 33-40.

Langguth, A. (1975). Ecology and evolution in the South American canids. In (M. W. Fox, Ed) *The Wild Canids: Their Systematics, Behavioral Ecology and Evolution*. Malabar, FL: Krieger Publishing Co., pp. 192-206.

Macdonald, D. W., Sillero-Zubiri, C. & Tattersall, F. H. (1995). Habitat selection and daily activity of giant molerats *Tachyoryctes macrocephalus*, significance to the Ethiopian wolf *Canis simensis* in the Afroalpine ecosystem. *Biological Conservation* **72,** 77-84.

Nesbitt, W. H. (1975). Ecology of a feral dog pack on a wildlife refuge. In (M. W. Fox, Ed) *The Wild Canids: Their Systematics, Behavioral Ecology and Evolution*. Malabar, FL: Krieger Publishing Co., pp. 391-395.

Peters, R. (1978). Communication, cognitive mapping, and strategy in wolves and hominids. In (R.Hall & H. S. Sharp, Eds) *Wolf and Man: Evolution in Parallel*. New York: Academic Press, pp. 95-108.

Peterson, R. O., Page, R. E. & Dodge, K. M. (1984). Wolves, moose, and the allometry of population cycles. *Science* **224,** 1350.

Robertshaw, J. D. & Harden, R. H. (1985). The ecology of the dingo in north-eastern New South Wales. II. Diet. *Australian Wildlife Research* **12,** 39-50.

Rubin, H. D. & Beck, A. M. (1982). Ecological behavior of free-ranging urban dogs. *Applied Animal Ethology* **8,** 161-168.

Schenkel, R. (1967). Submission: its features and functions in the wolf and dog. *American Zoologist* **7,** 319-331.

Scott, J. P. (1950). The social behavior of dogs and wolves: an illustration of sociobiological systematics. *Annals of the New York Academy of Science* **51,** 1009-1021.

Scott, M. D. & Causey, K. (1973). Ecology of feral dogs in Alabama. *Journal Wildlife Management* **37,** 253-265.

Scott, J. P. & Fuller, J. L. (1965). *Genetics and the Social Behavior of the Dog*. Chicago: University of Chicago Press.

Servin, J. (1991). Some aspects of the social behaviour of the Mexican wolf (*Canis lupus baileyi*) in captivity. *Acta Zoologica Mexicana Nueva Serie* **45,** 1-28.

Sperry, C. C. (1939). Food habits of peg-leg coyotes. *Journal of Mammalogy* **20,** 190-194.

Whitehouse, S. J. O. (1977). The diet of the dingo in Western Australia. *Australian Wildlife Research* **4,** 145-150.

**Biochemical identification studies**

Aguirre, G.D., Ray, K. & Acland, G. M. (1999). Overview of the International Workshop on Canine Genetics. *Journal of Heredity* **90,** 1-2.

Asa, C. S., Mech, L. D., Seal, U. S. & Plotka, E. D. (1990). The influence of social and endocrine factors on urine-marking by captive wolves (*Canis lupus*). *Hormones and Behavior* **24,** 497-509.

Braend, M. (1988). Polymorphism of a serum protease inhibitor system in Alaskan wolves. *Hereditas* **108,** 269-270.

Brownlow, C. A. (1996). Molecular taxonomy and the conservation of the red wolf and other endangered carnivores. *Conservation Biology* **10,** 390-396.

Burbidge, M. J. (1998). *Polyphyletic origins of extinct and extant domestic dogs*. Masters Thesis, University of Victoria, British Columbia.

Chiarelli, A. B. (1975). The chromosomes of the Canidae. In (M. W. Fox, Ed) *The Wild Canids: Their Systematics,*

*Behavioral Ecology and Evolution*. Malabar, FL: Krieger Publishing Co., pp. 40-53.
Clark, P., Ryan, G. E., & Czuppon, A. B. (1975). Biochemical genetic markers in the family Canidae. *Australian Journal of Zoology* **23**, 411-417.
Dresler, S. L., Runkel, D., Stenzel, P., Brimhall, B., & Jones, R. T. (1974). Multiplicity of the hemoglobin alpha-chain in dogs and variations among related species. *New York Academy Science Annals* **241**, 411-415.
Ellegren, H., Savolainen, P. & Rosen, B. (1996). The genetical history of an isolated population of the endangered grey wolf *Canis lupus*: A study of nuclear and mitochondrial polymorphisms. *Philosophical Transactions of the Royal Society of London, B* **351**, 1661-1669.
Elliot, D. & Wong, M. (1973). Catalase and hemoglobin in the Canidae studied by starch-gel electrophoresis. *Zoologischer Anzeiger* **191**, 178-181.
Elliot, O. & Wong, M. (1972). Acid phosphatase: handy enzyme that separates the dog from the wolf. *Acta Biologica et Medica Germanica* **28**, 957-962.
Fisher, R. A., Putt, W. & Hackel, E. (1976). An investigation of the products of 53 gene loci in three species of wild canidae: *C. lupus*, *C. latrans* and *C. familiaris*. *Biochemistry & Genetics* **14**, 963-974.
Garcia-Moreno, J., Matocq, M. D., Roy, M. S., Geffen, E. & Wayne, R. K. (1996). Relationships and genetic purity of the endangered Mexican wolf based on analysis of microsatellite loci. *Conservation Biology* **10**, 376-389.
Girman, D. J., Kat, P. W., Mills, M. G. L., Ginsberg, J. R., Borner, M., Wilson, V., Fanshawe, J. H., Fitzgibbon, C., Lau, L. M. & Wayne, R. K. (1993). Molecular genetic and morphological analyses of the African wild dog (*Lycaon pictus*). *Journal of Heredity* **84**, 450-459.
Gotelli, D., Sillero-Zubiri, C., Applebaum, G. D., Roy, M. S., Girman, D. J., Garcia-Moreno, J., Ostranders, E. A. & Wayne, R. K. (1994). Molecular genetics of the most endangered canid: The Ethiopian wolf *Canis simensis*. *Molecular Ecology* **3**, 301-312.
Hillis, D. M. (1987). Molecular versus morphological approaches to systematics. *Annual Review of Ecology and Systematics* **18**, 23-42.
Hong, L. & Li-Ming, S. (1996). The mitochondrial DNA evolution of four species of Canidae. *Acta Theriologica Sinica* **42**, 87-95 [Chinese, English summary].
Hungerford, D. A. & Snyder, R. L. (1966). Chromosomes of a European wolf (*Canis lupus*) and of a bactrian camel (*Camelus bactrianus*). *Mammalian Chromosome Newsletter* **20**, 72.
Johnson, M. R., Boyd, D. K. & Pletscher, D. H. (1994). Serologic investigations of canine parvovirus and canine distemper in relation to wolf (*Canis lupus*) pup mortalities. *Journal of Wildlife Diseases* **30**, 270-273.
Laikre, L., Ryman, N. & Thompson, E. A. (1993). Hereditary blindness in a captive wolf (*Canis lupus*) population: Frequency reduction of a deleterious allele in relation to gene conservation. *Conservation Biology* **7**, 592-601.
Laikre, L., Tegelstrom, H. & Gelter, H. P. (1992). DNA fingerprinting of captive wolves (*Canis lupus*). *Hereditas* **117**, 293-296.
Lehman, N., Clarkson, P., Mech, L. D., Meier, T. J. & Wayne, R. K.. (1992). A study of the genetic relationships within and among wolf packs using DNA fingerprinting and mitochondrial DNA. *Behavioral Ecology and Sociobiology* **30**, 83-94.
Lorenzini, R. & Fico, R. (1995). A genetic investigation of enzyme polymorphisms shared by wolf and dog: suggestions for conservation of the wolf in Italy. *Acta Theriologica* **40 (Suppl. 3)**, 101-110.
Mercure, A., Ralls, K., Koepfli, K. P. & Wayne, R. K. (1993). Genetic subdivision among small canids: mitochondrial DNA differentiation of swift, kit, and arctic foxes. *Evolution* **47**, 1313-1328.
Meyer, W. & Tsukise, A. (1995). Lectin histochemistry of snout skin and foot pads in the wolf and the domesticated dog (Mammalia: Canidae). *Annals of Anatomy* **177**, 39-49.
Okumura, N., Ishiguro, N., Nakano, M. Matsui, A. & Sahara, M. (1996). Intra- and interbreed genetic variation of mitochondrial DNA major non-coding regions in Japanese native dog breeds (*Canis familiaris*). *Animal Genetics* **27**, 397-405.
Randi, E., Lucchini, V. & Francisci, F. (1993). Allozyme variability in the Italian wolf (*Canis lupus*) population. *Heredity* **7**, 516-522.
Randi, E., Francisci, F. & Lucchini, V. (1995). Mitochondrial DNA restriction-fragment-length monomorphism in the Italian wolf (*Canis lupus*) population. *Journal of Zoological Systematics and Evolutionary Research* **33**, 97-100.
Richards, M., Smalley, K. Sykes, B. & Hedges, R. (1993). Archaeology and genetics: analyzing DNA from skeletal remains. *World Archaeology* **25**, 18-28.
Richards, M. B., Sykes, B. C. & Hedges, R. E. M. (1995). Authenticating DNA extracted from ancient skeletal remains. *Journal of Archaeological Science* **22**, 291-299.
Roy, M. S., Geffen, E., Smith, D. & Wayne, R. K. (1996). Molecular genetics of pre-1940 red wolves. *Conservation Biology* **10**, 1413-1424.
Roy, M. S., Girman, D. J., Taylor, A. C. & Wayne, R. K. (1994). The use of museum specimens to reconstruct the genetic variability and relationships of extinct populations. *Experientia* **50**, 551-557.
Seal, U. S. (1987). Molecular approaches to taxonomic problems in the Canidae. In (M. W. Fox, Ed) *The Wild Canids: Their Systematics, Behavioral Ecology and Evolution*. Malabar, FL: Krieger Publishing Co., pp. 27-39.
Simonsen, V. (1976). Electrophoretic studies on the blood proteins of domestic dogs and other Canidae. *Heridatas* **82**, 7-18.
Tanabe, Y., Ota, K., Ito, S., Hashimoto, Y., Sung, Y. Y., Ryu, R. J. & Faruque, M. O. (1991). Biochemical-genetic relationships among Asian and European dogs and the ancestry of the Japanese native dog. *Journal of Animal Breeding and Genetics* **108**, 455-478.
Templeton, J. W. (1990). Canine DNA fingerprinting: Can it identify breeds? *Journal of the American Veterinary Medical Association* **196**, 1357, 1359, 1365.
Thomas, P. A. (1994). Dosimetry of 210Po in humans, caribou, and wolves in northern Canada. *Health Physics* **66**, 678-690.
Todd, N. B. (1970). Karyotypic fissioning and canid phylogeny. *Journal of Theoretical Biology* **26**, 445-480.
Tsuda, K., Kikkawa, Y., Yonekawa, H. & Tanabe, Y. (1997). Extensive interbreeding occurred among multiple matriarchal ancestors during the domestication of dogs: evidence from inter- and intraspecies polymorphisms in the D-loop region of mitochondrial DNA between dogs and wolves. *Genes and Genetic Systems* **72**, 229-238.
Vilà, C., Savolainen, P., Maldonado, J. E., Amorim, I. R., Rice, J. E., Honeycutt, R. L., Crandall, K. A., Lundeberg, J. &

Wayne, R. K. (1997). Multiple and ancient origins of the domestic dog. *Science* **276**, 1687-89.

Vilà, C., Maldonado, J. E. & Wayne, R. K. (1999). Phylogenetic relationships, evolution, and genetic diversity of the domestic dog. *Journal of Heredity* **90**, 71-77.

Wayne, R. K. (1992). On the use of morphologic and molecular genetic characters to investigate species status. *Conservation Biology* **6**, 590-592.

Wayne, R. K., Gilbert, D. A., Lehman, N., Hansen, K., Eisenhawer, A., Girman, D., Mech, L. D., Gogan, P. J. P., Seal, U. S. & Krumenaker, R. J. (1991). Conservation genetics of the endangered Isle Royale gray wolf. *Conservation Biology* **5**, 41-51.

Wayne, R. K., Lehman, N., Allard, M. W. & Honeycutt, R. L. (1992). Mitochondrial DNA variability of the gray wolf: Genetic consequences of population decline and habitat fragmentation. *Conservation Biology* **6**, 559-569.

Wayne, R. K., Meyer, A., Lehman, N., Van Valdenburgh, B., Kat, P. W., Fuller T. K., Girman, D. & O'Brien, S. J. (1990). Large sequence divergence among mitochondrial DNA genotypes within populations of eastern African black-backed jackals. *Proceedings of the National Academy of Science* **87**, 1772-1776.

Wayne, R. K., Nash, W. G. & O'Brien, S. J. (1987). Chromosomal evolution of the Canidae. I. Species with high diploid numbers. *Cytogenetics and Cell Genetics* **44**, 134-41.

Wayne, R. K. & O'Brien, S. J. (1987). Allozyme divergence within the Canidae. *Systematic Zoology* **36**, 339-55.

Wilton, A. N., Steward, D. J. & Zafiris, K. (1999). Microsatellite variation in the Australian dingo. *Journal of Heredity* **90**, 108-111.

Wurster-Hill, D. H., Ward, O. G., Davis, B. H., Park, J. P., Moyzis, R. K. & Meyne, J. (1988). Fragile sites, telomeric DNA sequences, B chromosomes, and DNA content in raccoon dogs, *Nyctereutes procyonoides,* with comparative notes on foxes, coyotes, wolf, and raccoon. *Cytogenetics and Cell Genetics* **49**, 278-281.

Zajc, I. & Sampson, J. (1999). Utility of canine microsatellites in revealing the relationships of pure bred dogs. *Journal of Heredity* **90**, 104-107.

**Wild canids**

Anderson, R. M. (1943). Summary of the large wolves of Canada, with description of three new Arctic races. *Journal of Mammalogy* **24,** 386-393.

Anonymous. (1993). Identity of canine shot in Yellowstone still a puzzle. *International Wolf* **3**, 24.

Aylsworth, L. (1997). International wolf center bibliography of wolf resources. On the Internet: http://www.wolf.org/bib/index.html

Bailey, V. (1940). The home life of the big wolves. *Natural History* **46**, 120-122.

Barker, B. C. W. & A. Macintosh. (1979). The Dingo - a review. *Oceania* **14**, 27-53.

Bauer, E. A. (1994). *Wild Dogs: The Wolves, Coyotes, and Foxes of North America*. San Francisco: Chronicle Books.

Bekoff, M. (1977). Canis latrans. *Mammalian Species* **79**.

Bekoff, M. (1978). *Coyotes: Biology, Behavior, and Management*. New York: Academic Press.

Bergerud, A. T. (1988). Caribou, wolves and man. *Trends in Ecology and Evolution* **3**, 68-72.

Berta, A. (1987). Origins, diversification, and zoogeography of the South American Canidae. In (B. D. Patterson & R. M. Timm, Eds) *Studies in Neotropical Mammalogy: Essays in Honor of Philip Hershkovitz, Fieldiana Zoology New Series* **39**, 455-471.

Berta, A. (1988). *Quaternary evolution and biogeography of the large South American Canidae (Mammalia: Carnivora).* University of California Geological Sciences **132**.

Bharos, A. M. K. (1996). Occurrence of the wolf, *Canis lupus* Linnaeus, in Rewa District, M.P. *Journal of the Bombay Natural History Society* **93**, 81.

Bibikov, D. I. (1980). Wolves in the USSR. *Natural History* **89**, 58-63.

Billings, E. (1856). The natural history of the wolf (*Canis lupus*), and its varieties. *The Canadian Naturalist and Geologist* **1**, 209-215.

Bogan, M. A. & P. Mehlop. (1983). Systematic relationships of gray wolves (*Canis lupus*) in southwestern North America. *Occasional Papers of the Museum of Southwest Biology* **1**, 1-21.

Bueler, L. E. (1973). *Wild Dogs of the World*. Briarcliff Manor, NY: Stein & Day and Publishers, Scarborough House.

Cabrera, A. (1931). On some South American canine genera. *Journal of Mammalogy* **12**, 54-67.

Carbyn, L. N., Fritts, S. H. & Seip, D. R. (1995). *Ecology and Conservation of Wolves in a Changing World.* Edmonton: Canadian Circumpolar Institute Occasional Publication **35**.

Coues, E. (1873). The prairie wolf or coyote, *Canis latrans. American Naturalist* 7, 384-389.

DeBlieu, J. (1992). Could the red wolf be a mutt? *New York Times Magazine* 14 June, 30-36.

Dixon, J. S. (1916). The timber wolf in California. *California Fish and Game* **2**, 125-129.

Elder, W. H. & Hayden, C. M. (1977). Use of discriminant function in taxonomic determination of canids from Missouri. *Journal of Mammalogy* 58, 17-24.

Fiennes, R. (1976). *The Order of Wolves*. Indianapolis & New York: Bobbs-Merrill.

Flower, W. H. (1880). On the bush dog (*Icticyon venaticus*, Lund). *Proceedings of the Zoological Society of London* 1880, **70-76**.

Fox, M. W. (1975, 1983). *The Wild Canids: Their Systematics, Behavioral Ecology and Evolution*. Malabar, FL: Krieger Publishing Co.

Fox, M. W. (1984). *The Whistling Hunters: Field Studies of the Asiatic Wild Dog (Cuon alpinus)*. Albany: State University of New York Press.

Friis, L. K. (1985). *An investigation of subspecies relationships of the grey wolf,* Canis lupus, *in British Columbia.* Masters Thesis. University of Victoria, British Columbia.

Gipson, P. S. (1978). Coyotes and related *Canis* in the southeastern United States with a comment on Mexican and Central American *Canis*. In (Marc Bekoff, Ed) *Coyotes: Biology, Behavior, and Management,* New York: Academic Press, pp.191-209.

Gipson, P. S., Sealander, J. A. & Dunn, J. E. (1974). The taxonomic status of wild *Canis* in Arkansas. *Systematic Zoology* **23**, 1-11.
Goldman, E. A. (1937). The wolves of North America. *Journal of Mammalogy* **18**, 37-45.
Gray, R.(1970). *Gray Wolf: The Natural Life of North American Wolves*. New York: W. W. Norton & Co.
Harrington, F. H. & Paquet, P. C. (1982). *Wolves of the World: Perspectives of Behavior, Ecology, and Conservation*. Park Ridge, NJ: Noyes Publications.
Hershkovitz, P. (1957). A synopsis of the wild dogs of Colombia. *Novedades Colombianas Contribuciones Cient ficas del Museo de Historia Natural de la Universidad del Cauca* **3**, 157-161.
Hershkovitz, P. (1961). On the South American small-eared zoro *Atelocynus microtis* Sclater (Canidae). *Fieldiana Zoology* **39**, 505-523.
Hilton, H. (1978). Systematics and ecology of the eastern coyotes. In (M. Bekoff, Ed) *Coyotes: Biology, Behavior and Management*. New York: Academic Press, pp. 209-228.
ITIS (Integrated Taxonomic Information System). (1998). Integrated Taxonomic Information System. On the Internet: http://www.itis.usda.gov/itis/
Klinghammer, E., Sloan, M. & Klein, D. (1990). *Wolf Literature References*. Battle Ground, IN: Institute of Ethology, North American Wildlife Park Foundation, Inc.
Knee, M. (1995). Wolves: a bibliography and guide to the literature (1968-1987). On the Internet: http://www.albany.edu/~knee/wolf.html
Knight, J. (1997). On the extinction of the Japanese wolf. *Asian Folklore Studies* **56**, 129-159.
Luomala, K. (1960). The native dog in the Polynesian system of values. In (S. Diamond, Ed) *Essays in Honor of Paul Radin*. New York: Columbia University Press, pp.190-240.
McCarley, H. (1962). The taxonomic status of wild *Canis* (Canidae) in the south central United States. *Southwestern Naturalist* **7**, 227-235.
Mader, T. R. (1988). *Wolf Reintroduction in the Yellowstone National Park: A Historical Perspective*. Gillette, WY: Common Man Institute.
Mech, L. D. (1966). *The Wolves of Isle Royale*. Fauna of the National Parks of the United States **7**.
Mech, L. D. (1970). *The Wolf*. Garden City, NY: Natural History Press.
Mech, L. D. (1974). Canis lupus. *Mammalian Species* **37**.
Mech, L. D. (1981). *The Wolf: The Ecology and Behavior of an Endangered Species*. Minneapolis: University of Minnesota Press [reprint].
Mivart, St. G. (1890). *Dogs, Jackals, Wolves and Foxes: Monograph of the Canidae*. London.
Murie, A. (1940). *Ecology of the coyote in the Yellowstone*. Fauna of the National Parks of the United States **4**.
Murie, Adolph. (1944). *The wolves of Mount McKinley*. Fauna of the National Parks of the United States **5**.
Murie, A. (1985). *The Wolves of Mount McKinley*. Seattle: University of Washington Press [reprint].
Nowak, R. M. (1967). The red wolf in Louisiana. *Defenders of Wildlife News* **42**, 60-70.
Nowak, R. M. (1972). The mysterious wolf of the South. *Natural History* **81**, 50-53, 74-77.
Nowak, R. M. (1974). Red wolf: our most endangered mammal. *National Parks and Conservation Magazine* **48**, 9-12.
Nowak, R. M. (1978). Evolution and taxonomy of coyotes and related *Canis*. In (M. Bekoff, Ed) *Coyotes: Biology, Behavior and Management*. New York: Academic Press, pp. 3-16.
Nowak, R. M. (1983). A perspective on taxonomy of wolves in North America. *Canadian Wildlife Service Report Series* **45**, 10-19.
Nowak, R. M. (1991). Dogs, wolves, coyotes, and jackals. In *Walker's Mammals of the World, Fifth edition, Volume II*. Baltimore: John Hopkins University Press, pp. 1065-1080.
Nowak, R. M. (1992). The red wolf is not a hybrid. *Conservation Biology* **6**, 593-595.
Obee, B. (1994). *Wolf: Wild Hunter of North America*. Stamford, CT: Longmeadow Press.
Osgood, W. H. (1934). The genera and subgenera of South American canids. *Journal of Mammalogy* **15**, 45-50.
Paradiso, J. L. (1967). Recent records of coyotes, *Canis latrans*, from the southeastern United States. *Southwestern Naturalist* **11**, 500-501.
Paradiso, J. L. (1968). Canids recently collected in East Texas, with comments on the taxonomy of the red wolf. *American Midland Naturalist* **80**, 529-534.
Paradiso, J. L. (1971). A report on the taxonomic status and distribution of the red Wolf. *United States Fish and Wildlife Service Wildlife* **145**, 1-36.
Paradiso, J. L. (1973). New data on the red wolf in Alabama. *Journal of Mammalogy* **54**, 506-509.
Paradiso, J. L. & Nowak, R. M. (1971). A report on the taxonomic status and distribution of the red wolf. *United States Department of the Interior Special Scientific Report on Wildlife* **145**, 1-36.
Paradiso, J. L. & Nowak, R. M. (1972). Canis rufus. *Mammalian Species* **22**.
Perzanowski, K. (1995). The wolf: its ecology and interactions with other species. *Honyurui Kagaku* **35**, 29-33. [Japanese].
Peterson, R. O. (1977). *Wolf Ecology and Prey Relationships on Isle Royale*. National Park Service Scientific Monograph Series **11**.
Peterson, R. O. (1995). *The Wolves of Isle Royale*. Minocqua, WI: Willow Creek Press.
Peterson, R. O. & Allen, D. L. (1976). *Ecological Studies of the Wolf on Isle Royale, 18th Annual Report on Isle Royale Studies*. Washington, DC: Government Printing Office.
Peterson, R. O. & Page, R. E. (1988). The rise and fall of Isle Royale wolves, 1975-1986. *Journal of Mammalogy* **69**, 89-99.
Pimlott, D. H. & Joslin, P. W. (1968). The status and distribution of the red wolf. *Transactions of the North American Wildlife Natural Research Conference* **33**, 373-388.
Potvin, F., Jolicoeur, H. & Huot, J. (1988). Wolf diet and prey selectivity during two periods for deer in Quebec: decline versus expansion. *Canadian Journal of Zoology* **66**, 1274-1279.
Pulliainen, E. (1965). Studies of the wolf (*Canis lupus* L.) in Finland. *Annales Zoological Fennici* **2**, 215-259.
Russel, D. & Shaw, J. H. (1972). Distribution and relative density of the red wolf in Texas. *Proceedings of the Annual Conference of the Southeastern Association of Game and Fish Commissions* **25**, 131-137.
Saito, H. (1964). *Japanese Dogs and Wolves*. Tokyo: SekkaSha. [Japanese].
Schmitz, O. J. & Kolenosky, G. B. (1985). Wolves and coyotes in Ontario: morphological relationships and origins. *Canadian Journal of Zoology* **63**, 1130-1137.

Searching Wolf. (1998a). Recent wolf technical journal articles: 1988 to present. On the Internet: http://www.iup.edu/~wolf/wtechart.htmlx

Searching Wolf. (1998b). Recent magazine articles. On the Internet: http://www.iup.edu/~wolf/wpopart.htmlx

Searching Wolf. (1998c). A bibliography of European wolf literature. On the Internet: http://www.iup.edu/~wolf/

Shigerhara, N., Matsura, S., Nakamura, T. & Kondo, M. (1993). First discovery of the ancient dingo-type dog in Polynesia (Pukapuka, Cook Islands). *International Journal of Osteoarchaeology* **3**, 315-320.

Smithsonian Institute. (1993). Selected references on wolves. On the Internet: http://www.si.edu/resource/faq/nmnh/wolves.htm

Snyder, L. M. (1988). A controlled feeding study involving gray wolf (*Canis lupus*) and whitetailed deer (*Odocoileus virginianus*). *University of Tennessee Anthropological Association Newsletter* **13**, 1-9.

Sullivan, J. O. (1978). Variability in the wolf, a group hunter. In (R. Hall & H. S. Sharp, Eds) *Wolf and Man: Evolution in Parallel.* New York: Academic Press, pp. 31-40.

Stains, H. J. (1975). Distribution and taxonomy of the Canidae. In (M. W. Fox, Ed) *The Wild Canids: Their Systematics, Behavioral Ecology and Evolution.* Malabar, FL: Krieger Publishing Co., pp. 3-26.

Steinhart, P. (1995). *The Company of Wolves.* New York: Vintage Books.

Stromberg, M. R. & Boyce, M. S. (1986). Systematics and conservation of the swift fox, *Vulpes velox*, in North America. *Biological Conservation* **35**, 97-110.

University of Wyoming. (1997). Wolf reintroduction: A selected bibliography. On the Internet: http://www.uwyo.edu.lib/wolf.htm

Waithman, J. & Roest, A. (1977). A taxonomic study of the kit fox, *Vulpes macrotis*. *Journal of Mammalogy* **58**, 157-164.

Wayne, R. K. & Gittleman, J. L. (1995). The problematic red wolf. *Scientific American* **273**, 36-39.

Wilson, D. E. & Reeder, D. M. (1993). *Mammal Species of the World: A Taxonomic and Geographic Reference, 2nd ed.* Washington, DC: Smithsonian Institution Press. Also on the Internet: http://nmnhwww.si.edu/msw/

Woodhouse, S. W. (1852). The North American jackal - *Canis frustror*. *Journal Academy Natural Science* **2**, 87-88.

Wright, N. P. (1970). *A Guide to Mexican Mammals and Reptiles*. Mexico City: Minutiae Mexicana.

Young, S. P. & Jackson, H. H. T. (1951). *The Clever Coyote.* Washington, DC: The Stackpole Co., & Harrisburg: The Wildlife Management Institute.

Young, S. P. & Goldman, E. A. (1964). *The Wolves of North America.* New York: Dover Publication.

www.ingramcontent.com/pod-product-compliance
Ingram Content Group UK Ltd.
Pitfield, Milton Keynes, MK11 3LW, UK
UKHW061844041026
15591UKWH00015B/906